ENVIRONMENTAL
SCIENCE

FOR A CHANGING WORLD

WITH EXTENDED COVERAGE

ENVIRONMENTAL SCIENCE

FOR A CHANGING WORLD

WITH EXTENDED COVERAGE

SUSAN KARR
Carson-Newman College

ANNE HOUTMAN
Rochester Institute of Technology

JENEEN INTERLANDI
Science Writer

With contributions by Teri Balser

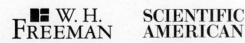

W. H. FREEMAN   SCIENTIFIC AMERICAN

Publisher: Susan Winslow
Senior Acquisitions Editor: Jerry Correa
Developmental Editor: Andrea Gawrylewski
Project Manager: Karen Misler
Associate Director of Marketing: Debbie Clare
Associate Media Editor: Betsye Mullaney
Associate Editor: Anna Bristow
Editorial Assistant: Jane Taylor
Art Direction, Cover, Design, and Illustrations:
MGMT. design
Photo Editor and Researcher: Stephanie Heimann
Proofreader: Anna Paganelli
Senior Illustration Coordinator: Bill Page
Director of Production: Ellen Cash
Printing and Binding: RR Donnelley

Library of Congress Control Number: 2012952131
ISBN-13: 978-1-4292-40307
ISBN-10: 1-4292-4030-X

Printed in the United States of America

First printing

W. H. Freeman and Company
41 Madison Avenue
New York, NY 10010
Houndmills, Basingstoke RG21 6XS, England
www.whfreeman.com

Front cover: A nighttime peek inside the California Academy of Science's rainforest exhibit through the "Living Roof." The 197,000-square-foot roof is planted with nine plant species native to California and selected to thrive in San Francisco's Golden Gate Park, home of the academy. The plants cut down on the urban heat island effect, keep the building 10x cooler than a normal roof would, and capture rainwater for slower absorption.

BRIEF CONTENTS

UNIT 4
ENERGY: A WICKED PROBLEM WITH MANY CONSEQUENCES

UNIT 5
ENERGY: TOWARD A SUSTAINABLE FUTURE

DETAILED CONTENTS

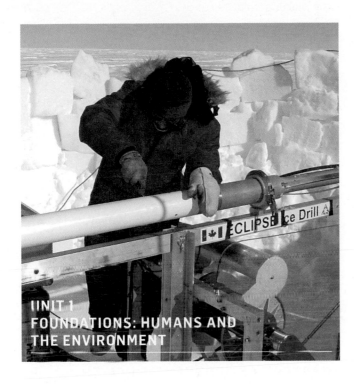

UNIT 1
FOUNDATIONS: HUMANS AND
THE ENVIRONMENT

**UNIT 2
ECOLOGY, PATTERNS, AND PROCESSES**

**UNIT 3
EARTH'S RESOURCES, CURRENT
CHALLENGES, AND SUSTAINABLE OPTIONS**

**UNIT 4
ENERGY: A WICKED PROBLEM WITH MANY CONSEQUENCES**

**UNIT 5
TOWARD A SUSTAINABLE FUTURE**

AUTHORS

SUSAN KARR, MS, is an Instructor in the biology department of Carson-Newman College in Jefferson City, Tennessee, and has been teaching for more than 15 years. She has served on campus and community environmental sustainability groups and helps produce an annual "State of the Environment" report on the environmental health of her county. In addition to teaching non-majors courses in environmental science and human biology, she teaches an upper-level course in animal behavior where she and her students train dogs from the local animal shelter in a program that improves the animals' chances of adoption. She received degrees in animal behavior and forestry from the University of Georgia.

ANNE HOUTMAN, PhD, is Professor and Head of the School of Life Sciences at Rochester Institute of Technology, which includes programs in environmental and biological sciences. Her research interests are in the behavioral ecology of birds. Currently, research in her laboratory focuses on the ecology and evolution of hummingbird song. She also has an active research program in science pedagogy. Anne received her doctorate in zoology from the University of Oxford and conducted postdoctoral research at the University of Toronto.

JENEEN INTERLANDI, MA, MS, is a science writer who contributes to *Scientific American* and *The New York Times Magazine*. Previously, she spent four years as a staff writer for *Newsweek*, where she covered health, science, and the environment. She was a 2009 Kaiser Foundation Fellow for global health reporting and she is a 2013 Nieman Fellow, studying at Harvard University. Jeneen has worked as a researcher at both Harvard Medical School and Lamont Doherty Earth Observatory. She holds Master's degrees in environmental science and journalism, both from Columbia University in New York.

Dear Reader,

For more than 15 years as an environmental science and biology instructor I've found that "stories" capture the imagination of my students. Students are genuinely interested in environmental issues—and using stories to teach these issues makes the science more relevant and meaningful to them. Many leave the class with an understanding that what they do really matters, and they feel a willingness to act on that knowledge. This is why I am enthusiastic about our new textbook, *Environmental Science for a Changing World*, presented here with extended coverage, including six additional chapters to give instructors more flexibility in what they teach.

Each chapter will keep students engaged and reading to find out "what happens next." At the same time, explanations of science are woven into the narrative and illustrated in vivid infographics that give additional detail without slowing down the story.

We've broken some topics down into multiple chapters—for instance, we present separate chapters on coal, petroleum, and nuclear power rather than the traditional single chapter on conventional energy. This gives instructors the flexibility to focus on discrete topics if they choose. This extended edition will be especially useful for instructors who are looking for stand-alone chapters on topics of environmental health, mineral resources and geology, and environmental policy. A chapter on world hunger sets the stage for our two chapters on agriculture, giving context to the pressing question of how we will feed the world's growing population today and in the future. The expanded coverage of biodiversity issues also provides a more in-depth look at this important environmental science topic, presenting a separate chapter on preserving biodiversity.

Based on instructor feedback, it was important to focus on the competencies that a general education science course should instill in the non-major: environmental literacy, science literacy, and information literacy. The text provides the necessary background to build those skills and plenty of opportunities to practice them.

Environmental Literacy—Many of the biggest challenges facing society are environmental issues. Throughout the book we examine the scientific, social, political, and economic facets of contemporary environmental issues. We strive to give a balanced presentation, especially for controversial topics, focus on the underlying science concepts, and discuss sustainable solutions.

Science Literacy—Educated citizens need to understand how scientific investigations are done and how results and conclusions are reported in order to make informed decisions in their own lives. In each chapter we've included experimental evidence, graphical representations of data, and the work of scientists in the field, giving students practice in evaluating evidence and understanding the process of science.

Information Literacy—Students must be able to find information and then evaluate its quality. We explain how to search for and find scientific information and how to critically analyze the veracity of that information.

Every person involved in this book—the writers, illustrators, editors, and fellow instructors—has one sincere objective: to help students become informed citizens who are able to analyze issues, evaluate arguments, discuss solutions, and recognize trade-offs as they make up their own minds about our most pressing environmental challenges.

Sincerely,

Susan Karr

Susan Karr

WRITTEN FOR TODAY'S STUDENT

"AS GLOBAL FISHERIES COLLAPSE, ONE RESEARCHER RAISES SCHOOLS OF MARINE FISH IN A BALTIMORE WAREHOUSE."

"THE BARK OF AN ORDINARY TREE IN SAMOA COULD HOLD THE CURE FOR CANCER."

Environmental Science for a Changing World with Extended Coverage captivates students with real-world stories while exploring the science concepts in context. Engaging stories plus vivid photos and infographics make the content relevant and visually enticing. The result is a text that emphasizes environmental, scientific, and information literacies in a way that engages students.

ENGAGING STORIES ILLUMINATE PRESSING ENVIRONMENTAL ISSUES

It was his mother's death in the fall of 1984, from a particularly aggressive form of breast cancer, that drove Paul Cox back to the Samoan rainforest. Cox had first visited the South Pacific island in 1973, through an undergraduate research program with Brigham Young University, where he was majoring in botany. Since then, the Utah native had earned a Ph.D. from Harvard and made a career studying plant physiology in the United States.

◉ WHERE IS SAMOA?

SAMOA
APIA
PAPUA NEW GUINEA
SAMOA
AUSTRALIA
NEW ZEALAND

A RUNNING MARGIN GLOSSARY MEANS STUDENTS ENCOUNTER IMPORTANT TERMS IN THE CONTEXT OF THE STORY

biodiversity The variety of life on Earth; it includes species, genetic, and ecological diversity.
species diversity The variety of species, including how many are present (richness) and their abundance relative to each other (evenness).

...moa, 1986.

...tern medicine failed to cure ...ered the Samoans—their rich ...d the profound influence that ...' health and well-being. Surely, he thought, somewhere in the lush abundance that had long sustained this impoverished island nation must be a compound powerful enough to obliterate the most insidious of tumors. "I told my friends at the National

biodiversity The variety of life on Earth; it includes species, genetic, and ecological diversity.
species diversity The variety of species, including how many are present (richness) and their abundance relative to each other (evenness).
genetic diversity The heritable variation among individuals of a single population or within the species as a whole.
ecological diversity The variety within an ecosystem's structure, including many communities, habitats, niches, and trophic levels.

Cancer Institute, 'If there is even a 1% chance of finding something, it's worth taking a look,'" he recalls. "They said, 'We think there is like a 3% chance.' So I went." Six months after his mother's funeral, with his wife and four young children in tow, Cox returned to Samoa.

Biodiversity benefits humans and other species.

Why did Cox suspect that Samoa might hold the key to a cancer cure? Tropical regions—warm, lush, close to the equator—contain the greatest concentration and variety of plant and animal life forms on Earth. This variety is called **biodiversity**. Ecosystems of this region have both high **species diversity** (see Chapter 9) and high **genetic diversity**. They also usually have high **ecological diversity**, a wide variety of communities and ecosystems with

SCIENCE IS TAUGHT IN THE CONTEXT OF THE STORY

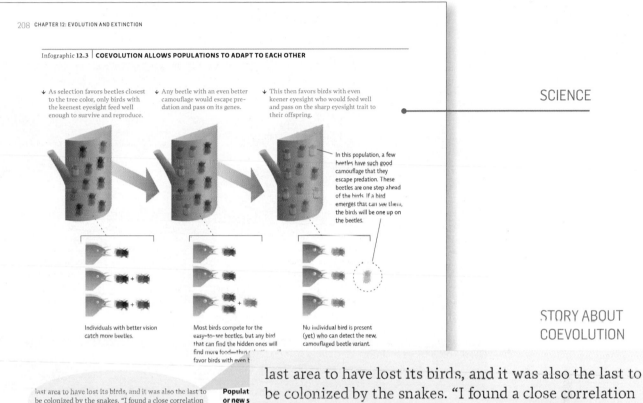

208   CHAPTER 12: EVOLUTION AND EXTINCTION

Infographic 12.3 | **COEVOLUTION ALLOWS POPULATIONS TO ADAPT TO EACH OTHER**

↓ As selection favors beetles closest to the tree color, only birds with the keenest eyesight feed well enough to survive and reproduce.

↓ Any beetle with an even better camouflage would escape predation and pass on its genes.

↓ This then favors birds with even keener eyesight who would feed well and pass on the sharp eyesight trait to their offspring.

SCIENCE

In this population, a few beetles have such good camouflage that they escape predation. These beetles are one step ahead of the birds. If a bird emerges that can see them, the birds will be one up on the beetles.

Individuals with better vision catch more beetles.

Most birds compete for the easy-to-see beetles, but any bird that can find the hidden ones will find more food—thus selection favor birds with even b

No individual bird is present (yet) who can detect the new, camouflaged beetle variant.

STORY ABOUT COEVOLUTION

last area to have lost its birds, and it was also the last to be colonized by the snakes. "I found a close correlation between the bird decline and the expansion of brown tree snakes around the island," she says, which suggested to her that brown tree snakes really might be the culprit. [INFOGRAPHIC 12.4]

Some of Guam's bird species went extinct sooner than others. For instance, the **endemic** bridled white-eye, the gregarious bird species that was extinguished first, happens to be very small, raising the possibility that the small size of these birds might have put them at a disadvantage. Larger species like flycatchers survived longer, though they, too, eventually disappeared. Other bird species experienced **extirpation**; the Guam rail, for instance, is gone from Guam but other populations still live on the nearby island of Rota.

Populat or new s

If even a extinct s due to a l predator that coul been sur have bee brown tree snake.

When populations diverge because of isolation, food availability, new predators, or habitat fragmentation such that their members can no longer freely interbreed, new species may arise (*speciation*). This increases species diversity in a community and sometimes produces

last area to have lost its birds, and it was also the last to be colonized by the snakes. "I found a close correlation between the bird decline and the expansion of brown tree snakes around the island," she says, which suggested to her that brown tree snakes really might be the culprit.

HOW WE SELECTED THE STORIES:

We polled more than 250 environmental science instructors over the course of 3 years to find out their favorite case studies for teaching environmental science. Our authors and team of advisors considered the results and chose chapter stories that could most effectively teach the science, while remaining fresh and engaging. Finally, 20 focus group participants gave us feedback on the strength of our story selection for each chapter.

REAL VOICES, CHARACTERS, AND VIVID PHOTOS BRING THE ISSUES TO LIFE

PHOTOJOURNALISTIC
IMAGES DRAW
STUDENTS IN

↑ An overcrowded train approaches as other passengers wait to board at a railway station in Dhaka, Bangladesh.

REAL PEOPLE AND
REAL VOICES BRING
CONCEPTS TO LIFE

and she already had a son. "People from the Communist Brigade said I couldn't have a second baby," she told National Public Radio (NPR) in a 2005 interview. "But I was determined. I needed a son to work our farm." So

the rural Sichuan Province, found that women whose ... likely to die ... re sanctioned ... f all married ... ed with ... countries— ... ich type of ... cted by the ... f women ... method their ... all, IUDs and sterilizations account for more than 90% of contraceptive methods used since the mid-1980s.

Still, there are signs of improvement. The number of sterilizations has declined considerably since the early 1990s. And in 2002, facing strong international pressure, the Chinese government finally outlawed the use of physical force to compel a woman to have an abortion or be sterilized. (Though experts say that prohibition is not entirely enforced. "The policy is largely exercised at the local level," says Greenhalgh. "And in many provinces, local governments still demand abortions.")

In any case, by 2011, China's fertility rate had plummeted to 1.54, spurring China's leaders to proclaim that the one-child policy had succeeded in preventing some 400 million births and all the calamities those births would have brought—disease, more famine, overtaxed social services, and so on.

But while the drop in fertility was plain, not everyone would agree that one-child deserved the credit. Critics argue that by the time the first generation of "Little Emperors" reached adulthood, many other forces had reshaped their society and the world.

Factors that decrease the death rate can also decrease overall population growth rates.

When Qin Yijao's mom was pregnant with him back in 1981, the one-child policy was being strictly enforced, and she already had a son. "People from the Communist Brigade said I couldn't have a second baby," she told National Public Radio (NPR) in a 2005 interview. "But I was determined. I needed a son to work our farm." So taking care to evade the birth police, she snuck off and

PRACTICE IN ANALYZING SCIENCE AND EVALUATING NEW INFORMATION

EACH CHAPTER HAS A CORE
TAKE-AWAY MESSAGE

CORE MESSAGE
The variety of life on Earth is tremendous. This biodiversity provides vital ecological services that support life; we depend on these same services for things like food, medicine, and economic development. Biodiversity faces many threats today, jeopardizing the ability of ecosystems to provide these services. One of the first steps in protecting biodiversity is simply trying to determine what is here.

ANALYZING THE SCIENCE

The following graph depicts the relationship between numbers of extinctions and human population size since the 19th century.

INTERPRETATION

1. Describe what is happening to:
   a. the extinction rate over time.
   b. human population growth over time.

2. The two curves have been graphed together. What is the implication of presenting these data in this manner?

Species extinction and human population

DATA-BASED PROBLEMS
ENHANCE SCIENCE
LITERACY SKILLS

EVALUATING NEW INFORMATION

Life on Earth as we know it is a result of millions of years of evolutionary processes. These processes are not immune to changes in the environment; changes in evolutionary processes will ultimately result in changes in biodiversity, which will necessarily affect life on Earth, including humans.

Go to the website: www.actionbioscience.org/evolution/myers_knoll.html and read the article on the effects of the sixth mass extinction on evolution.

Evaluate the website and work with the information to answer the following questions:

1. Is this a reliable information source? Does it have a clear and transparent agenda?
   a. Who runs this website? Does this person's/group's credentials

a. What changes in particular do they think will impact future evolution?
b. What types of evidence do they provide to support their arguments? Give specific examples.

3. The sixth mass extinction is the result of human activities. Read "The First Human Caused Extinction: The Dodo" (http://catherineowen. suite101.com/the-first-human-caused-extinction-the-dodo a85161).
   a. Why did the dodo go extinct? Be specific—there are direct and indirect consequences discussed in the article.
   b. On Mauritius, what happened as a consequence of the dodo's extinction?
   c. Do you think humans today could have a similar impact on a species? Consider both direct and indirect human impacts. Which types of species might be vulnerable? Find an example on the Conservation of Nature (IUCN) "Species ... www.iucnredlist.org/species-of-the-day/

Go to the website www.actionbioscience.org/evolution/myers_knoll.html
and read the article on the effects of the sixth mass extinction on evolution.

WEB-BASED PROBLEMS
DEVELOP INFORMATION
LITERACY SKILLS

MAKING CONNECTIONS

Gyps indicus: **A SPECIES WORTH SAVING?**

Background: Vultures are a type of bird that feeds on dead organisms. They are found worldwide and provide an important ecosystem service. Many are large (with wingspans of over 4 feet) and depend on large animal carcasses for food. Some of the larger species, including the California condor (*Gymnogyps californianus*), are critically endangered due to habitat loss and human hunting. Loss of vultures has the potential to impact ecosystems, so conservation groups have worked to save them.

Case: You are a member of a philanthropic organization devoted to saving species from extinction. Your organization has a limited amount of funds, so it is imperative that the funds be given to worthwhile projects. The Indian government has approached you for funds to help save the critically endangered Indian vulture *Gyps indicus*. Before your organization can make a decision, you need to gather information:

1. Research *Gyps indicus* (the IUCN Red List is a good place to start). Consider the following questions:
   a. Why i
   b. What
   c. Where

d. Does its habitat still exist? If it were brought back from the brink of extinction, is there habitat for it to live in and food for it to eat?
e. What plans are currently in place to save it?

2. Do some further research to answer the following questions.
   a. Why did the condor decline? Consider both human impacts and the condor's biology.
   b. How much money has been spent to date on the rescue of the California condor?
   c. What is the status of condors that currently live in the wild?
   d. What is the outlook for the future of this species?
   e. Answer questions a. – d. for *G. indicus* and compare your results to those for the California condor.

3. Based on your research about *G. indicus* and California condors, write a report recommending a course of action to your organization. Should you give money to save the *G. indicus* species? Why or why not? Specifically, you should address the issue of whether it is:

2. Do some further research to answer the following questions.
   a. Why did the condor decline? Consider both human impacts and the condor's biology.
   b. How much money has been spent to date on the rescue of the California condor?
   c. What is the status of condors that currently live in the wild?

CRITICAL THINKING PROBLEMS BUILD
ENVIRONMENTAL LITERACY SKILLS

SOLUTIONS TO ENVIRONMENTAL PROBLEMS INSPIRE STUDENTS

ENVIRONMENTAL SOLUTIONS ARE DISCUSSED IN EVERY CHAPTER

BIOMIMICRY IS A FREQUENT THEME IN THE CHAPTERS—USING NATURE'S MODELS AND STRATEGIES TO SOLVE HUMAN PROBLEMS—SUSTAINABLY

Taking our cues from nature, we can learn to use rangelands sustainably.

Factory

Some parts are reprocessed into new products.

owned Horse Creek, will also suffer.

Scientists around the world have spent decades trying to prevent or even reverse desertification, to little avail. These days, most experts tend to agree that beyond a certain point, recovering grasslands that have swirled into deserts is impossible. "Most of our efforts to reverse desertification have failed dismally," says Dr. Richard Teague, a research ecologist working with ranchers in West Texas to restore degraded rangeland. "But a number of ranchers here are having success with protocols developed halfway around the world."

Taking our cues from nature, we can learn to use rangelands sustainably.

As the sun rises over Zimbabwe (formerly Rhodesia) in southern Africa, herds of antelope and zebra traverse a patchwork of temperate and tropical grasslands, feeding steadily on reedy stalks and short, fat shrubs; elephants and wildebeest splash around a precious watering hole, well fed and content. The animals may not realize it, but they have stumbled upon the African Center for Holistic Management (ACHM), 6500 acres of thriving rangeland in the heart of an otherwise parched and ailing prairie. [INFOGRAPHIC 14.5]

Perhaps nowhere else on Earth is such an oasis more urgently needed. Because the region is too arid to support much else, ranching provides the only livelihood for most of the people living there; about 75% of all land is used to graze domestic herds of cattle, goat, and sheep, and even that has not been enough. With population, and thus the number of mouths in need of food, rising steadily, farmers have crowded more and more livestock onto lands that grow sparser and drier by the day. Already stressed by climate change, that land is crumbling quickly into desert. And as viable pastures become increasingly difficult to find, neighboring tribes have descended further into violent conflict—sometimes killing each other over a few stalks of grass.

The ACHM was established in 1992 by Allan Savory, a Rhodesian-born scientist-turned-rancher. Before then, the land had been so thoroughly desertified that neither wild nor domestic herds bothered to graze there. But in the nearly two decades since, the picture has changed dramatically. Both plants and wild herds have rebounded with surprising speed; even during the dry season, water is plentiful enough to sustain fish and water lilies. And if that's not enough, livestock has increased by 400%. Ranchers come from as far as Texas to marvel at the

Infographic 14.5 | **SOUTHERN AFRICA MAP**

AFRICA

Zimbabwe

Southern Africa

↑ Biologist Allan Savory, squatting beside a patch of dry grasses in the desert where he teaches holistic land management.

turnaround and to seek counsel from Savory, who is widely credited for the dramatic recovery.

Savory came by his expertise in a circuitous way. He spent the 1950s working as a research biologist and game ranger for the British Colonial Service. At the time, the British government was culling thousands of wild herds in an effort to create more land for farming. "Zimbabwe

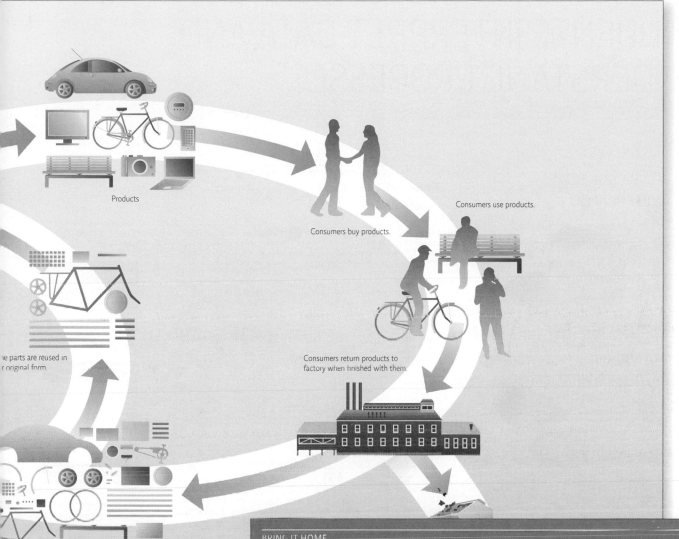

Products

Consumers buy products.

Consumers use products.

Consumers return products to factory when finished with them.

...e parts are reused in ...r original form.

Products are disassembled.

BRING IT HOME OFFERS WAYS THAT STUDENTS CAN ADDRESS ENVIRON- MENTAL ISSUES ON THE INDIVIDUAL, GROUP, AND POLITICAL LEVEL

BRING IT HOME

⊘ PERSONAL CHOICES THAT HELP

Species and habitats provide numerous benefits to people, including water and air purification, food sources, recreation, and medicine. Unfortunately, many species are facing threats at ever-increasing levels. The good news is that we as a society have a direct impact on these threats and can make changes to ensure the survival of many of our at-risk species.

Individual Steps
→ Don't buy products made from wild animal parts such as horns, fur, shells, or bones. Only buy captive-bred tropical aquarium fish, not wild-caught fish.
→ Research the policies of a not-for-profit organization that protects biodiversity, such as World Wildlife Fund (WWF), Ocean Conservancy, Defenders of Wildlife, or the African Wildlife Foundation. Is it worth donating money to their cause?

→ Make your backyard friendly to wildlife, using suggestions from www.nwf.org.
→ Install an Audubon Guide app on your smartphone, or buy a field guide to learn the plant and animal species in your area.

Group Action
→ Work with faculty and other students to organize a bioblitz for a protected area in your region. A bioblitz, which is an intensive survey of all the biodiversity in the area, can generate a large amount of data to be used for habitat management and species protection.
→ Join a citizen science program monitoring wildlife. Many regional conservation groups have monitoring opportunities and provide training. For national programs, see the Cornell Lab of Ornithology website at www.birds.cornell.edu and the Izaak Walton League website at www.iwla.org.

Policy Change
→ The Endangered Species Act was the first U.S. legislation established to protect species diversity. To learn more about current challenges and updates to the program, visit epa.gov/espp.

ENGAGING GRAPHICS HELP STUDENTS INTERPRET DATA AND UNDERSTAND PROCESSES

INFORMATION + GRAPHIC = INFOGRAPHIC

MAPS APPEAR IN EVERY CHAPTER TO ANCHOR THE STORY TO A GEOGRAPHICAL LOCATION AND PRESENT NEW DATA, HELPING STUDENTS DEVELOP SKILLS IN GEOGRAPHY AND MAP READING

Infographic 11.5 | **TRACKING POACHERS BY USING CONSERVATION GENETICS**

↓ Because elephant populations from different regions of Africa are genetically distinct, scientists can match the DNA from a confiscated tusk to the region in Africa where it originated, helping law enforcement officials track down poachers and identify regions of high poaching activity.

1 DNA microsatellites
Researchers track DNA microsatellites—regions on the DNA molecule where small sequences of DNA subunits called nucleotides (shown here by letter designations) repeat over and over. Microsatellites are not a functional part of a gene and are not used produce a protein.

Microsatellite

2 DNA fingerprinting
Because they are not functional, micro-satellites can accumulate changes without harming the individual. They are passed on to offspring and the microsatellites in different breeding populations of African elephants are distinct enough to serve as a fingerprint of that population.

3 Tracking ivory
Wasser and associates have created a reference map of elephant populations in Africa based on microsatellite analysis. The microsatellite pattern in ivory can be compared to this reference map, identifying poaching "hotspots."

Dung sampling locations of reference populations

Each colored band represents a different microsatellite region—banding patterns differ for different populations.

DNA fingerprint of dung from known location

DNA fingerprint of dung from known location

Compare the banding pattern for the unknown sample to the 2 shown above. Which population did this sample come from?

DNA fingerprint of tusk from unknown location

Infographic 24.3 | **OIL RESERVES**

CANADA (179 bbl)

RUSSIA (60 bbl)

U.S. (22 bbl)

CHINA (18 bbl)

76 Middle East
63 Europe & Eurasia
15 Asia Pacific
15 Africa
9 North America
7 South & Central America

MEXICO (15 bbl)

LIBYA (39 bbl)
NIGERIA (35 bbl)

VENEZUELA (77 bbl)

IRAN
IRAQ
KUWAIT
QATAR
SAUDI ARABIA
UAE
(815 bbl)

15
30
200
800

PROVED GAS RESERVES 2008
(trillion cubic meters)

BIGGEST OIL RESERVES 2005
(billions of barrels (bbl))

↑ Different regions of the world have differing amounts of conventional oil and natural gas reserves. The Middle East has the largest reserves of both oil and natural gas.

DATA AND NUMBERS AREN'T INTIMIDATING

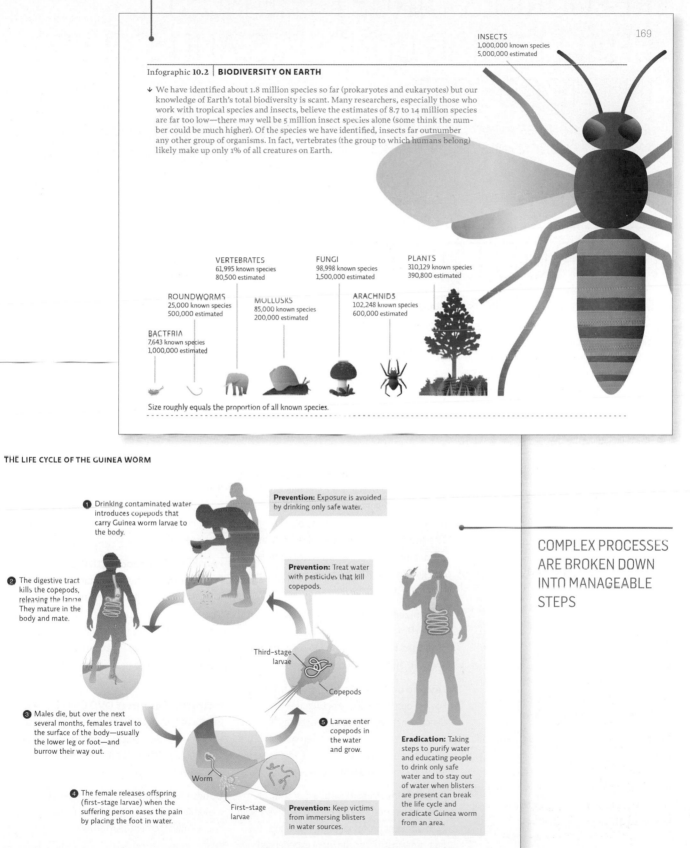

169

Infographic 10.2 | **BIODIVERSITY ON EARTH**

↓ We have identified about 1.8 million species so far (prokaryotes and eukaryotes) but our knowledge of Earth's total biodiversity is scant. Many researchers, especially those who work with tropical species and insects, believe the estimates of 8.7 to 14 million species are far too low—there may well be 5 million insect species alone (some think the number could be much higher). Of the species we have identified, insects far outnumber any other group of organisms. In fact, vertebrates (the group to which humans belong) likely make up only 1% of all creatures on Earth.

INSECTS
1,000,000 known species
5,000,000 estimated

VERTEBRATES
61,995 known species
80,500 estimated

FUNGI
98,998 known species
1,500,000 estimated

PLANTS
310,129 known species
390,800 estimated

ROUNDWORMS
25,000 known species
500,000 estimated

MOLLUSKS
85,000 known species
200,000 estimated

ARACHNIDS
102,248 known species
600,000 estimated

BACTERIA
7,643 known species
1,000,000 estimated

Size roughly equals the proportion of all known species.

THE LIFE CYCLE OF THE GUINEA WORM

❶ Drinking contaminated water introduces copepods that carry Guinea worm larvae to the body.

Prevention: Exposure is avoided by drinking only safe water.

Prevention: Treat water with pesticides that kill copepods.

❷ The digestive tract kills the copepods, releasing the larvae. They mature in the body and mate.

Third-stage larvae

Copepods

❸ Males die, but over the next several months, females travel to the surface of the body—usually the lower leg or foot—and burrow their way out.

❺ Larvae enter copepods in the water and grow.

Worm

First-stage larvae

❹ The female releases offspring (first-stage larvae) when the suffering person eases the pain by placing the foot in water.

Prevention: Keep victims from immersing blisters in water sources.

Eradication: Taking steps to purify water and educating people to drink only safe water and to stay out of water when blisters are present can break the life cycle and eradicate Guinea worm from an area.

COMPLEX PROCESSES ARE BROKEN DOWN INTO MANAGEABLE STEPS

RESOURCES TARGET THE MOST CHALLENGING CONCEPTS AND SKILLS IN THE COURSE

CLASSROOM ACTIVITIES, ANIMATIONS, GRAPHING TUTORIALS, AND ASSESSMENT MATERIALS ARE ALL BUILT AROUND THE CONCEPTS AND SKILLS THAT ARE MOST DIFFICULT FOR STUDENTS TO MASTER

INSTRUCTOR RESOURCES

EnviroPortal An online learning space to help instructors administer their courses by combining our fully customizable e-Book with a robust set of instructor resources. These include news feeds, tutorials, interactive infographics, the Test Bank, and homework management tools.

LearningCurve LearningCurve uses a game-like interface to guide students through a series of questions tailored to their individual level of understanding. A personalized study plan is generated based upon their quiz results. LearningCurve is available to students in the EnviroPortal.

Interactive e-Book A complete online version of the textbook fully integrated with links right where you need them. You can even personalize the e-Book just like you would a printed textbook, with highlighting, bookmarking, and notes.

Story Abstracts The abstracts offer a brief story synopsis, providing interesting details relevant to the chapter and to the online resources not found in the book.

Chapter Topic Overviews Overviews list specific lesson outcomes for each Guiding Question, page numbers in the chapter where each Guiding Question is covered and answered, and a list of the key terms associated with each Guiding Question.

Video Case Studies Assignable videos from an array of trusted sources bring the stories of the book to life. Critical thinking questions allow students to apply their environmental, scientific, and information literacy skills. Also includes PPTs for classroom activities.

Clicker Questions Designed as interactive in-class exercises, these questions reinforce core concepts and uncover misconceptions.

Test Bank/Computerized Test Bank A collection of over a thousand questions, organized by chapter and Guiding Question, presented in a sortable, searchable platform. The Test Bank features multiple-choice and short-answer questions, and uses infographics and graphs from the book. Available through our Portal and on the IR DVD/CTB discs.

Optimized Art (JPEGs and layered PowerPoint slides) Infographics are optimized for projection in large lecture halls and split apart for effective presentation.

Layered or Active PowerPoint Slides PowerPoint slides for select figures deconstruct key concepts, sequences, and processes in a step-by-step format, allowing instructors to present complex ideas in clear, manageable chunks.

PowerPoint Presentations Our suite of lecture presentations for PowerPoint were built with different professor profiles in mind: the first-time instructor, the more experienced teacher, and the professor who wants to take an active learning approach.

• Lecture Outlines for PowerPoint for New Instructors Adjunct professors and instructors who are new to the discipline will appreciate these detailed companion lectures, perfect for walking students through the key ideas in each chapter. These rich prebuilt lectures make it easy for instructors to transition to the book.

• Enhanced Lectures for PowerPoint Using dynamic, layered art taken directly from the text, these lectures are intended for the more experienced professor who feels comfortable with the material.

• Author's Classroom Activities for PowerPoint Developed by author Susan Karr, these classroom activities use proven active-learning techniques to engage students in the material and inspire critical thinking. These lectures are intended for an instructor who is interested in taking an active learning approach to the course.

Instructor Resources DVD Combines a variety of Instructor Resources—Optimized Art files, Lecture Presentations for PowerPoint, Test Bank questions, and more—in one convenient package.

Course Management System e-packs available for Blackboard, WebCT, and other course management platforms.

→ ## ORGANIZED BY GUIDING QUESTIONS

STUDENT AND INSTRUCTOR RESOURCES ARE ARRANGED TO SUPPORT THE LEARNING GOALS OF EACH CHAPTER

→ ## SUPPORT ENVIRONMENTAL, SCIENCE, AND INFORMATION LITERACY

SUPPLEMENTARY MATERIALS EXPLORE TIMELY ENVIRONMENTAL ISSUES, DEVELOP CRUCIAL SCIENCE LITERACY SKILLS SUCH AS DATA ANALYSIS AND GRAPH INTERPRETATION, AND PROVIDE PRACTICE IN EVALUATING SOURCES OF INFORMATION

STUDENT RESOURCES

EnviroPortal An online learning space combining a fully customizable e-Book and student resources. Students can access a variety of study tools, chapter quizzes, flashcards, animated interactive infographics, and a lecture art notebook.

LearningCurve LearningCurve is a set of formative assessment activities that uses a game-like interface to guide students through a series of questions tailored to their individual level of understanding. A personalized study plan is generated based upon their quiz results. LearningCurve is available to students in the EnviroPortal.

Online Study Guide Organized by the chapter's Guiding Questions, the study guide elaborates on each infographic in the book and provides questions to promote the students' critical thinking.

Tutorials Our tutorials give students the opportunity to explore how science is done by analyzing data and drawing conclusions. Students build critical thinking skills with a variety of media tools, including:

- **Evaluating Sources of Information** These tutorials encourage students to examine real-world environmental issues and think critically about the opposing sides.

- **Graphing** Graphing tutorials let students build and analyze graphs, using their critical thinking skills to predict trends, identify bias, and make cause-and-effect connections.

Video Case Studies Videos from an array of trusted sources bring the stories of the book to life and allow students to apply their environmental, scientific, and information literacy skills. Each video includes questions that engage students in the critical thinking process.

Interactive Infographics/Animations All infographics in the text include an animated interactive tutorial or an infographic activity.

Interactive e-Book A complete online version of the textbook and all of its media resources, fully integrated with links right where you need them. You can even personalize the e-Book just like you would a printed textbook, with highlighting, bookmarking, and notes.

Key Term Flashcards Students can drill and learn the most important terms in each chapter using interactive flashcards.

Lecture Art Notebook The infographics for each chapter are available as PDF files that students can download and print before lectures.

Quizzing with Feedback Student response-specific feedback helps explain concepts and correct student misunderstandings.

Free Book Companion Website Features most student resources in an online format.

WHAT IS **BIOMIMICRY**?

Many of the environmental problems that humans face—from needing clean water to searching for a more efficient biofuel—have already been solved by nature. In *Environmental Science for a Changing World with Extended Coverage,* we feature many solutions to environmental problems that use nature as a model. Finding solutions that mimic nature (biomimicry) shows that often the best answers are right under our noses! Below, check out a few examples of biomimicry...

CHAPTER 4

In **Chapter 4,** we profile the eco-friendly carpet company Interface Carpet. The company created a new carpet product called TacTiles, **whose design is based on the physics that explains how a gecko lizard clings to walls and ceilings**. The microscopic hairs on a gecko's foot bond to the molecular layer of water that's present on nearly every surface, allowing its feet to cling. Interface used this as inspiration to develop tiles that bond to one another rather than the floor—eliminating the need for toxic glues. (See page 58.)

CHAPTER 14

Though the majority of the world's grasslands are rapidly drying up into desert, some brave ranchers are trying a different solution for grazing their animals. In **Chapter 14** we meet Allan Savory, a Rhodesian-born scientist turned rancher. Savory proposes that, instead of grazing livestock on the same patch of prairie in too many or too few numbers, thereby decimating the vegetation and soil, **we mimic the grazing patterns that nature has already devised:** wild herds graze in tightly packed bunches that move from one patch to another, fertilizing the soil with their droppings as they go. (See page 240.)

CHAPTER 32

City living can be both an environmental blessing and a curse; tall buildings concentrate a lot of people over a much smaller land area than in the suburbs, but more building means more hard surfaces and less vegetation and soil. As a result, polluted waters can run quickly over hard rooftops and streets, flooding waterways and wrecking water nutrient levels. In **Chapter 32**, residents of the Bronx are improving water runoff by installing green roofs and rain gardens that **take advantage of the natural filtering and absorbing abilities of vegetation and soil.** (See page 578.)

...and many more!

ACKNOWLEDGEMENTS

From Susan Karr...

Working on this book has been a journey unlike any I've been on in my professional career. I've discovered that an undertaking this big is truly a collaborative effort. It is amazing what you can accomplish when you work with talented and highly skilled people. I want to thank W. H. Freeman acquisitions editor Jerry Correa for his vision for the book and for his personal support; and developmental editors Andrea Gawrylewski and Beth Howe for their insights and outstanding editorial skills in crafting each chapter. I also want to thank Jeneen Interlandi, the accomplished and gifted writer who has made these chapters such a pleasure to read; and the fabulously detail-oriented project managers, copyeditors, and proofreaders, Karen Misler, Anna Paganelli, and Jane Taylor who helped polish up the final work. Outstanding writing was also done by Alison McCook and Melinda Wenner Moyer. Kelly Cartwright, Shamili Sandiford, JodyLee Estrada Duek, Michelle Cawthorn, and James Dauray made excellent contributions to the Bring it Home feature and end-of-chapter materials. Thanks also go to the entire team at MGMT. design for their skills, vision, and patience, and to Ellen Cash who expertly coordinated the production of the book. Shannon Howard, Donna Brodman, Lindsey Veautour, and Debbie Clare worked tirelessly to involve hundreds of environmental science instructors and spread the word about this exciting text.

I've had tremendous support from my biology department colleagues at Carson-Newman College and from focus group participants, who offered advice, answered questions, and helped track down elusive information—thanks for sharing your expertise. In particular, Teri Balser, Shamili Sandiford, and Lissa Leege helped develop a roadmap for the book with their insights from years of teaching environmental science. I also owe a debt of gratitude to my environmental science students over the years for their questions, interests, demands, and passion for learning that have always challenged and inspired me. Finally, I want to thank my husband, Steve, for supporting me in so many different ways it is impossible to count them all.

From Jeneen Interlandi...

Each of these chapters is a story—of scientists and everyday people, often doing extraordinary things. It has been my great pleasure to tell those stories here. For that, I thank each and every one of my sources. Their time and patience is what made this book possible.

I would also like to thank Susan, with whom it has been an honor to work, and the entire W. H. Freeman team for their tireless efforts.

REVIEWERS

We would like to extend our deep appreciation to the following instructors who reviewed, tested, and advised on the book manuscript at various stages of development.

CHAPTER REVIEWERS

Matthew Abbott, *Des Moines Area Community College*

David Aborn, *University of Tennessee at Chattanooga*

Michael Adams, *Pasco–Hernando Community College*

Shamim Ahsan, *Metropolitan State College of Denver*

Keith Allen, *Bluegrass Community and Technical College*

John Anderson, *Georgia Perimeter College*

Clay Arango, *Central Washington University*

Walter Arenstein, *San Jose University*

Deniz Ballero, *Georgia Perimeter College*

Marcin Baranowski, *Passaic County Community College*

Brad Basehore, *Harrisburg Area Community College*

Damon Bassett, *Missouri State University*

Sean Beckmann, *Rockford College*

Stacy Bennetts, *Augusta State University*

David Berg, *Miami University*

Joe Beuchel, *Triton College*

Aaron Binns, *Florida State University*

Karen Blair, *Pennsylvania State University*

Barbara Blonder, *Flagler College*

Steve Blumenshine, *California State University, Fresno*

Ralph Bonati, *Pima Community College*

Polly Bouker, *Georgia Perimeter College*

Richard Bowden, *Allegheny College*

Anne Bower, *Philadelphia University*

Jennifer Boyd, *University of Tennessee at Chattanooga*

Scott Brame, *Clemson University*

Allison Breedveld, *Shasta College*

Mary Brown, *Lansing Community College*

Robert Bruck, *North Carolina State University*

Brett Burkett, *Collin College*

Alan Cady, *Miami University*

Elena Cainas, *Broward College*

Deborah Carr, *Texas Tech University*

Mary Kay Cassani, *Florida Gulf Coast University*

Niccole Cerveny, *Mesa Community College*

Karen Champ, *College of Central Florida*

Lu Anne Clark, *Lansing Community College*

Reggie Cobb, *Nash Community College*

Richard Clements, *Chattanooga State University*

Jennifer Cole, *Northeastern University*

Eric Compas, *University of Wisconsin, Whitewater*

Jason Crean, *Saint Xavier University & Moraine Valley Community College*

Michael Dann, *Pennsylvania State University*

James Dauray, *College of Lake County*

Robert Dennison, *Heartland Community College*

Michael Denniston, *Georgia Perimeter College*

Robert Dill, *Bergen Community College*

Craig Dilley, *Des Moines Area Community College*

JodyLee Estrada Duek, *Pima Community College*

Don Duke, *Florida Gulf Coast University*

James Dunn, *Grand Valley State University*

John Dunning, *Purdue University*

Kathy Evans, *Reading Area Community College*

Brad Fiero, *Pima Community College*

Linda Fitzhugh, *Gulf Coast Community College*

Steven Forman, *University of Illinois at Chicago*

Paul Gier, *Huntingdon College*

Michael Golden, *Grossmont College*

Sherri Graves, *Sacramento City College*

Michelle Groves, *Oakton Community College*

Myra Carmen Hall, *Georgia Perimeter College*

Sally Harms, *Wayne State College*

Cornelia Harris, *Marist College*

Stephanie Hart, *Lansing Community College*

Wendy Hartman, *Palm Beach State College*

Alan Harvey, *Georgia Southern University*

Keith Hench, *Kirkwood Community College*

Robert Hollister, *Grand Valley State University*

Tara Holmberg, *Northwestern Connecticut Community College*

Jodee Hunt, *Grand Valley State University*

Meshagae Hunte-Brown, *Drexel University*

Kristin Jacobson, *Illinois Central College*

Jason Janke, *Metropolitan State College of Denver*

David Jeffrey, *Georgia Perimeter College*

Thomas Jurik, *Iowa State University*

Charles Kaminski, *Middlesex Community College*

Michael Kaplan, *College of Lake County*

Richard Keen, *University of Colorado, Boulder*

John Keller, *College of Southern Nevada*

Myung-Hoon Kim, *Georgia Perimeter College*

Elroy Klaviter, *Lansing Community College*

Paul Klerks, *University of Louisiana at Lafayette*

Janet Kotash, *Moraine Valley Community College*

Jean Kowal, *University of Wisconsin, Whitewater*

John Krolak, *Georgia Perimeter College*

James Kubicki, *Pennsylvania State University*

Diane LaCole, *Georgia Perimeter College*

Katherine LaCommare, *Lansing Community College*

Andrew Lapinski, *Reading Area Community College*

Jennifer Latimer, *Indiana State University*

Stephen Lewis, *California State University, Fresno*

Robert Loney, *Trent University*

Eric Lovely, *Arkansas Tech University*

Kristine Lowe, *University of Texas, Pan American*

Marvin Lowery, *Lone Star College System*

Steve Luzkow, *Lansing Community College*

Steve Mackie, *Pima Community College*

Nilo Marin, *Broward College*

Eric Maurer, *University of Cincinnati*

Costa Mazidji, *Collin College*

DeWayne McAllister, *Johnson County Community College*

Vicki Medland, *University of Wisconsin, Green Bay*

Alberto Mestas-Nunez, *Texas A&M University, Corpus Christi*

Chris Migliaccio, *Miami Dade College*

Jessica Miles, *Palm Beach State College*

Dale Miller, *University of Colorado, Boulder*

Kiran Misra, *Edinboro University of Pennsylvania*

Scott Mittman, *Essex County College*

Edward Mondor, *Georgia Southern University*

Brian Mooney, *Johnson & Wales University*

Zia Nisani, *Antelope Valley College*

Ken Nolte, *Shasta College*

Kathleen Nuckolls, *University of Kansas*

Segun Ogunjemiyo, *California State University*

Bruce Olszewski, *San José State University*

Jeff Onsted, *Florida International University*

Nancy Ostiguy, *Pennsylvania State University*

Daniel Pavuk, *Bowling Green State University*

Barry Perlmutter, *College of Southern Nevada*

Craig Phelps, *Rutgers, The State University of New Jersey*

Neal Phillip, *Bronx Community College*

David Polcyn, *California State University, San Bernardino*

Keith Putirka, *California State University, Fresno*

Jodie Ramsay, *Northern State University*

Bob Remedi, *College of Lake County*

Erin Rempala, *San Diego City College*

Angel Rodriquez, *Broward College*

Deanne Roquet, *Lake Superior College*

Dennis Ruez, *University of Illinois at Springfield*

Robert Ruliffson, *Minneapolis Community and Technical College*

Christopher Sacchi, *Kutztown University of Pennsylvania*

Melanie Sadeghpour, *Des Moines Area Community College*

Seema Sah, *Florida International University*

Jay Sah, *Florida International University*

Shamili Sandiford, *College of DuPage*

Waweise Schmidt, *Palm Beach State College*

ON THE ROAD TO COLLAPSE

What lessons can we learn from a vanished Viking society?

The remains of Hvalsey, a
Viking settlement church,
in southern Greenland.

CORE MESSAGE

Humans are a part of the natural world and are dependent on a healthy, functioning planet. We put pressure on the planet in a variety of ways, but our choices can help us move toward sustainability.

GUIDING QUESTIONS

After reading this chapter, you should be able to answer the following questions:

→ What constitutes the "environment" and what fields of study collaborate under the umbrella of environmental science?

→ What are some of the environmental dilemmas that humans face and why are many of these considered "wicked problems"?

→ What challenges does humanity face in dealing with environmental issues and how can environmental literacy help us make more informed decisions?

→ What does it mean to be sustainable and what are the characteristics of a sustainable ecosystem?

→ What can human societies and individuals do to develop sustainably?

Although not much of a tourist destination, Greenland offers some spectacular sights—colossal ice sheets, a lively seascape, rare and precious wildlife (whales, seals, polar bears, eagles). But on his umpteenth trip to the island, Thomas McGovern was not interested in any of that. What he wanted to see was the garbage—specifically, the ancient, fossilized garbage that Viking settlers had left behind some seven centuries ago.

McGovern, an archaeologist at the City University of New York, had been on countless expeditions to Greenland over the past 40 years. Digging through layers of peat and permafrost, he and his team had unearthed a museum's worth of artifacts that, when pieced together, told the story of the Greenland Vikings. But as thorough as their expeditions had been, that story was still maddeningly incomplete.

Here's what they knew so far: A thousand or so years ago, an infamous Viking by the name of Erik the Red led a small group of followers across the ocean from Norway, to a vast expanse of snow and ice that he had dubbed Greenland. Most of Greenland was not green. In fact, it was a forbidding place marked by harsh winds and sparse vegetation. But tucked between two fjords along the southwestern coast, protected from the elements by jagged, imposing cliffs, the Vikings found a string of verdant meadows, brimming with wildflowers. They quickly set up camp here, and proceeded to build a society similar to the one they had left behind in Norway. They farmed, hunted, and raised livestock. They also built barns and churches as elaborate as the ones back home. They established an economy and a legal system, traded goods with mainland Europe, and at their peak, reached a population of 5,000 (a large number in those days).

And then, after 450 years of prosperity, they disappeared—seemingly into thin air—leaving little more than the beautiful, tragic ruins of a handful of barns and churches in their wake.

The how and why of this vanishing act remained a tantalizing mystery, one that has drawn hundreds of scientists—McGovern among them—to Greenland each summer. Recently, some of McGovern's colleagues had begun to suspect that disturbances in the natural environment—a cooling climate, loss of soil, problems with the food supply—may have been the deciding factors.

While other researchers probed ice sheets and soil deposits in search of clues, McGovern stuck to the garbage heaps, or *middens*, as Vikings called them. Every farmstead had one, and every generation of the farmstead's owners threw their waste into it. The result was an archeological treasure trove: fine-grain details about what people ate, how they dressed, and the kinds of objects they filled their homes with. It gave McGovern and his team a clear picture of how they lived.

If they dug deep enough, McGovern thought, it might also explain how they died.

WHERE IS THE VIKING SETTLEMENT IN GREENLAND?

GREENLAND

NORSAQ

CANADIAN TERRITORY

ICELAND

Environmental science is all encompassing.

From a modern developed society like the United States, it can be difficult to imagine a time and place when the natural world held such sway over our fate. Our food comes from a grocery store, our water from a tap; even our air is artificially heated and cooled to our liking. These days, it seems more logical to consider societal conflict, or

environment The biological and physical surroundings in which any given living organism exists.
environmental science An interdisciplinary field of research that draws on the natural and social sciences and the humanities in order to understand the natural world and our relationship to it.

Infographic **1.1** | **ENVIRONMENTAL SCIENCE IS HIGHLY INTERDISCIPLINARY**

ENVIRONMENTAL SCIENCE CAN HELP US ANSWER QUESTIONS LIKE "WHY DID THE VIKINGS DISAPPEAR FROM THIS REGION IN GREENLAND?"
NATURAL SCIENCES
• What is the climate like?
• Which plants and animals live here?
• Which crops or animals can be raised here?
• How can soil erosion be prevented?
• What energy sources are available and how do they impact the environment?
SOCIAL SCIENCES
• How have indigenous people lived here?
• What environmental policies would best fit this culture and place?
• Will residents accept changes to their lifestyle that might benefit the environment?
• Which energy sources are most cost effective?
HUMANITIES
• How can people express their love, fears, and hopes for their homeland (literature, theater, music)?
• How do religion and tradition influence choices?

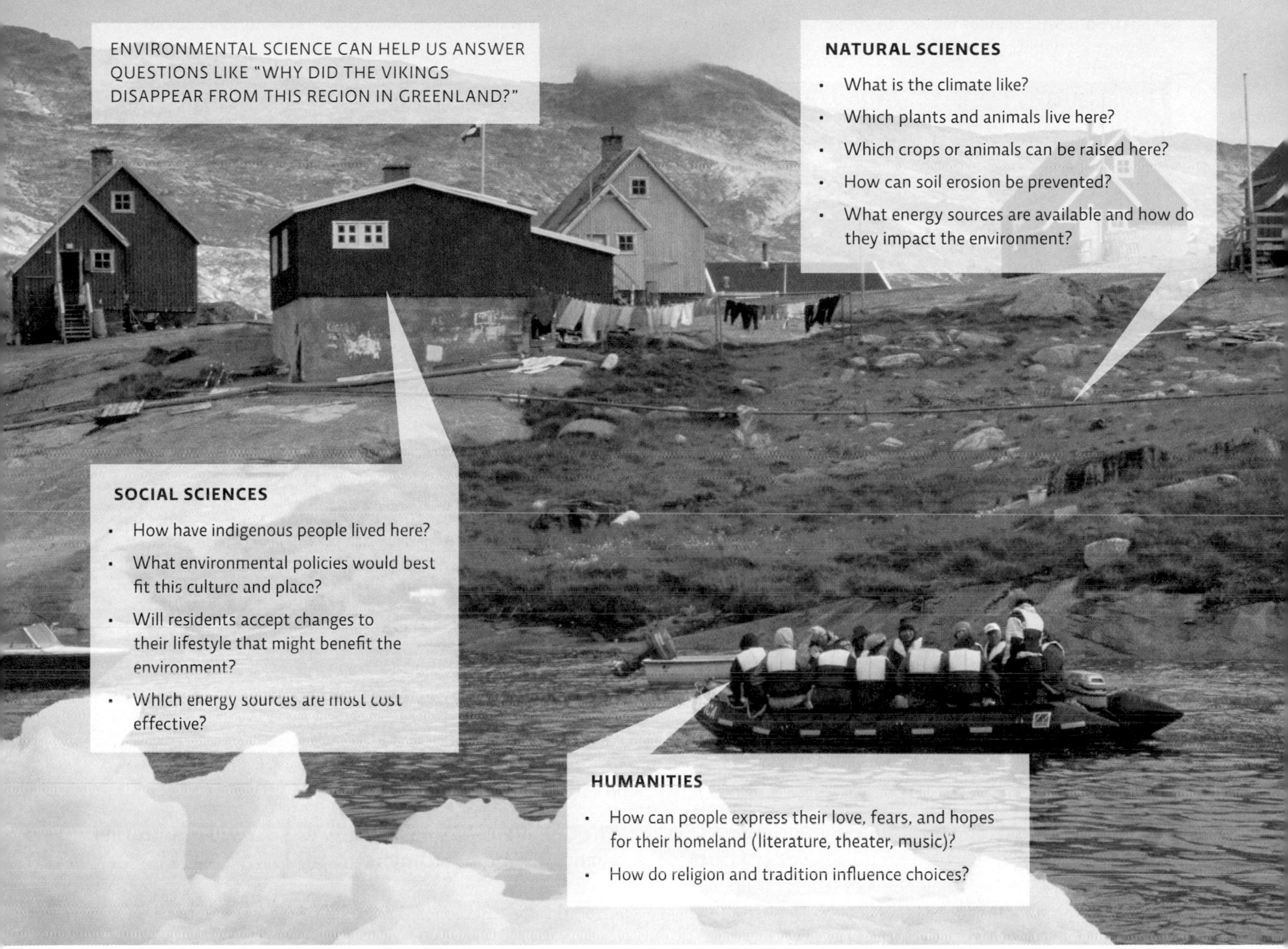

↑ Environmental science studies the natural world and how humans interact with and impact it. We must look to the natural and social sciences as well as to the humanities to help us understand our world and effectively address environmental issues and environmental questions such as, "Why did the Vikings disappear from this region in Greenland and how do humans live now in such a harsh environment?"

even collapse, through the lens of politics or economics. But, as we will see time and again throughout this book, the natural environment—and how we interact with it—plays a leading role in the sagas that shape human history; this is as true today as it was in the time of the Vikings.

Environment is a broad term that describes the surroundings or conditions (including living and nonliving components) in which any given organism exists. **Environmental science**—a field of research that is used to understand the natural world and our relationship to it—is extremely interdisciplinary. It relies on a range of natural and applied sciences (such as ecology, geology, chemistry, and engineering) to unlock the mysteries of the natural world, and to look at the role and impact of humans in the world. It also draws on social sciences (such as anthropology, psychology, and economics) and the humanities (such as art, literature, and music) to understand the ways that humans interact with, and thus impact, the ecosystems around them. [INFOGRAPHIC 1.1]

Infographic **1.2** | **DIFFERENT APPROACHES TO SCIENCE HAVE DIFFERENT GOALS AND OUTCOMES**

↓ Environmental science is used to systematically collect and analyze data to draw conclusions and use these conclusions to propose reasonable courses of action.

EMPIRICAL SCIENCE IS USED TO INVESTIGATE THE NATURAL WORLD

↑ Through observation, glaciologists study and record the rate of glacier melt in Greenland; it is increasing dramatically in some places.

IN APPLIED SCIENCE, KNOWLEDGE IS USED TO ADDRESS PROBLEMS OR NEEDS

↑ Engineers harness the power of water by diverting extra glacial meltwater to produce hydroelectric power. This power can be converted to hydrogen fuel and can provide energy to remote areas.

Environmental science is an **empirical science**: It scientifically investigates the natural world through systematic observation and experimentation. It is also an **applied science**: We use its findings to inform our actions and, in the best cases, to bring about positive change. [INFOGRAPHIC 1.2]

The ability to understand environmental problems is referred to as **environmental literacy**. Such literacy is crucial to helping us become better stewards of Earth. Environmental problems can be extremely complicated and tend to have multiple causes, each one difficult to address. We must also understand that because of their complexity, any given response to an environmental problem involves significant **trade-offs** and no one response is likely to present the ultimate solution. Scientists refer to such problems as "wicked problems." In confronting them, we must consider not only their environmental but also their economic and social causes and consequences. Scientists refer to this trifecta as the **triple bottom line**. [INFOGRAPHIC 1.3]

In his book *Collapse*, University of California at Los Angeles biologist Jared Diamond details how wicked problems can lead to a society's ultimate demise. He identifies five factors in particular that determine whether any given society will succeed or fail: natural climate change; failure to properly respond to environmental changes; self-inflicted environmental damage; hostile neighbors; and loss of friendly neighbors. According

to Diamond, the relative impact of each factor varies by society.

The situation of the Greenland Vikings was a rare case. It turns out that, like a perfect storm, all five of these factors conspired together.

The Greenland Vikings' demise was caused by natural events and human choices.

Greenland's interior is covered by vast ice sheets that stretch toward the horizon—3,000 meters (10,000 feet) thick and more than 250,000 years old. To residents of the hard land, these ice sheets are not good for much— they create harsh winds and brutal cold—but to climate scientists, they're a treasure trove. As snow falls, it absorbs various particles from the atmosphere and lands on the ice sheets. As time passes, the snow and particles

empirical science A scientific approach that investigates the natural world through systematic observation and experimentation.
applied science Research whose findings are used to help solve practical problems.
environmental literacy A basic understanding of how ecosystems function and of the impact of our choices on the environment.
trade-offs The imperfect and sometimes problematic responses that we must at times choose between when addressing complex problems.
triple bottom line The combination of the environmental, social, and economic impacts of our choices.

Infographic **1.3** | **WICKED PROBLEMS**

↓ Wicked problems are difficult to address because, in many cases, each stakeholder hopes for a different solution. Solutions that address wicked problems usually involve trade-offs, so there is no clear "winner." One example of a wicked problem is climate change. There are many causes of the current climate change we are experiencing, both natural and anthropogenic (caused by human actions) and the effects of climate change will be varied for different species and people depending on where they live and their ability to adapt to the changes.

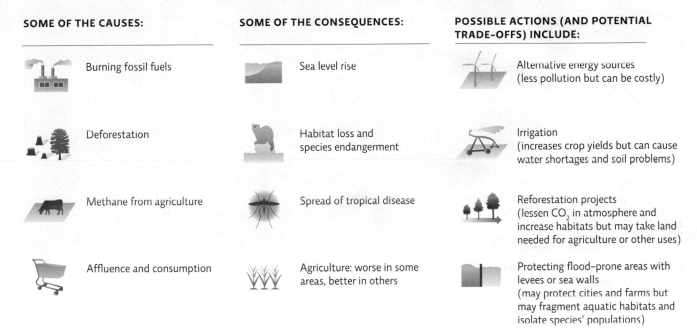

SOME OF THE CAUSES:

Burning fossil fuels

Deforestation

Methane from agriculture

Affluence and consumption

SOME OF THE CONSEQUENCES:

Sea level rise

Habitat loss and species endangerment

Spread of tropical disease

Agriculture: worse in some areas, better in others

POSSIBLE ACTIONS (AND POTENTIAL TRADE-OFFS) INCLUDE:

Alternative energy sources (less pollution but can be costly)

Irrigation (increases crop yields but can cause water shortages and soil problems)

Reforestation projects (lessen CO_2 in atmosphere and increase habitats but may take land needed for agriculture or other uses)

Protecting flood-prone areas with levees or sea walls (may protect cities and farms but may fragment aquatic habitats and isolate species' populations)

---

compact into ice, freezing in time perfect samples of the atmosphere as it existed when that snow first fell. By analyzing those ice-trapped particles—dust, gases, chemicals, even the water molecules themselves—scientists can get a pretty good idea of what was happening to the climate at any given time. "It's like perfectly preserved slices of atmosphere from the past," says Lisa Barlow, a geologist and climate researcher at the University of Colorado at Boulder. "It gives us additional clues as to what was going on."

To uncover those clues, a team of scientists and engineers picked an accessible segment of ice sheet, not far from the Viking settlements, drilled from the surface all the way down to the bedrock below, and extracted a 12-centimeter-wide (5-inch-wide), 3,000-meter-long cylinder of ice, which they then divided and dispersed among a handful of labs across the globe, including Jim White's light stable isotope lab—also at the University of Colorado at Boulder. Analysis of thin sequential segments of ice showed that when the Vikings first arrived in Greenland the temperature was anomalously higher than average for the last

1,000 years—atmospheric temperature can be deduced from the amount of oxygen-18 or Deuterium (heavy Hydrogen) present in the sample. By the time the Vikings had vanished, temperatures had lowered so much that scientists qualified the period as a mini ice-age, overlain on seasonal changes. "It's no wonder they didn't make it," says Barlow. "With lower temperatures, livestock would have starved for lack of hay over the long winter, and self-inflicted environmental damage made the situation worse."

Indeed, in addition to natural climate change (Diamond's first factor), the Vikings also suffered from self-inflicted environmental damage (Diamond's second factor).

In addition to that single ice core, scientists have analyzed hundreds of mud cores taken from lake beds around the Viking settlements. These mud samples—which contain large amounts of soil that was blown into the lakes during Viking times—indicate that soil erosion had become a significant problem long before the region descended into a mini ice age. "This wasn't a climate problem," says Bent

↑ Scientist from the National Snow and Ice Data Center working with an ice core drill.

Fredskild, a Danish scientist who extracted and studied many of the mud cores. "This was self-inflicted. It happened the same way that soil erosion happens today— they overgrazed the land, and once it was denuded, there was nothing to anchor the soil in place. So the wind carried it away."

Overgrazing wasn't their only mistake. The Greenland Vikings also used grassland to insulate their houses against the cold of winter; typical insulation consisted of 6-foot-thick slabs of turf, and a typical home took about 10 acres of grassland to insulate. On top of that, they chopped down the forests, harvesting enough timber to not only provide fuel and build houses but also to make the innumerable wooden objects to which they had become accustomed back in Norway.

Greenland's ecosystem was far too fragile to endure such pressure, especially as the settlement grew from a few hundred to a few thousand. The short, cool growing season meant that plants developed slowly, which in turn

meant that the land could not recover quickly enough from the various assaults to protect the soil.

As climate cooling and overharvesting conspired to destroy pasturelands, summer hay yields shrank. When scientists counted the fossilized remains of insects that lived in the fields and haylofts of Viking Greenland, they found that their numbers fell dramatically in the settlement's final years. "The falloff in insects tells us that hay production dwindled to the point of crisis," says Fredskild. Without hay, livestock could not survive the ever-colder, ever-longer winters. And without livestock, the Vikings themselves went hungry. As scientists soon discovered, they needn't have.

Responding to environmental problems and working with neighbors help a society cope with changes.

Back in his Manhattan lab, McGovern sorts through hundreds of animal bones collected from various Greenland middens. By examining the bones and making careful note of which layers they were retrieved from, McGovern can tell what the people ate and how their diets changed over time. "This is a pretty typical set of remains for these people from this region and time period," he says,

sustainable development Development that meets present needs without compromising the ability of future generations to do the same.

leaning over a shiny metal tray of neatly arranged bone fragments. Some are the bones of cattle imported from Europe. Others are the remains of sheep and goats; still others of local wildlife such as caribou. Conspicuously absent, McGovern says, are fish of any kind.

"If we look at a comparable pile of bones from [Norwegian settlers of] the same time period, from Iceland, we see something very different," McGovern explains. "We have fish bones and bird bones and little fragments of whale bones. Most of it, in fact, is fish—including a lot of cod." It turns out that while the Greenland Vikings were guilty of Diamond's third factor, failure to respond to the natural environment, their Icelandic cousins were not.

Like their cousins in Greenland, the Vikings who settled Iceland at about the same time were initially fooled into thinking that their newly discovered land could sustain their cow-farming, wood-dependent ways: Plants and animals looked similar to those back in Norway, and grasslands seemed lush and abundant. They cleared about 80% of Iceland's forests and allowed their cows, sheep, and goats to chew the region's grasslands down to nothing before finally noticing how profound the differences between Iceland and Norway actually were: Growing seasons in Iceland were shorter, both soil and vegetation were much more fragile, and because the land could not rebound quickly, cow farming was unsustainable.

But once they saw that their old ways would not work in this new country, the Icelandic Vikings made changes. Not only did they switch from beef to fish, they also began conserving their wood and abandoned the highlands, where soil was especially fragile. And, as a result of these and other adaptations, they survived and prospered. The Icelanders responded to the limitations of their natural environment in a way that allowed them to meet present needs without compromising the ability of future generations to do the same—an approach known today as **sustainable development**.

Some of the most telling clues to the mystery of the Greenland Viking's demise come not from the Viking colonies but from another group of people who lived nearby: the Inuit. The Inuit are an indigenous tribe who arrived in the Arctic centuries before the Vikings. They were expert hunters of ringed seal—an exceedingly difficult-to-catch but very abundant food source. They knew how to heat and light their homes with seal blubber (instead of firewood). And they loved to fish.

Fishing is not nearly as labor-intensive as raising cattle, and in the lakes and fjords of Greenland, it provides an easy, reliable source of protein. A comparison of Inuit and Viking middens shows that even as the Greenland Vikings were scraping off every last bit of meat and marrow from their cattle bones, the Inuit had more food than they could eat.

The Vikings might have learned from their indigenous neighbors; by adapting some of their customs, they might have survived the Little Ice Age and gone on to prosper as the Icelandic Vikings did. But excavations show that virtually no Inuit artifacts made their way into Viking settlements. And according to written records, the Norse detested the Inuit who, on at least one occasion, attacked the Greenland colony; they called the Inuit *skraelings*, which is Norse for "wretches," considered them inferior, and refused to seek their friendship or their counsel, In addition to these hostile neighbors (Diamond's fourth factor), the Greenland Vikings also suffered a loss of friendly neighbors (Diamond's fifth factor).

As the productivity of the Viking colonies declined, so did visits from European ships. As time wore on, it became apparent that the Greenland Vikings could expect very little in the way of trade; royal and private ships that had visited every year came less and less often. After a while, they did not come at all. For the Greenland Vikings, who depended on the Europeans for iron, timber, and other essential supplies, this loss proved devastating. Among other things, it meant that, as the weather grew colder, and food supplies dwindled, they had no one to turn to for help.

Humans are an environmental force that impacts Earth's ecosystems.

When it comes to the environment, modern societies are not as different from the Vikings as one might assume. Vikings chose livestock and farming methods that were ill-suited to Greenland's climate and natural environment. We too use farming practices that strip away topsoil and diminish the land's fertility. We have overharvested our forests, and in so doing have triggered a cascade of environmental consequences: loss of vital habitat and biodiversity, soil erosion, and water pollution. We have overfished and overhunted and have allowed invasive species to devastate some of our most valuable ecosystems.

In part, these problems stem from a disconnect between our actions and their environmental consequences. For example, unless they live close by, many people in the United States don't realize that entire mountains are being leveled to produce their electricity, destroying thousands of acres of habitat and miles of streams and rivers to access coal seams deep beneath the surface of West Virginia and Wyoming. We are slow to make the connection between the burning of that coal and mercury-contaminated fish or increased asthma rates.

> In part, these problems stem from a disconnect between our actions and their environmental consequences.

We also face a suite of new problems that did not trouble the Vikings. Chief among them is population growth; as we will discuss in subsequent chapters, global population is poised to top 9 billion come 2050. The sheer volume of people will strain Earth's resources like never before. This is relevant because every environment has a **carrying capacity**—the population size that an area can support indefinitely—but some of our actions are decreasing carrying capacity, even as our population swells (see Chapter 5 for more on human populations and carrying capacity). In addition, we generate more pollution than the Vikings did, and much of what we generate is more toxic.

Environmental scientists evaluate the impact any population has on its environment—due to the resources it takes and the waste it produces—by calculating its **ecological footprint** (see Chapter 4). Some analysts feel we have already surpassed the carrying capacity of Earth; our collective footprint already surpasses what Earth can support over a long period of time.

Another serious consequence of larger populations, increasing affluence, and more sophisticated technology is **anthropogenic** climate change. While the Vikings had to contend with periodic warming and cooling periods that were part of the natural climate cycle, the vast majority of scientists today conclude that modern humans are faced with rapidly warming temperatures caused largely by our own use of greenhouse gas-emitting fossil fuels.

Scientists have coined the term "ecocide"—willful destruction of the natural environment—to describe this constellation of forces, which, as Diamond writes, have "come to overshadow nuclear war and emerging diseases as the biggest threat to global civilization."

We have something else in common with the Vikings of Greenland: Our attitudes frequently prevent us from responding effectively to environmental changes. According to Diamond, the Vikings were stymied by their own sense of superiority. They considered themselves masters of their surroundings and so they did not notice signs that they were causing irreparable harm to their environment, nor did they bother to learn the ways of their Inuit neighbors who had managed to survive in the same region for centuries before their arrival.

MANY ENVIRONMENTAL PROBLEMS CAN BE TRACED TO THREE UNDERLYING CAUSES

↑ Population Size: There are more than 7 billion people on the planet. Though some have more impact than others, our sheer numbers contribute to many environmental problems.

↑ Resource Use: We use resources faster than they are replaced, and convert matter into forms that don't readily decompose.

↑ Pollution: We generate pollution that compromises our own health and that of ecosystems.

Infographic **1.4** | **OUR ATTITUDES AFFECT HOW WE RESPOND TO PROBLEMS**

↓ Attitudes affect how we respond to problems; some attitudes cause problems because they prevent us from responding to problems.

CLIMATE CHANGE
Changing global temperatures

TECHNOLOGICAL FIX
Human ingenuity & science will be able to solve problems we run into (someone else will fix it).

EVALUATE AND RESPOND
Move forward based on the best available evidence (we have options).

GLOOM AND DOOM
There is nothing we can do (give up).

ROSY OPTIMISM
Don't worry—it will all work out (ignore the problem).

FRONTIER
Resources are here for us to use and we'll find others when needed (nature will provide the solutions).

Our own attitudes can also help or hurt our ability to respond to changes. These attitudes tend to fall into one of a few groups, and range from pessimistic to more optimistic to pragmatic. [INFOGRAPHIC 1.4]

The United Nations' Millennium Ecosystem Assessment looks at how environmental problems affect humans and makes recommendations about addressing those problems. According to its 2005 assessment, a consensus report of more than 1,500 scientists, human actions are so straining the environment that the ability of the planet's ecosystems to sustain future generations is gravely imperiled. But there is hope: If we act now, the report's authors write, we can still reverse much of the damage. Some of the best lessons about how we can do this come from the natural environment itself.

Human societies can become more sustainable.

Unlike their counterparts in Greenland, the Icelandic Vikings responded to their environment and adopted more **sustainable** practices. They didn't have to look far for a model of sustainability: natural ecosystems are sustainable. This means they use resources—namely, energy and matter—in a way that ensures that those resources continue to be available.

To survive, organisms need a constant, dependable source of energy. But as energy passes from one part of an ecosystem to the next, the usable amount declines; therefore, new inputs of energy are always needed. A sustainable ecosystem is one that makes the most of **renewable**

energy—energy that comes from an infinitely available or easily replenished source. For almost all natural ecosystems, that energy source is the sun. Photosynthetic organisms such as plants trap solar energy and convert it to a form that they can readily use, or that can be passed up the food chain to other organisms.

Unlike energy, matter (anything that has mass and takes up space) can be recycled and reused indefinitely; the key is not using it faster than it is recycled. Naturally sustainable ecosystems waste nothing; they recycle matter so that the waste from one organism ultimately becomes a resource for another. They also keep populations in check so that the resources are not overused and there is enough food, water, and shelter for all.

Lastly, sustainable ecosystems depend on local **biodiversity** (the variety of species present) to perform many of the jobs just mentioned; different species have different ways of trapping and using energy and matter, the net result of which boosts productivity and efficiency (see Chapter 9). And predators, parasites, and competitors serve as natural population checks.

carrying capacity The population size that a particular environment can support indefinitely.
ecological footprint The land needed to provide the resources and assimilate the waste of a person or population.
anthropogenic Caused by or related to human action.
sustainable A method of using resources in such a way that we can continue to use them indefinitely.
renewable energy Energy that comes from an infinitely available or easily replenished source.
biodiversity The variety of species on Earth.

Thus, natural ecosystems live within their means and each organism contributes to the ecosytem's overall function. This is not to imply that natural ecosystems are perfect places of total harmony, but those that are sustainable meet all four of these characteristics.

Human ecosystems are another story. Humans tend to rely on **nonrenewable resources**—those whose supply is finite or is not replenished in a timely fashion. The most obvious example of this is our reliance on fossil fuels such as coal, natural gas, and petroleum, culled from deep within Earth, to power our society. Fossil fuels are replenished only over vast geologic time—far too slowly to keep pace with our rampant consumption of them. On top of that, we have a hard time keeping our population under control, despite (or maybe because of) all the advances of modern technology. We also generate volumes of waste, much of it toxic, and have yet to fully master the art of recycling.

Increasingly, however, we humans are looking to nature to help us learn how to change our ways. In Janine Benyus' book, *Biomimicry* (the term used to describe the strategy of mimicking nature), Benyus explains that biomimicry involves using nature as a *model, mentor,* and *measure* for our own systems. Emulating nature (nature as model) gives us an example of *what* to do; it can also teach us how

to do it (nature as mentor) and the level of response that is appropriate (nature as measure). As we'll see in later chapters, scientists are using biomimicry to design more sustainable methods of growing crops and livestock for human consumption, and of trapping and using energy. [INFOGRAPHIC 1.5]

Despite such efforts, many critics say that modern global societies are not acting nearly as quickly as they could or should. Once again, the charge echoes those archaeologists have leveled at the Vikings of Greenland. To understand what prevents us from changing our ways even in the face of brewing calamity, it helps to understand why the Greenland Vikings failed to do the same.

Humanity faces some challenges in dealing with environmental issues.

Experts agree that if the Vikings had simply switched from cows to fish they might have avoided their own demise. Such a switch would have saved at least some of the pastureland, not to mention the tremendous amount of time and labor it took to raise cattle in such an ill-suited environment. But cows were a status symbol in Europe, beef was a coveted delicacy, and for reasons that still elude researchers, the Vikings had a cultural taboo

---

Infographic **1.5** | **SUSTAINABLE ECOSYSTEMS CAN BE A USEFUL MODEL FOR HUMAN SOCIETIES**

SUSTAINABLE ECOSYSTEMS:	WE CAN MIMIC THIS BY:
RELY ON RENEWABLE ENERGY	**USING SUSTAINABLE ENERGY SOURCES** We can move away from nonrenewable fuels such as fossil fuels by turning to sustainable energy sources such as solar, wind, geothermal, and biomass (harvested at sustainable rates).
USE MATTER SUSTAINABLY	**USING MATTER CONSERVATIVELY AND SUSTAINABLY** We can reduce our waste by recovering, reusing, and recycling matter that we do use; we would also benefit from minimizing the toxins we create or release into the environment that degrade our natural resources.
HAVE POPULATION CONTROL	**GETTING HUMAN POPULATION GROWTH UNDER CONTROL** While predation controls many natural populations, there are many ways to reduce human birth rates without increasing the death rate through war or disease.
DEPEND ON LOCAL BIODIVERSITY TO MEET THE FIRST THREE REQUIREMENTS	**DEPENDING ON LOCAL HUMAN CONTRIBUTIONS AND BIODIVERSITY** Protecting biodiversity will help us achieve the three above goals; we can emulate nature by using a variety of local energy sources, building materials, and crops, and by exploring the many ideas and innovations that come from a diverse human community.

↑ A replica of the first church built in Brattahlid, Greenland—a Viking settlement founded by Erik the Red more than 1,000 years ago.

against fish. It was a taboo they clung to, even as they starved to death.

"It's clear that the Viking's decisions made them especially vulnerable to the climatic and environmental changes that descended upon them," says McGovern. "When you're building wooden cathedrals in a land without trees, when you create a society reliant on imports in a situation where it's difficult to travel back and forth between the homeland, when you absolutely refuse to collaborate with your neighbors in such a harsh, unforgiving environment, you are setting yourself up for trouble." Ultimately, he says, the decisions this society made played as much of a role in its demise as the actual environmental changes did.

nonrenewable resources Resources whose supply is finite or not replenished in a timely fashion.
social traps Decisions by individuals or groups that seem good at the time and produce a short-term benefit, but that hurt society in the long run.
tragedy of the commons The tendency of an individual to abuse commonly held resources in order to maximize his or her personal interest.
time delay Actions that produce a benefit today set into motion events that cause problems later on.
sliding reinforcer Actions that are beneficial at first but that change conditions such that their benefit declines over time.

The problem, McGovern says, is not that the Greenland Vikings made so many mistakes in their early days, but that they were unwilling to change later on. "Their conservatism and rigidity, which we can see in many different aspects, seems to have kept them on the same path, maybe even prodded them to try even harder—build bigger churches, etc.—instead of trying to adapt."

Decisions by individuals or groups that seem good at the time and produce a short-term benefit, but that hurt society in the long run, are called **social traps**. The **tragedy of the commons** is a social trap, first described by ecologist Garrett Hardin, that often emerges when many people are using a commonly held resource such as water or public land. Each person will act in a way to maximize his or her own benefit but as everyone does this, the resource becomes overused or damaged. Herders might put more animals on a common pasture because they are driven by the idea that "if I don't use it, someone else will." We do the same thing today as we overharvest forests and oceans or release toxins into the air and water. Other social traps include the **time delay** and **sliding reinforcer** traps—actions that, like the tragedy of the commons, have a negative effect later on. [INFOGRAPHIC 1.6]

Infographic **1.6** | **SOCIAL TRAPS**   ↓ Social traps are decisions that seem good at the time and produce a short-term benefit but that hurt society (usually in the long run).

TRAGEDY OF THE COMMONS   ↓ When resources aren't "owned" by anyone (they are commonly held), individuals who try to maximize their own benefit end up harming the resource itself.

This common resource (pasture) can support four animals sustainably. Each of the four farmers benefits equally.

If one farmer adds two more cows, the commons becomes degraded. All four farmers share in the degradation (less milk produced per cow), but the farmer with three cows gets all the benefit from adding the cows.

The other farmers also must add cows to return to the same level of production as before. **Over time, the commons degrades even more and at some point will no longer support any animals.**

TIME DELAY   ↓ Actions that produce a benefit today set into motion events that cause problems later on.

Modern fishing techniques that use giant nets can harvest large numbers of fish. Fishers try to get the most fish possible in the short term.

If more fish are taken over a number of years than are replaced naturally through fish reproduction, the population decreases. Other populations also decline as the fish they depend on are depleted.

After a decade or so, overfishing may so deplete the population that fishers cannot catch enough to meet needs. **The effects of decisions about how many fish to harvest are not felt until later.**

SLIDING REINFORCER   ↓ Actions that are beneficial at first may change conditions such that their benefit declines over time.

First generation: Most pests will be killed by the pesticide.

Diverse pest population

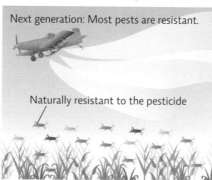

Next generation: Most pests are resistant.

Naturally resistant to the pesticide

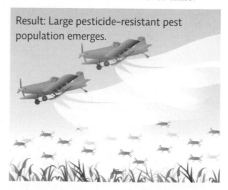

Result: Large pesticide–resistant pest population emerges.

Pesticide application can reduce pest numbers on a crop, but a few pests might survive.

The surviving pests reproduce. The pesticide is only helpful if more is applied or a more toxic pesticide is used. This can be harmful to other organisms and to humans.

In addition to becoming resistant to the pesticide, the pest population may even become bigger if the pest's predators are also killed. **What was helpful at first is no longer helpful and is even harmful over time.**

Infographic **1.7** | **WEALTH INEQUALITY**

↓ As there was in the Greenland Viking colony, there is disparity in wealth worldwide. The World Bank estimates that about 40% of the world's population lives on less than $2 a day; almost half of the 2.2 billion children worldwide live in poverty. A small percentage of our population (perhaps 20%) controls 80% of all resources. Actions by the wealthy today that degrade the environment degrade it for all sooner or later.

Percentage of the population
living on less than $2 per day

- Under 2%
- 2–5%
- 6–20%
- 21–40%
- 41–60%
- 61–80%
- Over 80%
- No data

Education is our best hope for avoiding such traps. When people are aware of the consequences of their decisions, they are more likely to examine the trade-offs to determine whether long-term costs are worth the short-term gains and, hopefully, to make different choices when necessary. In the United States, environmental laws, such as the Clean Air and Clean Water acts and the Endangered Species Act, have gone a long way toward protecting our natural environment from various social traps, but enforcing those laws is a constant challenge.

Social traps are not the only challenge societies face. Another obstacle to sustainable growth is wealth inequality. In the Greenland Viking colony, wealth at first insulated the people in power from the environmental problems and they didn't feel the strain of the decline until it was too late. Today, wealthier nations are less affected by resource availability, while 2 billion or more people lack adequate resources to meet their needs. In fact, just 20% of the population controls roughly 80% of all the world's resources. On one hand, the affluent minority (of which the United States is a part) uses more than its share. Deep pockets allow us to exploit resources for wants, not just needs, and to exploit them all over the world, so that we can spare our own natural environments at the expense of someone else's. For example, our demand for mahogany furniture drives deforestation in Central and South America where the tree flourishes. Because we are far away from the trees, the environmental fall-out is easy to ignore at the moment, but as it

did for the Vikings, it may reach a point where it affects everyone.

On the other hand, the underprivileged also exploit the environment in an unsustainable way. With limited access to external resources, they are often forced to over-exploit their immediate surroundings just to survive. The above example applies here, too. For an impoverished landowner in Costa Rica or Brazil, chopping down trees may be the only way she can feed her family—to her, harvesting the forest is more valuable than preserving it.

There are societal costs, as well, to this inequality. A growing gap between the haves and have-nots exacerbates tension and strife all over the world. In fact, fighting over resources has long been one of the contributing factors to societal decline and collapse. [INFOGRAPHIC 1.7]

Conflicting worldviews are another challenge to sustainable living. Because our **worldviews**—the windows through which we view our world and existence—are influenced by cultural, religious, and personal experiences, they vary across countries and geographic regions, even within a society. People's worldviews determine their **environmental ethic**, or how they interact with their natural environment; they also impact how they

worldviews The window through which one views one's world and existence.
environmental ethic The personal philosophy that influences how a person interacts with his or her natural environment and thus affects how one responds to environmental problems.

UNDERSTANDING THE ISSUE

CHECK YOUR UNDERSTANDING

1. **Environmental science is:**
   a. the physical surroundings or conditions in which any given organism exists or operates.
   b. an interdisciplinary field of research that seeks to understand the natural world and our relationship to it.
   c. the ability to understand environmental problems.
   d. a problem that is extremely complicated and tends to have multiple causes, each of which can be difficult to address.

2. **According to the United Nations' Millennium Ecosystem Assessment:**
   a. human actions are straining the planet's ecosystems, imperiling the ability to sustain future generations.
   b. sustainable development now ensures that the planet's ecosystems will sustain future generations in perpetuity.
   c. human actions have irreversibly damaged the planet's ecosystems; they will not sustain future generations.
   d. developing technological fixes to reverse the damage caused to the planet's ecosystems is the only way to assure sustaining future generations.

3. **Which of the following demonstrates a lack of environmental literacy?**
   a. A cooling climate caused summer hay yields to shrink, which lowered livestock survival over the longer, colder winters.
   b. The Greenland Viking middens show no evidence of the Vikings eating locally available fish—instead they continued to eat beef, scraping off every last bit of meat and marrow from the bones.
   c. The Inuit were expert hunters of ringed seal, which were an abundant food source in the Arctic.
   d. The Icelandic Vikings switched to eating fish when they realized that grazing cattle were not suited to Iceland's shorter growing season.

4. **Fishing while allowing sufficient numbers to be left behind for the fish population to regenerate reflects:**
   a. gloom and doom thinking.
   b. a frontier attitude.
   c. sustainable resource use.
   d. tragedy of the commons.

5. **Which of the following is NOT an example of a social trap?**
   a. Sliding reinforcer trap
   b. Time delay trap
   c. The tragedy of the commons
   d. Education trap

6. **Jack is opposed to the selective killing of deer that have vastly overpopulated the local forest. Jill argues that the ecosystem will be destroyed if some are not removed. While Jack is _____ as he sees deer as having intrinsic value, Jill is _____ as she values not just the species but the ecosystem processes as well.**
   a. biocentric; ecocentric
   b. ecocentric; biocentric
   c. anthropocentric; ecocentric
   d. biocentric; anthropocentric

WORK WITH IDEAS

1. Why are environmental problems considered "wicked problems"? What factors make addressing these wicked problems a challenge?

2. According to Jared Diamond, what are the five factors that determine whether any given society will succeed or fail? How did these factors play out for the Vikings in Greenland versus those in Iceland, and what were the consequences for each society? Provide evidence to support your conclusions.

3. In what ways is contemporary society both similar to and different from the Greenland Viking society? What can we learn from that society's mistakes to ensure that we do not suffer the same fate?

4. How can human societies become more sustainable? What are the challenges to achieving sustainability?

ANALYZING THE SCIENCE

The following data show the number of different types of livestock grazed in the world and in Nigeria from 1961–2008. Use these graphs and this table to answer the next 5 questions.

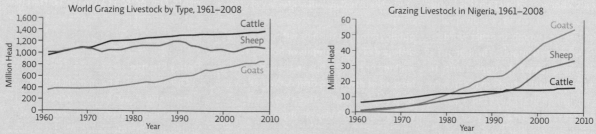

Grazing Livestock Population (in millions)

Type of livestock	World in 1961	World in 2008	% change	Nigeria in 1961	Nigeria in 2008	% change
Cattle	942	1,372		6	16.3	
Goats	349	864		0.6	53.8	
Sheep	994	1,086		1	33.9	

INTERPRETATION

1. How many different livestock types are included in the graphs and what are the trends in the numbers?

2. Which animals constituted the bulk of the grazing herds around 2008 on a worldwide basis? In Nigeria? Was this true in 1961 (the first year for which data are reported)? Provide the data to explain your responses.

3. Using the data from the table, calculate the percent change for each type of animal. According to your calculations which animal changed the most in each case (i.e., worldwide and in Nigeria)?

Hint: percent change $= \dfrac{(\text{2nd value} - \text{1st value})}{\text{1st value}} \times 100$

ADVANCE YOUR THINKING

4. Unlike cattle and sheep, goats are much more flexible in what they eat, but their sharp hooves also pulverize soil more easily. The data show that the growth in goat populations is particularly dramatic in a developing country like Nigeria. What might explain this pattern? And what might be some potential consequences?

5. How might the increase in the size of the goat herd be a potential social trap? What are some ways that we could avoid this trap?

EVALUATING NEW INFORMATION

The Lorax, a children's book by Dr. Seuss, tells the story of the Lorax, a fictional character who speaks for the trees against the greedy Once-ler who represents industry. Written in 1971, *The Lorax* was banned in parts of the United States for being an allegorical political commentary. Today the book is used for educating children about environmental concerns (see http://www.seussville.com/loraxproject/). Even so, some people consider the book inappropriate for young children due to its "doom and gloom" environmentalism.

The book *The Truax*, by Terri Birkett, involves a forest industry representative offering a logging-friendly perspective to an anthropomorphic tree, known as the Guardbark. This story was criticized for containing skewed arguments, and in particular a nonchalant attitude toward endangered species. About 400,000 copies of the book have been distributed to elementary schools nationwide.

Read both books. You can find *The Lorax* at your local public library and *The Truax* can be downloaded as a pdf from http://woodfloors.org/truax.pdf.

Evaluate the stories and work with the information to answer the following questions:

1. What are the credentials of the author of each book? In each case, do the person's credentials make him or her reliable/unreliable as a storyteller? Explain.

2. Connect each story to the key concepts in the chapter.
   a. What are the underlying attitudes and worldviews of each story?
   b. Does each story reflect social traps, and if so in what way?
   c. How might each story contribute to environmental literacy? Explain.
   d. What does each story have to say about sustainability? Explain.

3. What supporting evidence can you find for the main message in each story? In the story itself? From doing some research?

4. What is your response to each story? What do you agree and disagree with in each case? Explain.

MAKING CONNECTIONS

THE ARCTIC AND THE FUTURE OF A GLOBAL COMMONS

Background: According to a 2008 United States Geological Survey, the Arctic has 90 billion barrels of oil—13% of the undiscovered oil in the world. Now, evidence suggests that warming in the Arctic has already resulted in earlier break-up of ice in the spring and thinner ice year round. Computer models further suggest that by 2080 Arctic sea ice will completely disappear during the summer months. Ironically, the disappearance of ice will make exploration of oil resources in the Arctic more feasible, while the burning of oil and other fossil fuels have been blamed for the loss of ice in the first place.

As temperatures increase and sea ice declines, many arctic species will find it difficult to survive. Meanwhile, the need to build a massive infrastructure for oil drilling will impact ecologically intact areas. And a single oil spill can have serious impacts, especially given that there is no effective method for containing and cleaning up an oil spill in ice conditions.

Although no country owns the North Pole, there is growing concern among the indigenous Inuit that new measures are needed to preserve and sustainably use the Arctic's resources. Though not opposed to development, the Inuit want to ensure that their communities benefit without compromising the environment.

Case: The Inuit have selected you to help them develop a vision for the future of the Arctic. Based on your research and analysis, write a position paper on how Arctic resources should be utilized and how this region should be governed to ensure sustainability and consideration of the "triple-bottom line."

In your report include the following:

a. An assessment of the importance of the Arctic region as a global commons and the challenges of governing it as such.
b. An analysis of how the situation in the Arctic presents a "wicked problem."
c. A discussion of how we can prevent "ecocide" in the Arctic—specific strategies to minimize self-inflicted environmental damage, appropriately respond to on-going environmental changes, and ensure global cooperation to sustainably utilize resources in the Arctic.
d. The potential challenges to developing and implementing a sustainable Arctic policy and how these challenges could be addressed.

Setup for launch of the high-altitude research balloon for ozone testing. McMurdo Station, Antarctica.

SCIENCE AND THE SKY

Solving the mystery of disappearing ozone

CORE MESSAGE

Depletion of the stratospheric ozone caused by synthetic chemicals allows more dangerous solar radiation to reach Earth's surface, threatening the health and well-being of many plants and animals, including humans. Researchers employed the scientific process to collect physical evidence, analyze it, and report their findings to the scientific community. This helped us understand what was causing the depletion, and guided wise policy decisions at the national and international level to address this tragedy of the commons.

GUIDING QUESTIONS

After reading this chapter you should be able to answer the following questions:

→ What kinds of questions are under the purview of science and why is science limited in this way? Why do we say science is a "process" and that conclusions are always open to further study?

→ How are scientific hypotheses generated and tested? What are the two main types of scientific studies and why do we need both types?

→ Why is ozone in the stratosphere beneficial to life on Earth and how is it being depleted? What is the evidence that ozone depletion might be harmful to the health of humans or other organisms?

→ How did scientists use the process of science to help us understand what was happening with the ozone layer and what unknowns still exist?

→ How did scientists, policy makers, and world leaders take the science about ozone depletion and turn it into policy?

There's a point of no return halfway into the 9-hour flight from New Zealand to the Antarctic. Once that 4.5-hour mark passes, if something goes wrong with the plane, there's nowhere to stop in the Southern Ocean for repairs.

Susan Solomon is very familiar with that trip's all-important midway point. During her very first flight to the southernmost continent, the pilot told the passengers that their plane was not working properly: The front ski at the nose of their C-130 was frozen and couldn't be lowered into position, so landing on the packed snow at the Antarctic research station would be impossible. They had to turn around.

It was August, 1986—late winter in the Antarctic—and the atmospheric chemist was on her way to the southern continent to investigate a mystery: Why was the stratospheric ozone above the South Pole disappearing? Suddenly, the remoteness of where she was going hit home.

"I remember, as we were flying back to New Zealand, thinking, 'Wow, I really am going to the Antarctic,'" Solomon says.

The next night, the atmospheric scientist and her team from various research institutions, including the National Oceanic and Atmospheric Administration (NOAA), managed to make it to Antarctica, landing at McMurdo Station. On her very first excursion to the Antarctic, Solomon and her colleagues collected the initial data that would eventually grab the world's attention and settle a long-standing, hard-fought scientific debate that was taking place on an international stage.

Science gives us tools to observe the natural world.

Solomon and the team had decided to fly to the other end of the world after reading a scientific paper published the year before, in which Joe Farman of the British Antarctic Survey and his colleagues showed that, since the late 1970s, the ozone layer had thinned by about a third during the Antarctic spring.

The British team had collected nearly three decades of data in Antarctica, starting in 1957 with on-the-ground instruments. Like all good scientists, Farman and his team depended on **observations** (information detected with the senses or with equipment that extends our senses) of the natural world. His team collected data on the atmosphere's composition (*lower than normal ozone levels*) and then used these observations to draw conclusions or make **inferences**—explanations of what else might be true or what might have caused the observed phenomenon.

◉ WHERE IS McMURDO STATION?

AFRICA

SOUTH AMERICA

ANTARCTICA

McMURDO STATION

AUSTRALIA

NEW ZEALAND

observations Information detected with the senses—or with equipment that extends our senses.

inferences Conclusions we draw based on observations.

atmosphere Blanket of gases that surrounds Earth and other planets.

troposphere Region of the atmosphere that starts at ground level and extends upward about 7 miles.

stratosphere Region of the atmosphere that starts at the top of the troposphere and extends up to about 31 miles; contains the ozone layer.

ozone Molecule with 3 oxygen atoms that absorbs UV radiation in the stratosphere.

Farman's group also connected their results to studies by other researchers that had shown higher concentrations of an important compound: chlorofluorocarbons (CFCs), which in turn produce atmospheric chlorine (Cl). The concentrations of these chemicals seemed to increase at a rate matching the disappearance of ozone. The observation of a decrease in ozone did not come from just a few readings, but represented data collected at two different sites over more than a dozen seasons. *Replication* within a study (multiple test subjects or measurements) and between studies (independent tests that collect the same data, preferably conducted by other researchers) increase the reliability of the data. When replicates produce similar results it is less likely that the original data was "a fluke" or an unusual response. Farman's research exemplified this hallmark of good science—in this case multiple data points at two different testing sites. His team's results have subsequently been confirmed by other researchers.

Farman's team inferred that the ozone depletion in the Antarctic was somehow connected to the increased presence of chlorine compounds in the atmosphere.

It was a serious proposition: Without ozone, the world as we know it would not exist. Ozone is a key element of the **atmosphere**, the blanket of gases surrounding our planet that is made up of discernable layers, which

↑ Top: The McMurdo Station, the largest human settlement in the Antarctic. Bottom: Susan Solomon at the McMurdo Station in 1987.

differ in temperature, density, and gas composition. The lowest level, the **troposphere**, extends about 7 miles up. This level is familiar to us—it is the air we breathe and where our weather occurs. The next level in the atmospheric blanket, the **stratosphere**, rises to 31 miles above Earth's surface. The stratosphere is much less dense than the troposphere but contains a "layer" of **ozone**

Infographic **2.1** | **THE ATMOSPHERE AND UV RADIATION**

↓ The atmosphere is composed of vertical layers that differ in temperature and chemical composition. There is not much ozone in the atmosphere, but most of what ozone there is occurs in a "layer" in the stratosphere. Ozone is important because it prevents some UV-radiation from reaching Earth's surface.

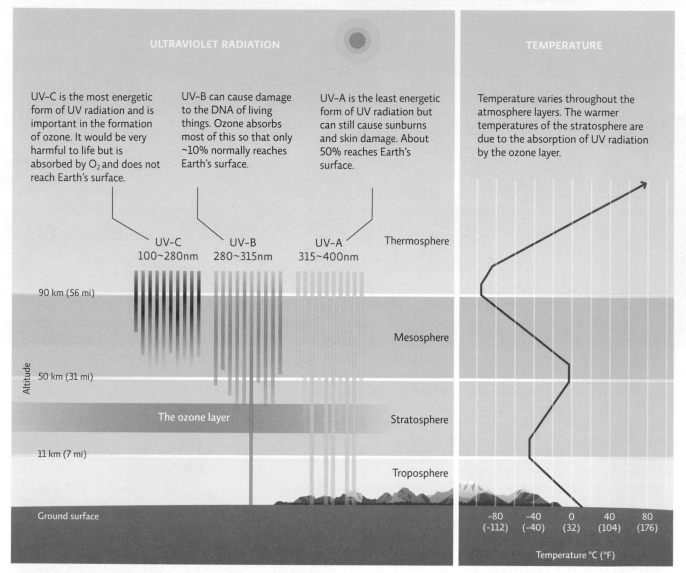

ULTRAVIOLET RADIATION

UV-C is the most energetic form of UV radiation and is important in the formation of ozone. It would be very harmful to life but is absorbed by O_2 and does not reach Earth's surface.

UV-B can cause damage to the DNA of living things. Ozone absorbs most of this so that only ~10% normally reaches Earth's surface.

UV-A is the least energetic form of UV radiation but can still cause sunburns and skin damage. About 50% reaches Earth's surface.

TEMPERATURE

Temperature varies throughout the atmosphere layers. The warmer temperatures of the stratosphere are due to the absorption of UV radiation by the ozone layer.

UV-C
100~280nm

UV-B
280~315nm

UV-A
315~400nm

Thermosphere

90 km (56 mi)

Mesosphere

Altitude

50 km (31 mi)

The ozone layer

Stratosphere

11 km (7 mi)

Troposphere

Ground surface

-80
(-112)
-40
(-40)
0
(32)
40
(104)
80
(176)

Temperature °C (°F)

(abbreviated as O_3, because it contains 3 oxygen atoms), a region where most of the atmosphere's ozone is found.

Without ozone, the world as we know it would not exist.

Solar radiation enters the atmosphere every day, including three forms of **ultraviolet (UV) radiation**: UV-A, UV-B, and UV-C. Ozone molecules in the stratosphere absorb much of the UV-B, a vital service since UV-B can damage cells and biological molecules like DNA. In humans, exposure to UV-B radiation increases the risk of cataracts, skin damage, and cancer. Fortunately, ozone prevents

most of the UV-B from reaching Earth's surface, where it can harm organisms. This stratospheric (good) ozone should not be confused with ground-level (bad) ozone found in the troposphere. Ground-level ozone is a component of smog and is harmful to living things (see Chapter 25). [INFOGRAPHIC 2.1]

Susan Solomon became a scientist because she was curious about the natural world around her. **Science** is both a body of knowledge (facts and explanations) and the process used to get that knowledge. Understanding the process is more important than the "facts," since facts may change as more information is collected through the scientific process. This process is a powerful tool that allows us to gather evidence to test our ideas, *and* to evaluate the quality of that evidence.

Science, however, is limited to asking questions about the *natural* world—not all questions are open to science. Scientific investigation, in both the natural and social sciences, is based on data gathered through **empirical evidence**, or observations. Only physical phenomena that can be objectively observed—meaning data that could be collected by *anyone* in the same place, using the same equipment, etc.—are fair game for science. Scientists can gather empirical evidence about the environment and living things using a wide variety of tools, including natural tools, such as their eyes, ears, and other senses. Phenomena that are not objectively observable (What is my dog thinking? Do ghosts exist?) and ethical or religious questions (Is the death penalty wrong? What is the meaning of life?) cannot be empirically studied, and therefore are not under the purview of science.

Arriving at McMurdo Station, Solomon knew she had to apply the scientific process to understand why the ozone layer was thinning in the Antarctic. And she already had a culprit in mind.

The scientific view of CFCs did not change overnight.

Throughout Solomon's childhood, she was surrounded by items that contained the synthetic molecules known as CFCs. The compounds were first developed in the 1930s as a commercial coolant, to replace more toxic ammonia and sulfur dioxides, and were therefore included in refrigerators, air conditioners, and other household and industrial items. By the time Solomon reached graduate school at the University of California, Berkeley, in the late 1970s, CFCs were being used in everything from hairspray to foam containers for fast food.

CFCs contain atoms of carbon, fluorine, and chlorine. For example, CFC_{12}, a common refrigerant, has the molecular formula CCl_2Fl_2—one carbon atom (C) is bound to two chlorine (Cl) and two fluorine (Fl) atoms. By tweaking the

balance of these atoms, chemists synthesized CFCs that they believed were stable, nonflammable, and harmless to people and the environment—qualities that led to their broad acceptance.

Over time, these items emitted CFC molecules into the air; in 1971, British scientist James Lovelock detected CFCs in the atmosphere over England. Since CFCs were considered safe, the news was not a cause for concern. But it caught the attention of scientist Sherwood Rowland of the University of California, Irvine, who wondered what would happen to this completely synthetic substance in the atmosphere. He and his colleague Mario Molina set out to look for answers.

In 1974, they proposed that CFCs were not entirely harmless in the atmosphere. The compounds were specifically designed by chemists to be stable, and Molina and Rowland realized that CFCs would stay aloft for a remarkably long time in the atmosphere, residing there and accumulating for 100 years or more. And once in the stratosphere, the molecules would be exposed to UV light so intense that it would break them apart. The process would release solitary chlorine atoms, which previous research had shown could chemically react with—and destroy—ozone, aligning with Farman's observations from Antarctica. The two chemists calculated that at the rate CFCs were being produced in 1972, the chemicals could destroy 6% of the ozone layer. And manufacturers were making more CFCs every year.

But the scientific view of CFCs didn't change overnight—because all conclusions in science are considered open to revision (because our understanding of a concept or process will change as scientists learn more), more evidence was needed to overturn the prevailing conclusion that CFCs were safe.

The scientific method systematically rules out explanations.

Farman's data was compelling, but would it hold up to scrutiny? Less than a year after his research was published in 1985, in the journal *Nature*, NASA scientists published their own report, verifying what Farman had observed. This corroboration strengthened Farman's conclusions. That ozone was depleting much faster than earlier projections set off alarm bells. These data provided a **correlation** between the presence of CFCs and ozone depletion—both occurred together. It suggested that CFCs may be related to ozone decline, but did not establish a **cause-and-effect relationship.** The two trends could

---

ultraviolet (UV) radiation Short-wavelength electromagnetic energy emitted by the Sun.

science A body of knowledge (facts and explanations) about the natural world, and the process used to get that knowledge.

empirical evidence Information gathered via observation of physical phenomena.

correlation Two things occur together—but it doesn't necessarily mean that one caused the other.

cause-and-effect relationship An association between two variables that identifies one (the effect) occurring as a result of or in response to the other (the cause).

occur together by coincidence, or something else entirely could be causing both to occur.

To get a better picture of what was going on, scientists needed to further apply the **scientific method**, in which they would work logically and systematically to design studies specific to the question being asked. The fact that the ozone layer was thinning above the Antarctic triggered great debate among scientists, resulting in more than one possible explanation, known as **hypotheses**, for what was occurring. Some researchers thought that Antarctic air was mixing and lifting lower, low-ozone air into the stratosphere, changing the concentration. Other researchers believed that solar activity was creating nitrogen oxides (NO_x) which could be destroying ozone, as the amount of NO_x fluctuated with sunlight in the Antarctic. But Susan Solomon had a different idea.

Solomon was a young scientist when she visited Antarctica in 1986 and 1987, the beginning of her long relationship with the southern, icy world. While trying to understand why ozone was disappearing over the region, Solomon kept thinking about temperature. October is spring in Antarctica, and scientists knew that cold spring winds would swirl around, producing a cyclone of air in the atmosphere (a polar vortex), keeping cold air in place over the poles and leading to the formation of polar clouds in the stratosphere. Solomon proposed the hypothesis that cloud particles in the polar stratospheric clouds were providing surfaces for the reactions that would free chlorine molecules (Cl_2) from CFCs. In sunlight, the chlorine molecules would then break up into chlorine atoms. These isolated chlorine atoms destroyed ozone—particularly in the Antarctic spring, when sunlight streamed in.

Scientific hypotheses must be **testable** (generate **predictions** about what we could objectively observe if we conducted the test). In turn, these predictions must be **falsifiable**, meaning that it would be possible to produce evidence that shows the prediction is wrong. (Predictions based on untestable ideas—such as, "reincarnation exists"—are not falsifiable and therefore are not considered suitable for science.) In this way, if that falsifying evidence does not appear, it may be reasonable to conclude that the evidence supports the prediction.

Notice we do not claim that the hypothesis is *proven*, only that it is supported (or confirmed). This is a hallmark of the tentative nature of science. "Proven" suggests we have the final answer; science, however, is open ended, and no matter how much evidence accumulates, there are always new questions to ask and new studies to conduct that could alter our conclusions. But this does not mean that all possible explanations are equally valid. This is precisely the reason why hypotheses must be tested again and again and in different ways. As evidence mounts in support of a hypothesis, the probability that it is wrong lessens and it becomes unreasonable to reject the hypothesis in favor of another, less supported explanation. But,

scientific method Procedure scientists use to empirically test a hypothesis.
hypothesis A possible explanation for what we have observed that is based on some previous knowledge.
testable A possible explanation that generates predictions for which empirical evidence can be collected to verify or refute the hypothesis.
prediction A statement that identifies what is expected to happen in a given situation.
falsifiable An idea or a prediction that can be proved wrong by evidence.

Infographic 2.2 | **CERTAINTY IN SCIENCE**

↓ There are degrees of certainty in science—we know some ideas are better than others. The more evidence we have in support of an idea, especially when the evidence comes from different lines of inquiry, the more certain we are that we are on the right track. But since all scientific information is open to further evaluation, we do not expect or require "absolute" proof.

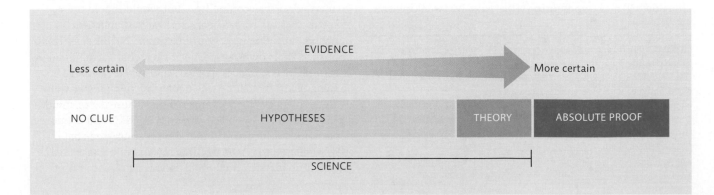

Infographic **2.3** | **SCIENTIFIC PROCESS**

↓ Scientists work from previous knowledge and observation to ask new questions and pose possible explanations (hypotheses) for what they observe. They then design a study to gather evidence to test predictions made from their hypotheses. The scientific method is not really a linear sequence. Rather, it is more of a cycle that scientists move through in whatever order of "steps" best suits their needs.

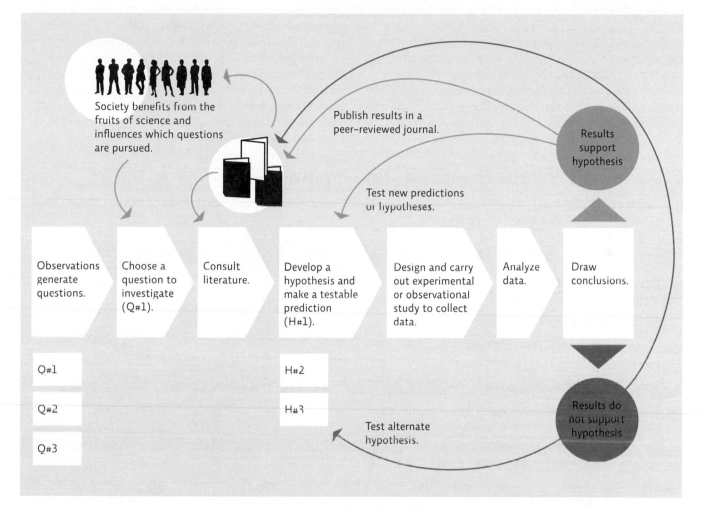

in keeping with the tentative nature of science, even the best-supported hypotheses will be revised or even abandoned if new data strongly supports a new conclusion. [INFOGRAPHIC 2.2]

Once a study is conducted and data gathered, the data are evaluated to determine whether they confirm or fail to confirm the hypothesis, but the scientific process doesn't end there. If a hypothesis is rejected, alternative hypotheses can be tested. If a hypothesis is confirmed, the researchers should repeat the study to validate the data. They can also generate new predictions that test the same hypothesis from different angles. As evidence mounts from replicate studies and from multiple

predictions, we become more confident in our data and conclusions. [INFOGRAPHIC 2.3].

Different types of studies amass a body of evidence.

Solomon's hypothesis generated the prediction that the stratosphere would contain high levels of chlorine monoxide, or ClO. If polar clouds and sunlight were causing chlorine to react with ozone, then the atmosphere should contain many molecules of chlorine bound to individual oxygen atoms. This prediction was falsifiable, since Solomon might not find high levels of ClO.

On both of her trips to the Antarctic, her team raised balloons into the air which measured the composition of the atmosphere where the ozone hole was found, and simultaneously analyzed light reaching the ground to determine whether it was changing in the region with less ozone and more ClO. They came back with measurements of ClO that would turn out to be their so-called smoking gun.

At the time, NASA ozone-modeler Paul Newman believed the loss of ozone in the Antarctic spring was due to excess solar activity. But when the ClO measurements from Solomon's team streamed out of a fax machine in NASA's Goddard Research Center, he knew the evidence supported Solomon's polar cloud hypothesis. [INFOGRAPHIC 2.4]

Solomon's experiment is an example of an **observational study**—collecting data in the real world without intentionally manipulating the subject of study. In these types of studies, researchers may simply be gathering data to learn about a system or phenomenon, or they may be comparing different groups or conditions found in nature. Often, researchers can conduct observational studies that take advantage of natural changes in the environment such as collecting "before and after" data in an area to examine the effect of a natural perturbation such as a flood or volcanic eruption, or in this case comparing areas with different levels of ozone depletion. These types of opportunities represent valuable "natural experiments" that can allow us to test cause-and-effect hypotheses,

---

Infographic **2.4** | **THE CHEMISTRY OF OZONE FORMATION AND BREAKDOWN**

↓ Ozone is naturally formed and broken down in the stratosphere, but ozone-depleting substances like CFCs catalyze additional ozone breakdown, leading to ozone depletion.

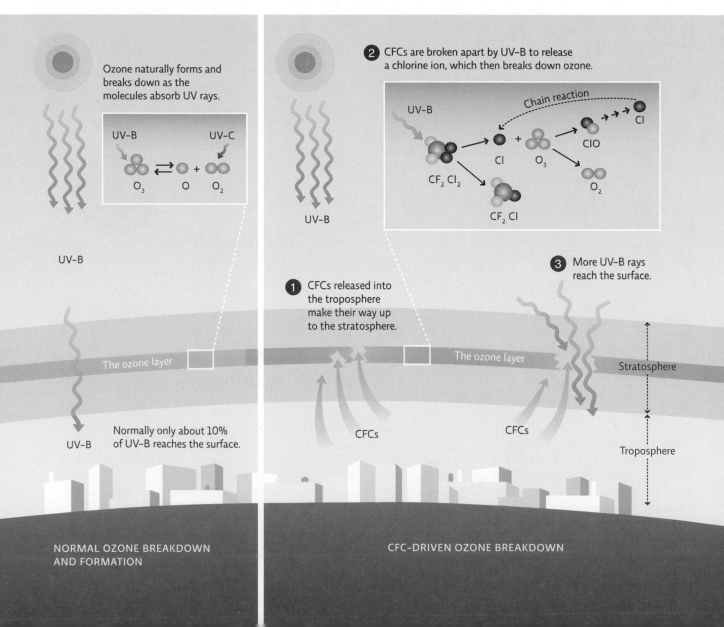

NORMAL OZONE BREAKDOWN
AND FORMATION

CFC-DRIVEN OZONE BREAKDOWN

↑ Researchers at the Finnish Ultraviolet International Research Center in the Arctic Circle research the effects of increased UV-B radiation on plant growth and biology. UV-B radiation at high latitudes damages living organisms, and can alter their DNA (genetic material).

just as a controlled **experimental study**, in which the researcher intentionally manipulates the conditions of the experiment in a lab or field setting, allows us to do. For instance, much research is underway on the effects on living organisms of ozone depletion and elevated UV exposures. Previous studies have shown that UV exposure can increase the incidence of skin cancer. A reasonable question would be: Is ozone depletion causing more skin cancer? This could be used to generate an experimental hypothesis such as: Lower ozone levels will lead to more cases of skin cancer. This hypothesis generates many predictions; some would be tested with observational studies and others experimentally.

In 2002, Chilean researchers Jaime Abarca and Claudio Casiccia investigated this hypothesis by evaluating skin cancer incidence in residents of Punta Arenas, Chile, a region exposed to higher UV-B due to its proximity to the ozone hole. Because of the ozone hole, global air circulation patterns have actually shifted; the southern hemisphere jet stream now circulates at a latitude closer to the South Pole, taking the ozone-depleted air over the more populated regions of the Southern Hemisphere like

southern Chile. They predicted that more cases of skin cancer would be seen there in years when ozone depletion was high. To test this they calculated skin cancer rates of the population between 1987 and 1993, and compared those to rates from the same population between 1994 and 2000. The results of their study showed that non-melanoma skin cancer rates were significantly higher in times when ozone depletion was higher. This evidence supports their hypothesis and correlates living close to the Antarctic ozone hole with one's chance of developing skin cancer. This observational study does not manipulate any variable—no people were exposed to higher or lower levels of UV-B or intentionally relocated to areas of higher or lower ozone depletion. The researchers simply collected data that was available in the population and then evaluated it to see if the two variables—ozone depletion and skin cancer incidence—were correlated.

---

observational study Research that gathers data in a real-world setting without intentionally manipulating any variable.

experimental study Research that manipulates a variable in a test group and compares the response to that of a control group that was not exposed to the same variable.

We can't ethically manipulate people to further test this hypothesis but we can conduct an experimental study on cells or model organisms such as mice. Australian researcher Scott Menzies and his colleagues did just that in the early 1990s by testing the prediction that mice exposed to high UV-B would develop more skin cancer than mice exposed to normal levels of UV-B. This experiment compared two groups—the **control group** of mice was exposed to normal levels of UV-B and the **test group** was exposed to the same amount of UV-B received in areas where ozone depletion has been observed. The two groups are identical in every way except for the test variable (in this case, the amount of UV-B radiation exposure). The inclusion of a control group is key because it allows researchers to attribute any differences seen between the two groups to the single test variable that was altered. If researchers only looked at the test group, they would have no way to determine whether the incidence of skin cancer was higher than normal.

In an experimental study, we have both an **independent variable** and a **dependent variable**. We manipulate the independent variable (in this case, the amount of UV-B radiation), and measure the dependent variable (the incidence of skin cancer) to see if it is affected. In other words, if development of skin cancer is *dependent* on UV-B radiation, then we should see skin cancer incidence change as UV-B exposure changes. Scientists often represent their data on a graph, on which the x-axis (horizontal axis) displays the independent variable and the y-axis (vertical axis) shows the response (dependent variable). Menzies' data showed a clear difference between his control group (none developed skin cancer) and his test group (100% developed skin cancer). It is rare to see such an absolute difference and, as with any study, we would like to see these results replicated, but a well-designed and well-conducted study increases our confidence that the results are valid. [INFOGRAPHIC 2.5]

Both observational and experimental studies gather data systematically to produce scientifically valid evidence. In contrast, anecdotal accounts (individual occurrences or observations) represent data that was not systematically collected or tested and cannot be compared to any control (many uncontrolled variables exist). While anecdotes are not considered acceptable scientific evidence, a claim based on anecdotal accounts could be tested to see if a correlation or cause-and-effect relationship exists.

Solomon and her team reported their evidence to the international scientific community by publishing their results in *Nature* and the *Journal of Geophysical Research* in 1987. These are **peer-reviewed** journals—meaning that before results are published, they are reviewed by a group of outside experts. Studies that are not well-designed

or well-conducted are not accepted for publication. Therefore, peer-reviewed published research represents high-quality scholarship in the field. In this case, the reviewers already knew that CFCs were present in the stratosphere, and lab studies showed they were capable of destroying ozone. Now, Solomon and her team were presenting evidence that, at the time of year when ozone was dropping, the Antarctic stratosphere contained high levels of ClO, demonstrating that free chlorine atoms were reacting with ozone.

The evidence was amassing that CFCs were contributing to ozone depletion, but that didn't mean the other hypotheses would be immediately abandoned. Research on these alternative hypotheses continued but scientists were not finding evidence to support the other hypotheses' predictions. If lower, ozone-poor air was lifting and mixing with the stratosphere, then researchers should observe gases moving upward in the atmosphere; instead, they saw the opposite: Air seemed to be flowing downward. If solar activity was creating nitrogen oxides (NO_x) that were destroying ozone, as NASA modeler Paul Newman had once believed, then scientists should have observed increases in NO_x in the South Pole. But when they took measurements, they found that NO_x levels were actually decreasing. This observation provided further support for Solomon's hypothesis, which predicted that NO_x levels should be low, not high—because otherwise NO_x would combine with ClO molecules and prevent them from interacting with ozone.

The multiple lines of evidence are so compelling that they have elevated the "CFC hypothesis" to the status of **theory**—a widely accepted explanation that has been extensively and rigorously tested. (But just like well-supported hypotheses, we do not claim that a theory is *proven*. Even well-substantiated theories are always open to further study.) This differs from the casual meaning of theory, which suggests a speculative idea without much substance. To discount any scientific theory as "just a

---

control group The group in an experimental study that the test group's results are compared to; ideally, the control group will differ from the test group in only one way.

test group The group in an experimental study that is manipulated somehow such that it differs from the control group in only one way.

independent variable The variable in an experiment that the researcher manipulates or changes to see if it produces an effect.

dependent variable The variable in an experiment that is evaluated to see if it changes due to the conditions of the experiment.

peer-reviewed Researchers submit a report of their work to a group of outside experts who evaluate the study's design and results of the study to determine whether it is of high-enough quality to publish.

theory A widely accepted explanation of a natural phenomenon that has been extensively and rigorously tested scientifically.

BACKGROUND KNOWLEDGE
UV light can cause
skin cancer.

QUESTION
Is ozone depletion causing
more skin cancer?

HYPOTHESIS
Lower ozone levels will lead
to more cases of skin cancer.

← Scientists collect evidence to test
ideas. Experimental studies are
used when the test subjects can be
intentionally manipulated while
observational studies allow a look
at entire ecosystems or other complex systems.

Scientifically test the hypothesis.

OBSERVATIONAL STUDY (Abarca, *et al.*)
Prediction: The incidence of skin cancer in an area where
ozone depletion is occurring will be higher than it was in times
of less ozone depletion.
Procedure: Collect data on skin cancer incidence.

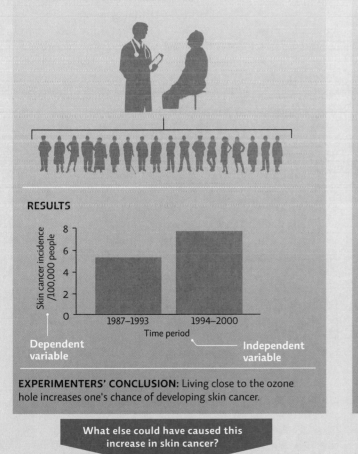

RESULTS

Dependent
variable

Independent
variable

EXPERIMENTERS' CONCLUSION: Living close to the ozone
hole increases one's chance of developing skin cancer.

What else could have caused this
increase in skin cancer?

EXPERIMENTAL STUDY (Menzies, *et al.*)
Prediction: Mice exposed to UV–B levels comparable to that seen
in areas where ozone depletion is observed will develop more skin
cancer than mice exposed to normal levels of UV–B.
Procedure: Expose mice to different UV–B levels and monitor
them for skin cancer.

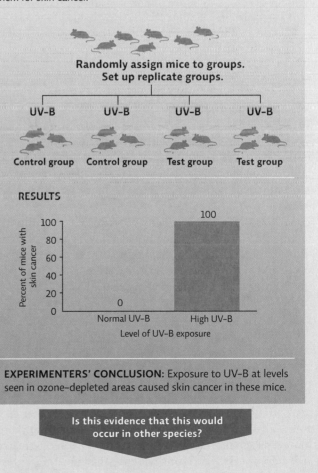

Randomly assign mice to groups.
Set up replicate groups.

UV–B   UV–B   UV–B   UV–B

Control group   Control group   Test group   Test group

RESULTS

EXPERIMENTERS' CONCLUSION: Exposure to UV–B at levels
seen in ozone-depleted areas caused skin cancer in these mice.

Is this evidence that this would
occur in other species?

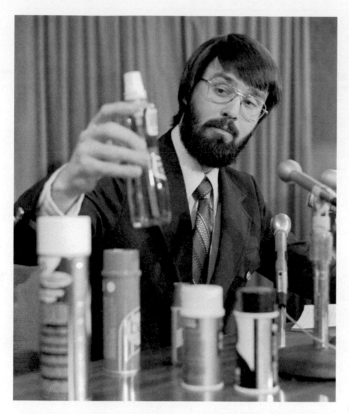

↑ In 1975, Representative Perry Bullard of Michigan introduced legislation to outlaw aerosol cans, 12 years before the Montréal Protocol was signed. This is an example of applying precautionary principles.

theory" represents a serious flaw in one's understanding of what a scientific theory really is.

As mentioned earlier, in science there are *degrees of certainty*; we know some things better than others. The more evidence we have in support of an idea, especially from different types of experiments, the more certain we are that we are on the right track. These degrees of certainty are expressed mathematically in terms of probabilities using **statistics**. A mathematical analysis of the data is done to determine the probability that the occurrence of a phenomenon (in this case the experiment's result) is a random event, rather than being caused by the variable being investigated.

This type of analysis of the data allows us to quantitatively assign a level of certainty to our conclusions. In statistics, this probability is expressed as a *p-value* that represents the likelihood our conclusions are wrong. Scientists generally require a high probability (at least 95%) that their conclusions are correct ($p \leq 0.05$ represents a level of certainty of 95%). This means that there is only a 5% chance we have incorrectly accepted or rejected our hypothesis. The more evidence that accumulates in favor of a particular conclusion, the more certain we are that it is likely to be correct and can be used to inform our decisions. Just because we don't know exactly what will happen if we continue to release CFCs into the atmosphere (including how fast the ozone layer will deplete, or when it could jeopardize human health) doesn't mean we don't know enough to take action. [For more on statistics and experimental design, see Appendix 3.]

Indeed, the loss of ozone was starting to have the effects that the studies in Chile and Australia were showing. Other observational studies revealed that UV-B levels were increasing in high- and midlatitude regions. Research by NASA scientists showed that the amount of UV-B reaching the ground in the midlatitudes in the mid-1990s had increased over 1979 levels. Other research showed that UV-B levels were 45% higher than normal in the spring of 1990 at the southern tip of Argentina.

In the meantime, skin cancer incidence has increased in human populations (as have other skin disorders and eye problems such as cataracts) and other organisms are also affected by extra UV-B exposure. Photosynthesis rates in marine organisms such as the tiny phytoplankton of the Antarctic sea are lower than normal in the spring because of increased UV-B exposure—a troublesome fact since phytoplankton forms the base of the Antarctic food chain. Decreases in photosynthesis are also troubling because they could lead to lower agricultural productivity and reduce the value of these areas as "carbon sinks"—areas where carbon is stored, keeping it out of the atmosphere (see Chapter 26 on climate change).

The international community got together to meet the problem head on.

Even while scientists were still trying to understand why the ozone layer was depleting, they knew they had to do something about it. In 1985, a group of experts from around the world met in Vienna to discuss ways to research and solve the problem. This set the stage for the international community to come together in Montréal, Canada, in 1987, to produce a plan to deal with the problem of ozone depletion. It would involve sacrifices—notably, phasing out dangerous chemicals such as CFCs—and only two dozen governments signed on to the **Montréal Protocol** that September. By 2009, the protocol (treaty) was eventually ratified by all 196 countries in the world when East Timor signed on.

The Montréal Protocol, administered by the United Nations, outlined a series of deadlines over the next decade for cutting back production of CFCs. Governments

Infographic **2.6** | **THE MONTRÉAL PROTOCOL AND ITS AMENDMENTS HAVE BEEN EFFECTIVE**

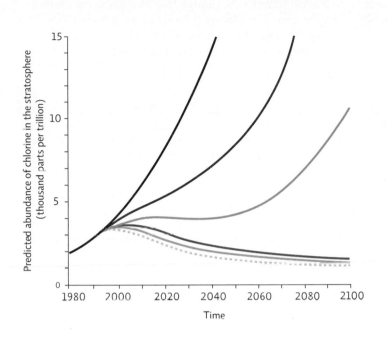

← Actual and projected change over time for total global emissions of ozone-depleting substances (ODS) with and without the Montréal Protocol and its amendments. Adjustments to the phase-out schedule of various ODSs in the form of amendments represent the success of adaptive management in dealing with complex environmental issues.

— No protocol
— Montréal Protocol 1987
— London Amendment 1990
— Copenhagen Amendment 1992
— Beijing Amendment 1999
····· Zero emissions

would have to put in place their own plans for achieving a desired outcome, or **policy**, for reducing CFCs.

Policy has been described as *translating our values into action*. Science provides information we can use to make informed decisions to protect our health, safety, or environment. Society then decides what "ought" to be addressed (do we address ozone depletion?) and hopefully uses the best scientific information available to set reasonable policies that take into account the ecological, economic, social, and political issues at stake.

Concerns raised by researchers in the 1970s had already led to a ban on CFCs for use as a propellant in hairsprays and other products in some countries, including the United States. As the evidence mounted, industry, the public, and governments began to realize the seriousness of this problem.

The U.S. government required that CFCs be phased out starting in the early 1990s, through regulations from the U.S. Environmental Protection Agency, the federal agency responsible for seeing that environmental laws are followed.

Note that the 1987 Montréal Protocol and the international commitment to address CFCs and ozone depletion came before Susan Solomon's definitive studies were published in 1988. This is an example of applying the **precautionary principle**—acting in the face of uncertainty when there is a chance that serious consequences might occur. By 1987, though we didn't know all the details, we knew enough to take action. As more information poured in, it quickly became apparent that the Montréal Protocol targets would not be sufficient to stop ozone depletion. Amendments are still proposed and negotiated in annual meetings which strengthen the response and adjust the target dates to phase out harmful compounds. This is an ongoing process, and an example of **adaptive management**—allowing room for altering strategies as new information comes in or the situation itself changes. This same approach is needed to address other global environmental issues such as climate change. We cannot know the exact details of future climate change impacts before they occur but we do have a good idea of what we are facing. By acting now, we can lessen the severity of those impacts and adjust our response as we go along. [INFOGRAPHIC 2.6]

statistics The mathematical evaluation of experimental data to determine how likely it is that any difference observed is due to the variable being tested.
Montréal Protocol International treaty that laid out plans to phase out ozone depleting chemicals like CFC.
policy A formalized plan that addresses a desired outcome or goal.
precautionary principle: Acting in a way that leaves a safety margin when the data is uncertain or severe consequences are possible.
adaptive management Plan that allows room for altering strategies as new information comes in or the situation itself changes.

Fortunately, companies that produced CFCs, such as Dupont, were already researching alternative chemicals in the lab that could serve the same purposes with less damage. If companies could sell a CFC alternative just as easily, then cutting back on CFCs would not hurt them financially. Still, some of the replacements, particularly some hydrochlorofluorocarbons (HCFCs) used in refrigerants,

eventually would also prove to be detrimental to ozone. In addition, illegal stockpiles and old refrigerators continue to release some CFCs, though that amount is decreasing rapidly under the Montréal Protocol. Pockets of CFCs remain because of continued legal "essential" uses, for medicines and nuclear power, for example.

Infographic **2.7** | **OZONE DEPLETION AND CFC LEVELS**

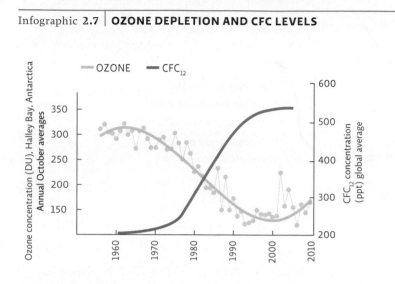

← This graph shows a correlation between the levels of CFC_{12} and ozone over Antarctica. As CFC_{12} increased, ozone declined (a negative correlation). When CFC_{12} amounts leveled off, around the year 2000, the ozone hole over Antarctica started to recover. (Ozone concentration is expressed in Dobson units, DU, a measure of how thick the air sample would be if it were compressed at 1 atmosphere of pressure at 0°C; 1 DU = 1 mm thick.)

↓ These images show the size of the ozone hole over Antarctica in 1979 (the 1st year ozone satellite images were available), 1993 (a "mid" year), and 2006 (the worst ozone hole ever recorded). Though the 1979 image might not be "normal," since CFC-driven depletion had already begun, ozone levels are considerably higher than in subsequent years. The 1993 image shows a much larger ozone hole than that in 1979, but the record ozone hole of 2006 is wider and deeper (indicated by the purple color—lowest ozone levels ever recorded).

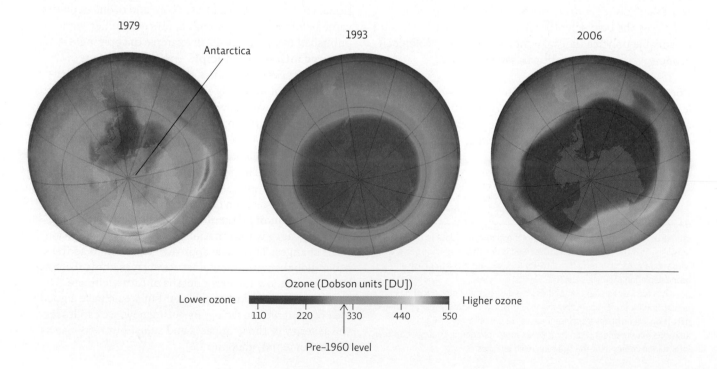

Today, ozone-depleting compounds in the stratosphere are decreasing at a rate consistent with the policy changes of the Montréal Protocol. Outside of polar regions, ozone levels declined through the 1990s but seem to be holding steady more recently. The Montréal Protocol has been hailed as the most successful international environmental agreement in history.

There is still much we don't understand about the atmospheric chemistry of ozone depletion. Just when it seemed ozone was beginning to recover, the winter of 2004–2005 in the Arctic was unusually cold and ozone loss was very high that year. This was eclipsed by a record 40% loss of ozone over the Arctic in spring of 2011. And Antarctica set two new records for ozone depletion in 2006—the widest and deepest hole ever observed, with some areas near 90% depletion. What this tells us is that even though CFCs are declining, very cold winters will still give us years of higher-than-expected ozone depletion. [INFOGRAPHIC 2.7]

Future projections estimate that midlatitude areas should be back to pre-1980 ozone levels by 2050; polar regions should be back by 2075, though the next 15 years should show periodic large declines and produce large "ozone holes" in very cold years like those seen in 2006 and 2011. Although she retired from NOAA in 2011, Solomon continues her research on CFCs, and is both hopeful and realistic. "It's clear that ozone will ultimately recover but it's also clear that it will take many decades to do so," she says. "It has been a real privilege to work on such an interesting problem and to feel that the world found it useful in making choices about the Montréal Protocol."◉

Research articles referenced in this chapter:
Abarca, J.F., & Casiccia, C.C. 2002. *Photodermatology, Photoimmunology & Photomedicine*, 18: 294–302.
Farman, J., *et al.* 1985. *Nature*, 315: 207–210.
Menzies, S.W., *et al.* 1991. *Cancer Research*, 51: 2773–2779.
Molina, M.J., & Rowland, F.S. 1974. *Nature*, 249: 810–812.
Solomon, S., *et al.* 1988. *Science*, 242: 550–555.

BRING IT HOME

⊘ PERSONAL CHOICES THAT HELP

The depletion of the ozone layer is a great example of how science documented a problem and its cause, and public action confronted the problem. All of the environmental changes we face, from rising levels of greenhouse gases to loss of biodiversity, can be addressed using science. How scientific information is or is not put into action has far-reaching consequences, making science literacy a matter of importance for every individual.

Individual Steps
→ Practice thinking like a scientist. Go outside for 10 minutes and observe the world around you. Make observations of what you see or hear. What predictions could you make from your observations; how could you test them?
→ Stay informed. Read or watch a science-related article or show once a month.

Group Action
→ Demonstrate the importance of scientific literacy with your friends and family. Develop three additional questions from the material in the chapter and discuss them over dinner.
→ Support science education. Find out about public lectures and programs in your area and attend one with your friends.

Policy Change
→ Attend a city council or county board meeting to see how policy issues are addressed in your area.
→ Make knowledgeable voting decisions on ballot initiatives.
→ Serve on local civic committees that address environmental issues in your community.

UNDERSTANDING THE ISSUE

CHECK YOUR UNDERSTANDING

1. **The scientific process includes:**
   a. observational studies that collect data in the real world without manipulating the subject of study.
   b. both testable and untestable explanations for what we have observed.
   c. only provable cause–and–effect relationships.
   d. just the information that can be detected with our senses: sight, hearing, taste, and smell.

2. **Which of the following is NOT true about the Montréal Protocol?**
   a. It is a treaty that was eventually ratified by all 196 countries in the world and is considered the most successful international environmental agreement in history.
   b. It is administered by the United Nations and outlines a series of deadlines for cutting back production of CFCs.
   c. It is a legally binding agreement and national governments of ratifying nations have to work in lock step with each other to reduce CFC use and production.
   d. Amendments to the protocol are still proposed and negotiated in annual meetings, which strengthen the response to ozone depletion and speed up target dates to phase out the harmful compounds.

3. **The statement "It is wrong to inflict pain on animals" is not scientifically testable because it is not _____ by any demonstration of empirical evidence.**
   a. provable
   b. falsifiable
   c. predictable
   d. correlational

4. **What is the relationship between CFCs and ozone in the stratosphere?**
   a. Ozone naturally breaks down in the stratosphere but substances like CFCs regulate its re-formation so that less harmful UV radiation reaches Earth's surface.
   b. Ozone from the stratosphere migrates down to the troposphere where it reacts with chemicals like CFCs to produce more oxygen.
   c. Ozone in the stratosphere is broken-down by chemicals like CFCs, but as CFCs themselves are broken apart by UV radiation, ozone depletion slows.
   d. Ozone is naturally formed and broken down in the stratosphere but substances like CFCs catalyze additional ozone breakdown that results in increased UV radiation reaching Earth's surface.

5. **Data gathered via objective observation of physical phenomena is:**
   a. scientific proof.
   b. an inference.
   c. empirical evidence.
   d. a hypothesis.

6. **In an experiment to study the effects of UV–B on leopard frogs, an effective control is:**
   a. a group of leopard frogs with skin lesions treated in the same way, but exposed to normal levels of UV–B.
   b. a group of leopard frogs treated in the same way, but given a lower exposure of UV–B radiation.
   c. a group of leopard frogs treated in the same way, but exposed to normal levels of UV–B.
   d. a group of red-spotted toads treated in the same way, but given a lower exposure of UV–B radiation.

WORK WITH IDEAS

1. What is science? Describe the key elements of science and apply this framework to show how the story of CFCs and stratospheric ozone depletion is science.

2. What is a correlation in science? How is it different from a cause-and-effect relationship? Explain how both types of data are used to understand the connection between CFCs and stratospheric ozone.

3. What are the differences between observational and experimental studies? Under what circumstances is each of these types of studies useful?

4. Define scientific hypothesis and explain how it differs from a scientific theory and the more general use of the word theory by lay-people.

5. Describe the chemistry of ozone formation and breakdown in the troposphere and explain how chemicals like CFCs interfere with this process. What is it about the chemistry of CFCs that makes a seemingly harmless chemical so dangerous?

6. Why is the Montréal Protocol considered the most successful international environmental agreement in history? Explain what makes this treaty so successful and discuss the possible underlying reasons for this success.

ANALYZING THE SCIENCE

The graph on the following page comes from a study exploring the correlation between chlorine monoxide (ClO) and ozone. The data was collected by placing the measurement instrument on the wing of a plane and flying through the Antarctic ozone hole.

INTERPRETATION

1. What does this graph show? Describe the pattern that you see.

2. Looking at the levels of ozone relative to the levels of chlorine monoxide, which side of the graph do you think most likely represents the inside of the ozone hole? On what basis do you think this?

3. Compare this graph to the graph in Infographic 2.7 (Ozone Depletion and CFC Levels). Although both graphs show a similar pattern, they are about two different aspects of the CFC-ozone story. Compare and contrast the information presented in the two graphs.

ADVANCE YOUR THINKING

HINT: If you want to access the actual article, you can find it at: www.undsci.berkeley.edu/article/ozone_depletion_01

4. Do the data in this graph show a correlation or a cause-and-effect relationship? Do they support the hypothesis that chlorine is the cause of ozone depletion? Explain your responses.

5. How are these data similar to and/or different from the data collected by Susan Solomon's team? And what do these data mean for the hypothesis that Solomon proposed?

6. How does this research study support the idea that science is a community enterprise?

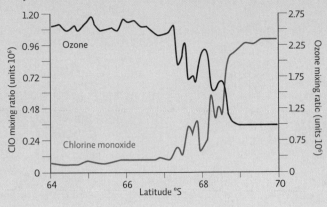

A plot of chlorine monoxide and ozone concentrations from data collected by the aircraft.

EVALUATING NEW INFORMATION

You can use your understanding of the nature of science to discover the real meaning of news stories on health, science, and other national issues.

For instance, how does the Montréal Protocol affect asthma sufferers? CFCs were used in metered dose inhalers (MDIs) to deliver medication to asthma patients. Initially the Food and Drug Administration (FDA) requested—and was granted—an essential-use exemption for MDIs using CFC propellants. But as alternative therapies became available, the FDA started removing CFC-based MDIs from circulation, with complete phase-out in 2010.

But while the Asthma and Allergy Foundation of America (www.is.gd/pags2h) and the American College of Allergy, Asthma, and Immunology (www.is.gd/p154hB) supported the new alternatives, the National Campaign to Save CFC Asthma Inhalers did not (www.is.gd/gOaiuB).

Evaluate the websites and work with the information to answer the following questions:

1. Who runs each website? Do their credentials make the information presented about non-CFC MDIs reliable or unreliable? Explain.

2. What is the mission of each website? What are the underlying values? How do you know this?

3. What claim does each website make about CFCs, the ozone hole, and MDIs? How do the websites compare in providing scientific evidence in support of their position on non-CFC MDIs? Is the information provided accurate and reliable? Explain.

4. What sorts of links do the websites provide for additional information? Are these supporting links reliable and representative of scientific knowledge? Explain.

5. Whose position on non-CFC MDIs do you agree with? Explain why.

MAKING CONNECTIONS

METHYL BROMIDE AND OZONE DEPLETION

Background: In 2004, the Montréal Protocol banned widespread use of another chemical called methyl bromide (MeBr), which also destroys ozone. While some 27–42% of MeBr comes from natural sources such as oceans and soil, additional MeBr is released as a result of human activity. In the United States, MeBr is used primarily as a broad-spectrum pesticide.

Chemical companies insist that making effective substitutes for all uses of MeBr is not currently possible, and farmers maintain that relying on less effective alternatives will pose economic hardships. These groups also point out that MeBr has a much shorter atmospheric lifetime (0.7 years compared to 100 years for CFCs), a lower ozone depletion potential, and the production of MeBr has been steadily declining anyway (from 12,994 metric tons in 2004 to 1.803 metric tons in 2010). In contrast, environmental advocates want a complete ban on MeBr.

Case: You have been assigned the task of developing a policy for the future use of MeBr. Your team must evaluate various options such as:

1. A complete ban on all anthropogenic sources of MeBr and switching to United States EPA approved alternatives (see www.is.gd/hAwJPp) even if this switch comes at economic cost.

2. A continuance of the current ban on non-essential uses with exemptions for critical uses approved on an annual basis. (See www.is.gd/sZfh5m for current exemptions.)

Research these (and possibly other) options and write a report recommending a course of action. In your report include the following:

a. An analysis of the pros and cons of each proposal including: a discussion of the consequences of each choice for stratospheric ozone depletion, human health, food availability and prices, and the economics of farming.

b. Based on the information at hand, what is the best course of action for the future of MeBr? Who should be involved in making this decision? Provide justifications for your proposal.

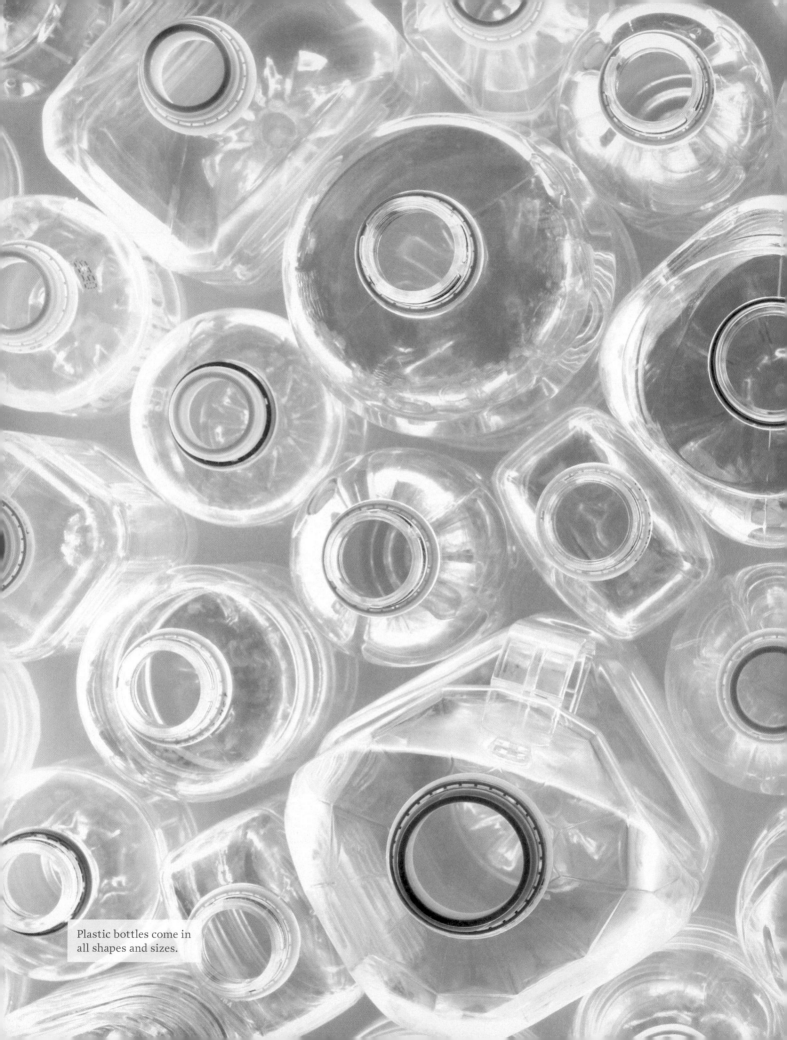

Plastic bottles come in all shapes and sizes.

TOXIC BOTTLES?

On the trail of chemicals in our everyday lives

CORE MESSAGE

Our understanding of the natural world changes almost daily as new evidence is gathered. Unfortunately, misinformation abounds in the popular press and on the Internet about the latest "science." For example, we live in an environment full of natural and synthetic chemicals, some of which are toxic, but knowing how to respond to the latest toxic scare is often difficult as we receive conflicting messages about their safety or risk. Developing information literacy skills enables us to better evaluate the usefulness and trustworthiness of various sources of information and then use the highest-quality information we can find to make reasoned decisions about how to respond.

GUIDING QUESTIONS

After reading this chapter, you should be able to answer the following questions:

→ What is information literacy and why is it important?

→ What common logical fallacies are used in presenting arguments? How can we use critical thinking skills to logically evaluate the quality of information and its source?

→ What factors influence whether or not a particular chemical is toxic and how toxic it is?

→ What kind of scientific studies allow us to assess the potential hazard of a particular chemical and how do we determine "safe" exposures?

→ What are endocrine disruptors and why can small amounts of exposure produce big effects in an individual?

In 2008, after a decade of study and contentious debate, a U.S. government-appointed panel of scientists known as the National Toxicology Program (NTP) finally arrived at a tentative consensus: Based upon data they had accumulated up to that point, they wrote in a report that would splash across headlines, they had "some concern for effects on the brain, behavior, and prostate gland in fetuses, infants, and children at current human exposures to bisphenol A."

Bisphenol A, or *BPA* as it is more commonly known, is a synthetic chemical. Since the late 1940s, it has been a staple ingredient in the linings of metal food cans, and plastic products of every kind, including food and beverage containers and plastic baby bottles. But in the two preceding decades, a mountain of scientific studies had implicated the unassuming compound in a rash of serious medical conditions, from impaired neurological and sexual development to cancer. Both Canada and the European Union were considering banning the use of BPA in baby bottles and baby food cans. Most of the NTP panel's scientists felt that the data were still too uncertain to warrant such a drastic step. But, they thought, it would be prudent for industries that used the chemical to start looking for a replacement. "The panel raised important research questions and public health concerns," says Sarah Vogel, a public health historian with the Johnson Family Foundation who has closely tracked the case of BPA. "For the first time ever, the government was suggesting that BPA might not be safe." The report incited a frenzy.

The plastic industry decried the report's conclusions. In a firestorm of press releases, in newspapers across the country and on cable news, industry spokespeople insisted that their own data showed BPA to be perfectly safe.

Parents everywhere were torn. On the one hand, it was hard to believe that something as commonplace as BPA could be so dangerous; if at least some studies were showing it to be safe, maybe it was. On the other hand, if there was even a chance that this chemical could harm their children, shouldn't the government take every possible precaution? BPA-free products were exceedingly difficult to find, and without some sort of federal mandate, that was unlikely to change. What, on Earth, were average consumers to do?

We live in an environment full of toxins.

Toxins are chemicals that cause direct damage upon exposure. They fall into two broad categories: synthetic

and natural. Natural toxins are not to be taken lightly: Arsenic, a basic element that sometimes leaches into groundwater, can cause cancer and nervous system damage in humans.

But synthetic toxins are a particular problem because there are quite a lot of them and many are **persistent chemicals** (meaning they don't readily degrade over time). According to the U.S. **Environmental Protection Agency (EPA)**, more than 80,000 chemicals are used in the United States alone. And some 1,000–2,000 new chemicals enter the consumer market each year.

The debate over how to regulate these chemicals—how to determine what quantity of any particular compound is safe for humans, and then how to ensure that our exposure levels stay well below those quantities—begins in 1962 with a book called *Silent Spring*.

In this book, legendary environmental activist Rachel Carson asked her readers to imagine a world without the sounds of spring, a world in which the birds, frogs, and crickets had all been poisoned to death by toxic chemicals. Just 20 years had passed since the widespread introduction of herbicides and pesticides (like DDT), she explained, but in that relatively short time, they had thoroughly permeated our society.

These chemicals were obviously great for killing off weeds and pests; they had done an amazing job conquering mosquito-borne diseases like malaria during World War II, and combating world hunger by boosting global food production until the 1980s. But no one seemed terribly concerned about the effects they might have on nontarget species, or on their (or our) ecosystems. After all, they were designed to kill living things. Wasn't it at least possible that what killed one organism might also kill others?

Carson went on to identify three specific concerns that were being overlooked at the time: Some chemicals can have large effects at small doses; certain stages of human development are especially vulnerable to these effects; and mixtures of different chemicals can have unexpected

↑ A group of men from Todd Shipyards Corporation run their first public test of an insecticidal fogging machine at Jones Beach State Park, New York, in July 1945. As part of the testing, a 6.5 kilometer (4 mile area) was blanketed with the DDT fog.

impacts. The book created an uproar, which led to much stricter regulations for chemical pesticides in general, and, in the United States, a complete ban on DDT in particular. But half a century later, we are still struggling to effectively regulate the chemicals in our world.

Regulation happens even in the face of change.

Regulation begins with **risk assessment**—a careful weighing of the risks and benefits associated with any given chemical. In an ideal world, unbiased, professional regulators would assess the safety of every new chemical before it entered our lives. They would discern all the potential consequences of excessive or continued long-term exposure, and in so doing would protect us from any slow, unwitting poisoning. In reality of course, a variety of factors—practicality, economic forces, sheer need—make such thorough precautions nearly impossible to implement.

To be sure, federal agencies (namely the Food and Drug Administration (FDA) and EPA) are mandated with protecting us from harmful chemicals; they have the authority to heavily regulate or ban outright those deemed to be dangerous. For substances where the data are uncertain and where the substances may cause unexpected or unpredictable effects, they can employ a "better safe than sorry" strategy known as the **precautionary principle**. This rule of thumb calls for leaving a wide safety margin when setting the *exposure limit*—the maximum quantity

humans can safely be exposed to. The width of that margin depends on the severity of the potential health effects and environmental damage.

The precautionary principle is becoming a favored tactic in the European Union. But in the United States, for the vast majority of chemicals, we take a different approach: "innocent until proven guilty." Rather than thoroughly testing each individual compound, regulators make educated guesses about safety, based on how other, similar compounds have fared. As a result, toxic products are often discovered only after (sometimes long after) reaching the marketplace—usually when some person or group of people suffer the effects. Rather than preventing these products from reaching store shelves in the first place, we recall them after the fact. This ad hoc regulation puts on the public the burden of proving that a chemical is actually more dangerous than expected.

And as the case of BPA shows, even when concerns about safety emerge, deciding which precautions to take can seem like an impossible task.

---

toxins Chemicals that cause direct damage upon exposure.
persistent chemicals Chemicals that don't readily degrade over time.
Environmental Protection Agency (EPA) The federal agency responsible for setting policy and enforcing U.S. environmental laws.
risk assessment Weighing the risks and benefits of a particular action in order to decide how to proceed.
precautionary principle A rule of thumb that calls for leaving a safety margin when the data about a particular substance's potential for harm are uncertain and where the substance may cause unexpected or unpredictable effects.

Part of the problem is that, even in our era of warp-speed communication, information is a slippery thing; this is especially true when it comes to science. We are constantly uncovering new information and gleaning new insights about the environment and our relationship to it. And as our understanding grows and changes, existing information often becomes obsolete. In fact, much of what we learn in science class today will be outdated 5 years from now—not because we are wholly ignorant in the present, but because we will know so much more in the future.

In this rapidly moving current of knowledge floats a seemingly endless array of information sources: newspapers and magazines, websites, scientific journals, and so on. Not all of these sources are equal. While some are carefully vetted for accuracy, others are incomplete or deliberately misleading.

Scientific information can generally be divided into two types: primary and secondary.

The ability to distinguish between reliable and unreliable sources is referred to as **information literacy**. It's the key to drawing reasonable, evidence-based conclusions about any given issue or topic, and it is especially important in cases like that of BPA, because when it comes to scientific issues, hyperbole and misinformation abound.

We are constantly uncovering new information and gleaning new insights about the environment and our relationship to it. And as our understanding grows and changes, existing information often becomes obsolete.

Primary sources are those that present new and original data or information, including novel scientific experiments and first-hand accounts of any given observation. Scientific journals are primary sources; they contain original reports of scientific studies. Almost all of these reports, or papers, are rigorously evaluated through **peer review**—a process whereby experts in the field (a panel of the author's "peers") assess the quality of the study's design, data, and statistical analysis, as well as the soundness of the paper's conclusions. Good studies are published; bad ones are rejected.

↓ Professor Heather Patisaul of North Carolina State University looks over brain scans of mice exposed to BPA.

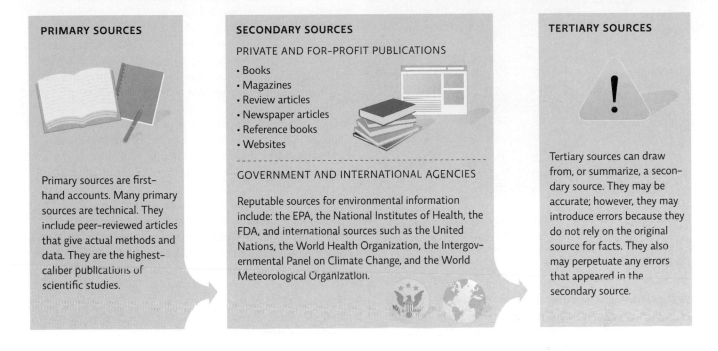

PRIMARY SOURCES

Primary sources are first–hand accounts. Many primary sources are technical. They include peer-reviewed articles that give actual methods and data. They are the highest-caliber publications of scientific studies.

SECONDARY SOURCES

PRIVATE AND FOR-PROFIT PUBLICATIONS

- Books
- Magazines
- Review articles
- Newspaper articles
- Reference books
- Websites

GOVERNMENT AND INTERNATIONAL AGENCIES

Reputable sources for environmental information include: the EPA, the National Institutes of Health, the FDA, and international sources such as the United Nations, the World Health Organization, the Intergovernmental Panel on Climate Change, and the World Meteorological Organization.

TERTIARY SOURCES

Tertiary sources can draw from, or summarize, a secondary source. They may be accurate; however, they may introduce errors because they do not rely on the original source for facts. They also may perpetuate any errors that appeared in the secondary source.

Secondary sources present and interpret the information from primary sources. In science, anything that hasn't been peer reviewed is a secondary source; this includes all reports from the popular press. **Tertiary sources** are those that present and interpret information from secondary sources. Many blogs, websites, and even news shows qualify as tertiary sources: They provide additional commentary on, or foster debate over, reports from the popular press. [INFOGRAPHIC 3.1]

Because the popular (nonscientific) press runs on catchy sound bites and easily digestible bits of information, news outlets (both secondary and tertiary) tend to oversimplify the results of individual studies, or present them as if they provided definitive answers. But there are rarely easy answers to environmental questions, and science is almost never as straightforward as we would like it to be. In fact, by its very nature, science is incremental; each study is just one small piece of a much larger puzzle, and existing hypotheses are subject to endless revision and qualification as new bits of data trickle in.

What are the dangers presented by toxins and how do we determine safe exposure levels?

In the wake of the National Toxicology Program's 2008 report, a team of scientists at the Centers for Disease Control and Prevention analyzed more than 2,000 urine samples collected from a statistically representative cross section of the American population. More than 90% of those samples tested positive for BPA; the average concentration was 2.6 parts per billion (ppb), though the top 5% of samples had an average concentration of almost 16 ppb. That's well past the amount of BPA known to cause harm in rodents. Concentrations seemed to decrease with age, so that children had more BPA in their systems than adolescents, and adolescents had more than adults. "This study really laid to rest any doubts about whether or not BPA was in fact leaching from our food containers into our bodies," says Vogel. But did that necessarily mean that BPA was dangerous in humans?

information literacy The ability to find and evaluate the quality of information.

primary sources Sources that present new and original data or information, including novel scientific experiments or observations and first–hand accounts of any given event.

peer review A process where researchers submit a report of their work to outside experts who evaluate the study's design and results to determine if it is of a high–enough quality to publish.

secondary sources Sources that present and interpret information from primary sources. Secondary sources include newspapers, magazines, books, and most information from the Internet.

tertiary sources Sources that present and interpret information from secondary sources.

↑ A warning sign with suggested consumption levels of fish greets anglers to the pier in Berkeley, California. A state water pollution board has unveiled a plan to rid San Francisco and Suisun Bays of PCBs.

Any given toxin has several characteristics we must consider when evaluating its safety. One such characteristic is its **persistence**—how long it takes the substance to break down in the environment. Chemicals with low persistence tend to break down quickly in the presence of sunlight. Chemicals with high persistence tend to linger for a long time and can affect ecosystems well after their initial release.

Another trait that must be considered is the toxin's **solubility**—its ability to dissolve in liquid, particularly water. In some cases, *water-soluble* chemicals are safe for humans but not so good for the environment. Because we can excrete them in our urine, they don't linger in our bodies for very long (of course, at high enough doses these chemicals can still prove toxic; and at low but continual doses they can cause kidney damage). But because water-soluble chemicals are easily taken up by aquatic organisms, they can wreak slow havoc on aquatic environments, and by extension, on the ecosystems that surround them.

Fat-soluble chemicals present an extra level of complexity. Because they pass easily through cell membranes, our cells can readily absorb these chemicals. Once they're inside, our bodies have a hard time expelling fat-soluble chemicals. In some cases, the liver can covert a fat-soluble molecule into a water-soluble one, so that it can be broken down and excreted in urine. But when our livers can't work this magic, fat-soluble chemicals are stored in our fatty tissue, where they can pile up in a process known as bioaccumulation.

Bioaccumulation refers to the buildup of fat-soluble substances in the tissue of an organism over the course of its lifetime. **Biomagnification** describes a consequence of bioaccumulation; it's what happens when animals that are higher up on the food chain eat other animals that have bioaccumulated toxins: they consume their prey's entire lifetime dose of those toxins. Biomagnification means that animals higher on the food chain accumulate far more toxins than do those lower on the chain. The best example in our human diets is tuna or swordfish. These fish are large predators, and thus are high up on the ocean food chain. So when we eat them, we consume all the

Infographic **3.2** | **BIOACCUMULATION AND BIOMAGNIFICATION**

↓ When animals eat other animals that have bioaccumulated toxins, they acquire those toxins. Fat-soluble toxins build up in the tissue of the animal over the animal's lifetime if the animal has continued exposure to the toxins; the lifetime accumulation is stored in fatty tissue.

BIOACCUMULATION

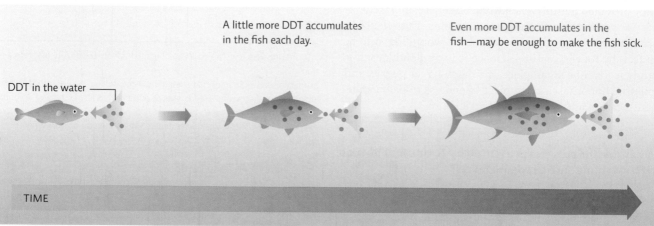

A little more DDT accumulates in the fish each day.

Even more DDT accumulates in the fish—may be enough to make the fish sick.

DDT in the water

TIME

45

toxins—of special concern is mercury—that they have picked up from preying on smaller fish. [INFOGRAPHIC 3.2]

As for BPA, it has a low persistence, meaning that it breaks down rapidly in the environment. Although it is fat soluble, liver and gut cells can readily convert it to a water-soluble form, so that it's easily excreted in urine. This means it should not bioaccumulate or biomagnify.

So how then was it present in more than 90% of human urine samples? BPA is so commonplace and people are exposed to it so continuously that it remains ever present in our systems. Even as we are breaking down and excreting some bits of BPA, we are ingesting more. Scientists have spent the past decade trying to determine what such exposure might mean for human health.

Figuring out the cause-and-effect relationships between our bodies and the chemicals that enter them is tricky work. **Epidemiologists**—the scientists charged with this type of research—can't just give a test group of humans a toxin to see what effects it has. They must do a bit of detective work. They can start by looking for health problems in specific populations and work their way backward to find the culprit. Or they can look at groups that have been exposed to a given toxin and see if any common health problems emerge or have already emerged. This latter approach is the one researchers took for BPA. Epidemiologists looked at the health profiles of hundreds of individuals who had BPA in their urine; in one study of 1,455 such people, they found a correlation between BPA concentrations and cardiovascular disease.

The task of determining exactly how BPA might go about wreaking such havoc inside actual human bodies, falls to toxicologists. **Toxicologists** concern themselves with determining the specific properties of any given potential toxin and how it affects cells or tissues. They do this by testing lab animals through ***in vivo*** (*in vivo* means "in the body") studies, or by testing cells in petri dishes in what scientists call ***in vitro*** studies (*in vitro* means "in glass"). Toxicologists use these data to determine how toxic a substance is and what its effects on living organisms are. [INFOGRAPHIC 3.3]

Toxicity can be affected by a host of factors. Individual susceptibility varies with genetics, age, and underlying health status. When exposed to a toxin, the type and amount of chemicals already in a person's system are

persistence The length of time it takes a substance to break down in the environment.

solubility The ability of a substance to dissolve in a liquid or gas

bioaccumulation The buildup of fat-soluble substances in the tissue of an organism over the course of its lifetime.

biomagnification The increased levels of fat-soluble substances in the tissue of predatory animals that have consumed organisms that have bioaccumulated toxins.

epidemiologist A scientist who studies the cause and patterns of disease in human populations.

toxicologists Scientists who study the specific properties of any given potential toxin.

in vivo study Research that studies the effects of an experimental treatment in intact organisms.

in vitro study Research that studies the effects of experimental treatment cells in culture dishes rather than in intact organisms.

↓ Fat-soluble substances such as mercury and DDT are not easily broken down and are stored in fatty tissue. Top predators living in ecosystems contaminated with mercury may contain high levels in their flesh. Fish advisories often reflect this danger.

BIOMAGNIFICATION

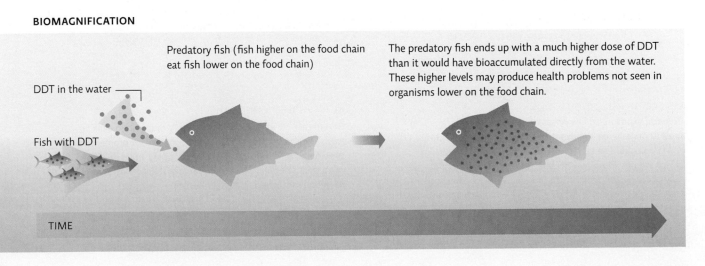

DDT in the water

Fish with DDT

Predatory fish (fish higher on the food chain eat fish lower on the food chain)

The predatory fish ends up with a much higher dose of DDT than it would have bioaccumulated directly from the water. These higher levels may produce health problems not seen in organisms lower on the food chain.

TIME

↓ Both *in vivo* and *in vitro* studies are used to experimentally study the effects of BPA. Note that none of the researchers discussed below concludes that they have definitive evidence that BPA is harmful, but each bit of research presents a piece of the puzzle. It is the accumulation of evidence that will be most helpful in drawing conclusions about the danger or safety of BPA in our food containers and other plastic goods.

TOXICOLOGICAL STUDIES

In vivo studies

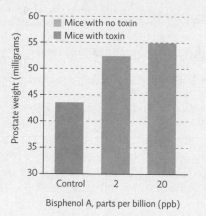

In vitro studies

DATA COLLECTED AND ANALYZED

Researchers Nagel and vom Saal studied the effect of low-dose BPA on the prostate size of male mice pups whose mothers were fed one of two doses of BPA when pregnant, compared to males whose mothers were not fed BPA.

DATA COLLECTED AND ANALYZED

Researchers Ishido and Suzuki placed droplets of rat neural stem cells onto a dish treated with various doses of BPA. They then monitored the cells with a microscope. As the stem cells migrated out of the spherical droplets across the dish, the researchers measured, on average, how far the cells migrated in both the test and control groups.

RESULT

At both doses, male mice had significantly larger (30–35% larger) prostate glands (p < 0.01) but did not vary from controls in overall body size.

RESULT

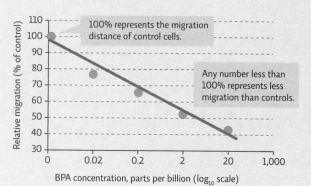

BPA had a negative effect on cell migration compared to control cells; as BPA concentration went up, cell migration (and survival) went down.

RESEARCHERS' CONCLUSIONS

Prenatal exposure to low doses of BPA alters prostate growth in male mice and is thus biologically active at doses seen in nature. The effect was greater than predicted from *in vivo* studies. Perhaps BPA acts synergistically with naturally occuring estrogens in the animal feed or is converted to a more active product in the body. Further studies on the processing of BPA in the body are needed.

RESEARCHERS' CONCLUSIONS

BPA negatively affects migration ability of rat brain cells *in vitro*. Since successful migration is necessary for proper brain development, low doses of BPA may impact brain development and function in these animals. Further testing is needed to determine if the same effect is seen *in vivo*.

↓ Epidemiological studies look at human populations to see if any BPA effects emerge in particular groups, such as those with a diagnosis of disease. The Lang study shown here compared the amount of BPA in the urine of subjects with and without various common health conditions to see if high BPA correlated with any of the illnesses.

EPIDEMIOLOGICAL STUDIES

DATA COLLECTED AND ANALYZED

The Lang study evaluated 1,455 subjects' urine BPA levels. Subjects were interviewed about whether they had been diagnosed with various health conditions. The data were adjusted for age and sex, and statistically analyzed to see if any groups (those with or without a certain health condition) differed in their levels of BPA.

RESULTS

■ Subjects diagnosed with the condition
■ Subjects who have not been diagnosed with the condition

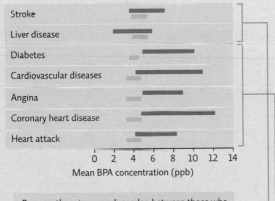

Mean BPA concentration (ppb)

Because there is so much overlap between those who have had a stroke or liver disease and those who have not, BPA does not seem to be implicated as a cause.

For diabetes and all forms of cardiovascular disease reported, there is a significant difference in BPA urine levels (higher in those with the disease). Notice that the bars overlap very little, if at all.

RESEARCHERS' CONCLUSIONS

Concentrations of BPA were associated with an increased prevalence of cardiovascular disease and diabetes. These findings add to the evidence that suggests there are adverse effects of low-dose BPA in animals. Independent replication and follow-up studies are needed to confirm these findings and to provide evidence about whether the associations are causal.

important. Some chemicals in the body might increase overall toxicity (**additive effects**); other chemicals may reduce toxicity due to interactions between the toxins that "cancel each other out" or at least lessen the effect (**antagonistic effects**). Still other chemicals may work together to produce an even bigger effect than expected (**synergistic effects**). Route of exposure (for example, inhalation, injection, or skin contact) and the dose at the time of exposure also play a role (large doses can cause immediate effects that differ from those caused by lower doses acquired over a longer time period). [INFOGRAPHIC 3.4]

But in general, toxicologists like to say that "the dose makes the poison." This means that almost anything can be tolerated in low enough doses; conversely, anything—even water—can be toxic if the dose is big enough. And in most cases, as the dose increases, so does the severity of the effect. This idea—that higher doses of something harmful are worse for you than lower doses—makes obvious sense. It guides both regulatory efforts and modern medicine. And it applies to almost every chemical you can think of—except for the class of which BPA is a part: endocrine disruptors.

Endocrine disruptors cause big problems at small doses.

As their name suggests, **endocrine disruptors** interfere with the endocrine system, typically by mimicking a **hormone**, or preventing a hormone from having an effect. BPA is an estrogen mimic (estrogen is a hormone that plays many roles in the body; its main task is to guide reproduction and development in both males and females); it binds to the body's cellular estrogen **receptors** and triggers the same effects that actual estrogen would trigger. In the wild, chemicals like this have been shown to cause feminization of males (even sex changes), as evidenced by lower sperm counts and the production of egg proteins normally only produced by females.

additive effects Exposure to two or more chemicals has an effect equivalent to the sum of their individual effects.

antagonistic effects Exposure to two or more chemicals has a lesser effect than the sum of their individual effects would predict.

synergistic effects Exposure to two or more chemicals has a greater effect than the sum of their individual effects would predict.

endocrine disruptor A molecule that interferes with the endocrine system, typically by mimicking a hormone or preventing a hormone from having an effect.

hormone A molecule released by the body that directs cellular activity and produces changes in how the body functions.

receptor A structure on or inside a cell that binds a hormone, thus allowing the hormone to affect the cell.

Infographic **3.4** | **FACTORS THAT AFFECT TOXICITY**

↓ Some chemicals are more toxic than others due to their mode of action. Other factors also affect how toxic a particular chemical will be for an individual.

EXPOSURE
Whether a toxin is inhaled, ingested, or contacts the skin may affect how much of a problem it causes. Dose is also important, since some chemicals have a threshold of toxicity. Frequency also matters—a single exposure may be tolerable at a given dose but repeated exposure can cause problems.

INDIVIDUAL FACTORS
Factors related to the individual may affect how toxic a chemical is. Some toxins are more of a problem for the very young or very old, or for those who are ill. In some cases, genetic differences make a person more or less vulnerable to a given toxin.

CHEMICAL INTERACTIONS
We are never exposed to just one chemical. The fact that chemicals can interact in ways that increase or decrease their toxic effects complicates our efforts to determine a "safe dose." For instance, suppose there are two chemicals (A and B); each raises one's temperature 1 degree at a given dose. If we are exposed to both chemicals at the same time, one of three interactions is possible:

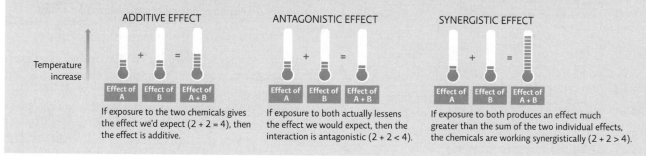

Temperature increase

ADDITIVE EFFECT
Effect of A   Effect of B   Effect of A + B
If exposure to the two chemicals gives the effect we'd expect (2 + 2 = 4), then the effect is additive.

ANTAGONISTIC EFFECT
Effect of A   Effect of B   Effect of A + B
If exposure to both actually lessens the effect we would expect, then the interaction is antagonistic (2 + 2 < 4).

SYNERGISTIC EFFECT
Effect of A   Effect of B   Effect of A + B
If exposure to both produces an effect much greater than the sum of the two individual effects, the chemicals are working synergistically (2 + 2 > 4).

In humans, we know that sperm counts are down among men and that puberty onset is earlier in both boys and girls than it has ever been before. We also know that 95% of the U.S. population has trace amounts of BPA in their urine, and since there are no known natural sources of BPA, this suggests that it is leaching from our food and beverage containers into our bodies. No one can say for sure whether one fact (changes in sexual development) is related to the other (BPA in our systems), but because hormones control the development of body organs, scientists are especially concerned about the exposure of developing fetuses, newborns, and infants to endocrine disruptors.

Endocrine disruptors are curiously different from most other chemicals, where the relationship between dose and effect is linear (the more you ingest, the sicker you get—"the dose makes the poison"). They can have one set of effects at a very low dose, and then no effects (or much different effects) at higher doses.

To track the effects of a dose of a chemical, toxicologists use data from *in vitro* and *in vivo* studies to create a **dose-response curve**, from which they can calculate an **LD50** (lethal dose 50%), the dose that would kill 50% of the population. The lower the LD50, the more toxic the substance. [INFOGRAPHIC 3.5]

Because endocrine disruptors like BPA have different effects at low and high doses, LD50s and dose-response curves are much trickier to calculate, and a "safe dose" is much tougher to determine. "We can't just find the threshold dose, where effects are first seen, and test higher doses from there to see what the impact will be,"

Infographic **3.5** | **DOSE-RESPONSE CURVES**

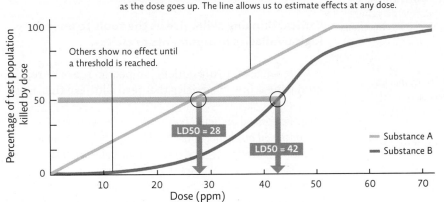

Some substances show effects at even the lowest dose, with increasing effects as the dose goes up. The line allows us to estimate effects at any dose.

Others show no effect until a threshold is reached.

LD50 = 28

LD50 = 42

Substance A
Substance B

Percentage of test population killed by dose

Dose (ppm)

← Dose-response studies evaluate the effect of a toxin at various doses (on animals or cells). Knowing the LD50 allows us to compare toxins directly— i.e., which are more toxic than others? Charting the actual change in effect as the dose increases (the "dose response") allows us to predict effects at doses other than those tested. We can also determine if there is a threshold of exposure that must be reached before harmful effects are seen.

"SAFE DOSE"
Because there is uncertainty in the determination of "safe dose," regulatory agencies factor in a safety margin of 100 or even 1,000 to be on the safe side. Example: If we have evidence that says the toxin is safe at 100 ppm, we might set the environmental standard (what is allowed) at 1 ppm (or even 0.1 ppm if children are at risk).

says Frederick vom Saal, a biologist at the University of Missouri who has spent two decades studying low-dose effects of BPA. "For endocrine disruptors, we have to test the effects of high doses and low doses separately."

Early claims of BPA safety did not take this detail into account. In the late 1980s, when EPA and FDA scientists were testing the chemical, they started with very high doses, given to rats, lowered the dose until they saw no adverse effects, and then stopped. As is the usual practice, a "safe dose" was established by dividing that "no effect" level by 1,000, to account for the possibility that humans might be more sensitive than lab animals, or that some people, such as children and the elderly or sick, might be more sensitive than others.

This high-dose assessment is the basis for statements like the one Steven Hentges, spokesperson for the American Plastics Council, made to the press when the 2008 National Toxicology Program (NTP) report came out. "At the rate of migration into food," he told innumerable reporters, "a consumer would have to ingest more than 1,300 pounds of food and beverage in contact with BPA every day for a lifetime to exceed the EPA safety limit." Like the scientists whose data he was citing, Hentges was not considering the possibility of separate low-dose effects.

But it turns out that, thanks to one unfortunate chapter in our chemical history, we already have a very good idea of what low doses of estrogen mimics can do to humans.

Between 1938 and 1971, doctors used the estrogen mimic diethylstilbesterol (DES) to prevent premature labor in expectant mothers. Thirty years later, when a higher incidence of reproductive abnormalities started showing up in the population, epidemiologists traced it back to DES given to pregnant mothers.

The DES research gave scientists an important tool for future estrogen mimic studies. In the course of confirming the chemical's toxicity, toxicologists discovered a particular strain of rat that, when given DES, or any other estrogen mimic for that matter, responded exactly as humans did. "It's a rare human–animal corroboration," says vom Saal. "It gives us an unprecedented degree of certainty about how well these mice studies relate to humans."

Back in 1997, when vom Saal tested low doses of BPA on these rodents, he found a litany of effects: increased postnatal growth in both males and females, early onset of sexual maturation in females, decreased testosterone and increased prostate size in males, altered immune function and increased mortality of embryos. Over the years, as other scientists replicated these data, more and more of them began to worry that, at low doses—such as those

dose-response curve A graph of the effects of a substance at different concentrations or levels of exposures.
LD50 (lethal dose 50%) The dose of a substance that would kill 50% of the test population.

Infographic **3.6** | **HOW HORMONES WORK**

↳ Steroid hormones like estrogen (and its mimics) pass into cells, where they bind to DNA and actually "turn on" genes. These effects can be far reaching since the activated genes may affect other genes and many cell processes, having many later effects.

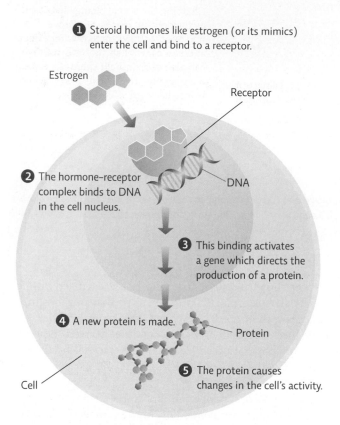

1 Steroid hormones like estrogen (or its mimics) enter the cell and bind to a receptor.

Estrogen

Receptor

2 The hormone-receptor complex binds to DNA in the cell nucleus.

DNA

3 This binding activates a gene which directs the production of a protein.

4 A new protein is made.

Protein

5 The protein causes changes in the cell's activity.

Cell

quantities that leach from plastic containers into food and drinks—BPA might be particularly dangerous for pregnant mothers, developing fetuses, and young children.

An equal number of researchers thought that developing fetuses were safe from any potential effects. They reasoned that BPA would be broken down in the mother's gut and excreted in her urine long before it had a chance to reach the womb. Even if the gut failed, they thought, BPA would never get past the placenta.

According to one 2002 study of 37 pregnant women, they were mistaken. In that study, researchers found BPA not only in maternal blood, but also in fetal blood and the placenta. Although no one could say for certain whether the quantities that reached the fetus were enough to cause harm, this study and others like it were enough to convince some scientists that BPA ought to be more tightly regulated. [INFOGRAPHIC 3.6]

But when the 2008 report came out saying as much, a firestorm erupted.

Critical thinking skills give us the tools to uncover logical fallacies in arguments or claims.

A wide range of media outlets jumped on the NTP report, producing scores of articles that read a lot like this one:

Tests performed on liquid baby formula found that they all contained bisphenol A (BPA). This leaching, hormone-mimicking chemical is used by all major baby formula manufacturers in the linings of the metal cans in which baby formula is sold. BPA has been found to cause hyperactivity, reproductive abnormalities and pediatric brain cancer in lab animals. Increasingly, scientists suspect that BPA might be linked to several medical problems in humans, including breast and testicular cancer.

While the article is factually correct, it commits several errors of omission. By failing to explain the high degree of uncertainty (similar studies reached different conclusions, scientists had yet to reach any sort of consensus about what the risks might be) the author creates the impression that the risks are far more certain, and dire, than they actually are. Hundreds of articles in dozens of publications followed a similar tack, and before long, the story was shortened, in the public's mind, to one simple statement: BPA causes cancer.

Almost immediately, environmental and consumer groups began calling for a nationwide ban on the chemical. These groups frequently attacked the plastics industry, suggesting that because large powerful companies profited handsomely from BPA-made products, they had an incentive to downplay or even suppress troubling data. Such *red herring* and *ad hominem* attacks succeeded in stirring up the public's fears, but told them nothing about whether or not BPA was actually safe. [INFOGRAPHIC 3.7]

The plastics industry countered by repeatedly reminding the public that neither the EPA nor the FDA had moved to ban BPA from use, and that the recent NTP report cited only "some concern" about safety. This was also true. But it ignored the fact that a small army of National Institutes of Health-funded scientists had long expressed concerns about the purported dangers of low-dose BPA exposure,

---

critical thinking Skills that enable individuals to logically assess the information they find, reflect on that information, and reach their own conclusions.
logical fallacies Arguments which attempt to sway the reader without using actual evidence.

and were now urging regulatory agencies to at least set a new, lower exposure limit.

Average consumers were left to figure out for themselves which side to trust: Were environmental activists exaggerating, or was BPA truly dangerous? Why did some scientific studies show deleterious effects from the chemical, while others did not? More importantly, what precautions could, or should, individuals take?

Critical thinking is the antidote to **logical fallacies**; it enables individuals to logically assess and reflect on information and reach their own conclusions about it. The skill set can be broken down into a handful of tenets, or measures.

Be skeptical. Just like a good scientist, one should not accept claims without evidence, even from an expert. This doesn't mean refusing to believe anything; it simply means requiring evidence before accepting a claim as reasonable. For example, a *Science News* article, "Receipts a Large and Little-Known Source of BPA," quoted a researcher as saying that he believed BPA from paper cash register receipts "penetrated deeply into the skin, perhaps as far as the bloodstream." The actual research documented the amount of BPA in the receipts and recovered from the surface of fingers that held a receipt, not the amount in the bloodstream of individuals who handled the receipts. A skeptical reader would therefore question the inference that BPA from receipts enters the body, since no evidence was presented regarding its uptake, only its presence on the skin's surface. Perhaps BPA does enter the bloodstream through receipts, but we need evidence in order to view this claim as reasonable.

Evaluate the evidence. Is the claim being made derived from anecdotal evidence (unscientific observations, usually relayed as secondhand stories) or from actual scientific studies? If it is based on actual studies, how relevant are those studies to the claim? Were they done on primates? Rodents? Cells in a Petri dish? Or did the researchers look at human populations? Most of the BPA studies were done

---

Infographic 3.7 | **DON'T BE MISLED BY LOGICAL FALLACIES**

↓ Logical fallacies are literary devices used to confuse or sway the reader to accept a claim/position in the absence of evidence.

COMMON LOGICAL FALLACIES	EXAMPLE
Hasty generalization: Drawing a broad conclusion on too little evidence.	A study might show that BPA is present in the urine of babies who drink from plastic bottles; however, this is not evidence that babies are being poisoned by their bottles. All we could conclude would be that babies who drink from plastic bottles have more BPA in their urine than babies who do not.
Red herring: Presents extraneous information that does not directly support the claim but that might confuse the reader/listener.	An argument that the buildup of toxins in modern humans is significant and that many of these buildups have led to problems may be true but tells us nothing about the safety of BPA.
Ad hominem attack: Attacks the person/group presenting the opposite view rather than addressing the evidence.	Opposing the use of BPA on the grounds that the chemical industry is untrustworthy (because they profit from plastics) does not address the safety of BPA.
Appeal to authority: Does not present evidence directly but instead makes the case that "experts" agree with the position/claim.	Claiming that BPA is a health hazard (or not) because a noted toxicologist or scientific group has drawn that conclusion is not evidence in itself. If the experts agree, there should be a reason, and that evidence should be presented.
Appeal to ignorance: A statement or implication that the issue is too complex and we are not capable of understanding it.	One might claim there is no way to know the effect of BPA since humans are exposed to so many toxins in order to justify doing nothing about the use of BPA. We may not know everything about the effects of BPA, but we do know some things and it is on that evidence that we should draw our conclusions.
False dichotomy: The argument sets up an either/or choice that is not valid. Issues in environmental science are rarely black and white, so easy answers (it is "this" or "that") are rarely accurate.	The claims that BPA must be *completely avoided* or is *totally safe for everyone* is a false dichotomy. The evidence so far suggests that the effects of BPA may be negligible for healthy adults but problematic for fetuses and very young children. Recommendations differ for different groups.

on rodents. As we'll see, some of those rodents were very good models for human health effects and some were not.

Be open minded. Try to identify your own biases or preconceived notions (most chemicals are dangerous; most people overreact to things like this, etc.) and be willing to follow the evidence wherever it takes you.

Watch out for biases. Does the author of the study or person making the claim have a position they are trying to promote? Are they financially tied to one conclusion or another? Are they trying to use evidence to support a predetermined conclusion?

In their defense of BPA, the plastics industry frequently cites a 2004 report by the Harvard Center for Risk Analysis, which had concluded that the evidence for adverse effects from low-dose BPA was weak. Vom Saal criticized this report because it looked at only 19 of the more than 115 published studies available at the time. The report was funded by the American Plastics Council and vom Saal charged that the authors only chose to examine those studies that gave the desired "no effect" results. When he himself evaluated all 115 published studies, he found that results (effects versus no effects) were strongly correlated with the source of funding. None of the 11 industry-funded studies found adverse effects from BPA, whereas 90% (94 out of 104) of the government-funded studies showed significant effects of one kind or another.

The report's authors argued that they excluded studies where rats had been injected with BPA rather than force-fed food pellets laden with the chemical. They reasoned that because ingestion was the most common route of exposure for humans, it made more sense to feed the rodents than to inject them. They felt that effects seen in the injection studies were not relevant to humans because in humans, BPA would not enter directly into the bloodstream. Vom Saal and others disagreed. "Fetuses and newborns—the ones we are most concerned about—haven't yet acquired the gut microbes that adults have," he said. "So we can't expect their digestive systems to break BPA down as efficiently as an adult digestive system might. That's as true for rodents as it is for humans."

As it turned out, there was another technical reason that some of the industry studies did not find low-dose effects. It had to do with the type of rodent those researchers were using. Rather than use the strain of rats that had been so instrumental in cracking the DES case, industry scientists used a strain of rat known to be highly resistant to estrogen. This was akin to stacking the deck in favor of BPA: If the rats weren't sensitive to estrogen, there was virtually no chance that they'd be affected by

an estrogen mimic like BPA. The panel that composed the 2008 report chose not to include these studies in its assessment.

Risk assessments help determine safe exposure levels.

By the time that panel drew its conclusions, BPA was already a $3 billion industry unto itself; it was used in an infinite variety of products, from baby bottles and food and beverage cans to dental sealants and football helmets. And, in some cases at least, it had no obvious substitute. That made employing the precautionary principle, or resetting the exposure limits, almost impossible. "Makers of plastic bottles had other things they could substitute BPA with," says Aaron L. Brody, a food scientist at the University of Georgia. "But what about cans?" Food manufacturers have long used BPA to create epoxy linings for steel cans. This coating makes the cans extremely durable; it protects the contents of canned goods without affecting taste, and, most importantly, the coatings are integral to reducing the incidence of bacterial contamination and foodborne illness. "BPA-based plastic is among the few materials so far that can withstand the high temperatures and pressures used to kill bacteria," says Brody. "It's not a perfect plastic, but without it, you'd probably have a lot more food preservation and safety issues."

As scientists debated, and regulatory agencies struggled, people were left to decide for themselves whether the risk posed by continued exposure to BPA warranted the hassle of trying to purge it from their lives, or at least from their baby bottles. "Sometimes, if there's enough momentum from the public, regulatory change will follow," says Vogel. "But more often than not, because of the way the system works, it really has to start with individual consumers deciding for themselves that they won't purchase or use certain items."

To make such decisions, we need to know two things: whether a given chemical has the potential to harm us, and how great that harm might be. It's crucial to consider both. Say, for example, a new study reports that BPA exposure doubles one's chances of developing a particular disease. Before hitting the panic button, we must first ask what the chances are of getting the disease in the first place, without BPA exposure. If it's a rare condition, with a 1% probability, then BPA increases our chances to a mere 2%. If it is more common, with, say, a 20% probability, then BPA doubles our chances to 40%—a significantly greater risk. How important that increase is, and thus, what we should do in light of it, depends on the seriousness of the disease.

So, what to do about BPA? Let's take a look at what we've learned so far: We know that chemicals like BPA can have effects at low doses. We know that developing fetuses and young mammals are particularly susceptible to these low-dose effects. We know that BPA is, in fact, leaching into our systems from food and beverage containers, and that it can indeed cross the human placental barrier.

But we also know that BPA is water-soluble and can be excreted in our urine. This means that, in most cases, it may not be building up, or bioaccumulating, in our tissues.

What we don't know—what we may never know—is how well the effects seen in mice correspond to the risks faced by humans. And because of all the additive and synergistic effects BPA is likely to have with all of the countless other chemicals we encounter, it will be difficult to tell, even in the future, how much of any given health condition can be specifically attributed to BPA.

The bane of risk assessment is that we can't really wait for those facts to come in.

Over the past few years, the public has employed a de facto precautionary principle. With the scientific jury still out, consumer and environmental groups took matters into their own hands. They issued dozens of public statements advising consumers to avoid using food and beverage containers made with BPA. They also lobbied

Congress for a full, nationwide ban. The public response to those campaigns led Walmart to say that it would no longer stock any bottles or containers made with BPA on its shelves. Facing hundreds of millions of dollars in lost sales, Nalgene, a leading plastic bottle maker, finally conceded, promising that it would phase BPA out over the next 5 years. Other companies quickly followed suit.

By 2011, both Canada and the European Union have banned BPA, several individual U.S. cities and states (including Chicago and Minnesota) have banned the sale of baby bottles made with BPA, and more and more companies have stopped using it. In fact, many plastic bottles now tout their BPA-free status.

A century hence, such drastic measures may prove unwarranted. Even vom Saal concedes that it is possible the long-term studies on BPA will not yield conclusions nearly as horrendous as those on, say, DES. It's possible that BPA will prove to be harmless. But in the meantime, he insists, we're much better safe than sorry.◉

Research articles referenced in this chapter:
Inhido, M. and Suzuki, J. 2010. *Journal of Health Science*, 56: 175–181.
Lang, I.A., et al. 2008. *Journal of the American Medical Association*, 300: 1303–1310.
Nagel, S.C., et al. 1997. *Environmental Health Perspectives*, 105: 70–76.
Schönfelder, G., et al. 2002. *Environmental Health Perspectives*, 110: 703–707.
Vom Saal, F.S., and Hughes, C. 2005. *Environmental Health Perspectives*, 113: 926–933.

BRING IT HOME

❂ PERSONAL CHOICES THAT HELP

What do some air fresheners, nail polish, and plastic storage containers have in common? They are all potential sources of chemicals that people are exposed to every day. Many of the products designed to improve our lives actually contain chemicals that may harm us in the long run. With just a few changes, you can dramatically reduce your long-term chemical exposure.

Individual Steps
→ Check to see if the body products you use contain potentially harmful chemicals at www.ewg.org/skindeep.
→ Avoid storing or microwaving food in plastic containers; chemicals such as BPA can leach into your food when heated.

Always use microwave-safe glass or ceramic containers when microwaving food.
→ Check with your city's solid waste agency for guidelines about how to correctly dispose of household wastes including paint, medication, cleaning products, and yard chemicals.

Group Action
→ Talk to your friends and family, especially those you live with, to see if you can switch to products with fewer harmful chemicals.

Policy Change
→ Research the policy points of the "Safe Chemical Acts of 2011," the proposed

overhaul of the 35-year-old Toxic Substances Control Act. Chronicle its progress through Congress and contact your legislator to voice an opinion about the policy.

UNDERSTANDING THE ISSUE

CHECK YOUR UNDERSTANDING

1. **What are toxins?**
    a. Natural or human-made chemicals that cause direct damage upon exposure to them
    b. Human-made chemicals like BPA that have been implicated in causing human health problems
    c. Chemicals that interfere with the endocrine system, typically by mimicking a hormone or preventing a hormone from having an effect
    d. Chemicals released by the body that direct cellular activity and change how the body functions

2. **The test data for a new pesticide provided mixed results about its effects on mice. The pesticide seemed harmless on its own, but when the new pesticide was combined with another pesticide some mice got sick. However, this new pesticide was very effective against an invasive insect ravaging local trees. In deciding to approve the pesticide for sale, the chemists were applying:**
    a. the precautionary principle.
    b. logical fallacies.
    c. information literacy.
    d. risk assessment.

3. **Pesticides are useful to increase agricultural yields but some people oppose their use because of their inherent toxicity. An argument against pesticide use that attacks the pesticide maker on the grounds that they are simply profit driven is:**
    a. an ad hominem attack.
    b. an appeal to authority.
    c. an appeal to ignorance.
    d. a false dichotomy.

4. **All of the following are secondary sources of information except:**
    a. an encyclopedia.
    b. a newspaper article.
    c. a peer-reviewed article.
    d. a government website.

5. **Pesticides such as DDT are known to biologically magnify in food chains. This means that:**
    a. organisms in lower trophic levels accumulate lethal doses of toxins.
    b. organisms at higher trophic levels have more concentrated levels of toxins.
    c. toxins build up in the tissue of an organism over the course of its lifetime.
    d. the environment has higher concentrations of toxins than organisms in the food chain.

6. **The toxicity of a chemical is evaluated using animal models by creating a(n) _____; a high LD50 indicates _____.**
    a. *in vitro* study; a safe dose
    b. dose analysis; a threshold dose
    c. *in vivo* study; high toxicity
    d. dose-response curve; low toxicity

WORK WITH IDEAS

1. What potential dangers do toxins present and what characteristics of a toxin make it more or less harmful?

2. What are the differences between *in vitro* and *in vivo* studies? How are these types of studies useful in toxicology?

3. Why is the debate over the safety of synthetic chemicals contentious?

4. What is the precautionary principle? In what way does the story of BPA reflect the precautionary principle?

5. What factors influence the toxicity of a given chemical? How do regulators setting "safe exposure" standards deal with the uncertainty associated with these factors?

ANALYZING THE SCIENCE

The graphs on the following page come from a 2005 study by the Toxic-Free Legacy Coalition and the Washington Toxics Coalition to identify chemical residues in the human body (http://pollutioninpeople.org/results). Samples of hair, blood, and urine of ten Washingtonians were tested for a variety of toxins including mercury and DDT. Data for each individual (identified by his or her initials) are shown for p,p' DDE (a breakdown product of DDT) and for mercury.

INTERPRETATION

1. Identify the people with the highest and lowest levels of mercury and DDT.

2. Compare the levels of mercury and DDT in each study participant to the national median. Which study participants have levels of DDT and mercury that are at least twice the national median?

3. The EPA "safe dose" for mercury is 1,100 ppb (parts per billion) for women in their childbearing years, as mercury levels above this value may impair neurological development in the fetus. Which study participant is above the safe level and by how much?

ADVANCE YOUR THINKING

4. Who conducted this study and for what purpose? What type of study do these data represent? Are the data reliable? Explain your responses.

5. The pesticide DDT has been banned in the United States since 1972. How do you explain the presence of DDT in 8 of the 10 study participants?

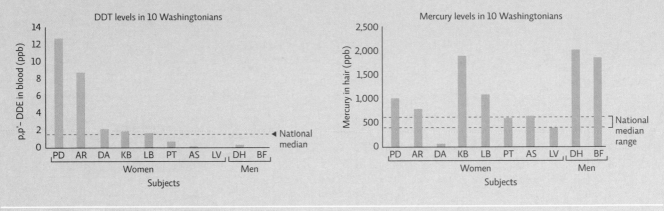

EVALUATING NEW INFORMATION

Some sources estimate that there are more than 12,000 chemical ingredients used in cosmetics, of which less than 20% have been reviewed for safety. There are no laws in the United States to regulate chemicals in cosmetic products; it is up to us individually to decide on the safety of a particular product. The Environmental Working Group's Skin Deep cosmetics database (www.ewg.org/skindeep) has ratings for more than 68,000 products from approximately 3,000 brands. Each product is rated on a scale of 0–10 for the level of hazard (low, moderate, high) and data availability for the ingredients.

Evaluate the website and work with the information to answer the following questions:

1. Is this a reliable information source? Does it have a clear and transparent agenda?
    a. Who runs this website? Do the credentials of the organization running the site make the information presented reliable or unreliable? Explain.
    b. What is the mission of this website? What are its underlying values? How do you know this?
    c. What data sources does EWG rely on and what methodology does the organization employ in constructing its databases? Are the sources EWG uses reliable?

d. Identify a claim made on this website. Is there sufficient evidence provided in support of this claim? Are there any logical fallacies used? Explain.
e. Do you agree with EWG's assessment of the problems with and concerns about cosmetics and other personal care products? What about its solutions? Explain.

2. Select one of your favorite personal care products that is also on the EWG website:

    a. Check out the company website for that product. What sort of information about the product and its ingredients does the company website offer? Does the company employ any logical fallacies in talking about the product or its ingredients? Is the information provided useful to you as a consumer? Explain your responses.
    b. How do the two websites (Skin Deep and the company website) compare in providing useful and reliable information about your particular product? Explain.
    c. Search the Skin Deep website for alternative brands that have a safer rating than your personal care product. Are there other choices? Would you consider switching your brand? Explain.

MAKING CONNECTIONS

TRICLOSAN—A TOXIN OF EMERGING CONCERN?

Background: Triclosan (2,4,4'–trichloro-2'-hydroxydiphenyl ether) is a synthetic chlorinated aromatic compound used as a broad-spectrum antimicrobial agent in a variety of consumer products such as toothpaste, antibacterial hand soaps, deodorants, and detergents. It has also been incorporated into plastics such as children's toys and kitchen utensils, as well as many other industrial and household items. Triclosan was first registered as a pesticide in 1969. According to one estimate, the annual use of triclosan is more than 300,000 kg/year.

Triclosan's efficacy and value are under question. The American Medical Association does not consider triclosan either necessary or efficacious in personal care products (see www.ama-assn.org/resources/doc/csaph/csaa-00.pdf). Physicians are particularly concerned about the development of antibiotic resistance, among other health concerns (triclosan has been shown to be an endocrine disruptor and allergan). There are also concerns about the impact of triclosan and its breakdown products on the environment.

Case: You have been assigned the task of critically evaluating the research on triclosan and developing a position paper for the future use and regulation of this chemical. In your report, include the following:

a. An assessment of the costs versus the benefits of triclosan in various products. Be sure to discuss health, economic, and environmental costs and benefits.
b. An analysis of any logical fallacies in how stakeholders (both those for and those against the use of triclosan) present their arguments.
c. Based on the information at hand, what is the best course of action for the future of triclosan? Who should be involved in making this decision? Provide justifications for your proposal.
d. Based on the story of BPA and triclosan, what kind of balance should we strike between the economic well-being of chemical companies and the health of humans and the environment? Develop a set of guiding principles for the development and use of chemicals in our society. Discuss the principles you develop and explain the underlying rationale.

Factory

Products

Some parts are reprocessed into new products.

Some parts are reused in their original form.

→ In cradle-to-cradle management, the manufacturer is responsible for the product from its production (cradle) to its final disposition after the consumer is finished with it. If the item were merely disposed of, it would be sent to its "grave" and those resources wasted. If, however, the item is disassembled and the parts reused, these parts become raw material again for a new product—a new "cradle." This provides an incentive to produce the product in a way that uses durable, reuseable parts and that minimizes toxins, since the manufacturer is responsible for dealing with those toxins.

Products are disassembled.

WALL TO WALL, CRADLE TO CRADLE

A leading carpet company takes a chance on going green

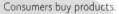

Consumers buy products.

Consumers use products.

Consumers return products to the factory when finished with them.

Waste is minimized because many components are reclaimed.

CORE MESSAGE

Human impact on Earth can be measured in terms of our ecological footprint which increases as our population increases, but is also closely tied to the way we use resources. Our economic choices, both corporate and individual, tend to focus on short term gain rather than long term sustainability, but we can make better and more informed decisions by taking all the costs—economic, social, and environmental—of a given action into account. Using nature as a model can help us make more sustainable choices while still supporting a viable economy.

GUIDING QUESTIONS

After reading this chapter, you should be able to answer the following questions:

→ What are ecosystem services? How can it be useful to place a monetary value on these services even if we know it will not be accurate?

→ What is the concept of an ecological footprint, and how does it relate to our use of natural interest and natural capital?

→ What factors influence how much human actions impact the environment, and how can we reduce that impact?

→ What are externalities and internalities in the business world and how do these concepts relate to the concept of true costs?

→ How does environmentally based economics differ from mainstream economics, and how might ideas from environmental economics help industry and consumers make better choices?

It was the summer of 1994, and Ray Anderson was feeling pretty good about things. His Atlanta-based company, Interface Carpet, was the world's leading seller of carpet tiles—small square pieces of carpet that are easier to install and replace than rolled carpet—raking in more than a billion dollars per year. One day, though, an associate from his research division approached him with a question. Some customers apparently wanted to know what Interface was doing for the environment. One potential customer had told Interface's West Coast sales manager that environmentally speaking, Interface "just didn't get it."

Anderson was dumbfounded. The carpet industry was not generally an eco-conscious industry; after all, synthetic carpet is made from petroleum in a toxic process that releases significant amounts of air and water pollution, along with solid waste. Indeed, Interface used more than a billion pounds of oil-derived raw materials each year, and its plant in LaGrange, Georgia, released 6 tons of carpet trimming waste to landfills each day. "I could not think of what to say, other than 'we obey the law, we comply,'" he recalls—in other words, his company did things by the book, in terms of the environment. That wasn't enough? His research associate suggested that the company launch a task force to create a companywide environmental vision. Anderson agreed, albeit reluctantly.

Desperate for inspiration, Anderson began leafing through *The Ecology of Commerce*, a book by environmental activist, entrepreneur, and writer Paul Hawken, which one of his sales managers had lent him. The book told the story of a small island in Alaska to which the U.S. Fish and Wildlife Service had introduced a population of reindeer in World War II. Although the reindeer thrived for a time on the available plants, eventually the population exploded beyond what their environment could support. They ultimately died out because, as Anderson explains, "you can't go on consuming more than your environment is able to renew." Yet that, he suddenly realized, was precisely what Interface was doing—using more resources than it could possibly renew. "As I read the book, it

↑ Ray Anderson, founder of Interface. The background displays sample pieces of his carpet tiles.

◉ **WHERE IS LAGRANGE, GEORGIA?**

ATLANTA
◎

● LAGRANGE

GA

GEORGIA

became clear that, God almighty, we're on the wrong side of history, and we've got to do something."

Anderson realized that he had to make changes to Interface—he needed to build it into a **sustainable**, environmentally sound business. "I didn't know what it would cost, and I didn't know what our customers would pay, so it was a leap of faith," Anderson recalls. "I knew we had to do this, but it was like stepping off of a cliff and not knowing where your foot was going to come down."

By virtue of our sheer numbers, humans have a substantial impact on the environment. But the choices businesses (and by extension, consumers) make can increase or decrease that impact. The amount and type of energy and water they use, the way they handle the waste they produce, the raw materials they use—these decisions affect not only business operations themselves, but also Earth as a whole, especially considering the magnitude of the resources and waste that some large businesses use and produce.

Businesses that are environmentally mindful don't just try to minimize their impact on nature—they actually

look to nature as an economic model from which to learn and model their choices. After all, **economics**—the social science that deals with how we allocate scarce resources—is not just about money. Most of the resources on which we depend actually come from the environment. Environmental resources like timber and water can be considered ecosystem goods. And ecological processes like water purification, pollination, climate regulation, and nutrient cycling are essential **ecosystem services** that also deserve to be considered in an economic accounting of value. Some of these resources are priceless because there are no substitutes—consider the oxygen produced by green plants, which we need in order to survive. [INFOGRAPHIC 4.1]

As explained in Chapter 1, when ecosystems are intact, they are naturally sustainable: They rely on renewable

---

sustainable Capable of being continued without degrading the environment.
economics The social science that deals with how we allocate scarce resources
ecosystem services Essential ecological processes that make life on Earth possible.

Infographic 4.1 | **VALUE OF ECOSYSTEM SERVICES**

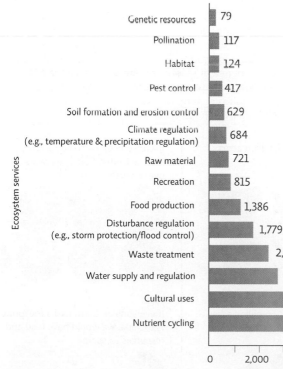

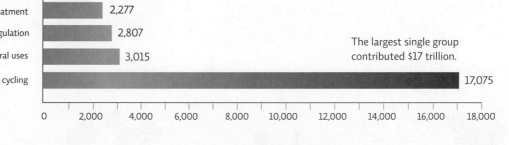

↓ Robert Constanza, an ecological economist at Portland State University, evaluated ecosystem services and quantified their values to be more than $33 trillion (1994 dollars). The estimate in 2009 dollars is $44 trillion. Though the figures are considered to be gross underestimates, especially for entities hard to quantify like habitat and genetic resources, they point out that ecosystems provide us with valuable, sometimes irreplaceable services, but that they can continue to do this only to the degree that human impact will allow. When we degrade ecosystems, we reduce their ability to provide these services.

Ecosystem services

Ecosystem service	Value
Genetic resources	79
Pollination	117
Habitat	124
Pest control	417
Soil formation and erosion control	629
Climate regulation (e.g., temperature & precipitation regulation)	684
Raw material	721
Recreation	815
Food production	1,386
Disturbance regulation (e.g., storm protection/flood control)	1,779
Waste treatment	2,277
Water supply and regulation	2,807
Cultural uses	3,015
Nutrient cycling	17,075

The largest single group contributed $17 trillion.

Billions of dollars per year (1994 dollars)

Infographic **4.2** | **ECOLOGICAL FOOTPRINT** ↓ The ecological footprint is the land area needed to provide the resources for and assimilate the waste of a person or population and may extend far beyond the actual land occupied by the person or population; it is usually expressed as a per capita value (hectares or acres/person). The current world footprint would require about 1.5 Earths to maintain, but obviously we just have one to work with.

Raw materials used in the city are imported from elsewhere.

Some waste is assimilated by areas outside the city.

Physical footprint of the city

Ecological footprint

BOTH PER CAPITA IMPACT AND POPULATION SIZE AFFECT HOW MUCH ENVIRONMENTAL IMPACT A NATION HAS

↓ The United States doesn't come close to having enough land area to meet its current ecological footprint; China, a more populous nation, just barely misses the mark, due to its lower per capita footprint. But as China develops economically and more of its people enter the middle class, per capita consumption will likely increase, along with China's overall impact.

U.S. POPULATION 300 MILLION

Footprint = 24 acres/person

U.S. uses 7.2 billion acres.

If everyone on Earth had a footprint like the United States, we would need over 6 Earths.

~2 BILLION ACRES

👤 =

CHINA'S POPULATION 1.3 BILLION

Footprint = 2 acres/person

China uses 2.6 billion acres.

If everyone on Earth had a footprint like China, we would have land and resources to spare.

~2 BILLION ACRES

👤 = ●

resources and also provide services that help to renew and recycle these resources. But ecosystems will only be able to provide us with their valuable goods and services as long as we let them. As Anderson came to realize, when we degrade ecosystems by using more from them than we replenish, we threaten our planet's ability to provide the services we need, and this ultimately threatens our own future. By using nature as a model, businesses can lessen their impact on the environment and still make choices that support a viable industrial economy.

Businesses and individuals impact the environment with their economic decisions.

Like many businesses, Interface Carpet has a large **ecological footprint**, that is, the land needed to provide its resources and assimilate its waste (typically expressed as hectares (ha) or acres (ac) per person or population). The ecological footprint is a value that businesses, individuals, and populations use to quantify their impact on the environment.

The ecological footprint of a population is influenced by both its sheer size (see Chapter 5 for more on human population growth) and its per capita (per person) use of resources. This means a small population of people, each using many resources and generating a lot of waste, can have an impact as big, or bigger, than a much larger population of people with a lower per capita impact.

The United States, for instance, has a particularly high per capita footprint, in that it requires much more land area to support each person than it actually possesses—the country is forced to import resources from other countries, and even export some waste. In fact, if the 7 billion people who currently populate the planet all lived like the average person in the United States, we would need the landmass of six or more Earths to sustain everyone. According to the World Wildlife Federation's 2010 *Living Planet* report, humans currently use 30% more resources than is ultimately sustainable. This means that unless we stop using them so quickly, we are going to run out. [INFOGRAPHIC 4.2]

What kinds of essential resources does Earth provide us? Considered in financial terms, our **natural capital** includes the natural resources we consume, like oxygen, trees, and fish, as well as the natural systems—forests, wetlands, and oceans—that produce some of these resources. Our **natural interest** is what is produced from this capital, over time—more trees and oxygen, for example—just like the interest you earn with a bank account. Natural interest represents the amount of readily produced resources that we *could* use and still

leave enough behind to, in time, replace what we took. Natural interest might be represented by an increase in a fish population, for instance, or new growth in a forest—basically, the extra that is added in a given time frame. [INFOGRAPHIC 4.3]

If we only withdraw resources equivalent to (or less than) the natural interest, we will leave behind enough natural capital to replace what we took. When Anderson spoke to his employees in the summer of 1994 about his new plan for sustainability, he stressed that his goal was to begin putting back more than the company took from the planet—in other words, he wanted Interface to be what he calls a "restorative enterprise." Up to that point, the company was using up far more natural capital in the form of resources like petroleum and water than was ultimately sustainable. Anderson realized that if we take more than is replaced, capital will shrink and therefore produce less the next year. By continually taking more than the equivalent of natural interest, we diminish resources and can potentially eliminate them. This can be an especially big problem with commonly held resources like water: Once we remove it from wells or rivers, we have to wait for the next rainfall to replenish it. When many users are accessing the resource, it can quickly become degraded if they do not work together to manage it—a tragedy of the commons (see Chapter 1).

When we dip into our natural capital as humans are doing today, using 30% more resources than is sustainable, we are decreasing future interest potential. Essentially, we are taking resources away from the future, in what eco-architect Bill McDonough calls *intergenerational tyranny*. By liquidating our natural capital more quickly than it can be replaced and calling that "income," the question becomes this: Where will future income come from?

From 1994 to 2006, Interface made major changes in order to achieve its new goals. The company cut the amount of energy it derived from fossil fuels by 55% and reduced its total energy use by 43%. It did this in part by maximizing energy efficiency in its facilities, installing skylights and solar tubes to replace artificial, electricity-dependent lighting, and installing more energy-efficient heating, ventilation, and air conditioning systems. In one of its factories, Interface also installed a real-time

---

ecological footprint The land needed to provide the resources for and assimilate the waste of a person or population.

natural capital The wealth of resources on Earth.

natural interest Readily produced resources that we could use and still leave enough natural capital behind to replace what we took.

Infographic **4.3** | **CAPITAL AND INTEREST**

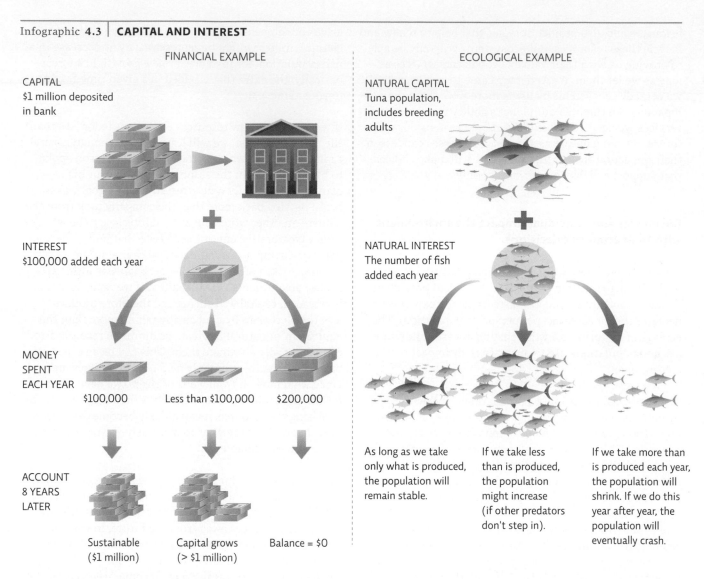

FINANCIAL EXAMPLE

CAPITAL
$1 million deposited
in bank

+

INTEREST
$100,000 added each year

**MONEY
SPENT
EACH YEAR**

$100,000   Less than $100,000 $200,000

**ACCOUNT
8 YEARS
LATER**

Sustainable   Capital grows   Balance = $0
($1 million)   (> $1 million)

ECOLOGICAL EXAMPLE

NATURAL CAPITAL
Tuna population,
includes breeding
adults

+

NATURAL INTEREST
The number of fish
added each year

As long as we take   If we take less   If we take more than
only what is produced,   than is produced,   is produced each year,
the population will   the population   the population will
remain stable.   might increase   shrink. If we do this
                            (if other predators   year after year, the
                            don't step in).   population will
                                                     eventually crash.

↑ Natural resources can be compared to the financial concepts of capital and interest. Natural capital is the wealth of resources on Earth and includes all the natural resources we use as well as the natural systems that produce some of those resources (forests, wetlands, oceans, etc.). Natural interest is the amount produced regularly that we could use and still leave enough natural capital behind to replace what we took.

energy tracker that displays energy use prominently for its employees to see, inspiring them to think of new ways to conserve energy. Although they declined to reveal how much money they invested in such improvements and technology, the company has ultimately recouped its costs in energy savings, according to a company spokesperson.

Researchers often use the **IPAT model** to estimate the size of a population's ecological footprint, or impact (I), based on three factors: population (P), affluence (A), and technology (T). The premise is that as population size increases, so does impact. More affluent and technology-dependent populations use more resources and generate more waste than do less affluent and technology-dependent populations (technology allows us to build more

things, dig deeper, and fly higher, all of which drain the environment). [INFOGRAPHIC 4.4]

One caveat with regard to this model is that technology can have the opposite effect: It can decrease, rather than increase, environmental impact. In 2006, for instance, after deciding to become sustainable, Interface invented a new technology called TacTiles— 2.5 × 2.5-inch squares of adhesive tape that join carpet tiles together. The adhesive is made from the same plastic used to make soda bottles.

IPAT model An equation (I = P x A x T) that identifies 3 factors that increase human impact (I) directly: population size (P), affluence (A), and technology (T).

Infographic 4.4 | **THE IPAT EQUATION**

↓ Population (P), affluence (A), and technology (T) all affect how much of an impact an individual or population has on the environment. As any or each of these factors increase, so does the population's overall impact, as indicated by the model's equation: I = P × A × T. The right kind of technology can actually lower overall impact, in which case the equation becomes I= (P × A)/T.

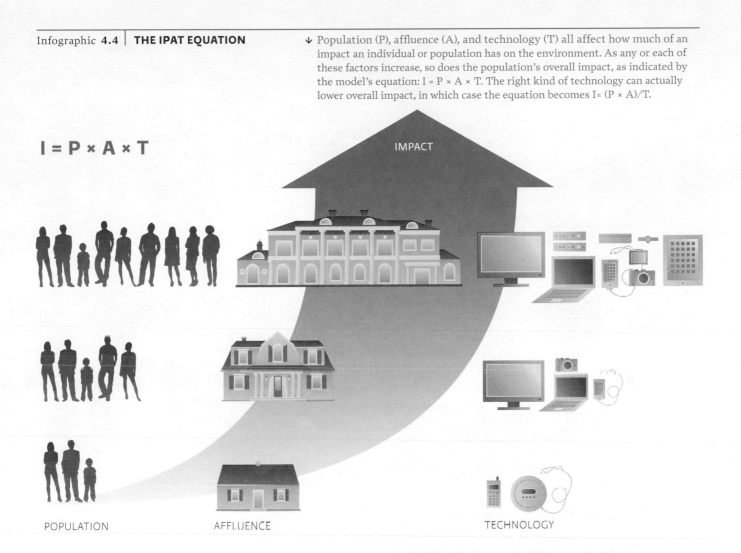

$$I = P \times A \times T$$

IMPACT

POPULATION   AFFLUENCE   TECHNOLOGY

In contrast with traditional "spread on the floor" adhesives, Interface's new tape does not contain any volatile organic compounds, which the U.S. Environmental Protection Agency recognizes as a health risk. TacTiles also make it possible for customers to replace single carpet tiles easily, when, for instance, there has been a spill. Although resources are still required to make TacTiles, they reduce health risks and waste compared to other approaches. When technologies such as TacTiles reduce the environmental footprint rather than increase it, the equation used to describe their impact changes to: I = (P × A)/T.

In 2007, to further reduce its impact, Interface launched a major carpet-recycling initiative called ReEntry 2.0. More than 5 billion pounds of carpet are pulled up and discarded globally each year, and less than 5% of that has historically been reused or recycled. With ReEntry 2.0, Interface developed a way to recycle carpets—both its own and those made by its competitors—to make new carpet, using only a small amount of virgin materials to

↑ Interface's process that reclaims old carpet and converts it into recycled raw materials diverted 28 million pounds of carpet from landfills in 2010. Since 1995, Interface has diverted a cumulative total of 228 million pounds of carpet and carpet scraps.

↑ The Sierra SunTower solar power plant in Lancaster, California, is a 20-acre array of 24,000 mirrors that focuses sunlight and creates a high-temperature steam that runs a traditional turbine generator. The facility provides energy for 4,000 homes in Southern California.

do so. With ReEntry 2.0, Interface has diverted 100,000 tons of material from landfills. Interface has promised to eliminate *any* negative impact it has on the environment by 2020, in a plan it calls "Mission Zero."

Mainstream economics supports some actions that are not sustainable.

One of the limitations of mainstream economics is that it doesn't take into account *all* potential costs. For instance, a carpet tile might require a certain amount of material that has a particular monetary cost—but what about the environmental costs associated with drilling enough oil to make that material in the first place, or the costs associated with cleaning up the pollution it creates? The direct cost of the material is an **internal cost**—a cost that is accounted for when a product or service is priced—but it is often incomplete. There can also be **external costs**, such as the health costs associated with the waste produced by making the carpet tiles or the environmental

damage caused by pollution. Historically, economists have regarded these as external to the business cost (the business doesn't pay for them) and they aren't reflected in the price the consumer pays for the good or service. But if the business doesn't pay for the costs and pass those costs on to the consumer, who does pay? Other people, present and future, and other species do. They pay in the form of degraded health, ecosystems, or opportunities.

An assessment of the cost of a good or service should include more than just the economic costs but also the social and environmental costs—the **triple-bottom line**. By ignoring the external costs, economies create a false idea of the true and complete costs of particular choices. A customer may pay $8 for every 50 square centimeter carpet tile, but the **true cost** for that piece of carpet would be much higher if it included the cost of greenhouse gas emissions and, for instance, the cost of treating people for asthma who might fall ill as a result of the particulate matter released during the carpet's production. The inadequate valuation of a product could

Infographic **4.5** | **TRUE COST ACCOUNTING**

What does it take to produce the paper you use every day?

↓ Many environmental and health costs of our goods and services are externalized (not included in the price the consumer pays). But if consumers don't pay all the costs to produce a product, such as paper, who does?

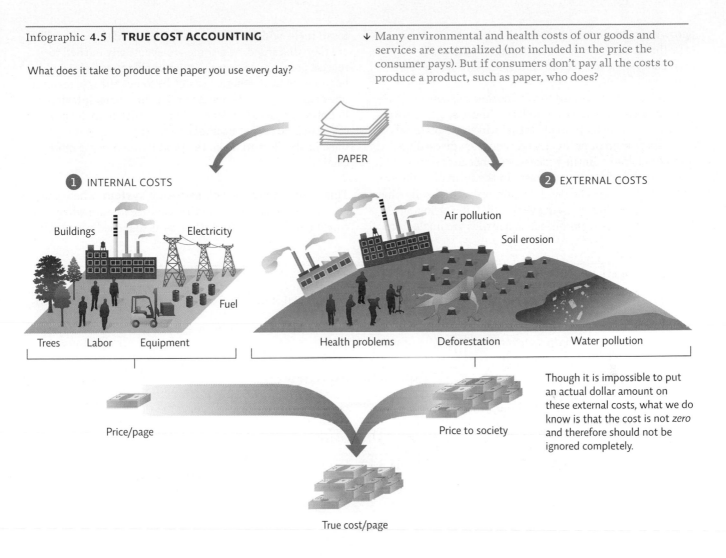

PAPER

1 INTERNAL COSTS

Buildings   Electricity

Fuel

Trees   Labor   Equipment

Price/page

2 EXTERNAL COSTS

Air pollution

Soil erosion

Health problems   Deforestation   Water pollution

Price to society

Though it is impossible to put an actual dollar amount on these external costs, what we do know is that the cost is not *zero* and therefore should not be ignored completely.

True cost/page

eventually lead to the exploitation or overuse of resources needed to produce it, an example of market failure. When external costs are internalized, on the other hand, people (or species) who don't benefit from the transaction do not pay for it. In this case, the product or service can be more appropriately priced or valued; this new price more accurately reflects its true cost.

Because we are so accustomed to not paying true costs, we would most likely be appalled at how much goods and services would really cost if all externalities were internalized. Although it sounds discouraging, any time

we purchase products that were made in a more environmentally or socially sound manner, we come a little closer to bearing the consequences of our choices. We also create a demand for these products in the marketplace. [INFOGRAPHIC 4.5]

Another inaccurate assumption of mainstream economics is that natural and human resources are either infinite or that substitutes can be found if needed. This is true for some but not all resources. For instance, fossil fuels are finite and will run out, even with technological advances that allow us to access more of the fuel that is left. It remains to be seen if we can replace fossil fuels with sustainable alternatives at current levels of use. Additionally, our actions can degrade air and water resources faster than nature can restore them; and crop productivity has its limits.

Mainstream economics also assumes that economic growth will go on forever. Since there are inherent limits to what Earth can provide, resource-dependent unlimited

internal cost Those costs—such as manufacturing costs, labor, taxes, utilities, insurance, and rent—that are accounted for when a product or service is evaluated for pricing.

external cost Costs that are not taken into account when a price is assigned to a product or service.

triple-bottom line Considering the environmental, social, and economic impacts of our choices.

true cost Including both internal and external costs when setting a price for a good or service.

growth is not, in fact, possible. We have to work within the limits of available resources in ways that allow essential ecosystem services to continue.

These assumptions lead to yet another misconception—that models of production follow a linear sequence: Raw materials come in, humans transform those materials into some kind of product, and then they discard the waste generated in the process. But, because some resources are finite—and waste in the form of pollution can damage natural capital like air, water, and soil—linear models of production will eventually fail. For instance, most traditionally produced carpet tiles are made from

fossil fuels, which are not sustainable. Old unwanted tiles are then discarded, and some are eventually burned to release greenhouse gases. A sustainable approach would be more cyclical, where "waste" becomes the raw material once again and can be used to make products. Interface's ReEntry 2.0 program uses old carpet tiles to make new ones, and old carpet backings to make new carpet backings, in an effort to make the production process more cyclical.

This is an example of a **closed-loop system**, where the product is folded back into the resource stream when consumers are finished with it, or is disposed of in such a

---

Infographic **4.6** | **ECONOMIC MODELS**

↓ Mainstream economics assumes that resources will always be available and that waste can be disposed of in a linear (one-way) system. Environmental economics recognizes that natural ecosystems provide our resources and assimilate our wastes. If companies could fold "waste" back into production or make sure it can be decomposed by nature, we could reduce our extraction costs, and be operating in a sustainable closed-loop system.

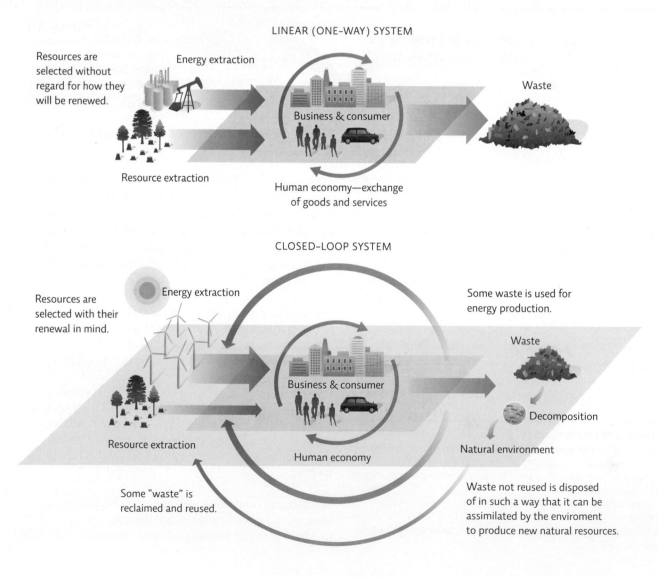

LINEAR (ONE-WAY) SYSTEM

Resources are selected without regard for how they will be renewed.

Energy extraction

Business & consumer

Resource extraction

Human economy—exchange of goods and services

Waste

CLOSED-LOOP SYSTEM

Resources are selected with their renewal in mind.

Energy extraction

Business & consumer

Resource extraction

Human economy

Some "waste" is reclaimed and reused.

Some waste is used for energy production.

Waste

Decomposition

Natural environment

Waste not reused is disposed of in such a way that it can be assimilated by the enviroment to produce new natural resources.

way that nature can decompose it. By leasing carpet rather than selling it, Interface manages its product in a **cradle-to-cradle** fashion—the carpet needs to be durable and recyclable, because Interface is responsible for the impact of its use at every stage of the process (See the diagram that opens this chapter on pages 56-57). [INFOGRAPHIC 4.6]

Another problem with traditional economics is that it **discounts future value**: It tends to give more weight to short-term benefits and costs than it does long-term ones. In other words, something that benefits or harms us today is considered more important than something that might do so tomorrow. For instance, we value the tuna we can harvest today more highly than tuna we might harvest 10 years from now, so the value of taking a large harvest of tuna today outweighs the benefits of taking less now to ensure there is still some later. If the money we could earn by using the resource now is higher than sustainable harvesting yields, modern economics tells us it is more profitable to use it now and invest the resulting money in another venture. But this investment approach doesn't take into account where those other ventures might come from or whether they are in any way diminished by the elimination of the first resource. How might the loss of tuna affect the ecosystem and other populations? Will there always be another fish population to harvest?

A sustainable approach would be more cyclical, where "waste" becomes the raw material once again and can be used to make products.

Two schools of thought are emerging regarding how to incorporate environmental considerations into economic decisions. Both **environmental** and **ecological economics** are disciplines that consider the long-term impact of our choices on human society and the environment. Environmental economists feel we can apply the tools of modern economics to solve environmental problems and still enjoy unlimited economic growth. They support actions such as improving technology to increase production efficiency and reduce waste; valuing resources as realistically as possible; moving away from dependence on nonrenewable resources; and shifting away from a product-oriented economy. In other words, the problems are real but they are solvable within current economic systems.

Ecological economists look to natural ecosystems as models for resource use and, like environmental economists,

stress recycling matter, only using renewable resources, and living within the limits of nature. However, they believe that our ingenuity will only take us so far and that economic growth does have limits; therefore, we must significantly change the way we do things in order to become sustainable as a society. While there are differences between these two approaches, both agree that some of the underlying assumptions of mainstream economics fall short with regard to how we view natural resources and our impact on the planet and people.

What about consumers? We can all decrease our impact by making more sustainable choices and by consuming less. This doesn't necessarily mean "doing without," but it does mean being mindful of our choices and opting for sustainable or low-impact choices whenever possible. However, this requires transparency from the industries that produce and sell us goods and services. That is hard to come by with current business models, often because the businesses themselves don't know all the external costs associated with their products. In order for their new plan to be a success, Interface was counting on their customers to make more sustainable choices as well. And make them they did. ReEntry 2.0 drew many new customers to Interface, including the Georgia state legislature, which purchased 13,000 yards of new carpets.

Businesses can learn a great deal about how to be sustainable from nature.

Anderson vowed in 1994 that Interface Carpet would become the world's first sustainable business. But what does that mean, exactly? By definition, **sustainable development** must meet present needs without preventing future generations from meeting their needs. It should enhance the quality of life without damaging the environment that helps meet those needs. In short, Anderson says, sustainability means "Take nothing. Do no harm." Some people have said that the term "sustainable development" is an oxymoron because development

closed-loop system A production system in which the product is folded back into the resource stream when consumers are finished with it, or is disposed of in such a way that nature can decompose it.
cradle to cradle Management of a resource that considers the impact of its use at every stage of the process.
discount future value To give more weight to short-term benefits and costs than to long-term ones.
environmental/ecological economics New theories of economics that consider the long-term impact of our choices on people and the environment.
sustainable development Economic and social development that meets present needs without preventing future generations from meeting their needs.

↑ Yumi Someya is head of U Corporation, a Japanese company which collects around 100 tons of used cooking oil from 5,400 Tokyo restaurants and 100 collecting stations each month. The company converts the oil into biodiesel to fuel city buses, service vehicles, and the company's own trucks. Someya hopes that by 2017, her company will be recycling all the used cooking oil in Tokyo—some 200,000 tons each year.

implies constant upward progress and, at some point, resource restriction will prevent further development. However, this would be true if development could only be considered a physical process, dependent on resource extraction. In reality, development can also be abstract: Some, for instance, consider improved quality of life and happiness to be development, even if it is not tied to physical resource use.

When he first vowed to make Interface sustainable, Anderson did not know whether his business would thrive or suffer as a result. "I was very apprehensive about it," he recalls. But he, along with other entrepreneurs who have followed suit, are finding that **green business**—doing business in a way that is good for people and the environment—is also profitable. It can provide a competitive advantage either because the consumer is willing to support the company's efforts or because green actions end up saving money.

Since 1994, Interface Carpet has reduced greenhouse gas emissions from its manufacturing facilities by 35%, slashed total energy use by 43% per unit of production,

and now relies on recycled or bio-based ingredients for 40% of its raw materials. During this same time, the company has increased sales by two-thirds and doubled its earnings. Some of this extra money was the direct result of its efforts—by reducing the amount of waste it produces, for instance, the company has saved $438 million dollars in waste elimination costs since 1994. But Interface has also won many new customer contracts as a result of the changes it has made. At one point, for instance, Interface was in competition with two other carpet companies over a $20 million contract at the University of California. After Interface filled out a 200-page questionnaire about how the company was addressing various environmental issues, one of the university's representatives turned to a colleague of Anderson's and exclaimed, "This is *real*."

How does a company find inspiration to become sustainable? One way is to look to natural ecosystems—a perfect example of sustainable resource use and waste minimization (biomimicry—see Chapter 1).

At Interface, Anderson was strongly inspired by biomimicry. For instance, the TacTiles technology that the company developed to replace glue was based on the physics that explains how a gecko lizard clings to walls and ceilings. The microscopic hairs on a gecko's foot bond to the molecular layer of water that's present on nearly every surface, allowing its feet to cling. Interface used this information to develop tiles that bond to one another rather than to the floor, making a kind of "floating" carpet that stays in place due to gravity rather than being actually glued to the floor. This makes carpet installation and removal much faster and easier. Interface also revolutionized its operations by considering itself part of a **service economy**; it focuses on selling a *service* rather than a *product*. The idea is simple: A customer pays for the service, and the vendor makes sure that the service is always available. The service might be the ability to photocopy pages, walk on comfortable carpet, or keep refrigerated food cold. Interface, for instance, sells the service of carpet—its color, texture, durability, and comfort—rather than the product itself. The customer pays a monthly fee to "lease" the carpet, and Interface maintains it and replaces it as needed. This encourages Interface to produce carpet that is durable and recyclable and also easily replaceable. [INFOGRAPHIC 4.7]

---

green business Doing business in a way that is good for people and the environment.

service economy A business model whose focus is on leasing and caring for a product in the customer's possession rather than on selling the product itself (selling the *service* that the product provides).

Infographic **4.7** | **PRODUCT VERSUS SERVICE ECONOMY**

↓ A service economy that focuses on providing the consumer the service desired, rather than a product, decreases resource drain and lessens waste while still potentially providing a profit for the seller.

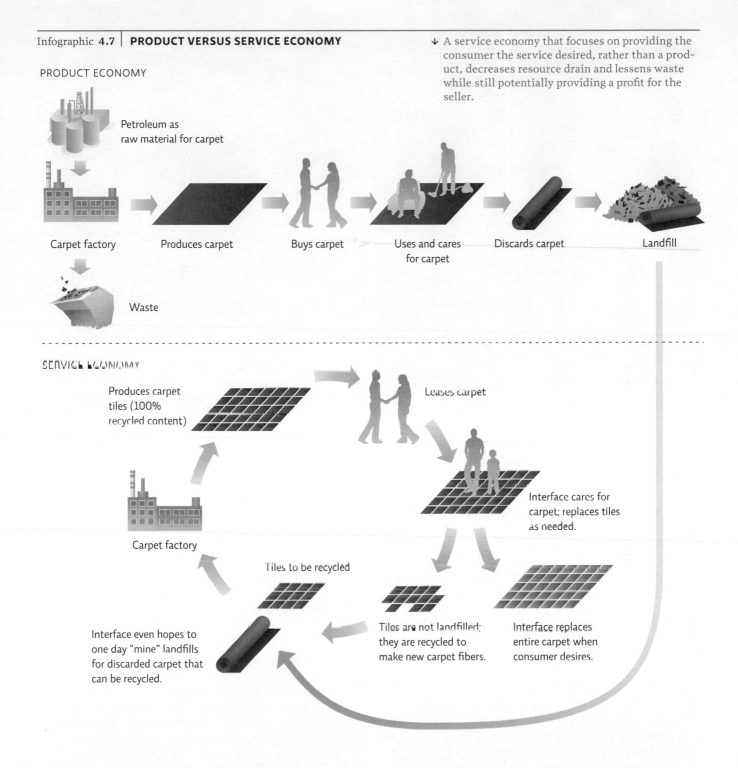

PRODUCT ECONOMY

Petroleum as raw material for carpet

Carpet factory   Produces carpet   Buys carpet   Uses and cares for carpet   Discards carpet   Landfill

Waste

SERVICE ECONOMY

Produces carpet tiles (100% recycled content)

Leases carpet

Interface cares for carpet; replaces tiles as needed.

Carpet factory

Tiles to be recycled

Interface even hopes to one day "mine" landfills for discarded carpet that can be recycled.

Tiles are not landfilled; they are recycled to make new carpet fibers.

Interface replaces entire carpet when consumer desires.

Another sustainable business practice involves *take-back programs*, particularly for products with a defined life span, such as electronics: Customers return the product to the producer when they are finished or when they need an upgrade. This provides an incentive to the producer to make a durable, high-quality product that can be reused or recycled.

There are tactics for achieving sustainability.

Changing the way we do business is not going to be easy. Even Interface has progressed slowly, despite its strong desire to become sustainable. Start-up or upgrade costs can be substantial, and even though they may pay for themselves in the long run, many businesses simply do not have the capital to fund the improvements. Plus,

↑ Craig Martineau, Brandon Sargent, and Dan Blake (from left) dropped out of Brigham Young University to pursue their green business, EcoScraps. Founded in 2010, the company collects roughly 20 tons of food waste a day from more than 70 grocers, produce wholesalers, and Costco stores across Utah and Arizona. Then it composts the waste into potting soil, which retails for up to $8.50 a bag in nurseries.

they may find themselves at a competitive disadvantage with businesses who are not trying to internalize costs. Consumers also have a role to play. Recycled paper's higher cost may more closely reflect the true cost of paper, but if consumers are not willing to put their buying dollars behind their environmental ideals, businesses that make and sell paper from trees will still be more successful. In other words, it will take changes by both consumers and producers to put business and industry on the path to sustainability. But there are things that can be done to level the playing field.

One way governments can encourage sustainability is by providing incentives for businesses to account for true costs rather than just internal costs. This could be accomplished by taxing companies based on how much pollution they generate, subsidizing environmentally friendly processes, or *cap and trade* programs that give out pollution "permits" that companies could sell if they release less pollution than they are allowed (see Chapters 25 and 31). For instance, if there is a pollution tax, it will be passed on to the consumer, who then decides whether or not to buy the product. The manufacturer that minimizes waste production and relies on fewer fossil fuels than its competitors would be able to meet the regulations at the lowest cost, have lower prices, and as a result, have a major market advantage. While promising, these policies can be complex to implement, because external costs are hard to quantify (How is a pollution tax fairly assessed?), and because they may put a burden on smaller manufacturers who have less ability to absorb the cost of upgrades; passing these costs on to the consumer also stresses low- income households.

Another important aspect of becoming sustainable requires that a business handle resources cyclically rather than in a linear fashion. A company that adopts cradle-to-cradle resource management is responsible for the resource or impact of its use at every stage of the process. It can do this by considering not just the immediate impact of using a product but also the upstream and downstream impacts—a *life-cycle analysis*. This can lead to better material choices (less toxic, more sustainable) and better process choices (reusable materials, less waste and pollution).

ecolabeling Providing information about how a product is made and where it comes from. Allows consumers to make more sustainable choices and support sustainable products and the businesses that produce them.

Anderson stresses that the consumer needs to know how a good or service is made—what the cradle-to-grave environmental impact of that product is. "You lay out for the consumer everything that goes into that product, and you lay it out for your competitors too—it's a totally transparent revelation of how you made that product, and what that footprint is at every step," he explains. This is hard to achieve when we buy so many products made in faraway places and shipped over long distances.

One way to communicate this information is through **ecolabeling**. But consumers have to be wary of labels because as "green" products become more attractive to consumers, more companies engage in *greenwashing*—claiming environmental benefits for a product when they are minor or nonexistent.

Fair trade items are, however, more likely to be produced with less environmental impact. In addition, for a product to be considered fair trade, workers must be paid a fair wage and work in reasonable conditions to produce the goods or services. *Share programs* are another useful option for items that people need infrequently, such as a car for those who live in a large city. Rather than buying, owning, and then storing the product for a large part of the time, consumers share ownership and only use the product when they need it.

Although Interface has come a long way since 1994, it is still working hard to achieve its sustainability goals. In June 2011, the company began producing its first 100% nonvirgin fiber carpet tiles, made from reclaimed carpet, fiber derived from salvaged commercial fishnets, and postindustrial waste. In addition to substantial energy savings, Interface has reduced waste some 76% since 1996 and has several LEED certified facilities. (LEED is an internationally recognized green building certification system; the acronym stands for Leadership in Energy and Environmental Design.) Practices like intercepting industrial waste destined for landfills have a positive effect, while other efforts lessen the company's overall negative impact: less toxic glues, less carpet waste. All the while, Interface is still the world's leading manufacturer of commercial carpet tiles, and its 2010 operating income increased 47% from 2009. "I think we're on the right track and we'll keep on going," says Anderson, who stepped down as the company's CEO in 2001 but still played the role of the company's conscience until his death in 2011 at the age of 77. "We'll get to the top of that mountain." ◉

Research article referenced in this chapter:
Constanza, R. *et al.* 1997. *Nature*, 387: 253–260.

UNDERSTANDING THE ISSUE

CHECK YOUR UNDERSTANDING

1. **Water would be an example of _____ , while the water cycle would be an example of _____.**
   a. an ecological footprint; the IPAT model
   b. an ecosystem good; an ecosystem service
   c. natural capital; natural interest
   d. an external cost; an internal cost

2. **The land needed to provide the resources for and assimilate the waste of a person or population is referred to as:**
   a. an ecological footprint.
   b. natural interest.
   c. sustainable development.
   d. true cost accounting.

3. **What does the term "cradle-to-cradle" mean when talking about product management?**
   a. Product materials must be tracked from production to disposal.
   b. The product is potentially more dangerous to children.
   c. Current legislation is too restrictive on new product development and causes the early demise of new businesses.
   d. Production is cyclical: "Waste" becomes the raw material once again and can be reused.

4. **"Natural interest" refers to:**
   a. activities of human society that can be continued without degrading the environment.
   b. the rules that deal with how we allocate scarce resources.
   c. essential ecological processes that make life on Earth possible.
   d. readily produced resources that can be used and still leave enough natural capital behind to replace what we took.

5. **Which of the following statements about sustainability is FALSE?**
   a. We could make all our industrial processes sustainable if we could transform linear processes into circular ones.

   b. The U.S. rate of consumption is not sustainable; if the world population consumed as much as the average U.S. citizen, we would need over six Earths.
   c. In its current form, mainstream economics is the optimal model for building sustainability because external costs are built in.
   d. Technology can be used to promote sustainability and decrease human environmental impact.

6. **Which of the following would NOT be part of an ecological economist's goals?**
   a. To identify points at which waste products of manufacturing can be landfilled
   b. To design human economies to function within natural ecosystems
   c. To increase resource use efficiency by altering manufacturing processes
   d. To eliminate the use of products that are harmful to the environment

WORK WITH IDEAS

1. What is the IPAT model? How is the equation $I = P \times A \times T$ similar to and/or different from the equation $I = (P \times A)/T$?

2. What are the differences between internal and external costs? How do these types of costs relate to true cost accounting?

3. What are the limitations of mainstream economics that make it an unsustainable model? How are environmental economics and ecological economics possible responses to these limitations?

4. Using biomimicry, conducting a life–cycle analysis, restructuring as a service economy, and ecolabeling are all methods that can be used to make businesses sustainable. Using the story of Interface, explain how employing these methods helps us to achieve sustainability in the way we manufacture and use products.

ANALYZING THE SCIENCE

The table on the following page comes from the Global Footprint Network's report, "Ecological Footprint Atlas 2010" (includes 2012 updates). Note: One hectare = 2.47 acres = 10,000 sq. meters or 107,600 sq. feet. By comparison, a standard U.S. football field is 57,600 sq. feet, or close to half a hectare.

INTERPRETATION

1. What percent of the world's population is found in low-, middle-, and high-income countries?

2. According to the table, which group of countries (by income level) does not have an ecological footprint that exceeds its total biocapacity?

3. Which regions of the world have the highest deficit—that is, which use more land than is available to them in their own country (implying that they must rely on imports from other countries)?

ADVANCE YOUR THINKING

Hint: Go to www.footprintnetwork.org/en/index.php/GFN/page/footprint_for_nations/. At the bottom of the page, click on "2011 Data Tables." Note: The income groups in the third column are: UI: Upper income; UM: upper middle; LM: lower middle; LI: lower income.

4. Compare a high-income, moderate-population country such as the United Arab Emirates with a high-income, fairly high-population country such as Japan. What patterns do you see?

5. Now compare a high-income, high-population country such as the United States with a low-income, high-population country such as Pakistan. What patterns do you see?

6. Go to www.footprintnetwork.org, click on "Footprint for You" at the bottom of the page, and calculate your own footprint. Put in the detailed information, not the basic information. Discuss your results, including some of the things the site suggests at the end that you could do to improve your footprint.

Country/region	Population (millions)	Ecological Footprint 2008 (global hectares per person)				Biocapacity 2008 (global hectares per person)			Biocapacity (deficit) or reserve
		Food production footprint	Forest footprint	Carbon footprint	Total ecological footprint	Food production	Forest land	Total biocapacity	
World	6,739.6	0.8	0.3	1.5	2.6	0.8	0.8	1.6	(1.0)
High-income countries	1,037.0	1.3	0.6	3.4	5.3	1.3	1.2	2.5	(2.8)
Middle-income countries	4,394.1	0.7	0.2	0.8	1.7	0.7	0.8	1.5	(0.2)
Low-income countries	1,297.5	0.6	0.2	0.2	1.0	0.7	0.3	1.0	(0.0)
Africa	938.4	0.7	0.3	0.3	1.3	0.8	0.5	1.3	(0.0)
Middle East/Central Asia	382.6	0.8	0.1	1.4	2.3	0.6	0.1	0.7	(1.6)
Asia-Pacific	3,725.2	0.6	0.1	0.8	1.5	0.5	0.2	0.7	(0.8)
South America	390.1	1.5	0.4	0.6	2.5	2.1	4.9	7.0	4.5
Central America/Caribbean	66.8	0.6	0.3	0.6	1.5	0.5	0.3	0.8	(0.7)
North America	448.9	1.3	0.7	4.0	6.0	1.7	1.8	3.5	(2.5)
EU	495.1	1.4	0.5	2.4	4.3	1.0	0.8	1.8	(2.5)
Other Europe	238.1	1.3	0.4	2.2	3.9	1.3	2.8	4.1	0.2

EVALUATING NEW INFORMATION

As consumers, we can rely on ecolabeling to some extent, but we need to remain vigilant to greenwashing, given that there are no clear labeling guidelines. One helpful source is the Good Guide, which has ratings for over 120,000 consumer products. Each product is given a summary score on a scale of 0–10, which is compiled from three sub-scores that address the product's health, environmental, and social impacts.

Explore the Good Guide website (www.goodguide.com).

Evaluate the website and work with the information to answer the following questions:

1. Is this a reliable information source? Does it have a clear and transparent agenda?
   a. Who runs this website? Do the credentials of the individual or organization make the information presented reliable/unreliable? Explain.
   b. What is the mission of this website? What are its underlying values? How do you know this?
   c. What data sources does the Good Guide rely on and what methodology does it employ in calculating its ratings? Are its sources reliable?

d. Do you agree with the Good Guide's assessment of the problems with and concerns about consumer products? Explain.
e. Do you agree with the Good Guide's solution (that is, its rating system)? Do you think the criteria it uses for rating products are sufficient and reasonable? Which criteria are most important to you as a consumer? Explain.

2. Select one of your favorite products that is also on the Good Guide website.
   a. How is your product rated by the Good Guide? Discuss both the overall score as well as the details of the three sub-scores.
   b. Check out the company website for your product. What sort of information about the product does the company website offer? How does it compare to the information on the Good Guide? Which information source is more useful to you as a consumer? Explain your responses.
   c. On the Good Guide website, search for alternative brands that have a better rating than your preferred product. Are there other choices? Would you consider switching your brand? Explain.

MAKING CONNECTIONS

THE EARTH CHARTER: A VISION FOR THE FUTURE

Background: The Earth Charter Initiative is an international effort whose stated mission is "to promote the transition to sustainable ways of living and a global society founded on a shared ethical framework that includes respect and care for the community of life, ecological integrity, universal human rights, respect for diversity, economic justice, democracy, and a culture of peace." The Earth Charter document (download it at www.earthcharterinaction.org) that is the basis of this initiative is promoted as a global consensus statement on what sustainability should mean and the principles by which sustainable development should be achieved at all levels of society.

Case: You have been invited to participate in the Earth Charter Initiative—that is, you are being asked to encourage people in your school, community, or workplace to endorse the Earth Charter and to apply its principles as part of their daily operations so they can join in the effort toward developing a sustainable society.

Write a letter to the people in your school, community, or workplace to describe the Earth Charter and explain why they should become part of this initiative. In your letter, include the following:

a. An assessment of the importance of sustainability and the challenges to achieving it.
b. An analysis of the Earth Charter as a framework for reaching sustainability—that is, how can the Earth Charter be used to make choices that take all the costs (economic, social, and environmental) of a given action into account and allow a school, community, or workplace to strike a balance between the current and future well-being of humans and the environment?
c. Propose some specific sustainability initiatives that your school, community, or workplace could engage in that reflect the Earth Charter principles.
d. Identify the potential challenges to implementing the Earth Charter and discuss how these challenges could be addressed.

CORE MESSAGE

A variety of factors influence whether and how fast a population grows. Our human population has grown explosively in the recent past and is still growing, especially in developing nations, leading to overpopulation in many of those areas. Though a population can have a substantial impact on its environment because of its high per capita use of resources, sheer numbers can also have a major impact. We can pursue a variety of ways to reduce population growth and stabilize population size; many of these methods focus on issues of social justice.

GUIDING QUESTIONS

After reading this chapter, you should be able to answer the following questions:

→ How and why have human population size and growth rate changed over time? What is the size and distribution of today's population?

→ What cultural and demographic factors influence population growth in a given country? How do they differ between developed and developing nations?

→ What is the "demographic transition" and why is it important?

→ What do population growth rates look like today and how can we achieve zero population growth?

→ What impact does our current world population have on the Earth and can Earth support us all?

A child looks out of a train at the Huaibei Railway Station during the travel rush period in Huaibei in eastern China's Anhui Province.

ONE-CHILD CHINA GROWS UP

A country faces the outcomes of radical population control

In most countries, including the United States, it's not hard to imagine a life without siblings—plenty of families have just one child. But what about a life without cousins or aunts and uncles? Imagine not just a single family, but an entire country of single children and you'll begin to get a sense of what China is like for the generation now entering adulthood.

Their elders call them the "Little Emperors"—a title that is meant to reflect the spoiled life and haughty temperament we sometimes associate with only children; they are the result of a colossal social experiment set in motion by the Chinese government 30-some years ago in an effort to curb population growth: one child per family. For 30 years.

Human populations grew slowly at first and then at a much faster rate in recent years.

At least twice in the course of human history, global human population has hit a dramatic growth spurt. The first came around 10,000 years ago, when the Agricultural Revolution dramatically increased the amount of food we could grow, and thus the number of mouths we could feed. As more food translated into healthier people, lower death rates, and longer life spans, the population **growth rate** increased and global population raced toward the 100 million mark. The next growth spurt came in the 1700s, when the Industrial Revolution ushered in a rapid succession of advances in both sanitation and health care—including vaccines, cleaner water, and better nutrition. Once again, death rates fell, life expectancy rose, and the population swelled. [INFOGRAPHIC 5.1]

In China, this cycle played out most dramatically during the latter half of the 20th century, when the country experienced a spectacular improvement in its standard of living. Between the 1950s and 1970s, **life expectancy** increased from 45 to 60. But as the **crude death rate** fell,

growth rate The percent increase of population size over time; affected by births, deaths, and the number of people moving into or out of a regional population.

life expectancy The number of years an individual is expected to live.

crude death rate The number of deaths per 1,000 individuals per year.

crude birth rate The number of offspring per 1,000 individuals per year.

population density The number of people per unit area.

overpopulation More people living in an area than its natural and human resources can support.

population momentum The tendency of a young population to continue to grow even after birth rates drop to "replacement rates" (2 children per couple)

Infographic **5.1** │ **HUMAN POPULATION THROUGH HISTORY**

↓ For most of human history, there have been fewer than 1 billion people on the planet.

Agricultural Revolution
10,000–8000 BCE

OLD STONE AGE	NEW STONE AGE	BRONZE AGE	

| 9000 | 8000 | 3000 | 2000 |

the **crude birth rate** held steady. By 1949, China was home to 500 million people—triple the United States' population at that time—making it the most populous country on Earth. By 1970, population had swelled to almost 900 million: roughly a quarter of the world's people crammed onto just 7% of the planet's arable land.

Today, China is the most populous nation, with more than 1.3 billion people, but it will soon be surpassed by India, which is projected to have 1.5 billion by 2030. Like many nations around the world, more and more of China's people live in big cities, with many urban areas reaching record **population densities**. [INFOGRAPHIC 5.2]

In China, such growth became a political issue. In the late 1950s, a famine claimed 30 million lives; and during the 1970s, a severe shortage in consumer goods—soap, eggs, sugar, and cotton—led to strict rationing. The government blamed the country's woes on **overpopulation**. Today, experts say that the real culprit was botched state planning. "China was not unlike other parts of the world at that time, in terms of higher life expectancies and growing population," says Wang Feng, director of the Brookings-Tsinghua Center for Public Policy. "We could have managed, as other countries did, with better agricultural and economic policies."

Maybe so. But in the late 1970s, with two-thirds of the population under the age of 30, and the baby boomers of the '50s and '60s just entering their reproductive years, China's leaders had good reason to panic. A population with a lot of young women has **population momentum**; it will continue to increase for another generation, even if each couple only has two children, just enough to replace

→ It took 130 years to go from 1 billion to 2 billion people. Since 1974, it has taken about 12 years to add each additional billion. In 2011 we passed the 7 billion mark.

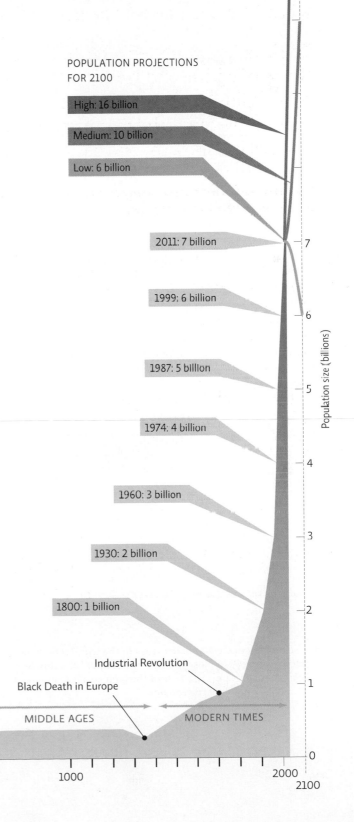

↓ Population size in 2050 and beyond will largely depend on future growth rates and may stabilize around 9 billion or, if we maintain our current growth rate, will exceed 16 billion by 2100.

POPULATION PROJECTIONS FOR 2100

High: 16 billion
Medium: 10 billion
Low: 6 billion

2011: 7 billion
1999: 6 billion
1987: 5 billion
1974: 4 billion
1960: 3 billion
1930: 2 billion
1800: 1 billion

Industrial Revolution
Black Death in Europe

IRON AGE
MIDDLE AGES
MODERN TIMES

Population size (billions)

1000
BCE (before common era)
CE (common era)
1000
2000
2100

Infographic **5.2** | **POPULATION DISTRIBUTION**

↓ The 10 most populous countries are not necessarily those with the largest land mass. China and the United States have roughly the same area—so China has a higher population density than the United States. Bangladesh has one of the highest population densities in the world, at almost 1,000 people per square kilometer.

TEN MOST POPULOUS COUNTRIES

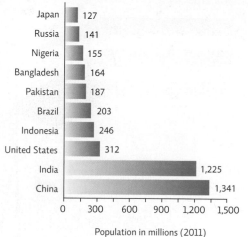

Country	Population (millions)
Japan	127
Russia	141
Nigeria	155
Bangladesh	164
Pakistan	187
Brazil	203
Indonesia	246
United States	312
India	1,225
China	1,341

Population in millions (2011)

POPULATION DENSITY

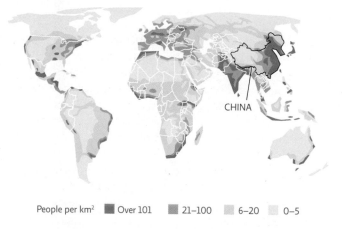

CHINA

People per km²   ■ Over 101   ■ 21–100   ■ 6–20   □ 0–5

↑ The most densely populated areas in the world tend to be in coastal areas or close to major waterways. Roughly 90% of all people live on just 10% of the land surface, and most people live north of the equator.

them when they die. If the government could not feed the mouths it already had, how in the world would it feed any more?

So in 1979, the country's leaders issued a mandatory decree: No family could have more than one child. The government vowed to deny state-funded health care and education to all but the firstborn. Parents who didn't comply also risked losing their jobs and faced severe fines (often several times their annual income) and penalties (including a loss of government subsidies for baby formula and other foodstuffs).

Dramatic as it sounded, the idea of strictly enforced population control was not entirely new. In fact, modern societies have fretted over our swelling ranks as far back as the Industrial Revolution (that second growth spurt); that's when an English priest by the name of Thomas Malthus first noticed that food supplies were not increasing in tandem with population. Unless something changed, he warned, we would soon have more mouths to feed than food to feed them with. And when that happened, disease, famine, and war would surely follow. According to Susan Greenhalgh, Harvard University anthropologist and author of the book, *Just One Child: Science and Policy in Deng's China*, such catastrophic thinking was a big part of how the Chinese government arrived at its one-child policy.

When China's leaders enacted the policy, they promised that the policy would be a short-term one, set to expire in 30 years. By then, they said, Chinese society would have adapted to a small-family culture, and with hundreds of millions of births prevented, the quality of life would have improved substantially.

Fertility rates are affected by a variety of factors.

China was once a culture built around large, extended families, and shaped by a constellation of *pronatalist pressures*—cultural and economic forces that encourage women to have more children. Parents and grandparents lived under the same roof; aunts, uncles, and cousins remained close by, usually in the same village. And couples had as many children as possible so that there would be enough hands to work the family farm, tend to household chores, and most importantly, care for parents as they aged.

Desired family size (how many children a couple wants) is one of the best predictors of actual fertility (how many children a couple has); therefore, any factors that increase or decrease one's desire to have children will significantly impact population growth rates.

为四化只生一个孩子好

THE 4 MODERNIZATION REQUIRES "ONE CHILD FAMILY"

北京计划生育宣传教育中心
BEIJING CENTER OF COMMUNICATION AND
EDUCATION FOR FAMILY PLANNING

↑ A poster promoting China's one-child policy.

The need for labor is a common pronatalist pressure in agrarian societies; so are the status and prestige associated with large families, and a lack of other options for women (in many countries, women's rights and freedoms are still strictly limited). In many developing countries, especially in Africa, a high **infant mortality rate**—the number of infants who die in their first year of life, per every thousand births—also contributes to higher fertility. Under these conditions, couples tend to have more children, which increases the odds that at least some will survive to adulthood.

infant mortality rate The number of infants who die in their first year of life per every 1,000 live births in that year.
total fertility rate (TFR) The number of children the average woman has in her lifetime.
demographic factors Population characteristics such as birth rate or life expectancy that influence how a population changes in size and composition.
developed country A country that has a moderate to high standard of living on average and an established market economy.
developing country A country that has a lower standard of living than a developed country, and has a weak economy; may have high poverty.

For centuries—right up until the 1970s, in fact—China's **total fertility rate (TFR)**, or the average number of children a woman has in the course of her lifetime, hovered between five and six. Ironically enough, the most dramatic decline in that rate took place in the decade before China implemented its one-child policy. Between 1970 and 1979, a largely voluntary mandate known as "late, few, long" managed to cut the total fertility rate in half—from 5.9 to 2.9—just by encouraging couples to marry later, delay childbearing, and space conceptions as far apart as possible.

Around the world we see big differences in the kinds of factors that influence how a population changes in size and composition. **Demographic factors** such as health (especially that of children), education, economic conditions, and cultural influences are, on average, quite different for **developed** and **developing countries**. [INFOGRAPHIC 5.3]

In China, cutting the average family size from six children down to three seemed easy. But making the leap from three children down to one proved considerably

Infographic **5.3** | **WE LIVE IN TWO DEMOGRAPHIC WORLDS**

↓ Values for demographic factors differ tremendously between developing nations like many African countries and developed nations like the United States. Most of the world's population, as well as most population growth, is in developing nations, but most wealth is in developed nations. The higher death rate in developed nations is due to an aging population; the much higher infant mortality rate in developing nations reveals the differences in quality of life and health care between the two categories.

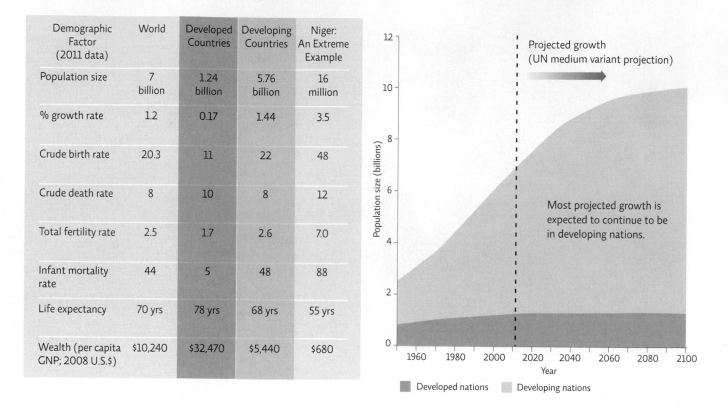

Demographic Factor (2011 data)	World	Developed Countries	Developing Countries	Niger: An Extreme Example
Population size	7 billion	1.24 billion	5.76 billion	16 million
% growth rate	1.2	0.17	1.44	3.5
Crude birth rate	20.3	11	22	48
Crude death rate	8	10	8	12
Total fertility rate	2.5	1.7	2.6	7.0
Infant mortality rate	44	5	48	88
Life expectancy	70 yrs	78 yrs	68 yrs	55 yrs
Wealth (per capita GNP; 2008 U.S.$)	$10,240	$32,470	$5,440	$680

Projected growth (UN medium variant projection)

Most projected growth is expected to continue to be in developing nations.

Population size (billions)

Year

■ Developed nations   ■ Developing nations

more difficult. One child meant no siblings, and in subsequent generations, no cousins or aunts and uncles, either. Who would help tend the farms? If a couple's child died or was disabled, who would care for them when they aged? "Many Chinese families in the countryside found it unthinkable at the time to have just one child," says Greenhalgh. So much so, that even with the penalties they faced, they resisted.

The policy's most horrendous consequences remain the source of great speculation and worry. Many experts have said that in the early years, enforcers carried out numerous sterilizations and abortions—some estimates range as high as the tens of millions—many of them by force. But because official numbers are so difficult to come by, it's nearly impossible to say whether such figures are grossly overestimating or underestimating the problem. Ultimately, the true numbers may be unknowable.

In 1998, however, a former Chinese family planning official shed some light on the inner workings of one-child, when she testified before the U.S. Congress that

women as far along as 8.5 months were forced to abort by injection of saline solution, and women in their ninth month of pregnancy, or even women in labor, were having their babies killed in the birth canal or immediately after birth. The policy's defenders say that such practices have abated, but as recently as 2001, the British newspaper the *Telegraph* reported that a single county in Guangdong Province was tasked with an annual quota of 20,000 abortions and sterilizations, after some residents disregarded the one-child policy.

Chinese officials have argued that their focus on population control has actually improved women's access to health care—both contraception and prenatal classes are free to all—and dramatically reduced the incidence of childbirth-related deaths and injuries.

But critics say that the human rights violations engendered by the one-child policy far outweigh any possible health gains and that, at any rate, health care is wildly imbalanced. One study, conducted by public health researchers from Oregon State University in 1990 in

↑ An overcrowded train approaches as other passengers wait to board at a railway station in Dhaka, Bangladesh.

the rural Sichuan Province, found that women whose pregnancies were unapproved were twice as likely to die in childbirth as those whose pregnancies were sanctioned by the government. And while nearly 90% of all married women in China use contraception—compared with roughly one-third in most other developing countries—most women are not given a choice about which type of contraceptive to use; in a 2001 survey conducted by the Chinese family planning commission, 80% of women said they were compelled to accept whatever method their family planning worker chose for them. Overall, IUDs and sterilizations account for more than 90% of contraceptive methods used since the mid-1980s.

Still, there are signs of improvement. The number of sterilizations has declined considerably since the early 1990s. And in 2002, facing strong international pressure, the Chinese government finally outlawed the use of physical force to compel a woman to have an abortion or be sterilized. (Though experts say that prohibition is not entirely enforced. "The policy is largely exercised at the local level," says Greenhalgh. "And in many provinces, local governments still demand abortions.")

In any case, by 2011, China's fertility rate had plummeted to 1.54, spurring China's leaders to proclaim that the one-child policy had succeeded in preventing some 400 million births and all the calamities those births would have brought—disease, more famine, overtaxed social services, and so on.

But while the drop in fertility was plain, not everyone would agree that one-child deserved the credit. Critics argue that by the time the first generation of "Little Emperors" reached adulthood, many other forces had reshaped their society and the world.

Factors that decrease the death rate can also decrease overall population growth rates.

When Qin Yijao's mom was pregnant with him back in 1981, the one-child policy was being strictly enforced, and she already had a son. "People from the Communist Brigade said I couldn't have a second baby," she told National Public Radio (NPR) in a 2005 interview. "But I was determined. I needed a son to work our farm." So taking care to evade the birth police, she snuck off and delivered her son in a cave. When she was found out, she was penalized. "I had to pay double the price for grain, and they confiscated our television and sewing machine." In the end, she got to keep and raise her son. But at 30, he has never worked a single day on the farm. "I don't even know where my mom's farm plot is," Qin also told NPR, adding that while he appreciates the risks his mother took, he and his wife have no desire for a second child themselves. The anecdote underscores a key point: Chinese culture has changed dramatically in the years since one-child was enacted, becoming much more urban and thus lessening the need for farmhand children. Many demographers say that such social changes may be more responsible for smaller families than the policy itself.

The **demographic transition** holds that as a country's economy changes from preindustrial to postindustrial, low birth and death rates replace high birth and death rates. The reasons are cultural as much as economic. As health care improves, infant mortality declines, and with it, the need for couples to have many children just to ensure that some make it to adulthood. And as cities grow and jobs materialize, the need for children to work the farm decreases. Meanwhile, as women find themselves with more opportunities, better health care, and greater access to education, they come to want fewer children. [INFOGRAPHIC 5.4]

Developed countries, including the United States and those of the European Union, have already undergone the demographic transition. Demographers are split over whether or not other countries are likely to do the same. But that uncertainty has not stopped world leaders from trying to cull lessons from the demographic transition model. Around the world, countries from Bangladesh to Brazil are working to reduce pronatalist pressures, increase gender equality, and combat poverty and infant mortality, all in an effort to improve standards of living, which in turn should lower fertility rates. The ultimate goal is to achieve **zero population growth**, when population size is stable, neither increasing nor decreasing. This occurs when the number of people born equals the number of people dying; in other words, **replacement fertility rate** is reached.

In the United States, the replacement rate is 2.1 (rather than the expected 2.0), since a few children die before maturity and not every female has children. In countries with high infant mortality rates, the replacement rate is higher.

There are at least a handful of clues suggesting that the demographic transition is as responsible for reductions in total fertility rates as the one-child policy is. Clue number one: The most significant fertility declines in the policy's

Infographic **5.4** | **DEMOGRAPHIC TRANSITION**

↓ Some industrialized nations have gone through the demographic transition from high birth and death rates to low birth and death rates, giving them a stable population. It is unclear whether industrialization will produce the same pattern in all developing countries. Steps to improve the quality of life and decrease death rates are important worldwide, even if other measures are needed to slow birth rates.

history coincided with the largest gains in the Chinese economy, not with periods of strict enforcement. Clue number two: Many other countries have had substantial declines in fertility during the past 25 years without the use of strict population-control policies, including Japan, Singapore, and Thailand, whose fertility rates mirror those of China's almost exactly—falling from 5 children per family in 1970 to 1.8 children per family in 2011.

"The parallels are hard to ignore," says Therese Hesketh, a global health professor at University College London. "It suggests that China could have expected a continued reduction in its fertility rate just from continued economic development." Wang Feng puts it more bluntly. "The claim that one-child prevented 400 million births is bogus for sure," he says. "Many other countries experienced nearly identical fertility declines over the same time frame. So unless you say Chinese people are necessarily different from other populations, it's very hard to argue the decline is due to government policy."

Worldwide, population growth rates are declining, but overall the number is still positive, so we are still increasing. The United Nations estimates future population size leveling off somewhere between 6 and 16 billion people by 2100, depending on how quickly we reduce growth rates (see Infographic 5.1). [INFOGRAPHIC 5.5]

But if the predominant causes of declining fertility remain a source of debate, the consequences are becoming clearer and more stark by the day. Because the reasons for population growth are likely to be different in different regions, reaching zero population growth requires that the pronatalist pressures in each region be addressed. A variety of measures that don't include government-restricted family size have proven successful.

Programs that address the needs of a given population and work within its cultural and religious traditions are likely to be most successful. Many demographers believe that addressing the social justice issues associated with overpopulation (poverty, high infant mortality, lack of education and opportunities for women) will prove the most instrumental in enabling countries with high TFRs to confront what lies ahead. [INFOGRAPHIC 5.6]

demographic transition Theoretical model that describes the expected drop in once-high population growth rates as economic conditions improve the quality of life in a population.
zero population growth The absence of population growth; occurs when birth rates equal death rates.
replacement fertility rate The rate at which people must be born to replace those dying in the population.

Infographic **5.5** | **DECLINING POPULATION GROWTH RATES**

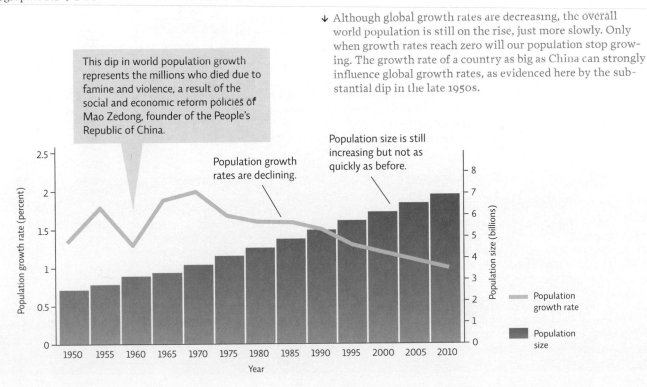

↓ Although global growth rates are decreasing, the overall world population is still on the rise, just more slowly. Only when growth rates reach zero will our population stop growing. The growth rate of a country as big as China can strongly influence global growth rates, as evidenced here by the substantial dip in the late 1950s.

This dip in world population growth represents the millions who died due to famine and violence, a result of the social and economic reform policies of Mao Zedong, founder of the People's Republic of China.

Population growth rates are declining.

Population size is still increasing but not as quickly as before.

Population growth rate

Population size

↓ Reaching zero population growth takes two steps: identifying why birth rates are high (what are the pronatalist pressures?) and then taking steps (education, birth control, social campaigns) to address those pronatalist pressures and reduce birth rates.

STEP 1: IDENTIFY PRONATALIST PRESSURES

INFANT MORTALITY RATE IMPACTS BIRTH RATE

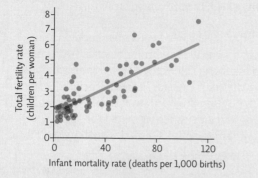

High infant mortality is closely correlated with poverty: Total fertility rate (TFR) goes up as infant mortality and poverty increase.

DESIRED FERTILITY AND ACTUAL TOTAL FERTILITY

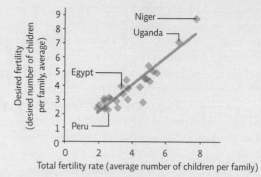

TFR closely aligns with the desired fertility of families—families usually have the number of children they would like to have.

STEP 2: STRATEGIES TO REDUCE TFR

AS EDUCATION OF WOMEN INCREASES, TFR DECREASES

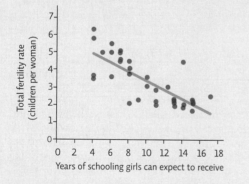

Fertility declines as educational opportunities for girls and women increase. This means that funding for education and job opportunities for women may be more effective at lowering TFR than other approaches. It also improves the survival of children, another factor that leads to a reduced TFR.

FAMILY PLANNING: THERE IS STILL UNMET NEED FOR ACCESS TO CONTRACEPTIVES

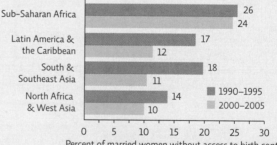

Family planning programs have been effective in many areas of the world. However, many women still have "unmet need" (they would like to use contraceptives but do not have access to them), which is as high as 39% in rural areas of some African nations.

DIFFERENT SOLUTIONS FOR DIFFERENT REGIONS

Thailand's very successful family planning program started with better maternal and infant health care, family planning programs, and access to contraceptives, but also reflected the culture's love of humor—from condom distribution at public gatherings to free vasectomies on the king's birthday, to "ads" painted on the sides of water buffaloes.

Kerala, India, a state in the world's most populous country, adopted a population-control plan centered around the three "e's": education, employment, and equality. As a result, Kerala's school system boasts more than a 90% literacy rate, which is almost identical

between boys and girls. Educated women join the workforce before deciding to have children, and some 63% (compared to 48% in India as a whole) of women there use contraceptives. The state's birth rate has stabilized at around 2.0.

In Brazil, the dropping birth rate over the past 40 years is attributed in part to pop culture. Families in popular TV shows are small; this coupled with rising opportunities for women and crowded living spaces in large cities, has produced a generation of young women whose desired family size is only one or two children.

Infographic 5.7 | AGE STRUCTURE AFFECTS FUTURE POPULATION GROWTH

↓ The fastest-growing regions are those with a youthful or very young population.

Very young
Youthful
Transitional
Mature
No data

↓ Age structure diagrams show the distribution of males and females of a population in various age classes. The width of a bar shows the percent of the total population that is in each gender and age class. The more young people in a population, the more population momentum it has—it will continue to grow for some time. The more people of reproductive age, the higher the growth rate, but this measure is also influenced by income—growth rates decline as income increases.

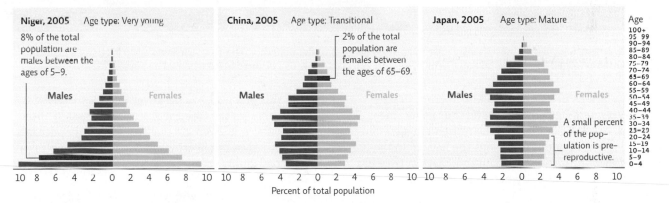

Niger, 2005 Age type: Very young

8% of the total population are males between the ages of 5–9.

Males | Females

China, 2005 Age type: Transitional

2% of the total population are females between the ages of 65–69.

Males | Females

Japan, 2005 Age type: Mature

Males | Females

A small percent of the population is pre-reproductive.

Age
100+
95–99
90–94
85–89
80–84
75–79
70–74
65–69
60–64
55–59
50–54
45–49
40–44
35–39
30–34
25–29
20–24
15–19
10–14
5–9
0–4

Percent of total population

In Niger's very young population, most people are under 30 and there is a very high capacity for growth (high population momentum).

China's transitional population is growing more slowly than Niger's because China's pre-reproductive and reproductive age cohorts are smaller. The slight skew toward males is noticeable.

Japan's mature population has a fairly even distribution among age classes, with two slight bulges seen around 30 and 60. This population is fairly stable or may even be decreasing slowly as deaths start to outnumber births.

The age and gender composition of a population affects its potential for growth.

Demographers create a diagram that displays a population's size as well as its **age structure** (the percentage of the population that is distributed into various age groups, called cohorts) and **sex ratio**. These **age structure diagrams** give insight into the future growth potential of a population. [INFOGRAPHIC 5.7]

When the majority of the people in a population are young, the diagram looks very much like a pyramid; there are nearly equal portions of men and women and fewer and fewer people at each higher age group. Because so

many people are young, and many have not yet reached reproductive age, the population will continue to grow rapidly over time.

Currently, many industrialized nations—including the United States and many European countries—have top-heavy age structure diagrams with many older people. In many cases, the struggle to care for these rapidly

age structure The part of a population pyramid that shows what percentage of the population is distributed into various age groups of males and females.
sex ratio The relative number of males to females in a population; calculated by dividing the number of males by the number of females.
age structure diagram A graphic that displays the size of various age groups, with males shown on one side of the graphic and females on the other.

↑ Multiple generations of a Chinese family pose at their home in an undated photograph from the last century. Historically, sons were so highly valued that a second son might be sent to join the family of a close relative who had no sons.

aging populations has turned prickly. In 2010, in order to relieve retirement-age problems (and thus government pension payouts), France raised the retirement age from 60 to 62, sparking riots throughout the nation. In the United States, the prospect of baby-boomers' retirement bankrupting Social Security has spurred intense and vitriolic debate.

China's age structure diagram has lost the bottom-heavy shape, reflective of its changing population age structure. Falling birth rates and rising life expectancy have tipped the balance between old and young. In 1982, just 5% of China's population was older than 65. In 2004, that percentage had climbed to 7.5; by 2050, it is expected to exceed 15%. These figures are lower than those in most industrialized countries (especially Japan, where the proportion of people over the age of 65 years is 20%). But because elderly people are still dependent on their children for support, this situation could spell disaster.

Demographers call it the 4-2-1 conundrum: As they settle into middle age, the members of each successive generation of only children will find themselves responsible for two aging parents and four grandparents. If that single child fails, family elders would be left with virtually no options—no second children, or nieces and nephews, or even friends and neighbors who could help out, since most other families would be in an equally precarious situation. In a country without an extensive pension

program, this prospect has the government especially worried. Although the Chinese government said in 2008 that the policy would remain in effect for another decade, in 2011 they began considering allowing select couples to have a second child.

There's another problem with China's evolving age structure: the workforce. Economists predict that between 2010 and 2020, the annual size of the labor force aged 20—24 will shrink by 50%. This is a forecast with global implications. "The deep fertility declines have plunged China into a demographic watershed," says Feng. "It means the days of cheap, abundant Chinese labor are over." For countries like the United States, which have come to depend on a steady influx of cheap goods from China, that's not such good news. But for the only-children of China, it almost certainly means higher wages, better working conditions, and more jobs to choose from in the coming decades.

An out-of-whack age structure isn't the only unintended consequence of the one-child policy. The sex ratio of men to women has also grown alarmingly high. It's true that almost everywhere in the world, there are slightly more males than females (though this differs according to age; because they tend to live longer, there are more women than men in older age groups), but in most industrial-ized countries, it tends to hover close to 1.05 (which, coincidentally is the exact sex ratio of the United States).

In China, the ratio has soared from 1.06 in 1979 to 1.17 in 2011; in some rural provinces, it's as high as 1.3. What does that mean in actual numbers? The State Population and Family Planning Commission estimates that come 2020, there will be 30 million more men than women in China.

How did this happen?

Like most Asian countries, and many African ones, China has a long tradition of preference for sons. The reasons are easy enough to see. For one thing, boys are often perceived as being better suited to the rigors of manual labor that drive rural, and even industrial, economies. And as they grow into men, they become the main financial providers for retired and aging parents. Daughters, on the other hand, are tasked with housework and the care of younger siblings and aging grandparents. When they marry, daughters traditionally become part of the groom's family, and a son's parents are generally taken better care of than his wife's.

Sex-selective abortions—terminating a female pregnancy—have been outlawed in China, but most experts say they are still common, thanks in large part to private-sector health care and the ability of a growing number of Chinese citizens to pay the high cost of a gender-determining ultrasound.

> ## "The deep fertility declines have plunged China into a demographic watershed."—Wang Feng

A recent study by Chinese researchers from Zhejiang Normal University shows that the sex ratio is high (favoring males) in urban areas where only one child is allowed but in rural areas where a second child is allowed, if the first is a girl, sex ratios are even higher for the second child—as high as 160 males for every 100 females. The researchers attribute this high male bias to sex-selective abortion. Although most demographers agree that outright infanticide is increasingly rare, subtler forms of **gendercide**—the systematic killing of a specific gender—are known to occur. If a female infant falls ill, for example, she might be treated less aggressively than a male infant would be.

These days, however, it is much more common for girl babies to be put up for adoption. Officially registered adoptions increased 30-fold, from about 2,000 in 1992 to 55,000 in 2001. The trend was especially pronounced in the United States, where families adopted nearly 8,000 Chinese babies in 2005 alone (compared with just 200 or

so in 1992). The practice is not necessarily as benign as it sounds. In 2009, the *Los Angeles Times* reported that many babies put up for adoption had not been abandoned by their parents, but confiscated by Chinese family planning officials.

The result of this skewed sex ratio, 30 years out, seems to be lots of prospective husbands in want of wives. Recent census data shows that in a growing portion of rural Chinese provinces, one in four men are still single at 40. "The marriage market is already getting more intense and competitive," says Feng. "Men of lower social ranks—who make up a very significant chunk of the population—are being left out. And it's only going to get worse."

A growing number of social scientists say that the dearth of women will ultimately threaten the country's very stability. Experts worry that such large numbers of young men who can't find partners will prove a recipe for disaster. "The scarcity of females has resulted in kidnapping and trafficking of women for marriage and increased numbers of commercial sex workers," writes Therese Hesketh, "with a potential rise in human immunodeficiency virus infection and other sexually transmitted diseases."

Carrying capacity: Is zero population growth enough?

The population size of a given area is not just determined by its birth and death rates but also by the migration of people in and out of its borders. Migration can be over large or small distances, across or within national boundaries, permanent or temporary (e.g., seasonal workers or refugees), and in large groups or small. **Immigration** refers to the movement of people into a given population; **emigration** refers to their movement out of a given population. Taken together, annual migration data and birth and death rates can tell us how quickly a population is changing in size. In China today, migration is mostly internal—from rural areas to urban ones, expanding those burgeoning populations and stressing the ability of the cities to meet the needs of the residents.

Every given environment has a **carrying capacity**—the maximum population size that the area can support. How large that population can be is determined by a range of forces, including the supply of nonrenewable resources, the

---

gendercide The systematic killing of a specific gender (male or female).
immigration The movement of people into a given population.
emigration The movement of people out of a given population.
carrying capacity The population size that an area can support for the long term; it depends on resource availability and the rate of per capita resource use by the population.

rate of replenishment for renewable resources on which we depend, and the impact each individual person has on the environment (a degraded ecosystem can sustain far fewer people than a well-maintained, healthy one).

There are roughly 7 billion people on the planet today. Whether we stabilize at 9 or 10 billion or more depends on how quickly we lower TFR to achieve replacement fertility worldwide. Even at 7 billion people, we may have already exceeded the carrying capacity of Earth. One problem we face is our dependence on nonrenewable energy sources; they will not last indefinitely. But we are also overusing our biological resources. There are differences in how well different regions of the world live within the *biocapacity* (the ability of the ecosystem's living components to produce and recycle resources, and assimilate wastes) of their own environments. North America, and the United States in particular, uses far more resources than its land area can provide, whereas Latin America and the Caribbean use less than half of the biocapacity of their region. This means nations like the United States and many others are importing resources from other regions.

The problem, then, is twofold: On one hand, we are increasing, rapidly, in sheer numbers. On the other, we are consuming more resources per person than ever before. [INFOGRAPHIC 5.8]

In general, population increase is a problem of the developing world—the highest fertility rates tend to be in the least developed countries, those that have yet to complete the demographic transition. Overconsumption, meanwhile, is a problem of more developed countries like the United States, where a single person might consume more food, fuel, and other resources (and thus place more strain on the environment) than a whole group of people in a less developed country like Sudan or Bangladesh. The impact humans have on Earth is thus due to a combination of factors—population size, affluence, and how we use resources, especially our use of technology, which tends to increase our overall use of resources. This "IPAT model" is discussed more fully in Chapter 4.

In the years since one-child was enacted, China has found itself careening from one end of this spectrum to the other. The problems that come with overpopulation—slums, epidemics, overwhelmed social services, and the production of high volumes of waste, for example—have been mitigated by a reduction in the fertility rate. But as fertility has declined, relative affluence has increased, and the Chinese are now straining the environment in a different way: by overconsuming.

It's important to understand that overconsumption in one region or country can strain carrying capacity in other, disparate regions. In Bangladesh, for example, rising sea levels, spurred by global climate change, are pushing that nation's carrying capacity to the brink. Most experts agree that a major driver of current climate change is rampant fossil fuel consumption, not by Bangladesh itself, but by developing nations, many of which are on the other side of the world. China now leads the way in the release of fossil fuel-based carbon emissions linked to climate

Infographic **5.8** | **HUMAN IMPACT AND CARRYING CAPACITY**

↓ Many ecologists and economists think that, at our current rate of consumption, human population size has already surpassed Earth's long-term carrying capacity. For long-term sustainability, we should leave behind enough resources so that nature can replenish what we harvest. If we had one-and-a-half "Earths," the resources left behind could replace what we take—but of course we do not.

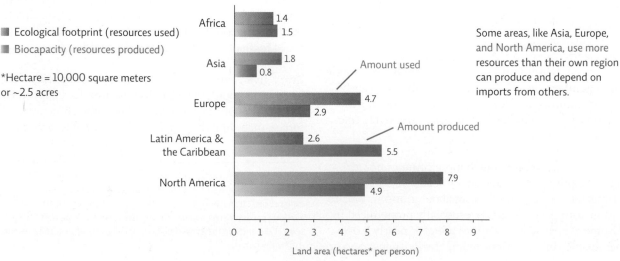

■ Ecological footprint (resources used)
■ Biocapacity (resources produced)

*Hectare = 10,000 square meters or ~2.5 acres

Some areas, like Asia, Europe, and North America, use more resources than their own region can produce and depend on imports from others.

Africa: 1.4 / 1.5
Asia: 1.8 / 0.8
Europe: 4.7 / 2.9
Latin America & the Caribbean: 2.6 / 5.5
North America: 7.9 / 4.9

Amount used
Amount produced

Land area (hectares* per person)

change—because of its huge population and its growing affluence.

Ultimately, the question of how many people the planet can support may not be the right one. Trying to pinpoint whether Earth's carrying capacity is 7 billion or 9 billion or even 16 billion is meaningless unless we identify what an acceptable quality of life is. And while overpopulation may be an issue with global significance and causes, it is one that ultimately plays out at the regional level: As land becomes degraded and water supplies depleted or polluted, people (and other organisms) live in an increasingly impoverished environment.

What awaits China's generation of Little Emperors?

Evidence suggests that despite the plight of the latest generation, China is indeed becoming a small-family culture. The National Family Planning and Reproductive Health Survey found that 35% of women prefer having only one child; 57% prefer having two children; and only 5.8% want more than two. True to the demographic transition's predictions, educated, urban women wanted fewer children than their rural, farm-dwelling counterparts. That's good news for those striving to bring China toward zero population growth: In the last three decades, the federal government has allowed hundreds of millions of Chinese people to move to cities in search of work. Those families will no longer need lots of children to work on the farm.

In recent years, the government has softened its stance and added several significant exceptions to the one-child rule: Couples made up of two only-children, rural families with land to farm, and several groups of ethnic minorities, are all allowed to have more than one child if they so choose.

Still, it's unlikely that the one-child policy will officially expire any time soon. In a 2008 interview with the state newspaper, *China Daily*, the country's minister of State Population and Family Planning, Zhang Weiqing, said that loosening the one-child policy would unleash a new baby boom. "Given such a large population base, there would be major fluctuations in population growth if we abandoned the one-child rule now. It would cause serious problems and add extra pressure on social and economic development."

For better or worse, China is now a country of small families, and of ever-fewer children. That may mean more resources per child, but it also means fewer workers, taxpayers, and innovators. Ultimately, it means that even as they bask in the spoils of one-child, the "Little Emperors" are burdened with the hopes and fears of an entire country.◉

Research articles referenced in this chapter:

Ewing, B., *et al.* 2010. *Ecological Footprint Atlas 2010*. Oakland, CA: Global Footprint Network.

Greenhalgh, S. 2008. *Just One Child: Science and Policy in Deng's China*. Berkeley: University of California Press.

Hesketh, T., *et al.* 2005. *New England Journal of Medicine*, 353: 1171–1176.

Ni, H., and A.M. Rossignol. 1994. *Epidemiology*, 5: 490–494.

Pritchett, L. 1994. *Population and Development Review*, 20: 1–55.

Sedgh, G., *et al.* 2007. *Women with an Unmet Need for Contraception in Developing Countries and Their Reasons for Not Using a Method*. Guttmacher Institute.

Yin, Q. 2003. *Theses Collection of 2001 National Family Planning and Reproductive Health Survey*. Beijing: China Population Publishing House. 116–26.

Zhu, W., *et al.* 2009. *British Medical Journal*, 338: b1211.

BRING IT HOME

◉ PERSONAL CHOICES THAT HELP

The impact of humans on the planet is created by a combination of population size and our resource use. The issue of population and carrying capacity is complex. We cannot have a truly sustainable society until key components such as poverty, lack of education, and basic human rights are addressed.

Individual Steps
→ Buy Fair Trade-Certified products. These products provide a livable wage to workers and are often linked to education and community development.

→ Research the products you buy to make sure that you are not supporting child slave labor or sweatshop facilities.

Group Action
→ Raise money and invest it in a socially responsible project. Kiva.org is a non-profit organization that provides micro-loans to help people start small businesses in less developed regions.

→ Join an organization such as Habitat for Humanity, which builds housing for low-income families.

Policy Change
→ We all know that our tendency as humans is to build up and out; however, revitalizing our current older downtown areas is important to prevent new habitat destruction as well as to make use of current infrastructure. Urge your local government to keep shopping and dining establishments in historic downtowns. If needed, work to develop a community partnership that starts clean-up programs and community gardens in abandoned lots.

UNDERSTANDING THE ISSUE

CHECK YOUR UNDERSTANDING

1. _____ is the number of children a couple must have to insure the population neither increases nor decreases, but the most useful measure for projecting future population change is _____.
   a. Replacement fertility; total fertility rate
   b. Infant mortality rate; pronatalist pressures
   c. Crude death rate; crude birth rate
   d. Demographic transition; population density

2. **Population momentum inherent in the populations of developing countries refers to the fact that:**
   a. people in these countries are living longer.
   b. infant mortality rates in these nations are on the decline.
   c. populations will continue to grow even at replacement fertility rates.
   d. economic growth in these nations is unstoppable.

3. **An age structure diagram can be used to:**
   a. measure potential population migration to cities.
   b. see the historical growth of a population.
   c. predict the future growth of a population.
   d. measure the quality of life of a population.

4. **The term "demographic transition" refers to:**
   a. slower growth as the population size approaches carrying capacity.
   b. the decline in death rates and then birth rates as a country becomes industrialized.
   c. the requirement for a population to reach a specific size before it becomes stable.
   d. migration from the overpopulated countryside to urban centers.

5. **Which of the following is NOT a pronatalist pressure?**
   a. The child is part of the family labor pool.
   b. There is a high infant mortality rate.
   c. Contraceptives are not available.
   d. Women have many opportunities to participate in the work force.

6. **Which statement illustrates human overconsumption?**
   a. Fertility rate of 1.4 in Japan is below replacement-level fertility.
   b. Average per capita daily water usage in the United States is 575 liters, while the amount needed for a reasonable quality of life is 80 liters.
   c. At 176 deaths per 1,000 live births, Angola has the highest infant mortality rates in the world.
   d. Per capita use of fossil fuels is decreasing in Denmark as a result of the carbon tax.

WORK WITH IDEAS

1. How has human population changed over time? What factors account for this change?

2. Compare and contrast different tactics to reduce population growth. What are the methods, effectiveness, drawbacks, and cultural constraints (why a particular method might not be appropriate everywhere) of each?

3. What is a demographic transition and why is it considered important for human populations? Where have demographic transitions yet to occur and what can we predict about whether these countries will make this transition?

4. Some countries like Japan are dealing with large aging populations, while others like Niger are dealing with large numbers of youth. Why does such a demographic divide exist and what problems arise in populations that are either "very young" or "mature"?

ANALYZING THE SCIENCE

The graphs on the following page come from the United Nations Population Division and are used in its population projections.

INTERPRETATION

1. What are the population projections for each fertility variant in 2050? In 2100? What do you predict the population will be in 2150 for the constant fertility variant?

2. What is the trend in the numbers in graph A for each line?

3. How does the information in graph A relate to that in graph B? Explain.

ADVANCE YOUR THINKING

4. The three projections in graph A are based on different fertility variants. The medium fertility variant assumes global fertility rate will drop to replacement (2.1); the high fertility variant assumes global fertility rate will be 2.6; and the low fertility variant assumes global fertility rate will be 1.6. Explain why the UN might have three population projections based on different fertility rates.

5. Why is Africa projected to have such large growth if population projections assume that global fertility rates will be at replacement? Use information from the chapter to support your conclusion.

6. The following table shows 2010 population data for various world regions. Using this information as well as the data from graph B, calculate the total population in each region in 2100 as well as the distribution of the world population in each region for 2010 and 2100. According to your calculations, what are the significant changes in the distribution of the world population from 2010 to 2100?

Region	Population in 2010 (in millions)	Population in 2100 (in millions)	Population change 2010–2100	% of world population in 2010	% of world population in 2100
Africa	1,022				
Asia	4,164				
Europe	738				
Latin America & Caribbean	590				
North America	345				
Oceania	37				
World total	**6,896**				

GRAPH A Estimated and projected world population according to different variants

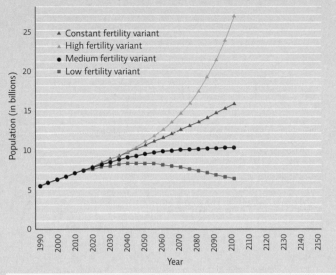

GRAPH B Projected population change between 2010 and 2100 by major region

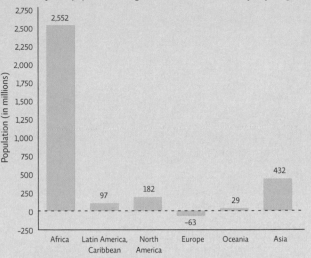

EVALUATING NEW INFORMATION

Providing education and job opportunities for women is considered important by many, not just for their effects on fertility, but also as a matter of social justice. One organization that works on this is the Foundation for International Community Assistance (FINCA), which provides microfinance services to low-income people, mostly women. Explore the FINCA website (www.finca.org).

Evaluate the website and work with the information to answer the following questions:

1. Is this a trustworthy organization? Does it have a clear and transparent agenda?
    a. Who runs this organization? Do his or her credentials make the work of the organization and the information presented reliable/ unreliable? Explain.
    b. What is the mission and vision of this organization? What are its underlying values? How do you know this?
    c. Do you agree with FINCA's assessment of the problems and concerns about poverty? Explain.

2. Explore the "About FINCA" and "Microfinance Programs" links.
    a. What are microfinance and village banking? Does the website provide supporting evidence for FINCA's claims about how these programs can help the poor? Is the evidence reliable?
    b. From the "Microfinance" menu, select "Client Protection." What is the Smart Microfinance Campaign and why is it important to microfinance clients?
    c. Do you think FINCA's business model is valid and effective? Explain.

3. Select the "Get Involved" link.
    a. Who can be a part of FINCA's solution? How can they get involved?
    b. Identify a specific solution and the strategy to accomplish it. Is the solution sufficient and the strategy reasonable? Explain.

4. How might the FINCA model influence cultural, economic, and demographic factors that influence population growth?

MAKING CONNECTIONS

CAN THE UN MILLENNIUM DEVELOPMENT GOALS ENABLE A DEMOGRAPHIC TRANSITION?

Background: The world population reached 7 billion in October 2011. This last billion was added in record time—12 years—as was the billion before it. It is also very likely that the next billion will be added in another 12 years, making this the most rapid period of human population expansion in history. This reality seems counterintuitive, given that fertility rates have been falling more rapidly than ever before. In fact, 40% of the world population lives in nations where birth rates are below replacement level. Most population growth is concentrated in the world's poorest nations.

Case: You have been given the task of analyzing the United Nations Millennium Development Goals (MDG)—eight goals for addressing the various dimensions of extreme poverty—in the context of the demographic transition. There are two basic questions:

1. Are the MDGs sufficient to enable developing countries to make the demographic transition?

2. Based on what is achieved by 2015, what should the next step be?

In your report, include the following:
a. A description of the UN MDGs, why they were selected, and what they are intended to achieve.
b. An analysis of the progress toward the MDGs. Will they meet their intended targets by 2015? Why or why not?
c. There has been much criticism of the MDGs, not only based on the assessments of progress toward their achievement, but also regarding how and why they were chosen in the first place. Who are some critics (discuss two) and what are their concerns? Discuss whether you agree with their assessment.
d. In what ways can such global goal setting be useful despite its limitations? Provide some specific examples of how such goals have been/can be used to address population matters.

Agostina Sunday, left, and Monica Emanuel get water from the sacred pond called Ogi in Ogi, Nigeria. The pond was contaminated with the water fleas that were infected with the Guinea worm larvae. Villagers, holding to traditional beliefs, initially tried to dissuade health officials from treating the water. Now, thanks to a relentless 20-year campaign led by former U.S. President Jimmy Carter, Guinea worm is poised to become the first disease since smallpox to be pushed into oblivion.

ERADICATING A PARASITIC NIGHTMARE

Human health is intricately linked to the environment

CORE MESSAGE

Human health is impacted by the environment. Our choices can alter the environment to either facilitate disease or to reduce its transmission. Infectious diseases are the leading cause of death in developing nations, while water contamination, land and waterway alteration, deforestation, and climate change are contributing to these and other health problems. Health can be improved significantly with better sanitation, access to clean air and water, and public health programs that reach people in affected areas.

GUIDING QUESTIONS

After reading this chapter, you should be able to answer the following questions:

→ How do the fields of public health and environmental health help improve the health of human populations?

→ What environmental hazards impact the health of people?

→ Which environmental factors facilitate the spread of Guinea worm disease and other diseases, and what steps are needed to eradicate these diseases?

→ How do factors that affect human health differ between developed and developing nations?

→ What can be done to reduce environmentally mediated health problems?

↑ Guinea worm disease is a major impediment to a farmer's ability to work. A health volunteer in Ghana educates children on how to use pipe filters when they go to the fields with their families. Pipe filters, individual filtration devices worn around the neck, work similarly to a straw, allowing people to filter their water to avoid contracting Guinea worm disease while away from home.

◉ **WHERE IS NIGERIA?**

Ernesto Ruiz-Tiben shook the tube of water and held it up to the sunlight so that the women who were gathered around him could see the tiny black flecks that had settled out. There was a soft, collective gasp at the spectacle. The black flecks—tiny water fleas known as copepods—offered the Nigerian women the first visible proof of what Ruiz-Tiben, director of the Carter Center's Guinea Worm Eradication Program, had been trying to convince them: The water they drank and bathed in was contaminated with tiny bugs, and these bugs were solely responsible for the searing worm infections that seemed to sweep through the village every year or so, usually right around harvest time.

For those unlucky villagers, the infection began when a person ingested contaminated water that contained copepods infected with Guinea worm larvae. Once digested, the copepods released the larvae, which burrowed into the victim's abdominal tissue. There the larvae matured into adult males and females who then mated. Males died but the females continued to grow—up to 100 centimeters (3 feet) long—all the while migrating through the victim's tissue. About a year after infection, the full-grown females forced their way out by releasing acid just beneath the skin, which in turn created a blister. When the fiery pain of that blister drove the victim to plunge into water, the female squirted out a dense cloud of milky white larvae, starting the cycle over again.

The disease is not fatal, but recovery is both very slow and very debilitating. Even after she releases her larvae, the mother worm can take as long as 3 months to fully emerge from the skin, during which time the victim is often completely laid up—unable to work for the entire harvest—or even to walk or move much, depending on which limbs are infected and with how many worms. Worse yet, there are no medications or vaccines for dracunuliasis, or Guinea worm disease (GWD) as it is more commonly known, and the infection itself does not confer immunity; that means the same people can fall prey to the worms over and over again.

Efforts to reduce the incidence of GWD began as part of an international program to provide safe drinking water to all people. According to the World Health Organization (WHO), some one billion people per year fall victim to **waterborne diseases**—those acquired by consuming contaminated water. In fact, water- and **vector-borne diseases** are the main **infectious disease** threats to human health. (Vectors are organisms that transmit a **pathogen** from one host to another; in the case of Guinea worm, copepods serve as the vector). Guinea worm is just one of countless such diseases, but in the course of eliminating it, Ruiz-Tiben and his colleagues hoped, they might also eliminate others.

They were aided in their quest by a simple fact: Unlike some of those other bugs, Guinea worms are utterly dependent on humans to complete their life-cycle—humans are the only known reservoir for adult worms (larvae can only survive in copepods, and only for a few weeks). That meant Ruiz-Tubin and his colleagues could break the Guinea worm's life cycle—and thus obliterate the disease—simply by changing human behavior.

Convincing villagers to filter their water before drinking it would stop new infections. Getting the villagers to apply a mild pesticide would decontaminate area ponds. And teaching people to avoid communal swimming or bathing while worms were emerging from the skin, and to treat new infections as soon as blisters emerge, would break the worm's life cycle once and for all. Their goal was total eradication of GWD; if they succeeded, it would be only the second disease in human history (after smallpox) to be completely wiped off the face of the Earth.

As straightforward as it all sounded, Ruiz-Tiben had not had much luck so far. Despite his careful detailing of the science, most of the villagers he had spoken with still believed—rather fiercely—that the sickness was delivered by angry gods as punishment for various misdeeds. And if that explanation seemed ludicrous to Ruiz-Tiben, well, his

waterborne disease An infectious disease acquired through contact with contaminated water.
vector-borne disease An infectious disease acquired from organisms that transmit a pathogen from one host to another.
infectious diseases Illnesses caused by an invading pathogen such as a bacterium or virus.
pathogen An infectious agent that causes illness or disease.

counter explanation—that the worms actually came from water the villagers had consumed a year prior—seemed equally ludicrous to them. He thought maybe showing them the dead bugs in the water would help him make some headway.

Environmental hazards impact human health.

Public health is a field that deals with the health of human populations as a whole: Public health **epidemiologists** like Ruiz-Tiben work to gauge the overall health status of a population or even of a nation. They use statistical analysis (e.g., rates of infant mortality, incidence of various diseases) to identify specific health threats to groups of people, then they recommend ways to mitigate those threats. Devising a plan of action is often the trickiest part; it requires the study of a whole host of interacting variables—from cultural and social forces (which influence things like diet and smoking habits) to economic stability or instability (which determines a given population's access to resources) to environmental factors like water cleanliness or changes in habitat that affect disease transmission.

Environmental health is a branch of public health that focuses on potential health hazards in the natural world and the human-built environment; such hazards include not only things like contaminated water, air, and soil, but also human behaviors—hand washing and water drinking, for example—that help determine whether those natural factors become hazards or not. According to the WHO, environmental hazards are responsible for about 25% of disease and deaths worldwide.

Fortunately, many of them are *modifiable*, that is, we can take action to change them. Like GWD, they can be mitigated (indeed, some 13 million deaths could be prevented each year) through reasonable measures. It is these modifiable hazards—like contaminated drinking water—that environmental public health workers like Ruiz-Tiben focus their efforts on. [INFOGRAPHIC 6.1]

In general, these can be divided into three broad categories: physical hazards (see Chapter 2), chemical hazards (see Chapter 3), and biological hazards (the focus of this chapter). *Biological hazard* typically refers to infectious diseases—illnesses caused by an invading pathogen such as a bacterium or virus. [INFOGRAPHIC 6.2]

Infographic **6.1** | **PUBLIC HEALTH PROGRAMS SEEK TO IMPROVE HEALTH OF THE POPULATION AS A WHOLE**

↓ The goal of public health programs is to improve the health of human populations through prevention and treatment of disease at the community level.

PUBLIC HEALTH

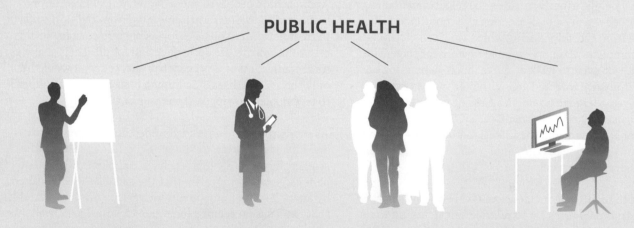

EDUCATES
Public health professionals provide information and health care advice to communities. Changing behaviors can be a critical part of improving public health.

PROVIDES HEALTH CARE
Public health care workers provide needed preventative medical care and treatment.

PROPOSES ACTIONS
Once risks are identified, public health professionals make recommendations to improve health in specific groups and in the population as a whole.

CONDUCTS RISK ASSESSMENTS
Epidemiologists analyze statistics related to a population's health to determine risk for various groups in the population (e.g., children, the elderly, the chronically ill) to a variety of environmental factors (e.g., sanitation, pollution, climate change).

Infographic **6.2** | **TYPES OF ENVIRONMENTAL HAZARDS**

↓ The WHO recognizes that modifiable environmental factors contribute significantly to disease, injury, and death worldwide. Environmental hazard categories can be broken down in a variety of ways, but here is one simple breakdown: physical, chemical, and biological hazards. All of these hazards can interact, sometimes in unpredictable ways, which increases the complexity of public health issues. Of these 3 groups, biological hazards remain the major risk in low-income nations, the areas where death rates are the highest.

BIOLOGICAL HAZARDS
These types of hazards include infectious agents (pathogens), which can be bacteria, viruses, protozoa, fungi, and worms. Pathogens can be transmitted through vectors like insects and ticks, through direct person-to-person contact, and from contaminated food and water.

Someone whose health is impacted by pollution may be more vulnerable to infection by a biological hazard; and those weakened by infection may be more affected by air pollution.

Floodwaters are likely to be contaminated with pathogens or provide a habitat for mosquitoes (biological hazards). Additionally, people who are ill or malnourished may be less able to cope with such natural disasters.

CHEMICAL HAZARDS
This type of hazard includes environmental pollution and exposure to hazardous chemicals in the home or workplace, or that occur naturally in the environment (such as arsenic in water). The largest contributor in this group is smoke from burning solid fuels indoors and smoking, including secondhand smoke.

Natural disasters can increase the risk of exposure to toxic chemicals. For example, chlorofluorocarbons released in air pollution decreased stratospheric ozone, leading to increased incidence of skin cancer in affected areas.

PHYSICAL HAZARDS
Natural disasters, extreme weather events, exposure to ultraviolet radiation, and even traffic accidents are examples of physical hazards.

To be sure, infectious diseases account for less of the global burden of disease than **noncommunicable diseases (NCDs)** like cardiovascular illness and diabetes. (Also known as lifestyle diseases, NCDs are largely determined by choices about things like diet and exercise.) In fact, NCDs cause the most deaths globally. But infectious diseases are still a major problem. They account for about 26% of deaths worldwide each year, the vast majority of these in the developing world.

The main environmentally mediated infectious diseases are diarrheal diseases (due to environmental factors like unsafe drinking water or poor hygiene and sanitation),

lung infections (linked to air pollution), and mosquito-transmitted diseases like malaria and dengue fever. [INFOGRAPHIC 6.3]

public health The science that deals with the health of human populations.
epidemiologist A scientist who studies the causes and patterns of disease in human populations.
environmental health The branch of public health that focuses on factors in the natural world and the human-built environment that impact the health of populations.
noncommunicable diseases (NCDs) Illnesses that are not transmissible between people; not infectious.

Infographic 6.3 | **ENVIRONMENTAL FACTORS CONTRIBUTE TO THE GLOBAL BURDEN OF DISEASE**

↓ Modifiable environmental factors contribute to some diseases more than others. This is a result of how a particular disease is transmitted (for example, if it is a waterborne disease) and how well a particular region is able to address environmental conditions (for example, whether there is clean water).

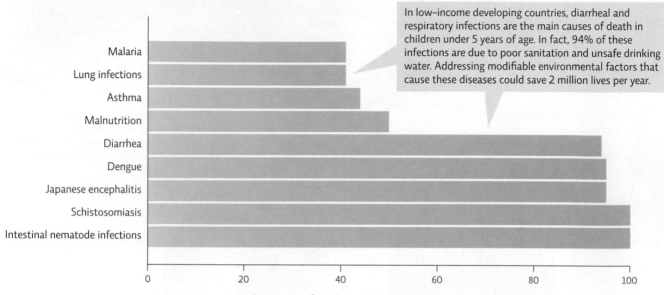

DISEASES WITH A HIGH ENVIRONMENTAL BURDEN

In low-income developing countries, diarrheal and respiratory infections are the main causes of death in children under 5 years of age. In fact, 94% of these infections are due to poor sanitation and unsafe drinking water. Addressing modifiable environmental factors that cause these diseases could save 2 million lives per year.

Percentage of incidence attributable to environmental factors

Unlike GWD, for which humans are the only host for the worm, many of these diseases are **zoonotic**—that is, they can spread between infected animals and humans. In fact, 75% of all **emerging infectious diseases**—those that are new to humans or have rapidly increased their range or incidence in recent years—are zoonotic; this includes new strains of influenza, one of the most common zoonotic diseases. About 72% of zoonotic diseases come from wildlife, a number that is increasing due to human encroachment into formally wild areas, and the African *bushmeat trade*—killing wild animals like primates for food. Many pathogens that infect wild primates can also infect humans, and the hunting and consumption of those animals increases the chances that we will be exposed to these pathogens.

Environmental changes are believed to play a role in many of these emerging infectious diseases. In the United States, recent increases in cases of West Nile virus may be linked to climate change—specifically extreme heat and drought. West Nile virus first showed up in the United States in 1999 in New York City; since then it has rapidly spread from coast to coast. High temperatures increase the ability of the mosquito vector to pick up the virus from its bird hosts. Though mosquitoes require water to breed, droughts tend to increase urban mosquito populations by increasing standing water in drains that would normally be flushed out by rains. With the hottest summer on record in Texas and other areas in the United States, 2012 is on track to break the record for U.S. cases of West Nile virus set back in 2007. Such outbreaks may become more common as a warming global climate allows vectors like mosquitoes to expand their range and thrive in areas where they were formally scarce or nonexistent. [INFOGRAPHIC 6.4]

As previous chapters have taught us, we are never exposed to any hazard—be it physical, chemical, or biological—in isolation. Rather, different types of hazards interact with one another, and with other elements of the human environment, in ways that can make it tricky for epidemiologists to map cause-and-effect relationships. For example, someone negatively affected by a chemical hazard like air pollution may be more susceptible to a biological hazard such as a lung infection.

zoonotic disease A disease that is spread between infected animals (not merely a vector that transmits the pathogen but another host that harbors the pathogen through its life cycle) and humans.
emerging infectious diseases Infectious diseases that are new to humans or that have recently increased significantly in incidence, in some cases by spreading to new ranges.

Infographic 6.4 | **A VARIETY OF PATHOGENS CAUSE DISEASE**

↓ A wide variety of infectious agents can cause disease in humans and other organisms. The main pathogenic threats to human health are waterborne and vector-borne diseases, with the mosquito being the number one vector of pathogens. Public health officials follow emerging infectious diseases closely; many of these diseases are zoonotic and their recent rise is usually related to environmental factors that increase the spread of the pathogen or its vector.

PATHOGEN CLASS	VIRUSES	BACTERIA	PROTOZOA	FUNGI	WORMS
	Dengue, Japanese encephalitis, and West Nile virus are emerging viral diseases that are spread by mosquitoes. Increased habitat, international human travel, and climate change are all implicated in these diseases. Other viruses are spread by direct contact or by contaminated water.	Bacterial pathogens can be spread by contaminated water (typhoid fever) or through vectors (Lyme disease). The 2010–2011 cholera outbreak in Haiti (caused by the *Vibrio* bacterium) killed more than 7,000 people.	Protozoa are single-celled organisms. Some are pathogenic and cause diseases such as giardiasis, contracted by drinking contaminated water, and African sleeping sickness. Malaria, caused by the *Plasmodium* protozoan, kills as many as 1 million people annually, mostly African children.	Many fungal infections plague humans. Ringworm is not a worm at all but is caused by a fungus. It is most common in areas with inadequate water for washing. Candidiasis is another common fungal infection; outbreaks in hospitals are linked to the overuse of antibiotics. (The fungus can thrive when bacteria are killed.)	A variety of worms can infect humans and other animals via ingestion or direct contact with the worm or its vector. The Guinea worm is a nematode. Other nematode species cause trichinosis, elephantitis, and hookworm.

EXAMPLES					
Pathogen	Dengue virus	*Vibrio cholerae*	*Plasmodium*	*Tinea corporis*	*Dracunculus medinenisis*
Transmission	Vector: Mosquito	Contaminated water	Vector: Mosquito	Contaminated water or direct contact	Vector: Copepods
Disease	Dengue fever	Cholera	Malaria	Ringworm	Guinea worm disease

One of the biggest influences on the seriousness of any given hazard is the health of the natural environment, something we've had unprecedented influence over in recent centuries. "We humans have been remarkably effective at rearranging the natural world to meet our own needs," said Sam Myers, a scientist at the Harvard University School of Public Health, in a recent article in *Annual Review of Environment and Resources*. In the article, he says that between one-third and one-half of global resources produced by ecosystem functions are now diverted to human uses. In the past 300 years alone, we have completely deforested between 7 and 11 million square kilometers of land—an area the size of the continental United States—and converted some 40 percent of the planet's ice-free land surface to cropland or pasture. An additional 2 million square kilometers of forest have

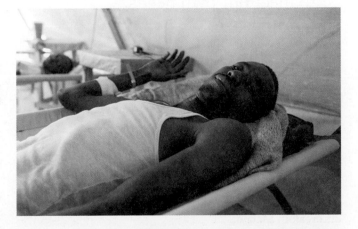

↑ A man who is recovering from a severe case of cholera at the Doctors Without Borders clinic in Haiti.

been converted into highly managed plantations with significantly less biodiversity. We are already using roughly half of the planet's accessible surface freshwater, and fishing three-quarters of monitored fisheries at or beyond their sustainable limits. And with population rising, these numbers are only likely to increase.

Though this type of manipulation has made life better for many, Myers and others worry that by so dramatically altering the ecosystems around us, we have not only imperiled our access to some of the most basic components of human health—namely adequate nutrition, safe water, and clean air—but have also increased our exposure and vulnerability to both natural disasters (see Chapter 13) and infectious diseases.

In the northeastern United States, for example, a cascade of interrelated environmental changes have helped drive the incidence of Lyme disease upward. Habitat fragmentation (a consequence of land development) has increased the edge habitats where field mice thrive. The mice are the main host reservoir for the bacterium that causes Lyme disease—it thrives in them and can be passed from host to host via ticks that bite the mice and then bite another animal. Biodiversity loss has also played a role. Unlike field mice, squirrels and possums are usually able to kill and remove the ticks before becoming infected. As biodiversity decreases, and populations of these less vulnerable hosts diminish, the ticks increasingly end up on the mice that more readily transmit the disease, building the infected tick population and increasing the chance that one will bite a person and transmit the disease.

> "We humans have been remarkably effective at rearranging the natural world to meet our own needs."
> —Sam Myers

Meanwhile, in the Amazon Basin, deforestation has increased the breeding habitat for mosquitoes that transmit malaria; in Cameroon, it has altered aquatic habitats in ways that favor a schistosomiasis-carrying snail.

In Asia and South America, monsoon rains and agricultural runoff have conspired to alter the salt and nutrient levels of coastal waters in ways that favor cholera population explosions. Floods that washed away latrines, emptying sewage into nearby waters, also sparked a 2012 cholera outbreak in Malawi.

And throughout the developing world, both dam building and urbanization have increased the incidence of a wide range of waterborne pathogens, including but not limited to Guinea worm. In cities, not only do people live in much closer quarters, but an infinite variety of humanmade objects—namely old tires and discarded plastic food containers—find second life as vessels for rainwater that collects during wet seasons, bakes in sunlight, and grows dank and stagnant over time. This water provides an excellent habitat for a whole suite of vectors that transmit a host of diseases: mosquitoes that carry dengue and malaria, black flies that carry worms which cause debilitating diseases like elephantiasis and river blindness (onchocerciasis), and snails that carry schistosomiasis. Dams—especially those in tropical regions—do something similar; they create large bodies of standing water that have been associated with an uptick in the same cadre of diseases.

Biological hazards contribute to the global burden of disease.

In the Nigerian village where Ruiz-Tiben was working, however, the main problem was not urbanization, or habitat fragmentation, or even deforestation—but a normal human trait: resistance to change. Even after the villagers saw the bugs in the water (which he had killed with a dash of the pesticide Abate), they continued to resist both the water filters and the use of pesticides. When Ruiz-Tiben discovered a hidden pond infested with copepods, the women of the village formed a human shield around it so that his team could not treat it with Abate. The pond was a sacred ancestral pool, they insisted. To douse it with chemicals would only invite the wrath of their gods.

The belief was anchored in an ancient history. In fact, GWD is itself ancient; it has been traced all the way back through the Old Testament to the Egyptian mummies. Some say the very symbol of modern medicine—a serpent coiled around a staff (known as a caduceus)—derives from our treatment of GWD: To prevent the worm from breaking off at the blister, it is wound slowly around a stick as it emerges from the body—a practice that has not changed for thousands of years.

Both the worms and the copepods that carry them are native to Africa and Asia; both evolved across human history to exploit human hosts and human water sources. The copepods thrive in open, stagnant water sources—like ponds and dams—whose availability varies by season and region. In the Sahelian zone, transmission generally occurs in the rainy season (from May to August) when shallow ponds grow deep enough to bathe in. In the

Infographic **6.5** | **GUINEA WORM INFECTION AND ERADICATION PROGRAMS**

↓ The Guinea worm is a parasite that spends part of its life cycle inside copepods (water fleas) and part in a human host. Humans are exposed to the parasite when they drink water contaminated with the water fleas. Because there is no other animal host in which the worm can complete its life cycle, if we can prevent infection at the source, we can eradicate this parasite.

THE LIFE CYCLE OF THE GUINEA WORM

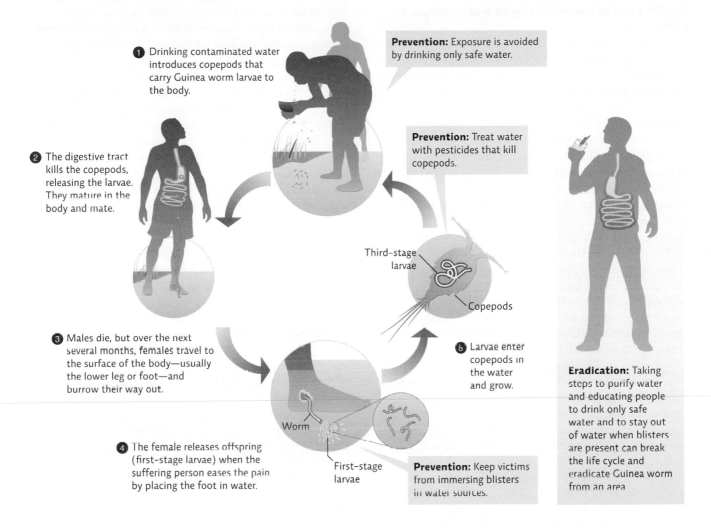

1 Drinking contaminated water introduces copepods that carry Guinea worm larvae to the body.

Prevention: Exposure is avoided by drinking only safe water.

Prevention: Treat water with pesticides that kill copepods.

2 The digestive tract kills the copepods, releasing the larvae. They mature in the body and mate.

Third-stage larvae

Copepods

3 Males die, but over the next several months, females travel to the surface of the body—usually the lower leg or foot—and burrow their way out.

5 Larvae enter copepods in the water and grow.

Eradication: Taking steps to purify water and educating people to drink only safe water and to stay out of water when blisters are present can break the life cycle and eradicate Guinea worm from an area.

Worm

4 The female releases offspring (first-stage larvae) when the suffering person eases the pain by placing the foot in water.

First-stage larvae

Prevention: Keep victims from immersing blisters in water sources.

humid savanna and forest zones, infections peak during the dry season (from September to January) when water holes shrivel and grow still, inviting copepods to multiply.

The only humans who come in contact with copepods are those who drink unpurified water from these sources. Like most such people, the villagers Ruiz-Tiben was working with were poor and lived in a naturally dry area; that meant that for generations, they had had no choice but to drink and bathe in whatever water they could find. Consistent water sources—those that do not dry up when the rains pass, or evaporate during particularly hot years—were sacred. And the women were understandably

leery of polluting their most treasured supply with a foreign chemical that had been made in a distant land.

It was not until a revered general from the region intervened on Ruiz-Tiben's behalf that the women finally relented. The general, a former president of Nigeria, assured the women that the pesticide would not harm their fish, but would instead keep their families from getting sick. Their ancestors, he said, would not want them to be sick. Reluctantly, the human shield broke up, and Ruiz-Tiben and his team were able to treat the pond with Abate, ridding it of copepods. [INFOGRAPHIC 6.5]

But as the Carter Center's army of public health workers would soon discover, Nigeria's challenges paled in comparison to the obstacles faced by other Guinea worm—infested countries.

The factors that affect human health differ significantly between developed and developing nations.

Back in 1995, when Sudan was still one nation with two warring factions—north vs. south—Nabil Aziz Mikhail bore witness to a quiet sort of miracle.

The country's Ministry of Health had long reported Guinea worm cases in the low thousands. But Mikhail, who had just assumed the role of Guinea Worm Eradication Coordinator, had quickly discovered that the number was much, much higher than that: 100,000-plus cases was the better estimate. He knew that simple measures, such as those the Carter Center was employing elsewhere, could stop the disease in its tracks. But he also knew that no such measures could be employed in Sudan, marred as it was by poverty and extreme violence. The most afflicted areas were simply too dangerous to venture into. Even if they could be reached, eradication programs—community education, latrine building, even pesticide application—would be impossible to implement under the circumstances.

Here's where the miracle comes in: Desperate to make a dent in the problem, Mikhail and his colleagues called former U.S. President Jimmy Carter himself, inviting him to host a conference on GWD in their war-torn home. Carter did them one better—not only did he come to Sudan, but he quickly negotiated what would later be called the "Guinea Worm Cease Fire"—a laying down of arms that lasted 6 full months that allowed public health workers to secure unprecedented gains in the region. Infected water sources were detected and decontaminated, filters were distributed, and active infections treated, on a scale the country had never seen before. "It was a dream," Mikhail says now. "I've never heard of a health activity that brought any sort of cease-fire."

It was also a lesson, Mikhail says: If there's one human behavior that favors the Guinea worm even more than bathing in infested water, it's war. Plain and simple.

In fact, war is just one reason that the death rate from environmentally mediated diseases is much greater (12 times greater, in 2006) in developing nations than it is in developed ones. Poverty is another; poor basic nutrition is another, still.

As we discuss elsewhere in this book, the differences between developed nations and developing ones are vast. In developed nations, cardiovascular disease and cancer represent the bulk of disease burden. (These are caused by lifestyle choices and industry and vehicle pollution.) In developing nations, while NCDs are ticking upward, infectious diseases, caused by all the environmental factors we've already discussed (lack of clean water, poor sanitation, burning solid fuels indoors for heat and energy), are still the leading cause of death. [INFOGRAPHIC 6.6]

Today, though Sudan's civil war is waning, violence persists and has left the country one of the few still plagued by GWD; in 2009, 81% of the world's remaining cases were found here. "Once the violence stopped, progress was imminent," says Ruiz-Tiben. "The Sudanese health workers and foreign nongovernment organizations surprised themselves with what they were able to accomplish in those 6 months. But when the cease-fire ended, the infection rates climbed right back up."

Environmentally mediated diseases can be mitigated with funding, support, and education.

In Ezza Nkwubor, in southeastern Nigeria, a team of elderly men—all local villagers—stand guard over a large, silent pond. The water has been treated with pesticides, and the men's job is to ensure that nobody with emerging worms comes into contact with it. Elsewhere, the sick have been quarantined—their wounds carefully tended and water and free food is brought to them, for the entire month that it takes the worm to emerge. Young boys—travelers and hunters—carry whistle-shaped cylinders tied to strings around their necks. The cylinders serve as portable filters so that the boys can drink directly from an environmental water source and still protect themselves from infection when they're out hunting. And season after season, women teach their protégés the importance of filtering water before giving it to their families. [INFOGRAPHIC 6.7]

It's the picture of success that Ruiz-Tiben and his colleagues have envisioned for decades. "It just shows what you can accomplish when the support is there," Ruiz-Tiben says, underscoring what has been a key lesson of the GWD eradication campaign: Even the simplest technologies require financial and political support to implement—not only from the developed world, or the international community, but from the countries themselves.

Infographic **6.6** | **DEATH RATES AND LEADING CAUSES OF DEATH DIFFER AMONG NATIONS**

↓ Death rates due to modifiable environmental factors are highest in developing countries; almost half of those deaths are in children. Poverty that restricts access to medical care, clean water, and an adequate diet is perhaps the leading "health risk" worldwide. This is nowhere more evident than in the low-income nations of Africa. Over 70% of people in high-income countries will live to at least age 70; in middle-income nations, that number drops to 40% and in low-income countries only 17% will make it to age 70.

ENVIRONMENTAL DISEASE BURDEN, 2004

Deaths 100,000 population

- <150
- 151–200
- 201–350
- 351–500
- 501–1,000
- No data

TOP 10 CAUSES OF DEATH WORLDWIDE, RANKED BY NUMBER OF DEATHS, 2008

LOW-INCOME COUNTRIES (163 OF 1,000)

Lower respiratory infections	Diarrheal diseases	HIV/AIDS	Ischemic heart disease	Malaria	Stroke	Tuberculosis	Prematurity and low birth weight	Birth asphyxia and birth trauma	Neonatal infections
18	13	13	10	8	8	7	5	5	4

MIDDLE-INCOME COUNTRIES (677 OF 1,000)

Ischemic heart disease	Stroke	Chronic obstructive pulmonary disease	Lower respiratory infections	Diarrheal diseases	HIV/AIDS	Road traffic accidents	Tuberculosis	Diabetes	Hypertensive heart disease
93	86	49	36	30	18	17	16	15	15

HIGH-INCOME COUNTRIES (159 OF 1,000)

Ischemic heart disease	Stroke	Respiratory system cancers	Alzheimer's and other dementias	Lower respiratory infections	Chronic obstructive pulmonary disease	Colon and rectal cancers	Diabetes	Hypertensive heart disease	Breast cancer
25	14	9	7	6	6	5	4	4	3

↑ Consider a group of 1,000 people representing the diverse make-up of the world's population. Of these 1,000, 163 would be from low-income countries, 677 from middle-income countries, and 159 from high-income countries.* The main causes of death vary when comparing these 3 groups and the top 10 causes of death are shown here. In low-income nations, infectious diseases are the leading cause of death (36.8% of all deaths), almost 10 times the rate seen in high-income nations (0.38%). Many cases of infectious diseases are due to modifiable environmental causes such as poor sanitation. In high-income nations, many modifiable environmental risks have been successfully addressed. In those nations, lifestyle choices associated with diet and exercise have become the major risk factors and death is usually due to chronic disease. Why do you think middle-income nations suffer both from lifestyle factors and from infectious diseases like diarrheal diseases and tuberculosis?

* WHO income brackets are based on World Bank groupings using income from 2009: Low-income nations (those with gross national incomes of $1,025 or less); middle-income nations (those with gross national incomes greater than $1,025 but less than $12,475); high-income nations (with gross national incomes greater than $12,475).

Infographic **6.7** | **ERADICATING GUINEA WORM DISEASE**

↴ Eradication programs have been in effect since 1980 and have made great progress. Prevention methods include purifying water, and education about how to acquire safe water sources and handle infections to avoid reintroducing larvae into the water. Donald Hopkins, the Carter Center Vice President of Health Programs, wrote in a 2011 report that the last cases are always the hardest to wipe out, but doing so is not a matter of "if" but "when."

1986
3.5 million cases in 20 countries

2010
About 1,800 cases in 4 countries

Mali
Ghana

Sudan
Ethiopia

Infographic **6.8** | **REDUCING ENVIRONMENTAL HEALTH HAZARDS**

Provide access to clean water	Digging wells and filtering surface water, including by using personal filters such as the LifeStraw, are two common methods to purify water, but these methods cost money and many low-income countries need financial assistance to make access to clean water a reality.
Improve sanitation and hygiene	Proper disposal of human and animal fecal waste includes methods—such as building latrines—to keep sewage out of surface waters, planting streamside vegetation to reduce runoff, and keeping animals out of water sources.
Reduce vector exposure	Removing vector habitat (like standing water for mosquitoes), providing barriers like mosquito netting, and applying pesticides can significantly reduce infection. Vaccinating pets can also reduce exposure to zoonotic diseases.
Reduce air pollution	Using cleaner-burning fuels and better-constructed and ventilated indoor stoves reduces indoor air pollution. Solar ovens can eliminate smoke and are inexpensive and effective. Adopting and enforcing air quality standards can reduce outdoor air pollution.
Education	Public education programs are vital in areas where simple behavioral steps can reduce exposure to hazards. Educational campaigns can teach people how to avoid exposure to pathogens, and how to protect themselves from infection and hazardous chemicals.
Effective public policy	Government support is needed to reduce environmental hazards and improve health care. (Governments can accept foreign aid if they cannot provide support directly.) Unfortunately, not all areas are politically stable and armed conflict or disinterested leaders can disrupt progress.

Yes, it is relatively cheap and technologically simple to build pit latrines and septic systems that make proper waste disposal possible, or to build fences that can keep animals out of human water supplies, or to plant vegetative buffers that can soak up runoff before it pollutes area streams. And yes, when combined, such a roster of straightforward measures might dramatically reduce the incidence of any number of life-threatening diseases (including, for example, the diarrheal diseases that kill so many children each year). But without the money to buy wood (for fences) or plants (for buffer zones), and without know-how and "local buy-in," such projects would never get off the ground.

Even education campaigns that help people understand how to avoid exposure to certain infectious agents—a measure that requires almost no material support—still take a concerted effort, and thus a well-trained workforce. "It's a lot of work to overcome preconceived notions," says Ruiz-Tiben. "You have to present the information in a way they can relate and respond to." [INFOGRAPHIC 6.8]

To generate those things—international support, local support, and so on—environmental health workers from the developed world need, urgently, to consider the perspectives of their developing-world brethren.

Consider the case of DDT: The pesticide can go a long way toward keeping insect vectors like mosquitoes from human hosts. But when scientists in the developed world linked it to a roster of poor health and poor environmental outcomes (see Chapter 3), the chemical was banned in many regions, including places where malaria is common. Leaders in those countries were, by many accounts, responding to pressure from the developed world. Today, world leaders are reconsidering: In some malaria hotspots—certainly in places where mosquito-borne diseases kill tens of thousands of people every year—it turns out that DDT still provides the best mitigation strategy, and may yet prove to be worth those risks.

As with most environmental problems, there are no easy solutions. To conquer GWD, environmental health workers like Ruiz-Tiben have had to battle indifference, poverty, human stubbornness, and more. In the end, those battles have paid off: In 2009, Nigeria became the 15th African country to rid itself of the ancient worm. By then, it was estimated that just 3,500 or so cases remained throughout the entire continent, and those numbers were dwindling rapidly. Indeed, some three decades after they began their quest, the Carter Center is finally closing in on its ultimate goal: eradicating Guinea worm disease, everywhere, once and for all.◉

Research articles referenced in this chapter:
Guimarães, R.M., et al. 2007. Cadernos de Saúde Pública [Online], 23: 2835–2844. doi: 10.1590/S0102-311X2007001200004.
LoGiudice, K., et al. 2003. Proceedings of the National Academy of Sciences, 100: 567–571.
Myers, S.S. and Patz, J.A. 2009. Annual Review of Environment and Resources, 34: 223–252.

BRING IT HOME

◉PERSONAL CHOICES THAT HELP

People in developed countries typically do not experience the same types and prevalence of infectious disease as do those in the developing world. However, outbreaks of illnesses like whooping cough, West Nile virus, bacterial food poisoning, and antibiotic-resistant bacterial infections do occur in the developed world and are largely preventable.

Individual Steps
→ Many diseases are spread by contaminated hands. The most effective way to remove infectious bacteria and viruses is with 20 seconds or more of washing with soap. This is even more effective than using hand sanitizer.
→ Reduce the likelihood that antibiotic-resistant bacteria will emerge by taking the entire prescription of any antibiotic you are prescribed.

Group Action
→ Mosquitoes are responsible for spreading many diseases. Organize your neighbors to take preventative steps to reduce mosquito breeding, including removing containers that might trap rainwater, draining areas of standing water, and cleaning out rain gutters.

→ Organize a fund-raising campaign to help purchase pipe filters such as the LifeStraw or finance well digging in areas that need access to clean water.

Policy Change
→ While the Environmental Protection Agency is required by the Safe Drinking Water Act to test public water supplies for contaminants and bacteria, the Food and Drug Administration is not empowered to require the same level of testing of bottled water. Ask your legislators what steps could be taken to hold bottled water to similar standards.

UNDERSTANDING THE ISSUE

CHECK YOUR UNDERSTANDING

1. **Which of the following is NOT true of environmental health?**
   a. It is a branch of public health.
   b. It focuses only on how natural disasters impact health.
   c. It focuses on health risks from the natural and human-built environment.
   d. It focuses on human health, not on the health of other species in the environment.

2. **Which of the following is the main focus of public health officials who address environmental hazards?**
   a. Biological hazards
   b. Human-caused hazards
   c. Modifiable hazards
   d. Waterborne hazards

3. **Which action below would help reduce the health problem that has the largest percentage of its occurrence attributed to environmental conditions?**
   a. Provide solar ovens to low income areas.
   b. Provide better sanitation and clean water.
   c. Improve diets to help people lose weight.
   d. Use cleaner burning fuels indoors.

4. **Guinea worm disease:**
   a. infects people who drink water contaminated with copepods carrying worm larvae.
   b. is spread by people with active Guinea worm infections entering communal water.
   c. can be controlled by filtering drinking water and stopping infected people from entering shared water sources.
   d. All of the above.

5. **Which factor below reduces our ability to address many environmental health problems?**
   a. Scientists don't understand how most infectious diseases are spread.
   b. Most environmental health problems are in rural areas, far from health professionals.
   c. People in developed countries don't care about the health of those in developing countries.
   d. Affected regions may not have the political will or financial means to address environmental problems.

6. **Compared to those in high-income nations, people in low-income developing nations:**
   a. are more likely to die from infectious diseases.
   b. develop more genetically based diseases.
   c. do not develop "lifestyle" diseases.
   d. are less affected by environmental hazards.

7. **Zoonotic diseases:**
   a. are increasing in frequency worldwide.
   b. come mainly from domesticated animal species.
   c. are diseases that spread between humans and plant species.
   d. All of the above:

8. **Infectious agents that cause disease are called:**
   a. vectors.
   b. physical hazards.
   c. pathogens.
   d. toxins.

WORK WITH IDEAS

1. How does the focus and goal of public health programs differ from that of individual health care providers?

2. Environmental hazards can be divided into physical, chemical, and biological hazards. Give an example of each of these hazards. Describe a scenario in which exposure to one of the hazards could make a person more vulnerable to another type of hazard.

3. What is malaria and how is it transmitted? What are the costs and benefits of using DDT to kill the malaria vector? What criteria should be used to determine whether to use DDT for mosquito control?

4. Guinea worm disease is still found in the Sudan. What can be done to eradicate it and why has this not already been done?

5. Compare the leading causes of death in low-, middle-, and high-income countries. Why do these differences exist?

6. Identify possible interventions that could decrease health issues associated with air pollution. (Smoking, burning solid fuels indoors without adequate ventilation, and vehicle exhaust are the most common air pollutants that impair health.)

ANALYZING THE SCIENCE

Accurate estimates of deaths due to malaria are important for many reasons, including decisions regarding vector control and health interventions, and also to direct charitable donations effectively. A recent analysis estimated much higher annual death rates from malaria than did previous studies, with the majority of the increase coming from victims over the age of 5. The table on the right shows the annual number of deaths from malaria in different world regions, estimated by the Institute for Health Metrics and Evaluation (IHME) and the World Health Organization (WHO).

Comparison of IHME and WHO estimates of malaria deaths by WHO region

WHO Region	IHME	WHO	Difference
Number of malaria deaths			
Africa (AFR)	1,098,818	596,000	502,818
Americas (AMR)	986	1,000	–14
Eastern Mediterranean (EMR)	47,499	15,000	32,499
Europe (EUR)	3	—	3
Southeast Asia (SEAR)	84,573	38,000	46,573
Western Pacific (WPR)	5,596	5,000	596
Total	**1,237,475**	**655,000**	**582,475**
% malaria deaths under age 5	**58%**	**86%**	

INTERPRETATION

1. Describe in one sentence what the table shows about the total number of deaths due to malaria.

2. What region of the world has the most mortality due to malaria? What proportion of the world total comes from this region?

3. What percentage and what number of malaria deaths do IHME and WHO estimate are of children under the age of 5?

ADVANCE YOUR THINKING

4. In which region are the new mortality estimates most enlarged? Why do you think this might be the case?

5. It is often extremely difficult to estimate the cause of death in a developing country; health workers rely on a "verbal autopsy," in which they ask surviving family members a series of questions about the deceased's symptoms. If it were possible to confirm that a person who died was or was not infected by malaria, rather than relying on verbal autopsies, would you predict that the estimates of mortality from malaria would increase or decrease? Explain your reasoning.

EVALUATING NEW INFORMATION

Humanitarian organizations are increasingly interested in ensuring that their time and energy is invested in the most efficient and effective manner possible. The Bill and Melinda Gates Foundation has led the way with this approach, and has donated a large amount of money to the Institute for Health Metrics and Evaluation (IHME) with this goal in mind.

Go to the IHME website (www.healthmetricsandevaluation.org/) and review the home page and, under the "About IHME" tab, read about its history.

Evaluate the website and work with the information to answer the following questions:

1. Is the IHME website a reliable information source?
   a. Does the IHME give supporting evidence for its claims?
   b. Does the organization give sources for its evidence?
   c. What is the mission of IHME? How is it funded?

2. Select the "Data Visualization" link under the "Tools" menu, and choose "Interact with this tool" for the figure "Deaths due to malaria by age, region, country, and year (Global), 1980–2010."
   a. Run your mouse over some of the bubbles. What do they represent?
   b. Describe in one sentence what the bubble graph shows. If you are having difficulty interpreting it, click on the upper right tab for a bar graph or a line graph. These show the same data in alternative formats. See "Terms defined" below the graph if you need more help with interpretation.
   c. Click the Play button in the bottom right corner to see the data from 1980 through 2010. What is the general trend shown?
   d. Choose one country by clicking on the dot representing it, click the "trails" option to the right of the graph just under the box containing the list of countries, then click the bottom black arrow (located to the left of "1980"). Describe what you see about your chosen country.
   e. IHME claims that death rates from malaria decreased dramatically worldwide beginning in 2004. What evidence does the organization provide to support this claim? Is it sufficient? Where would you attempt to find more evidence to support or refute this claim?

3. Select the "News & Events" link under the "News" menu, and choose "Owning insecticide-treated bed nets lowers child mortality by 23%." What is the main finding of the study? What policy recommendation would you make on the basis of the study? Going back to the graph in #2, above—which country do you think would most benefit from your policy recommendation and why?

4. IHME's slogan is "Accelerating global health progress through sound measurement and accountable science." Do you agree that it is worthwhile to invest time and money (that could otherwise go directly to humanitarian aid) in collecting and evaluating data more accurately? Justify your answer.

MAKING CONNECTIONS

CHOLERA RETURNS TO HAITI

Background: Cholera is a bacterial disease that spreads through water that has been contaminated with the feces of an infected person. When others bathe or drink in the contaminated water, they become infected. Infection results in diarrhea and vomiting, and without aggressive treatment, up to a quarter of those infected may die.

Case: In January 2010, a magnitude 7.0 earthquake struck Haiti, killing more than 200,000 people. Ten months later, a different kind of environmental disaster struck—cholera, which had not been reported in Haiti for over 100 years. Could this outbreak have been prevented or dealt with more quickly and efficiently?

Research Haiti and the Haitian cholera outbreak that began in 2010 and write a report that includes the following:

a. A description of the demographics of Haiti: health statistics (such as infant mortality, life expectancy, infectious disease rates, leading causes of death) and the percent of the population with access to clean water, sanitation facilities, and health care.
b. A description of the extent of the cholera outbreak, its demographics (which populations were most vulnerable), and the possible origin of the outbreak.
c. An explanation of why Haiti was vulnerable to an outbreak like this after the earthquake.
d. A brief synopsis of the response, including a list of the main groups that came to help, and a description of problems that arose in trying to combat the outbreak.
e. Your recommendation for an intervention that would better prepare responders to prevent and/or deal with a waterborne disease outbreak like this in Haiti in the future.
   - This can be an educational, a political, or an environmental intervention.
   - Address the potential benefits (including the economic, social, and ecological benefits) of the intervention.
   - Address the possible costs and consequences, including unintended and secondary consequences.

CORE MESSAGE

Ecosystems are complex assemblages of many interacting, living and nonliving components. Living organisms play irreplaceable roles in nature, supporting life and allowing ecosystems to function over the long term. It is important that we protect and work to restore ecosystems in nature to keep these connections intact so that we and other species can continue to live and thrive on this planet.

GUIDING QUESTIONS

After reading this chapter, you should be able to answer the following questions:

→ What is the hierarchy of organization recognized by ecologists and why might it be useful to recognize such distinctions?

→ What are biomes and how do environmental factors affect their distribution and makeup?

→ What are tolerance limits and how do they affect the distribution of a species within its ecosystem?

→ How do important nutrients like carbon, nitrogen, and phosphorus cycle through ecosystems?

→ What factors must be considered in order to create or restore an ecosystem?

ENGINEERING EARTH

A desert experiment gone awry

Biosphere 2 glows in the Arizona dusk.

On September 26, 1993, with their first mission complete, four men and four women emerged from Biosphere 2—a hulking dome of custom-made glass and steel—back into the Arizona desert, where throngs of spectators stood cheering. They had been sealed inside the facility, along with 3,000 other plant and animal species, for exactly 2 years and 20 minutes; it was the longest anyone had ever survived in an enclosed structure.

The feat was part of a grand experiment, the goals of which were twofold. First, scientists wanted to prove that an entirely self-contained, humanmade system—the kind they might one day use to colonize the Moon or Mars—could sustain life. Second, they hoped that by studying this mini-earth, which could be controlled and manipulated in ways the real Earth could not, they might better understand our own planet's delicate balance and how best to protect it.

Despite the fanfare surrounding the biospherians' emergence, it was tough to say whether the mission had been a success or a failure. More than one-third of the flora and fauna had become extinct, including most of the vertebrates and all of the pollinating insects. Morning glory vines had overrun other plants, including food crops. Cockroaches and "crazy ants" were thriving. Too little wind had prevented trees from developing stress wood—wood that grows in response to mechanical stress and helps trunks and branches shift into an optimal position; without stress wood, the trees were brittle and prone to collapse. And too many sweet potatoes had turned the biospherians themselves bright orange (a string of plant diseases had decimated other crops).

On top of that, nitrous oxide (laughing gas) had grown concentrated enough to "reduce vitamin B12 synthesis to a level that could impair or damage the brain," according to one interim report. And oxygen levels had plummeted from 21% (roughly the same as Earth's atmosphere) to 14% (just barely enough to sustain human life). To fix this, project engineers had been forced to pump in 600,000 cubic feet of outside air, violating the facility's sanctity as a closed system.

Worst of all, missteps and course corrections had been mired in secrecy—each one leaked to the press only months after the fact. Rumors had begun to circulate that the eight people sealed inside—not to mention the ones they took their orders from—were more interested in creating a futuristic utopia than in conducting rigorous scientific research. As evidence for this theory mounted, the scientific community grew suspicious. Was Biosphere 2 legitimate science, a publicity stunt, or some bizarre mix of the two?

To be sure, the eight biospherians had survived, and many experts agreed that in principle at least, the facility still held enormous potential as a scientific tool. But before that potential could be realized, the scientific community and the public at large would need to know exactly what had happened inside the desert dome.

To answer that question, we need to answer a few others first: What exactly is a biosphere, and just how did Biosphere 2's creators set about building one?

Organisms and their habitats form complex systems.

The term **biosphere** refers to the total area on Earth where living things are found—the sum total of all its ecosystems. An **ecosystem** includes all the organisms in a given area plus the nonliving components of the

◉WHERE IS BIOSPHERE 2?

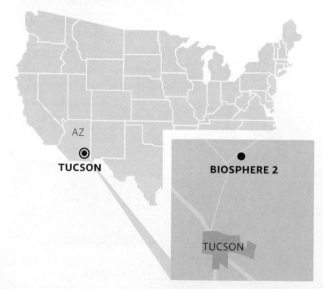

AZ

◉
TUCSON

●
BIOSPHERE 2

TUCSON

---

biosphere The sum total of all of Earth's ecosystems.
ecosystem All of the organisms in a given area plus the physical environment in which they interact.

← The eight biospherians emerge from Biosphere 2 after living there for two years.

↓ Visitors can tour the facility. Here they view the desert biome.

Infographic **7.1** | **ORGANIZATION OF LIFE: FROM BIOSPHERE TO INDIVIDUAL**

BIOSPHERE
The total area on Earth (air, land, or water) where living things are found

ECOSYSTEM
A specific portion of a biome consisting of the living (biotic) and nonliving (abiotic) environmental components that interact

BIOME
A portion of the biosphere characterized by a distinct climate and a particular assemblage of plants and animals adapted to it

COMMUNITY
All the populations (plants, animals, and other species) living and interacting in an area. Communities represent the "living" portion of the ecosystem.

← Ecologists recognize a nesting hierarchy of organization from the biosphere down to the individual organism. Each category is made up of the smaller ones. Ecologists often focus their study on populations, communities, and ecosystems, and the interactions between organisms and their surroundings. In contrast, a zoologist or botanist might focus on individual animals or plants.

POPULATION
A group of individuals of the same species living and interacting in the same region

INDIVIDUAL
A single member of the population

Infographic **7.2** | **HABITAT AND NICHE** ↓ Species depend on suitable habitats in which to live. Species fill a specific niche in their community.

SPECIES
A group of plants or animals that have a high degree of similarity and can generally only interbreed among themselves

HABITAT
The physical environment in which individuals of a particular species can be found

NICHE
The role a species plays in its community, including how it gets its energy and nutrients, its habitat requirements, and what other species and parts of the ecosystem it interacts with

physical environment in which they interact. This physical environment is often referred to as an organism's **habitat**. Ecologists study how ecosystems function by focusing on **species**' interactions with their surroundings and with other species in their communities. They also study interactions between members of the same species within a population. Individuals of each species in a community occupy a specific ecological **niche** shared by no other species in that community. In the natural world, ecosystems assume a range of shapes and sizes—a single, simple tide pool qualifies as an ecosystem; so does the entire Mojave Desert. [INFOGRAPHIC 7.1, 7.2]

All ecosystems function through two fundamental processes collectively referred to as ecosystem processes, namely nutrient cycling and **energy flow**. The term **nutrient cycles**, or biogeochemical cycles, refers to the movement of life's essential chemicals or nutrients through an ecosystem. Energy, on the other hand, enters as solar radiation and is passed along from organism to organism, some released as heat, until there is no more usable energy left. Therefore we can say that matter *cycles* but energy *flows*—a one-way trip.

Earth—or "Biosphere 1," as the creators of Biosphere 2 liked to call it—is materially closed but energetically open. [INFOGRAPHIC 7.3] In other words, the plants and other organic material that make up an ecosystem, called **biomass**, cannot enter or leave the system, but energy can: some leaves as heat or light and new energy is absorbed from outside. In fact, plant biomass is produced with energy from the Sun through photosynthesis.

Biomes are specific portions of the biosphere determined by climate and identified by the predominant vegetation and organisms adapted to live there. Biomes can be divided into three broad categories—marine, freshwater, and terrestrial. Within those three categories are several narrower groups, and within those, a variety of

Infographic 7.3 | **EARTH IS A CLOSED SYSTEM FOR MATTER BUT NOT FOR ENERGY**

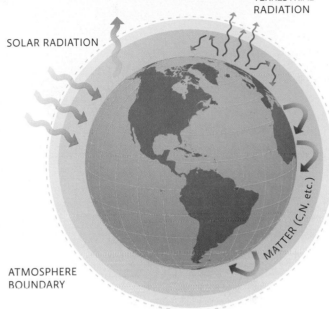

↑ Energy can enter and leave Earth as light (solar radiation) and heat (terrestrial radiation) but matter stays in the biosphere, cycling in and out of organisms and environmental components.

subgroups. An entire biome itself may be considered an ecosystem, as are the smaller groups and subgroups. For example, forests, deserts, and grasslands are the three main types of terrestrial biomes. Within the forest biome category are different types of forests, such as tropical, temperate, or boreal forest, and within each of those groups are subgroups (for example, dry tropical forest and tropical rain forest.) [INFOGRAPHIC 7.4]

When ecologists study entire ecosystems, they are limited to making observations and trying to discern cause and effect from those observations. This is no small challenge; even the simplest phenomenon is impacted by a myriad of factors. Precise, systemwide measurements are exceedingly difficult to come by. And unlike laboratory science, field research doesn't often accommodate rigorous controls.

Biosphere 2 would offer ecologists an unprecedented research tool: a mini-planet where a variety of environmental variables—from temperature and water availability to the relative proportions of oxygen and carbon dioxide (CO_2) at any given moment—could be tightly controlled and precisely measured. "Manipulating these variables and tracking the outcomes could greatly

---

habitat The physical environment in which individuals of a particular species can be found.

species A group of plants or animals that have a high degree of similarity and can generally only interbreed among themselves.

niche The role a species plays in its community, including how it gets its energy and nutrients, what habitat requirements it has, and what other species and parts of the ecosystem it interacts with.

energy flow The one-way passage of energy through an ecosystem.

nutrient cycles Movement of life's essential chemicals or nutrients through an ecosystem (also known as biogeochemical cycles).

biomass The sum of all organic material—plant and animal matter—that make up an ecosystem.

biome One of many distinctive types of ecosystems determined by climate and identified by the predominant vegetation and organisms that have adapted to live there.

↓ Biomes are specific types of terrestrial ecosystems with characteristic temperature and precipitation conditions. Temperature varies with latitude (decreasing as one moves away from the equator) and altitude (decreasing as elevation increases); thus a cold climate can be found above 60° north and south latitudes, as well as on an equatorial mountaintop. Latitude also affects precipitation, with wet areas occurring at the equator and at 30° north and south and dry areas occurring at 60° north and south (due to global air circulation patterns).

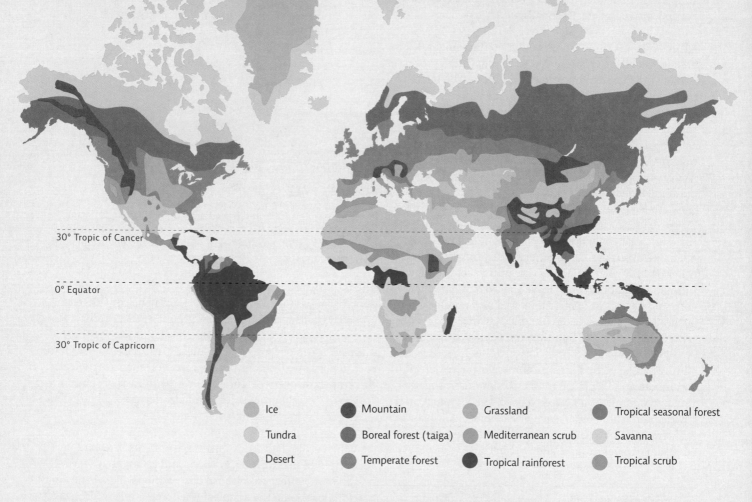

30° Tropic of Cancer

0° Equator

30° Tropic of Capricorn

Ice

Tundra

Desert

Mountain

Boreal forest (taiga)

Temperate forest

Grassland

Mediterranean scrub

Tropical rainforest

Tropical seasonal forest

Savanna

Tropical scrub

TUNDRA

TROPICAL RAINFOREST

BOREAL FOREST

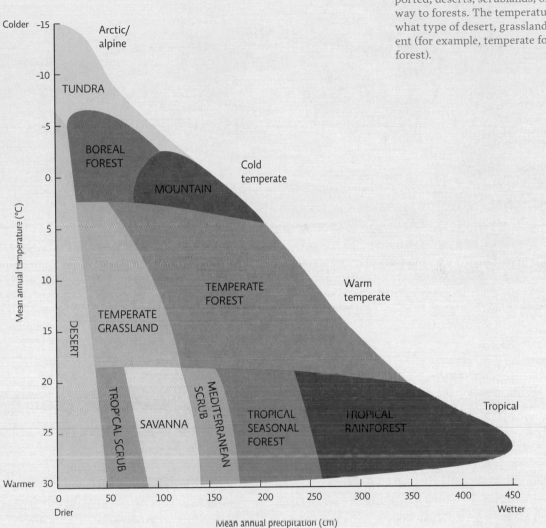

This biome climograph shows the approximate distribution of biomes with regard to annual precipitation and temperature. As precipitation increases and more plant life is able to be supported, deserts, scrublands, or grasslands give way to forests. The temperature also influences what type of desert, grassland, or forest is present (for example, temperate forest versus boreal forest).

Mean annual temperature (°C)

Colder -15
-10
-5
0
5
10
15
20
25
Warmer 30

Arctic/alpine

TUNDRA

BOREAL FOREST

MOUNTAIN

Cold temperate

DESERT

TEMPERATE GRASSLAND

TEMPERATE FOREST

Warm temperate

TROPICAL SCRUB

SAVANNA

MEDITERRANEAN SCRUB

TROPICAL SEASONAL FOREST

TROPICAL RAINFOREST

Tropical

0   50   100   150   200   250   300   350   400   450
Drier   Wetter

Mean annual precipitation (cm)

MEDITERRANEAN SCRUB

DESERT

SAVANNA

Infographic **7.5** | **MAP OF BIOSPHERE 2**

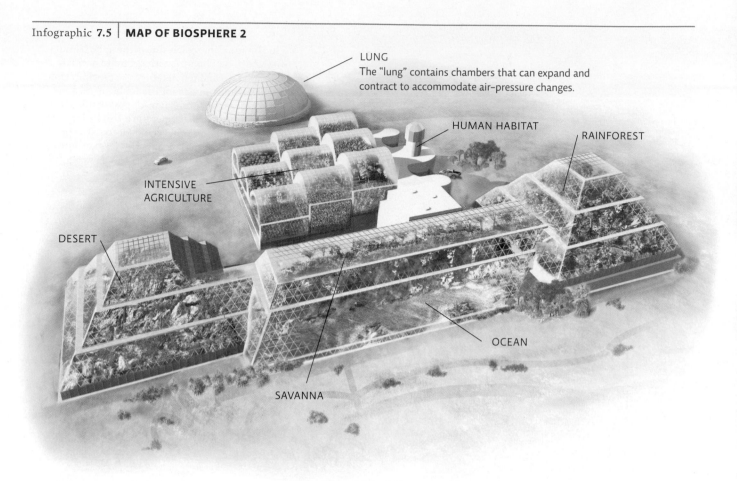

LUNG
The "lung" contains chambers that can expand and
contract to accommodate air-pressure changes.

HUMAN HABITAT

RAINFOREST

INTENSIVE
AGRICULTURE

DESERT

OCEAN

SAVANNA

↑ Biosphere 2 houses several biomes under one roof, each contributing to overall function. One of the challenges faced by designers
was how to include a variety of biomes in the close quarters of the 3-acre Biosphere 2 structure. For example, in nature, a tropical
rainforest would not be next to an arid desert. To deal with this, an ocean was placed between the desert and rainforest to serve as
a temperature buffer.

advance our understanding of natural ecosystems and all
the minute, complex interactions that make them work,"
says Kevin Griffin, a Columbia University plant ecologist
who conducted research at the Biosphere 2 facility. "The
plan was to use that knowledge to figure out how to repair
degraded ecosystems in the real world, so that they continue to provide the services so essential to our survival."

On top of all that, proving humans could survive in a
completely enclosed, manufactured system would take us
one giant step closer towards colonizing space.

But would it work?

The concept of enclosed ecosystems was not a new one.
Since before they put a person on the moon, in fact,
astronauts and engineers had been tinkering with their
own artificial ecosystems—systems they hoped could
one day be used to colonize space. The earliest versions
of this technology were developed in the 1960s and 1970s

by Russian and American scientists who, in the spirit of
their times, had pitted themselves against one another
in a mad dash to the finish line. The ecosystems they
designed ranged from small, crude structures in which a
single person might last for a single day, to larger, more
sophisticated enterprises that could sustain a few people
for a few months.

Biosphere 2 is by far the most elaborate. At 2½ football
fields in diameter, it remains the largest enclosed ecosystem ever created and the only one to house several biomes
under one roof. [INFOGRAPHIC 7.5] From a mountain under
the dome's 91-foot zenith, a stream rushes down through
a tropical rainforest before snaking southward into a
savanna. From there, the stream wends its way through
a mangrove swamp into a million-gallon ocean complete
with a coral reef. On the other side of the ocean lies a
desert. Biosphere 2 also includes a human habitat and an
agricultural biome.

Infographic **7.6** | **RANGE OF TOLERANCE FOR LIFE**

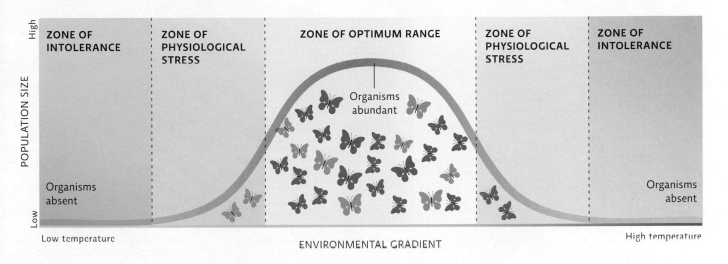

↑ Populations have a range of tolerance for a given environmental factor (such as temperature). Every species has an upper and a lower limit beyond which it cannot survive (in this example, temperatures that are too cool or too hot). Most individuals, like the butterflies in this population, can be found around the optimum temperature, though what is "optimum" for each individual may differ slightly because of genetic variability. Some individuals may find themselves in areas of the habitat that are warmer or colder than the optimum. They can tolerate these conditions but may be physiologically stressed and not grow well or successfully breed. Genetic differences that allow some individuals to tolerate or even thrive at the edges of the population's tolerance offer the population a chance to adapt to changing conditions (such as a warmer climate) if needed. The more narrow the range of tolerance and the less genetically diverse the population, the less likely it will survive a change in conditions.

Living things survive within a specific range of environmental conditions.

Each biome required a mind-boggling array of considerations—not only how diverse plant and animal species would interact within and across biomes, but also the nutrient requirements of each organism they planned to include. Termites, for example, would need enough dead wood at the beginning of closure to sustain them until some of the larger plants began dying off. Termites live in the soil, stirring it and allowing air to penetrate soil particles. If the termites ran out of dead wood and starved to death, organisms living in the soil would not get enough oxygen and the entire desert would be jeopardized. Hummingbirds, on the other hand, would need nectar-filled flowers. "Try figuring out how many flowers a day a hummingbird needs," says Tony Burgess, a University of Arizona ecologist who helped design the biomes in Biosphere 2 and remained involved until 2004. "From there you need to know what the blooming season is, and then what the nectar load per flower is. And then you have to translate all of that into units of hummingbird support. Now imagine doing that sort of thing about 3,000 times."

Biome-level planning was equally complicated. A rainforest would need consistently warm temperatures, but most desert biomes fluctuate wildly between day and night. In the summer, the grassland biome might literally starve the rainforest by gobbling up all the CO_2 needed for photosynthesis.

In this context, dead wood, flowers, and CO_2 are all examples of **limiting factors**—resources so critical that their availability controls the distribution of species and thus of biomes. The principle of limiting factors states that the critical resource in least supply is what determines the survival, growth, and reproduction of a given species in a given biome. Living things can only survive and reproduce within a certain range (between an upper and lower limit that they can tolerate) for a given critical resource or environmental condition, referred to as their **range of tolerance**. [INFOGRAPHIC 7.6]

limiting factor The critical resource whose supply determines the population size of a given species in a given biome.
range of tolerance The range, within upper and lower limits, of a limiting factor that allows a species to survive and reproduce.

↑ The Biosphere 2 ocean, shown here, is still used for marine research on the effect of increasing CO_2 levels on coral.

Ecologists routinely monitor a variety of limiting factors as a way of assessing ecosystem health, but anticipating what each individual organism would need to survive before the fact proved daunting.

An all-star team of scientists—oceanographers, forest ecologists, and plant physiologists—spent 2 years sorting through these challenges. Drawing on their combined expertise, they set about choosing the combination of soils, plants, and animals that seemed most capable of working together to recreate the delicate balances that had made Biosphere 1 such a spectacular success. A summer-dormant desert, like the ones found in Baja California, was chosen because it would reduce the desert's CO_2 demands when the savanna's productivity was at its highest. The ocean was situated between the desert and rainforest so that it could serve as a temperature buffer between the two. And each biome was created from a carefully selected array of species: The marsh biome was composed of intact chunks of swampland harvested from the Florida Everglades, and the savanna was composed of grasses from Australia, South America, and Africa. Well water mixed with aquarium salt filled the ocean, as did coral reefs culled from the Caribbean.

But it wasn't long before the rigor and pragmatism of good science began to clash with the idealism of Biosphere 2

financiers. And when that happened, critics say, science lost out.

Some scientists worried that the ocean wouldn't get enough sunlight to support plant and animal life. Others opposed the use of soil high in organic matter; soil microbes decompose the soil's organic carbon and release it into the air as CO_2. Although this organic soil might eliminate the need for chemical fertilizers, there were concerns that it would provide too much fuel for the soil microbes, and would thus send atmospheric CO_2 concentrations through the roof. Despite these concerns, scientific advisors to the project were overruled.

The first few months of Mission 1 went smoothly enough, but eventually, plants and animals started dying. Humans grew hungry and mysteriously sleepy. And before long, they turned on each other.

Like Earth, Biosphere 2 was designed as a materially closed and energetically open system: Plants would conduct photosynthesis with sunlight that streamed through the glass, but no biomass would enter or leave. Temperature, wind, rain, and ocean waves would be controlled mechanically. "But the facility would have to be self-sustaining," says Burgess. "Everything would die, unless the biota met its most fundamental purpose—using energy flow for biomass production." Humans would have to survive exclusively on what they could grow or catch under the dome. The system as a whole would have to continuously recycle every last bit of nutrient that was in the soil on day one.

At first, the carefully constructed agricultural biome seemed well suited to the challenge. Carrots, broccoli, peanuts, kale, lettuce, and sweet potatoes were grown on broad half-acre terraces that sat adjacent to the sprawling six-story human habitat; a bevy of domestic animals also provided sustenance—goats for milk, chickens for eggs, and pigs for pork. Indeed, eating only what they could grow made the biospherians healthier. Everyone lost weight. Bad cholesterol and blood pressure went down; so did white blood cell counts. And slowly, the biospherians say, their relationship to food changed. "Inside, I knew exactly where my food came from, and I totally understood my place in the biosphere and how it impacted the food I ate," says Jayne Poynter, the biospherian responsible for tending the farm. "When I breathed out, my CO_2 fed the sweet potatoes that I was growing. When I first got out, I lost sense of that. I would stand for hours in the aisles of shops, reading all the names on all the food things and think 'where does this stuff come from?' People must have thought I was nuts."

But the glazed glass of the dome admitted less sunlight than had been anticipated. Less sunlight meant less biomass production. And that meant less food. Mites and disease also cut crop production. Biosphere 2's size precluded the use of pesticides and herbicides: In that small an atmosphere, the toxins would build up rapidly and could have had a deleterious impact on air quality and human health.

Now, a few months in, food was starting to run out. And that wasn't the only problem.

The humans were so tired they couldn't work. Nobody knew why, but scientists on the outside suspected it had something to do with nutrient cycles.

Nutrients such as carbon cycle through ecosystems.

On Earth, nutrients cycle through both **biotic** and **abiotic** components of an ecosystem—organisms, air, land, and water. They are stored in abiotic or biotic parts of the environment called **reservoirs**, or sinks, and linger in each for various lengths of time, known as *residence times*. Organisms acquire nutrients from the reservoir, the chemical cycles through the food chain, and eventually the chemical is returned to the reservoir. For carbon, the atmosphere—where carbon is stored as CO_2—is the most

---

Infographic **7.7** | **CARBON CYCLES VIA PHOTOSYNTHESIS AND CELLULAR RESPIRATION**

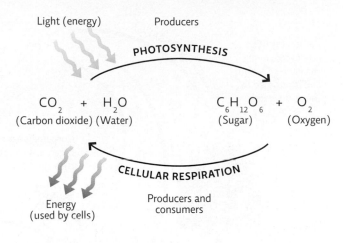

↑ In photosynthesis, producers use solar energy to combine CO_2 and H_2O to make sugar, releasing O_2 in the process. When producers (or any consumer who eats another organism) need energy, they break apart the sugar via the reverse reaction, cellular respiration. Oxygen is required for this reaction (which is why it is called "respiration").

---

important reservoir. (Oceans and soil are also abiotic reservoirs for carbon. Oceans absorb CO_2 directly from the atmosphere and soils accumulate it during decomposition.) Plants and other photosynthesizers use carbon molecules from atmospheric CO_2 to build sugar, releasing oxygen in the process. Because they "produce" sugar, an organic molecule, from inorganic atmospheric CO_2, they are called **producers**.

This sugar molecule represents stored chemical energy that the producer can use. A **consumer**, the organism that eats the plant (or that eats the organism that eats the plant), also uses the chemical energy of sugar. This energy is released to the cell via the process of **cellular respiration**. [INFOGRAPHIC 7.7] All organisms—producers and consumers—perform cellular respiration. (For

---

biotic The living (organic) components of an ecosystem, such as the plants and animals and their waste (dead leaves, feces).

abiotic The nonliving components of an ecosystem, such as rainfall and mineral composition of the soil.

reservoirs (or sinks) Abiotic or biotic component of the environment that serves as a storage place for cycling nutrients.

producer An organism that converts solar energy to chemical energy via photosynthesis.

consumer An organism that obtains energy and nutrients by feeding on another organism.

cellular respiration The process in which all organisms break down sugar to release its energy, using oxygen and giving off CO_2 as a waste product.

Infographic **7.8** | **THE CARBON CYCLE**

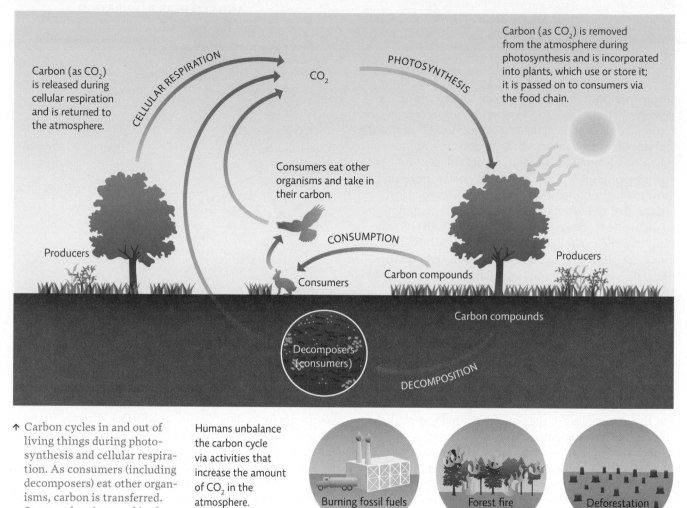

Carbon (as CO_2) is released during cellular respiration and is returned to the atmosphere.

CELLULAR RESPIRATION

CO_2

PHOTOSYNTHESIS

Carbon (as CO_2) is removed from the atmosphere during photosynthesis and is incorporated into plants, which use or store it; it is passed on to consumers via the food chain.

Consumers eat other organisms and take in their carbon.

Producers

CONSUMPTION

Consumers

Carbon compounds

Producers

Carbon compounds

Decomposers (consumers)

DECOMPOSITION

↑ Carbon cycles in and out of living things during photosynthesis and cellular respiration. As consumers (including decomposers) eat other organisms, carbon is transferred. Some carbon is stored in the bodies of organisms and in soil, but over the long term, the carbon cycle is balanced between photosynthesis and respiration.

Humans unbalance the carbon cycle via activities that increase the amount of CO_2 in the atmosphere.

Burning fossil fuels

Forest fire

Deforestation

more information on producers, consumers, and the food chain, see Chapter 9.)

From its initial incorporation into living tissue via photosynthesis, to its ultimate return to the atmosphere through respiration or through the burning of carbon-based fuels, carbon cycles in and out of various molecular forms and in and out of living things as it moves through the **carbon cycle**. [INFOGRAPHIC 7.8] Even though a single carbon atom might cycle in a few weeks or years, decades can pass before changes in a Brazilian rainforest impact a farm in Iowa. Inside Biosphere 2, the same cycle took approximately 3 days, which meant that changes in one biome could be felt in another biome much more quickly than on Earth.

Still, Biosphere 2's carbon cycle was not that different from Earth's—carbon moved from living tissue to the atmosphere and back in the same predictable manner. Or at least it should have. As the biospherians' energy waned, it became clear that something had gone terribly wrong.

It turned out that oxygen levels had fallen steadily—from 21% down to 14%; at such low concentrations, the biospherians were unable to convert the food they consumed into usable energy. "We were just dragging ourselves around the place," Poynter says. "And we had sleep apnea at night. So we'd wake up gasping for air because our blood chemistry had changed."

Infographic 7.9 | **OXYGEN DEPLETION IN BIOSPHERE 2**

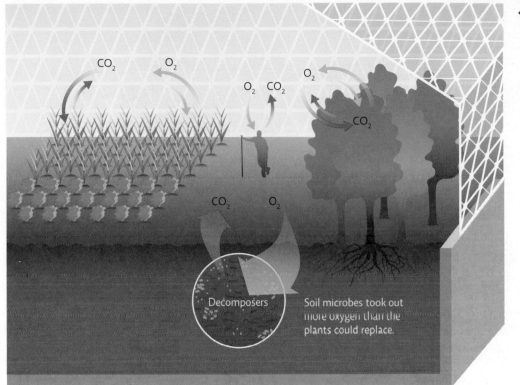

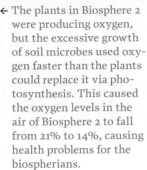

← The plants in Biosphere 2 were producing oxygen, but the excessive growth of soil microbes used oxygen faster than the plants could replace it via photosynthesis. This caused the oxygen levels in the air of Biosphere 2 to fall from 21% to 14%, causing health problems for the biospherians.

In just a few months, some 7 tons of oxygen—enough to keep six people breathing for 6 months—had gone missing. As scientists from Columbia University later discovered, soil microbes were gobbling up all that O_2 and converting it into CO_2 as they decomposed the organic matter in the soil. [INFOGRAPHIC 7.9]

The biospherians responded by filling all unused planting areas with morning glory vines, a pretty and fast-growing (but as it turned out, invasive) species they hoped would maximize the amount of CO_2 converted back into O_2 by photosynthesis. But even with an abundance of plants and enough CO_2, photosynthesis was still limited by the availability of sunlight; not even morning glories could keep up with the soil microbes in their warm, well-watered, highly organic soil.

carbon cycle Movement of carbon through biotic and abiotic parts of an ecosystem. Carbon cycles via photosynthesis and cellular respiration as well as in and out of other reservoirs such as the oceans and soil. It is also released by human actions such as fossil fuel burning.

nitrogen cycle Continuous series of natural processes by which nitrogen passes from the air to the soil, to organisms, and then returns back to the air or soil through decomposition or denitrification.

Biosphere 2 is not alone with regard to a disrupted carbon cycle. Human activity has greatly altered carbon amounts in Earth's atmosphere. Many of our actions (such as burning fossil fuels) increase the amount of carbon normally released into the atmosphere or degrade natural ecosystems so that less carbon is removed from the atmosphere (as in the case of deforestation). Just like with Biosphere 2, this extra carbon causes problems such as global climate change, acidification of oceans, and alterations of communities worldwide.

Adding to the confusion, concrete used to build parts of Biosphere 2 was absorbing some of the CO_2 and converting it into calcium carbonate, trapping some of the carbon and oxygen in this unexpected sink.

Besides carbon, two other chemicals essential for life nitrogen and phosphorus—cycle through ecosystems. Nitrogen, the most abundant element in Earth's atmosphere, is needed to make proteins and nucleic acids, but plants cannot utilize nitrogen in its atmospheric form (N_2). All plant life, and ultimately all animal life too, depends on microbes (bacteria) to convert atmospheric nitrogen into usable forms as part of the **nitrogen cycle**.

Infographic **7.10** | **THE NITROGEN CYCLE**

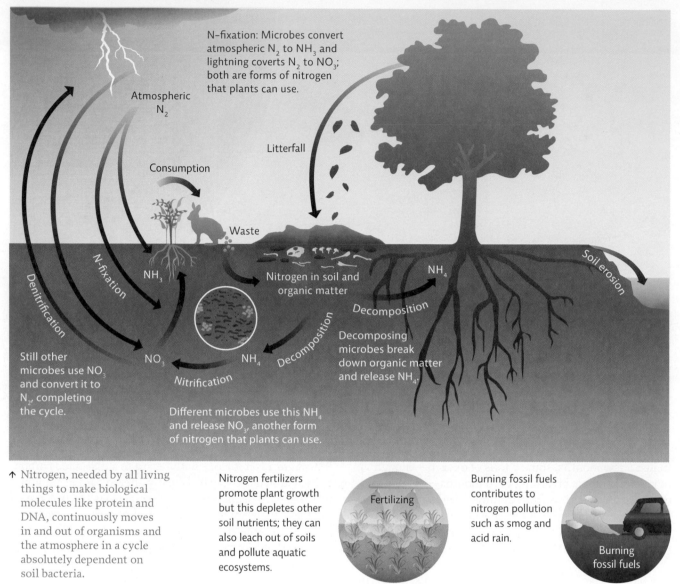

N-fixation: Microbes convert atmospheric N_2 to NH_3 and lightning coverts N_2 to NO_3; both are forms of nitrogen that plants can use.

Atmospheric N_2

Litterfall

Consumption

Waste

Soil erosion

Denitrification

N-fixation

NH_3

Nitrogen in soil and organic matter

NH_4

Decomposition

Decomposing microbes break down organic matter and release NH_4.

Still other microbes use NO_3 and convert it to N_2, completing the cycle.

NO_3

NH_4

Decomposition

Nitrification

Different microbes use this NH_4 and release NO_3, another form of nitrogen that plants can use.

↑ Nitrogen, needed by all living things to make biological molecules like protein and DNA, continuously moves in and out of organisms and the atmosphere in a cycle absolutely dependent on soil bacteria.

Nitrogen fertilizers promote plant growth but this depletes other soil nutrients; they can also leach out of soils and pollute aquatic ecosystems.

Fertilizing

Burning fossil fuels contributes to nitrogen pollution such as smog and acid rain.

Burning fossil fuels

In a process called **nitrogen fixation**, atmospheric nitrogen (N_2) is converted by bacteria into ammonia (NH_3) which plants take up through their roots; consumers take in nitrogen via their diet. A small amount of N_2 is fixed by lightning, producing nitrate (NO_3). In other steps of the nitrogen cycle (decomposition, nitrification, and denitrification) various types of bacteria feed on nitrogen compounds in organic matter or the soil, eventually returning it as N_2 to the atmosphere. [INFOGRAPHIC 7.10] The nitrous oxide (N_2O), or laughing gas, that gave the biospherians trouble is a by-product of denitrification that normally exists in trace amounts in Earth's atmosphere (it is also

becoming a dangerous greenhouse gas, as it is produced in ever-higher concentrations by some human activities).

Unlike nitrogen and carbon, phosphorus—which is needed to make DNA and RNA—is found only in solid or liquid form on Earth, so the **phosphorus cycle** does not move through the atmosphere, but passes from inorganic to organic form through a series of interactions with water and organisms. [INFOGRAPHIC 7.11]

In Biosphere 2, the nitrogen and phosphorus cycles were disrupted. Thanks to an overabundance of soil microbes, nitrous oxide reached levels high enough to interfere with

Infographic **7.11** | **THE PHOSPHORUS CYCLE**

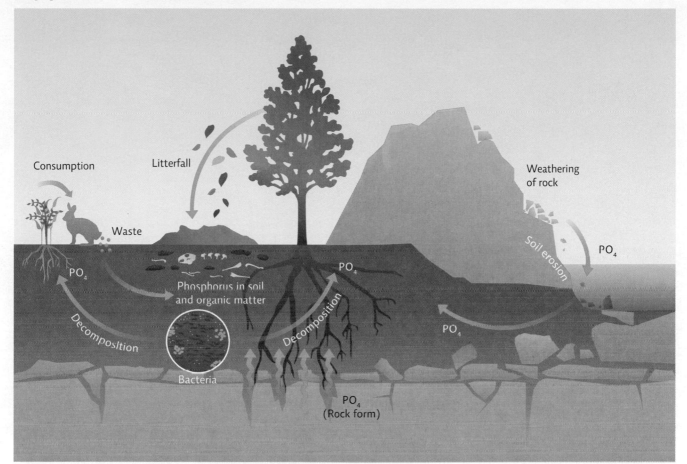

↑ Phosphorus, needed by all organisms to make DNA, cycles very slowly. It has no atmospheric component but instead depends on the weathering of rock to release new supplies of phosphate (PO₄) into bodies of water or the soil, where it dissolves in water and can be taken up by organisms. Microbes also play a role when they break down organic material and release the phosphate to the soil.

Dust released through mining or in eroded areas can introduce phosphorus into the environment much more quickly than it would normally enter.

Open-pit mining

Fertilizers and animal waste (including sewage) can alter plant growth and nutrient cycling, especially in aquatic ecosystems where phosphorus is usually a limiting nutrient.

Animal waste

the metabolism of vitamin B12, which is essential to the brain and nervous system.

Phosphorus got trapped in the water system, polluting aquatic habitats. The underwater and terrestrial plants of Biosphere 2 were dying off too quickly to complete this

cycle. Biospherians removed excess nutrients from their water supply by passing the water over algal mats that would absorb the nutrients and could then be harvested, dried, and stored.

As food reserves dwindled, the eight biospherians split into two factions. One group felt that scientific research was the top priority, and wanted to import food so that they would have enough energy to continue with their experiments. The other group felt that maintaining a truly closed system—one where no biomass was allowed to enter or leave—was the project's most important goal;

nitrogen fixation Conversion of atmospheric nitrogen into a biologically usable form, carried out by bacteria found in soil or via lightning.
phosphorus cycle Series of natural processes by which the nutrient phosphorus moves from rock to soil or water, to living organisms, and back to the soil.

↑ Justin Peterson, a Biosphere 2 undergraduate intern who assists PhD candidate Henry Adams with his Pinon Pine Tree Drought Experiment. The experiment's goal is to predict the effects of climate shifts on the trees.

proving that humans could survive exclusively on what the dome provided would be essential to one day colonizing the Moon or Mars. To them, importing food would amount to a mission failure. "It was a heartbreaking split," Poynter says. "Just 6 months into the mission, and two people on the other side of the divide had been my closest friends going in." Eventually, Poynter snuck food in. That wasn't the only breech. To solve the various nutrient cycle conundrums, the project's engineers had installed a CO_2 scrubber, and pumped in 600,000 cubic feet of oxygen.

Biosphere officials hid these actions from all but a few key people. When one reporter finally broke the story, the public was outraged. Spectators of every ilk—seasoned scientists, skeptical reporters, and casual observers alike—became convinced that other data was also being fudged. "Secrets are like kryptonite to the scientific process," says Griffin. "Once you find out something has been deliberately overhyped or downplayed, or just plain lied

about, all the data from that research becomes suspect. And data that can't be trusted has no scientific value."

Three years after the first mission was completed, the editors of the respected journal *Science* deemed the entire project a failure. "Isolating small pieces of large biomes and juxtaposing them in an artificial enclosure changed their functioning and interactions, rather than creating a small working earth as originally intended," they wrote. For the $200 million dome to survive as a scientific enterprise, they concluded, it would need dramatic retooling.

Ecosystems are complicated, but learning how they function will help us restore degraded ones.

Biosphere 2 taught scientists that Earth is far too complex, that ecosystem components intertwine in far too many complicated ways, for humans to recreate. Each is governed by a countless array of interacting factors and a change in one can set off a whole chain of events that degrade the system's capacity to sustain life.

> Three years after the first mission was completed, the editors of the respected journal *Science* deemed the entire project a failure.

From a scientific point of view, the fact that Biosphere 2 was not able to sustain life as hoped does not mean the project was of no value. Negative results can be just as informative as positive ones—in some cases even more so because they uncover gaps in our knowledge and help us decide how to move forward. In fact, Biosphere 2's greatest liability—its skyrocketing CO_2 levels—proved to be its most valuable asset. "Now it's like a time machine," says Griffin who points out that it is allowing us a look at the consequences of elevated atmospheric CO_2 levels, the main contributor to climate change today. Recent research by Griffin has uncovered some of the complexities of carbon cycling. His group saw unexpected fluctuations in carbon release at various levels in the tree canopy, telling him there is much we still don't know about how carbon cycles—data that could only be gathered in an enclosed forest such as that found in Biosphere 2. Today, scientists from all over the world still use the facility to study the effects of an atmosphere loaded with carbon dioxide. Ultimately, though, the most valuable lesson Biosphere 2 has provided is how irreplaceable Biosphere 1 is.◉

Research articles referenced in this chapter:
Cohen, J.E., and Tilman, D. 1996. *Science*, 274: 1150–1151.
Griffin, K.L., *et al.* 2002. *New Phytologist*, 154: 609–619.

BRING IT HOME

◎ PERSONAL CHOICES THAT HELP

Nutrient cycling is critical for maintaining Earth's ecosystems, but we interfere with nutrient cycles through our daily activities. Driving a car interferes with the carbon cycle by releasing carbon from fossil fuel reservoirs. Applying synthetic chemical fertilizers to food crops interferes with the nitrogen cycle by adding soluble nitrogen compounds to aquatic ecosystems through runoff. The challenge is to figure out ways to work with nutrient cycles rather than against them—in other words, to return nutrients to the reservoirs from which they come. How might this be done in our daily lives and in our own communities?

Individual Steps

→ Reduce your fossil fuel use to curtail carbon, nitrogen, and sulfur emissions. Take public transportation, walk, ride a bike, and drive a fuel-efficient vehicle.
→ Compost food and yard waste. Then use this material to fertilize flowerbeds, trees, and garden plots. Composting will reduce or eliminate the need for inorganic fertilizers in your yard.

Group Action

→ Participate in or organize an event to plant trees or native grasses. By doing so, you can help recapture the carbon put into the atmosphere by driving a car.
→ Many urban areas welcome individuals who are willing to "adopt a median" and plant and maintain flowerbeds and trees in small plots along roadways.

Policy Change

→ Public policy currently prevents large-scale composting of municipal wastes in most areas. Working to change these policies will extend the life of our landfills and make use of valuable nutrient-rich materials.
→ Support legislation to increase fuel efficiency of vehicles and subsidies for clean, renewable energy. More-efficient cars and cleaner energy sources will reduce our carbon outputs from fossil fuel use.

UNDERSTANDING THE ISSUE

CHECK YOUR UNDERSTANDING

1. **Biosphere 2:**
   a. wasted taxpayers' money.
   b. was poor science and is ignored.
   c. was a utopian dream that did not work.
   d. was a learning experience for the scientific community.

2. **Which of the following is the correct organization of life?**
   a. Individual, community, biome, ecosystem, population, biosphere
   b. Individual, population, community, ecosystem, biome, biosphere
   c. Biosphere, community, biome, ecosystem, population, individual
   d. Biome, biosphere, population, ecosystem, community, individual

3. **Which biome description is correct?**
   a. Grasslands receive less rainfall than forests, but more than deserts.
   b. Forests have freezing temperatures regularly.
   c. Grasslands are much warmer annually than forests.
   d. Deserts are hot year round, much hotter than forests.

4. **The range of tolerance for an organism might be:**
   a. for a field mouse, how cold it gets.
   b. for a robin, the competing species in the area.
   c. for an earthworm, the amount of rainfall.
   d. for a new wolf pack, the number of other wolf dens in the area.

5. **Carbon dioxide in the atmosphere is:**
   a. captured by plants during photosynthesis.
   b. released by plants during photosynthesis.
   c. captured between water molecules.
   d. captured by plants during cellular respiration.

6. **In order for nitrogen to be usable for other life it must first be:**
   a. taken up by plants through photosynthesis.
   b. taken up by soil bacteria through nitrogen fixation.
   c. taken up by animals through respiration.
   d. dissolved in water and taken up through plant roots.

WORK WITH IDEAS

1. Using an example such as spring wildflowers and rainfall, explain the terms *environmental gradient* and *range of tolerance*.

2. List the steps involved in the nitrogen cycle. What would happen if a wildfire burned so hot that it sterilized the soil, killing all the microbes?

3. You are designing a habitat for living on another planet, about the same size as Biosphere 2. Do you want several small biomes or one large biome? Justify your answer, using terms such as *habitat*, *nutrient cycles*, and *biomass*.

4. You are studying a small island with the following characteristics. It has one type of grass and one type of tree. It has only four types of small invertebrates, one type of lizard, and a seed-eating mouse. It has a detritivore and some soil bacteria for decomposers. Draw your island biome, showing how carbon will move.

5. A large hawk reaches the island described in question 4. Given your biome's size, and the amount of carbon-based biomass, is there enough carbon available for the hawk to live on the island? Explain.

6. You are studying a pocket mouse which lives in an underground burrow where it stores seeds. Make a list of biotic and abiotic things that might be limiting factors to its survival.

ANALYZING THE SCIENCE

The following table shows six tropical forests, arranged from least to most fertile soils, estimated from total soil nitrogen.

Parameter	Lower montane rainforest, Puerto Rico	Amazon caatinga, Venezuela	Oxisol forest, Venezuela	Evergreen forest, Ivory Coast	Dipterocarp forest, Malaysia	Lowland rainforest, Costa Rica
Above-ground biomass (tons/ha)	228	268	264	513	475	382
Root biomass (tons/ha)	72.3	132	56	49	20.5	14.4
Total soil nitrogen (kg/ha)		785	1,697	6,500	6,752	20,000
Turnover time of leaves (years)	2	2.2	1.7		1.3	

INTERPRETATION

1. Which two areas have the poorest soils, in terms of nitrogen availability?

2. Look at the root biomass in the two areas from Question 1. What is the relationship between root biomass and poor soils? Propose an explanation.

3. Can you determine soil nutrient levels by looking at the above-ground biomass? Explain.

ADVANCE YOUR THINKING

4. Notice how long it takes leaves to break down ("Turnover time of leaves"). Speculate why trees in poorer soils produce leaves that last longer.

5. If you removed the trees through logging or burning, would these areas be good for agriculture? (Hint: Cleared areas are usually used for small farms that do not have funds to purchase farm equipment, fertilizers, or other chemicals.)

EVALUATING NEW INFORMATION

Biosphere 2 covers a mere 3 acres with finite resources. Although Earth is vastly larger, its resources are also finite. People have to decide, for each biome, whether to leave some untouched, or use all of an area and its resources for human purposes. One example would be plowing under an entire prairie and growing wheat, displacing all of the native plants and animals. The rainforests are another such biome. Left alone, they produce huge amounts of oxygen for the whole planet, while supporting millions of species. Many humans want to use the trees for lumber and the land to grow crops or raise cattle.

Some groups advocate using rainforests in sustainable ways: Manufacturing herbal products, or raising shade-grown coffee under the canopy. Will these tactics help preserve the biome?

Go to the Smithsonian Institution's website and read about Bird Friendly® coffee (nationalzoo.si.edu/scbi/migratorybirds/coffee/default.cfm).

Evaluate the website and work with the information to answer the following questions:

1. Is this a reliable information source?
   a. Does the organization give supporting evidence for its claims?
   b. Does it give sources for its evidence?
   c. What is the mission of this organization?

2. Notice the list of scientific articles at the bottom of the Bird Friendly® coffee page. Also be sure to see the guidelines (www.nationalzoo.si.edu/scbi/migratorybirds/coffee/certification/Norms-English_1.pdf).
   a. Do you agree with the assessment of the need for Bird Friendly® coffee?
   b. Identify a claim the organization makes and the evidence it gives in support of this claim. Is it convincing?

3. The Global Canopy Association is a group that asks companies to disclose their "forest footprint" (www.forestdisclosure.com). Should companies be required to participate?
   a. Should some intergovernmental group investigate companies and estimate impacts?
   b. Is it important to allow a "free market" for goods and services, assuming businesses will take care of the lands they use but do not own?

MAKING CONNECTIONS

DEAD ZONES IN THE GULF OF MEXICO

Background: In the northern Gulf of Mexico, there are areas where organisms die every summer. This was first noticed in the mid-1980s; by the early years of the new century, the areas had grown so large that they were impacting the commercial harvesting of fish and shrimp in what has traditionally been one of the primary areas for obtaining seafood from U.S. coastal waters. Investigations showed that these "dead zones" were hypoxic—severely lacking in dissolved oxygen that needs to be available for fish, shellfish, plants, algae, and all the other organisms that live there. There were also reports of high nitrogen levels in the Mississippi River, which empties into the Gulf of Mexico, and of vast algal blooms in the Gulf preceding the hypoxia. These events were followed by the appearance of the dead zones each summer.

Case: The National Ocean Service (NOS) is very concerned about the continued growth in size of the dead zones each year, and the threat they pose to the entire gulf ecosystem. The company where you work wants to apply for a grant from the NOS to investigate possible solutions to the annual dead zones.

Research the issues and write a report that includes:

1. An introduction explaining the causes of the dead zones. Be sure to take into account the following factors which can affect the nitrogen cycle in the Gulf of Mexico:
   a. Farms in Kentucky and Kansas
   b. Large cities, such as Chicago
   c. Manufacturing in Ohio and Michigan
   d. Concentrated Animal Feeding Operations (CAFOs) in Nebraska and Iowa

2. A possible course of action you think your company might want to pursue, such as greatly slowing or eliminating the pollution generated by one of the four sources listed above.

THE WOLF WATCHERS

Endangered gray wolves return to the American West

CORE MESSAGE

Population size and makeup can fluctuate or remain stable. The stability of populations is often dependent on regulation by the environment or by other species like predators. If human impact has killed or removed predators, we may have to step in to fill their role.

GUIDING QUESTIONS

After reading this chapter, you should be able to answer the following questions:

→ What is a population and why is it an ecologically significant concept? What factors do ecologists use to describe and monitor natural populations?

→ What types of population growth patterns are seen and what factors affect population growth?

→ What are the reproductive strategies of r- and K-selected species and how do these relate to population growth patterns and potential?

→ What is carrying capacity and why do some populations slowly approach it and then stabilize, whereas others overshoot it and then crash?

→ What role do predators play in regulating populations and how can the presence or absence of predators affect entire communities?

A gray wolf on the prowl at Yellowstone National Park.

At least a half a dozen times each winter, Doug Smith climbs into a helicopter, gun in tow, and hunts wolves in Yellowstone National Park. He's not looking to kill them—just to put them to sleep for a little while so that he can outfit them with radio collars to track the size of their packs, what they eat, and where they go over the course of the following year. Smith, a biologist, spends about 200 hours per year on these "hunting" expeditions, which are part of the Yellowstone Gray Wolf Restoration Project that he leads. The project has been responsible for reintroducing 41 wolves to Yellowstone since 1994, when their numbers had dwindled as a result of predator control programs implemented by the U.S. government in the early 20th century.

Sometimes, though, things go awry on Smith's radio collar missions. For instance, the tranquilizer dart doesn't fully sedate big wolves—some of which can reach 80 kilograms (175 pounds). Smith is forced to approach the animals while they're still awake. "I have to grab them on the scruff of the neck and manhandle them until I get the collar on," he explains. "They're typically not dangerous then—they've had enough drug to be kind of out of it—but they're still able to walk around. It's a wild experience." A few times, the wolves—who are typically frightened of the helicopter—instead try to attack it while it's hovering with the doors open a few feet above the ground. "I've had two females turn and run at the helicopter, teeth gnashing, jumping up trying to get me," Smith recalls. "I'm hovering above it, going back and forth, and I can't get a shot because all I'm seeing is face and teeth." Despite these adventures, Smith says a wolf has never actually bitten him—and he has tranquilized and collared more than 300 of them.

Understanding how wolf populations interact with biotic and abiotic forces in their environment, through programs like the Wolf Restoration Project, is key to preventing this vital species from disappearing from the Park—and the country—forever. That's because so many factors influence the livelihood of a species and its ability to survive and reproduce. In the early 20th century, humans not only threatened wolves by hunting them; they also destroyed the animals' natural habitat when they cut down forests to build farms, and they starved the animals by hunting elk, deer, and bison, wolves' usual sources of food. At the time, people didn't think that this combination of changes would very nearly cause wolves to go extinct. But now that scientists like Smith have spent years watching how the animals live, they have a much better understanding of what needs to be done to keep

◉ WHERE IS YELLOWSTONE?

them alive and maintain them at their natural population numbers.

Before humans started killing wolves in the early 1900s as part of a government-sponsored program, the exact numbers of gray wolves living in Yellowstone were unknown, but estimates range from 300 to 400. Chances are, their numbers fluctuated over time. A **population**—all the individuals of a species that live in the same geographic area and are able to interact and interbreed—frequently fluctuates in size and distribution and for many reasons. They change in size as a result of ease of access to food, water, and nesting sites, the presence of predators (including humans, who affect population stability through hunting and fishing), as well as the population's innate characteristics. Ecologists who study changes in population size and makeup (**population dynamics**) find that the population size of some species increases and decreases rather predictably (barring a catastrophic

↑ Doug Smith and Yellowstone Delta wolf 487M, the largest wolf collared in the winter of 2005.

event), while others tend to fluctuate more randomly, affected by a variety of factors.

Population biologists like Smith study a population's features in order to understand how best to sustain it and promote its survival. Today, such efforts are more crucial than ever, with more than 18,000 plant and animal species categorized as threatened on the International Union for Conservation of Nature's (IUCN's) Red List.

But even without human interference, the population size of any species can change in a variety of ways. It can increase or decrease, go through cyclic changes or fluctuate wildly, or remain the same. The size of a population in a given geographic area is determined by the interplay of factors that simultaneously increase the number of individuals in a population (births and immigration) and those that decrease numbers (deaths and emigration). An understanding of these factors helps us predict population size at any given time.

Populations fluctuate in size and have varied distributions.

In the early 20th century, the U.S. government began its Predator Control Program when Congress allocated $125,000 for predator and rodent control through poisoning—and later through hunting—to eradicate or control wild animals that might harm crops or livestock. Wolves were one of the targeted species because wolves preyed upon livestock. Wolf eradication was also supported because it boosted deer populations for human hunting. The National Park Service herded wolves out of parks to be hunted by rangers and other Park Service employees; moving them outside made them more accessible to hunters and made the park more welcoming for visitors. "It was park policy to kill all predators, and wolves were their biggest objective," Smith explains. Between 1914 and 1926, at least 136 wolves were killed in Yellowstone. In 1944, the last known wolf in the Yellowstone area was killed.

Every population has a **minimum viable population**, or the smallest number of individuals that would still allow a population to be able to persist or grow, ensuring long-term survival. This is an important concept when considering how to conserve endangered or threatened species. A population that is too small may fail to recover for a variety of reasons. For example, some species' courtship rituals require a minimum number of individuals for success. Other activities that depend on numbers—like flocking, schooling, and foraging—fail below certain

---

population All the individuals of a species that live in the same geographic area and are able to interact and interbreed.
population dynamics The changes over time of population size and composition.
minimum viable population The smallest number of individuals that would still allow a population to be able to persist or grow, ensuring long-term survival.

population sizes. Genetic diversity (inherited variety between individuals in a population) is also important: a population with little genetic diversity is less able to adapt to changes and is therefore more vulnerable to environmental change. A small population is also subject to inbreeding, which allows harmful genetic traits to spread and weaken the population.

Critics began raising concerns about the Predator Control Program in the 1920s, and wolves in Montana and Wyoming became protected under the Endangered Species Act in 1973, but it wasn't until 1987 that the U.S. Fish and Wildlife Service proposed reintroducing an "experimental population" of wolves into Yellowstone in its Northern Rocky Mountain Wolf Recovery Plan.

Not everyone was in favor of reintroducing wolves to the landscape of the American West, and opposition remains today. Ranchers in particular were and still are wary of the damage wolves might do to livestock. In some years, wolves were responsible for thousands of herd deaths—although not nearly as many as are killed by coyotes, says the United States Fish and Wildlife Service. Current programs compensate ranchers for lost livestock but they must prove the animal was killed by a wolf, which is not always easy to do. Sport hunters too worry that wolves will decimate game species like deer and elk; however, these animals succumb to many factors, including bad weather and other predators—not just wolves—so it is unclear how much pressure wolves actually put on these wild herds.

After the Fish and Wildlife Service unveiled its plan, Congress funded the organization to prepare an **environmental impact statement** on what, exactly, would happen if wolves were reintroduced to Yellowstone. By considering all aspects of wolf ecology, scientists were able to organize the reintroduction to maximize survival of the species. For instance, scientists believed that it would be best to "softly" release wolves in the park by holding them temporarily in packs in areas that would be suitable for them to live, instead of doing a "hard" release in which the wolves could immediately disperse anywhere in the park. The soft release curtailed the wolves' movement and helped them survive and acclimate to the move. In addition, it allowed the park staff, who brought road-killed deer, elk, moose, and bison to the wolves twice a week, to become familiar with the wolves so that they could identify them once they were fully released.

After several decades of tracking wolves in Yellowstone, Smith has a good idea of where to look for the wolf packs. And the helicopter hovers a mere 50 feet off the ground, making it easy to spot roaming groups of wolves.

The number of wolves distributed over the acreage of Yellowstone is their **population density**—the number of individuals per unit area. Population density is an important feature that varies enormously among species, or even among populations of the same species in different ecosystems. If a population's density is too low, individuals may have difficulty finding mates, or the only potential mates may be closely related individuals, which can lead to inbreeding, loss of genetic variability, and, ultimately, extinction. Density that is too high can also cause problems, such as increased competition, fighting, and spread of disease. Deer, elk, and moose populations in the United States, whose density has increased in recent years because of exploding numbers combined with shrinking habitats, now frequently suffer from an infectious disease known as chronic wasting disease.

In addition to size and density, another important feature is **population distribution**, or the location and spacing of individuals within their range. A number of factors affect distribution, including species characteristics, topography, and habitat makeup. Ecologists typically speak of three types of distribution. In a **clumped distribution**, individuals are found in groups or patches within the habitat. Yellowstone examples include social species like prairie dogs or beaver that are clustered around a necessary resource, like water. Wolves travel in packs and therefore have a clumped distribution. Elk, one of their prey species, congregate as well; living in herds offers some protection against the wolves.

In a **random distribution**, individuals are spread out in the environment irregularly with no discernible pattern. Random distributions are sometimes seen in homogeneous environments, in part because no particular spot is considered better than another. Species that disperse through random means and germinate where they fall—like wind-blown seeds—also often distribute randomly. **Uniform distributions**, rare in nature, include individuals that are spaced evenly, perhaps due to territorial behavior or mechanisms for suppressing growth of nearby individuals (seen in some plant species). [INFOGRAPHIC 8.1]

environmental impact statement An evaluation of the positive and negative impacts of a proposed environmental action, including alternative actions that could be pursued.

population density The number of individuals per unit area.

population distribution The location and spacing of individuals within their range.

clumped distribution Individuals are found in groups or patches within the habitat.

random distribution Individuals are spread out over the environment irregularly with no discernable pattern.

uniform distribution Individuals are spaced evenly, perhaps due to territorial behavior or mechanisms for suppressing the growth of nearby individuals.

Infographic **8.1** | **POPULATION DISTRIBUTION PATTERNS** ↓ The distribution of individuals in populations varies from species to species and is influenced by biotic and/or abiotic factors.

↑ Elk stay in herds, which offers some protection against predators.

↑ The seeds of many Yellowstone flower species are distributed randomly and germinate where they fall.

↑ The creosote bush of the desert Southwest produces toxins that prevent other bushes from growing close by.

Populations display various patterns of growth.

The researchers in the wolf restoration project could observe these factors—distribution, genetic diversity, initial population size, and density as well as factors like prey availability and habitat structure and see how they influenced the size and dynamics of a population in Yellowstone. However, in order to predict population dynamics, scientists use some simple mathematical models that describe population growth over time. The annual **population growth rate** is determined by **birth rate** (the number of births per 1,000 individuals per year) minus the population **death rate** (number of deaths per 1,000 individuals per year), or by taking a simple census

of population size at the two time points in question. Between 2009 and 2010, the wolf population went from 320 to 343, an increase of 23 animals. This is a growth rate of 7% (23/320=0.07).

When there are no environmental limits to survival or reproduction, a population will reach its maximum per capita rate of increase (r), called its **biotic potential**. This occurs, theoretically, when every female reproduces to her maximal potential and every offspring survives. A population increasing in this manner will quickly grow to fill its environment. This period of growth, which can't go on indefinitely, is referred to as **exponential growth**, named for the mathematical function it represents.

Populations that have a high biotic potential have high fecundity (females typically produce lots of offspring, reach reproductive maturity quickly, and produce many "clutches" per year). The higher the biotic potential, the faster the population of a given species will grow under ideal conditions. Yellowstone species such as deer mice and the invasive weed known as spotted knapweed have higher biotic potential than species such as grizzly bears and spruce trees.

Exponential growth is typically seen when a species first enters a new environment, or if there is an influx of new resources. The population must have a high birthrate—most individuals must have access to enough food, water, and habitat in which to reproduce—and a low death rate. The loss of predators can also lead to exponential growth among their prey species. For instance, elk numbers in Yellowstone doubled between 1914 and 1932, after the Predator Control Program had been implemented.

A population that is growing exponentially will have a *J-shaped curve* if plotted on a graph with time on the *x* axis and population size on the *y* axis. The J curve shows a slight lag at first and then a rapid increase. This is due to the fact that the larger the population, the faster it grows, even at the same growth rate. Think of it this way: Doubling a small number yields a number that is still small. Doubling a large number, on the other hand, produces a very large number.

As an example of how profound exponential growth can be, imagine if someone offered to give you a penny one day and then, each subsequent day for a month, doubled the amount given to you the previous day. On day 1 you would have 1 cent; on day 2 you would be given 2 cents; on day 3 you'd be given 4 cents, and on day 4, you would be given 8 cents. By day 31, you would have over $21 million. Exponential growth can create large populations quickly. [INFOGRAPHIC 8.2]

Infographic 8.2 | **EXPONENTIAL GROWTH OCCURS WHEN THERE ARE NO LIMITS TO GROWTH**

↓ Because deer mice have a high biotic potential, even a single pair could produce more than 31,000 descendants in their lifetime.

BIOTIC POTENTIAL OF DEER MICE

Assume each pair produces 10 pups/litter and none die: *r* = 5 (*r* is expressed as surviving offspring per adult; we divide the litter size by 2 since there are 2 parents per litter).

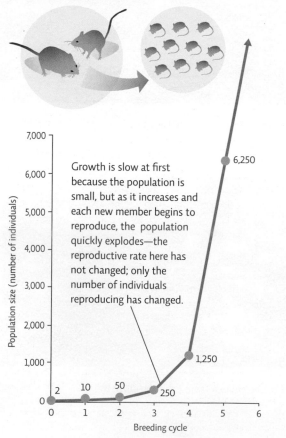

Growth is slow at first because the population is small, but as it increases and each new member begins to reproduce, the population quickly explodes—the reproductive rate here has not changed; only the number of individuals reproducing has changed.

Population size (number of individuals) — vertical axis: 0, 1,000, 2,000, 3,000, 4,000, 5,000, 6,000, 7,000

Breeding cycle — horizontal axis: 0, 1, 2, 3, 4, 5, 6

Data points: 2, 10, 50, 250, 1,250, 6,250

Infographic **8.3** | **LOGISTIC POPULATION GROWTH**

→ Exponential growth turns into logistic growth (S curve) as population size approaches carrying capacity (*K*) and resistance factors begin to limit survival.

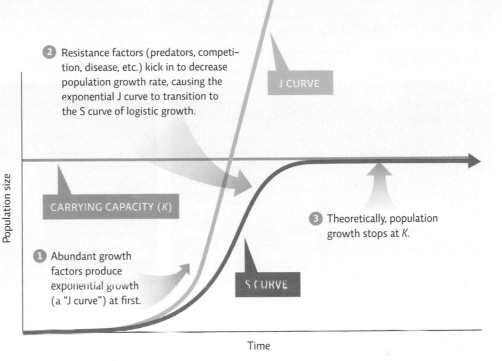

2 Resistance factors (predators, competition, disease, etc.) kick in to decrease population growth rate, causing the exponential J curve to transition to the S curve of logistic growth.

J CURVE

CARRYING CAPACITY (*K*)

1 Abundant growth factors produce exponential growth (a "J curve") at first.

3 Theoretically, population growth stops at *K*.

S CURVE

Population size

Time

Exponential growth can't last forever, however; as a population begins to fill its environment, its growth rate decreases, because as more individuals use the available resources, the resources become scarce. Some individuals starve or are unable to find habitat in which to reproduce, and crowding may also bring about an increase in disease and aggression. Predation pressure may increase as the more numerous prey is easier to track and capture or simply because the predator population itself has increased. This kind of growth—in which as population size increases, growth rate decreases—is called **logistic growth**. A population that grows logistically will produce an *S-shaped curve* if plotted on a graph with time on the *x* axis and population size on the *y* axis. The S is created by the initial growth of the J-shaped curve of exponential growth, followed by decelerating growth as the species approaches its maximum sustainable population size, where it levels off.

The population size that a particular environment can support indefinitely—without long-term damage to the environment—is called its **carrying capacity**, signified as *K* in population models. Carrying capacity depends on the presence of *growth factors* (resources needed to survive or reproduce) and varies between species; the same environment can support many more mice than wolves, for example. Over time, a population's carrying capacity can change. If resources are diminished at a faster rate than they are replenished, the carrying capacity will

drop. If, on the other hand, new resources are added or become available, perhaps due to the loss of a competitor, the carrying capacity for a given species will rise. [INFOGRAPHIC 8.3]

A variety of factors affect population growth.

Populations will grow as long as growth factors are available, but as the population gets larger, resources start to decline. The *limiting factor*, that resource that is most scarce, tends to determine carrying capacity. *Resistance factors*, which include predators, competitors, and disease, also control population size and growth. The effects that predators have on populations can vary widely, in part because predators, along with disease

---

population growth rate The change in population size over time (births minus deaths over a specific time period).
birth rate The number of births per 1,000 individuals per year.
death rate The number of deaths per 1,000 individuals per year.
biotic potential (*r*) The maximum rate at which the population can grow due to births if each member of the population survives and reproduces.
exponential growth Population size becomes progressively larger each breeding cycle; produces a J curve when plotted over time.
logistic growth The kind of growth in which population size increases rapidly at first but then slows down as the population becomes larger; produces an S curve when plotted over time.
carrying capacity (*K*) The population size that a particular environment can support indefinitely without long-term damage to the environment.

Infographic 8.4 | **DENSITY-DEPENDENT AND DENSITY-INDEPENDENT FACTORS AFFECT POPULATION SIZE**

↓ Density-dependent factors exert more of an effect as population size increases. On the other hand, density-independent factors have the same effect regardless of population size.

DENSITY DEPENDENT

COMPETITION

↑ Mule deer compete with elk for food; the larger the deer population, the more competition.

DISEASE

↑ Infectious diseases, such as chronic wasting disease, which weakens and eventually kills the animal, spread more easily in large populations of elk, deer, or moose.

PREDATORS

↑ Predators are more successful at capturing prey from larger populations than from smaller ones.

DENSITY INDEPENDENT

FIRE/FLOOD

↑ Forest fire

STORMS

↑ Row of trees flooded by a river

AVALANCHE AND OTHER NATURAL DISASTERS

↑ Snow avalanche

Infographic **8.5** | **REPRODUCTIVE STRATEGIES** ↳ Different species have different potentials for population growth, known as reproductive strategies. A species' biology may place it anywhere along a continuum between two extremes—the *r*- and *K*-selected species.

CHARACTERISTICS OF *r*-SELECTED SPECIES

1. Short life
2. Rapid growth of individual
3. Early maturity
4. Many, small offspring
5. Little parental care
6. Adapted to unstable environment
7. Prey
8. Niche generalists

CHARACTERISTICS OF *K*-SELECTED SPECIES

1. Long life
2. Slower growth of individual
3. Late maturity
4. Few, large offspring
5. High parental care
6. Adapted to stable environment
7. Predators
8. Niche specialists

r-SELECTED SPECIES ———————————————————————————— **K-SELECTED SPECIES**

DANDELIONS

DEER MICE

ELK

BEARS

SPOTTED KNAPWEED

SPRUCE TREES

and competition, are **density-dependent** factors—their effects all increase as population size goes up. On the other hand, some factors affect a population no matter how large or small it is, such as droughts, storms, and floods. These **density-independent** factors don't necessarily regulate population size, but they can certainly decrease it. [INFOGRAPHIC 8.4]

A species' biology also affects its growth potential. For instance, ecologists recognize a continuum of **reproductive strategies** among species. Those that reproduce quickly, with high biotic potential, are known as

r-selected species—so named because of their high rate of population increase (*r*). Yellowstone *r*-selected species, such as deer mice and spotted knapweed, are well adapted to exploit unpredictable environments, and are able to increase quickly if resources suddenly become available.

On the other hand, **K-selected species**, which in Yellowstone include bears, wolves, and slow-growing trees like spruce, are found at the other end of the continuum and have much lower reproductive rates. Because they reproduce slowly, their population growth rates are responsive to environmental conditions; they decrease or increase slowly if resources decrease or increase in availability. This responsiveness tends to keep population sizes close to carrying capacity (*K*). [INFOGRAPHIC 8.5]

K-species and *r*-species often experience different types of population change. For instance, population sizes tend to be stable, especially for *K*-species, in undisturbed, mature areas. On the other hand, *r*-species with rapid reproductive potential sometimes have sudden, rapid population growth, characterized by occasional surges to high peaks, which may overshoot carrying capacity,

density dependent Factors, such as predation or disease, whose impact on the population increases as population size goes up.
density independent Factors, such as a storm or an avalanche, whose impact on the population is not related to population size.
reproductive strategies How quickly a population can potentially increase, reflecting the biology of the species (life span, fecundity, maturity rate, etc.).
r-selected species Species that have a high biotic potential and that share other characteristics such as short lifespan, early maturity, and high fecundity.
K-selected species Species that have a low biotic potential and that share characteristics such as long lifespan, late maturity, and low fecundity; generally show logistic population growth.

Infographic **8.6** | **SOME POPULATIONS FLUCTUATE IN SIZE OVER TIME**

OVERSHOOT AND CRASH

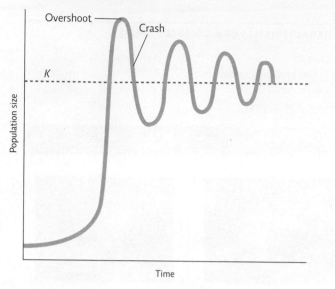

↑ Not all populations show logistic growth that levels off nicely at carrying capacity. Some populations might overshoot carrying capacity, drop below it, and increase and overshoot it again until they settle down close to carrying capacity.

PREDATOR–PREY POPULATION CYCLES

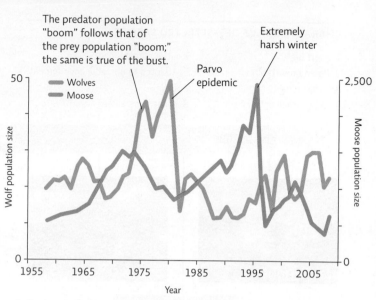

↑ Some predator–prey populations, like that of the moose and wolves of Isle Royale, Michigan, regularly go through boom-and-bust cycles as part of their natural population cycle.

followed by sudden crashes, especially in response to seasonal availability of food or temperature changes; their high rate of reproduction does not allow the population the time to adjust and produce fewer offspring as resources become scarce. When this occurs, the population that exceeds carrying capacity will drop below carrying capacity and then increase again; some populations will eventually level off close to carrying capacity while others continue to overshoot and crash.

Predator and prey also respond to each other; predator populations increase as their prey population increases. But more predators eventually reduce the prey population. Fewer prey means less food, and eventually decreases the number of predators, allowing the prey species to recover. In some cases, the fluctuations in population size are large enough to result in **boom-and-bust cycles**, with the predator population peaks lagging behind those of its prey or food source. A classic example comes from an island called Isle Royale in Michigan. Wolves first appeared on the island in 1949 and preyed upon moose there, causing the moose population numbers to drop. After the moose population dropped, so did that of the wolf, which allowed the moose population to increase in number once again. The population size of one affected the survival and thus the population size of

the other, but other density-dependent and density-independent factors also played a role: Winters in the 1970s were particularly harsh and killed off many moose; and in the early 1980s, the introduction of parvovirus dropped wolf numbers to their lowest point. [INFOGRAPHIC 8.6]

The loss of the wolf emphasized the importance of an ecosystem's top predator.

Thanks in part to Smith's determination, the Yellowstone Gray Wolf Restoration project has been incredibly successful. There are now more than 5,000 gray wolves living in the lower 48 states, approximately 100 of which are in Yellowstone National Park. But one chief lesson hammered home by observing wolves in Yellowstone is that populations do not exist in isolation. "Work in central Yellowstone clearly demonstrated that the addition of a keystone species, such as wolves, can result in observable changes in the behavior of its prey," says Claire Gower, a wildlife biologist at Montana Fish, Wildlife, and Parks. "These consequences may not stop at the prey individual, but may have cascading effects on other community-level processes."

↑ The blackened areas of the trunks of these aspen trees in Wind Cave National Park, South Dakota, show the winter browse line from herbivores such as elk and deer, which strip off the soft bark during winter. Black scar tissue forms on the tree trunks in place of bark.

After wolves were reintroduced to Yellowstone, scientists noticed that coyote populations in the area shrank, in part because the wolves were hunting them. This has had a cascade of effects on the other smaller mammals upon which coyotes typically prey. For instance, scientists speculate that the elusive red fox—whose nocturnal behavior makes these foxes difficult to track—has probably become more common in the park, because fewer coyotes are around to hunt them, and because foxes do not share any prey with wolves.

The elk population has also been affected. According to the National Park Service, the elk population in Yellowstone doubled between 1914 and 1932, a direct result of wolf **extirpation** (local extinction). These elk then began exerting pressure on their food sources. Willow trees were overgrazed because elk like to eat the tender young shoots; this removed an important resource for other animals like birds (habitat) and beaver (food and building material). The decline of the beaver population was especially significant because beaver dams change the flow of water, creating lakes that support a wide variety of fish, amphibian, and plant species that would not frequent the faster flowing stream. The loss of these dams allowed streams to return to their original flow, changing the community that lived in the area.

"Work in central Yellowstone clearly demonstrated that the addition of a keystone species, such as wolves, can result in observable changes in the behavior of its prey."—Claire Gower

As the wolf population increased again after reintroduction, the elk spent less time in willow stands, preferring to be out in the open where they could see their predators. As the willows recovered, so did songbird populations. Beavers returned as well, whose dam building restored lakes and fish populations. Damming the river also slowed the flow of water enough to increase the amount that soaked into the ground, recharging groundwater supplies and providing more water for the deep roots of trees. Populations interact with one another in complex and often unpredictable ways in their ecological community. [INFOGRAPHIC 8.7]

boom-and-bust cycles Fluctuations in population size that produce a very large population followed by a crash that lowers the population size drastically, followed again by an increase to a large size and a subsequent crash.
extirpation Local extinction of a species.

Infographic **8.7** | **THE PRESENCE OR ABSENCE OF WOLVES AFFECTS THE ENTIRE ECOSYSTEM**

↓ Wolves impact the entire ecosystem by keeping elk populations in check. Whether elk are low or high in number affects the entire plant community, which then affects which animal species are present.

WITHOUT WOLVES

↑ When willow thickets are abundant, beavers thrive and build dams that create lakes and ponds. With no wolves to look out for, elk stay in willow thickets and overgraze willow needed by beavers.

↑ Without willow thickets, the beavers leave. Dams are not built or maintained and water flow changes, reducing bird and other wildlife populations.

WITH WOLVES

↑ Willows regrow because elk now feed in meadows to watch for wolves.

↑ With more willows, the beavers return and build dams. The area floods again and wildlife populations recover.

American ecologist and environmentalist Aldo Leopold, who lived in the early 20th century, was one of the first American conservationists concerned with the effect of eliminating the wolf from the American West. He recognized that the problems resulting from interference with ecological communities could be far reaching. "I now suspect that just as a deer herd lives in mortal fear of its wolves, so does a mountain live in mortal fear of its deer," he wrote in his book, *A Sand County Almanac*. "And perhaps with better cause, for while a buck pulled down by wolves can be replaced in two or three years, a range pulled down by too many deer may fail of replacement in as many decades."

Many populations of animals other than wolves are declining worldwide due to human impact, especially from habitat loss, the introduction of non-native species, and predator removals. In fact, the number one reason that species become endangered today is habitat destruction. People remove habitats or resources and break needed connections within ecosystems, and populations respond: Some species may benefit from the change and their population could increase (spotted knapweed, a non-native plant in Yellowstone, spreads quickly in disturbed areas), while other species may decline in number because they are displaced by others, or because needed conditions for growth are no longer present (songbirds may lose habitat if the woodland patches they nest in are cut down). Understanding community interactions and population dynamics helps managers monitor, protect, and restore populations.

Often these dynamics are complex. For instance, while many people believe that Yellowstone's Wolf Restoration Project is the sole cause of the Park's recent elk population decline—the reintroduced wolves, they say, are simply eating all the elk—in reality, many factors have played a role in their decline. "Wolves have come back, cougars have come back, the state was managing for fewer elk by hunting them when they left the park, and grizzly and black bears increased in number, which eat elk calves too," Smith explains. "There are many reasons why they've declined." Some studies suggest that winter weather may be one of the main factors that control elk and deer population sizes. To ensure that a population survives, it is crucial that managers assess all of the possible reasons for a population's changes.

The success of the reintroduction program has led to the wolf being "delisted" in Montana and Idaho (its status has been changed from "endangered" to "threatened"), though not without opposition from some conservation groups. This delisting has the potential to allow wolf

hunts with quotas set by the Fish and Wildlife Service. To Smith, the management of wolves includes the protection of some packs but policies that allow hunting of others. He reasons that allowing wolf hunting to keep wolves out of areas where conflict with humans is greatest may be one of the best ways to protect wolves in wild places like Yellowstone. If people know they can protect their animals and livelihoods, they may be more amenable to allowing the wolves to remain in wilderness areas.

To ensure that the wolves reintroduced to Yellowstone are given the chance to really flourish, Smith and his colleagues diligently stay on the wolves' trail, studying their population dynamics. "We want to know their population size, what they are eating, where they are denning, how many pups they have and how many survive, and how the wolves interact with each other," he explains. Why is it so important to ensure that the wolves do well? Simply put: They were here first, he says. "We want to restore the original inhabitants to the Park."◉

Research articles referenced in this chapter:
Beyer, H.L., *et al.* 2007. *Ecological Applications*, 17: 1563–1571.
Vucetich, J.A. and Peterson, R.O. 2004. *Proceedings of the Royal Society London Biological Sciences*, 271: 183–189.

BRING IT HOME

◉ PERSONAL CHOICES THAT HELP

Understanding the factors that influence how populations change can help us manage species that are either becoming invasive in nature or are facing extinction. How people view species and their connection to our world has a large impact on how management plays out.

Individual Steps

→ Learn more about wolves at the International Wolf Center (www.wolf.org).
→ Use the Internet and books on wildlife to research what your area might have been like prior to human settlement. Which species have been extirpated; which ones have been introduced? How have wildlife populations changed as a result of human action?
→ See if you can recognize distribution patterns in the wild. Do some flowers or trees grow in clumps or patches? Can you

find species that appear to have a random or uniform distribution?

Group Action

→ Explore organizations that support predator preservation, such as Defenders of Wildlife and Keystone Conservation for suggestions on how you can help educate others about the importance of predators.
→ Join a local, regional, or national group that works to monitor, protect, and restore wildlife habitats, such as the Defenders of Wildlife Volunteer Corps.
→ Investigate predator compensation funds such as the Defenders of Wildlife Wolf Compensation Trust and the Maasailand Preservation Trust (for livestock losses due to lion predation). Do you feel this is a worthwhile approach?

Policy Change

→ Check the U.S. Fish and Wildlife Service website, www.fws.gov, for updates on wolf management and protection status.
→ Write a letter to the editor of your newspaper in support of or in opposition to the decision to delist the wolf in parts of the American West.

UNDERSTANDING THE ISSUE

CHECK YOUR UNDERSTANDING

1. **Consider the roach. Why does its population size never reach its biotic potential?**
   a. Females produce few offspring at a time.
   b. Its tolerance limits are too broad.
   c. Overharvesting by humans
   d. Resistance factors limit population size.

2. **The concept of minimum viable population:**
   a. explains how many individuals can fit into a habitat.
   b. describes the potential number of individuals if there are no predators.
   c. describes the potential number of individuals if there is no competition.
   d. describes the smallest number of individuals needed to ensure the long-term continuation of a particular population.

3. **Population density is:**
   a. the number of individuals there are in a given area, such as an acre or a square mile.
   b. the number of members there are in each pack or group of animals.
   c. complicated by the home range of each organism.
   d. the minimum number of individuals that must be in an area in order to ensure long-term continuation of that population.

4. **Groups of fish called blue tang swim in large schools around coral reefs in the ocean, while barracuda, a fierce, predatory fish, prowl the waters alone, looking for prey throughout the area. Regarding distribution:**
   a. blue tang are random and barracuda are solitary.
   b. blue tang are random and barracuda are uniform.
   c. blue tang are clumped and barracuda are random.
   d. blue tang are uniform and barracuda are random.

5. **Exponential growth of a population:**
   a. is often seen if a population reaches a new environment that is favorable.
   b. is a J-shaped curve on a population graph.
   c. occurs when every female reproduces at every opportunity and every offspring survives.
   d. All of the above are true.

6. **Carrying capacity, or *K*, is:**
   a. determined primarily by the number of predators in the area.
   b. the number of individuals that can live in a sustainable fashion in a particular environment.
   c. the total number of individuals that could possibly live in a particular environment for a short time during an exponential growth phase.
   d. determined only by the potential birth rate of a population.

WORK WITH IDEAS

1. Compare the reproductive strategy of a deer mouse with that of a bear.

2. Kangaroo rats eat seeds and are eaten by coyotes. Under what conditions might the kangaroo rat population increase exponentially? Logistically? Identify at least two factors that could lead to a decrease in the kangaroo rat population size.

3. Mountain gorillas live in extended families and search daily for favorite tree types that have tasty leaves. Contrast the *population distribution* of mountain gorillas with that of the trees they eat.

4. Lesser goldfinches are small, seed-eating birds. In cities, they are eaten by wild hawks and by domestic cats. Discuss several density-dependent and density-independent factors, including both critical factors and resistance factors, that could affect their carrying capacity.

5. Beavers in some areas of Michigan and Minnesota live almost exclusively on slow-growing aspen trees and can harvest them faster than the trees can reproduce. Describe how these two populations (beaver and aspen trees) undergo population fluctuations, boom-and-bust cycles.

6. Ranchers in your area want to impose a bounty on wolves and mountain lions; they fear the predators are eating young calves. As a wildlife biologist, explain to the ranchers why this may or may not be a good idea.

ANALYZING THE SCIENCE

The following graph shows the numbers of Kaibab deer on the isolated Kaibab Plateau in northern Arizona. President Theodore Roosevelt declared it a National Game Preserve in 1906 and predator removal was encouraged to protect the deer. Between 1907 and 1917, more than 600 mountain lions were removed; 200 more were removed over the next 20 years. Wolves were exterminated by 1926 and more than 7,000 coyotes and 500 bobcats were also removed. The deer population increased, and by 1915 there was significant damage to the grasses, shrubs, and trees that were being eaten by the deer.

Kaibab deer population, Kaibab Plateau, Arizona

Deer population (in thousands) vs. Year

- 100,000 — Seven successive warnings
- First fawn starved
- Damage seen; first warning given
- 60% of herd starved in two winters
- 40,000
- Probable capacity if herd reduced in 1918
- 30,000
- 25,000
- 20,000
- 10,000
- *K*

INTERPRETATION

1. With the initial level of predators, what was the probable carrying capacity (K) of the Kaibab Plateau for deer at the start of this story?

2. Once the population had passed its probable K, would it be able to sustain itself at the higher numbers? Explain your answer.

3. In what 2 to 3 year period did the deer population have the highest reproductive growth rate?

ADVANCE YOUR THINKING

4. Explain the factors—critical, resistance, density-dependent, density independent—that accounted for the changes in population numbers for the deer in 1905, 1915, and 1930.

5. Decide if Kaibab deer behaved like an r-selected species or a K-selected species, and justify your answer.

6. Compare the probable look of the habitat of the Kaibab Plateau (727,000 acres, 9,000 feet elevation) in 1905 with the way it probably looked in 1935 after 30 years of damage.

EVALUATING NEW INFORMATION

For over 100 years, up until the 1960s, wolves were extirpated in most of the United States. Over the last few decades there have been studies indicating that wolves may have been more important to the functioning of an ecosystem than humans realized. Many groups have worked to reintroduce wolves in states such as Colorado, Montana, Idaho, North Carolina, Washington, and New Mexico. These groups sometimes encounter fierce resistance from other groups concerned with human safety and with possible financial impacts of wolves preying on livestock.

Read the article "Reintroducing the Gray Wolf in the U.S.," at www.actionbioscience.org/biodiversity/johnson.html. Then go to the main website for the group, at www.actionbioscience.org, and investigate the organization.

Evaluate the websites and work with the information to answer the following questions:

1. Is this a reliable information source?
   a. Does the organization give supporting evidence for its claims?
   b. Does it give sources for its evidence?
   c. What is the mission of this organization?

2. Now visit the American Farm Bureau's website, www.fb.org. In the site's search engine, enter the word "wolves" and read at least two articles. Answer all the items under Question 1 for this website.

3. Finally, visit the Defenders of Wildlife website, www.defenders.org. In the site's search engine, enter the word "wolves" and read at least two articles. Answer all the items under Question 1 for this website.

4. Do you believe that wolves should be reintroduced in a few, isolated areas; in many areas, including those where human contact is frequent; or not at all? Justify your decision.

MAKING CONNECTIONS

PRAIRIE DOGS AND THE GRASSLANDS OF SOUTHERN ARIZONA

Background: Early in the 20th century, the valleys of southern Arizona were filled with tall grasses, which attracted cattle ranchers. Summer rainfall on the mountain ranges washed down into the valleys and permeated the soil, nourishing the grasslands. In the late 1800s, the ranchers had steadily killed off large predators, such as wolves, mountain lions, coyotes, and eagles. By the early 1900s, hundreds of thousands of prairie dogs lived in vast underground "towns" (interconnected tunnels and burrows) throughout the grasslands. Prairie dogs are social, very active animals, constantly digging and moving their tunnels. They kept the soil soft and porous, and filled with numerous holes. Grazing cattle would fall in the holes, break a leg, and die, leading to the ranchers' collective decision to remove the prairie dogs.

Ralph Morrow was a game warden who received permission and assistance from the federal government to trap and poison the prairie dogs east of the Chiricahua Mountains. Within 10 years, he had removed the prairie dogs from this area. By the 1930s, the grasslands were gone as well, replaced by desert scrub vegetation. Summer floods were eroding

large gullies, and the topsoil was gone, leaving hard-packed sandy clay with little organic matter.

Case: A group in Idaho is interested in eliminating prairie dogs because it is trying to farm buffalo commercially. The group members have come to your agency, which does environmental research and recommends strategies for remediation of local problems. Write a report detailing the ecological events relating to the extermination of prairie dogs east of the Chiricahua Mountains, and explaining what happened to change the area so dramatically. Make sure to consider the following in your report:

a. predator removal
b. prairie dog populations
c. prairie dog removal and its effects on soils, rainfall runoff, and erosion
d. ecology and effects on the grasslands of prairie dog removal and soil changes
e. effects of cattle (and buffalo) on soils
f. your advice regarding the group's wish to eliminate prairie dogs

CORE MESSAGE

Ecological communities are complex assemblages of all the different species that can potentially interact in an area. All the pieces of the ecological community are connected; change one thing and many others are affected. This means ecosystems are often negatively affected by human impact. Understanding the interconnections within the communities may allow us to better protect and even help restore damaged ecosystems.

GUIDING QUESTIONS

After reading this chapter, you should be able to answer the following questions:

→ How does matter and energy move through ecological communities?

→ How do biotic and abiotic factors affect community composition, structure, and function?

→ How do species interactions contribute to the overall viability of the community?

→ In what ways do human actions affect ecological communities and how can we take steps to help restore damaged ecosystems?

→ How do ecosystems change over time through ecological succession? How can we use this knowledge to assist in ecosystem restoration?

Wood Storks in Florida.

WHAT THE STORK SAYS

A bird species in the Everglades reveals the intricacies
of a threatened ecosystem

James Rodgers steered his canoe toward a large cypress tree as sunlight trickled through the dizzy pattern of leaves overhead. The tree had several Wood Stork nests in it, and he and his assistant wanted to get a closer look at all of them. They were in the thick of a dense swamp near the northwestern edge of the Florida Everglades, and it was the height of breeding season for the storks—eggs had hatched and nestlings everywhere were crying, loudly, for food. Rodgers was silent. He knew from experience that alligators patrolled the waters surrounding stork nests, and that too much human disturbance could "flush" the adult storks—forcing them to flee in a hurry, which would leave their babies vulnerable to aerial predators.

The Wood Stork is an unassuming sort-of bird: more than 3 feet tall, yes. But also covered with a mottled black-and-white coat of feathers—bland compared to some of its tropical neighbors. Despite the lack of majesty of the Wood Storks, however, Rodgers and others at the Florida Fish and Wildlife Service keep close tabs on their ranks.

Here's why: In the late 1970s, the number of nesting pairs of the bird plummeted to an all-time low of 4,500 or so. By the early 1980s, the bird earned a spot on the endangered species list. It was then that Rodgers and his colleagues were first tasked with determining which of several factors (Reduced nesting habitat? Health of females? Damaged feeding grounds?) was most responsible for the decline of this particular bird. And it was through those research efforts—focused intently on the Wood Stork—that they found an entire ecosystem on the brink.

⦿ **WHERE ARE THE FLORIDA EVERGLADES?**

EVERGLADES NATIONAL PARK

GULF OF MEXICO

FL

EVERGLADES

The well-being of a species depends on the health of its ecosystem.

Community ecology is the study of how a given ecosystem functions—how space is structured, why certain species thrive in certain areas, and how individual species in the same community interact with one another.

This includes understanding how various species contribute to ecosystem services like pollination, water purification, and nutrient cycling (see Chapter 4). Wetlands such as the Florida Everglades provide extremely important water management services, including the recharge of groundwater and the capture of contaminants and excess nutrients, enabling them to be stored or converted to safer forms. This process prevents those nutrients from reaching downstream fresh- and saltwater ecosystems. Most important to us humans, however, is the wetlands' contribution to flood control. By capturing and storing large amounts of precipitation and runoff (overland flow of water), and then releasing it downstream slowly over time, they significantly reduce peak flood levels during major rain events.

Community ecology also includes understanding the myriad ways in which we humans have altered various ecosystems, and in so doing have changed the ways they function. In the Florida Everglades, which have been heavily developed over the past half-century, infrastructure like roadways and canals have dramatically reordered the physical landscape. Meanwhile, all the pollutants that come with modern living—solid waste, agricultural chemicals, etc.—have upended the delicate balance of chemical and physical reactions that make this natural world function.

↑ The inlets of Everglades National Park contain a unique mix of tropical and temperate plants and animals, including more than 700 plant and 300 bird species.

For the Wood Stork—a tall wading bird that weighs up to 3 kilograms (7 pounds)—it's all about food. During mating season alone, the birds consume an estimated 45 kilograms (100 pounds) of food—each. One captive bird ate more than 650 small fish in just 35 minutes. Indeed, their feeding habits alone make for great spectacle. They hunt almost exclusively in shallow, muddy, plant-filled water—just 15-50 centimeters (5-20 inches) deep, but so cloudy that fish cannot be seen. They inch their way through these waters at a steady two-steps-per-second clip, sweeping their long, narrow bills—which are kept precisely 8 centimeters (3 inches) agape, and submerged all the way up to the breathing passage—side to side in a relentless hunt for food. When the bill's methodical searching meets the sensation of a wriggling fish (or crayfish), it snaps shut with spectacular speed—in just 25 milliseconds, to be exact. It's the fastest reflex known to all vertebrates, and it enables the Wood Stork to capture prey that no other wading birds can access.

---

community ecology The study of all the populations (plants, animals, and other species) living and interacting in an area.

indicator species The species that are particularly vulnerable to ecosystem perturbations, and that, when we monitor them, can give us advance warning of a problem.

ecosystems All of the organisms in a given area plus the physical environment in which they interact.

But for this tactile (or non-visual) feeding method to work, the prey must be densely concentrated. That means Wood Storks need seasonally drying wetlands to forage. And lots of them. Even a small drop-off in feeding success can impact the ability of these colossal birds' to successfully rear their young.

It's this sensitivity that makes the Wood Stork such a good **indicator species** for the Everglades. An indicator species is one that's particularly vulnerable to ecosystem perturbations. Because even minor environmental changes can affect them dramatically, they can warn ecologists of a problem before it grows. "It is much easier to follow one or two species than to try and monitor an entire ecosystem," says Rodgers, who is a Wood Stork specialist. "So if an indicator species can be identified, this makes it much easier to keep tabs on the health of the ecosystem."

Human alterations have changed the face of the Everglades.

South Florida once provided an ideal breeding ground for these amazing but picky birds. Before giving way to a hodgepodge of resorts, sugar plantations, and dense urban centers, the region was defined by an uninterrupted web of natural **ecosystems**, collectively known

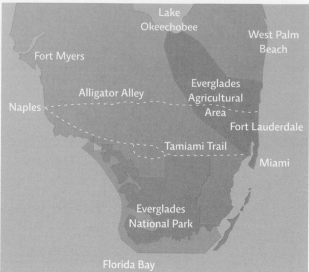

↑ A residential community in Weston, Florida, built where Everglades once existed.

↑ Map of South Florida.

as the Everglades. Marshes, prairies, swamps, and forests stretched across some 4,000 square miles of land. Each distinct community was connected by the same slow-moving water, which began at the southern edge of Lake Okeechobee and flowed south for 100 miles before empty-ing into the Florida Bay. The glacial pace of this water (it can take months or even years for a given eddy to travel from lake to bay) across such a broad, shallow expanse (60 miles wide, and in some places just a few inches deep) is known as *sheet flow*. The Everglades' nickname, "River of Grass," comes from the image of sheet flow through the region's iconic sawgrass marshes.

It was here that Wood Storks flourished. In the 1930s, an estimated 15,000 to 20,000 pairs nested throughout the southeastern United States—largely in South Florida, where foraging grounds were ideal. And because their plain black-and-white plumage was not lovely enough to attract bird hunters collecting feathers for fashionable ladies' hats, the storks thrived even as other wading bird stocks were decimated.

But it was not long after they first discovered the Everglades that American explorers began hatching plans to drain and then develop them. Swamps and muddy rivers choked with grass were seen as having no inher-ent value. "From the middle of the 19th century to the middle of the 20th, the United States went through a period in which wetland removal was not questioned," says University of Florida geographer and historian Christopher Meindl. "In fact, it was considered the proper thing to do." The Central and Southern Florida Project, authorized by Congress in 1948, set about to

systematically drain the Everglades. And over time, a vicious new cycle was established—humans would drain the swamps and replace them with towns and cities; a rash of floods would prompt more complete drainage (under the rubric of flood control and prevention), which would in turn be followed by even more development (on the newly drained land).

As the human population swelled in the region, water that once fed swamps and marshes was rerouted to the faucets of burgeoning developments. And as water levels changed, becoming deeper in some areas and completely disappearing in others, the total wading bird population plummeted—by 90% between the 1930s and 1990s.

Matter and energy move through a community via the food web.

As ecologists would soon discover, the loss of even one species can disrupt an entire ecosystem—from the health of giant wading birds, right down to the movement of matter and energy.

Energy is the foundation of every ecosystem; it is captured by photosynthetic organisms and then passed from organism to organism via the **food chain**—a simple, linear path that shows what eats what. Any given ecosys-tem might have dozens of individual food chains. Linked together they create a **food web**, which shows all the many connections in the community. Both food chains and webs help ecologists track energy and matter through a given community. They can vary greatly in length and

Infographic **9.1** | **EVERGLADES FOOD WEB**

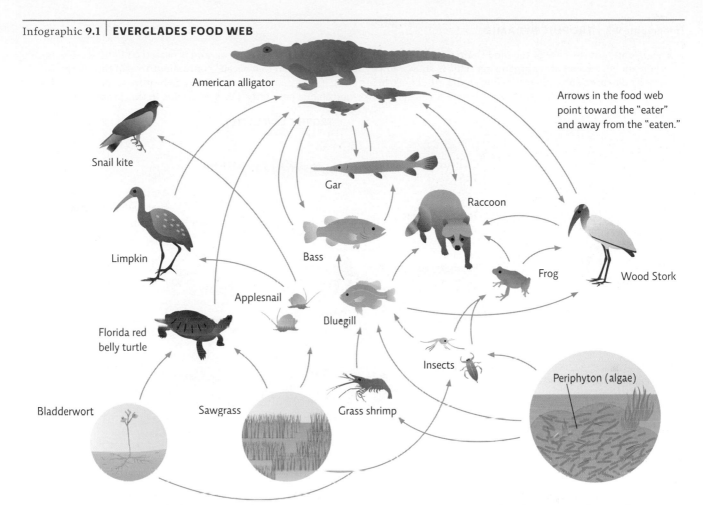

↑ The food web of the Everglades is very complex and varies among the different ecosystems found there. Periphyton algae mats form the base of the food web and may be the most important producers in the ecosystem. The American alligator is the main apex predator, though when young is also prey to various birds, fish, mammals, and even other alligators.

complexity between different types of ecosystems. But most share a few common features and all are made up of the same basic building blocks—namely, producers and consumers. [INFOGRAPHIC 9.1]

Florida Wood Storks sit near the top of a food chain that begins with sawgrass, other plants like cypress and mangrove trees, and a mix of algae and bacteria known as periphyton. These photosynthetic organisms are all known as **producers**. Producers capture energy directly from the sun and convert it to food (sugar) via photosynthesis. They are then eaten by a wide range of **consumers**—organisms that gain energy and nutrients by eating other organisms. Animals, fungi, and most bacteria are consumers. These different feeding levels are known as **trophic levels**. Consumers are organized into trophic levels based on what they eat. *Primary consumers* eat producers; *secondary consumers* eat primary consumers; *tertiary consumers* eat secondary consumers, and so on, ending with the apex predators in the last trophic

level. Of course, some organisms feed at more than one trophic level: Wood Storks eating crayfish are feeding at trophic level 3 but when they eat small fish like bluegill, they are feeding at trophic level 4.

In an ecosystem as diverse and complex as the Everglades, there are dozens of different organisms at each trophic level, making for a wide variety of food chains. For example, periphyton might be eaten by grass shrimp (a primary consumer), which in turn might be eaten by bluegill (a secondary consumer), who fall prey to raccoons

food chain A simple, linear path starting with a plant (or other photosynthetic organism) that identifies what each organism in the path eats.
food web A linkage of all the food chains together that shows the many connections in the community.
producer A photosynthetic organism that captures solar energy directly and uses it to produce its own food (sugar).
consumer An organism that eats other organisms to gain energy and nutrients; includes animals, fungi, most bacteria.
trophic levels Feeding levels in a food chain.

Infographic **9.2** | **TROPHIC PYRAMID**

↓ Energy enters at the base of the food chain in the first trophic level (TL) via photosynthesis and is passed on to higher levels as consumers feed on other organisms. This is shown as a pyramid (smaller on top) because only about 10% of the energy is passed on to each subsequent level, with the other 90% being "lost" to the environment (usually as heat from the energy that the organism burns in day-to-day life before it is eaten). Most food chains have only 4 or 5 levels due to this progressive loss of energy.

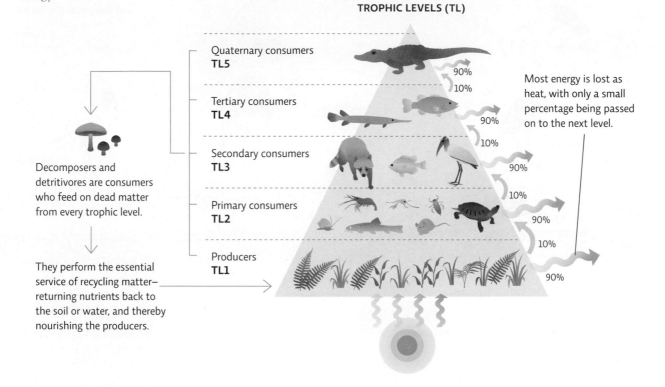

TROPHIC LEVELS (TL)

Quaternary consumers
TL5

Tertiary consumers
TL4

Secondary consumers
TL3

Primary consumers
TL2

Producers
TL1

90%
10%
90%
10%
90%
10%
90%
10%
90%

Most energy is lost as heat, with only a small percentage being passed on to the next level.

Decomposers and detritivores are consumers who feed on dead matter from every trophic level.

They perform the essential service of recycling matter—returning nutrients back to the soil or water, and thereby nourishing the producers.

(tertiary consumers), who might be eaten by alligators (quaternary consumers). Alligators eat a wide variety of animals—turtles, fish, birds, even mammals like the raccoon—making them the apex predator in many Everglades food chains.

When any of these organisms die, they are eaten by an army of consumers known as **detritivores**—animals like worms, insects, and crabs, that feed on dead plants and animals—and **decomposers**—organisms like bacteria and fungi that break decomposing organic matter all the way down into its constituent atoms and molecules.

As one moves up the food chain, energy and *biomass* (all the organisms at that level) decrease, creating a trophic pyramid. The reason is simple: Nearly every organism uses up the majority of its energy and matter in the complicated act of living. So when it is killed and consumed by a predator, it only passes on about 10% (some species pass on more or less than others) of all the energy and matter it consumed during its lifetime. [INFOGRAPHIC 9.2]

A pyramid's ultimate size is determined by its first trophic level—the one made up of photosynthesizing

producers, namely plants. More plant growth means more food for primary consumers, which in turn means more food for those organisms above them, and so on up the food chain. The end result is a larger pyramid with more and larger trophic levels.

Productivity—the amount of energy trapped by producers and converted into organic molecules like sugar—is limited by sunlight and nutrient availability. **Gross primary productivity** is a measure of total photosynthesis. But plants only use a portion (actually less than 50%) of this energy to fuel their daily needs. Scientists use the term **net primary productivity (NPP)** to describe how much energy is left over after that. The "net" is a measure of energy available to higher trophic levels—in other words, the amount that's converted to new growth.

The Everglades are blessed with long summer days that favor plant and algal photosynthesis. But in many other ecosystems, winter months cast the downside of sunlight dependence into stark relief. In most temperate and boreal forests, for example, less sunlight and cooler temperatures limit photosynthesis during these months, causing plant productivity to slow, or shut down completely,

↑ Great White Egret and Wood Stork feed in shallow ponds and channels (called sloughs) in a glade in Everglades National Park.

resulting in less new growth, and thus less food for other organisms. This is why bears hibernate and birds fly south for the winter: The lack of productivity drives them to these extremes. NPP can be a window into the health of an ecosystem. If it rises or falls unexpectedly, ecologists can look for the cause of the change, which might be an invasion by a non-native plant or a sudden drop in a producer population. Anything that alters NPP can potentially affect organisms at every other level of the trophic pyramid.

As researchers discovered back in the 1980s, it was a kink in the food chain that hurt the Wood Storks. While they feed on many things, they prefer fish—and not just any fish, but those between 2 and 15 centimeters (1 and 6 inches) long. Most fish need more than a single season to grow this big; in fact, they need wetlands that are flooded for longer than a year and only very rarely go completely dry. It turns out that as humans altered water cycles in South Florida, there were fewer and fewer such areas, and thus fewer fish for the storks to feed their young.

The Everglades are shaped by biotic and abiotic factors.

Much of Florida is made up of low, flat land that floods from June through September. During this period, Florida

typically receives about 75% of its annual rainfall; water forms a vast sheet, covering thousands of acres, and small changes in depth can amount to large changes in surface area. Many wetland fish grow and reproduce in this expanding habitat. Then, as rains taper off, water begins to recede. And where they were once spread out, fish become concentrated in small ponds and *sloughs* (free-flowing channels of water that develop in between sawgrass prairies). Foraging storks follow these receding waters, which ecologists like to call "dry down," from upland ponds to lowland coastal areas, feeding on fish as they become concentrated. They are so dependent on this water cycle, studies show, that their breeding cycle is regulated by water levels. Such profound connectedness—between landscape and life—is common in the Everglades.

detritivores Consumers (including worms, insects, crabs, etc.) who eat dead organic material.

decomposers Organisms such as bacteria and fungi that break down organic matter all the way down to constituent atoms or molecules in a form that plants can take back up.

gross primary productivity A measure of the total amount of energy captured via photosynthesis and transferred to organic molecules in an ecosystem.

net primary productivity (NPP) A measure of the amount of energy captured via photosynthesis and actually stored in the photosynthetic organism.

Infographic **9.3** | **MEASURING SPECIES DIVERSITY**

↓ The species diversity in an area is a measure of species richness (the total number of species) and species evenness (a comparison of the population size of each species). To survey birds in the state, the Florida Fish and Wildlife Conservation Commission divides the state into more than 1,000 "blocks" and records the number of bird species seen in each; their 1998 data is shown here. Lower diversity is seen in areas with more homogeneous habitats such as sand pine forests or agricultural areas, as well as in areas that are more developed.

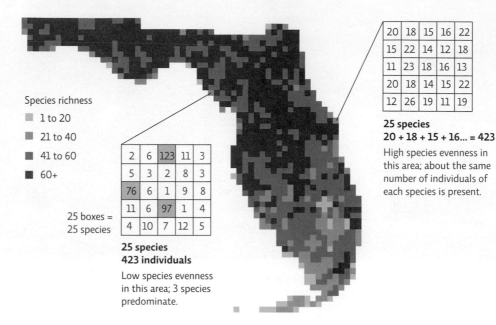

In a hypothetical comparison of two of those blocks, one on the west coast and one on the east coast, let's say both blocks have 25 bird species. The number of individuals seen for each species is shown in a 25-box grid. We see that both areas have the same number of individuals overall (423); however, in our example, the east coast block has much greater species evenness than the west coast block. Therefore the east coast block is considered to have greater overall species diversity than the west coast block.

Species richness

■ 1 to 20
■ 21 to 40
■ 41 to 60
■ 60+

25 boxes = 25 species

2	6	123	11	3
5	3	2	8	3
76	6	1	9	8
11	6	97	1	4
4	10	7	12	5

25 species
423 individuals
Low species evenness in this area; 3 species predominate.

20	18	15	16	22
15	22	14	12	18
11	23	18	16	13
20	18	14	15	22
12	26	19	11	19

25 species
20 + 18 + 15 + 16... = 423
High species evenness in this area; about the same number of individuals of each species is present.

Such connections among species, and between species and their environment, give rise to *ecosystem complexity*. Ecosystem complexity is a measure of the number of species at each trophic level, as well as the total number of trophic levels and available niches. Each species occupies a unique **niche**—that is, a unique role and set of interactions in the community, how it gets its energy and nutrients, and its preferred **habitat**. If two different species tried to occupy the same exact niche, one would outcompete the other. The less successful species has three choices: leave the area, switch niches, or die out. Greater ecosystem complexity means more niches and thus more ways for matter and energy to be accessed and exchanged. This generally increases a community's **resilience**—its ability to adjust to changes in the environment and return to its original state rather quickly.

Species diversity, which refers to the variety of species in an area, is measured in two different ways: species richness, and species evenness. **Species richness** refers to the total number of different species in a community. **Species evenness** refers to the relative abundance of each individual species. In general, organisms at a higher trophic level will have fewer members than those at a lower trophic level, but organisms within the same trophic level should have relatively similar numbers. If they do, the community is said to have high species evenness. If, on the other hand, one or two species dominate any given trophic level, and there are few members of other species, then the community is said to have low species evenness. In such uneven communities, the less abundant species is at a greater risk of dying out.

Both richness and evenness have an impact on diversity. In general, higher species richness and evenness makes for a more complex community and a more intricate food web. It enables more matter and energy to be brought into the system and also makes the community less likely to collapse in the face of calamity. [INFOGRAPHIC 9.3]

A community's composition, and thus complexity, is also heavily influenced by its physical features. As physical features like temperature and moisture change, so does community composition. This often happens in **ecotones**, places where two different ecosystems meet—like the edge between a forest and field, or river and shore. The different physical makeup of these edges creates different conditions, known as **edge effects**, which either attract or repel certain species. For example, it is drier, warmer, and more open at the edge between a forest and field than it is further into the forest. This difference produces conditions favorable to some species but not others.

Infographic 9.4 | **MANGROVE EDGES**

↓ The mangrove—seagrass ecotone provides an example of an edge effect. Fish such as immature gray snapper and bluestriped grunt "commute" between the mangrove trees and the seagrass beds. The proximity of these two areas is vital to provide both the protection during the day and feeding opportunities at night that these young fish need.

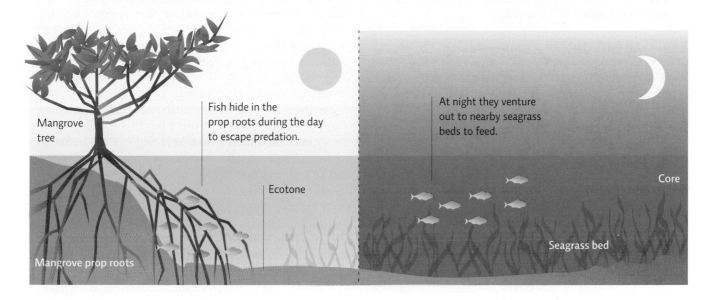

Ecotones may also attract some species that use different aspects of the two adjacent communities; fish such as young snapper or grunts, for example, prefer to live in areas where seagrass beds are fairly close to a shoreline populated by mangrove trees. The mangrove "prop" roots, which anchor the trees into the wet, sandy ground below, offer the fish safety from predators during the day but are close enough to the seagrass beds where the snapper and grunts feed at night for easy "commuting." These fish are not found in coastal areas without the combination of protective coastal mangrove trees and close-by, offshore seagrass beds. [INFOGRAPHIC 9.4]

Species that thrive in edge habitats like this are called **edge species**. Other species, those that can only be found deep within the core of a given habitat, are called **core species**. Some of the many species who find food and refuge in the seagrass, such as crustaceans, sea urchins, and worms, prefer to stay in core areas where they are better hidden and protected from wave action or can make use of deeper sediment buildup in these inner areas. Because they will not venture out across the edge in search of new habitat, core species are easily trapped by habitat fragmentation; we may eliminate the species altogether if we don't leave enough core area behind. (See Chapter 12 for more on habitat fragmentation and core species.)

Changing community structure changes community composition.

Wood Storks are spectacular fliers. From a perch, they spring their giant bodies into air in a single motion, then extend their necks and legs fully as they take flight. They can reach altitudes as high as 1,500 meters (5,000 feet), and can glide for miles without flapping their wings (a feat accomplished by riding vertical air currents: The currents support their weight and allow the storks to

niche The role a species plays in its community, including things like how it gets its energy and nutrients, what habitat requirements it has, and which other species and parts of the ecosystem it interacts with.
habitat The physical environment in which individuals of a particular species can be found.
resilience The ability of an ecosystem to recover when it is damaged or perturbed.
species diversity The variety of species in an area; includes measures of species richness and evenness.
species richness The total number of different species in a community.
species evenness The relative abundance of each species in a community.
ecotones Regions of distinctly different physical areas that serve as boundaries between different communities.
edge effects The different physical makeup of the ecotone which creates different conditions that either attract or repel certain species (for instance, it is drier, warmer, and more open at the edge of a forest and field than it is further in the forest).
edge species Species that prefer to live close to the edges of two different habitats (ecotone areas).
core species Species that prefer core areas of a habitat—areas deep within the habitat, away from the edge.

spiral upward). When foraging grounds dry up, or flood, or are converted into human developments, these aerial skills are pushed to the limit. Surveys showed some Wood Storks flying farther and farther from their nesting habitat in search of foraging grounds—as much as 120 kilometers (75 miles) in some cases.

But they weren't the only ones to struggle in the newly developed region. As the Florida Everglades' natural landscape was modified—as cities replaced swamps and roads replaced rivers—so too were the species interactions and thus the composition of the natural communities that remained.

Dams, dikes, and bridges installed to give humans total control over water levels—in any given portion of the Everglades at any given time—disrupted the flow of water like never before—causing some portions of the Everglades to stay too wet for too long, and others to stay far too dry. As sloughs ran dry, key detritivores and decomposers like worms, grass shrimp, and microbial communities that had thrived there were decimated. This then led to the decline of the snakes, fish, alligators, turtles, and wading birds that fed on them.

Meanwhile, agricultural lands—sugar plantations, in particular—were doing as much damage as flood control efforts. The Everglades are a nutrient-poor ecosystem, especially low in phosphorus. This is fine for the plants and animals that live there; they have evolved and adapted to such conditions and are very effective at moving nutrients through the food chain. But, as early developers discovered, it's not so good for agriculture. To grow crops, farmers must add large quantities of synthetic nutrients to the mucky wetland soil. The runoff from those nutrients has created vast algal blooms, from Lake Okeechobee and elsewhere, which have in turn choked off plant and animal life. As scientists recently discovered, phosphorus runoff is flushed into the canals, then pumped into the lake. When the lake drains, the phosphorus enters marshes and trickles through other ecosystems, changing nutrient levels, and plant species composition along with it.

Sawgrass, typically the dominant species in the marsh, is well adapted to obtain phosphorus from the normally nutrient-poor waters. Cattails, another Everglades producer species with unique flowering spikes, normally prefer the marshes' edges, its growth limited by the lack of phosphorus. However, Danish ecologist Hans Brix and his colleagues have shown that when phosphorus levels increase, cattails can quickly outcompete sawgrass. In the experiment, both plants took in more phosphorus from nutrient-enriched waters. But cattails increased their phosphorus uptake ten-fold, whereas sawgrass increased theirs only five-fold. This allows the cattails to grow more quickly than sawgrass. In areas with nutrient enrichment, they have pushed beyond their natural habitat, through ecotones and into neighboring communities, where they now grow in such dense mats that they're outcompeting sawgrass, choking off native invertebrates on the bottom of the food chain, and physically preventing birds and alligators from nesting. (See Chapter 7 for more on nutrient cycles.)

Replacing mangrove forests with oceanfront resorts has also proven problematic. It turns out that mangrove trees are a **keystone species**—one that impacts its community more than its mere abundance would predict. It's a species that many other species depend on, and one whose loss creates a substantial ripple effect, disrupting interactions for many other species, and ultimately, altering food webs. From their natural habitat at the water's edge, mangrove "prop roots" stabilize the shoreline and provide shelter for a wide variety of fish. So when the mangrove forests are cleared, many other species suffer: the fish that hide among their roots, the fish that feed on those fish, and so on.

Alligators are also a keystone species in the Everglades, one that a great many species depend on during the dry season. As the waters recede, depressions made by alligators (gator holes) are some of the few places that still hold standing water. These holes become refuges for fish, invertebrates, and aquatic plants; they also become very attractive to the animals who feed on these aquatic creatures. Without gator holes, many species would not survive the dry season.

Wood Storks also depend on the presence of alligators, but not just for the dry season gator holes. Back in the 1980s, Rodgers and his colleagues embarked on a comprehensive study of stork nests in an effort to see which types of trees the storks preferred to nest in, and whether or not the availability of those trees was impacting their ability to breed. "We went to 20 stork colonies," Rodgers remembers. "We measured every tree, recorded its species, size, cored it for age, noted its branching structure." The results, arrived at after 5 years of painstaking work, can be summed up in a single sentence, Rogers says: Wood Storks will nest in just about anything, as long as it's surrounded by water that is patrolled by alligators. "Without the alligators, raccoons swim across, and climb up and destroy everything," Rogers says. "Without the alligators, when predators get in, we've seen them abandon entire colonies." [INFOGRAPHIC 9.5]

Infographic **9.5** │ **KEYSTONE SPECIES SUPPORT ENTIRE ECOSYSTEMS**

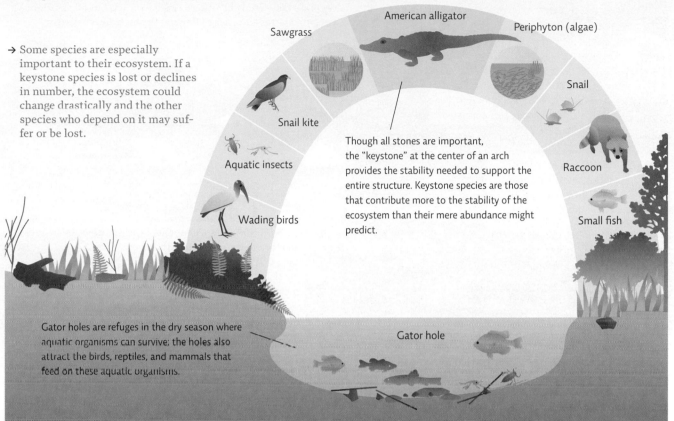

→ Some species are especially important to their ecosystem. If a keystone species is lost or declines in number, the ecosystem could change drastically and the other species who depend on it may suffer or be lost.

Though all stones are important, the "keystone" at the center of an arch provides the stability needed to support the entire structure. Keystone species are those that contribute more to the stability of the ecosystem than their mere abundance might predict.

American alligator

Periphyton (algae)

Sawgrass

Snail kite

Aquatic insects

Wading birds

Snail

Raccoon

Small fish

Gator holes are refuges in the dry season where aquatic organisms can survive; the holes also attract the birds, reptiles, and mammals that feed on these aquatic organisms.

Gator hole

Species interactions are extremely important for community viability.

Communities are all about relationships. Successful communities are those where a certain balance has evolved between all the organisms living there. Species interactions serve many purposes; for example, they control populations and affect carrying capacity. Biodiversity (lots of species, lots of variety within a species) is important, because more diversity means more ways to capture, store, and exchange energy and matter. But it is not sheer numbers that matter most; it is all the connections between species—how they help or hurt one another—that determine how and how well an ecosystem works. Each species is unique and thus interacts in its own unique way with all the species around it.

Many of the species have adaptations that bind them to others or that allow them to coexist. In the Everglades, for example, alligators have adaptations—like sharp teeth and powerful jaws—that allow them to stalk and capture prey while most of the fish that they prey on have adaptations—like camouflage and a wary nature—that help them avoid capture.

Competition—the vying between organisms for limited resources—is another way that species interact. In general, it is subtle, rather than outright fighting. **Intraspecific competition** (that which occurs between members of the same species) is generally stronger than **interspecific competition** (that which occurs between members of different species). This is because members of the same species share the exact same niche and thus compete for all resources in that niche, whereas members of different species may compete for only a single resource, like water.

Other neighbors—those who prey on the same food, or inhabit similar niches—find a way to partition resources. That is, they divvy up the goods in a way that reduces competition and allows several species to coexist. For

keystone species A species that impacts its community more than its mere abundance would predict.
competition Species interaction in which individuals are vying for limited resources.
intraspecific competition Competition between members of the same species.
interspecific competition Competition between individuals of different species.

↓ The heart of a functioning community is its species interactions. Some interactions are beneficial and others cause conflict, but all are important in keeping matter and energy flowing through an ecosystem.

MUTUALISM Both species benefit: The moth gains nutrition while the flower gets pollinated.

PARASITISM One species benefits and the other is harmed: An animal that has too many leeches will be weakened from the loss of blood.

COMMENSALISM One species benefits and the other is unaffected: The heron can catch twice as many fish when foraging alongside the ibis; this doesn't impact the ibis's ability to forage.

example, limpkins and snail kites (two Everglade birds that feed almost exclusively on apple snails) hunt in different regions of the Everglades. This strategy—known as **resource partitioning**—increases the ecosystem's overall capture of matter and energy and thus benefits the entire community.

There are other strategies, too, that keep an ecosystem functioning and strong. Some of these interactions show a tremendous interdependency on the part of the participants. Known as **symbiosis**, these relationships can take one of three forms: **mutualism**, where both parties benefit; **commensalism**, where one benefits from the relationship but the other is unaffected; and **parasitism**, where one benefits from the relationship and the other is negatively affected (this is actually a form of predation). [INFOGRAPHIC 9.6]

resource partitioning When different species use different parts or aspects of a resource, rather than competing directly for exactly the same resource.
symbiosis A close biological or ecological relationship between two species.
mutualism A symbiotic relationship between individuals of two species in which both parties benefit.
commensalism A symbiotic relationship between individuals of two species in which one benefits from the presence of the other but the other is unaffected.
parasitism A symbiotic relationship between individuals of two species in which one benefits and the other is negatively affected (a form of predation).
restoration ecology The science that deals with the repair of damaged or disturbed ecosystems.

By ensuring that all populations persist, even as individuals die, these delicate checks and balances allow more energy to be captured and exchanged, and thus increases the amount of biomass the ecosystem is able to produce.

As we've seen, the Wood Storks rely on several relationships to survive. Their ability to feed depends not only on very specific wetlands hydrology, but also on the health and size of the fish they feed upon. And their ability to raise and fledge young depends not only on the availability of "moated" trees—those surrounded by water—but also on a healthy population of alligators that can patrol those moats for raccoons. When individual species are lost, or when a landscape is physically altered, the balance is tipped. And when that happens, things can fall apart. Fast.

Ecologists and engineers help repair ecosystems.

The federal 1992 Water Resources Development Act enlisted the U.S. Army Corps of Engineers to investigate the damage to the Everglades that resulted from nearly 50 years of unchecked expansion. The final report, published in 1999, acknowledged that the original Everglades (as we found them upon first exploration in the late 1800s) had been reduced by 50%. Constructed canals and levees had dramatically altered water levels, leaving some areas parched and others flooded. And poorly timed water releases were further starving ecosystems that had already been affected by hypersalinity, excessive

PREDATION Alligators prey on a variety of animals and are prey themselves when young.

COMPETITION Organisms that vie for the same resource are in competition with each other. The greatest competition for an individual is from a member of its own species, since both individuals are competing for all the same resources.

RESOURCE PARTITIONING Snail kites and limpkins don't directly compete for their food source, apple snails, because each predator feeds in a different region of the Everglades.

nutrients (from agricultural runoff), and an ever-growing list of non-native species.

Ignoring these problems any longer could greatly imperil the 10 million people that had made their home in the region. "What folks finally realized when we reexamined the area was that the wetlands were this essential filter—they cleaned the water of pollutants," says Kim Taplin, a restoration ecologist who works for the U.S. Army Corps of Engineers restoring the Florida wetlands. "So as the ecosystems have suffered, water quality has declined considerably. We're going to have millions of people with no clean water, unless we fix it." Fixing it is the work of restoration ecologists. **Restoration ecology** is the science that deals with the repair of damaged or disturbed ecosystems. It requires a special blend of skills—not only biology and chemistry, but also engineering and a heavy dose of politics.

> "We're going to have millions of people with no clean water, unless we fix it." —Kim Taplin

In 2000, the U.S. Congress enacted the most comprehensive—and expensive—ecological repair project in history. The Comprehensive Everglades Restoration Plan, or CERP, included more than 60 construction projects to be completed over a 30-year period. The idea was to restore some of the natural flow of water through the Everglades and to capture a portion of the water that now flows to the ocean for South Florida cities and farms.

One of the Army Corps' biggest challenges has been to take down at least part of the Tamiami Trail, a 240-kilometer (150-mile) stretch of U.S. highway that connects the Southern Florida cities of Tampa and Miami. The road, which was built in the 1920s, has proven itself one of the most serious barriers to freshwater flow in the region. It's also a heavily traveled, essential piece of human infrastructure that connects two major cities. That means the U.S. Army Corps must not only tear the road down, but also build something in its place. "Most of our restoration projects involve building even more structures," says Tim Brown, project manager for the U.S. Army Corps of Engineers' Tamiami Trail project. "It's a delicate balance. We of course want to restore as much of the natural system as possible. But we are also charged with protecting lives and property, and in this case, that means building bridges."

Tamiami is not the only plan in the works. In 2008, the state of Florida agreed to buy U.S. Sugar Corporation and all of its manufacturing and production facilities in the Everglades Agricultural Area south of Lake Okeechobee for roughly $1.7 billion. State officials declared that they would allow U.S. Sugar to operate for 6 more years before shuttering facilities and beginning the work of restoration. After that, water flow from Lake Okeechobee would be funneled through a series of holding and treatment ponds that would release clean water into the

Infographic **9.7** | **THE COMPREHENSIVE EVERGLADES RESTORATION PLAN**

↓ Darker-green areas represent wetland areas or river floodplains; white arrows show overland water flow.

HISTORIC FLOW

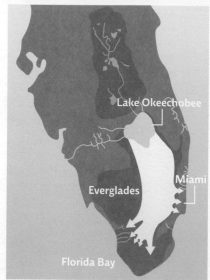

CURRENT FLOW

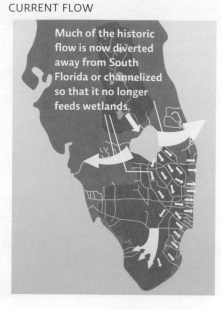

Much of the historic flow is now diverted away from South Florida or channelized so that it no longer feeds wetlands.

PROPOSED FLOW UNDER CERP

↑ Historically, the Everglades covered most of South Florida—more than 10,000 square kilometers (4,000 square miles).

↑ Projects to drain the wetlands and divert water to agricultural lands disrupted normal flow, drained about half of the wetlands and ended up resulting in water shortages for the downstream ecosystems and for people as well.

↑ The goal of the Comprehensive Everglades Restoration Plan (CERP) is to restore the flow of water back to some historic wetland areas through the removal of some canals and levees, as well as to capture some of the freshwater that drains into the ocean, benefiting both the ecosystems and the residents of South Florida.

↑ Michael Korvela, of the South Florida Water Management District, with vegetation from an artificial marsh in Palm Beach, Florida, that filters pollutants from water bound for the Everglades.

Everglades, rehabilitating some 187,000 acres of land. But the agreement has been revised several times since then as the economy has fluctuated and state officials and sugar executives have adjusted and readjusted exactly how many acres would be bought for exactly how many dollars. The proposed purchase was hotly debated, as many feared the cost would take money away from other Everglades restoration projects. In October 2010, just under 27,000 acres were purchased for $197 million, with a 10-year option to acquire another 153,000 acres. A year later, U.S. Sugar was still farming the land, leasing it back from the state.

Each facet of CERP has brought its own fresh round of debate over how best to balance the needs of a swelling human population against the importance of restoring and protecting a heavily degraded ecosystem. Of course, no one knows for certain what will work and what won't. The Everglades landscape has changed dramatically, in ways that not even the best scientists can reverse; decades of development will do that. To plan effective restoration efforts, then, scientists and engineers must be flexible; they must be willing to experiment and respond as conditions change—an *adaptive management* approach similar to that used to address stratospheric ozone depletion (see Chapter 2). [INFOGRAPHIC 9.7]

"The bottom line, though, is that there's only so much we can do," says Rodgers. "It's a lot just to figure out what the baseline was or should be. Some plant species have probably gone extinct, and some non-natives are virtually impossible to remove. What we can do is figure out what some of the big obstacles to recovery are, remove them, and after that, let nature take its course." For his part, Rodgers says he can't imagine that the great 1,000-breeding-pair Wood Stork colonies that early settlers described will ever return to South Florida. The landscape has been too dramatically altered, he says. Though sometimes, of course, nature can surprise us.

Community composition changes over time as the physical features of the ecosystem itself change.

Though the changes the Everglades have experienced are extreme, changes to ecological communities are really the norm—nature is not static. Predictable transitions can sometimes be observed in which one community replaces another, a process known as **ecological succession**. **Primary succession** begins when **pioneer species** move into new areas that have not yet been colonized. In terrestrial ecosystems, these pioneer species are usually lichens—a symbiotic combination of algae and fungus. Lichens can tolerate the barren conditions. As time goes

by and they live, die, and decompose, they produce soil. As soil accumulates, other small plants move in—typically sun-tolerant annual plants that live one year, produce seed, and then die—and the plant community grows. Gradually, the plant growth itself changes the physical conditions of the area—covering sun-drenched regions with broad, shady leaves, for example. Since these conditions are no longer suitable for the plants that created them, new species move in and those changes beget even more changes until the pioneers have been completely replaced by a succession of new species and communities.

Secondary succession describes a similar process that occurs in an area that once held life but has been damaged somehow; the level of damage the ecosystem has suffered determines what stage of plant community moves in. For example, a forest completely obliterated by fire may start close to the beginning with pioneering lichens, whereas one that has suffered only moderate losses may start midway through the process with shrubs or sun-tolerant trees moving in. The stages are roughly the same for any terrestrial area that can support a forest: first annual species, then shrubs, then sun-tolerant trees, then shade-tolerant ones. Grasslands follow a similar pattern with different species of grasses and forbs (small leafy plants) moving in over time.

We tend to view this progression as a "repair" sequence. While we can certainly step in to assist in this natural progression to help a damaged ecosystem recover to a former state, the ecosystem is simply doing what comes naturally—responding to changing conditions.

Intact ecosystems have a better chance at recovering from, and thus surviving, perturbations. Ecosystems that recover quickly from minor perturbations are said to be resilient—they can bounce back. More complex communities tend to be more resilient than simpler ones with fewer species because it is less likely that the loss of one or two species will be felt by the community at large— even if some links in the food web are lost, other species are there to fill the void. Of course, if keystone species are lost, the community will feel the effect. The loss of the

ecological succession Progressive replacement of plant (and then animal) species in a community over time due to the changing conditions that the plants themselves create (more soil, shade, etc.).

primary succession Ecological succession that occurs in an area where no ecosystem existed before (for example, on bare rock with no soil).

pioneer species Plant species that move into an area during early stages of succession; these are often *r* species and may be annuals, species that live one year, leave behind seeds, and then die.

secondary succession Ecological succession that occurs in an ecosystem that has been disturbed; occurs more quickly than primary succession because soil is present.

Infographic **9.8** | **ECOLOGICAL SUCCESSION**

FOREST ECOLOGICAL SUCCESSION DEPENDS ON SOIL AND LIGHT AVAILABILITY

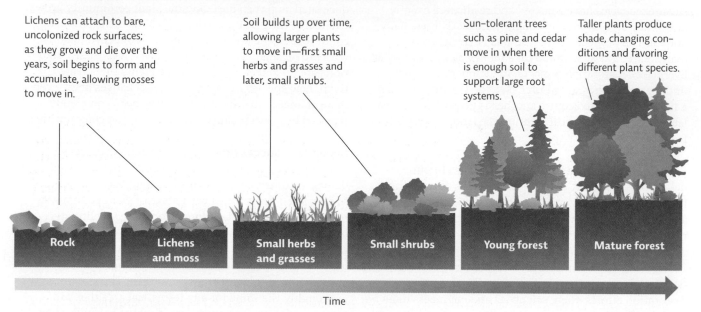

Lichens can attach to bare, uncolonized rock surfaces; as they grow and die over the years, soil begins to form and accumulate, allowing mosses to move in.

Soil builds up over time, allowing larger plants to move in—first small herbs and grasses and later, small shrubs.

Sun-tolerant trees such as pine and cedar move in when there is enough soil to support large root systems.

Taller plants produce shade, changing conditions and favoring different plant species.

Rock | **Lichens and moss** | **Small herbs and grasses** | **Small shrubs** | **Young forest** | **Mature forest**

Time

↑ In terrestrial ecosystems we see natural stages of succession occur whenever a new area is colonized or an established area is damaged. Sun-tolerant species give way to shade-tolerant ones as more soil is built up, supporting larger plant species.

EVERGLADES ECOLOGICAL SUCCESSION DEPENDS ON THE WATER LEVEL

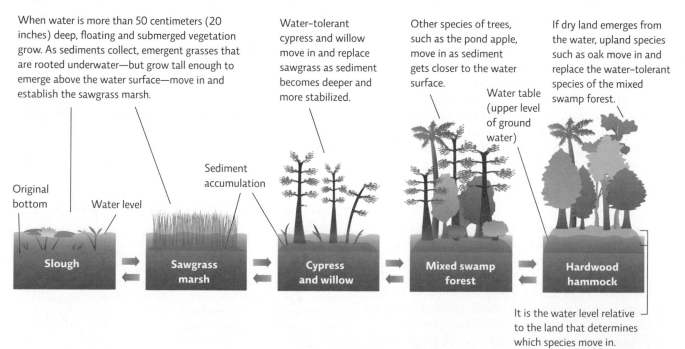

When water is more than 50 centimeters (20 inches) deep, floating and submerged vegetation grow. As sediments collect, emergent grasses that are rooted underwater—but grow tall enough to emerge above the water surface—move in and establish the sawgrass marsh.

Water-tolerant cypress and willow move in and replace sawgrass as sediment becomes deeper and more stabilized.

Other species of trees, such as the pond apple, move in as sediment gets closer to the water surface.

If dry land emerges from the water, upland species such as oak move in and replace the water-tolerant species of the mixed swamp forest.

Water table (upper level of ground water)

Original bottom

Water level

Sediment accumulation

Slough | **Sawgrass marsh** | **Cypress and willow** | **Mixed swamp forest** | **Hardwood hammock**

It is the water level relative to the land that determines which species move in.

↑ Ecological succession in the Everglades doesn't necessary follow the tidy predictable sequence seen in terrestrial ecosystems; in fact, periodic fires and the cyclic rainfall patterns may not support a predictable progression at all. Water levels are also important and influence which species move in. Succession in this area can actually go both ways: As ground level changes relative to the water level, an area might flood anew or sediment build-up might continue to raise the land relative to the water level. In the absence of disturbance, succession will progress to the hardwood hammock forest when sediment builds up enough to expose dry land.

alligator from the complex Everglades community would impact many species and change the face of the ecosystem.

Some ecosystems remain in a constant cycle of succession; others eventually reach an end-stage equilibrium where the conditions are well suited for the plants that created them—for example, trees whose seedlings can grow in shady habitat. These species, which can persist if their environment remains unchanged, are called **climax species**. End-stage **climax communities** can stay in place until disturbance restarts the process of succession—although there is debate among scientists over whether any community ever reaches an end point of succession, or continues to change and adapt.

Wetland areas also go through succession, responding to the presence of water and sediment depth. In the Everglades, each ecosystem is guided along this path by its own constellation of forces. Some, like the iconic sawgrass ecosystems, are fire adapted; fire returns them to early stages again and again, where the underwater roots of the emergent plants (those that are rooted underwater but grow above the waterline) such as sawgrass survive and quickly regrow. Others, if left undisturbed, would pass through successional stages of pioneers (grasses) to

shrubs or small trees to larger species of trees, depending on the deposition of soil and proximity of the water table to the surface (the top of the groundwater in the area). Others still are guided by the engineering changes of animals such as alligators, whose digging habits provide the foundation for an entire food chain. In each, though, the same general concept applies: As conditions change, other species better adapted to those conditions move in and displace previous residents. [INFOGRAPHIC 9.8]

However precarious their recovery might be, Wood Storks have indeed rebounded in recent years. Some say this rebound is the result of careful conservation efforts—including a restriction on development in certain areas—implemented under the Endangered Species Act. Others insist it is merely the result of above-average rainfall in recent years. For his part, Rogers sees another trend at work. Once again, he says, the storks are trying to tell us something. "They have shifted their center of distribution from South Florida to Central and North Florida," he says. "They're now spilling into Georgia and North Carolina—something we've never seen before." Rogers suspects the shift has something to do with the way climate is changing in the region, though he says much more research is needed before anyone can say for certain. "We're still trying to figure out what that means," he says. "But we know it's a clue to something."◉

climax species Species that move into an area at later stages of ecological succession.

climax community The end stage of ecological succession in which the conditions created by the climax species are suitable for the plants that created them so they can persist as long as their environment remains unchanged.

Research articles referenced in this chapter:
Brix, H., et al. 2010. BMC Plant Biology, 10: 23.
Rodgers, J.A., et al. 1996. Colonial Waterbirds, 19: 1–21.

BRING IT HOME

◔ PERSONAL CHOICES THAT HELP

The world is full of weird and wonderful species. Every year we discover new information about how intricate our biological communities are. By restoring habitats and increasing our understanding of the relationships between species, we can better ensure their long-term survival.

Individual Steps
→ Visit a park or nature preserve and watch for signs of species interactions. Do you hear animals or birds; can you see signs of predation or herbivory?
→ Buy a Duck Stamp. Usually purchased by waterfowl hunters for license purposes, nonhunters can purchase a stamp, which

supports wetland conservation in the National Wildlife Refuge System.

Group Action
→ The Everglades case study is an example of a very extensive restoration project. Call your local park district or nature preserve to see what restoration work is happening in your area and how you can become involved.

Policy Change
→ Follow the U.S. Fish and Wildlife Service Open Space blog to learn more about wildlife and issues facing conservation (http://www.fws.gov/news/blog).

UNDERSTANDING THE ISSUE

CHECK YOUR UNDERSTANDING

1. An indicator species:
a. helps predators keep tabs on the location of prey.
b. is an indication of the population size of a second species.
c. helps ecologists locate sensitive areas.
d. helps ecologists monitor the health of an area.

2. In ecological terms, a consumer is:
a. any plant.
b. any animal.
c. any organism that eats other organisms.
d. any animal that eats other animals.

3. Detritivores and decomposers:
a. might be earthworms and bacteria.
b. might be algae and fungi.
c. are usually insects such as flies.
d. are usually large predators.

4. Edge effects:
a. apply only to the largest and smallest members of a community.
b. occur in the areas where two or more habitats meet.
c. are beneficial for nearly all organisms.
d. are harmful for nearly all organisms.

5. In mutualism, both organisms receive an immediate benefit from the relationship. A good example of this is:
a. a dog and a flea.
b. an ant and a grasshopper.
c. a butterfly and a flowering plant.
d. a deer and a wolf.

6. An example of how secondary succession would occur in a particular area would be when:
a. one species has been outcompeted by another species, and is extirpated.
b. a flood has removed much of the vegetation.
c. hot ash from a volcano has completely burned and buried the area.
d. a disease reduces the top predator's population.

WORK WITH IDEAS

1. Which is more vulnerable to disturbances, a simple food web with only a few species, or a more complex one? Explain.

2. How do mangrove forests fit the definition of a *keystone species*?

3. Draw a simple food web for a natural area near you. Include producers and at least three levels of consumers, as well as detritivores, and decomposers.

4. Explain why both species richness and species evenness are important for a healthy ecosystem.

5. What happens to the net primary productivity and to the species diversity when humans disrupt wetlands by adding more nutrients, as in the example of agriculture in the Everglades?

6. Cowbirds are nest parasites—they lay their eggs in the nests of small forest birds such as bluebirds and leave them for the small birds to care for. The bluebirds then spend the next several weeks caring for the huge baby cowbird, which quickly kills the bluebird's own young. Cowbirds prefer open, disturbed areas and seldom venture far into a forest for any reason, even to lay eggs. Use the concept of edge effect to explain what happens to the populations of bluebirds when humans build roads, recreation areas, homes, and businesses in a large forest.

ANALYZING THE SCIENCE

The Mississippi River lies on three sides of the city of New Orleans, curving around it. To the fourth side of the city lies Lake Pontchartrain, 65 kilometers (40 miles) long and over 50 kilometers (30 miles) wide; the lake connects directly with the Gulf of Mexico. Until 140 years ago, much of southern Louisiana was swampland, lower than sea level and much lower than the level of the Mississippi River. As the swamps were drained, levees were built, beginning in the 1700s, speeding up after 1880, and continuing today.

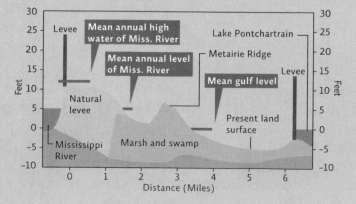

INTERPRETATION

1. What is the difference in mean annual water level, in feet, between Lake Pontchartrain and the Mississippi River at New Orleans? Why is this?

2. New Orleans lies between the levees that hold back the Mississippi River and Lake Pontchartrain. What are the highest and lowest elevations of the city itself, in feet?

3. What type of former ecosystem lies underneath most of the city?

ADVANCE YOUR THINKING

4. Without the levees throughout the southern half of the state, what would happen?

5. Why do southern Louisiana cemeteries feature above-ground tombs, crypts, and vaults, rather than graves?

6. In 2005, Hurricane Katrina devastated New Orleans and much of southern Louisiana. In its aftermath, researchers studied the area, trying to understand what happened. What steps could Louisiana take to avoid a repeat of this flooding and damage in the future?

EVALUATING NEW INFORMATION

Woodlands are some of the most important communities in the United States. Rich with diversity, we see them as permanent and unchanging places, filled with wildlife, which we wish to preserve. For over 70 years, Smokey the Bear has been telling us "Only YOU can prevent forest fires!" but the United States has had an active role in fighting and preventing forest fires for far longer. We began fighting wildfires in a systematic way around the turn of the last century; historic photographs show people building fire lines and shoveling dirt over smoldering spots in our woodlands. In the past 30 years, however, there have been increased discussions about the automatic response of immediately quelling all wildfires—perhaps some wildfires do more good than harm. Change from storms, fires, and floods is part of the natural cycle of an ecosystem.

The following articles about wildfires and ecology were written over the past 10 years:

- An overview of wildfires (geography.about.com/od/globalproblemsandissues/a/wildlandfire.htm)
- A history of humans and wildfires (www.foresthistory.org/Education/Curriculum/Activity/activ9/essay.htm)
- The ecology of fire (www.esa.org/education_diversity/pdfDocs/fireecology.pdf)
- Logging and fires (www.emagazine.com/magazine-archive/the-burning-west)
- Logging expansion (www.planetark.com/dailynewsstory.cfm/newsid/29330/story.htm)
- Spending on fires (www.emagazine.com/includes/print-article/magazine-archive/6631)

Evaluate the websites and work with the information to answer the following questions:

1. Are these reliable information sources?
   a. Do the authors of these articles give supporting evidence for their claims?
   b. Do they give sources for their evidence?
   c. What are the missions of these organizations?

2. Now go to the National Interagency Coordination Center for wildfires (www.nifc.gov/nicc). On the left-hand side of the page, click on *Incident Management Situation Report*.
   a. How many new fires are there in the United States?
   b. Large fires are wildfires of 100 acres or more occurring in timber, or a wildfire of 300 acres or more occurring in grass/sage. How many new large fires are there at the moment?
   c. How many uncontained large fires are there?

3. Finally, go to InciWeb (www.inciweb.org) and select your state in the upper right-hand corner of the page. Are there fires currently burning, or that have burned recently, in your state? If not, choose another state. How many acres have been burned in the past month in the state you're reviewing?

4. Based on the information you have found, explain your position:
   a. What should be done when a wildfire occurs?
   b. What, if anything, should be done in an area to prevent or lessen a possible wildlands fire?

MAKING CONNECTIONS

CORAL REEFS AND LOBSTER DINNER

Background: Belize is a tiny country in Central America composed of jungle, pine forests, limestone caves, Atlantic Ocean beaches, hundreds of small islands, and the major portion of the longest coral reef in this hemisphere—second only to the Great Barrier Reef in Australia. Coral reefs are found in warm, shallow waters and support huge numbers of organisms, estimated at one-fourth of all marine life. They are diverse, productive, and very fragile. The algae in coral reefs produce oxygen and the reef itself acts as a nursery for sealife, as shoreline protection from storms and erosion, and as an important feeding and breeding ground for thousands of species. Well aware of its precious resources, the Belizean government has declared over 40% of the country to be natural parks and reserves, including much of the coral reef and surrounding areas. The major businesses in Belize are tourism and commercial harvesting of fish, lobster, and conch. Can the delicate reef system withstand both uses? In July 2011, a group of 5 people were caught with nearly 300 tiny lobster tails, many weighing less than 1 ounce, and over 200 undersized Queen Conch that were also out of season. A month later, the senior marine conservationist declared that over 85% of the reef was dead, dying, or in serious difficulty.

Case: Investigate the following three important needs. How do we balance them?

1. The need for the preservation of an important world biome that gives us so many ecological services

2. The need for a productive harvest of fish, lobster, and conch each year to support thousands of fishers and to supply hundreds of thousands of consumers with seafood

3. The need for a tourism industry for the enjoyment of millions and for the kind of "clean" industry that can help a small country support itself sustainably

Write a report that focuses on one of the three needs as most important while preserving and maintaining the other two needs. In your report, be sure to use facts to justify what you consider to be the most important need and be sure to include the following:

a. The ecological services provided by a coral reef and an indication of what will happen if the reef is damaged or destroyed
b. The economics and ecology of the seafood industry in a small, rural, coastal country such as Belize
c. The damage, both to the economy and to the ecology, of poaching, and a possible solution
d. The costs and benefits of tourism in a small, rural country
e. Some possible solutions to balancing these issues

CORE MESSAGE

The variety of life on Earth is tremendous. This biodiversity provides vital ecological services that support life; we depend on these same services for things like food, medicine, and economic development. Biodiversity faces many threats today, jeopardizing the ability of ecosystems to provide these services. One of the first steps in protecting biodiversity is simply trying to determine what is here.

GUIDING QUESTIONS

After reading this chapter, you should be able to answer the following questions:

→ What is biodiversity and how do genetic, species, and ecosystem diversity each contribute to overall diversity?

→ How many species are estimated to live on Earth and which taxonomic groups have the most species? How sure are we of these numbers?

→ Why is biodiversity important?

→ What are biodiversity hotspots and why are they important?

→ What role does isolation play in making an area a potential biodiversity hotspot?

Medicinal plants.

NATURE'S MEDICINE CABINET

Will the bark of an ordinary tree in Samoa become a cure for cancer?

A man walks along the Samoan coastline carrying bundles of vegetation in baskets made of woven leaves.

It was his mother's death in the fall of 1984, from a particularly aggressive form of breast cancer, that drove Paul Cox back to the Samoan rainforest. Cox had first visited the South Pacific island in 1973, through an undergraduate research program with Brigham Young University, where he was majoring in botany. Since then, the Utah native had earned a Ph.D. from Harvard and made a career studying plant physiology in the United States.

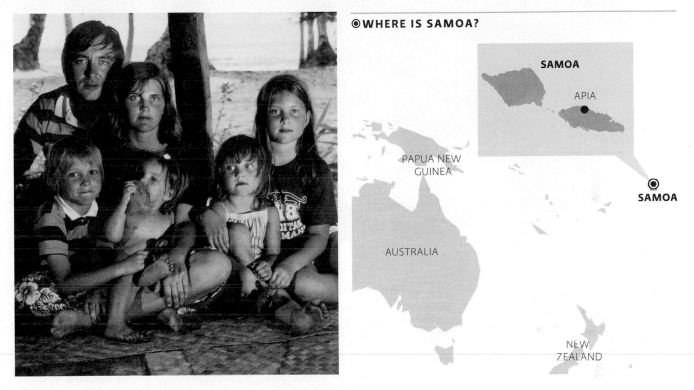

↑ Paul Cox and family in Samoa, 1986.

⊙ WHERE IS SAMOA?

But when the best of Western medicine failed to cure his mother, Cox remembered the Samoans—their rich folk healing traditions, and the profound influence that jungle plants had on their health and well-being. Surely, he thought, somewhere in the lush abundance that had long sustained this impoverished island nation must be a compound powerful enough to obliterate the most insidious of tumors. "I told my friends at the National Cancer Institute, 'If there is even a 1% chance of finding something, it's worth taking a look,'" he recalls. "They said, 'We think there is like a 3% chance.' So I went." Six months after his mother's funeral, with his wife and four young children in tow, Cox returned to Samoa.

Biodiversity benefits humans and other species.

Why did Cox suspect that Samoa might hold the key to a cancer cure? Tropical regions—warm, lush, close to the equator—contain the greatest concentration and variety of plant and animal life forms on Earth. This variety is called **biodiversity**. Ecosystems of this region have both high **species diversity** (see Chapter 9) and high **genetic diversity**. They also usually have high **ecological diversity**, a wide variety of communities and ecosystems with

---

biodiversity The variety of life on Earth; it includes species, genetic, and ecological diversity.
species diversity The variety of species, including how many are present (richness) and their abundance relative to each other (evenness).
genetic diversity The heritable variation among individuals of a single population or within the species as a whole.
ecological diversity The variety within an ecosystem's structure, including many communities, habitats, niches, and trophic levels.

Infographic **10.1** | **BIODIVERSITY INCLUDES GENETIC, SPECIES, AND ECOSYSTEM DIVERSITY**

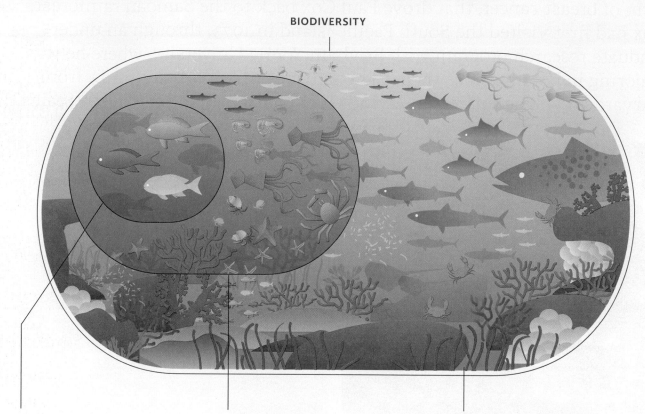

BIODIVERSITY

GENETIC DIVERSITY
Variations in the genes among individuals of the same species

↑ Though these squarespot anthias fish are members of the same species, *Pseudanthias pleurotaenia*, they show tremendous genetic diversity.

SPECIES DIVERSITY
The variety of species present in an area; includes the number of different species that are present as well as their relative abundance

↑ The area of Polynesia where Samoa is found has some of the highest coral reef species diversity in the world.

ECOLOGICAL DIVERSITY
The variety of habitats, niches, trophic levels, and community interactions

↑ Samoa's rainforest and coral reef ecosystems contain complex, three-dimensional structures with varied habitats and multiple niches that support high species diversity and a complex community.

different habitats, many trophic levels, and lots of niches. Higher levels of diversity tend to contribute to stability—genetically diverse populations are more likely to be able to adapt if faced with new environmental pressures; ecosystems with high species and ecological diversity may themselves be more stable and resilient in the face of environmental change. [INFOGRAPHIC 10.1]

It is virtually impossible to know just how many species exist on Earth. Indeed, the numbers vary widely—from 3 million to 100 million, depending on the source. What does seem certain is this: Of all the species now living on Earth, the vast majority—some 86%—have yet to be discovered or identified by humans. "The numbers are mind-boggling," says Jim Miller, a scientist at the New

York Botanical Garden in New York City. "Almost unfathomable." [INFOGRAPHIC 10.2]

What we do know for sure is that biodiversity is responsible for much more than the majesty of nature. In fact, it is essential to a vast array of **ecosystem services**: Photosynthetic organisms (plants on land; algae and phytoplankton in the sea) bring in energy, produce oxygen, and sequester carbon. Other organisms capture and pass along important nutrients like nitrogen and phosphorus.

ecosystem services Benefits provided by functional ecosystems that are important to all life (including humans); includes such things as nutrient cycles, air and water purification, and ecosystem goods such as food and fuel.

Infographic **10.2** | **BIODIVERSITY ON EARTH**

↓ We have identified about 1.8 million species so far (prokaryotes and eukaryotes) but our knowledge of Earth's total biodiversity is scant. Many researchers, especially those who work with tropical species and insects, believe the estimates of 8.7 to 14 million species are far too low—there may well be 5 million insect species alone (some think the number could be much higher). Of the species we have identified, insects far outnumber any other group of organisms. In fact, vertebrates (the group to which humans belong) likely make up only 1% of all creatures on Earth.

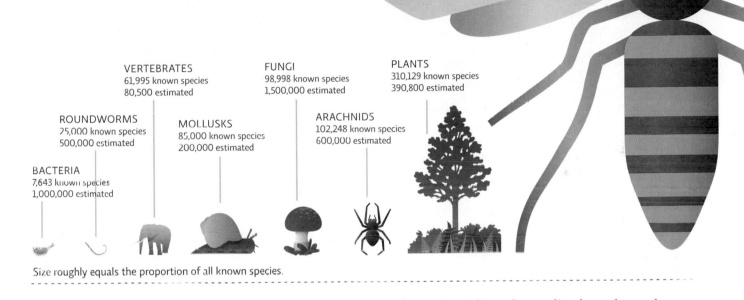

INSECTS
1,000,000 known species
5,000,000 estimated

VERTEBRATES
61,995 known species
80,500 estimated

FUNGI
98,998 known species
1,500,000 estimated

PLANTS
310,129 known species
390,800 estimated

ROUNDWORMS
25,000 known species
500,000 estimated

MOLLUSKS
85,000 known species
200,000 estimated

ARACHNIDS
102,248 known species
600,000 estimated

BACTERIA
7,643 known species
1,000,000 estimated

Size roughly equals the proportion of all known species.

↓ When we compare the estimated size of different groups of species to the number within each group listed as endangered or threatened, there is a clear bias toward species of interest to us (large animals and plants). However, we are discovering that the lesser-known and less appreciated groups, such as worms and insects, perform vital ecosystem services and also warrant our attention.

PERCENT OF SPECIES LISTED AS ENDANGERED COMPARED TO THE ESTIMATED SIZE OF THE GROUP

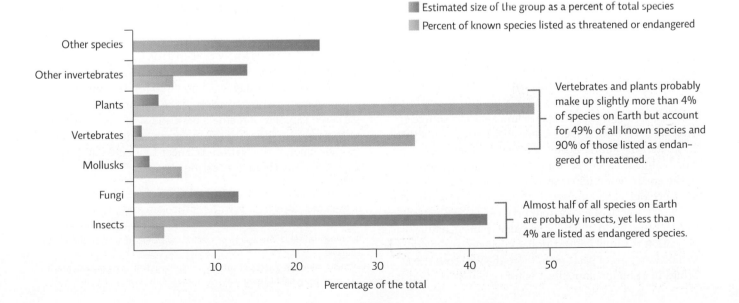

■ Estimated size of the group as a percent of total species
■ Percent of known species listed as threatened or endangered

Vertebrates and plants probably make up slightly more than 4% of species on Earth but account for 49% of all known species and 90% of those listed as endangered or threatened.

Almost half of all species on Earth are probably insects, yet less than 4% are listed as endangered species.

Percentage of the total

Others still help purify the air and water and eventually become food for other creatures. And populations are kept in check by predators and competitors so that no single species grows too populous or gobbles up too many needed resources. To take just one example from the Samoan rainforest, the tooth-billed pigeon plays an essential role in the propagation of mahogany, a tree species found throughout Polynesia. The pigeon may be the only native animal that can open the tree's seeds so new seedlings can take root. Meanwhile, adult mahogany trees provide habitat and food to a variety of insect, bird, mammal, and reptile species. They are also key players in carbon cycling and soil stabilization, and they serve as a windbreak in coastal areas. (See Chapter 4 for more on ecosystem services.)

Many societal traditions are also rooted in nature. In Samoa, native foods are an important part of traditional feasts; the Sunday meal typically features tropical foods like fresh fruit, root vegetables including taro, and lots of seafood. Kava, a traditional drink prepared from the roots of a native pepper plant, is often drunk before ceremonial events and has a mild tranquilizing effect. It is also purported to have analgesic effects.

Biodiversity can also provide direct protection against disease in various ways. The incidence of Lyme disease, for example, is often less in areas with higher biodiversity (especially areas with squirrels and possums, two species who effectively remove disease-carrying ticks and prevent the spread of the disease to other organisms). This is known as the *dilution effect*. Likewise, a study by John Swaddle and Stavros Calos showed that fewer human cases of West Nile virus occur in eastern U.S. counties with higher bird biodiversity, possibly because some species are less effective at transmitting the virus than others—the more species present, the less likely the virus will be transmitted to humans.

Biodiverse ecosystems have economic value, too. In Samoa, the forests provide not only food, fuel, and building material, but also pharmaceuticals. People use the chemicals they extract from plants both to attend to their individual human health issues and also as a source of income. Ever since Robert Constanza's 1997 study that estimated the annual value of ecosystem services to be worth almost twice the annual gross domestic product of the entire world (close to $44 trillion in today's dollars), ecological economists have set out to quantify the monetary worth of ecosystem services to bring attention to this often underappreciated value. Indeed, nature-based recreation is a multibillion-dollar business worldwide. Whale watching, a pastime in more than 80 countries, brings in $1 billion annually. Visitors spent $12 billion at

U.S. national parks in 2010. The U.S. Department of the Interior estimated the economic value of hunting, fishing, and wildlife viewing in 2001 to be $108 billion. In 2012, the UN adopted the System of Environmental-Economic Accounting (SEEA) as the international standard for assessing the value of environmental assets. Using this statistical system, the Australian Bureau of Statistics quantified Australia's ecological resources and services (in 2010) at almost $5 trillion—which is more than half of the total economic wealth of the nation. [INFOGRAPHIC 10.3]

The ecological, medical, and monetary benefits that a given species can provide for human communities are referred to as the species' **instrumental value**. But many scientists and laypeople agree that all species, great and small, also have **intrinsic value**—that is, they are inherently worth preserving, even aside from the services they offer us.

Once they arrived in Samoa, Cox and his family settled in a thatched hut on the far western edge of the island in the village of Falealupo—just a few steps from the sea and as far as they could get from the comparatively modern neighboring island of American Samoa. There, amid a tropical swirl of insects, humidity, and white sand, with no electricity or running water, Cox apprenticed himself to a mostly female cadre of healers. His primary teacher was Pela Lilo, an 82-year-old woman whose particular collection of plant-based remedies had been passed down to her through generations.

Lilo and her fellow healers shared hundreds of natural remedies with Cox—the bark of vavae (*Ceiba pentandra*) to treat asthma, leaves of the kuava tree (*Psidium guajava*) for diarrhea, and root of 'Ago (*Curcuma longa*) for rashes. It turned out that the Samoans did not have a word for breast cancer (one purported treatment for "lumpy breasts" did not prove effective against breast tumors). But they did have a treatment for "yellowing fever," a disease Westerners know as Hepatitis C; boiling the bark of the mamala tree (*Homalanthus nutans*) and drinking the extract twice a day was known to relieve symptoms. After seeing a demonstration of the potion's power, Cox gathered samples of mamala bark and sent them off (along with dozens of other promising roots, stems, and leaves) to colleagues at the National Cancer Institute (NCI), in Bethesda, Maryland, for testing.

---

instrumental value An object's or species' worth, based on its usefulness to humans.

intrinsic value An object's or species' worth, based on its mere existence; it has an inherent right to exist.

Infographic **10.3** | **ECOSYSTEM SERVICES**

↓ We depend on genetically diverse, species-rich communities to provide the goods and services we use every day. Impoverished ecosystems that have lost genetic, species, or ecological diversity cannot perform these tasks as well as highly diverse ecosystems can.

1. CULTURAL BENEFITS
Aesthetic
Spiritual
Educational
Recreational

2. HUMAN PROVISIONS
Food
Fiber products, such as cotton and wool
Fuel
Pharmaceuticals

3. ECOSYSTEM REGULATION AND SUPPORT
Nutrient cycling
Pollination and seed dispersal
Air and water purification
Flood control
Soil formation and erosion control
Climate regulation
Population control

2. Many forest products can be harvested

1. Many people enjoy spending time in nature.

3. Shoreline vegetation protects the banks from erosion.

But what he saw as a garden of medical promise, others saw as timber. By the time Cox shipped his samples over to scientists at the NCI, loggers had long since clear-cut (removed all the trees from) some 80% of Samoa's lowland forests. And now they wanted to do the same to the 30,000-acre forest where the mamala tree grew most abundantly. As it turned out, the villagers were being pressured by the government to build a school. And a logging company was offering to pay the villagers nearly $2 an acre for the mamala forest—just enough to cover construction costs. Cox knew that any health or financial benefits from the NCI research would be decades off—at least. But he also knew that if the loggers had their way, hundreds of other plants with potential medicinal properties might never be found.

Plants gain medicinal qualities as they adapt to other species.

The tropics are home to roughly 5.3 billion people, or 80% of the world's population. Because many of the countries in this region are poor, medicinal plants provide both a primary form of health care and a major source of income to local communities. Medicine is one of the most prominent facets of *ethnobotany*—the study of how different cultures make use of the plants that surround them.

According to the World Bank, the international trade in medicinal plants was worth nearly $100 billion in 2011. Roughly half of all prescription drugs—including some medicine cabinet staples like aspirin, codeine, and most hypertension drugs—were originally derived from plants. And by most accounts, there are many more medical treasures just waiting to be discovered: Less than 1% of the world's 265,000 known flowering species have been tested for their effectiveness against human diseases. "We are barely scratching the surface now," says Miller. [INFOGRAPHIC 10.4]

Despite that untapped potential, the role of plants in Western medicine has waxed and waned throughout history. For one thing, the odds are daunting. On average, only 1 in 15,000 compounds will demonstrate the potential to treat a human condition. For most of the modern medical era, finding that 1 compound took an average of 12 years and $300 million—enough to dissuade all but the most tenacious of scientists.

Roughly half of all prescription drugs—including some medicine cabinet staples like aspirin, codeine, and most hypertension drugs—were originally derived from plants.

Plants were once the biggest source of new medications in Western society. But advances in pharmacology quickly turned scientists' attention away from the forest and toward the test tube, where they could design molecules from scratch, rather than search for them like needles in a haystack. A few decades ago, scientists predicted that advances in biochemistry would eliminate the need to exploit fragile ecosystems for our own medical needs; they believed any molecule we discovered could easily be replicated in a test tube or Petri dish.

↑ Samoan healer, Pela Lilo, with her daughter-in-law, Fa'asaina Lamositele, treating a patient with traditional medicine.

Infographic **10.4** | **NATURALLY DERIVED MEDICINES**

Opium (flower)
Medicine: Morphine
Use: Pain

Foxglove (flower)
Medicine: Digoxin
Use: Heart disease

**Madagascar
periwinkle (flower)**
Medicine: Vinblastine
Use: Cancer (Hodgkin's lymphoma,
breast, testicular)

Ephedra (shrub)
Medicine: Salbutamol
Use: Asthma

Curarea (woody vine)
Medicine: Tubocurarine
Use: Muscle relaxant

Cobra (snake)
Medicine: Cobroxine
Use: Pain

Pacific yew tree
Medicine: Paclitaxel, Taxol
Use: Cancer (lung, breast, ovarian)

Penicillium (fungus)
Medicine: Penicillin, Metavastin
Use: Antibacterial, cholesterol lowering

Streptomyces (bacteria)
Medicine: Ivermectin
Use: Parasites

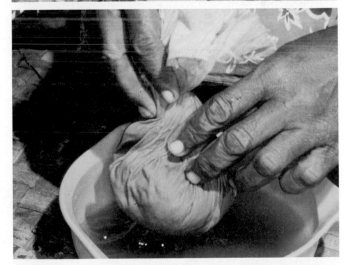

↑ Top: Leaves from the Samoan mamala tree *Homalanthus nutans* from which the anti-HIV drug candidate prostratin was derived. Middle: Samoan healer Ake Lilo preparing a branch from the mamala tree. Bottom: The bark of the mamala tree is boiled down into an extract the Samoans use to treat Hepatitis C.

But with the exception of a few remarkable successes, like the cancer drug paclitaxel that is now synthesized in the lab rather than being extracted from the bark of Pacific yew trees where it was first found, that prediction hasn't quite panned out. "An overwhelming number of plant-derived compounds have eluded all attempts to copy," says Sarah Oldfield, director of the nonprofit Botanic Garden Conservation Initiative. Because copying plant-derived molecules has proven so intractable, modern medicine continues to depend on their availability in nature. And by many accounts, evolution has bested the most apt chemists, resulting in molecules too complex to fathom, let alone replicate. "Mother Nature is the ultimate chemist," says Oldfield. "Modern chemistry has barely enabled us to copy her creations, so you can imagine the near impossibility of designing them from scratch."

It's little wonder. Plant chemicals evolve over eons, as individual plant species strive to fight off predators. Because they can't run or physically fight, the way animals can, plants employ two main antipredator strategies: physical defense structures like thorns, thick seed coats, or bark; and chemical weapons, like the ones that ultimately find their way into our medicine cabinets. In short, it's the diversity of life itself, challenging the plant kingdom in a mind-boggling variety of ways, that spurs such a broad range of protective chemicals, and in so doing, provides us with some precious lifesaving medicines. (See Chapter 12 for more on natural selection of adaptive traits.)

Desperate to save whatever such medicines might exist in Samoa from being lost to logging, Cox negotiated with the villagers, promising them he would raise the school

↑ A rare Hawaiian silversword.

money himself if they agreed to protect the forest from logging. Then he went knocking on the doors of nonprofits and conservation groups back in the United States. Before too many trees were felled, he had scrounged up $85,000—enough money to build the school and spare the forest. It was a rare victory. "It's common for indigenous people to have to choose between saving a forest or coral reef and building a school or medical clinic," says Cox. "Because they don't typically have the resources to do both, they must sacrifice one to pay for the other. And the loss when that happens is often incalculable."

Especially in the Pacific Islands like Samoa and Hawaii. The region has the highest percentage of threatened and endangered species (those at risk for extinction) of any region on Earth. These species are often specialists that occupy narrow niches. This region also contains a high proportion of **endemic** plants and animals—those that exist nowhere else on Earth. Areas like this that contain a large number of endemic but threatened species are known as **biodiversity hotspots**. Many hotspots are in tropical regions. [INFOGRAPHIC 10.5]

Isolated ecosystems like islands (or mountaintops or other remote ecosystems) tend to have a higher proportion of endemic species than larger areas such as continents. Species come to islands in various ways—they may be blown in on storms, arrive as lost migrators, or the island itself may have broken off from a larger landmass at some point. Because these are rare events, once a species arrives it is unlikely to be joined by other members of its species. This isolates the founding population and as it adapts over time to its new island home it may eventually evolve into a new species (see Chapter 12 for more on the founder effect). [INFOGRAPHIC 10.6]

The number of unique species (the degree of endemism) generally increases with isolation. On the Hawaiian Islands, almost 2,400 miles from the nearest mainland, 90% of native species are endemic. This isolation and high endemism makes remote islands particularly vulnerable to species loss. In fact, even though they make up a tiny percent of the land mass on Earth, islands have accounted for about half of all recorded extinctions in the last 400 years. In Hawaii alone, fully one half of the indigenous flora faces immediate extinction. Such a high extinction rate threatens the fragile tapestry of life on these islands, from the soil and freshwater supplies to the health and economic future of the islands' residents.

---

endemic A species that is native to a particular area and is not naturally found elsewhere.

biodiversity hotspot An area that contains a large number of endemic but threatened species.

Infographic **10.5** | **BIODIVERSITY HOTSPOTS**

↑ Biodiversity hotspots, areas with a high number of endemic but threatened or endangered species, cover a small percentage of land and water areas but hold more than 40%–50% of all plant and vertebrate endemic animal species. Most hotspots are located in tropical biomes or in isolated terrestrial ecosystems such as mountains or islands. Even small disturbances can threaten endemic species that populate specialized niches in these hotspots.

Samoa

TROPICAL ANDES 15,000 endemic plants, 487 threatened species, 2 extinct

GUINEAN FORESTS 1,800 endemic plants, 115 threatened species, 0 extinct

POLYNESIA–MICRONESIA 3,074 endemic plants, 99 threatened species, 43 extinct

INDO-MALAYAN ARCHIPELAGO 15,000 endemic plants, 162 threatened species, 4 extinct

↳ Isolation can increase the number of endemic species in an area because local populations do not "share" genes with other populations. Over time, an isolated population may diverge from its ancestral population as it becomes adapted to its immediate environment.

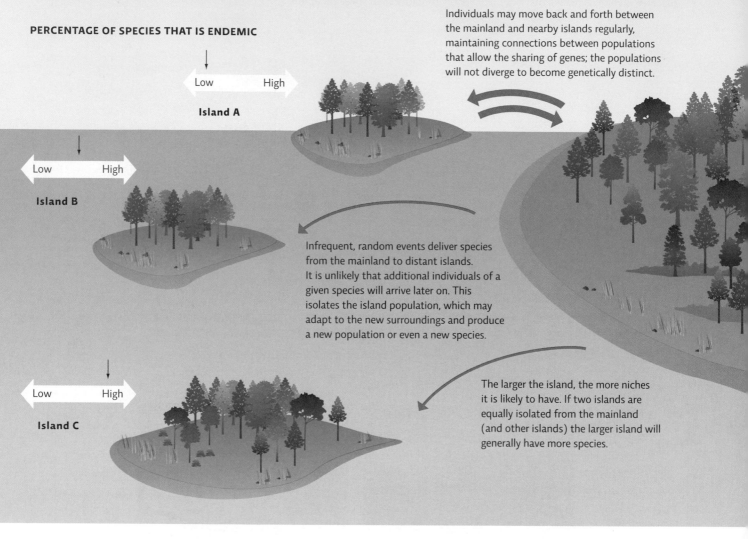

PERCENTAGE OF SPECIES THAT IS ENDEMIC

Low High
Island A

Low High
Island B

Low High
Island C

Individuals may move back and forth between the mainland and nearby islands regularly, maintaining connections between populations that allow the sharing of genes; the populations will not diverge to become genetically distinct.

Infrequent, random events deliver species from the mainland to distant islands. It is unlikely that additional individuals of a given species will arrive later on. This isolates the island population, which may adapt to the new surroundings and produce a new population or even a new species.

The larger the island, the more niches it is likely to have. If two islands are equally isolated from the mainland (and other islands) the larger island will generally have more species.

If a species is **extirpated**—that is, if it goes extinct in one ecosystem but still exists elsewhere—it is sometimes possible to reintroduce the species by bringing in individuals from a different population. But this is not as easy as it sounds. Local species are adapted to local conditions—they fill an *ecological niche*, or specific functional role in an ecosystem. Therefore, plants of the same species, but from different global locations, can occupy different niches.

The loss of even a few species can disrupt the connections that the ecological community depends on, triggering a cascade that threatens other species in the region. "Simple" ecological communities (ones with relatively few species) are much more affected by a single loss. But even "complex" ecosystems (those with many species and many filled niches) can be gravely imperiled if too many members are lost, especially if keystone species are lost. The loss of half of a native plant community is a loss that few, if any, ecosystems can endure without completely collapsing. Of course, once one ecosystem collapses, a new one may emerge. But such recovery takes a very long time. [INFOGRAPHIC 10.7]

In the meantime, scientists like Cox are racing against the clock, trying to find, catalogue, and ultimately preserve,

extirpated A species that is locally extinct in one or more areas but still has some individual members in other areas.

↓ Samoa has high biodiversity and is part of the diverse but threatened Polynesia/Micronesia hotspot. Some of the species that live in Samoan ecosystems are shown here.

TOOTH-BILLED PIGEON The endangered endemic tooth-billed pigeon (*Didunculus strigirostris*) specializes in eating mahogany seeds, using its sawtooth lower beak (like those seen in fossil birds) to open the hard seeds. Habitat loss due to deforestation threatens its survival.

FRUIT BAT Flying foxes (*Pteropus samoensis*) are fruit bats that pollinate large, tree-climbing Freycinetia flowers, the roots of which are used by Samoans to make fishing baskets. The bats are endangered due to habitat loss; as the bats have declined in number, so have the flowers and the ability of the Samoans to make the baskets.

SAMOAN MOORHEN The endemic Samoan moorhen (*Gallinula pacifica*) may be extinct (last sighted in 1984) because of habitat loss from deforestation.

RED GINGER (*Alpinia purpurata*, known as *teuila*) is the national flower of Samoa and a popular garden plant with its own national festival.

SAMOAN SWALLOWTAIL BUTTERFLY This endemic butterfly (*Papilio godeffroyi*) is critically endangered because of forest clearing for farming.

RACCOON BUTTERFLYFISH Samoan seas have at least 250 species of coral and almost 1,000 species of fish. These raccoon butterflyfish (*Chaetodon lunula*) are one of 30 or so species of butterflyfish found there.

↑ More than 90% of the native species on the Hawaiian Islands are found nowhere else, making these populations more vulnerable to extinction; if they are lost there, they are lost forever. The smaller the island, the smaller the population of a given species, another vunerability for extinction. Here, botanists work through the night setting up a protective fence around the last specimen of *Delissea undulata* found in the wild. This endangered plant was growing on the side of a collapsed lava tube, but had been knocked over by wind or animals and was dangling from its roots.

as many species as possible before they vanish from existence. To that end, they have deployed themselves throughout the Pacific—rappelling broad, steep cliffs that overlook the ocean, shimmying down into narrow valleys, sometimes even dangling from helicopters—to take a full census of island plants, and to collect seeds from those species with fewer than 100 remaining specimens. This careful accounting has provided the basis for a large-scale conservation effort that scientists are replicating in other biodiversity hotspots around the world. "If we know what we are losing and where we're losing it, we can plan a counter-offensive," says David Burney of the National Tropical Botanical Garden on the Hawaiian island of Kaua'i. "The census helps us to figure out what's due to habitat loss, and so on."

One such effort is the All Taxa Biodiversity Inventory (ATBI), a project of Discover Life in America, aimed at inventorying the estimated 100,000 species of living organisms in Great Smoky Mountains National Park, which spans North Carolina and Tennessee in the

southeastern United States. The project members are developing checklists, reports, maps, databases, and natural history profiles of the species found there, using an army of scientists and citizen volunteers. As of June 2012, they have discovered 922 species new to science and 7,391 new to the park.

This type of basic field research—evaluating an area to see which species are present, how healthy and genetically diverse that species' population is, and what threats the species face—is essential to protecting and maintaining biodiversity. After all, we can't save or protect what we don't even know exists or don't realize is in danger of extinction.

Biodiversity is proving invaluable in the search for cures.

For its part, the mamala tree turned out to be a good save. NCI researchers boiled down the bark and purified the remaining substance, which they named prostratin.

Although prostratin had no effect against cancer, as Cox had hoped, it did prove to be a valuable weapon against HIV, the virus that causes AIDS. The first human trials of prostratin began in 2008. Researchers expect the drug to be commercially available in a few years.

So far, mamala bark provides the only known source of prostratin. While small amounts of the compound have been enough to conduct preclinical animal studies, much larger quantities would be needed for large-scale drug manufacturing. Stripping the Samoan rainforest of its mamala trees would be environmentally destructive, not to mention inefficient. So biologists are looking for better, synthetic, options. By locating the genes responsible for prostratin synthesis and cloning them into a simple organism, like a bacterium, they hope to spare the mamala tree from overexploitation. The genetically altered bacteria would then serve as mini prostratin factories—quickly and efficiently producing mass quantities of the molecule, which could then be used to make medication. This approach has already been successfully used to scale up production of the plant-derived antimalaria drug artemisinin.

Cox is quick to point out that without the Samoans' help, he would never have found prostratin. The forest he scoured housed two varieties of the tree; only one contained prostratin and within that variety, only trees of a certain size seemed to work (thus emphasizing the value of genetic diversity). The Samoans freely shared this knowledge with Cox. In exchange, he negotiated a profit-sharing agreement with the people of Falealupo. The U.S. government has promised that half of all royalty income from prostratin will go back to them. This represents one of the first formal legal recognitions of indigenous intellectual property rights. In addition to the profit sharing, any commercialized drug developed from prostratin will be supplied to developing countries for free.

While Cox and his colleagues believe ardently in each species' intrinsic value, they also agree that ethnobotany offers one of the best chances to preserve as many plant species as possible. "There is a strong link between the health of forests and the health of humans," he says. "If people understand that a rainforest might contain the best cures for diseases that plague us, they will care a whole lot more about saving it." ◉

Research articles referenced in this chapter:
Australian Bureau of Statistics. 2012. *Completing the Picture—Environmental Accounting in Practice.*
Cox, P.A. 1993. *Journal of Ethnopharmacology*, 38: 181–188.
Mora, C., *et al.* 2011. *PLoS Biology* 9: e1001127 doi:10.1371/journal.pbio.1001127.
Swaddle, J.P. and Calos, S.E. 2008. *PLoS ONE* 3:e2488.doi:10.1371/journal.pone.0002488

BRING IT HOME

◔ PERSONAL CHOICES THAT HELP

Species and habitats provide numerous benefits to people, including water and air purification, food sources, recreation, and medicine. Unfortunately, many species are facing threats at ever-increasing levels. The good news is that we as a society have a direct impact on these threats and can make changes to ensure the survival of many of our at-risk species.

Individual Steps
→ Learn about the species that inhabit your own area by visiting a local park or wildlife sanctuary. Install an Audubon Guide app on your smartphone, or buy a field guide to identify the plant and animal species in your area.
→ Make your backyard friendly to wildlife, using suggestions from www.nwf.org.

Group Action
→ Work with faculty and other students to organize a bioblitz for a protected area in your region. A bioblitz, which is an intensive survey of all the biodiversity in the area, can generate a large amount of data to be used for habitat management and species protection.
→ Join a citizen science program that monitors wildlife. Many regional conservation groups have monitoring opportunities and provide training. For national programs, see the Cornell Lab of Ornithology website, at www.birds.cornell.edu and the Izaak Walton League website, at www.iwla.org.

Policy Change
→ The United States has not ratified the Convention on Biological Diversity, an international treaty that seeks to protect biodiversity worldwide. Learn about the convention at www.cbd.int/convention/ and contact your U.S. senators, asking them to push for the United States to ratify the treaty.

UNDERSTANDING THE ISSUE

CHECK YOUR UNDERSTANDING

1. **Which of the following explains why there are more endangered tropical species than temperate species?**
   a. More people live in the tropics.
   b. Tropical species are more likely to be generalists with broad niches.
   c. Global climate change is warming the tropics more rapidly.
   d. Both a. and c. are correct.

2. **Why are many of the biodiversity hotspots around the world on islands?**
   a. Islands accumulate species from many different areas.
   b. Populations of island species are isolated.
   c. Islands have more diverse habitats.
   d. There are more niches on islands.

3. **An example of species diversity might be:**
   a. the wide variety of coloration and tail size in guppies.
   b. the diverse habitat types and organisms inhabiting a deep lake, its edges, and the surrounding meadow and forest areas.
   c. the many different species inhabiting a swamp.
   d. None of the above.

4. **The total number of different species on Earth:**
   a. is unknown, but insects are the most numerous species.
   b. is a few million, mostly bacteria and fungi.
   c. is more than 20 million, with half of them being plants.
   d. is less than a million, mostly vertebrates.

5. **Whereas an individual with an anthropocentric, or human-centered, worldview would stress the ___ value of species as a reason to conserve biodiversity, most environmental scientists would stress their ___ value.**
   a. intrinsic; medicinal
   b. instrumental; intrinsic and ecological
   c. ecological; instrumental
   d. instrumental; cultural

6. **Why is biodiversity loss a concern?**
   a. It primarily occurs in the developed world, where most of the world's population lives.
   b. It increases the degree of endemism in an area.
   c. It decreases our ability to synthesize current medicines like taxol.
   d. It disrupts ecological connections, potentially diminishing ecosystem services.

WORK WITH IDEAS

1. Which is the most convincing reason for protecting endangered species: the intrinsic value of biodiversity, the possibility of finding medicines, or preserving ecosystem functions? Why?

2. What are the ethical issues involved in preserving biodiversity in tropical areas?

3. Some agricultural researchers are interested in finding the wild ancestors of our domesticated animals and plants. They are concerned that modern food production uses only a few varieties of a given species, although there are many varieties that could be used. How is the protection of genetic diversity within a species both similar and dissimilar to the conservation of threatened species?

4. Marine estuaries, on coastlines and the shores of islands, contain mangrove forests and thousands of organisms on the land and in the shallow waters of the estuary. Explain some of the ecosystem services provided by these estuarine environments.

5. Describe how biodiversity and national security may be connected. Why do some people make the argument that maintaining biodiversity is critical to national and international security?

6. How can you, as an individual, help maintain biodiversity worldwide? Choose at least two actions that you can take; justify your choices as they relate to biodiversity.

ANALYZING THE SCIENCE

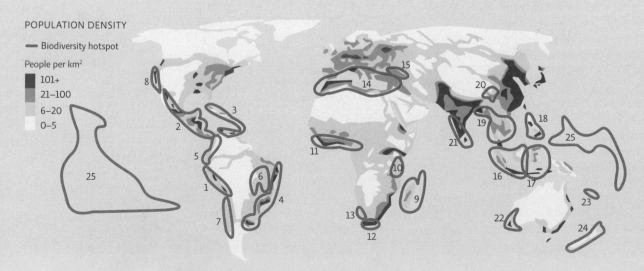

POPULATION DENSITY

— Biodiversity hotspot

People per km²
- 101+
- 21–100
- 6–20
- 0–5

Biodiversity hotspots are regions of the world that are of particular importance to global biodiversity both because of the total number of species they contain and because of the number of endemic species they have.

The map on the previous page shows both the location of biodiversity hotspots and human population density around the world.

INTERPRETATION

1. How many of the 25 hotspots include islands?

2. In one or two sentences, describe the relationship between population density and biodiversity.

3. Can you assume that biodiversity hotspots in highly populated areas are the most at risk? What other factors would you look for?

ADVANCE YOUR THINKING

4. Based on what you've already learned about human population growth, do you think that any hotspots that are not being threatened by development will become threatened in the next 50 years?

EVALUATING NEW INFORMATION

Scientists do not know how many species live on our planet. Our best estimate is that we have identified less than 20% of them. In the All Taxa Biodiversity Inventory (ATBI), scientists and citizen scientists are trying to catalogue all the species living in the Great Smoky Mountain National Park. Visit the Discover Life in America website (www.dlia.org/atbi/) to learn more about the ATBI project.

Evaluate the website and work with the information to answer the following questions:

1. Evaluate the organization and the ATBI project.
   a. What is the mission of Discover Life in America?
   b. What is your opinion of the ATBI project—is it a valid use of resources?
   c. Is this organization equipped to handle this challenge?

2. Who serves on the Science Advisory Panel? Choose two or three advisers and read about their accomplishments. Do you think these individuals are qualified for this role?

3. Open the "Science & Research" tab at the top of the page and select the "Taxa Tally" link. Look over the species list.
   a. Which groups have the most "New to the Park" species? Which have the most "New to Science" species?
   b. Does this surprise you or is it consistent with Infographic 10.2? Explain.

4. Open the "Education" tab at the top of the page and select the "Citizen Science" link.
   a. Would you consider participating in an ATBI activity if you were in the area? Does this page give you enough information to decide?
   b. What questions would you like to ask before committing to become a volunteer or donor? Where would you find answers to your questions?

MAKING CONNECTIONS

LET THERE BE DARK

Background: Sea turtles spend their adult lives at sea and there they face multiple threats, such as being caught in fishing nets or ingesting waste at sea. They are particularly vulnerable to things that threaten their ability to successfully reproduce. Females come ashore and lay eggs in pits dug in the sand; the eggs are covered in sand and the female returns to the sea. Hatchlings are on their own to dig out and make their way to the ocean. Most are eaten by predators before reaching the water or soon after, but the sheer numbers of hatchlings usually means that some make it to the sea and escape predation there. Though they hatch out at night, hatchlings orient toward light—moonlight reflecting on the water is likely to be the brightest spot in their vicinity. But human settlement is changing this. Bright lights from homes, streets, and parking lots attract the hatchlings, who scurry off in the wrong direction, usually succumbing to predators or exhaustion. Female turtles are also less likely to nest on beaches lit by artificial light, reducing the number of nesting beaches at their disposal.

Case: Your Florida beach town used to be relatively quiet but has recently seen an influx of tourists. The town is starting to develop; shops, restaurants, and rental beach homes are springing up close to the beach. You are concerned about the effect of increased lighting on the loggerhead and green sea turtles that nest each year on your beaches. The planning commission will be meeting soon and you have been invited to present a proposal on how to best deal with the lighting issue while still allowing the town to develop its tourist trade.

1. Research the two species of sea turtles that nest in your area. What are their requirements for successful nesting? Are these turtles endangered?

2. Research other communities that have successfully dealt with this issue. What steps have they taken to reduce the impact of lighting on nesting females and newly hatched turtles?

3. Prepare a written proposal and a PowerPoint presentation that outlines what your town could do to reduce the impact of lighting on sea turtles while still allowing development of the area. Include:

   a. An overview of the needs of the turtles and the problems created by beach lighting.
   b. The likely effect on the turtles and the community if no action is taken.
   c. A summary of what other towns have done to deal with the problem.
   d. A proposal of specific rules or procedures that could be enforced to prevent lighting issues.
   e. Your recommendations for educating the community members and the tourists who visit—what message will you share and how you will get the message across?
   f. Suggestions for ways to get local businesses and developers on board with your plan, and fund-raising options to finance the program.

CORE MESSAGE

Today, many species on Earth are in danger of becoming extinct predominately due to human impact. Protection plans can focus on individual species or entire ecosystems; both methods have proven successful. A combination of national and international efforts is helping address biodiversity loss.

GUIDING QUESTIONS

After reading this chapter, you should be able to answer the following questions:

→ What are the major causes of species endangerment and extinction today?

→ How do single-species conservation programs compare to ecosystem-based approaches?

→ What is conservation genetics and how can it contribute to the conservation of species?

→ What legal protections do endangered and threatened species have in the United States and internationally?

→ What options are available for financing conservation efforts and what role do consumers play in protecting biodiversity?

Elephants in Garamba National Park in the Democratic Republic of Congo. Conservation groups say poachers are wiping out tens of thousands of elephants a year in Africa, with the underground ivory trade becoming increasingly militarized.

A FOREST WITHOUT ELEPHANTS

Can we save one of Earth's iconic species?

Deep in the African rainforest, an elephant has just been killed. Her corpse is found by park rangers; the body has been poked through with spears, the cheeks and trunk have been eaten away by scavengers. The rest was taken by the poachers who killed her: the tail sliced off to be made into bracelets of black elephant hair, which are popular with tourists, the eyes to be sold to traditional Chinese medicine shops, where their purported medicinal powers fetch a high price. The biggest prize, though—indeed, the whole reason for the killing—were the tusks, two gleaming protrusions of pure ivory which have been hacked off with an ax and are now making their way, by looping, circuitous routes, to jewelry shops in Manhattan and Beijing.

Elsewhere in the forest, the victim's herd is grieving. Elephants exhibit complex behaviors and are known for both their elaborate communication networks and their uncannily human mourning rituals. Even when they are miles apart, the members of a family maintain regular contact, communicating with sounds too low for human ears. When a family member dies, they process past the body—marching in slow, single file (occasionally one elephant or another will place its foot over the heart of the deceased or prod the lifeless body with its trunk). They often remain with the deceased for days and may even cover the body with dirt and branches.

⊙ WHERE ARE THE ELEPHANT RANGES?

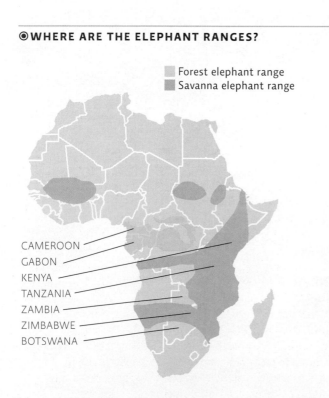

◼ Forest elephant range
◼ Savanna elephant range

CAMEROON
GABON
KENYA
TANZANIA
ZAMBIA
ZIMBABWE
BOTSWANA

Elephant poaching is at its highest level in decades, according to the Wildlife Conservation Society (WCS). In 2011 alone, authorities seized 14 large-scale ivory shipments totaling an estimated 24 tons of ivory—more than in any previous year. "Evidence is steadily mounting," says Tom Milliken, an elephant expert with WCS, and coauthor of a 2012 report detailing the crisis, that "African elephants are facing their most serious crisis since international commercial trade in ivory was first prohibited [in 1989]."

In some ways, the reasons for this uptick are obvious. Rising wealth and a growing market for elephant ivory in China (blocks made of ivory bearing a family's crest or signature are a popular status symbol throughout the country) and other nations are being abetted by a dramatic increase in road building in Africa (growing populations and rising consumption around the world have necessitated more and more resource extraction from previously untouched lands).

But the plight of forest elephants, and the saga of those racing to save them, is also part of a much larger, more complicated story—laced with grave dangers, daunting odds, and heroic efforts. It's a story in which the world's biodiversity—and thus life as we know it—hangs in the balance.

There are roughly five major causes of exinction and endangerment.

Conservation biology is the science of preserving biodiversity. Conservation biologists focus on protecting individual species and maintaining or restoring entire ecosystems. That means they also work intently to

↑ Congolese soldiers and rangers patrolling Garamba National Park discover a poached elephant in a remote area of the park.

understand the threats facing species and ecosystems. Today those threats are legion. As has been detailed in numerous chapters throughout this book, we humans are changing the environment with unprecedented speed—so much so that many populations cannot keep up. We're clearing more land and building more roads to access more natural resources than ever before. And we're killing off an alarming number of plants and animals in the process. The details for any given species may differ, but the basic story is often the same: As more and more individuals die off, the genetic diversity of that species dwindles and the population as a whole becomes increasingly vulnerable to extinction.

In general, conservation biologists agree that biodiversity the world over faces five main threats—threats tied to an expanding human population and increasing levels of affluence: overexploitation, pollution, climate change, invasive species, and—the number one cause—habitat destruction and fragmentation.

conservation biology The science concerned with preserving biodiversity.

"African elephants are facing their most serious crisis since international commercial trade in ivory was first prohibited."
—Tom Milliken

In recent decades, despite a rash of international attention to the problems of biodiversity loss, each of these factors has gotten worse. Overfishing has claimed some 85% of oyster reefs and as much as 90% of the world's populations of large predatory fish like tuna and cod (see Chapter 16); pesticides and mercury pollution have all but killed off the Mekong River dolphin; and climate change is threatening to bring the iconic polar bear (in the Arctic Circle), not to mention many other species around the world, to disastrous ends. Meanwhile, invasive species are running amok from Alabama to Zimbabwe, thanks to intentional and accidental introductions and human travel across the globe. And habitat fragmentation is isolating populations and limiting their ability to migrate

Habitat destruction and fragmentation

Humans change habitats on land and on river- and seabeds, both to harvest resources and to reclaim the area for agriculture or urban use. For example, in Southeast Asia, clearing forests for palm oil plantations has led to a loss of about 83% of bird and butterfly species. Habitat fragmentation splits up species' habitats, and many species have minimum space requirements. Forest elephants require large tracts of forest and are vulnerable to poachers in open, clear areas.

Climate change

A changing climate threatens species that cannot adapt or relocate to more suitable habitats. Species with specific habitat requirements (specialists) are particularly affected. For example, because ice cover on the Arctic Ocean has dropped as much as 34% in recent years, polar bears and ringed seals struggle to feed and reproduce.

Pollution

Pollution in the air and water is toxic to many species and damages habitats. For example, heavy pesticide use has brought the Mekong River dolphin to the brink of extinction.

Invasive species

Invasive species drive native species to extinction by outcompeting or preying on them. For example, the rapidly growing Japanese kudzu vine has spread over most of the American Southeast, blocking light and nutrients from native plants—some estimate that it is spreading by 150,000 acres a year.

Overexploitation

Humans overharvest and deplete species populations (on land and sea). For example, about 85% of oyster reefs have disappeared since the late 1800s due predominantly to overharvesting.

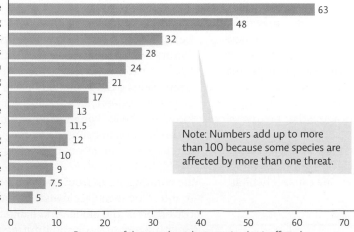

Threat	Percentage
Agriculture/aquaculture	63
Logging	48
Residential/commercial development	32
Invasive alien species	28
Pollution	24
Hunting/trapping	21
Climate change/severe weather	17
Change in fire regime	13
Dams/water management	11.5
Energy production/mining	12
Fisheries	10
Human disturbance	9
Transport/service corridors	7.5
Native species	5

Percentage of threatened vertebrate species that is affected

Note: Numbers add up to more than 100 because some species are affected by more than one threat.

← Conservation biologists agree there are 5 main threats to biodiversity. But each threat can be broken down into subparts. Here we show the subcategories of threats for vertebrates—the group that we know most about. This graph presents the percentage of threatened species according to specific threat.

or disperse to new areas. It is also making commercially valuable wildlife—not only elephants with their ivory tusks, but also tigers prized for their vibrant skins, and rhinos whose horns are used in traditional Chinese medicine—more vulnerable than ever to human predation. [INFOGRAPHIC 11.1]

To bring attention to the problem of **threatened species**—those that are at risk for extinction—the International Union for Conservation of Nature (IUCN) established the Red List of Threatened Species in 1963. The "Red List," as it is called, uses a series of designations to classify the seriousness of threat to a given species. Species are added to the list when they're at risk of becoming endangered, and removed when their status improves. The United States as a whole, its individual states, and other countries maintain their own lists of threatened species using the same designations. [INFO-GRAPHIC 11.2]

According to the United Nations Environmental Programme, one-third of all the plants and animals on Earth are now at risk of going extinct. And if we aren't worried about that, we should be. "Biodiversity... provide[s] a wide range of services to human societies," the WCS report's authors write. "Its continued loss, therefore, has major implications for current and future human well-being."

To avert the worst of those implications, we must figure out how best to preserve the biodiversity that's left. And to do that, we must first learn as much as possible about the ecosystems in question: What are the key requirements of its resident species? Are those requirements being met? If not, why not, and what other species are being threatened as a result? For conservation biologists working to save the forest elephants of Africa, answering those questions means, literally, following a trail of poop.

Human impact that threatens the forest elephant also puts its entire ecosystem at risk.

Steven Blake, an ecologist with WCS, followed several herds of forest elephants across several different wildlife refuges—in Congo, Cameroon, Gabon, and elsewhere—observing their behavior and collecting dung samples along the way for more than a decade. Here's what he learned:

threatened species Species that are at risk for extinction; various threat levels have been identified, ranging from "least concern" to "extinct."
keystone species A species that impacts its community more than its mere abundance would predict.

Infographic 11.2 | CONSERVATION DESIGNATIONS

↓ The International Union for Conservation of Nature (IUCN) maintains the Red List of Threatened Species, which identifies the conservation status of species worldwide, a process that helps conservationists focus efforts on the species most at risk. Status categories reflect extinction risk, and range from *extinct* to *least concern*; numbers indicate the number of species in that category as of October 2012.

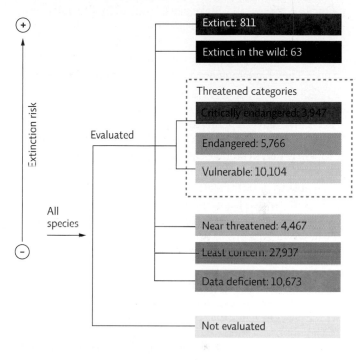

These elephants gobble up hundreds of pieces of fruit from a single tree. Then, as they walk on, they deposit the seeds of that fruit throughout the forest, with a generous helping of fertilizer. They can cross an entire national park in the space of 3 days, defecating roughly 50 times along the way. Each pile of dung contains thousands of individual seeds from more than a dozen different plant species. It may be gross, but it's no trivial matter. "Tropical forests are so diverse that a seed that lands near its parent plant has a suite of predators and pathogens waiting to nab it," Blake says. "So if you're a seed and you land under your parent, the probability of you surviving is almost zero." Without the elephants to disperse them, Blake says that plants with large fruits and seeds would eventually disappear. "The balance would be tipped in favor of those species that are dispersed by wind or other abiotic agents."

Of course, the impact elephants (or a lack of elephants) have on their ecosystem goes even further than that. In fact, the forest elephant is a **keystone species** (see Chapter 9): Not only does it disperse seeds and mobilize large amounts of nutrients with its feces, it also orders the physical structure of the forest—trampling vegetation, felling trees, and opening up the forest canopy in ways that facilitate plant growth and keep the ecosystem healthy. Savanna elephants also play a role as a keystone species. Without their grazing and trampling of young woody vegetation, the savanna grasslands that so many antelope and other species (and the predators who feed on them) depend on would quickly convert to shrublands. "Once they're gone," Blake says, "overall biodiversity will likely plummet."

But why exactly are elephants so imperiled in the first place? They have long been coveted for their ivory; why is poaching suddenly getting worse?

As it turns out, elephant feces provide researchers like Blake with some clues to that question as well. By diligently and systematically mapping the location of elephant dung across vast swaths of land, scientists have been able to detail the effects of newly created access points to various segments of forest. The elephants themselves proved elusive and difficult to count, but their dung provided a proxy census. The results of that census were a bit surprising.

"It's not the logging or mining that's doing the most damage," says Kate Evans, an animal behaviorist at the University of Bristol in the United Kingdom, and founder of the nonprofit Elephants for Africa. "It's the construction of roads." An endless network of new roads has cut its way into previously remote tracts of forest—slicing the landscape up into a patchwork of segregated fragments as they go. These roads have made elephant herds that were once protected by the dense maze of greenery easily accessible to poachers.

The elephants themselves are responding to the incursion. As Evans and others have documented, they travel 14 times faster than normal when crossing a road that is unprotected from poachers. Ultimately, as more and more roads crop up, their world shrinks. "If you put a 20-mile ring of death around your house, the chances are you won't want to go more than 15 miles from home," Blake

explains. "And if that ring closes in, you're going to feel besieged. You won't be able to go to the places you need to, you won't be able to see your friends, you will become imprisoned, and most likely the food will start to run out. It's just like that with forest elephants."

There are multiple approaches to species conservation.

While the causes of forest elephant endangerment (road building and ivory poaching) may be clear, the question of what to do about it is anything but. Indeed, experts have long debated how to best protect any given species or ecosystem. Early programs often took the **single-species approach**: They singled out well-known animals—known as **flagship species**—like pandas and condors, and focused on the specific threats those individual species faced, using a variety of methods, including captive breeding programs to increase population sizes and reintroducing the species to the wild. [INFOGRAPHIC 11.3]

In some ways, this approach has been a huge success: It has saved gray wolves, eagles, and even brown pelicans from complete decimation. And it's brought a great deal of attention and funding to conservation efforts. Indeed, the single-species approach is still employed today. Zoos around the world participate in *Species Survival Plans*, which use careful breeding programs to maximize genetic diversity. Among other things, this includes moving reproductive animals from zoo to zoo to introduce new genes into a breeding program or to minimize inbreeding.

The ultimate goal of these programs is to release animals back into the wild. But the tendency to focus mostly on those species (cute, furry, large) that are photogenic enough to capture the public's attention means that many, if not most, species in need of help fall through the cracks (people are more likely to donate money to "Save the panda!" than to "Save the white warty-back pearly mussel!").

As long as the main problem facing a species is low population size, reintroduction campaigns can be very effective, but this is insufficient if other threats—such as a degraded habitat or heavy poaching pressure—still exist. Indeed, such campaigns can give a false impression that we are saving many species, when in fact habitat destruction continues to decimate the vital habitat that supports innumerable species. For these reasons, many conservationists now support an **ecosystem approach**. This means identifying entire ecosystems—often *biodiversity hotspots*—that are at risk, and taking steps to restore or rehabilitate them. The goal of **ecosystem restoration** is to return the ecosystems to their original

---

single-species approach A conservation strategy that focuses on protecting one particular species.

flagship species The focus of public awareness campaigns aimed at generating interest in conservation in general; usually an interesting or charismatic species, such as the giant panda or tiger.

ecosystem approach A conservation strategy that focuses on protecting the ecosystem as a whole in an effort to protect the species that live there.

ecosystem restoration The repair of natural habitats back to (or close to) their original state.

remediation Restoration that focuses on the cleanup of pollution in a natural area.

Infographic **11.3** | **SINGLE-SPECIES APPROACH**

↓ Species Survival Plans, administered by the Association of Zoos and Aquariums, focus on increasing the population size and genetic diversity of threatened populations. Work goes on in zoos and aquariums as well as in the wild.

CAPTIVE BREEDING PROGRAMS

← Individuals for captive breeding are carefully chosen to maximize genetic diversity in the managed population. In 1987, the last 22 California condors were captured and became part of a captive breeding program. Chicks are fed with condor puppets so that they do not associate humans with food, to better facilitate their eventual release. As of 2012, there were 450 individuals, 266 in the wild.

FIELD CONSERVATION WORK

← Conservation groups work in the field to learn about and monitor species, and to protect or restore habitat. They also work with local residents to prevent human–wildlife conflicts. Here, a small antelope known as a dik-dik is released from a poacher's snare in Kenya.

REINTRODUCTION PROGRAMS

← Individuals from captive breeding programs or other wild populations can be introduced to an area if suitable habitat and protection is in place. The golden lion tamarin was downlisted from critically endangered to endangered thanks to the establishment of new populations from both reintroduction and relocation of animals into a new protected area in the coastal forests of Brazil.

states (or as closely to their original states as possible). It may involve reforestation projects, removal of non-native species, restoration of a natural river's flow patterns, or **remediation** (the cleanup of pollution); the restoration goal depends on the ecosystem in question.

In the Costa Rican rainforest, conservation biologist Daniel Janzen has worked for nearly 40 years restoring the area in and around Guanacaste National Park, forest land which had been converted to pastures and agricultural land. The largest forest restoration project of its kind, Janzen's plan involved protecting the area from invasive, non-native grasses and fire; replanting trees; and allowing pioneer species to move back into the cleared lands from neighboring forests. Through this process of secondary succession (Chapter 9), forest plant communities moved back into the area. All told, the project has restored more than 400 square miles of pasture land to forest and is now home to an estimated 235,000 species. While restoration programs like this are needed, efforts that prevent destruction in the first place are more likely to protect a greater number of species. The tropical dry forests of Guanacaste will take several hundred years to return to their original mature forest state, and we will never know how many species were lost on their way back to recovery.

Proponents point out that by focusing on the ecosystem as a whole, the entire community benefits—not just the species we knew were endangered, but also those that we didn't know were there. In order to implement an ecosystem approach to conservation, we must know how to identify when an ecosystem is in danger. Ecologists are developing metrics (factors that are measured) to assess ecosystem quality, such as *species richness* and *evenness* (see Chapter 9), soil health, water quality, plant community composition, and the abundance of non-native species.

Ecologists often monitor an **indicator species**—a species that is particularly vulnerable to ecosystem perturbations—to keep track of an ecosystem's health. A type of ecosystem conservation known as **landscape conservation** draws on this idea but instead of following one species, examines several indicator species in what is known as a *landscape species suite*. The species are specifically chosen to include a group that, together, uses all the vital areas within the ecosystem. The idea behind this approach is that if you monitor these species together and work to protect them as a group, you will simultaneously be protecting the entire ecosystem in which they reside, including all the species that live around them. [INFOGRAPHIC 11.4]

Either way, says Evans, the forest elephant is a good bet for conservationists. As a keystone species, it protects and facilitates an endless array of other species. A single-species approach that focuses on preserving the forest elephant will, by necessity, work to restore and protect its habitat and in doing so confer protection on the other

species that reside there. A similar strategy proved successful in the Pacific Ocean at the beginning of the 20th century. By the end of the 1800s, sea otter populations there had fallen drastically because of overhunting (otter pelts were highly valued); this led to an overpopulation of their prey, sea urchins, who feed on kelp. As sea urchin populations grew, the kelp forests were decimated, removing habitat and food for myriad species. After the International Fur Seal Treaty banned the hunting of sea otters in 1911, the otter populations recovered, which brought the sea urchin populations back in check, allowing the kelp to recover and benefitting all the species of the kelp forest.

As for the forest elephant—a highly intelligent mammal with a complex social system—its plight has captured human hearts, and thus shone a light on the oft-forgotten African rainforest. Still, to stop the ivory trade from obliterating this particular species, scientists will have to employ a range of tools, some of them much more complicated than dung.

National laws and international treaties aid in species conservation.

To Sam Wasser, a conservation biologist at the University of Washington, tracking the ivory trade is a bit like playing a shell game where one must guess which dish the bean is under. By sequencing ivory DNA from elephants all across Africa, Wasser's Seattle laboratory has mapped the ivory trade—from the slaughtered elephants in Africa to the curio shops in New York and Beijing. What Wasser's lab has found has astounded nearly everyone involved in elephant conservation. It turns out that countries regarded as veritable success stories—Gabon and Tanzania, for example—where poaching was thought to be all but eradicated, are actually poaching hotspots. "By shipping through all these completely not-intuitive routes," Wasser says, "the poachers evade detection, because you can't tell where it comes from." Moreover, virtually all of the ivory making its way to Wasser's lab originated in just a handful of places. "We used to think that so much ivory—even before the recent peak, authorities were seizing tons and tons of the stuff every year—we used to think it must be coming from all over the place, because how else could you possibly get that much ivory?" Wasser says. "But the DNA evidence showed that we were way wrong."

Wasser's work falls under the rubric of **conservation genetics**—the scientific field that relies on species' genetics to inform conservation efforts. Through DNA analysis, conservation biologists can determine the

---

indicator species The species that are particularly vulnerable to ecosystem perturbations, and that, when we monitor them, can give us advance warning of a problem.
landscape conservation An ecosystem conservation strategy that specifically identifies a suite of species, chosen because they use all the vital areas within an ecosystem; meeting the needs of these species will keep the ecosystem fully functional, thus meeting the needs of all species that live there.
conservation genetics Scientific field that relies on species' genetics to inform conservation efforts.
Convention on International Trade in Endangered Species of Wild Fauna and Flora (CITES) An international treaty that regulates the global trade of selected species.

↓ Protecting the entire ecosystem where an endangered species lives helps protect all the species who live there—even those that conservationists didn't know were endangered or threatened. One way to do this is to identify and meet the needs of a landscape species suite, a group of species that, between them, use most of the vital resources needed by other species in their ecosystem.

Forest elephants, chimpanzees, and mountain gorillas have been proposed as a landscape species suite for the northeastern forests of Congo.

Most of the other species in the ecosystem will benefit when elephants, gorillas, and chimpanzees—and their habitat—are protected.

Elephants and gorillas prefer slightly open forests with a wide variety of understory plants. Many other species also use these plants as food or habitat.

All three species avoid roads or rivers with heavy human use.

Elephants need the food, mineral, and water resources found near watering holes and wetlands. These areas also attract gorillas and other wildlife but not chimps.

Elephants need corridors wide enough to allow safe passage to other protected areas. This benefits gorillas, chimps, and other core species that will not use a narrow corridor.

Chimps prefer older, closed-canopy stands with many mature fruiting trees.

Elephants, gorillas, and other species that feed on fast-growing plants are attracted to logged areas that are regrowing. Chimps avoid these areas.

amount of genetic diversity within a population, or the kinship between separate groups—i.e., whether they are part of one extended population, or represent distinct populations that don't interbreed—or even whether a given population is part of an endangered species. [INFOGRAPHIC 11.5]

For example, scientists recently determined that the forest elephant and the savanna elephant of Africa are two different species. The former is several feet shorter than the latter, on average (8 feet versus 12 feet for full-grown males); eats much more fruit; has straighter, slimmer tusks and smoother skin that allow it to move easily through dense forests. The savanna elephant, on the other hand, being much larger and literally rougher around the edges, prefers wide-open grasslands. The two had been grouped together—by poachers, laypeople, and conservationists alike — for centuries. But it turns out they have no more genetic overlap than the Indian elephant and the now extinct woolly mammoth from which it evolved.

Such findings have huge implications for conservation biology. The **Convention on International Trade in Endangered Species of Wild Fauna and Flora (CITES)**—an international treaty that regulates global trade of selected species—banned the trade of African

Infographic 11.5 | TRACKING POACHERS BY USING CONSERVATION GENETICS

↓ Because elephant populations from different regions of Africa are genetically distinct, scientists can match the DNA from a confiscated tusk to the region in Africa where it originated, helping law enforcement officials track down poachers and identify regions of high poaching activity.

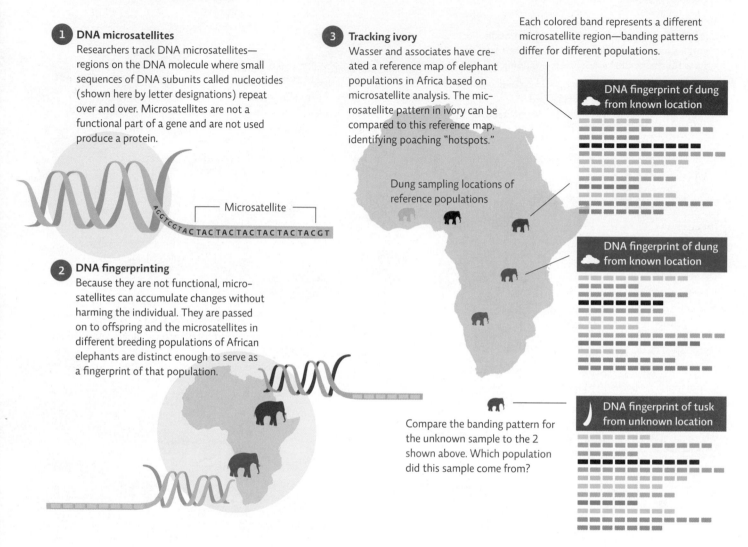

1 DNA microsatellites
Researchers track DNA microsatellites—regions on the DNA molecule where small sequences of DNA subunits called nucleotides (shown here by letter designations) repeat over and over. Microsatellites are not a functional part of a gene and are not used produce a protein.

Microsatellite

AGGTCGTACTACTACTACTACTACGT

2 DNA fingerprinting
Because they are not functional, microsatellites can accumulate changes without harming the individual. They are passed on to offspring and the microsatellites in different breeding populations of African elephants are distinct enough to serve as a fingerprint of that population.

3 Tracking ivory
Wasser and associates have created a reference map of elephant populations in Africa based on microsatellite analysis. The microsatellite pattern in ivory can be compared to this reference map, identifying poaching "hotspots."

Each colored band represents a different microsatellite region—banding patterns differ for different populations.

Dung sampling locations of reference populations

DNA fingerprint of dung from known location

DNA fingerprint of dung from known location

Compare the banding pattern for the unknown sample to the 2 shown above. Which population did this sample come from?

DNA fingerprint of tusk from unknown location

---

elephant ivory back in 1990. Since then, various governing bodies across Africa and Asia have been lobbying to ease those restrictions. So far, they have had little success, but with two distinct species, Blake says, they may finally have their way. "The forest elephant has been gravely imperiled by poaching," he says. "But the savanna elephant, by many accounts, is doing just fine. So, if they are two different species, CITES could now open the trade in savanna elephants." Demand would then rise, he says. And ivory prices would, too. "Rising demand and rising prices will make it even more profitable for black marketers to operate, and ultimately, the forest elephants will suffer even greater decimation."

There are also international treaties that protect species and ecosystems around the world. The 1992 **Convention on Biological Diversity (CBD)** attempts to reach even further than CITES, supporting conservation and sustainable use of all biological diversity, not just endangered species. But because CBD is less concrete than CITES—it does not outline any specific targets or provide a mechanism to reach its goals, leaving it up to each signatory to determine how best to proceed—it has been less successful.

Convention on Biological Diversity (CBD) An international treaty that promotes sustainable use of ecosystems and biodiversity.
Endangered Species Act (ESA) The primary federal law under which biodiversity is protected in the United States.

National laws also protect endangered species. The two main laws in the United States are the Marine Mammal Protection Act and the better-known **Endangered Species Act (ESA)**. Passed in 1973, the ESA mandates that listed species—those that have been officially declared threatened or endangered—be protected through a range of federally funded and scientifically proven strategies. Those strategies include the conservation of natural habitats as well as the breeding, relocation, and/or reintroduction of captive animals into the wild. The law has been controversial since its passage. To be sure, it's succeeded in bringing back several iconic species from the brink of extinction—the bald eagle and the American alligator, to name just two. But its effectiveness has also been perpetually hampered by a variety of constraints—including landowner disputes and what critics describe as colossal funding shortfalls. [INFOGRAPHIC 11.6]

Of course, CITES has troubles of its own—especially where the African forest elephant is concerned. Member nations meet every few years to review the status of various species and to modify their regulation if necessary, a process that critics say is as fraught as the ESA's with political maneuvering, budget woes, and other practical difficulties.

At a recent conference, Zambia and Tanzania petitioned CITES for permission to sell their stockpiled ivory—that which had come from elephants that died naturally, or that were killed before the ban was implemented. The ivory was just sitting there, the countries' representatives argued. The elephants were already dead, and the countries themselves could really use the money.

In the past, several African nations had, with CITES' permission, auctioned off their confiscated and stockpiled ivory (some of which had come from culled animals—those killed off deliberately when herds exceed park capacity). Selling this ivory was supposed to help elephant conservation efforts by reducing the market for illegally killed animals (and by increasing funds for conservation initiatives). But in the end, poaching only increased.

Opponents to the Zambian and Tanzanian request argued that such allowances would only bolster the illegal trade, which they said was already rampant in both the petitioning nations. Zambia and Tanzania denied the charge, insisting that poaching had been all but eliminated within their borders. The conference grew heated, and the two sides quickly reached an impasse.

And then Wasser presented his data: DNA analysis showing that much of the ivory seized that year—more than 60% of that which had been smuggled to ports in China and Thailand and New York—did indeed come from Tanzania and Zambia.

The petition was quashed.

Infographic 11.6 | **LEGAL PROTECTION FOR SPECIES**

→ A variety of national laws and international treaties (conventions) offer protection for species both inside and outside "protected areas." In the United States there are laws that specifically protect particular groups (e.g., wild horses or eagles) or specific habitats (e.g., wetlands), but the two broadest laws, both passed in the 1970s, are the main federal statutes that protect species. Internationally, there are many treaties, such as the ones mentioned here, that protect habitats or place restrictions on the use or harvesting of species.

U.S. NATIONAL LAWS	
Marine Mammals Protection Act (1972)	Protects all marine mammals (no killing, capture, or harassment without authorization)
Endangered Species Act (1973)	Mandates protection for "listed" species. Listing is a cumbersome process and many species don't make it to the list due to budgetary concerns rather than need.
INTERNATIONAL TREATIES	
Convention on International Trade in Endangered Species of Wild Fauna and Flora (1973)	Regulates the sale and trade of endangered or threatened species or products (175 signatory nations)
Convention on Biological Diversity (1992)	The 193 signatory nations agree to pursue goals of biodiversity conservation, sustainable use of biodiversity, and equitable sharing of genetic resources (crops and livestock).

Habitat protection requires local support.

Protected areas are clearly defined geographic spaces on land or at sea that are recognized, dedicated, and managed to achieve the long-term conservation of nature. Every nation in the world has protected areas. They include a range of designations. *National parks* are set aside primarily for human recreation. *Wildlife refuges* and *wilderness areas* are generally open to visitors but are not commercially developed (i.e., they have no restaurants, hotels, or other human accommodations). *Nature preserves* (also called nature or game reserves) are closed to hunting and fishing; their main goal is to protect wildlife.

Evidence is mounting that protected areas are helping. For example, after a serious coral bleaching event, Kanton Island reef in the South Pacific was designated a marine protected area. Its remarkable recovery (in less than a decade) is attributed to the protection of the entire ecological community (see Chapter 15). Likewise, the African white rhinoceros, a species once believed to be extinct, is thriving in protected areas; its conservation status has been upgraded to vulnerable. [INFOGRAPHIC 11.7]

As with legal safeguards like the ESA and CITES, protected areas have their limits. According to the WCS, elephant poaching is lower—and overall elephant populations are dramatically higher—in national parks and game reserves, especially in the eastern and southern regions of Africa where the savanna elephant is thriving. But most experts agree that protected areas are not enough to save much of anything in the long run. The physical condition of such reserves, and the type and amount of species that can survive inside them, are changing, says Evans. "Whether it's because of global warming, or road building, or other factors, we've already got huge percentages of elephants living outside these protected areas. And what happens then? When those protected areas become places that they can't live? We haven't yet thought about what we are going to do when that happens."

For its part, the WCS is urging private logging and mining companies to curb illegal hunting in their concessions, especially when those concessions are near protected areas. "We need to start moving away from that idea of distinct areas, towards a place of coexistence," Evans says. And that will never happen without the support of local communities.

Infographic **11.7** | **GLOBAL PROTECTED AREAS**

↓ Protected areas come in all shapes and sizes and vary according to what level of protection the area receives (e.g., a wildlife refuge that allows hunting of some species versus a nature preserve that does not allow any hunting). About 13% of land on Earth has some protected status (only 1.6% of the oceans are protected; see Chapters 15 and 16 for more on marine protected areas).

PERCENTAGE OF PROTECTED AREAS BY REGION (2011)

- Less than 10%
- 10%–30%
- 30%–50%
- More than 50%

Conservation plans should consider the needs of local human communities.

In Botswana's forests, peace seems to reign. Elephants, gathered in a clearing, play and groom and engage in all sorts of histrionics. According to Evans, the country boasts the largest population of forest elephants on the continent. To be sure, there are several reasons for this success: For one thing, the human population is much lower here than it is in countries like Kenya and Gabon, where poaching is epidemic. For another, Botswana's president has "led from the top," Evans says. "He's really made conservation a priority and that helps a great deal." But the main reason Botswana's forest elephants are doing so well is that the country's economy has come to rely on them. It turns out that **ecotourism**—low-impact travel to natural areas that contributes to the protection of the environment—is a big and thriving industry: Tourists will pay good money to see wildlife and well-preserved wild areas, especially if they know their money is helping conservation efforts. "The people of Botswana have made a very proactive decision," Evans says. "For their own long-term survival, they see wildlife as a crucial sustainable resource."

Therein lies the key. For conservation programs to work, they must protect not only the species in question, but also the humans who share its habitat.

A 2009 drought in Kenya, for example—one of the worst droughts in living memory—caused the region's tribes to lose many of their cows and crops; meanwhile, the price of ivory soared. Brokers just across the Tanzania border were paying around $20 a pound at the time for raw ivory—a deal too good for any starving family to ignore. The bushmeat trade in central Africa offers another example. The loss of agricultural land to environmental degradation and armed conflict has driven more area residents into natural areas to kill wild animals for meat, which in turn diminishes wildlife, increases access to the area, and spreads zoonotic diseases (see Chapter 6) as more people come into contact with wild animals. Steps to improve agriculture, increase access to nutritious food,

protected areas Geographic spaces on land or at sea recognized, dedicated, and managed to achieve the long-term conservation of nature.
ecotourism Low-impact travel to natural areas that contributes to the protection of the environment and respects the local people

↓ While the number of protected areas has increased dramatically in recent decades, from about 5 million km² in 1975 to almost 25 million km² in 2011, species extinction rates have not dropped; they are actually increasing as pressure outside of protected areas increases. Protection of the habitats of the most vulnerable species is a must if we want to effectively address species extinctions.

GLOBAL PROTECTED AREAS, 1911–2011

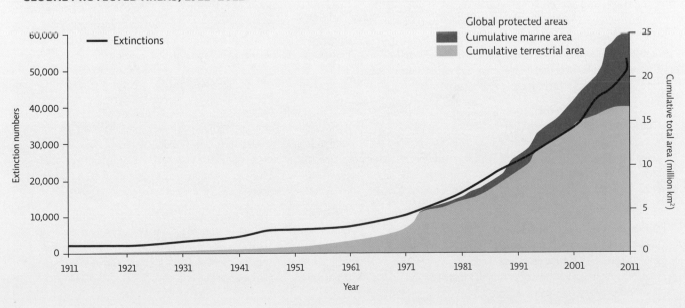

and resolve political disputes would do more to mitigate these problems than any law prohibiting bushmeat.

Ecotourism is not the only means to such ends. **Debt-for-nature swaps**—a wealthy nation forgives part of a developing nation's debt, in return for which the developing nation pledges to protect certain ecosystems—are another way to appeal to the needs of those living nearest the species and ecosystems that are so embattled (so far, almost $3 billion in debt has been forgiven, and tens, if not hundreds, of millions of acres of wilderness have been preserved through such programs). Nonprofit

---

Infographic **11.8** | **MANY ROUTES TO CONSERVATION**

↓ Conservation of biodiversity does not have to depend solely on regulations, laws, and treaties. There are other effective approaches to protecting species and their ecosystems. Some of these are market driven (there is a financial incentive to pursuing them), while others rely on individuals to voluntarily give of their time or money.

METHOD	EXAMPLE
Ecotourism Low-impact travel to natural areas that contributes to the protection of the environment and respects the local people.	A community, region, or nation may find that tourists who come to see intact ecosystems bring in more money than would be gained by harvesting resources. Ecotourism in Costa Rica brings in more than $3 billion annually.
Valuing ecosystem services Assessing the monetary value of an intact ecosystem to inform how best to use an area.	In Thailand, the value that a mangrove swamp provides in terms of local harvestable goods and value to offshore fisheries is about 6 times more than the economic value of shrimp farms that displace the mangroves. If the cost of replacing the mangroves is factored in, shrimp farming would cost money, not produce it.
Debt-for-nature swaps The preservation of natural areas is funded by an agreement in which a nation forgives part of the debt of a developing nation, or a nonprofit agency pays off the debt, in return for the developing nation's pledge to put a percentage of that saving toward conservation projects.	So far, more than $3 billion in debt has been forgiven in these programs, with almost $1 billion funding conservation in debtor countries. Between 1991 and 1998, the United States forgave $1.4 billion in debt for seven South American countries, generating around $170 million in conservation funding.
Nonprofit organizations Voluntary membership supports a large number of conservation-oriented organizations, which employ thousands of scientists, educators, legal advisors, and policy experts around the world.	The World Wildlife Fund (WWF) was the first wildlife-focused organization. In 2011, 85% of its donations went to conservation projects. The Sierra Club was founded in 1892 and has many college chapters; it seeks to protect American wilderness and promote environmental protection.
Land trusts and conservation easements Land trusts are nonprofit organizations set up to protect private land under their care. Conservation easements represent a legal agreement between a landowner and land trust or government that permanently limits land development and restricts use of that land to help preserve its conservation value (see Chapter 14).	Using the money from public donations, the nonprofit organization Nature Conservancy has purchased about 15 million acres in the United States, permanently protecting that land from future development. It also funds programs worldwide that protect another 100 million acres.
Consumer choices Consumers can influence actions taken by businesses and industries by purchasing products obtained in a way that does not harm species, or better yet, that helps conservation efforts. They can also vote with their dollars by choosing not to buy products whose acquisition harms biodiversity.	Avoid purchasing exotic pets (they may be wild-caught) or any items made from endangered species, such as ivory products, animal skins, and collectable specimens (e.g., butterflies under glass). Buying fair trade, shade-grown coffee and chocolate is another way to promote habitat protection and protect biodiversity in the regions where the crops are grown.
Opportunities for citizen scientists Volunteer work is an important part of keeping natural areas in good shape. There are a wide variety of educational, research, and wildlife/habitat management opportunities that allow individuals to contribute to conservation work.	The All Taxa Biodiversity Inventory of the Great Smoky Mountains National Park and the Chesapeake Bay Program, aimed at restoring the bay (see Chapter 18), are just two examples of robust programs that allow citizens to pitch in and help in conservation efforts.

organizations have also been major players in conserving species and their habitats. [INFOGRAPHIC 11.8]

For his money, Blake says better enforcement of existing laws, and a marketplace that reflects the true costs of natural resources, would go a long way toward curbing poaching. "Bad environmental companies need to be penalized and good ones rewarded," he says. Yes, this will drive the cost of certain goods up—low prices cannot be maintained if companies have to invest in good road planning, stop poaching, and other environmentally and socially friendly practices. But, as with any natural resource—from water to timber to coal and oil—higher prices might be the key to better conservation.

However, Evans says, education may be the most crucial element of all. "We want the communities living in close proximity to elephants to understand that they provide crucial ecosystem services," she says. "We also want people in developed countries to understand that even if they never set foot on the African continent, or see an elephant up close, their decisions have a direct impact on these ecosystems."

Because in the end, it's demand from these countries—not only for ivory products, but for timber and other resources, that's driving and facilitating the trade.

Ultimately, then, the story of the forest elephant ends not in a shrinking African forest, or even in a Chinese port city, but much closer to home: in a jewelry shop, in a bustling Manhattan neighborhood. That's where federal agents recently discovered $2 million worth of illegal ivory. In 2012, the shop's owners pled guilty to selling ivory without permits; under plea bargain, they agreed to forfeit the products and to pay $55,000 in donations to the WCS for use in elephant conservation and anti-poaching efforts. It was a rare victory for conservationists, one that underscored a crucial point for Wasser, Evans, Blake, and others. "It's a global problem that requires very local solutions," Evans says. In the end, preserving biodiversity—even in the jungles of Africa—begins right here at home.

References used in this chapter:
Baillie, J.E.M, et al. 2010. *Evolution Lost: Status and Trends of the World's Vertebrates.* London: Zoological Society of London.
Blake, S., et al. 2007. *PLoS Biology,* 5: e111. doi:10.1371/journal.pbio.0050111
Stokes, E.J., et al. 2010. *PLoS ONE,* 5: e10294. doi:10.1371/journal.pone.0010294
United Nations Environment Programme. 2012. *Global Environmental Outlook-5 Report,* http://unep.org/geo/pdfs/geo5/GEO5_report_full_en.pdf.
Wasser, S.K., et al. 2008. *Conservation Biology,* 22: 1065–1071.

debt-for-nature-swaps A wealthy nation forgives the debt of a developing nation in return for a pledge to protect natural areas in that developing nation.

BRING IT HOME

⊙ PERSONAL CHOICES THAT HELP

A recent study estimates that nearly 9 million different species are found on Earth. With such a stunning amount of biodiversity, it is hard to imagine that we could possibly be in danger of losing a significant number of these species to extinction. Yet the extinction rates will continue to grow if we do not consider serious changes.

Individual Steps
→ Gain an appreciation for the tremendous diversity we have by watching a documentary of an ecosystem far away from you. Make a list of all the species featured in the documentary, then check their status at www.iucnredlist.org/.
→ Avoid purchasing any pets that are listed as endangered species, even those bred in captivity, as this can increase the threat to the species in the wild. If you do choose to buy an exotic pet that is not an endangered species, request certification that it is captive bred.
→ Consider contributing to an organization that works to protect species or their habitat; check out options at www.fws.gov/endangered/what-we-do/ngo-programs.html.

Group Action
→ Identify county and state parks and preserves around you. Find out what specific actions they take to maintain the biodiversity of the ecosystems there, such as removing invasive species and cultivating native ones. Form a volunteer group to assist in these efforts.
→ Discover which endangered species live in your state at www.fws.gov/endangered/. Make flyers and posters with images and information about these species and their habitats. Share them around your school or community.

Policy Change
→ The Endangered Species Act was the first U.S. legislation established to protect species diversity. To learn more about current challenges and updates to the program, visit epa.gov/espp.

UNDERSTANDING THE ISSUE

CHECK YOUR UNDERSTANDING

1. **Which of the following is the leading cause of species extinction?**
   a. Invasive species
   b. Climate change
   c. Overexploitation
   d. Habitat destruction

2. **The number of elephants being slaughtered has increased because:**
   a. logging and mining roads within elephant habitats allow poachers easier access.
   b. there is high demand for ivory in China and other parts of the world.
   c. during droughts, when cattle and crops are lost, poaching is an alternative source of income.
   d. All of the above.

3. **How does an ecosystem approach to conservation differ from a single-species approach?**
   a. The ecosystem approach has been more successful at attracting the public's attention to conservation needs.
   b. The ecosystem approach focuses on a single "charismatic" species—large, furry, and photogenic.
   c. An ecosystem approach involves restoring and protecting an entire habitat and all the species within it.
   d. All of the above.

4. **Conservation genetics:**
   a. relies on analysis of species' DNA to make conservation decisions.
   b. is used only for captive breeding of endangered species.
   c. is relevant for single-species approaches to conservation, but not ecosystem approaches.
   d. is used by poachers to identify species.

5. **CITES:**
   a. regulates the hunting and/or gathering of endangered species.
   b. regulates international trade in endangered species.
   c. supports measures to protect biodiversity but doesn't specify what should be done.
   d. allows some international trade in elephant ivory each year.

6. **Which of the following is true of the Endangered Species Act?**
   a. It is very well funded by the government, due to the public's strong support for it.
   b. It has not been effective in saving any species from extinction.
   c. It is controversial due to its restrictions on land owners.
   d. It mainly protects endangered species outside the United States.

7. **Protecting a mangrove swamp rather than developing the area as a shrimp farm can actually be the better long-term financial choice. This is an example of:**
   a. ecotourism.
   b. debt-for-nature programs.
   c. a land trust.
   d. valuing ecosystem services.

WORK WITH IDEAS

1. Which of the five main threats to biodiversity identified in Infographic 11.1 are endangering the African elephant? Explain how the decline of the elephant affects other species who share its ecosystem.

2. If forest and savanna elephants are different species, how might this affect strategies to protect them from extinction? Would this have a positive or a negative effect on elephant populations? Explain your reasoning.

3. Local community members are protesting a proposed development because it would destroy a population of sunflowers they believe is a listed endangered species. Developers disagree and claim that this flower is found throughout the area and is not endangered. How would conservation genetics help settle this dispute?

4. How can CITES protect endangered species if it only regulates international trade? Do you think this is sufficient to protect endangered species? Why or why not?

5. Identify some actions you could personally take that would contribute to conservation.

ANALYZING THE SCIENCE

The International Union for Conservation of Nature (IUCN) has maintained an inventory of the extinction risk for the world's species since 1963, called the "Red List."

INTERPRETATION

1. Describe what the graph shows about the proportion of cycad species in each category.

2. Compare the proportion of threatened (critically endangered, endangered, vulnerable) amphibians to birds.

IUCN graph

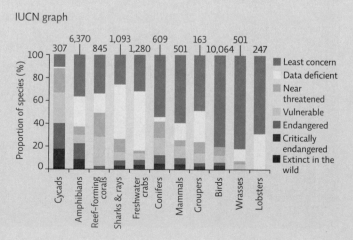

3 Compare the number of threatened amphibians to birds (Hint: The total number of species per group is at the top of the bar in the graph).

ADVANCE YOUR THINKING

4. Identify the three groups that contain the highest proportion of species that are "data deficient." Why do you think less is known about species in these particular groups? How would you predict that extinction risk estimates would change if more were known about

these species—that is, would a higher proportion be threatened or would more be of "least concern"?

5. Many North American species of birds are migratory, spending winters in tropical regions (such as Central or South America) and summers in North America. Habitat destruction from deforestation is one of the most common threats many of these species face. Suppose you spearheaded a program that returned adequate habitat to the migratory birds in your area. Why might this not be enough to allow the species to recover?

EVALUATING NEW INFORMATION

Everyone has limited time and money, and choosing how to invest that time and money to support biodiversity issues can be complex. Should you fund an organization that protects individual species or ecosystems? How much money is reasonable to spend on administration and fundraising? What conservation focus do you prefer?

Evaluate the website and work with the information:

1. Visit the World Wildlife Fund (WWF) website (www.worldwildlife.org) and the Nature Conservancy website (www.nature.org). For each, answer the following questions:
    a. What is the mission of the organization?
    b. Is the website up to date? Does it appear to be accurate? Reliable? Explain.
    c. Does the organization use a single-species conservation approach or an ecosystem approach? Support your answer.
    d. What does the organization claim about the percentage of its operating budget it spends on programs versus on other costs? What evidence does it provide to support this claim? Is it sufficient?

    e. How long has the organization been operating? What are some of the successes it claims?

2. Go to Charity Navigator (www.charitynavigator.org).
    a. Search for World Wildlife Fund. Does the information on the percentage of the operating budget spent on programs, fundraising, and administration match the claims of the WWF website? If there is a discrepancy, why might that be? Which source would you trust, and why?
    b. Search for Nature Conservancy. How does the budget information on the Charity Navigator site compare to that on the Nature Conservancy site?
    c. How does the Nature Conservancy compare to WWF in terms of money spent on programs versus on fundraising and administration? Be specific.
    d. Would these differences lead you to choose to donate to one organization over the other? Why or why not? What other information would you use to make a decision about which organization to support?

MAKING CONNECTIONS

ECOTOURISM: BIODIVERSITY SAVIOR OR BIODIVERSITY THREAT?

Background: Ecotourism is one of the fastest-growing sectors of the tourism industry. However, not all ecotourism is created equal. That "feel good" travel experience may carry a large carbon footprint (eco-tourists may travel thousands of miles by air and car), increase habitat destruction and pollution (with Western travelers' higher expectations for food, plumbing, and roads) and even introduce invasive species (ecotourists have been known to inadvertently distribute seeds on their hiking boots and gear).

Case: Is it possible to travel and still be ecologically responsible? Leading ecotourism organizations have proposed some guiding principles to help ensure that ecotourism is handled responsibly. Go to the International Ecotourism Society (TIES) website (www.ecotourism.org). Under the "About" drop-down menu, select "What is Ecotourism" and find their "six principles of ecotourism."

You will research and produce a "grade card" of 4 to 5 ecotourism destinations or tours. The guide should be for travelers embarking from the United States; include at least one destination in the continental U.S. and at least one on another continent. Your report should include:

1. A rubric for assigning grades for each of the six TIES ecosystem principles. The rubric should include a brief sentence describing the grade, as shown in this example for the first principle (minimize impact).

A	No measurable impact of the visit during travel to & from the site, nor of the activities on site
B	Minor impacts of the visit during travel to & from the site or of the activities on site
C	Minor impacts of the visit during both travel to & from the site and of the activities on site
D	Significant impacts of the visit during travel to & from the site or of the activities on site
F	Significant impacts of the visit during travel to & from the site and of the activities on site

2. For each destination or tour provide:
    a. A grade for each of the TIES principles, and a total grade for the overall destination or tour.
    b. A brief discussion of the highest grade received.
    c. A brief discussion of the lowest grade received.
    d. Any further information about conservation or social impacts of the destination/tour that may not have been covered by the TIES principles, including unintended and secondary consequences.
    e. Your recommendation: Is this trip an acceptable ecotourism destination or tour?
    f. Any suggestions you might have for mitigating the impact of the destination/tour (e.g., purchasing carbon offset credits).

3. Of the destinations/tours you researched, which would you most like to participate in? Explain.

CORE MESSAGE

The variety of life on Earth is a result of natural selection favoring those individuals within populations that are best able to survive in their particular environment. Given enough time, some populations may be able to adapt to environmental changes. Extinction is a natural part of this process, as less adapted populations are eliminated by better competitors or as populations are lost due to natural disasters. The pace of extinctions, however, has accelerated in recent decades and human impact may now be ushering in what many see as the sixth mass extinction.

GUIDING QUESTIONS

After reading this chapter, you should be able to answer the following questions:

→ What is evolution and how does natural selection allow populations to adapt to changing environments?

→ What factors other than natural selection influence the evolution of a population?

→ Why do some scientists say that we are currently in the middle of the "sixth mass extinction"?

→ How do humans mimic the mechanisms of natural selection for their own purposes?

→ What are some common misconceptions about evolution?

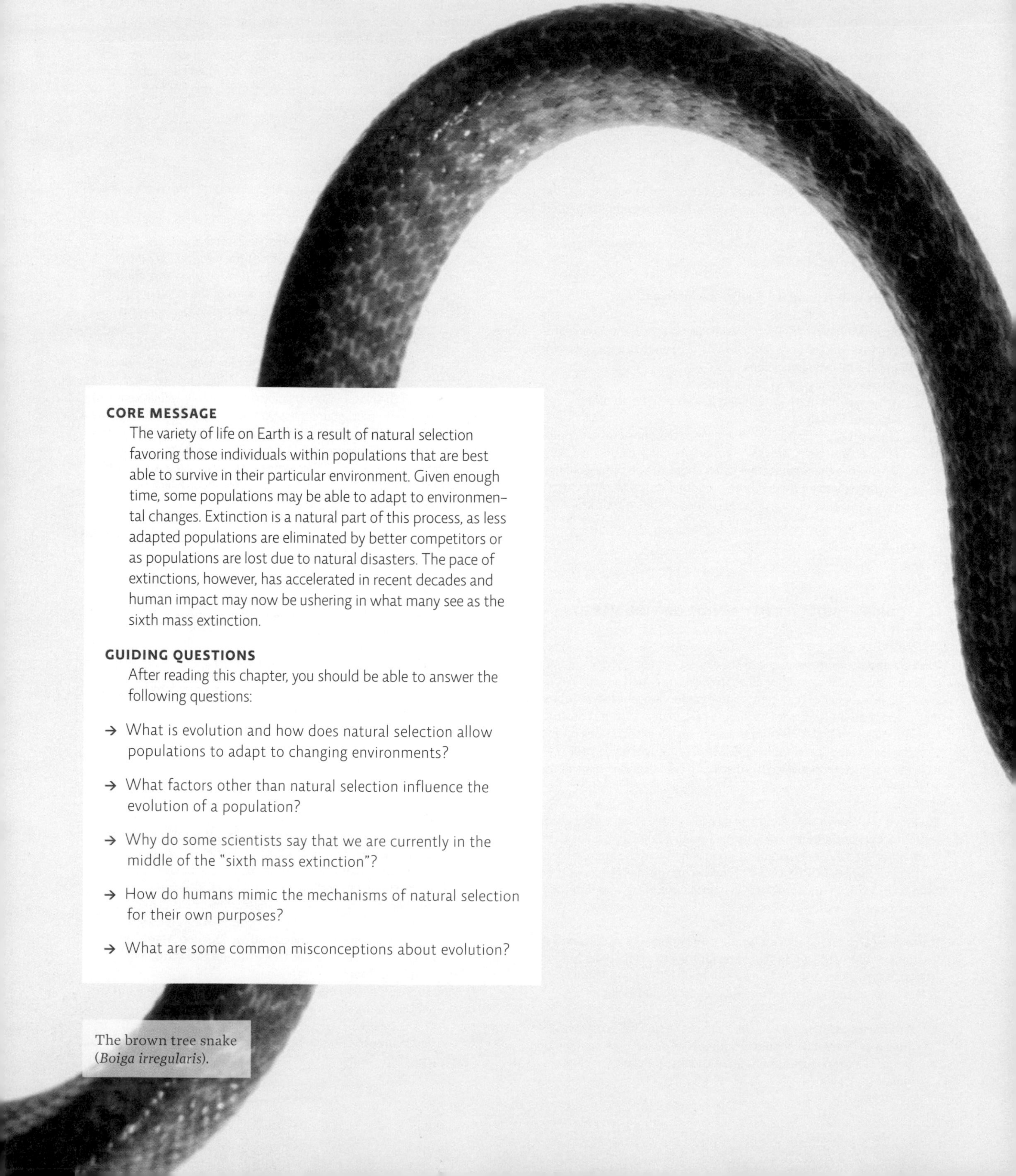

The brown tree snake
(*Boiga irregularis*).

CHAPTER 12 EVOLUTION AND EXTINCTION

A TROPICAL MURDER MYSTERY

Finding the missing birds of Guam

↑ A brown noddy (*Anodus stolidus*) guards an egg. The brown noddy nests in numbers on nearby Cocos Island but has not successfully nested on Guam since snake populations peaked in the 1970s and 1980s.

→ Dr. Julie Savidge holding a Mariana Fruit-Dove. This species only occurs on certain islands within the Mariana Islands and the last sighting on Guam was in 1985. This bird was caught as part of an early blood-sampling effort to see if exotic diseases might be causing the bird decline on Guam.

Something bizarre began happening on Guam, the southernmost island in the Mariana island chain nestled halfway between Japan and New Guinea, in the late 1960s. The island (a U.S. territory), where once some 18 native avian species filled the tropical forests with song, became eerily quiet as the nation began losing its bird populations. By the early 1980s, four species of birds had gone extinct; the first documented loss was the Guam bridled white-eye, a small bird that flocked in tree canopies, feeding on nectar and small insects. Ten others were in danger of extinction. Worse, researchers and wildlife experts did not have a clue why they were dying.

Back in the United States, in the winter of 1980, biologist Julie Savidge was gearing up to begin Ph.D. work at the University of Illinois when she by chance attended a lecture in Redding, California, organized by the Wildlife Society. A biologist from Guam had been invited to talk about the islands' bird disappearances, and he noted—rather grimly—that no one had yet solved the devastating mystery. Savidge was immediately enthralled. "I went up to him afterwards and said, 'This sounds fascinating. Is there any way that I might be able to get out to Guam?'" Savidge recalls.

Two summers later, after much back-and-forth between Savidge and the Guam Division of Aquatic and Wildlife Resources, Savidge was hired to investigate the bird disappearances as part of her Ph.D. dissertation.

At the time, Guam's scientists were convinced that diseases like blood parasites or pesticides were to blame for the avian losses. So Savidge began looking into those possibilities. But as soon as she started talking to Guam locals, she became skeptical of the idea. "When I would interview the natives, I'd be asking all these disease questions, and they'd say, 'Why are you asking about this? The real problem is the snakes,'" she explains.

The locals were certain that brown tree snakes (*Boiga irregularis*), 1- to 2-meter-long (3- to 6-foot-long), non-native snakes that had been accidentally introduced to the islands in military shipping cargo in the late 1940s and 1950s, were responsible for the birds' demise. Non-native species that cause ecological, economic, or human health problems and are hard to eradicate are considered

invasive species, and they can cause significant damage in areas they invade.

Oddly, Savidge's research was starting to show that, in fact, diseases weren't playing a factor in the bird deaths at all. When she and her colleagues sampled a variety of birds for bacteria, viruses, and parasites between 1982 and 1985, they found that the native birds were basically healthy. Curious about the tree snake hypothesis (an idea other scientists had dismissed), Savidge began to investigate, on her own time, whether these reptiles might be causing the **extinctions**. Surely it was something else—could a few snakes really obliterate a whole island of birds?

◉ **WHERE IS GUAM?**

---

invasive species A non-native species (a species outside of its range) whose introduction causes or is likely to cause economic or environmental harm or harm to human health.

extinct/extinction The complete loss of a species from an area; may be local (gone from an area) or global (gone for good).

Natural selection is the main mechanism by which populations adapt and evolve.

Before they started disappearing, the birds of Guam were a diverse and resplendent bunch. The island, about a fifth the size of Rhode Island, was home to 18 native species of birds, each specially suited to life on the island.

Populations usually contain individuals that are genetically different from one another. According to the evolutionary theory first put forth by Charles Darwin and Alfred Russel Wallace, and subsequently supported by a tremendous amount of evidence from a wide variety of scientific disciplines, **selective pressures** on the population—nonrandom influences affecting who survives or reproduces—favor individuals with certain inherited traits over others (such as better camouflage, tolerance for drought, or enhanced sense of smell). This can lead to *differential reproductive success* among members of the population if individuals who are better adapted leave more offspring than other individuals.

The traits that an environment favors are called **adaptations**, and the process by which organisms best adapted to the environment survive to pass on their traits is known as **natural selection**. Evolutionary biology helps us understand the diversity of life on Earth and how populations change over time. It is one of the pillars of biological science, and the vast amount of evidence in support of both the occurrence of evolution and the mechanisms by which it happens has elevated this explanation to the level of scientific theory (see Chapter 2).

For most populations, more offspring are born than can survive, since resources are limited and many species produce large numbers of young. Since only some individuals will survive, over time, the population will contain more and more of these better-adapted individuals and their offspring. Ultimately, this changes how common certain variants of **genes** (these variants are called *alleles*) are in the population—the frequency (percent in the population) of some alleles increases and that of others decreases. When this occurs, the population has experienced **evolution**, or changes in the **gene frequencies** within a population from one generation to the next. Natural selection may be *stabilizing*, *directional*, or *disruptive*, depending on which genetic traits are favored or selected against. [INFOGRAPHIC 12.1]

It is important to note that *individuals* are selected for, but *populations* evolve; individuals do not change their own genetic makeup to produce new necessary adaptations, such as bigger size or pesticide resistance. If they get the opportunity to reproduce, they pass their traits on to the next generation. If they cannot tolerate environmental changes, as was the case with the first bird species to disappear from Guam (the bridled white-eye), they die or fail to reproduce and do not pass on their genes. Individuals may be able to adjust their behavior to accommodate environmental changes, but if a trait is not genetically controlled, and therefore is not heritable, it will not influence the composition of the next generation.

Populations need genetic diversity to evolve.

The ability of a population to adapt is a reflection of its tolerance limits, which largely depend on **genetic diversity**—different individuals having different alleles. A population that is highly diverse (has individuals with many different traits) is likely to have wider tolerance limits, which increases the population's potential to adapt to changes. This means it is more likely that some individuals will exist that can withstand (or even thrive in) the changes, and that the population as a whole will survive. If a change occurs that produces a condition outside of the range in which individuals can survive and reproduce (for instance, if the climate becomes too warm or too dry), the population will die out. Similarly, if a new challenge is presented, such as the introduction of a new predator or competitor, the survival of the population will depend on whether there are any individuals in the population who can effectively deal with the new species. If the snakes on Guam were indeed responsible for killing off the birds, any birds that happened to have effective snake-avoidance behaviors would have had a greater chance of survival.

There are two main sources of variation that can increase genetic diversity in a population. The ultimate source of new variability is genetic *mutation*, a change in the DNA sequence in the sex cells that alters a gene, sometimes to the extent that it produces a new protein, and possibly a

---

selective pressure A nonrandom influence affecting who survives or reproduces.

adaptation A trait that helps an individual survive or reproduce.

natural selection The process by which organisms best adapted to the environment (the fittest) survive to reproduce, leaving more offspring than do less well-adapted individuals.

genes Stretches of DNA, the cell's hereditary material, that each direct the production of a particular molecule (usually a protein) and influence an individual's traits.

evolution Differences in the gene frequencies within a population from one generation to the next.

gene frequencies The assortment and abundance of particular variants of genes (alleles) relative to each other within a population.

genetic diversity The heritable variation among individuals of a single population or within the species as a whole.

Infographic **12.1** | **NATURAL SELECTION AT WORK**

↳ When the environment presents a selective force (a new predator, changing temperatures, change in food supply), natural selection is the primary force by which populations adapt. The survivors are those who were lucky enough to have genetic traits that allowed them to survive in their changing environment (others who did not possess the trait were not as likely to survive to reproduce). Because survivors pass on those adaptations to their offspring, the gene frequencies of the population change in the next generation, which means some traits are more common and others are less common than they used to be. When this happens, the population is said to have evolved.

Original population	Next generation	Later generation

Beetles resting on this tree vary in color.

Individuals with the less favorable trait (coloration that makes them stand out on a tree trunk) are more likely to be eaten.

Fewer dark individuals are born (though recombination might produce some from light or tan parents).

Over time, the population may be mostly or solely made up of tan individuals.

Genetic variation exists in the population: Individuals possess inherited differences.

Differential reproductive success: Not everyone will survive to reproduce.

Gene frequencies have changed: The population is evolving.

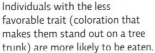

Number of individuals →

Population of beetles of different colors

— Original population
— Evolved population

← Natural selection can have different outcomes, depending on what varieties in the population the environment favors or selects against.

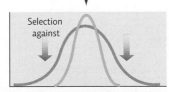

Selection against

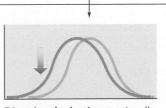

Stabilizing selection favors the norm and selects against extremes.

Directional selection continually favors a particular extreme of the trait (bigger, darker, etc.).

Disruptive selection favors the extremes but selects against the intermediate forms.

 All trees are tan; tan beetles are favored.

 In areas with trees darkened from pollution, darker beetles are favored.

 In a forest with light and dark trees but no tan trees, tan beetles (the intermediate color) are not favored.

new trait. Mutations are rare, but because DNA replication and repair occurs all the time, rare events do happen. When a mutation produces traits that are beneficial, they can quickly be passed on to the next generation, allowing the population to evolve to be better adapted to its environment. A second source of genetic variety occurs as eggs and sperm are made: *Genetic recombination* shuffles alleles around and sometimes produces individuals with new traits when a sperm fertilizes an egg.

The value of this genetic diversity is illustrated today in the example of the rock pocket mouse of the desert in the American Southwest. Animals of this species have coats that are either light tan or a darker color. It turns out that coat color corresponds to a population's environment—areas of light-colored rock contain populations with mostly tan mice, whereas darker mice inhabit black lava rock regions. Research by Michael Nachman and colleagues at the University of Arizona has shown that coat color is determined by a single gene that comes in

two "versions" (two different alleles). The dominant allele is designated by the letter D; the recessive allele by the lower case letter d. All individuals have two copies of the gene, and the color of their coat is determined by which two alleles they possess. Darker mice have at least one dominant allele (DD or Dd). Tan mice possess 2 recessive alleles (dd).

It is likely that coat color provides camouflage and protection from visual hunters, but only if the mouse is on a background of the same color. A study on deer mice (a similar species) showed that predatory owls are more successful at capturing mice on a contrasting background. This gives support to the conclusion that coat color is adaptive as camouflage and therefore is responsive to natural selection. [INFOGRAPHIC 12.2]

Another type of natural selection is known as **coevolution**. In coevolution, two species each provide the selective pressure that determines which of the other's

---

Infographic **12.2** | **EVOLUTION IN ACTION**

↳ Natural selection produces populations with different gene frequencies (more or less of a particular gene variant or allele). For this to occur, there must be genetic variation (more than one allele for a given trait) and a selective pressure (a reason one variant is better than another in a given situation).

Different color morphs of the rock pocket mouse (*Chaetodipus intermedius*) are found on different colored rocky outcroppings in the desert Southwest. An evaluation of the mice living on or near the Pinacate lava flow in southern Arizona represents the first documentation of the genetic basis (in this case, a single gene) for a naturally favored trait. The well-known peppered moth is another example in which different color variants are favored in different habitats but the genes responsible for that trait have not yet been identified.

Even though there is gene flow between dark and light populations that are close to one another, populations on tan rock have mostly tan individuals and populations on dark rock have mostly dark individuals, suggesting a strong selective pressure that favors one color over the other.

Predatory owls are likely the selective pressure that favors different coat colors in different habitats.

Gene flow

Tan mice (the recessive trait, dd) predominate in light-colored rocky outcroppings.

Darker mice (the dominant trait, DD or Dd) predominate on darker lava rocks.

traits is favored by natural selection. Predator and prey species usually evolve together, each exerting selective pressures that shape the other. As predators get better at catching prey, the only prey to survive are those a little better at escaping, and it is these individuals that reproduce and populate the next generation. This game of one-upmanship continues generation after generation, with each species affecting the differential survival and reproductive success of the other. The result can be a predator extremely well equipped to capture prey, and prey extremely well equipped to escape. [INFOGRAPHIC 12.3]

If the birds on Guam were indeed eradicated by the invasive snake species, it was because the speed at which the eradication happened prevented the bird populations from potentially coevolving survival strategies to deal with the new snake population. The brown tree snake was already well adapted to prey upon birds. But Guam's bird populations had never faced such a predator and had no natural defenses. It was an unfair fight. In fact,

that the birds on Guam disappeared so quickly made it difficult to tease out the cause of their extinction at all. Savidge had to work backward to solve the mystery, putting the missing pieces together, experiment by experiment. Her first step was to compare whether the distribution of the snakes on Guam matched up with the areas where birds had disappeared.

Based on what the residents told her, and what historical records like newspaper clippings reported, Savidge found that birds had begun to disappear first from southern Guam, and that their disappearances matched up perfectly with when the brown tree snakes had begun populating the area. Ritidian, an area on the very northernmost tip of Guam, on the other hand, was the

coevolution A process of evolution in which two species each provide the selective pressure that determines which traits are favored by natural selection in the other.

↓ A study done with dark and light varieties of deer mice revealed that owls caught twice as many opposite-colored mice (dark mice on a light background or tan mice on a dark background) as mice whose coloration matched their background, even in almost total darkness. Owl predation is therefore likely to be a strong selective pressure on coat color, driving directional selection that produces either light or dark populations of mice, depending on the background.

Infographic **12.3** | **COEVOLUTION ALLOWS POPULATIONS TO ADAPT TO EACH OTHER**

↓ As selection favors beetles closest to the tree color, only birds with the keenest eyesight feed well enough to survive and reproduce.

↓ Any beetle with an even better camouflage would escape predation and pass on its genes.

↓ This then favors birds with even keener eyesight who would feed well and pass on the sharp eyesight trait to their offspring.

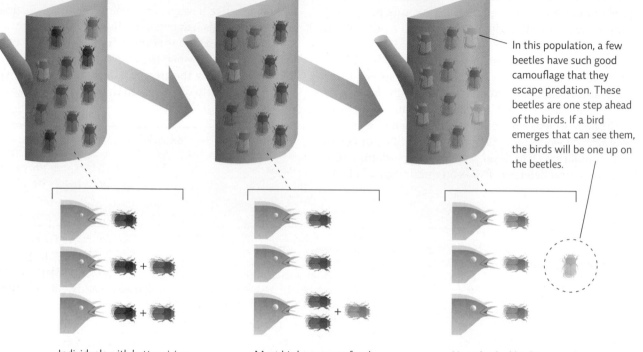

In this population, a few beetles have such good camouflage that they escape predation. These beetles are one step ahead of the birds. If a bird emerges that can see them, the birds will be one up on the beetles.

Individuals with better vision catch more beetles.

Most birds compete for the easy-to-see beetles, but any bird that can find the hidden ones will find more food—thus selection will favor birds with even better eyesight.

No individual bird is present (yet) who can detect the new, camouflaged beetle variant.

last area to have lost its birds, and it was also the last to be colonized by the snakes. "I found a close correlation between the bird decline and the expansion of brown tree snakes around the island," she says, which suggested to her that brown tree snakes really might be the culprit. [INFOGRAPHIC 12.4]

Some of Guam's bird species went extinct sooner than others. For instance, the **endemic** bridled white-eye, the gregarious bird species that was extinguished first, happens to be very small, raising the possibility that the small size of these birds might have put them at a disadvantage. Larger species like flycatchers survived longer, though they, too, eventually disappeared. Other bird species experienced **extirpation**; the Guam rail, for instance, is gone from Guam but other populations still live on the nearby island of Rota.

Populations can diverge into subpopulations or new species.

If even a few individuals of the bridled white-eye or other extinct species had been able to avoid the snake (perhaps due to a heritable trait that made them more wary of the predator), they might have given rise to new populations that could cohabitate with the snake. Or, had they not been surrounded by sea, some of the Guam birds may have been able to relocate and find a safe haven from the brown tree snake.

When populations diverge because of isolation, food availability, new predators, or habitat fragmentation such that their members can no longer freely interbreed, new species may arise (*speciation*). This increases species diversity in a community and sometimes produces

---

Infographic 12.4 | ENDANGERED AND EXTINCT BIRDS OF GUAM

↓ Of the 18 original native species on Guam, many are endangered or already extinct.

THE SPREAD OF THE SNAKE MAPPED WITH BIRD EXTINCTIONS OVER TIME

Last area to have all 18 native bird species — 1980s

All native species extinct in this area by 1984 — 1970s

GUAM

1960s

1950s

Bridled white-eyes and Guam flycatchers were last seen at this checkpoint in 1964.

Brown tree snake
Boiga irregularis
The snake's range expanded about 1.6 km (1 mi) per year.

Cocos Island
Boiga is absent from Cocos Island; all bird species are present.

↑ The loss of birds followed the spread of *Boiga* as predicted. The last area to have all 10 species of native birds was the last area *Boiga* invaded.

Extinct
Bridled white-eye
The endemic Guam bridled white-eye was last seen in 1983 and is now extinct.

Extinct
Rufous fantail
Once common all over Guam, the endemic subspecies of the Rufous fantail is now extinct.

Extinct
Guam flycatcher
The endemic Guam flycatcher was last seen on Guam in 1985 and is now extinct.

Only Captive
Guam kingfisher
By 1986, only seven Guam kingfishers existed in the wild; a captive breeding program is underway.

Only Captive
Guam rail
The endemic Guam rail is extinct in the wild; a captive breeding program is underway.

In other islands
Mariana Fruit-Dove
The Mariana Fruit-Dove is extirpated on Guam but is still found on other islands.

---

specialists who can exploit open niches. This separation may be physical (for example, geographic boundaries the individuals won't cross) or may arise when something prevents some individuals from choosing others as mates, as may happen when individuals spend their time in different parts of their habitat.

However, not all evolution is driven in this manner. Random events play a role, too, typically by decreasing genetic diversity rather than increasing it.

In any population, some individuals will survive and mate and others won't. For no other reason than pure chance, random mating may increase or decrease the frequency of a particular trait, a process known as **genetic drift**. The loss of this trait doesn't have anything to do with any inherent advantage or disadvantage the trait offers; it is merely the chance loss of some gene variants because the individuals with these genes did not happen

---

endemic A species that is native to a particular area and is not naturally found elsewhere.

extirpation Local extinction in one or more areas, though some individuals exist in other areas.

genetic drift The change in gene frequencies of a population over time due to random mating that results in the loss of some gene variants.

to mate. This can produce significant changes, especially in small populations where traits that may be present in just a few could be lost in a single generation if these individuals don't happen to reproduce.

Genetic drift can produce a new population that is different from the original population when only a subset of the original variants reproduce. The **bottleneck effect** occurs when a portion of the population dies, perhaps because of a natural disaster like a flood, or a strong new selective pressure like the introduction of a new predator. The survivors then produce a new generation, and any genes in the deceased individuals are lost from the population forever.

> "I found a close correlation between the bird decline and the expansion of brown tree snakes around the island."
> —Julie Savidge

The **founder effect** is seen when a small group that contains only some of the gene variants found in the original population becomes physically isolated from the rest. Even if these individuals do not have the most adaptive traits, natural selection takes over, favoring the best adaptations in the founding group, and producing a subsequent population that is likely to be different from the original population.

Today, human impact increases instances of both the founder effect and the bottleneck effect. Much of what we do isolates populations into smaller groups, forcing them into these situations. [INFOGRAPHIC 12.5]

The pace of evolution is generally slow but is responsive to selective pressures.

In addition to genetic diversity, the size of the population also makes a difference in how quickly natural selection can produce a change in a population: Beneficial traits can spread more quickly in smaller populations simply because it is more likely that the individuals with the trait will find each other and mate. Reproductive rate and generation time also influence how quickly a population can adapt to changes. Many problem species, like insect pests, are r-selected species and have fast generation times, which means they can often stay one step ahead of our efforts to control them. Many endangered species, on the other hand, are K-selected species, with longer generation times; therefore they take longer to recover if

population numbers fall (for more on r and K species, see Chapter 8).

The strength of the selective pressure also affects how quickly natural selection might produce a change in a population. One of the reasons the demise of birds in Guam was so stupefying was that it happened so quickly—particularly for the small birds, which were easiest for the snakes to eat. Larger birds disappeared later, when the snakes started eating their nestlings and eggs. While speciation can take thousands or millions of years, extinction can occur much more quickly if the rate of change exceeds the ability of the population to adapt.

To strengthen the case that the brown tree snake was indeed at the root of the bird extinction in Guam, Savidge had to perform a few more experiments. She needed to be sure, of course, that brown tree snakes actually *liked* eating birds. To do so, she set bird-baited traps all around the islands for the snakes and waited. "I put them in half a dozen locations throughout the island," she recalls. What she found shocked her: "In one area where the birds were extinct, 75% of my traps got hit within 4 nights," she recalls. On the other hand, in the baited traps she had set on Cocos Island, the island off the coast of Guam that is not populated by the snakes, all the birds survived.

Savidge knew, however, that birds weren't brown tree snakes' only prey. The reptiles also ate small mammals, so she checked to see whether these animals were also being adversely affected by the snakes' presence. In the 1960s, before the tree snakes had arrived, scientists on Guam had done a survey of small mammal density and had found, on average, 40 small mammals per hectare of land on the island. When Savidge did the same thing in the mid-1980s, she found only 2.8 animals per hectare. "My prediction was that I would see a decline in these rats and mice and shrews, and indeed, it was like a 94% decline," she explains. The findings suggested that brown tree snakes were devastating small mammal populations in addition to the birds. All in all, Savidge's three pieces of evidence—the fact that the geographic location of the snakes correlated strongly with the birds' disappearance, that brown tree snakes liked to eat birds, and that other small mammals also went missing after the snakes' arrival—convinced Savidge that she had finally solved the mystery of Guam's disappearing birds. Brown tree snakes, she concluded, were definitely the culprit.

---

bottleneck effect The effect of population size being drastically reduced; leads to the loss of some genetic variants and results in a less diverse population.
founder effect The effect when a small group with only a subset of the larger population's genetic diversity becomes isolated; the small group evolves into a different population, missing some of the traits of the original population.

Infographic **12.5** | **RANDOM EVENTS CAN ALTER POPULATIONS**

GENETIC DRIFT

↓ Random mating can eliminate some gene variants from the population not because these individuals were poorly adapted to their environment, or less attractive to the opposite sex, but because they were just unlucky and didn't mate.

A population contains a variety of individuals, but some gene variants are more common than others.

Random mating occurs, but some unlucky individuals don't find mates.

Subsequent generations may have different gene frequencies.

BOTTLENECK

↓ If something causes a large part of the population to die, leaving the survivors with only a portion of the original genetic diversity, the population may recover in size but will not be as genetically diverse as the original population.

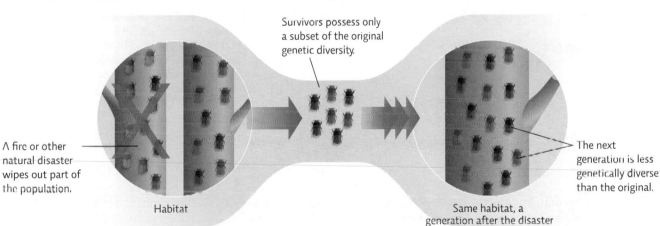

Survivors possess only a subset of the original genetic diversity.

A fire or other natural disaster wipes out part of the population.

Habitat

The next generation is less genetically diverse than the original.

Same habitat, a generation after the disaster

FOUNDER EFFECT

↓ If only a subset of a population colonizes a new area, and the subset becomes completely isolated from the original group, the new population will be less genetically diverse than the original.

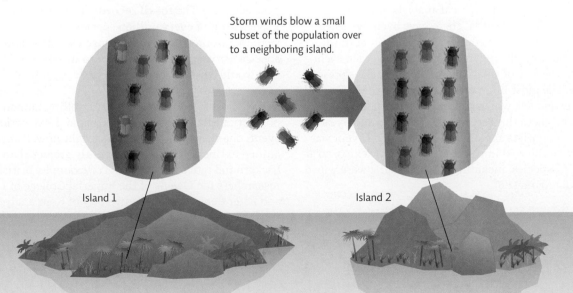

Storm winds blow a small subset of the population over to a neighboring island.

Island 1

Island 2

Infographic **12.6** | **EARTH'S MASS EXTINCTIONS**   ↓ There have been five mass extinctions in Earth's history. Many scientists believe the high rate of extinction and endangerment seen today is leading to a sixth mass extinction.

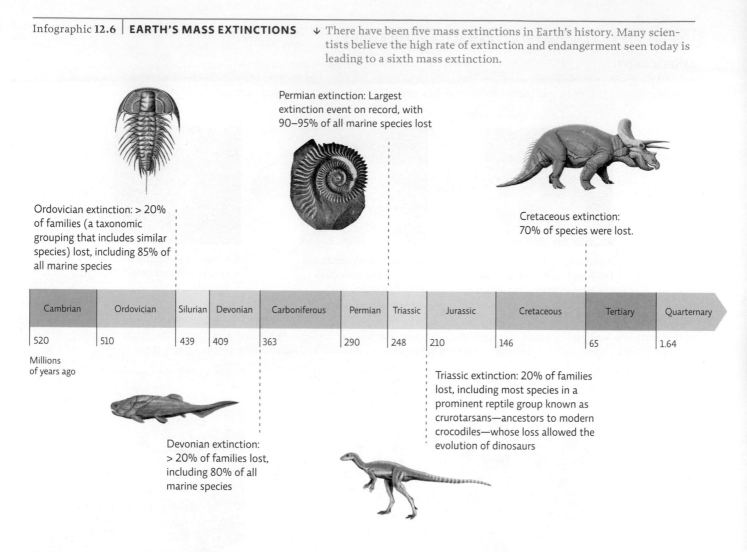

Permian extinction: Largest extinction event on record, with 90–95% of all marine species lost

Ordovician extinction: > 20% of families (a taxonomic grouping that includes similar species) lost, including 85% of all marine species

Cretaceous extinction: 70% of species were lost.

Cambrian	Ordovician	Silurian	Devonian	Carboniferous	Permian	Triassic	Jurassic	Cretaceous	Tertiary	Quarternary
520	510	439	409	363	290	248	210	146	65	1.64

Millions of years ago

Triassic extinction: 20% of families lost, including most species in a prominent reptile group known as crurotarsans—ancestors to modern crocodiles—whose loss allowed the evolution of dinosaurs

Devonian extinction: > 20% of families lost, including 80% of all marine species

Extinction is normal, but the rate at which it is currently occurring appears to be increasing.

Extinction is nothing new on Earth. It is as constant and as common as evolution; by most estimates, more than 99% of all species that have ever lived on the planet have gone extinct. Based on a critical analysis of the fossil record, scientists agree that there have been five *major extinction events*—when species went extinct at much greater rates than during intervening times, each event leading to the loss of 50% or more of the species present on Earth. The most infamous of these was the *K-T boundary mass extinction*, which occurred at the transition from the Cretaceous to the Tertiary period, 65 million years ago. Most scientists agree the K-T extinction event was set off by an asteroid impact in the Gulf of Mexico; 70% of all living species, including the dinosaurs, were wiped out. [INFOGRAPHIC 12.6]

But these kinds of events often lead to the emergence of new species, as other populations adapt to the open niches that are left. Cycles of extinction and evolution ultimately gave rise to the diversity of life we see on Earth today—which now is estimated to range from 3 to 100 million species. Throughout most of time, the **background rate of extinction**—the average rate of extinction that occurs between mass extinction events— has been slow. The **fossil record** tells us that, on average, 1 species out of every million species goes extinct each year. In a world with 3 million species, this would be 3 species per year; if Earth is home to 100 million species, that would be 100 species per year.

Today, most scientists agree that swelling human populations have triggered a sixth major extinction event, one that we are witnessing right now. Plant and animal extinction rates are currently greater than the background rate of extinction. For example, British researchers Ian Owens of the Imperial College of London and Peter Bennett of the University of Kent analyzed the extinction risk for 1,012 threatened bird species and found that habitat destruction was cited as a risk factor in 70% of the cases; and other human interventions, such as the introduction of non-native species or overharvesting, were implicated in 35% of the cases. In some areas with high endemic species diversity, such as tropical rain forests, the rate can be quite high: One estimate puts it as high as 27,000 extinctions per year in tropical rain forests that are being cut down. A look at the historic mammalian fossil record reveals that, on average, 1 mammal species has become extinct every 200 years, but in the last 400 years we've documented 89 mammalian species extinctions—that is almost 45 times faster than the background rate.

Estimates of just how rapidly we are losing species vary, in part because we don't know how many species exist and because it is hard to verify that a species actually is extinct. In an often-cited 1995 article published in the journal *Science*, researchers estimated that current rates of extinction range from 100 to 1,000 times greater than background rates. If all species currently threatened become extinct, this would raise the high-end estimate to 10,000 times greater. A recent report in *Nature* suggests that current methods overestimate extinction rates by as much as 160% but points out that even the low-end estimates are a cause for concern. And while scientists debate whether species extinctions are 100 times or 1,000 times faster than normal worldwide, it may be more useful to evaluate threats at a local level, like in Guam, and focus efforts on reducing species loss there. Especially on a small, isolated island, a rate that is even 10 times higher than normal is significant.

What we do know is that today's accelerated extinction is largely driven by human actions (see Chapter 11). As the human population increases, our impact is becoming much more devastating for other species. We remove the resources they need to survive, minimize their habitat ranges, introduce new predators or competitors, and strip them of their genetic diversity, all of which slowly eliminate them. In Guam, the near-total disappearance of birds between the 1960s and 1980s was a biological murder mystery of astounding proportions—one that illustrates just how vulnerable populations are to sudden changes.

Humans affect evolution in a number of ways.

The introduction of the brown tree snake to Guam was an accident: A snake hitchhiker crossed the ocean on a human tanker—unbeknownst to the crew—and landed in a veritable bird buffet. But humans also directly affect the evolution of a population through **artificial selection**. Artificial selection works the same way as natural selection but the difference is that the selective

↑ A Guam rail (*Rallus owstoni*), a critically endangered species whose decline was caused by predation by the introduced brown tree snake. The Guam rail is currently considered extinct in the wild, though reintroduction programs are underway.

pressure is us. For many animal species—from pets to farm animals and plant species—humans choose who breeds with whom, in an attempt to produce new individuals with the traits they desire. By doing this over many generations, people have accentuated certain plant and animal traits, sometimes to extremes. For instance, artificial selection created domestic dogs from their wolf ancestors. [INFOGRAPHIC 12.7]

But evolution is ever at work. Pesticide- and antibiotic-resistant populations can emerge as an inadvertent human-influenced selection. When we apply a chemical that kills a pest or pathogen, some individuals survive because of their natural genetic resistance (the individuals were already resistant even though they had never encountered the chemical). These survivors are then the only individuals who reproduce, producing the next generation that is also pesticide resistant, ultimately changing the frequency of resistant genes in the population (see Chapter 22).

---

background rate of extinction The average rate of extinction that occurred before the appearance of humans or occurs outside of mass extinction events.
fossil record The total collection of fossils (remains, impressions, traces of ancient organisms) found on Earth.
artificial selection Humans decide which individuals breed and which do not in an attempt to produce a population with desired traits.

Infographic **12.7** | **HUMANS USE ARTIFICIAL SELECTION TO PRODUCE PLANTS OR ANIMALS WITH DESIRED TRAITS**

→ All dogs (*Canis lupus familiaris*) are descendants of the wolf (*Canis lupus*). By only breeding those males and females with the traits desired (size, herding ability, protective instinct, etc.), humans have created more than 170 dog breeds.

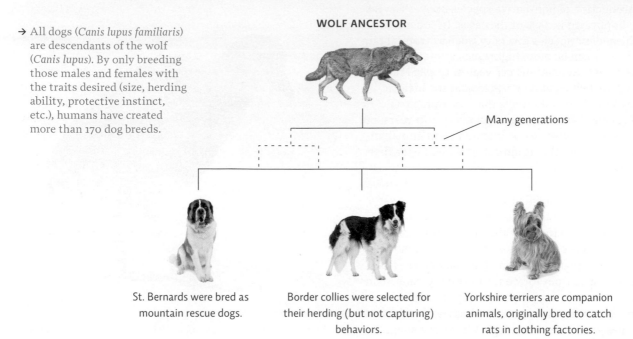

WOLF ANCESTOR

Many generations

St. Bernards were bred as mountain rescue dogs.

Border collies were selected for their herding (but not capturing) behaviors.

Yorkshire terriers are companion animals, originally bred to catch rats in clothing factories.

Infographic **12.8** | **COMMON MISCONCEPTIONS ABOUT EVOLUTION**

MISCONCEPTION	EVOLUTIONARY EXPLANATION
The study of evolution seeks to explain the creation of life	Evolution doesn't study creation or the origin of life. Evolution is the science that studies the "descent with modification" of life after it came into being.
Humans evolved from monkeys or apes	Humans did not evolve from monkeys or apes. Humans, monkeys, and apes share a common ancestor, who lived 5–8 million years ago, but each group followed different evolutionary paths to its current form.
Evolution is goal directed	Evolution is not working toward a particular final version or trait. It is unpredictable because the genes available to natural selection vary, and selective pressures (such as environmental conditions and competition) also vary.
Evolution proceeds from simple to complex	Simple is not always the ancestral condition. Parasites may be simple in design but many have evolved from nonparasitic, more complex forms.
Evolution produces perfect adaptations	New traits are rarely "perfect." Species have a long history of ancestral adaptations, each giving way to the next. Evolution works with whatever is available—it doesn't scrap the anatomy of an "arm" and reform a wing for a bat. It modifies existing structures for new functions. In addition, adaptations are often compromises between differing needs. For example, the flippers of a seal are better suited for swimming but make walking on land awkward.
Evolution is just a theory and many scientists don't accept it anyway	The claim that evolution is "just a theory" reveals a misunderstanding of the concept of a scientific theory. "Theories" represent our most well-accepted explanations in science—to attain the status of theory, an explanation must have substantial evidence, ideally from multiple lines of evidence. Multiple polls show that well over 95% of scientists accept evolutionary theory as the best explanation for the diversity of life and how populations adapt to their environments.

Evolution can be a contentious subject for some people who feel it conflicts with other views they hold. Whether or not you are convinced by the physical, scientific evidence that evolution has occurred or that natural selection is a mechanism by which it occurs, it is important that you base your criticisms of the theory on sound science and not on misconceptions of evolutionary theory. [INFOGRAPHIC 12.8]

Ultimately, by changing the environment rapidly, humans apply a number of new selective pressures on populations. Our changes have the capacity to be so great that natural selection simply cannot keep up—a new needed trait is not present in the population or it cannot spread quickly enough to prevent a population collapse. The accidental introduction of the predatory brown tree snake is one such example; it introduced a major change, basically overnight, that was able to eat its way through the vertebrate populations before those populations could adapt.

Now, however, things are looking up for birds in Guam. Thanks to efforts by the Guam Department of Agriculture and the U.S. Department of Agriculture's Wildlife Services, brown tree snakes are being controlled to allow bird populations to recover. "We are using traps in northern Guam to control snakes in areas where we would like to reintroduce native species," explains Diane Vice, a wildlife biologist with the Guam Department of Agriculture. Her organization is also working to prevent the snakes from infiltrating Cocos Island, the atoll off of southern Guam that is an important haven for nesting sea birds, Micronesian starlings, and sea turtles. "The hope is to create safe habitat," Vice says, so that these beautiful native species can once again thrive.◉

Research articles referenced in this chapter:

He, F., and Hubbell, S.P. 2011. *Nature*, 473: 368–371.

Nachman, M.W. 2005. *Genetica*, 123: 125–136.

Owens, I.P.F, and Bennett, P. 2000. *Proceedings of the National Academy of Sciences*, 97: 12144–12148.

Pimm, S., *et al.* 1995. *Science*, 269: 347–350.

Savidge, J.A. 1987. *Ecology*, 68: 660–668.

BRING IT HOME

⊘ PERSONAL CHOICES THAT HELP

The astonishing variety of life found on Earth is the result of natural selection favoring those individuals within populations that are best able to survive in their particular environment. Given enough time, some populations may be able to adapt to environmental changes. However, human activities may disrupt natural ecosystems so that organisms cannot adapt fast enough to survive, and may go extinct. Conservation activities can help protect vulnerable organisms and ecosystems.

Individual Steps
→ Live in an older, established area of your community. Suburban sprawl reduces habitat for wildlife, and reliance on cars causes greenhouse gas emissions that could result in species-threatening climate change.
→ Choose native plant species when landscaping your home.

→ If you own a cat, keep it indoors—pet cats kill millions of songbirds in the United States each year.

Group Action
→ With the help of officials at a local park or from your state wildlife agency, organize a work day to remove invasive plant species from a park or wildlife area.
→ Volunteer to give educational presentations aimed at stemming the introduction of invasive species in your area, such as educating boaters not to transport water, animals, or plants from one body of water to another or educating campers not to transport firewood from one area to another.

Policy Change
→ The U.S. Fish and Wildlife Service recently issued a rule under the Lacey Act that bans the importation and transport of 4 species of large snakes (3 species of African pythons and the yellow anaconda) popular as pets but that threaten native species if released into the wild. The service is considering a ban on other species of invasive snakes as well. Learn about this ongoing problem and voice your opinion to the service about whether to add other species to this ban.

UNDERSTANDING THE ISSUE

CHECK YOUR UNDERSTANDING

1. **Which of the following is TRUE of evolutionary processes?**
    a. They are goal driven.
    b. New genes evolve in response to environmental change.
    c. Evolution acts on existing genetic variation.
    d. Random events do not influence evolution.

2. **Ten thousand years ago, most members of Species A were killed by a series of volcanic eruptions. However, some members escaped; all modern members of Species A are descended from those 100 individuals. This is an example of:**
    a. genetic drift.
    b. artificial selection.
    c. mass extinction.
    d. the bottleneck effect.

3. **Why do most scientists think that we are in the midst of a sixth mass extinction?**
    a. Current extinction rates are greater than the background rate.
    b. The background extinction rate is approximately 20%.
    c. Most extinctions are occurring in the taiga and tundra.
    d. No new species are being discovered.

4. **Which of the following is an example of coevolution?**
    a. Moths that are preyed upon by bats can hear the ultrasonic sounds the bats use in hunting.
    b. Polar bears and Arctic foxes both are white for camouflage on snow.
    c. Dolphins and whales have flippers that are similar in shape and function to the fins of a fish, allowing them both to swim efficiently.
    d. Humans and chimpanzees share 98% of their DNA.

5. **In a population of butterflies, there used to be a wide range of sizes: small, medium, and large. But a non-native bird was introduced to the butterfly's habitat. It loves to eat butterflies, but it will only eat those that are medium size. Eventually, medium-sized butterflies became rare. This is an example of:**
    a. disruptive selection.
    b. stabilizing selection.
    c. artificial selection.
    d. directional selection.

6. **If sea levels rose 1 meter (3 feet) overnight, what would happen to coastal terrestrial organisms who could not retreat inland?**
    a. Their lungs would become gills, allowing the organisms to adapt to life in the water.
    b. They would all die.
    c. Individuals with adaptations that allowed them to live in water might survive, but the rest would die.
    d. A new gill-like structure would evolve, resulting in new species adapted to living in water.

WORK WITH IDEAS

1. Describe the forces that influence evolution. Why do evolutionary rates differ? Why can't all species adjust to changes in their environment and avoid extinction?

2. Describe the process of natural selection, using bacterial antibiotic resistance as an example. Apply the requirements for natural selection to the development of antibiotic resistance in your answer. Include examples of human behavior that have contributed to antibiotic resistance.

3. How might the genetic diversity within a population that, because of the founder effect, becomes established in a remote area compare to that of the original population? How would this affect the new population's ability to adapt to changes in its environment? Explain.

4. Suppose a native Guam bird species began to show evasive behaviors that allowed it to avoid the brown tree snake. Describe a potential coevolution scenario that might have allowed this adaptation to emerge in the population.

5. What kinds of predictions can you make about the effects of the current mass extinction on:
    a. communities?
    b. ecosystems?
    c. humans?

ANALYZING THE SCIENCE

The following graph depicts the relationship between numbers of extinctions and human population size since the 19th century.

INTERPRETATION

1. Describe what is happening to:
    a. the extinction rate over time.
    b. human population growth over time.

2. The two curves have been graphed together. What is the implication of presenting these data in this manner?

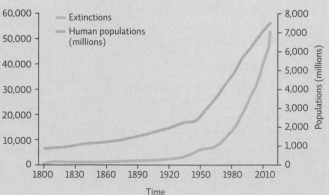

Species extinction and human population

ADVANCE YOUR THINKING

3. The y axis is labeled "Extinction numbers." What taxonomic units are being measured? What if the taxonomic unit being evaluated here had been genus (a taxonomic group that can contain more than one species) or family (a taxonomic group that can contain more than one genus)? Would you be more concerned about the trend of extinctions or less? Explain.

4. What type of relationship is suggested by the figure (correlational or causal)? What additional data would you like to see to support the graph's main point?

EVALUATING NEW INFORMATION

Life on Earth as we know it is a result of millions of years of evolutionary processes. These processes are not immune to changes in the environment; changes in evolutionary processes will ultimately result in changes in biodiversity, which will necessarily affect life on Earth, including humans.

Go to the website www.actionbioscience.org/evolution/myers_knoll.html and read the article on the effects of the sixth mass extinction on evolution.

Evaluate the website and work with the information to answer the following questions:

1. Is this a reliable information source? Does it have a clear and transparent agenda?
   a. Who runs this website? Do the credentials of the person or group make this source reliable/unreliable? Explain.
   b. Who are the authors? What are their credentials? Do they have the scientific background and expertise that lends credibility to the article?

2. In your own words, explain why the authors think that the current mass extinction will change evolutionary processes.

   a. What changes in particular do they think will impact future evolution?
   b. What types of evidence do they provide to support their arguments? Give specific examples.

3. The sixth mass extinction is the result of human activities. Read "The First Human Caused Extinction: The Dodo" (http://suite101.com/article/the-first-human-caused-extinction-the-dodo-a85161).
   a. Why did the dodo go extinct? Be specific—there are direct and indirect consequences discussed in the article.
   b. On Mauritius, what happened as a consequence of the dodo's extinction?
   c. Do you think humans today could have a similar impact on a species? Consider both direct and indirect human impacts. Which types of species might be vulnerable? Find an example on the International Union for Conservation of Nature (IUCN) "Species of the Day" website (www.iucnredlist.org/species-of-the-day/archives).

MAKING CONNECTIONS

Gyps indicus: A SPECIES WORTH SAVING?

Background: Vultures are a type of bird that feeds on dead organisms. They are found worldwide and provide an important ecosystem service. Many are large (with wingspans of over 1.2 meters (4 feet)) and depend on large animal carcasses for food. Some of the larger species, including the California condor (*Gymnogyps californianus*), are critically endangered due to habitat loss and human hunting. Loss of vultures has the potential to impact ecosystems, so conservation groups have worked to save them.

Case: You are a member of a philanthropic organization devoted to saving species from extinction. Your organization has a limited amount of funds, so it is imperative that the funds be given to worthwhile projects. The Indian government has approached you for funds to help save the critically endangered Indian vulture *Gyps indicus*. Before your organization can make a decision, you need to gather information:

1. Research *Gyps indicus* (the IUCN Red List is a good place to start). Consider the following questions:
   a. Why is this species considered critically endangered?
   b. What are the causes of its decline?
   c. Where does it live?

   d. Does its habitat still exist? If it were brought back from the brink of extinction, is there habitat for it to live in and food for it to eat?
   e. What plans are currently in place to save it?

2. Do some further research to answer the following questions.
   a. Why did the condor decline? Consider both human impacts and the condor's biology.
   b. How much money has been spent to date on the rescue of the California condor?
   c. What is the status of condors that currently live in the wild?
   d. What is the outlook for the future of this species?
   e. Answer questions a.–d. for *G. indicus* and compare your results to those for the California condor.

3. Based on your research about *G. indicus* and California condors, write a report recommending a course of action to your organization. Should you give money to save the *G. indicus* species? Why or why not? Specifically, you should address the issue of whether it is:
   a. biologically possible to save *G. indicus*.
   b. biologically desirable to save *G. indicus*.
   c. economically feasible to save *G. indicus*.

CORE MESSAGE

Forests have great economic value but we must balance that with the value of their ecosystem services and sociocultural benefits. Sustainable management practices allow us to harvest forest products without destroying the forest or its ability to provide ecosystem services and future products.

GUIDING QUESTIONS

After reading this chapter, you should be able to answer the following questions:

→ In general what are the characteristics of a forest biome? What are the three main types of forests and what influences which forest type is found in a given area?

→ What is the three-dimensional structure of a forest and how are the plant species found there adapted to their level of the forest?

→ What ecosystem services do forests provide? Why might it be useful to put a dollar value on these services?

→ What is the current state worldwide of forest resources? What threats exist?

→ What are ways that we can act to protect and sustainably manage forest resources? What trade-offs exist with our choices?

Teams of men and women pick their way across steep, rocky slopes as they plant trees in Mahotiere, Haiti.

RETURNING TREES TO HAITI

Repairing a forest ecosystem one tree at a time

When Jean Robert was a young boy, the mountains surrounding the Haitian city of Gonaïves (pronounced go-nah-eev) were still lush and green with trees. His family's hillside farm—small though it was—provided a steady crop of mangoes and avocados in the summer, coffee and cassava in the fall, and wood for charcoal throughout the year. But today, the forests are gone and the mountains are bare. Not only have crop yields shrunk—from enough to sell in the local markets, to barely enough to feed a family—but the mountain homes and the city below have been left defenseless against the onslaught of tropical storms that pound Haiti every summer. Unencumbered by trunks or roots or shrubs, water rushes freely downward, gathering into apocalyptic mudslides that destroy homes, crops, and livelihoods. In 2004, a single storm claimed more than 2,000 lives from this one city.

Now, instead of growing mangoes and cassava, Robert and his neighbors are trying to grow trees. Working as a team, moving in slow, deliberate, farm-by-farm patterns across the mountainside, the community work group, or kombit, plants a carefully planned mix of fruit and timber trees. By the end of the month, they say, each member will have his or her own saplings to tend to. If the saplings survive, and if the American ecologists they've been consulting with are right, their efforts will give rise to a new era of sustainable forestry, which strikes a balance between what people need from the forest and what the forest itself needs to survive.

It's a scene being replayed all over Haiti. When Europeans first arrived in the country some 500 years ago, two-thirds of the land was covered in forests so majestic that explorers dubbed this region "the Pearl of the Antilles." But as the centuries passed, the forests were cleared—sometimes with amazing breadth, to make way for coffee and sugar plantations; other times, in discrete chunks that Haitian peasants would then convert into subsistence farms. Many of the trees—along with the coffee and sugar crops that displaced them—were sold overseas. Most of the rest provided fuel for heating Haitian homes, and cooking Haitian meals.

Today, less than 2% of that original forest remains; 6% of the land has no soil left at all. In these places, dead tree stumps, bleached white by the sun, protrude from the ground like bones poking through decayed flesh. And Haiti has become both the poorest and the most environmentally degraded country in the Western Hemisphere. With few trees and little soil, drinking water has grown polluted, crops have dwindled, and the people have suffered.

Farmers like Jean Robert say that forests are the key to changing all of that. They have planted thousands of trees in these mountains in recent years, and hope to plant thousands more. "Almost all of the country's

◉ **WHERE ARE HAITI AND GONAÏVES?**

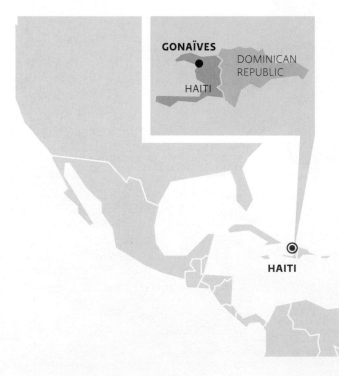

↑ South of Dajabón, Haiti's brown landscape on the left contrasts sharply with the rich forests of its neighbor, the Dominican Republic, on the right.

problems—natural disasters, food shortages, poverty—can be traced back to rampant **deforestation**," says Haitian ecologist Timote Georges, who is working with the American non-profit, Trees for the Future, to reforest the mountains around Gonaïves. "So if we want to fix the country, we have to put the forests back and then find a way to manage them better."

That's no small feat. Forests, after all, are one of our most contentious resources. Such is the range of economic benefits and ecosystem services they provide, that any one use must be weighed against a host of others, and immediate human needs are frequently pitted against long-term conservation goals. In Haiti, where most people live on less than $2 a day, trees provide food, energy, building material, and desperately needed income. To stop them from being chopped down, Georges and his fellow Haitians will have to find alternative sources of each, not to mention a farming method that doesn't require clearing the forests.

But to really understand how trees might help alleviate poverty, we must first consider just what forests are, what they do for us, and why so many of them are being chopped down in the first place.

Forest biomes are determined by temperature and precipitation.

Forests are biomes dominated by trees. They cover about 30% of the planet's landmass, but thanks to their sheer concentration of biodiversity, are home to more than 50% of Earth's terrestrial life, and more than 60% of its green, photosynthesizing leaves. There are many types of forest biomes around the globe, each determined by the temperature and amount of precipitation the area

---

deforestation Net loss of trees in a forested area.

Infographic **13.1** | **FORESTS OF THE WORLD**

↓ Forests are biomes whose dominant plant life is trees and other woody vegetation. There are 3 main types of forests, classified according to climate, and many subdivisions within these 3 types.

BOREAL FOREST

TEMPERATE FOREST

TROPICAL FOREST

Boreal forests (taiga) represent the largest terrestrial biome; they stretch from Canada to Siberia and are found at higher elevations in lower latitudes. They have a short growing season with little precipitation, most of it snow. Soils here are thin and acidic and the major tree species are evergreen conifers with needlelike leaves.

Temperate forests have distinct seasons. The soil is fertile, with a thick layer of decomposing leaf litter that supports the plant life. Depending on how much precipitation an area gets, the forest may be predominantly coniferous (evergreen) or broadleafed (deciduous).

Tropical forests have similar temperatures year round. Dry tropical forests have distinct wet and dry seasons, whereas tropical rain forests receive rain year round. The soils are thin, acidic, and low in nutrients. Rapid decomposition by fungi and bacteria supports the dense vegetation found in these soil-poor areas.

↑ Volunteers begin planting 25,000 donated trees in Mahotiere, Haiti, to combat soil erosion.

receives. **Boreal forests**, characterized by evergreen species like spruce and fir, cover vast tracks of land in the higher latitudes and altitudes and are characterized by low temperatures and precipitation levels; they represent some of the last expansive forests left on Earth. **Temperate forests**, which contain deciduous trees that lose their leaves in winter, like oak, hickory, and maple, are found in the wet areas of mid-latitudes. These forests are not as expansive as the boreal forests because they are found in latitudes with high human populations, but some areas have seen remarkable regrowth in the 20th century. For example, there is now more forest cover in the eastern United States than there was in the 19th century after heavy clearing. (This regrowth can be sustained because forests elsewhere are harvested.) **Tropical forests** contain a diverse mix of tree and undergrowth species and are found in tropical latitudes where temperatures do not vary much throughout the year. [INFOGRAPHIC 13.1]

Most forests consist of four distinct layers. The **canopy**, formed by the overlapping crowns of the tallest trees, makes up the ceiling of the forest. Some even taller trees may reach above the canopy to form an **emergent** layer. Beneath the canopy is the **understory** layer, where

boreal forests Coniferous forests found at high latitudes and altitudes characterized by low temperatures and low annual precipitation.

temperate forests Found in areas with four seasons and a moderate climate, receive 30–60 inches of precipitation per year, and may include conifers and/or hardwood deciduous trees (lose their leaves in the winter).

tropical forests Found in equatorial areas with warm temperatures year-round and high rainfall; some have distinct wet and dry seasons but none has a winter season.

canopy Upper layer of a forest formed where the crowns (tops) of the majority of the tallest trees meet.

emergent The region where a tree that is taller than the canopy trees rises above the canopy layer.

understory The smaller trees, shrubs, and saplings that live in the shade of the forest canopy.

Infographic **13.2** | **CROSS SECTION OF A FOREST**

↓ Forests are stratified, having four (or more) vertical layers, each of which contains species adapted to the level of sunlight and moisture available.

Emergent layer: A few trees grow above the general level of the forest canopy.

Canopy: The crowns of the dominant trees shade the layers below.

Understory: Trees and shrubs here are adapted to shade; saplings will grow rapidly when a spot in the canopy opens up.

Forest floor: This lowest level contains leaf litter, decomposing plant material, herbs, flowers, and seedlings.

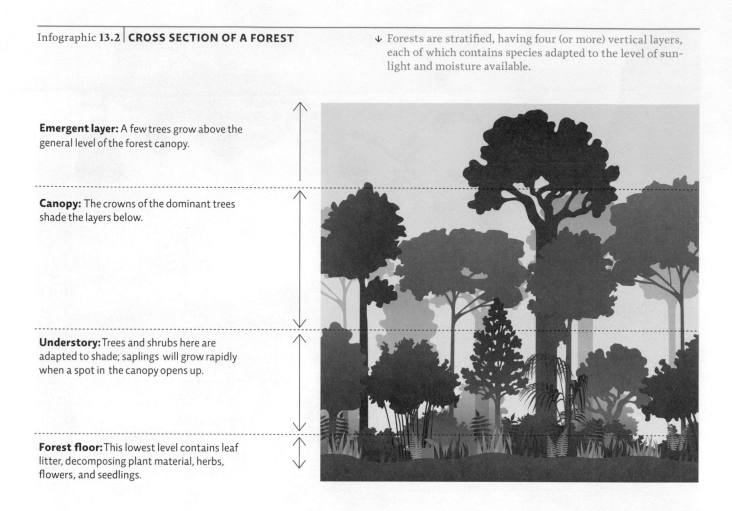

shade-tolerant shrubs, smaller trees, or the saplings of larger trees grow. Sometimes these trees are dense enough to form a lower canopy. The lowest level is the **forest floor**, which is typically made up of seedlings, herbs, wildflowers, and ferns. The forest floor also contains soil, which is composed of leaf litter and other debris—branches, logs, and stumps—that decomposes over time.

Within each forest layer is a range of species uniquely suited to the temperature, humidity, and amount of sunlight that layer receives, and well adapted to its particular neighbors. For instance, in a temperate deciduous forest, wildflowers on the forest floor will bloom early in the spring before the bigger trees "leaf out" and block the sun. Sunlight is a precious commodity on the forest floor and wildflowers compete for it. One wildflower species may bloom one week and another the next week. Disruption to one part of the forest (for instance, cutting down a tree that opens the canopy) can have a trickle-down effect that impacts each subsequent layer. [INFOGRAPHIC 13.2]

Most experts agree that the biggest modern contributors to Haitian deforestation are the food and fuel needs of

Haitians themselves. But throughout the nation's history, there were many other culprits—from 18th-century French colonizers' coffee and sugar plantations, to the timber industry of the 19th and 20th centuries. As the population grew—from 3 million in 1940 to 9 million in 2000—rural Haitians were forced to clear ever-larger swaths of mountainside to make room for subsistence crops. The trees themselves doubled as a source of fuel and cash for families who not only used the wood to cook with but also sold it as charcoal in Port-Au-Prince, the nation's densely populated, energy-starved capital city.

Charcoal—partially burned wood that ignites more easily and burns hotter than the original wood itself—is used in many developing countries that lack other reliable fuel sources. Charcoal is produced in a variety of ways but all include a slow, low-temperature "roasting" of the wood in a low-oxygen environment. This process releases both CO_2 and particulates (soot) into the atmosphere. Like Haiti, other areas dependent on charcoal as their major fuel have become severely deforested and plagued with air pollution. Many African nations including Mozambique, Malawi, Somalia, and Tanzania are

↑ Charcoal sellers in the Carrefour Feuilles district of Port-au-Prince, Haiti.

experiencing rampant deforestation largely to provide charcoal and fuel wood. In Mozambique, for example, charcoal is the main fuel for 80% of the population. More than 2.5 million trees are cut each year in Somalia—that is 10 trees per household, per month.

In time, Haiti's charcoal trade grew to account for 20% of the rural economy and 80% of the country's energy supply. Before long, 98% of the country's forests had been chopped down and Haitians were burning 30 million trees' worth of charcoal annually. "The trade itself became this incredibly destructive force," says Andrew Morton, a forest ecologist for the United Nations Environment Programme. "And the fact that it was based not on foreigners exploiting the land for profit, but on poor Haitians trying to earn money and feed their families and heat their homes, made it impossible to stop. There really was no other source of energy."

Of course, charcoal wasn't the only crucial service the trees provided.

Forests provide a range of goods and services and face a number of threats.

Though trees are the largest and most notable life-forms present, a forest is much more than just its trees. Together, the species inhabiting the different layers participate in a delicate symphony of chemical and physical cycles that produce an invaluable range of ecosystem services for the planet.

It starts with the soil, which the forest itself helps to form and maintain: leaves and branches die, fall to the ground, and decay, forming a thick brown layer of nutrients in which all future generations of plant life will take root.

While dead and decaying plants help form the soil, living ones hold it in place. During rain storms, soil anchored in by roots, especially by tree roots, can't flow as easily

---

forest floor The lowest level of the forest, containing herbaceous plants, fungi, leaf litter, and soil.

↑ The original forests of Haiti would have looked like this one on the Barahona Coast in neighboring Dominican Republic.

down hillsides into nearby surface waters (lakes, streams, oceans), where it would contaminate them with sediment and chemicals it picked up from the ground's surface. By slowing this flow of rainwater across the ground's surface (called **runoff**), a well-vegetated area allows more water to soak into the ground, recharging the groundwater supplies that provide area residents with their major source of drinking water. The soil also traps chemicals that might otherwise contaminate that drinking water. (See Chapter 18 for more on runoff.)

While plants, soil, and water are playing off one another in this manner, the forest is also conducting another important cycle: pulling carbon dioxide out of the atmosphere and replacing it with oxygen. In fact, forests as a whole store more carbon in their biomass, litter, and soil than all the carbon in the atmosphere, making this biome one of the world's largest **carbon sinks** (an area that stores more carbon than it releases, such as the standing timber in a forest or organic matter in soil). And forest leaves produce

so much oxygen that they are commonly referred to as "the lungs of the planet."

Last but not least, virtually every layer of forest provides food and habitat for a bevy of animals (vertebrates and invertebrates), fungi, and microbes; these creatures all do their part to contribute to the functioning of the forest ecosystem as a whole. It is difficult to overstate the importance of forests to the biosphere.

In addition to all these ecosystem services, forests also provide a range of economic benefits. In fact, humans have relied on forests for millennia for a litany of consumer goods. Wood products, including lumber, firewood, charcoal, paper pulp, and some medicines, account for some $100 billion in global trade every year. On top of that, the wildlife supported by forests provides humans with food and recreational hunting opportunities, both of which have economic value. [INFOGRAPHIC 13.3]

Infographic 13.3 | ECOLOGICAL AND ECONOMIC VALUE OF FORESTS

↓ All ecosystems, including forests, contribute to the ongoing functioning of the planet and the immediate well-being of humans. Some of these services we take for granted; others are recognized for their economic value. One calculation estimates the value of services provided free by ecosystems of the world at trillions of dollars per year, a value greater than the GNP of all nations of the world combined.

ECOSYSTEM SERVICES

Watershed services: Water purification and provision

Atmosphere and climate effects: A major sink for CO_2; increases rainfall in some areas; biggest contributor of oxygen to the atmosphere

Soil maintenance and protection: Soil production and recycling of nutrients; reduction of soil erosion

Disturbance regulation: Protection from storm damage, especially in coastal areas

Biodiversity and genetic resources: Food and habitat for biodiversity; a rich storehouse of genes that might prove useful to improve our crops or provide as yet undiscovered medicines

ECONOMIC VALUE

Goods: Provides many of the basic goods we depend on including:
- Food
- Fuel
- Building materials
- Other products such as rubber and cork
- Raw material for paper and other industrial products
- Medicines

Jobs: More than 10 million people make their living in and from forests

Recreation and ecotourism opportunities

SOCIOCULTURAL BENEFITS

The beauty of forests provides a place for spiritual renewal, artistic inspiration, and stress reduction. Ancient stands of trees provide a connection to the past; many indigenous people are an important part of their forest ecosystem, possessing ancestral knowledge of the forest and its inhabitants.

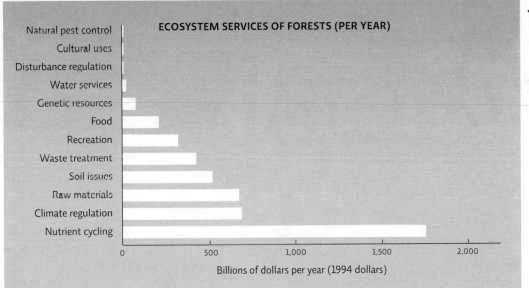

ECOSYSTEM SERVICES OF FORESTS (PER YEAR)

- Natural pest control
- Cultural uses
- Disturbance regulation
- Water services
- Genetic resources
- Food
- Recreation
- Waste treatment
- Soil issues
- Raw materials
- Climate regulation
- Nutrient cycling

Billions of dollars per year (1994 dollars)

← Robert Constanza and his colleagues evaluated ecosystem services of forests and quantified their value (in 1994 dollars) to be $4.7 trillion.

In Haiti, the cost of deforestation spun out over decades and proved catastrophic. Soil eroded down into streams, rivers, and gullies, clogging them with sediment and disrupting aquatic ecosystems. With nothing to absorb the water, floods became more severe and groundwater sources were quickly depleted as the water flowed away rather than soaking into the ground. Slowly, crop yields shrank. As the forest habitat was fragmented, biodiversity dwindled. And as the trees vanished, the people of Haiti suffered. Unlike wealthier countries they could not afford expensive water purification systems—a service the forests once provided for free. "It took some years before we could feel the other effects of deforestation," says Georges. "The floods got worse and we lost drinking

---

runoff Water that flows downhill across the land surface, usually after a rainfall.

carbon sinks Places such as forests, ocean sediments, and soil, where accumulated carbon does not readily reenter the carbon cycle.

↓ A variety of "drivers" are responsible for deforestation around the world.

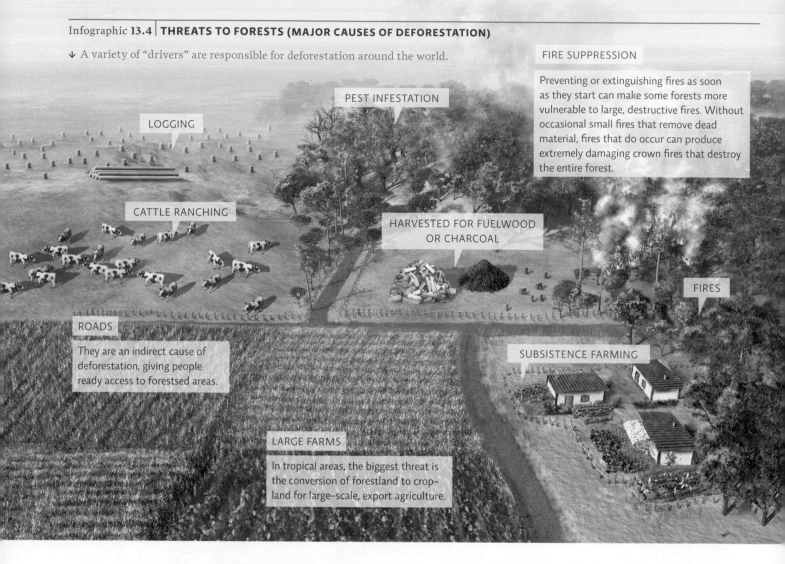

FIRE SUPPRESSION

Preventing or extinguishing fires as soon as they start can make some forests more vulnerable to large, destructive fires. Without occasional small fires that remove dead material, fires that do occur can produce extremely damaging crown fires that destroy the entire forest.

PEST INFESTATION

LOGGING

CATTLE RANCHING

HARVESTED FOR FUELWOOD OR CHARCOAL

FIRES

ROADS

They are an indirect cause of deforestation, giving people ready access to forestsed areas.

SUBSISTENCE FARMING

LARGE FARMS

In tropical areas, the biggest threat is the conversion of forestland to crop-land for large-scale, export agriculture.

NET CHANGE IN FOREST AREA BY COUNTRY, 2005–2010

← Today, most deforestation is occurring in developing countries. It must be noted, however, that the industrialization of developed countries, like those of the United States and Europe, was supported in large part by the harvesting of their own forests.

Net loss (hc/year)

■ More than 500,000
■ 250,000–500,000
■ 50,000–250,000

Small change (gain or loss)

■ Less than 50,000

Net gain (hc/year)

■ 50,000–250,000
■ 250,000–500,000
■ More than 500,000

water to runoff and pollution. And once those problems started, there was no easy way to fix them."

Eventually, everyone who could abandoned the countryside for the capital city. Even that did not stop the tide of deforestation. As the population swelled, so did the demand for fuel, which in Haiti, still comes almost exclusively from trees.

Today global deforestation has slowed considerably, from 9 million **hectares (ha)** per year (22 million acres [ac]) in the 1990s down to 5.2 million ha (13 million ac) in the year 2000, but the planet still has a net loss of forested land every year. There are three main culprits behind this trend: the harvesting of forests for wood and wood products, the conversion of forests into agricultural land, and urbanization. Management of fire is also a factor in forest destruction. Frequent fires remove deadwood and other flammable material but if we suppress these fires, the deadwood builds up so that when a fire does come through, it can burn so hot it catches the entire forest on fire. It turns out that fire is actually needed to maintain some forests whose trees are fire-adapted with seeds that only germinate when exposed to the heat of a fire. The nature and degree of each of these threats varies by country because forests are often used and managed differently in developing countries than they are in developed ones. [INFOGRAPHIC 13.4]

In general, deforestation occurs at a greater rate in developing countries like Haiti, where dire poverty and a lack of alternatives force people to harvest their forests or remove them for other land uses. They need wood and charcoal for fuel and housing; they need more space for agriculture; and they need commodities to sell in the market place. In most cases, developing countries also have far greater remaining forest stands than developed countries, but many of these stands are falling fast.

In fact, many if not most developed countries, including the United States and most European countries, became developed in part by harvesting their own forests. Today, these countries have fewer forest stands than many developing countries, but they also have much more stringent regulations in place to protect those that are left. These regulations—and the ability to enforce them—help keep deforestation in check.

Forests can be managed to protect or enhance their ecological and economic productivity.

Funding and technical expertise can also facilitate more effective forest management in developed nations. For

example, in 1905, the United States established the National Forest Service. Its first director, Gifford Pinchot, challenged the prevailing notion that U.S. forests were inexhaustible and introduced the idea of sustainable forestry—only taking what the forests could sustainably produce or replace. The focus in these early years was **maximum sustainable yield**: harvesting as much as sustainably possible (but no more) for the greatest economic benefit. In 1960, the United States expanded on Pinchot's ideas and enacted the **Multiple-Use Sustained-Yield Act,** which mandated that national forests be managed in a way that balances a variety of uses—outdoor recreation and timber interests as well as the health of watersheds, fish, and wildlife. No single use could predominate.

> "The floods got worse and we lost drinking water to runoff and pollution. And once those problems started, there was no easy way to fix them."
> —Timote Georges

Today the U.S. National Forest Service, administered by the U.S. Department of Agriculture, oversees 155 national forests (and 20 national grasslands) and provides guidance for the management of private forests and grasslands, both nationally and internationally. Its overriding objective is the management of these areas in an ecologically sustainable manner, and it promotes **Forest Ecosystem Management (FEM)** as the best way to meet this mandate. Rather than focus exclusively on timber harvests and maximum sustainable yields, FEM aims to manage the forest ecosystem as a whole. This includes a variety of techniques for timber harvesting, vegetation removal, and controlled burns to remove deadwood and stimulate seed germination in fire adapted species, as well as restoration of forested areas and research in forestry (ecology and economic uses).

Though some timber harvesting methods are very damaging to the integrity of the forest (such as large clear-cuts on steep slopes), methods are available that reduce disruption to the ecosystem while still providing

hectares (ha) Metric unit of measure for area; 1 ha = 2.5 acres (ac)
maximum sustainable yield Harvesting as much as sustainably possible for the greatest economic benefit.
Multiple–Use Sustained–Yield Act U.S. legislation (1960) mandating that national forests be managed in a way that balances a variety of uses.
Forest Ecosystem Management (FEM) Focuses on managing the forest as a whole, rather than for maximizing yields of a specific product.

Infographic **13.5** | **TIMBER-HARVESTING TECHNIQUES**

→ There are many ways to harvest trees from a forest, each with its own economic and ecological trade-offs. Here are variations of 4 techniques. To evaluate the impact of a particular method, consider what the area would look like 50 years after a harvest.

ORIGINAL FOREST

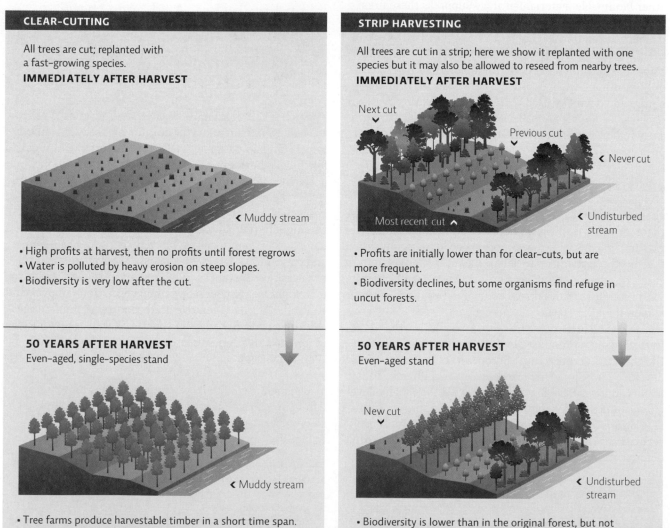

CLEAR-CUTTING

All trees are cut; replanted with a fast-growing species.

IMMEDIATELY AFTER HARVEST

‹ Muddy stream

• High profits at harvest, then no profits until forest regrows
• Water is polluted by heavy erosion on steep slopes.
• Biodiversity is very low after the cut.

50 YEARS AFTER HARVEST
Even-aged, single-species stand

‹ Muddy stream

• Tree farms produce harvestable timber in a short time span.
• Water may be polluted by runoff from the open stand, but to a lesser degree than when clear-cut.
• Tree farm has less biodiversity than the original forest.

STRIP HARVESTING

All trees are cut in a strip; here we show it replanted with one species but it may also be allowed to reseed from nearby trees.

IMMEDIATELY AFTER HARVEST

Next cut ⌄

Previous cut ⌄

‹ Never cut

Most recent cut ⌃

‹ Undisturbed stream

• Profits are initially lower than for clear-cuts, but are more frequent.
• Biodiversity declines, but some organisms find refuge in uncut forests.

50 YEARS AFTER HARVEST
Even-aged stand

New cut ⌄

‹ Undisturbed stream

• Biodiversity is lower than in the original forest, but not as low as clear-cut lands, since a forest remains standing at all times and still usable by some of the wildlife.

economically valuable forest products. Ideally, the stand of trees—its health, age, and species composition—and the slope of the land determine the harvesting method. Consideration is also given to the other species that reside there. When trees are harvested properly, they can provide immediate and long-term economic benefits without serious environmental damage. Even clear-cutting can be appropriate when it increases the overall health and viability of a future forest by removing unhealthy or genetically inferior trees and replacing them with better stock. Clear-cuts followed by planting of a fast-growing, high-value tree (a tree farm) can also reduce the pressure to cut other forests (though there must be enough native forests left in areas suitable for tree farms to avoid serious ecological damage due to biodiversity loss). [INFOGRAPHIC 13.5]

Forest management is not without its critics, however. Conflicting interests make it difficult to achieve a balance between multiple forest uses and ecosystem protection. For example, harvesting trees in the Pacific Northwest provides many jobs, good profits for timber companies, and useful products for our homes and businesses, but can reduce biodiversity and harm salmon runs, which are vital for ecosystem health, native cultures, and the tourism industry.

Debates are often oversimplified and shown as two competing options—owls versus jobs—however, a more accurate cost-accounting of potential forest uses requires that we factor in the economic and ecological value of ecosystem services. Forests are so important economically, ecologically, recreationally, and even spiritually that

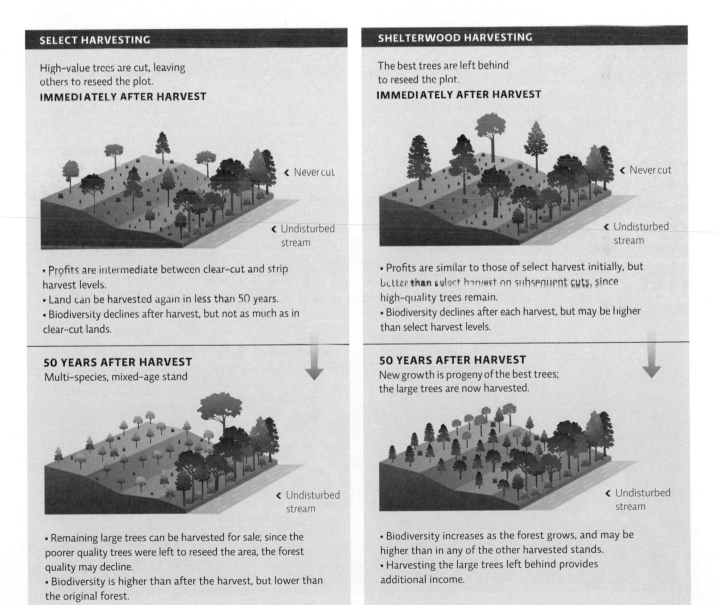

SELECT HARVESTING

High-value trees are cut, leaving others to reseed the plot.
IMMEDIATELY AFTER HARVEST

‹ Never cut

‹ Undisturbed stream

• Profits are intermediate between clear-cut and strip harvest levels.
• Land can be harvested again in less than 50 years.
• Biodiversity declines after harvest, but not as much as in clear-cut lands.

50 YEARS AFTER HARVEST
Multi-species, mixed-age stand

‹ Undisturbed stream

• Remaining large trees can be harvested for sale; since the poorer quality trees were left to reseed the area, the forest quality may decline.
• Biodiversity is higher than after the harvest, but lower than the original forest.

SHELTERWOOD HARVESTING

The best trees are left behind to reseed the plot.
IMMEDIATELY AFTER HARVEST

‹ Never cut

‹ Undisturbed stream

• Profits are similar to those of select harvest initially, but better than select harvest on subsequent cuts, since high-quality trees remain.
• Biodiversity declines after each harvest, but may be higher than select harvest levels.

50 YEARS AFTER HARVEST
New growth is progeny of the best trees; the large trees are now harvested.

‹ Undisturbed stream

• Biodiversity increases as the forest grows, and may be higher than in any of the other harvested stands.
• Harvesting the large trees left behind provides additional income.

no matter how well managed they are, they will always be a contentious resource.

Of course, any gains made by developed countries are still largely offset by deforestation in countries like Haiti. For example, deforestation in Mexico has largely offset new plantings and reforestation in the United States. Central America has lost more than 5 million ha (12 million ac) since 1990 and Europe has gained 12 million ha (30 million ac). "Industrial countries may be leading the way in conserving their own forests," says Morton. "But their demand for wood drives much of the deforestation elsewhere." Multinational corporations have simply moved deforestation operations to developing nations where regulation and enforcement are often lacking and people are desperate for income. They then export products to wealthier countries for sale.

Three months after the kombit visited his hillside farm, Jean Robert is harvesting Moringa leaves. They're tasty, packed with protein, and when harvested properly they regrow rather quickly. If everything goes as expected, he will eventually have a bevy of crops to see him through the year: mangoes in the summertime, coffee in the fall, and if he's lucky, oak and mahogany stands that will yield high prices in the timber market down the road. In the meantime, the Moringa trees provide his family with a sustainable supply of protein and fuel wood.

The trees Georges' team plants can be broken down into three main types. They start with fast-growing, multipurpose trees like the Moringa. Because it is a nitrogen-fixing plant, the Moringa helps refertilize the soil. And because it grows quickly, it can provide a sustainable source of food and fuel wood. Next, they plant fruit trees—mangoes, avocados, and citrus. These trees take longer—3 to 5 years—before they are ready to harvest, but they put down roots and thus stabilize the soil in just a few months' time. "Farmers are less likely to cut down fruit trees for charcoal, because they know it will provide the kind of food they can both eat and sell for profit," says Georges. "Fruit trees are worth something alive." The last thing the kombit will plant are the slow-growing timber trees, which may be sustainably and profitably harvested in the

↓ Women preparing fresh Moringa leaves for cooking. Nicknamed the "tree of life," the seeds, leaves, and roots are all edible. The nutritious leaves are high in protein, vitamins, and minerals, and contain enough iron to treat mild anemia; they can be harvested without killing the tree, eaten fresh, or dried for later use. Cuttings can be planted to start new trees.

← Wangari Maathai's Green Belt Movement of Kenya has spread to more than 30 other African nations. The first Kenyan woman to earn a Ph.D., Maathai received several environmental awards, including the Nobel Peace Prize. She passed away in 2011 but the Green Belt Movement lives on.

future, but don't provide any immediate benefits to the farmers. "The goal is to mix it up," says Georges. "You can create a whole stable system that's going to provide money and food throughout the seasons and across the years."

Of course, for any of this to work, urban-dwelling Haitians will need a new, sustainable energy source.

There are several ways to protect forests, but each comes with trade-offs.

The trade-offs associated with the use of forest resources are the subject of much debate and conflict because each decision impacts both human livelihood and the health of the environment. To be sure, the economic value of wood may be dwarfed by the ecosystem services lost if the area is overharvested, and protecting the ecosystem may actually prove more profitable, even in the short term. But people still need fuel, building material, and income. "Of course trees and forests are important for the environment," says Georges, "We know that they protect us from floods, and help keep drinking water clean and plentiful. But for many Haitians, selling those same trees

is the only way to feed a family. So how can you ask them not to?"

One solution, according to a growing number of experts, is to price the ecosystem services themselves. In Costa Rica, for example, higher utility bills offset the costs of maintaining rainforests that purify and replenish the water people use every day. This money is given directly to landowners who would otherwise have to chop the trees down to sell as fuel or timber or to convert the land to agriculture.

Other options include the promotion and increased availability of sustainable wood products. In developed and even in developing nations, certified sustainable wood products are becoming more available. The Forest Stewardship Council (FSC) certifies lumber and other timber products through a process that evaluates the forest itself and the timber-harvesting techniques in terms of wildlife, water, and soil quality. Worldwide, more than 6% of forests are certified by the FSC as sustainably managed. Other resources, like latex and tree nuts, can be sustainably harvested from standing forests. Alternatives for wood products also exist and can reduce the pressure

↑ Tourists on canopy walkway, Monteverde Cloud Forest Preserve, Costa Rica.

on forests. For example, lumber from old buildings can be salvaged to provide quality building materials. Paper can be made from old paper (recycled) or from fast-growing crops such as kenaf, jute, flax, and hemp. Reforestation projects are increasing in number around the world. For example, the Kenyan Green Belt Movement has planted and protected more than 30 million trees and has sparked similar movements worldwide.

For some, saving the forests may come down to recognizing their intrinsic value: their natural beauty, their inherent sacredness, and their right to exist as a living thing, regardless of what we humans might extract from them. Naturalists from John Muir to Wallace Stegner have argued for the protection and preservation of forests on these grounds alone. "We simply need that wild country available to us, even if we never do more than drive to its edge and look in," 20th-century American naturalist Wallace Stegner wrote in his famous *Wilderness Letter*. "For it can be a means of reassuring ourselves of our sanity as creatures, a part of the geography of hope."

This appreciation for the intrinsic value of a forest has been successfully translated into the economic enterprise known as **ecotourism.** Ecotourism is a viable option for many areas, especially less developed countries. These areas often possess high biodiversity precisely because they are less developed and are often found in tropical or subtropical areas with naturally high biodiversity. In the quest to develop economically, these regions may find that the highest economic value for their resources lies in keeping them intact. Ecotourism allows a way for funds to enter the country while protecting the natural areas at the same time.

But for many people, including those in Haiti, these options seem limited. Charcoal produced from wood is the only means people have to heat their homes, prepare their food, and in the cities, fuel their businesses. "We won't solve deforestation until we increase the market

---

ecotourism Low-impact travel to natural areas that contributes to the protection of the environment and respects the local people.

share of other energy sources," says Morton. "Right now everyone—bakers, rum makers, housewives, even factories—are dependent on charcoal." In fact, small businesses use the equivalent of 27 ac of wood each day.

"It's a vicious cycle," says Morton. "They cut down trees to make charcoal to cook and earn some money. This erodes the soil, which causes crops to shrink and floods to be more severe. Both of those render people even more impoverished, which in turn necessitates more tree-cutting." Charcoal, he says, is the fuel of the poor, and unless Haitians escape the poverty trap, they won't stop using it. In fact, declining soil fertility and falling commodity prices have led many rural Haitians to increase their charcoal production to generate more income.

In the countryside, Georges says, charcoal use will probably continue no matter how well reforestation efforts work. "It's not a matter of weaning people off charcoal," he says. "The use is ingrained into the culture. The real issue is sustainable practices." Georges notes that simple changes like harvesting only fast-growing trees for charcoal, pruning branches to encourage more growth and cutting trees 1 meter above the ground rather than at ground level for quicker regeneration could make a huge difference.

To meet intensive energy needs of the city, however, other measures will be needed. One potential alternative to charcoal that could be used in Haiti and elsewhere is Jatropha, a fast-growing plant whose oil-rich seeds have been hailed as a promising biofuel source. Even the material left after the oil is pressed out can be digested by bacteria to produce biogas, a fuel similar to natural gas. Another option is the production of composite briquettes from a variety of flammable materials such as grass, paper, or sawdust. Shredded plastic is added to make it more combustible (see Chapters 28 and 29 for more on sustainable energy).

There is a Haiti that people like Georges and Robert talk about in the quiet moments after a day's planting. It's a Haiti lush and green with trees, a country where families earn their living selling mangoes and Moringa leaves instead of charcoal and firewood. Whether this country resembles their past as much as their future will depend on an infinite number of variables—not just how well they manage their forests and reforestation efforts, but also whether they can find alternative building materials or establish a reliable energy sector based on something other than wood.

In parts of Haiti, where the top soil has long since given way to a barren landscape of rock, most experts agree that it is too late. But in much of the country there are still some trees left. And that means there is still hope.◉

Research article referenced in this chapter:
Constanza, R., et al. 1997. Nature, 387: 253–260.

BRING IT HOME

❯ PERSONAL CHOICES THAT HELP

Forests are a renewable resource that can be used sustainably for many years under proper management conditions. Using harvesting systems such as selective cutting or strip cutting can allow economic use of the forests without eliminating the ecological functions that forests provide.

Individual Steps
→ Buy paper products (toilet paper, facial tissue, and notebook paper) made of recycled content to decrease the unnecessary cutting of trees and to encourage recycling.
→ When purchasing lumber for projects, look for wood that has been certified as sustainably managed by the Forest Stewardship Council.

→ Avoid buying noncertified furniture made from tropical wood such as rosewood, teak, and ebony; harvesting these woods contributes to tropical deforestation.

Group Action
→ Organize a workday to clear invasive species such as common buckthorn and Japanese honeysuckle to ensure that our remaining forests provide high-quality habitat.

Policy Change
→ Support legislation that protects roadless areas and old-growth forests.
→ Ask your local grocery stores, restaurants, and college dining facilities to offer shade-grown coffee and chocolate, which can help encourage preservation of tropical forests.

UNDERSTANDING THE ISSUE

CHECK YOUR UNDERSTANDING

1. High-latitude and high-altitude forests characterized by a short growing season, a lot of snowfall, and thin, acidic soils are:
    a. temperate forests with trees that drop their leaves in winter, such as oak and maple.
    b. boreal forests with evergreen trees that bear needlelike leaves, such as spruce and fir.
    c. savanna forests with trees that have long taproots and thick, fire-resistant barks, such as acacia and eucalyptus.
    d. tropical forests with trees that are highly valued as timber for their rot-resistant wood, such as mahogany and teak.

2. The understory of a forest is made up of:
    a. trees that push through and grow above the level of the forest canopy.
    b. the seedlings, ferns, herbs, and wildflowers that grow on the forest floor.
    c. the overlapping crowns of the tallest trees that make up the roof of the forest.
    d. shade-tolerant shrubs or the saplings of larger trees that sometimes form a lower canopy.

3. Which of the following is NOT a cause of deforestation in Haiti?
    a. Palm oil plantations for biofuel production
    b. Coffee and sugar plantations for export crops
    c. Wood charcoal production for the domestic energy market
    d. Timber harvesting for commercial sale

4. How does deforestation contribute to loss of drinking water in Haiti?
    a. Without trees, the eroded soil clogs drinking water pipes.
    b. The loss of forest habitat fragments aquatic ecosystems that provide water to the local villages.
    c. Without the trees, the water rushes away rather than soaking into the ground to recharge groundwater supplies.
    d. Clearing the forest for crops means increased irrigation, which reduces drinking water supplies.

5. The timber-harvesting system that would be most likely to cause disruption to the ecosystem services provided by a forest is:
    a. shelterwood harvesting.
    b. strip harvesting.
    c. clear-cutting.
    d. select harvesting.

6. Which of the following activities will NOT simultaneously protect forests and provide for long-term economic well-being of local people in developing countries?
    a. Pricing ecosystem services provided by the forest and paying landowners to maintain their trees
    b. Clear-cutting the forest and planting fast-growing trees for charcoal production
    c. Promoting the harvesting and selling of wood that is certified by the Forest Stewardship Council
    d. Translating the intrinsic value of a forest into ecotourism enterprises

WORK WITH IDEAS

1. Tropical rain forests have thin, acidic soils yet they contain dense vegetation and high biodiversity. How can these tropical forests have poor soil but support such diverse arrays of plant and animal life?

2. What is an "ecosystem service"? Describe three such services provided by forests.

3. How is the current status of forests different in developing versus developed countries? What factors account for these differences?

4. How can we sustainably use forest resources?

5. What is the reforestation strategy employed by the Haitian kombit? Explain the rationale behind the approach and discuss whether it can be successful.

ANALYZING THE SCIENCE

The data in the following table comes from the most recent assessments of global forest resources conducted by the Food and Agriculture Organization of the United Nations. This table reports on the global, regional, and subregional trends in forest areas designated primarily for the protection of soil and water.

INTERPRETATION

1. How much total forest area was designated for protection of soil and water globally in 2000? In 2010? (Hint: Don't forget to account for the units of measure shown in the header.)

2. Which subregion of Asia had the most land designated for protection of soil and water in 1990? Was this the same in 2010? Provide the data to support your responses.

Region/subregion	Information availability		Area of forest designated for protection of soil and water (1,000 ha)		
	# of countries	% of total forest area	1990	2000	2010
Eastern and Southern Africa	21	80.9	14,033	13,311	12,611
Northern Africa	7	99.1	4,068	3,855	3,851
Western and Central Africa	22	52.5	2,639	3,281	3,079
Total Africa	**50**	**69.2**	**20,709**	**20,447**	**19,540**
East Asia	4	90.2	24,061	38,514	65,719
South and Southeast Asia	17	100.0	55,811	57,932	56,501
Western and Central Asia	23	99.7	12,222	13,059	13,669
Total Asia	**44**	**95.8**	**92,094**	**109,505**	**135,889**
Total Europe	**45**	**99.7**	**76,932**	**90,788**	**92,995**
Caribbean	11	53.1	869	1,106	1,428
Central America	3	36.9	124	114	90
North America	5	100.0	0	0	0
Total North and Central America	**19**	**97.8**	**994**	**1,220**	**1,517**
Total Oceania	**18**	**21.6**	**1,048**	**1,078**	**888**
Total South America	**10**	**85.1**	**48,656**	**48,661**	**48,549**
World	**186**	**86.9**	**240,433**	**271,699**	**299,378**

3. Summarize the trends in forest lands set aside for the protection of soil and water:
   a. Between 1990 and 2010, how much has the total (global) area grown or decreased?
   b. Between 1990 and 2010, which major region showed the most gain and how much did it gain?
   c. Between 1990 and 2010, which major region showed the most loss and how much did it lose?

ADVANCE YOUR THINKING

Hint: To answer some of the following questions it might help to access the actual report, at www.fao.org/docrep/013/i1757e/i1757e00.htm (see Chapter 6 of the report—Protective Functions of Forest Resources).

4. What might explain the patterns seen in the table?

5. Why does the United States not report any forest area designated for protection of soil and water? How might this affect how the data presented in the table is interpreted?

6. If the world population is 7 billlion, what is the per capita amount of forest land that is set aside for soil and water protection?

EVALUATING NEW INFORMATION

Many of the world's forests are severely degraded. Yet forests produce many consumer products that we depend on, such as food, medicine, building materials, and raw materials for industrial products like paper. So what should a conscientious consumer do?

Explore the Forest Stewardship Council (FSC) website (www.fsc.org).

Evaluate the website and work with the information to answer the following questions:

w
1. Is this a reliable information source? Does the FSC have a clear and transparent agenda?
   a. Who makes up this organization? Does its membership make the FSC reliable/unreliable? Explain.
   b. What is this organization's mission? What are its underlying values? How do you know this?

c. Does the FSC give supporting evidence for its claims about forest resources and its vision to address the problem? Does the website give sources for its evidence?
d. Identify a claim the FSC makes and the evidence it gives in support of this claim. Is it sufficient? Explain.
e. Do you agree with its assessment of the forest issues? Explain.

2. Select the link "Be part of the solution."
   a. Who can "be a part of the solution" and how?
   b. Identify a specific solution and the strategy to accomplish it. Is the solution sufficient and the strategy reasonable? Explain.
   c. Do you agree with the FSC's solutions? Explain

3. How can a consumer select FSC-certified products, according to this website? Is it easy for consumers to know if the wood products they are purchasing are FSC certified? Explain.

4. How might FSC certification help forests?

MAKING CONNECTIONS

INDONESIA'S PALM OIL DILEMMA

Background: Like those in Haiti, Indonesia's rainforests could potentially be a renewable energy source. But unlike Haiti with its charcoal production, Indonesia leads the global production of palm oil, a commodity for export as a biofuel, cooking oil, and ingredient in industrial products such as cosmetics. Palm trees can produce fruit for 30 years or more, and yield more oil per hectare than other oilseed crops. But the cost of these benefits is the loss of large stands of tropical forest to allow for plantations of palm trees. In Indonesia, the increasing demand for palm oil has accelerated seizure of indigenous peoples' land, dismantling local communities and cultures and threatening the habitat of endangered species like the orangutan.

A low-carbon alternative to fossil fuel–based gasoline, palm oil biodiesel was seen as a solution to climate change, but recent research revealed that palm oil development involving the clearing of intact tropical rainforest contributes more greenhouse gases than it helps to avoid. In fact, Indonesia already emits more greenhouse gases than any other nation besides China and the United States.

Case: You have been assigned to develop a policy for the future of the Indonesian rainforest. Your team must evaluate various options such as:

1. Protect Indonesian rainforests through the United Nations Reducing Emissions from Deforestation and Forest Degradation (REDD+) program, which offers financial incentives to developing countries to sustainably manage forests to sequester carbon. According to UN estimates, the financial value of greenhouse gas emission reductions through REDD+ could be US$30 billion a year.

2. Continue producing palm oil but seek a sustainable biofuel certification (similar to the FSC certification for wood). According to the Palm Oil Action Group, there are millions of hectares of degraded land in Indonesia that could be reclaimed for palm oil plantations.

Research these (and possibly other options) and write a report recommending a course of action. In your report, include the following:

a. An analysis of the pros and cons of each proposal, including: a discussion of the consequences of each choice on local human communities as well as on the rainforest ecosystem. What is the best option for Indonesian rainforests? Who should be involved in this selection? Provide justifications for your proposal.
b. What overall lessons can be learned from the stories of Haiti and Indonesia and applied to other forest ecosystems in jeopardy? Explain your suggestions.

At home on the range, bison graze the slopes of their refuge on the Fort Peck Reservation in Montana.

RESTORING THE RANGE

The key to recovering the world's grasslands
may be a surprising one

CORE MESSAGE

Grasslands are a critical resource, offering many important ecosystem services as well as being used for grazing of livestock and even production of biomass for biofuel energy. Grasslands are found all over the world, but are currently endangered by overuse and a changing climate. "Desertification" caused by overgrazing is the most common problem facing grasslands, but innovative practices in managing livestock can help protect the world's grasslands.

GUIDING QUESTIONS

After reading this chapter, you should be able to answer the following questions:

→ What is a grassland? What kinds of grasslands are there and where are they found?

→ Why are grasslands important?

→ How do overgrazing and undergrazing affect grasslands?

→ What is involved in the formation of soil and how does land use affect it?

→ How can we manage grasslands to lessen the threats they face while still using them productively?

When Jim Howell, an ecologist and fifth-generation cattle rancher, first announced his plans to revive Horse Creek Ranch in Butte County, South Dakota, friends said he was either crazy or foolish. Sure, the ranch had once encompassed some of the best grazing land in the country, but persistent drought and too many cattle had long since brought those days of plenty to a close. Like so many ranches around it, Horse Creek's pastures were too parched and degraded to sustain much of anything, let alone an entire herd of cattle.

Across the Great Plains, **rangeland** (land that humans use to graze livestock) is drying out and ranchers are growing desperate. In some places, the degradation is so bad that prairie grasslands are becoming deserts.

Normally this process, known as **desertification**, is both natural and slow. As climates shift over geologic time, grasslands morph into desert and deserts back into grassland in a cycle both never-ending and imperceptibly gradual. These days, it's occurring much more rapidly than that—10 feet per year in West Texas alone. Part of the problem is climate change. But most experts agree that the biggest culprit, by far, is **overgrazing**—when too many animals feed on a given patch of land.

Desertification is not unique to the Great Plains. Around the world, from Afghanistan to Zimbabwe, 70% of the planet's rangeland is threatened. This represents roughly one-third of the world's entire land surface. And while the phenomenon has not garnered as much media attention as, say, global climate change, the consequences are no less dire. In fact, from the Fertile Crescent of ancient Babylon to modern-day developing countries like Darfur, desertification has been the stuff of wars. The cascade is both simple and devastatingly comprehensive: plants die, soil erodes, prairies fall to dust, famine sets in, economies falter, societies fail.

Since the early 1980s, the U.S. government has paid hundreds of thousands of ranchers to quit ranching so that vast swaths of degraded grassland can have a chance to recover. But critics say that holding the land in such **conservation reserve programs** decimates local prairie economies. "There are real environmental benefits to placing the land in trust and saying it can't be grazed or farmed over anymore," says Howell. "But it leaves farmers dependent on government checks and does nothing to stimulate the local economy."

Starting with Horse Creek—a forgotten ranch, in a fading prairie town—Howell and his business partners, who took control of the ranch in 2008, are working on a different solution, one that aims to repair economy and environment in tandem. The bottom line, they say, is this: if you want to save the prairies, you've got to graze more cattle, not less.

⊙WHERE IS HORSE CREEK RANCH, SOUTH DAKOTA?

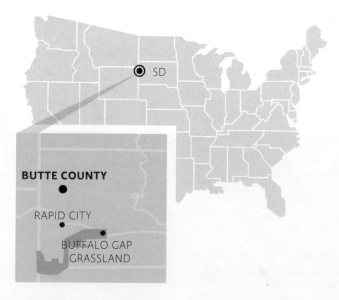

BUTTE COUNTY

RAPID CITY

BUFFALO GAP GRASSLAND

Grasslands provide a wide range of important goods and services.

Grasslands are biomes that receive enough rainfall to support grass and herbaceous plants, but not enough to support forests. They may also be found in regions where rainfall is plentiful but larger plants are kept in check by periodic fires or herds of grazing herbivores. Broadly, there are several different types: *tropical grasslands*, also known as savannas, occur in places that have both rainy and dry seasons, but are warm year-round. *Cold grasslands* like steppes are, as the name suggests, cold most of the

↑ Cattle and goats have pulverized the drought-prone Omo, Ethiopia, region into dust.

year, and are characterized by a very short growing season and ultrathin layers of soil. The Great Plains—which lies between the Mississippi River and the Rocky Mountains and stretches from the south of Texas into Canada—is *temperate grassland*. These grasslands, known as prairies, contain many species of plants, have thick soils, and have a truly seasonal climate with cold winters and hot summers. [INFOGRAPHIC 14.1]

Though we might most often think of grasslands in terms of their human uses (pasture, farmland), they provide extremely important ecosystem services such as nutrient cycling, soil formation and protection, carbon sequestration, protection of surface waters, and provision of habitat for both year-round and seasonal wildlife. But it is also true that for thousands of years, grasslands—which cover about 40% of Earth's surface and contain some of the richest soil in the world—have provided humans with a multitude of goods and services. In fact, most major cereal crops, including wheat, rye, barley, and millet, were originally derived, thousands of years ago, from grassland seedbeds. Today, scientists trying to develop disease-resistant crops continue to rely on the genetic material found in grasslands. Meanwhile, other scientists are

perfecting ways to harvest grassland plants as an energy source (see Chapter 29 for more on biofuels).

But the most widespread, human-centered use of grass lands is as food for large grazing herd animals.

In addition to supporting countless wild ungulate (hoofed animals) grazer herds, a little more than half of the world's grasslands—26% of the planet's land surface—are used to graze more than 3 billion domestic livestock (mostly cattle, goats, and sheep). This type of grazing provides the primary source of food and income for some 2 million of the world's poorest people. And while other livestock (like chicken and pigs) consume vast stores of grain that might otherwise go to humans, grazing animals

rangeland Grassland used for grazing of livestock.
desertification The process that transforms once-fertile land into desert.
overgrazing Too many herbivores feeding in an area, eating the plants faster than they can regrow.
conservation reserve program Farmers and ranchers are paid to keep damaged land out of production to promote recovery.
grasslands A biome that is predominately grasses, due to low rainfall, grazing animals and/or fire.

Infographic **14.1** | **GRASSLANDS OF THE WORLD**

↳ Grasslands are found on every continent except Antarctica. The examples of grasslands shown here vary based on climate.

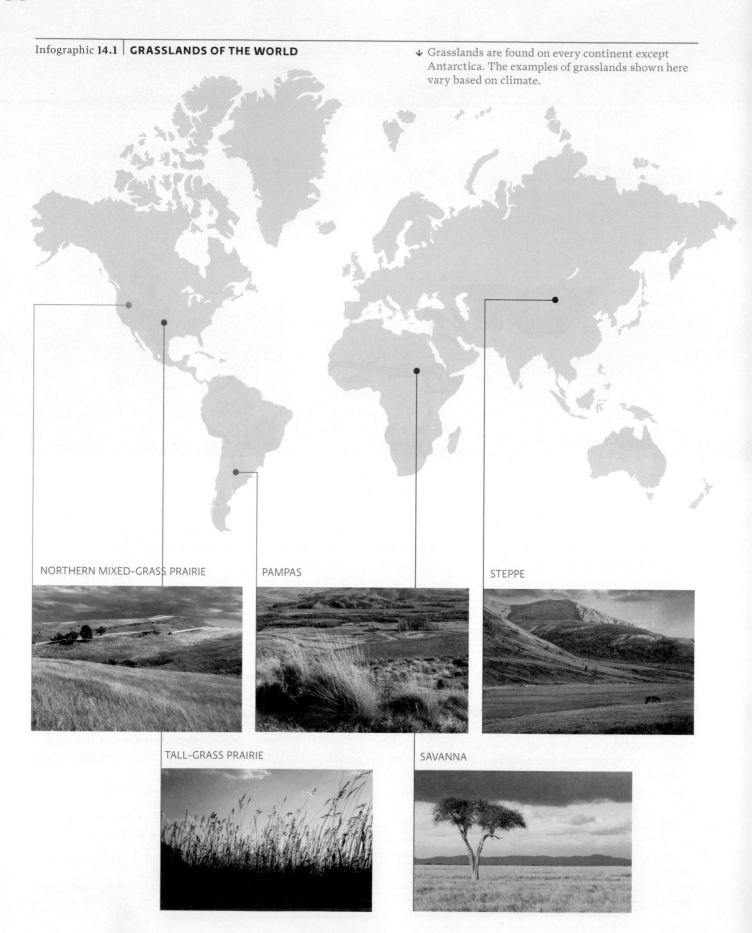

NORTHERN MIXED-GRASS PRAIRIE

PAMPAS

STEPPE

TALL-GRASS PRAIRIE

SAVANNA

are effectively converting food we cannot eat (grass) into food we can eat (meat and dairy). [INFOGRAPHIC 14.2]

Grasslands face a variety of human and natural threats.

A trifecta of forces now threatens all of these goods and services. First, global climate change: As temperatures rise, scientists expect that shifting rainfall patterns will help push many current grasslands—including vast swaths of the great American prairies—into desert. The second threat involves human land-use decisions: Rising global populations will force us to convert more land into cities and suburbs for living, and into croplands for food. Their wide, flat expanses and nutrient-rich soil make grasslands ideal for both. The third, and many say largest threat that grasslands face, is the one posed by overgrazing.

Ironically, grazing is normally very good for grasslands. Because wild **herbivores** evolved to subsist on grasslands alone, both grazers and grasses are well adapted to grazing. Grasses can grow from the base upward, so by clipping off the top part of the blade, herbivores expose new growth shoots to the sunlight, thus stimulating the plants' growth. By breaking up the soil with their hooves, they allow water to penetrate the ground and enable seeds to germinate and take root. And as they defecate and urinate, large grazing mammals return nutrients such as nitrogen and phosphorus to the soil in a form that plants can easily absorb.

But when grass is overgrazed, or chewed down to its roots, the growth area on the blade is destroyed, the blade can no longer regenerate, and the plant eventually dies. When too many plants die at once, the soil has nothing to hold it in place. And when too many large grazing animals stomp their hooves directly onto the soil, the soil becomes compacted, which makes it harder for water to penetrate, seeds to germinate, or seedlings to grow. [INFOGRAPHIC 14.3]

Together, plant loss and soil compaction increase the rate of **soil erosion**—a process in which soil is swept away by wind and rain down into streams, rivers, and gullies, faster than it can possibly be replenished. That's no small matter. Soil formation is a slow process that requires the weathering of rock and decomposition of organic material. Under the best conditions, it takes 1 year to generate just a millimeter of the precious brown gold in which our food

herbivore An animal that feeds on plants.
soil erosion The removal of soil by wind and water that exceeds the soil's natural replacement.

Infographic **14.2** | **GRASSLAND GOODS AND SERVICES**

WILDLIFE HABITAT

CONVERSION TO AGRICULTURE

GRAZING

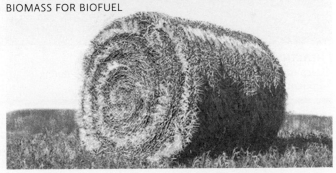

BIOMASS FOR BIOFUEL

243

Infographic **14.3** | **DESERTIFICATION**

↘ Every inhabited continent has grasslands vulnerable to desertification, especially in arid areas close to existing deserts.

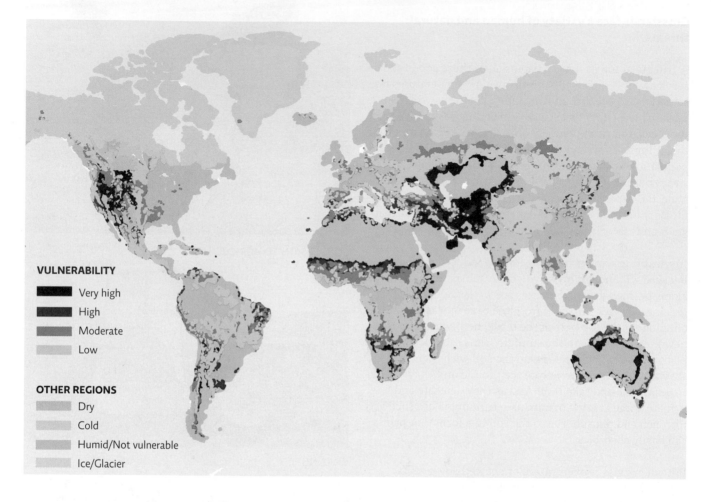

VULNERABILITY

- Very high
- High
- Moderate
- Low

OTHER REGIONS

- Dry
- Cold
- Humid/Not vulnerable
- Ice/Glacier

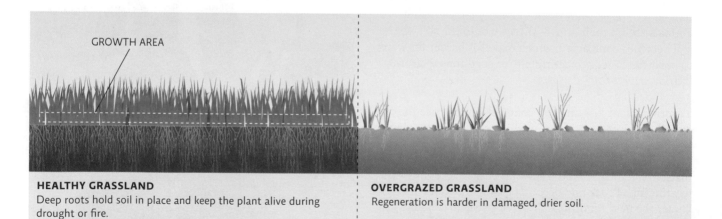

GROWTH AREA

HEALTHY GRASSLAND
Deep roots hold soil in place and keep the plant alive during drought or fire.

OVERGRAZED GRASSLAND
Regeneration is harder in damaged, drier soil.

↑ Grasses are adapted to grazing; cropping the grass stimulates the growth area at the base of the blade. Overgrazing may remove this growth area and kill grasses, increasing the potential for desertification.

Infographic **14.4** | **SOIL FORMATION**

↓ Soil is produced by the decay of organic material and the weathering of rock. Distinct layers are seen in healthy soils, with the topsoil (A horizon) being the most fertile for plant growth. Desertification will reduce or remove the O and A horizons and produce drier B and C horizons.

Desertification results in the loss of the O and A horizons.

↑ Native prairie grasses have deep roots (up to 5 meters [16 feet] long) which allow them to access deep water supplies and to weather droughts. The native grass roots also hold the soil in place much better than do shallow-rooted annual crops like wheat.

O horizon — Surface litter

A horizon — Topsoil: Contains decaying organic matter (humus) and living soil organisms

B horizon — Subsoil: Denser than A horizon, higher mineral content, lower fertility

C horizon — Contains rock in the process of being broken down (weathering) to produce new soil

R horizon — Solid rock that has not been broken down

grows. A closer look at a soil's profile sometimes reveals layers (called horizons) that reflect the formation process. As plants die out and soil erodes, the denuded landscape begins to reflect incoming sunlight rather than absorb it; this touches off a cascade that ultimately alters wind and temperature patterns. Before long, grasslands give way to deserts. [INFOGRAPHIC 14.4]

At least once in America's history, human activities so amplified the speed and scope of desertification that it triggered the largest human migration inside a decade our nation has ever seen. Beginning in 1934, clouds of dust so massive that ranchers named them "black blizzards" swirled relentlessly across the Great Plains, tearing the paint off of houses and cars and forcing some 2.5 million homesteaders from the land. History named this time and place the Great Dust Bowl, and it became a classic example of the tragedy of the commons. As each individual homesteader expanded farm and ranch for his own gain, the land as a whole fell victim to overgrazing. When the next drought cycled through, there was no grass left to hold the crumbling soil in place.

It took nearly three decades for the prairies to recover, but according to some critics, only half as long for us to forget the Dust Bowl's most important lessons. By the 1970s, with the market prices for agricultural goods rising steadily, Plains farmers were overplowing and overgrazing with the same pre–Dust Bowl fervor. Now, soil erosion is approaching Dust Bowl rates. The problem is worst in New Mexico and western Texas, where the Chihuahuan Desert claims 10 feet of grassland per year. But other swaths of prairie, including the ones in Butte County, have been plagued with drought, and are slowly being abandoned.

So far, grassland degradation has cost humans roughly 12% of global grain production, not to mention $23 billion per year in global GDP (gross domestic product). All told, the food supply of more than 1 billion people is threatened. The Food and Agricultural Organization (FAO) predicts that some 50 million people will be faced with displacement in the coming decades. The majority of these will be subsistence farmers who live in the world's poorest regions and depend solely on cattle ranching for their livelihoods; but U.S. ranchers, like the ones who owned Horse Creek, will also suffer.

Scientists around the world have spent decades trying to prevent or even reverse desertification, to little avail. These days, most experts tend to agree that beyond a certain point, recovering grasslands that have swirled into deserts is impossible. "Most of our efforts to reverse desertification have failed dismally," says Dr. Richard Teague, a research ecologist working with ranchers in West Texas to restore degraded rangeland. "But a number of ranchers here are having success with protocols developed halfway around the world."

Taking our cues from nature, we can learn to use rangelands sustainably.

As the sun rises over Zimbabwe (formerly Rhodesia) in southern Africa, herds of antelope and zebra traverse a patchwork of temperate and tropical grasslands, feeding steadily on reedy stalks and short, fat shrubs; elephants and wildebeest splash around a precious watering hole, well fed and content. The animals may not realize it, but they have stumbled upon the African Center for Holistic Management (ACHM), 6500 acres of thriving rangeland in the heart of an otherwise parched and ailing prairie. [INFOGRAPHIC 14.5]

Perhaps nowhere else on Earth is such an oasis more urgently needed. Because the region is too arid to support much else, ranching provides the only livelihood for most of the people living there; about 75% of all land is used to graze domestic herds of cattle, goat, and sheep, and even that has not been enough. With population, and thus the number of mouths in need of food, rising steadily, farmers have crowded more and more livestock onto lands that grow sparser and drier by the day. Already stressed by climate change, that land is crumbling quickly into desert. And as viable pastures become increasingly difficult to find, neighboring tribes have descended further into violent conflict—sometimes killing each other over a few stalks of grass.

The ACHM was established in 1992 by Allan Savory, a Rhodesian-born scientist-turned-rancher. Before then, the land had been so thoroughly desertified that neither wild nor domestic herds bothered to graze there. But in the nearly two decades since, the picture has changed dramatically. Both plants and wild herds have rebounded with surprising speed; even during the dry season, water is plentiful enough to sustain fish and water lilies. And if that's not enough, livestock has increased by 400%. Ranchers come from as far as Texas to marvel at the

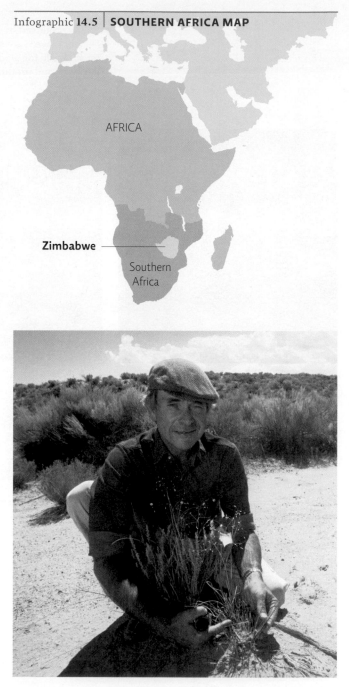

Infographic **14.5** | **SOUTHERN AFRICA MAP**

AFRICA

Zimbabwe ————

Southern Africa

↑ Biologist Allan Savory, squatting beside a patch of dry grasses in the desert where he teaches holistic land management.

turnaround and to seek counsel from Savory, who is widely credited for the dramatic recovery.

Savory came by his expertise in a circuitous way. He spent the 1950s working as a research biologist and game ranger for the British Colonial Service. At the time, the British government was culling thousands of wild herds in an effort to create more land for farming. "Zimbabwe

Infographic **14.6** | **HOW IT WORKS: GRAZING AND IMPACT ON GRASSES**

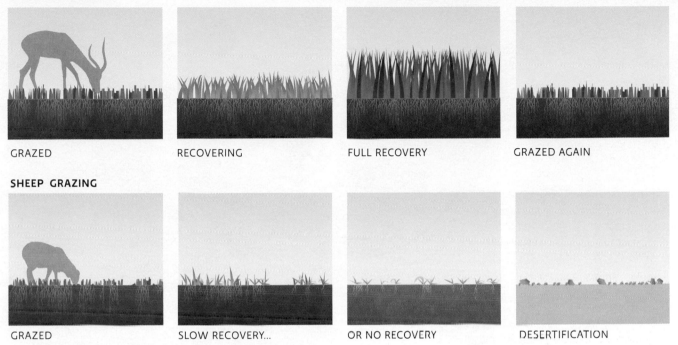

WILDLIFE GRAZING

GRAZED

RECOVERING

FULL RECOVERY

GRAZED AGAIN

SHEEP GRAZING

GRAZED

SLOW RECOVERY...

OR NO RECOVERY

DESERTIFICATION

↑ Savory noticed that wild herbivores grazed intensively and then moved on, allowing the grasses to recover. Livestock allowed to stay on a pasture too long will damage the plants, which slows or prevents recovery.

was infamous for the veterinary policy of shooting out all game over large areas," Savory says. "The unofficial slogan was, 'you cannot ranch in a zoo.'" That policy bred tension between ecologists like Savory who wanted to preserve the splendor of Africa's natural environment, and ranchers and government officials who wanted to farm right over it. Like many of his colleagues, Savory developed a healthy disdain for the cattle that were slowly replacing his beloved zebra, elephants, and antelope, and a firm belief that the land he loved was being destroyed by cattle ranchers who crowded too many animals onto too little land.

As he patrolled the vast terrain, Savory noticed that lands grazed by wild herds were healthier than those managed by cattle ranchers; plants were more abundant and diverse, rivers were cleaner and better stocked. "The wild herds were doing basically the same thing as the domestic herds," he says. "They were eating the grass. But they were having the exact opposite effect." He noticed that as they tried to protect themselves from predators and avoid feeding on their own feces, wild herds grazed in tightly bunched packs, and moved quickly from one patch of land to the next. They would stay just long enough to

fertilize the ground with their waste and agitate the soil without compacting it, and they would not return until the dung had been absorbed and the land was clean again. This meant that grasses were eaten down severely over a short period, but then left for a long time to recover. The result: animals fed and grasses regenerated in an endless and mutually beneficial cycle. [INFOGRAPHIC 14.6]

Once upon a time, **pastoralists**—individuals who herd and care for livestock as a way of life—mimicked the processes laid out by Mother Nature: grazing their stock in tight herds, moving them constantly across vast swaths of rangeland. But in the 19th and 20th centuries, ranchers began partitioning their grasslands into distinctly fenced-in pastures and dividing their livestock so as to control them more easily. This made ranching easier, to be sure. But as time wore on, farmers found that their lands were being overrun by shrubs and weeds. The cattle were selectively eating only the sweetest tasting grasses and leaving all the less palatable varieties to flourish. Season by season, the plant species composition shifted,

pastoralists Individuals who herd and care for livestock as a way of life.

↑ A pastoralist, a Rabari tribal shepherd herding the sheep home in Gujarat, India.

and as it did, the land became less productive. Less sweet grass meant less cattle feed, which in turn meant thinner cattle, less food and, ultimately, thinner profits. And as scientists like Savory began to notice this, a simple idea took root: The lands were being destroyed because too many animals were feeding off of them. They decided that trimming the populations of both domestic and wild herds would be the key to salvation.

But, as scientists—including Savory himself—soon realized, undergrazing presented its own set of problems. "Where we culled too many animals, there was nothing to eat the grasses," Savory says. "So they would die standing upright, and the detritus would prevent the sun from reaching the growth buds." This also meant that instead of being returned to the soil through animals' digestive tracts, nutrients would instead be processed by soil microbes—a much slower and thus less efficient process. And because there were fewer of them, animals could be more selective about which plants to eat; left untouched, unpalatable weeds and plants began to take over the pastures. [INFOGRAPHIC 14.7]

U.S. ranchers were experiencing similar problems, and by the 1970s the American farming industry had caught on. Land-grant universities in Texas and Arizona designed machines like the Dixon Imprinter that simulated the physical effects of large grazing herds—breaking soil crusts and laying down plant litter over vast swaths of pasture. Of course, those machines could not cycle nutrients the way animal digestive systems could, and so the colossal machine solved some problems (detritus no longer blocked the sun from growth shoots), but not others (soil quality still dropped because nutrients weren't cycled as effectively).

Around the same time back in Africa, after two decades of surveying the land and observing both ranching and wild grazing in action, Savory was certain that ranching practices were contributing heavily to land degradation. He was equally certain that humans could reverse the process by managing the land differently—but he didn't know exactly how. To figure it out, he would have to design and implement several different methods of ranching so that he could compare them to one another.

Infographic **14.7** | **INCREASING PROPORTION OF UNPALATABLE SPECIES**

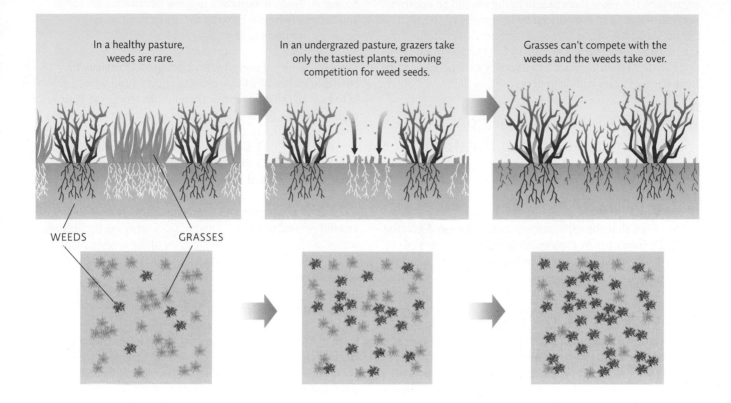

In a healthy pasture, weeds are rare.

In an undergrazed pasture, grazers take only the tastiest plants, removing competition for weed seeds.

Grasses can't compete with the weeds and the weeds take over.

WEEDS   GRASSES

↑ When lands are undergrazed, animals will only eat the tastiest plants, ignoring the less palatable weeds. This gives the weeds a competitive edge, allowing them to quickly take over the area.

And to do that, he would need to persuade the ranchers of Zimbabwe to let him experiment on their lands.

At first his imploring fell on deaf ears. Everyone agreed the lands were ailing, but government officials blamed the lack of rain, combined with overgrazing in some areas. Ranchers insisted they were already taking great care not to graze too many animals on too small a patch of land. What more could they do? "I went to the government scientists, and I tried talking to the ranchers themselves," he says. "Nobody wanted to hear that they weren't doing things in exactly the best way."

Then one night, a couple of ranchers paid him a visit. They'd been doing everything the government advised, they said, and both their ranches, which sat side by side on the open plains of Bulawayo, Zimbabwe, had enjoyed ample rainfall that season. But their grasses were not rebounding and their businesses were suffering. Could he help them? "I told them I didn't have any answers yet, just a sense that we were causing the land degradation ourselves, and that we could fix it by changing the way

we ranched," he says. "We agreed to be the blind leading the blind."

Savory's plan was based on biomimicry—grazing livestock the same way wild herbivores grazed. Borrowing some lessons from his time in the British Army, he devised a technique that called for controlling herds' movements with military precision. He used electric fencing to divide the land into small paddocks, and then pulled all the livestock together into just one such paddock. Then he allowed the animals a day or two to eat everything they possibly could before releasing them into the next paddock. By that time, the first paddock had been churned into lumps of soil, dung, and freshly exposed growth buds. And so the herd would move, from paddock to paddock, devouring and fertilizing each patch of land as it went. By the time they had moved through the last paddock, they would be ready for market. Each paddock would then have an entire season—roughly 180 days, depending on how the rains fell that year—to recover.

The concept was not entirely new. **Rotational grazing**—where animals are allowed to graze on a small section of pasture for a few days before being moved to another section—was first introduced by scientists of the 18th century and has since gained widespread acceptance by ranchers around the world. But so far, the technique had done little to restore plant biodiversity or stop the transformation of grassland into desert.

> Savory's plan was based on biomimicry—grazing livestock the same way wild herbivores grazed.

Savory says that the most widely used rotational grazing methods focus too heavily on limiting the number of animals allowed to graze and not enough on the amount of time they spend grazing any given patch of land. Overgrazing, he says, is a function of time, not numbers. In fact, he says, high stock densities are actually better for the land because they reduce animal selectivity (that is, hungry animals crowded together will eat whatever they can get their teeth on, including the less sweet-tasting varieties of grass that would normally be left to overtake the pasture) and thus help preserve biodiversity. "The results we are seeing at our center today are because of a 400% increase in the number of animals we graze, not despite it," he says. "I have to make that point to almost every scientist and rancher that passes through."

In 1960s Zimbabwe, degraded land gave way to starving people and civil unrest. Amid this turmoil, Savory and his wife were exiled, along with hundreds of others whom the government decided were activists or political dissidents. They took their work to the United States, where they found a surprisingly similar state of land-use affairs. Government agencies and land-grant colleges around the country scoffed at the idea of increasing the number of livestock as a means to combating desertification. But as individual ranchers caught wind of Savory's work, they began seeking him out.

Counteracting overgrazing requires careful planning.

One of these ranchers was Jim Howell. Howell had grown up on his family's ranch in Colorado, and was studying to be a veterinarian when he came across Savory's work in the mid-1990s. "I saw quickly that the way my family had managed the land, going back to when they first bought it in 1937, was hurting it," he says. "We were losing biodiversity, losing soil, and every year the land

was yielding up less profits than it had the year before." Convinced that being a good rancher meant being a good ecologist, Howell switched majors, became an ecologist and promptly began looking for opportunities to put Savory's methods to the test.

Eventually, he partnered with a group of investors to create Grasslands, LLC, a private equity fund whose aim is to buy up failing ranches throughout the Great Plains region and use Savory's holistic management techniques to resuscitate them. In 2010, the group made its first purchase: Horse Creek and a neighboring ranch that together hold 23 square miles of degraded South Dakota rangeland—roughly equal in size to the island of Manhattan.

With careful planning, Howell's ranching team can work around several seasonal constrictions at once, including water availability, the bloom cycles of poisonous plants, and the migration patterns of various wild animals. Not returning to the same small paddock for a full year also breaks the reproductive cycle of many parasites. And keeping animals tightly bunched for a set period of time ensures that all plants are grazed equally—not just the sweetest grasses, but the weeds, too. This actually gives the better-tasting plants an advantage: Because they don't waste as much energy on chemical defenses (which is what makes some plants so unpalatable in the first place), they regrow faster and thus colonize more area than the foul-tasting plants.

On top of that, planned grazing helps keep plant biomass levels within an ideal range, where plants capture a maximum amount of sunlight and thus grow exceedingly quickly. Below this range, plants have less leaf area, capture less solar energy, and grow much more slowly as a result; above it, excessive leaf growth blocks newer shoots from the sun, and old leaves die as fast as new ones are made. Grasses kept between these two extremes regenerate much more quickly and are thus ideal for grazing. [INFOGRAPHIC 14.8]

Howell says the end result of such careful planning is that land productivity is maximized and ranching becomes profitable once again. So far, the numbers support his claim. In one U.S.-based study, early adopters of planned grazing averaged 300% profit increases in the first 5 years. "You're raising twice as many animals in the same amount of space," explains Howell. "It's like getting a second ranch for free."

In fact, the financial benefits have proven so substantial, Howell and his partners are betting that the economy will rebound in tandem with the land. "Ultimately, we want to sell these places back to the surrounding communities,"

Infographic **14.8** | **PLANNED GRAZING**

↓ Livestock are allowed to graze intensively on a plot and then are moved to the next plot. In this example, each plot is grazed for 1 month. By the time the livestock return to a given plot, it will have recovered.

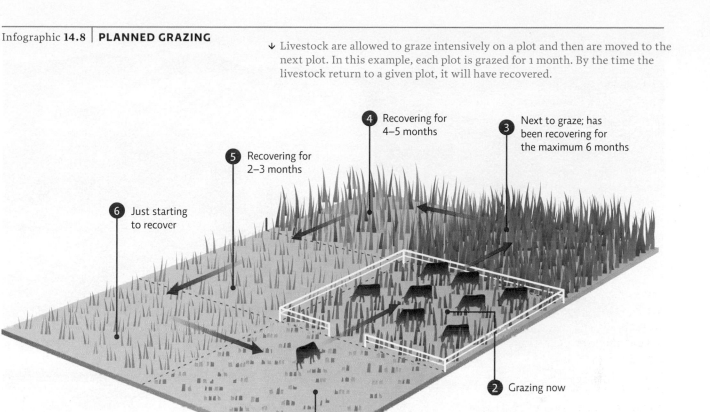

4 Recovering for 4–5 months

3 Next to graze; has been recovering for the maximum 6 months

5 Recovering for 2–3 months

6 Just starting to recover

2 Grazing now

1 Just grazed

says Howell. "At the very least, we hope to inject some energy into the region—provide jobs, attract young people."

Planned grazing is tricky work.

But planned grazing is tricky work, especially for ranchers who are set in their ways and are already anxious about the bottom line. "It's very hard to make a ranch profitable in the first place. So significant change can be very frightening," says Howell. In more traditional grazing methods, animals are left in the same pasture for long periods, sometimes for an entire growing season, which can run for as long as 180 days. With *planned grazing*—as Savory's method is called—animals are moved much more frequently. "It's really easy to screw things up when you switch from regular grazing to planned grazing," Howell, adds. "It's easy enough if you have 3 pastures and a 180-day growing season—each pasture gets about 60 days of grazing and 120 days of recovery," Howell says. "But say you have 45 pastures. That'd give you an average grazing period of only 4 days. If you're off by even a day, your

animals can suffer considerably." Leave them on a given pasture too long, and they will run out of food. Do that too often, and the animals' ability to gain weight, lactate, come into heat, and reproduce will all be compromised. And that can take a huge economic toll.

Not everyone agrees the risk is worth it. Most ranchers acknowledge that **sustainable grazing**—grazing that maintains the health of the ecosystem and allows the grasses to recover before the animals return—is essential to protecting grasslands the world over. But some argue that existing methods of continuous grazing work just as well, if not better, when done properly. "There is plenty of evidence showing comparable outcomes for biodiversity, soil erosion, etc.," says David Briske, a rangeland ecologist at Texas A&M University. "Effective management of grazed ecosystems is sufficiently dynamic and

rotational grazing Moving animals from one pasture to the next in a predetermined sequence to prevent overgrazing.
sustainable grazing Practices that allow animals to graze in a way that keeps pastures healthy and allows grasses to recover.

↑ A shelterbelt of trees may be able to hold back the Gobi Desert and keep this farmland fertile.

complex that it should not be envisioned to have any one correct solution."

Grasslands are not just here for human use and anything we can do to restore or protect them will benefit us and other species as well. Savory is showing that improved grazing techniques can help protect and even restore some degraded grasslands. Another land-management technique that also reduces soil erosion and protects grasslands from degradation is the planting of *shelterbelts*—a stand of trees that blocks the wind and thus decreases wind erosion. Shelterbelt programs helped the United States recover from the Dust Bowl and are being used today in areas facing desertification, such as Inner Mongolia and China (where the shelterbelt is referred to as the "Green Wall of China"). Other approaches, such as conservation easements, reserves, and protected parks also help protect our grasslands by limiting land uses and making it easier to keep the land in some level of protected status. [INFOGRAPHIC 14.9]

Nevertheless, a growing number of U.S. ranchers are following Howell's and Savory's lead. Success in Zimbabwe has been followed by successes in California, Texas,

Colorado, and New Mexico. "Change is tough," says Joe Morris, a rancher in central California who has been using Savory's methods with great success. "Ranching is a culture, steeped in tradition. But our lands are hurting and our communities are dying, and we know we've got to do something to fix that."

This past spring, Howell and his partners ran 2,300 yearling cattle out of Horse Creek Ranch—a nearly 100% increase from years past and a number they hope to double next season. "It's going to be a long road," says Howell. "But we've already made so much progress, in just one season." Meanwhile, back on the range, tiny buds of grass have begun to poke up everywhere.◉

Research articles referenced in this chapter:

Bailey, D.W., *et al.* 1996. *Journal of Range Management*, 49:386–400.

Briske, D. D., *et al.* 2011. *Rangeland Ecology and Management*, 64:325–334.

Briske, D. D. *et al.* 2008. *Rangeland Ecology and Management*, 61:3–17.

Derner, J.D., & Hart, R.H. 2007. *Rangeland Ecology and Management*, 60:270–276.

Diaz-Solis, H., *et al.* 2003. *Agricultural Systems*, 76:655–680.

Heitschmidt, R.K., *et al.* 2005. *Rangeland Ecology and Management*, 58:11–19.

Savory, A., & Parsons, S. 1980. *Rangelands*, 2:234–237.

Infographic **14.9** | **PROTECTING OUR GRASSLANDS**

MECHANISM	DESCRIPTION	BENEFITS
CONSERVATION EASEMENTS	Legal agreement that restricts how a landowner can develop and use land; allowable uses are typically agriculture, forestry, recreation, or natural preserves.	Land will be protected in perpetuity; reduced taxes to landowner
CONSERVATION RESERVE PROGRAMS	USDA program that assists farmers and ranchers in land management; farmers are paid to allow sensitive areas to be taken out of production and planted with grasses or trees.	Reduces soil erosion and water pollution, and increases wildlife habitat
PARK STATUS	Designation as a national or state park offers protection from overuse.	Areas can be managed in ways that protect the habitat and allow recreational use, a benefit to both nature and people.
SUSTAINABLE GRAZING	Grazing only the number of animals that the area can support; rotating animals from pasture to pasture is often used to maximize the number of animals that can be grazed.	Allows grassland to recover after grazing and to benefit from the rejuvenating effects of intensive short-term grazing
SHELTERBELTS	Trees are planted at the edge of a farm or grassland to reduce wind erosion.	Prevents or slows the encroachment of nearby deserts; also protects the crop or grasses from wind damage

BRING IT HOME

⊃ PERSONAL CHOICES THAT HELP

Because of overgrazing and rampant soil erosion, the grasslands have lost a larger proportion of their habitat than the tropical rainforests. But most people have never worried about preserving the grasslands, let alone are even aware of how little remains (only about 1–2%). Grassland habitats provide many ecosystem services—they capture CO_2 and are home to an immense range of biodiversity. They can be used for both economic growth as well as recreation, but only if they are managed properly.

Individual Steps
→ Visit a remnant or restored prairie. Such prairie preserves are scattered throughout the heart of North America, ranging from the Canadian provinces of Alberta, Saskatchewan, and Manitoba in the north, down through the Great Plains to southern Texas and Mexico, and from western Indiana westward to the Rocky Mountains. There are more than 20 national grasslands and several dozen more state and local prairie preserves.
→ Explore the difference between turf grass and native grasses. Find out what grasses are native to your area and plant some in your yard.
→ Purchase grass-fed beef or free-range bison meat as a way to support sustainable use of grassland ecosystems.

Group Action
→ If there are prairie restoration efforts underway in your community, participate in these efforts, which include seeding, brush cutting, removal of invasive species, and seed collecting.

Policy Change
→ Research national legislation designed to protect and restore native habitats like grasslands and write a letter to your state representative asking him or her to vote for the legislation.
→ Organize a fundraiser and donate the proceeds to research that examines the pros and cons of using grasslands for either biomass production or wind farms.

UNDERSTANDING THE ISSUE

CHECK YOUR UNDERSTANDING

1. **Which of the following statements describes a grassland biome?**
   a. Grasslands are biomes with abiotic conditions like fires that support the growth of grass and herbaceous plants, but not trees.
   b. Grassland biomes are found exclusively in semi-arid areas with very low levels of rainfall, most of which falls in the winter season.
   c. Grasslands have thick soils that develop due to the long, warm growing season characteristic of their tropical environment.
   d. Grasslands are only found in temperate regions that experience hot summers and cold winters, conditions that make it impossible to support tree growth.

2. **Temperate grasslands:**
   a. occur in places that have both rainy and dry seasons, but are warm year-round.
   b. have a seasonal climate with cold winters and hot summers.
   c. are cold most of the year and are characterized by very short growing seasons.
   d. are high-altitude grasslands that get all their precipitation in the form of snow.

3. **Which of the following is NOT a cause of desertification of rangeland?**
   a. Climate change, especially shifting rainfall patterns that will result in less precipitation
   b. Conversion of rangeland to urban and suburban development
   c. Soil erosion due to the loss of vegetative cover and the compaction of the soil column
   d. Overgrazing by domesticated livestock

4. **Pastoralists:**
   a. are herders who graze their livestock in tight herds, moving them constantly across vast swaths of rangeland.
   b. are landowners who plant fast-growing, nutrient-dense grasses to maximize livestock production in small pastures.
   c. partition their land into separate fenced-in pastures and divide their livestock between these fenced areas so as to control them more easily.
   d. are livestock owners who are paid to remove sensitive grassland areas out of production.

5. **A program that pays farmers to take sensitive areas out of production and plant them with grasses or trees instead is a:**
   a. conservation easement agreement.
   b. conservation reserve program.
   c. shelterbelt planting program.
   d. rotational grazing system.

6. **Shelterbelts are:**
   a. trees planted at the edge of a farm or grassland that protect crops or grasses from wind and prevent encroachment of nearby deserts.
   b. planned grazing systems where animals are allowed to graze on a small section of pasture for a few days before being moved to another section so as to prevent overgrazing.
   c. legal agreements, including tax breaks, that restrict how a landowner can develop and use land so that the land will be protected in perpetuity.
   d. areas managed in ways that protect the habitat but allow recreational use so as to benefit both nature and people.

WORK WITH IDEAS

1. What is a grassland? How and why do grasslands differ worldwide?

2. Describe the major threats to grasslands and explain the factors that underlie these threats.

3. How does grazing by native herbivores impact grassland ecosystems?

4. What are the differences between pastoralist grazers and ranchers? Which is better for the grassland and how do we know this?

5. What is overgrazing and how can it damage a grassland? Why is undergrazing just as problematic as overgrazing?

6. Planned grazing can make ranching activities mutually beneficial to both the livestock and the grassland ecosystem, yet it is a risky venture for ranchers in the United States. Why is this? Why are some farmers hesitant to change? What trade-offs do they face?

ANALYZING THE SCIENCE

The data in the table on the following page comes from a research document prepared by the Temperate Grasslands Conservation Initiative of the IUCN World Commission on Protected Areas. The document was prepared to build a case for the conservation and protection of temperate grassland ecosystems by evaluating their total economic value to human well-being.

INTERPRETATION

1. What does this figure show? Which of the *grassland* ecosystems is the most converted? Which one is the most protected?

2. The Conservation Risk Index (CRI) is a ratio of habitat converted to habitat protected for each biome. It is used to assess the level of threat to a particular biome.

   a. Calculate the CRI for each of the 13 biomes. Here is one example: CRI of deserts and xeric shrublands is 6.8 / 9.9 = 0.7.
   b. A CRI greater than 8 suggests a critically endangered biome (in other words 8 times as much land is being converted as is being protected). A CRI of 2 or less is considered a biome with a low risk of endangerment. According to the CRIs you calculated in part a, what is the status of each of the grasslands?

ADVANCE YOUR THINKING

3. What patterns do you see in the CRIs you calculated above? Describe the patterns and include data to support your explanation.

4. Which ecoregions are under the greatest threat?

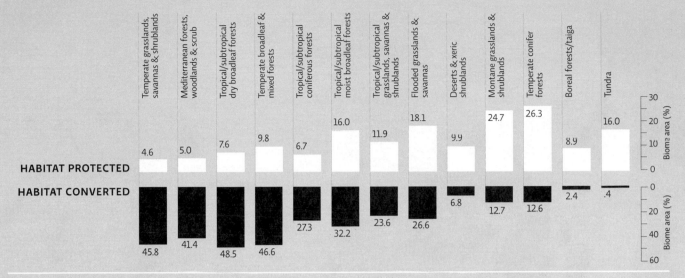

EVALUATING NEW INFORMATION

About 90% of grassland degradation is a result of intensive agriculture, including grazing practices and the conversion of croplands to grow livestock feed.

How can a conscientious consumer help protect the grasslands ecosystem? What choices can we make that are sensitive to the welfare of the ecosystem as well as the animals that we eat?

Explore the Eat Wild website (www.eatwild.com).

Evaluate the website and work with the information to answer the following questions:

1. Is this a reliable information source? Does it have a clear and transparent agenda?
   a. Who runs this website? Does this person's credentials make him or her reliable/unreliable? Explain.
   b. Does the website provide supporting evidence for its claims about grass-fed versus feedlot animal products? Does it give sources for its evidence?

2. Select the link "A direct link to local farms."
   a. What information does that link provide? How can you as a consumer use this information?
   b. Select the "Criteria" link. What are the criteria for listing a farm on the website? Do you think these criteria are sufficient and reasonable? Which criteria are most important to you as a consumer? Explain your responses.

3. Examine two other websites that certify animal products: Animal Welfare Approved (www.animalwelfareapproved.org) and American Grassfed (www.americangrassfed.org).
   a. How similar/different are the criteria each website uses for certifying animal products? Which website uses criteria that are most important to you as a consumer? Explain your response.

4. How might certifications for animal products help grasslands?

MAKING CONNECTIONS

THE BUFFALO COMMONS AND THE FUTURE OF THE GREAT PLAINS

Background: Allan Savory's planned grazing methods are one approach to reverse desertification in the Great Plains. But according to some, not only is farming and ranching ecologically unsustainable, it is economically impractical as well. Several researchers have proposed that cattle and grazing be replaced by wildlife refuges, where American bison (buffalo) could once again be allowed to roam freely. In 1987, Rutgers scientists Frank and Deborah Popper suggested converting about 139,000 square miles of the Great Plains, which they argued would most likely become depopulated anyway, into wildlife preserves called the Buffalo Commons.

Case: You have been assigned the task of determining the future of the Great Plains. Select between the following two options:

1. The federal government should use financial incentives to attract people to the Great Plains and should also offer subsidies to encourage sustainable grazing practices.

2. As the Great Plains become depopulated, the unoccupied land should be purchased and transformed into ecological reserves supporting activities such as ecotourism, wind farms, and free-range bison ranching.

Research these two alternatives and write a report recommending one of the two options above. In your report include the following:

a. An analysis of the pros and cons of each proposal including: a discussion of the consequences of each choice both for local human communities and for the grassland ecosystem, and a reflection on the values underlying each proposal.
b. Based on the information at hand, what is the best option for the future of the Great Plains and who should be involved in this decision? Provide justifications for your proposal.

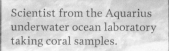

Scientist from the Aquarius
underwater ocean laboratory
taking coral samples.

SCIENCE UNDER THE SEA

Aquanauts explore an ecosystem on the brink

CORE MESSAGE

Marine ecosystems contain a huge diversity of life, though we know far less about ocean ecosystems than those on land. Many ocean ecosystems are suffering as a result of pollution, overfishing, misuse of the ecosystem's resources, and global climate change. Of particular concern is the change in ocean chemistry caused by the release of CO_2 from fossil fuel use. This has the potential to alter ocean ecosystems drastically, and some effects are already being seen. Choices we make today will influence the future of ocean ecosystems and the species that inhabit them.

GUIDING QUESTIONS

After reading this chapter, you should be able to answer the following questions:

→ What is contributing to ocean acidification and why is this a problem?

→ What environmental conditions determine the location and makeup of marine ecosystems?

→ Where are coral reefs found and what is the community makeup of these complex ecosystems?

→ What threats do coral reefs and other ocean communities face?

→ How can we reduce the threats to coral reefs and other ocean ecosystems?

↑ Two goliath groupers swim past a porthole on the Aquarius Reef Base.

→ A researcher swims past Aquarius Reef Base, the only undersea research station in the world. Scientists can spend up to 10 days on the station before they must return to the surface (to avoid decompression illness).

It was four o'clock in the morning and Marc Slattery could not sleep. He and his six crewmates had just settled into the Aquarius Reef Base—an artificial, undersea research station located on Conch Reef, 10 miles off shore from, and 50 feet below, Key Largo, Florida. Tomorrow they would begin an 8-day stretch of underwater experiments and data collection—all aimed at understanding how the physiology of various coral reef species was changing in response to changes in ocean conditions. Slattery, a scientist at the University of Mississippi and the team's principal investigator, was anxious to get started.

But it was not this eagerness keeping him awake; it was the two goliath groupers—each about 2.5 meters (8 feet) in length—gliding persistently past the viewport near his bunk. Sticking close to one another, they would swish up to the small circular window, stare at him briefly, and then swish away to complete another lap around the structure. Slattery was captivated. As a young kid, he had spent countless hours watching a collection of tropical fish glide around his home aquarium. Now, for the first time, he knew how the fish must have felt. "They seem to swim back and forth between the bunkroom and galley viewports looking for those fascinating creatures that walk on two legs, breathe air and eat constantly," he wrote in the expedition log. "They accept our odd behavior and even seem to enjoy our presence. Maybe they know we are here to help."

Whether they knew it or not, the groupers, and all of their underwater neighbors, were being threatened by an avalanche of forces—global climate change chief among them. Around the world, temperatures were rising, glaciers melting, and the ocean's chemistry changing in peculiar and disturbing ways. Scientists like Slattery had been on a quest to understand these changes from their land-based labs. Now Slattery wanted to dig for clues down below.

Acidification threatens life in the world's oceans.

From the beginning of the Industrial Revolution to the present day, we humans have burnt enough fossil fuels and clear-cut enough forests to release more than 500 billion tons of CO_2 into Earth's atmosphere, making it higher than at any point in the past 800,000 years. Even worse than these unprecedented levels is how fast they have risen—too fast, experts say, for many organisms to adapt. Much has been made of what heat trapping molecules do to terrestrial ecosystems. But, as scientists are now

⊙WHERE IS KEY LARGO, FLORIDA?

GULF OF MEXICO

KEY LARGO

AQUARIUS

FL

KEY LARGO

learning, their effect on aquatic environments is just as profound.

Ocean and atmosphere come into direct contact over 75% of Earth's surface, and they are constantly exchanging gases over that interface; anything emitted into one eventually ends up in the other—including CO_2. Winds quickly mix CO_2 into the top few hundred feet of water, and as years pass, currents pull it ever deeper into the ocean. Between the early 1990s and mid-2000s, scientists around the world collected and analyzed nearly 80,000 water samples from a range of ocean environments. By their estimates, some 30% of all the CO_2 released by humans in the last two centuries has been absorbed by the world's oceans. "For terrestrial ecosystems, it's a good thing," says Slattery. "Because it means that much

Infographic **15.1** | **pH AND OCEAN ACIDIFICATION**

↓ The pH scale is a measure of how acidic or basic a solution is. It is a measurement that compares the proportion of hydrogen (H^+) or hydroxide (OH^-) ions in a solution (ions are charged atoms or molecules). Water (H_2O) can dissociate into hydroxide ions (OH^-) and hydrogen ions (H^+). In pure water, there is always one H^+ for every OH^-. These solutions are neutral, with a pH of 7.0. Acids have a pH lower than 7.0 and have extra H^+ ions in the solution. Bases (alkaline solutions) have a pH higher than 7.0 and have extra OH^- ions in the solution.

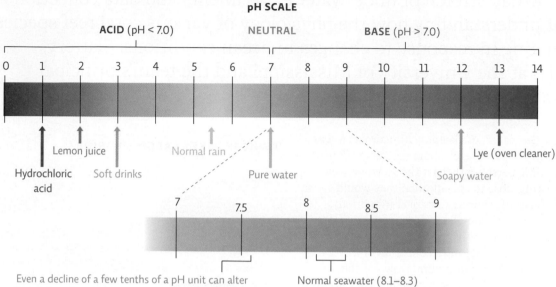

pH SCALE

ACID (pH < 7.0)   **NEUTRAL**   **BASE** (pH > 7.0)

0 1 2 3 4 5 6 7 8 9 10 11 12 13 14

Hydrochloric acid
Lemon juice
Soft drinks
Normal rain
Pure water
Lye (oven cleaner)
Soapy water

7   7.5   8   8.5   9

Even a decline of a few tenths of a pH unit can alter the ability of marine organisms to produce shells. At a pH of 7.6, most shells would dissolve quickly.

Normal seawater (8.1–8.3)
Seawater becomes more acidic as it absorbs CO_2 (it has dropped from an average of 8.2 to 8.1 since the Industrial Revolution).

OCEAN ACIDIFICATION HAS INCREASED OVER TIME

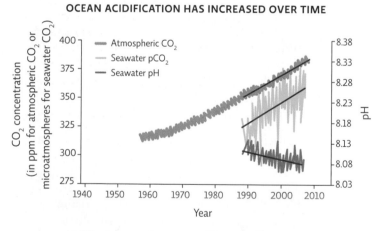

— Atmospheric CO_2
— Seawater pCO_2
— Seawater pH

y-axis (left): CO_2 concentration (in ppm for atmospheric CO_2 or microatmospheres for seawater CO_2)
y-axis (right): pH
x-axis: Year

↑ The increase in ocean acidity correlates well with the increases in CO_2 dissolved in seawater and atmospheric CO_2 concentrations.

→ Ocean pH will continue to drop significantly if we do not curtail fossil fuel use and our release of extra CO_2.

PREINDUSTRIAL

An estimate of ocean pH before the 1800s

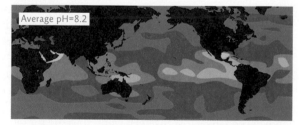

Average pH=8.2

YEAR 2100

Projected ocean pH in 2100, accounting for current CO_2 emissions levels.

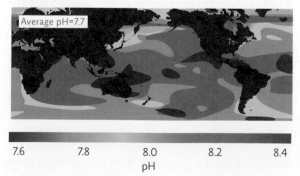

Average pH=7.7

7.6   7.8   8.0   8.2   8.4
pH

↑ A healthy reef area, like this one off the coast of Castello Aragonese near Naples, Italy, is full of life, made up of a variety of coral species and inhabited by a wide array of fish and invertebrates like the red sponges that dot the coral here. A well-camouflaged fish (a tompot blenny) hugs the coral just above the black sea urchin (bottom left).

↑ Just a few hundred yards away from the scene in the photo to the left, CO_2 can be seen bubbling out of volcanic vents in the sea floor, giving us a glimpse of what future ocean communities might look like if oceans continue to acidify. This CO_2 so acidifies the water that only sparse algal mats are found here.

less CO_2 is lingering in the atmosphere. But for oceans, it could be very bad."

Here's how: Normal seawater has an average pH of around 8.2, meaning it is slightly basic (alkaline). So far, CO_2 emissions have reduced the ocean's pH by about 0.1. That might not sound like much, but the pH scale is logarithmic, so even small numbers represent large effects. A pH drop of 0.1 corresponds to a 30% increase in ocean water acidity. If present trends continue, by 2100 the ocean's surface waters will be about 150% more acidic than they were in 1800. In 2003, scientists adopted the phrase "ocean **acidification**" to describe this coming catastrophe. [INFOGRAPHIC 15.1]

Scientists expect such a colossal reordering of ocean chemistry to have a huge impact on marine ecosystems. "Ocean acidification is global warming's equally evil twin," says Jane Lubchenco, a marine ecologist and the head of the National Oceanic and Atmospheric Administration.

One big concern is that as pH shifts, the availability of key nutrients like nitrogen and iron will change, plummeting in some areas, soaring in others, and threatening

the stability of virtually all marine ecosystems. In just one example, Michael Berman and his colleagues at the University of California have found that the rate of nitrification (a process that produces nitrate, a form of nitrogen that marine organisms need to grow) decreases in tandem with pH; as that happens, smaller species of plankton (which are more tolerant of nitrate declines), gain an advantage over larger ones. Were it pervasive enough, such a change in species composition could alter the food chain and decrease primary production throughout the oceans (see Chapter 9 for more on food chains and webs). Indeed, some researchers think this shift may already be occurring. Overall, plankton biomass may have decreased as much as 40% in the 20th century, they say, especially since 1950.

The consequences of this particular chain of events are, for now, anybody's guess. On one hand, there could be a **positive feedback** effect that amplifies ocean acidification and climate change: Less plankton means that less CO_2 is taken in by the organisms, and more is left behind

acidification The lowering of the pH of a solution.
positive feedback Changes caused by an initial event accentuate that original event (i.e., changes brought on by warming lead to even more warming).

to further acidify the water or reenter the atmosphere. On the other hand, there could be a **negative feedback** effect on climate change: Less nitrification means less N_2O (nitrous oxide—a potent greenhouse gas) is produced, and thus less is released into the atmosphere.

So far, the most well-documented effect of acidification seems to be on marine calcifiers—ocean organisms that make shells, plates, and exoskeletons from calcium minerals. There are thousands of different types of calcifiers—from snails to corals to plankton—dispersed widely throughout the ocean. Early research suggests that acidification may well threaten all of them. When dissolved in water, CO_2 forms carbonic acid, which not only eats away at existing calcium-based materials, but interferes with the chemical reactions by which new ones are made. [INFOGRAPHIC 15.2]

Scientists have already documented significant effects on pteropods—tiny swimming snails that are important food for whales, birds, and fish (including juvenile salmon, pollack, and other commercially important species) in the Arctic and Antarctic. Experiments show that pteropod shells grow more slowly—and even start to dissolve—in acidified seawater. At predicted ocean pHs of the near future, they will become too fragile to support all the life-forms that feed on them.

> If present trends continue, by 2100 the ocean's surface waters will be about 150% more acidic than they were in 1800.

To be sure, all calcifiers play an important part in the grand choreography of ocean life. But the most substantial of them are **coral reefs**, like the one Slattery and his team were studying near Aquarius.

Aquarius is the only facility of its kind: an 82-ton, double-lock pressure vessel, just 46 feet long and 10 feet wide. That's big enough to house six people for 8 days, sturdy enough to weather the violent storms that periodically shake the region, and unique enough that for the next week, Slattery and his five fellow aquanauts would be the only people on the planet living at the bottom of the sea.

---

negative feedback Changes caused by an initial event trigger events that then reverse the response (i.e., changes brought on by warming lead to cooling).

coral reef Colonies of tiny animals (coral) that produce a calcium carbonate exoskeleton that over time build up to form large underwater structures (the reef) in shallow, warm, tropical seas.

↑ Declines in plankton productivity lead to decreased carbon capture. This means less carbon, stored in the shells of dead plankton, sinks to the ocean bottom. This type of carbon capture is an important mechanism for locking away some of the extra CO_2 that enters the oceans.

To a marine biologist, the advantages of this particular perspective are innumerable. Without having to return to the surface every hour, or dive in shifts to complete a mere day's worth of work, individual team members would be able to take measurements in real time and could observe myriad reef species in all their splendor for hours on end. "Aquarius will enable us to observe and take measurements at much closer intervals than we otherwise might," Slattery says. "It will give us a much fuller picture of what's going on down there."

What he really wanted to get a picture of were the hidden crevices tucked deep within the reef. "These are areas within a coral reef landscape that are naturally acidified," Slattery explains. "They're packed with sponges and other ocean life, which means lots of respiration, and at the same time they have poor water circulation." Because CO_2 is released during cellular respiration, Slattery reasoned that CO_2 concentrations would be high and pH would be low (see Chapter 7 for more on cellular respiration). Slattery and his team hoped the creatures living in such crevices might provide some clues about how the larger reef would respond to an acidified ocean.

Once they found these acidified microhabitats, the team planned to measure individual cellular respiration rates for the creatures living in these crevices—a crucial detail that had yet to be ascertained by anyone. "Knowledge of the *in situ* rates of respiration processes and their impact on local pH has been virtually nonexistent," says Chris Martens, a marine biologist from the University of North Carolina, Chapel Hill, who has also studied acidification from the Aquarius Reef Base. "But it's hugely important." Without these data, researchers can't tell how much

↓ Acidic ocean water decreases the ability of aquatic organisms to form shells or exoskeletons, while also dissolving shells and coral that have already been formed.

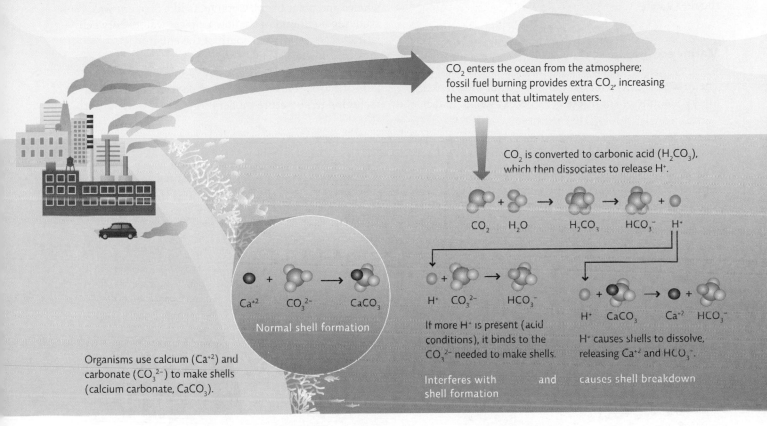

CO_2 enters the ocean from the atmosphere; fossil fuel burning provides extra CO_2, increasing the amount that ultimately enters.

CO_2 is converted to carbonic acid (H_2CO_3), which then dissociates to release H^+.

$$CO_2 + H_2O \rightarrow H_2CO_3 \rightarrow HCO_3^- + H^+$$

Normal shell formation

$$Ca^{+2} + CO_3^{2-} \rightarrow CaCO_3$$

Organisms use calcium (Ca^{+2}) and carbonate (CO_3^{2-}) to make shells (calcium carbonate, $CaCO_3$).

$$H^+ + CO_3^{2-} \rightarrow HCO_3^-$$

If more H^+ is present (acid conditions), it binds to the CO_3^{2-} needed to make shells.

Interferes with shell formation

$$H^+ + CaCO_3 \rightarrow Ca^{+2} + HCO_3^-$$

H^+ causes shells to dissolve, releasing Ca^{+2} and HCO_3^-.

and causes shell breakdown

↓ An experiment by Orr *et al.* showed the increased rate of dissolution (dissolving) of pteropod shells in water with less calcium carbonate. In this test, shells were exposed to water containing the lower levels of calcium carbonate we expect to see in 2100 if we continue to use fossil fuels at the same rate that we now do. (At lower pHs there will be less calcium carbonate in the ocean).

   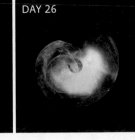 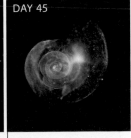

| Live healthy pteropod | Normal shell | DAY 16 — Shells become pitted (making them appear more opaque) as they begin to decalcify. | | After 45 days, major dissolution has occurred. |

Live healthy pteropod | Normal shell | Shells become pitted (making them appear more opaque) as they begin to decalcify. | After 45 days, major dissolution has occurred.

acidification is coming from CO_2 absorption from the atmosphere and how much is coming from CO_2 respired by the infinite array of ocean life in all the various micro-environments. "We need to know the rough contributions of each before we can possibly hope to develop effective mitigation strategies."

But first they had to find these acidified microhabitats on the reef itself.

Marine ecosystems are diverse.

The hidden crevices Slattery and his team were looking for are just one of countless ocean ecosystems. In fact, the world's oceans are a wonderland of diversity. They cover about 70% of Earth's surface, and house a greater variety of flora and fauna than all land masses combined. In shallow, temperate regions, forests of tall brown seaweed known as kelp provide both food and habitat to a wide variety of organisms. Meanwhile, algae that cling to the underbelly of sea ice in the Arctic and Antarctic oceans forms the base of a food chain that ultimately supports whales and polar bears. Coral reefs like Slattery was studying are found in shallow tropical waters between latitudes 30° north and south of the equator. [INFOGRAPHIC 15.3]

Lots of ecosystems means lots of ecosystem services, including temperature moderation (ocean water absorbs

---

Infographic **15.3** | **CORAL REEF DISTRIBUTION, THREATS, AND DESTRUCTION**

↓ Most reef-building corals are found in warm, tropical and subtropical waters between 30° N and 30° S latitudes, shown here as red dots. Coral reefs around the world are threatened by a variety of forces, and destruction of reefs is on the rise.

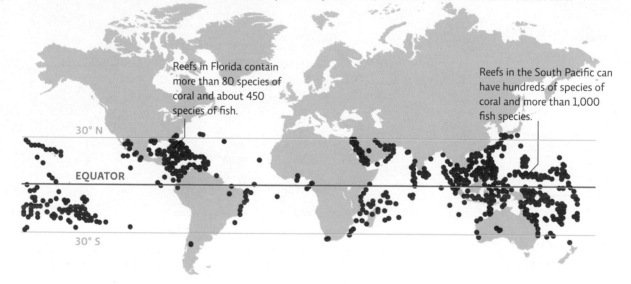

Reefs in Florida contain more than 80 species of coral and about 450 species of fish.

Reefs in the South Pacific can have hundreds of species of coral and more than 1,000 fish species.

30° N

EQUATOR

30° S

↓ NOAA estimates that 75% of coral reefs worldwide are threatened by human activities or environmental changes. The two graphs below show the percentage of coral reefs already destroyed and the threats that face those that remain. Overfishing alters the community makeup, disrupting important relationships that keep the reef alive and healthy; heavy nets can also directly damage coral reefs themselves. Coastal development can add sediment to water, making it cloudy; in some areas, coral reefs are actually mined for limestone building materials. Pollution, whether from land or sea (ships), can directly harm coral as well.

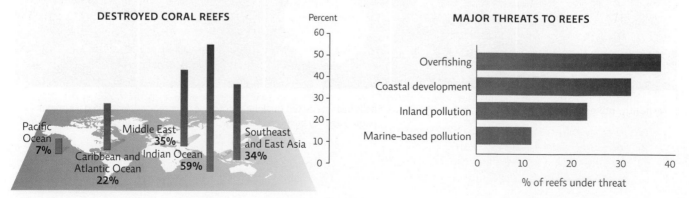

DESTROYED CORAL REEFS

Pacific Ocean
7%

Middle East
35%

Caribbean and Atlantic Ocean
22%

Indian Ocean
59%

Southeast and East Asia
34%

MAJOR THREATS TO REEFS

Overfishing
Coastal development
Inland pollution
Marine-based pollution

% of reefs under threat

↑ Researchers at the Aquarius Reef Base determine cellular respiration rates of a brown barrel sponge by measuring CO_2 production.

a lot of heat and releases it slowly), nutrient cycling, and support for large populations of commercially valuable fish. In particular, coral reefs are especially valuable for their services: protection of coastal areas from storms, purification of the water (many reef occupants are filter feeders), provision of recreational opportunities, and support of important commercial fisheries. They even serve as a source of current and potential medicines such as antibiotics and anticancer drugs. And oceans have absorbed a good bit of the CO_2 we have released due to fossil fuel burning, reducing the atmospheric warming that CO_2 would have caused if all of it had remained in the air.

Marine ecosystems, much like terrestrial ones, vary by location. In the ocean, depth is a key determinant of environmental conditions. About 80% of all sunlight is absorbed in the first 10 meters (33 feet) of the water column, even more than that in murky waters. Because sunlight supplies both heat and the essential energy for photosynthesis, upper layers of the ocean are much

warmer and more productive than lower layers. In fact, most life-forms are found in shallow waters where light can penetrate; both heat and productivity dwindle as we move down the water column to deeper depths.

Scientists have divided the oceans into zones, based on depth. Each region differs in the amount of available sunlight, which in turn affects rates of photosynthesis and thus strongly impacts the diversity and abundance of ocean life at that level. Ocean regions are also identified by their proximity to land; the closer to land and to rivers that empty into the ocean, the more nutrient rich the area, and thus productive, the ocean community. In fact, **estuaries**, regions where rivers empty into the ocean, are known as the nurseries of the sea because so many species come to these areas to spawn (see Chapter 18 for more on nutrient enrichment of estuaries). [INFOGRAPHIC 15.4]

estuary Region where rivers empty into the ocean.

↳ Different communities of species are found at different depths of the ocean and at different distances from shore. Most life is found on the continental shelf, the region off each shoreline that is relatively shallow compared to the deeper regions that begin at the edge of the shelf. Communities in deeper waters have life-forms adapted to little or no sunlight, whereas the most productive communities are found in relatively clear, shallow waters with ample sunlight.

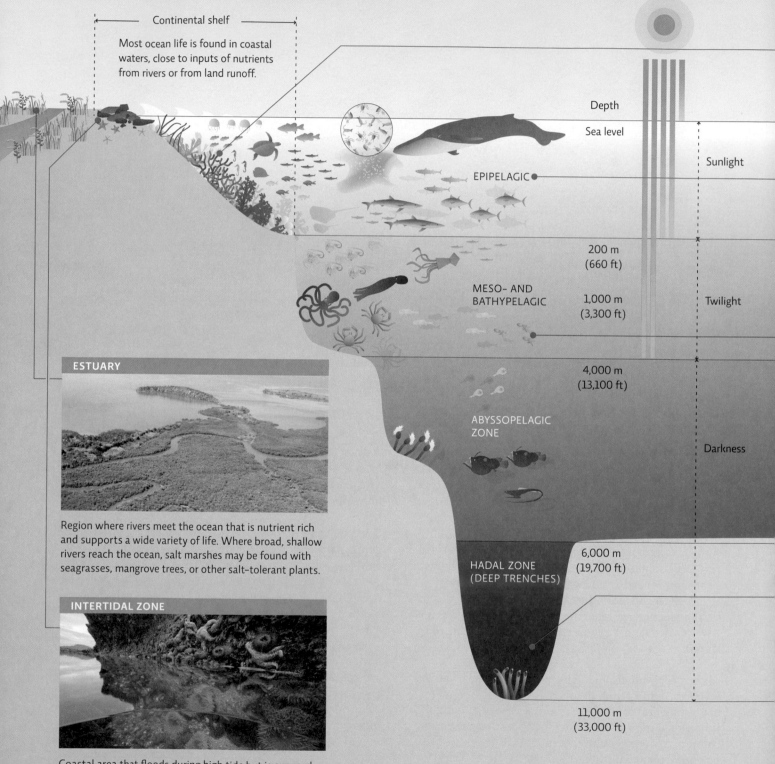

Continental shelf

Most ocean life is found in coastal waters, close to inputs of nutrients from rivers or from land runoff.

Depth

Sea level

EPIPELAGIC

Sunlight

200 m
(660 ft)

MESO- AND
BATHYPELAGIC

1,000 m
(3,300 ft)

Twilight

4,000 m
(13,100 ft)

ABYSSOPELAGIC
ZONE

Darkness

6,000 m
(19,700 ft)

HADAL ZONE
(DEEP TRENCHES)

11,000 m
(33,000 ft)

ESTUARY

Region where rivers meet the ocean that is nutrient rich and supports a wide variety of life. Where broad, shallow rivers reach the ocean, salt marshes may be found with seagrasses, mangrove trees, or other salt-tolerant plants.

INTERTIDAL ZONE

Coastal area that floods during high tide but is exposed to air at low tide. Organisms found here, like barnacles, starfish, mussels, and crabs, are adapted to withstand wave action and low-tide dry periods.

CORAL REEF

Found in clear, shallow water in tropical regions; in addition to the keystone coral species, the reef supports a wide variety of fish, crustaceans, sponges, and other marine organisms.

OPEN OCEAN

Not as densely populated by ocean life, but home to many species on the move, like sharks and whales

MESO-, BATHY-, AND ABYSSOPELAGIC ZONES

Deeper regions that are home to squid and strange luminescent fish like this anglerfish; plankton make up the base of this food chain.

HADAL ZONE COMMUNITIES

Organisms such as tube worms, sea stars, and even some fish living near thermal vents (areas where superheated water containing hydrogen sulfide and other chemicals is released from Earth's crust); these communities do not depend on photosynthesis but get their energy from chemicals released by the vents, captured by specialized bacteria, and passed up the food chain.

Can some organisms adapt to ocean acidification?

From high above, the Aquarius Reef Base looked something like an alien anthill. A small army of divers had descended on the structure to take care of some daily maintenance tasks, and all six aquanauts were shuttling samples and equipment back and forth from reef to station. Eventually, this would grow exhausting. "The pinnacle, where we sample, seems to get further away each day, particularly when the current gets ripping," wrote one aquanaut on the station's blog. "But then we get to ride the current back in, 'Finding Nemo' style!"

The team's main focus was the *Xestospongia muta*, a giant barrel sponge that thrives in both acidified and nonacidified microhabitats. By monitoring pH and CO_2 levels around each sponge, and taking tissue samples for later analysis in the lab, they hope to determine if those sponges that live in acidified waters have a different protein makeup that helps them adapt to, and even thrive in, these regions.

Some sponges lived in the crevices where pH was low; others lived out on the open reef, where pH was closer to normal. Slattery wanted to see whether the crevice dwellers had any special adaptations that their open-reef-dwelling cousins lacked. "It's just like humans and the common cold," he explains. "Some people go through the whole season without getting sick at all, and then others catch every little bug that's been flying around. We want to see if there's the same type of natural variation in sponges responding to acidification."

They also transplanted paired sponges—from acidified and nonacidified habitats—to sites facing additional stress, namely temperature increases. The idea was to see what impact prior exposures to low pH had on the organisms' health, and also to see what effects these environmental changes had on physiology.

Lastly, the team took a detailed census of the flora and fauna in various microhabitats, especially the acidified crevices, and collected samples of all the species they could. Like the sponge tissue, these samples would be assessed for protein expression to see if similar organisms behaved differently when they were exposed to lower pH.

Coral reefs are complex communities with lots of interspecific interactions.

Species of coral reef communities interact with and depend on each other in many ways (see Chapter 9). Clown fish and sea anemones offer each other both food

Infographic **15.5** | **CORAL BIOLOGY**

↓ Coral live in a mutualistic relationship with zooxanthellae ("zooks")—photosynthetic algae that live inside the epidermal cells of the coral. The two species share nutrients. Zooks also raise the pH of the coral cells slightly, which helps the coral lay down its coral skeleton. Though they come in a tremendous variety of shapes and colors due to species differences and even growing conditions, all coral have the same basic body plan: the polyp—a tube-like structure attached at the base to a substrate. At the top of the tube are stinging tentacles that trap food floating by, which is taken in by the mouth and sent to the digestive sac.

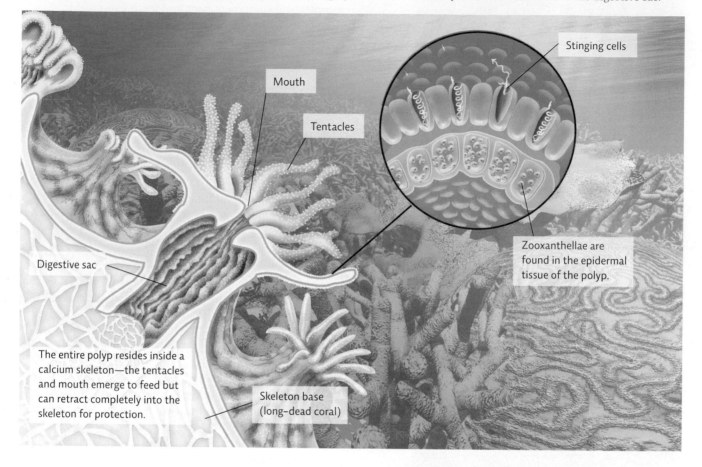

Mouth

Tentacles

Stinging cells

Zooxanthellae are found in the epidermal tissue of the polyp.

Digestive sac

The entire polyp resides inside a calcium skeleton—the tentacles and mouth emerge to feed but can retract completely into the skeleton for protection.

Skeleton base (long-dead coral)

and protection (mutualism); remora fish attach their suckers to manta rays and eat any bits of food that flow out of the ray's mouth (commensalism); and sea urchins act like lawn mowers, feeding on the reef itself in a way that enables new corals to attach and grow, thereby keeping the overall structure strong and healthy.

Like kelp, coral reefs attract and provide food and habitat for many other species. In fact, scientists estimate that 25% of all ocean species spend at least some portion of their life in a coral reef. This outsized role in marine ecology makes reefs an appealing target for oceanographers to study, and is but one of several reasons that the Aquarius station was positioned near a reef. "Coral reefs are the ultimate keystone species," says Martens. "Understanding them is the key to understanding everything else that's going wrong in the ocean right now."

Corals are tiny marine organisms that live in densely packed colonies of many individual polyps (basically a tube with one opening surrounded by tentacles). Coral larvae (immature stage) are free floating but must attach to a surface to survive. Established coral reefs actually attract the larvae via chemical cues, increasing the odds that the larvae will attach to and build on top of existing coral skeletons. Once attached to a surface, the larvae enter the polyp stage and secrete a calcium carbonate skeleton. A reef thus grows from corals building on top of other corals over the course of generations. Studies have also shown that lower pH leads to declines in fertilization, in larval development, and in a process called settlement—the dropping of coral larvae out of the water column, and their attaching to something solid—the necessary prerequisites to new coral colony formation.

Infographic **15.6** | **CORAL BLEACHING**

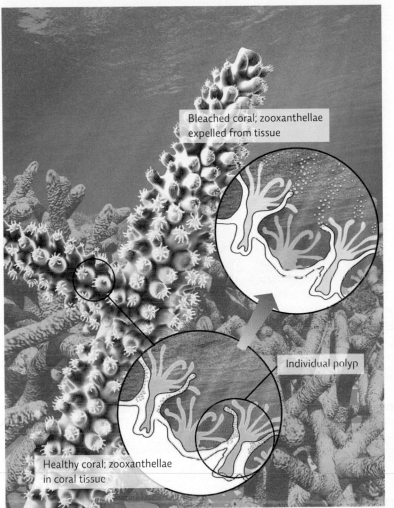

Bleached coral; zooxanthellae expelled from tissue

Individual polyp

Healthy coral; zooxanthellae in coral tissue

← Coral may expel their zooxanthellae if stressed, such as when water temperature gets too high (though the mechanism for how this expulsion occurs is poorly understood). Without their pigmented algae, the white coral skeleton can be seen through the translucent polyp—leading to the term "coral bleaching." Bleaching may be an adaptation that allows the coral to take up a different species of zooks, one that can tolerate the warmer temperatures, for instance. Coral may be able to survive for a few months without the zooks but if the polyps don't eventually take other algae back up, they will die. Multiple warming events or a warming event that persists for several weeks may stress the coral beyond its ability to survive. Pollution and even excess sun exposure are also linked to bleaching events.

↓ Bleached staghorn coral on a Fiji reef

Corals feed on decaying organic material and plankton—pretty much anything that floats by. But they live in nutrient-poor regions: Tropical waters tend to remain stratified, with warm water trapped above colder, deeper water. This stratification prevents the mixing of deep water and surface water that would normally bring up nutrients from below.

Thus the corals themselves could not survive without some help from their algal partners. They live in a mutualistic relationship with zooxanthellae (known affectionately to most marine biologists as "zooks")—a photosynthetic, dinoflagellate algae that lives inside, and shares nutrients with, the coral. Corals provide their resident zooks with nutrients (from the polyps' waste) and CO_2 for photosynthesis. Zooks, in turn, provide the coral polyps with food (sugars made during photosynthesis). The zooks also give the corals their beautiful colors. It's

a well-honed evolutionary collaboration, one that enables each species to thrive in a nutrient-poor environment. [INFOGRAPHIC 15.5]

Corals can also control the number of zooks they host in their cells by adjusting the amount of sunlight or nutrients they supply. Under extreme stress, corals have been known to expel the algae. This event is known as coral bleaching because when it happens, the coral turns bone white. Though the coral can survive for a short period of time without zooks, they will die if not recolonized. Since the many species of zooks are symbiotically tied to their coral hosts, any events that lead to coral death, such as bleaching or decalcification, could also lead to the loss of the zooxanthellae species that reside in those coral. [INFOGRAPHIC 15.6]

Scientists have found that in the past, corals have gone through natural death and birth cycles in response to environmental changes such as shifts in temperature or pH. However, many scientists feel that today's accelerating rates of acidification and bleaching are likely unprecedented, as is their global scope. But CO_2-related effects are by no means the only threats.

The world's oceans face many other threats.

Rising sea levels may decrease sunlight penetration—and thus reduce photosynthesis—in the deepening coral seas. And increased flooding is almost certain to bring more pollutants from land to sea.

On top of all that, oceans everywhere are now threatened by overfishing, pollution, and invasive species.

In the ocean at large, some 90% of predators in the top trophic-levels have been eliminated by overexploitation (a.k.a., fishing pressure). Such heavy losses disrupt the interdependent relationships needed to sustain each community. For example, without grazer fish to keep it in check, algae overgrows on coral. Fishing pressure inflicts other wounds as well: Bottom trawling can decimate sea beds, crushing or burying organisms that live close to the beds and uncovering those that need to remain buried (see Chapter 16). Cyanide sprays, used to stun fish for aquarium collection, kill most fish and coral that

encounter it. And dynamite, also used in fishing, physically destroys reefs and other ecosystems around it.

Meanwhile, sediments and high levels of nutrient runoff from agricultural areas are boosting algae production and creating algal blooms, which in turn smother corals and block sunlight, reducing photosynthesis. Trash from both land and sea, and petroleum from ships and boats are also polluting the marine environment at unprecedented rates.

Invasive species pose yet another threat; more than 80% of ocean harbors around the world now host at least one invasive. Whether they arrive in the ballast water of ships, escape from aquaculture pens, or are moved in the aquarium trade, the outcome is the same: They wreak havoc on aquatic ecosystems. "We don't even know how much damage they're doing," says Slattery, who has studied the impact of invasive lionfish on reefs near the Bahamas. "The fact is, we know much less about the oceans than we do about land. But if you think about it, that makes protection even more vital, because we need to err on the side of caution." [INFOGRAPHIC 15.7]

Scientists have been working to create marine protected areas (MPAs)—places where fishing and other human activities are restricted or completely prohibited. Evidence shows that in the right conditions, MPAs can significantly improve the marine ecosystems they encompass. Three years after a 2003 bleaching event that devastated the South Pacific coral reefs of Kanton

→ Lionfish (escapees from aquaria) are aggressive (and poisonous) predators and either kill native fish outright or outcompete them when foraging. When they are introduced to a coral reef, the survival of native fish can drop by as much as 80%. In some regions of Florida, divers who remove them are eligible for a $10,000 prize. In areas where groupers are protected, lionfish are not as big of a problem because the native groupers eat them—another reason to protect communities and try to keep them intact.

Infographic **15.7** | **THREATS TO OCEANS** ↓ Oceans are threatened by a variety of human activities.

FISHING PRESSURE

↑ Fishing with dynamite brings up lots of fish but severely damages the coral reefs and kills many nontarget species. (See Chapter 16 for information on overfishing.)

↑ Fish collectors spray cyanide on the reef. This stuns some fish, which are collected, but kills most of the fish and kills the coral as well.

↑ Many species of commercially valuable fish, such as bluefin tuna and Atlantic cod, are facing population crashes due to overfishing. (See Chapter 16 for information on commercial overfishing.)

POLLUTION

↑ Plumes of sediment pollution along coasts can introduce nutrients, toxins, pathogens, and solids that can smother and damage coral reef or other marine ecosystems. (See Chapter 18 for more on water pollution.)

↑ Debris like this plastic bag can choke ocean animals or quickly become entangled on a coral and smother it. (See Chapter 30 for more on trash in the oceans.)

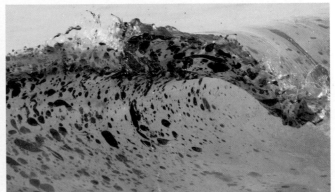

↑ Oil spills from drilling operations and from tanker transport damage coastal and offshore marine ecosystems. (See Chapter 24 for more on oil spills.)

Island, the area was declared an MPA where fishing was prohibited. By 2010, the reefs showed significant recovery. Credit is given to the community of fish: Because they were left as intact, undisturbed populations, the fish grazed on algae that would normally move in and prevent coral from recolonizing the area. By keeping the community connections intact, the reef appears to be well on its way to recovery, even after a serious bleaching event, showing the resiliency of intact ecosystems.

Of course, the best way to save the oceans may be to change the way we live on land. We humans now emit more than 30 billion tons of CO_2 each year. Steps taken now to decrease the amount of CO_2 released into the atmosphere will reduce the CO_2 available to be absorbed by the oceans. Unfortunately, even if we stopped completely, right now, it would take tens of thousands of years for ocean chemistry to return to its preindustrial state. "We can't reverse the tide at this point," says Slattery. "But if we act quickly, we can at least slow it down." [INFOGRAPHIC 15.8]

On the last day of the mission, Slattery and his team ascended back up to the sunlit surface. Slattery did not remember the sun being quite so bright. He was anxious to get back to his lab at the University of Mississippi, where he and his team would sort through the reams of data and samples, like detectives sorting through clues in an effort to solve a great mystery.

They already knew that their initial hypothesis would hold true: There were indeed major differences in pH across the reef landscape. But whether those differing conditions had any impact on the reef species' physiology remained to be seen. Back in the lab, they would measure the protein levels in each of the samples they had collected and try to determine whether species that had grown up in acidified microhabitats expressed different proteins, or responded differently to stress, than those that had grown up on the open reef.

They'd spent 9 days at the bottom of the ocean. Now the real work could begin.◉

Research articles referenced in this chapter:
Beman, J.M., et al. 2011. *Proceedings of the National Academy of Sciences*, 108: 208–213.
Orr, J., et al. 2005. *Nature*, 437: 681–686.

Infographic **15.8** | **REDUCING THE THREATS TO OCEAN ECOSYSTEMS**

STRATEGY	EFFECT
Designate vulnerable areas as marine protected areas.	Protecting ocean communities helps an area withstand or even recover from environmental damages, as evidenced by the astounding recovery of the Kanton Island reef.
Reduce use of fossil fuel.	Conservation efforts and transition to non–fossil fuel based energy sources will reduce the amount of anthropogenic CO_2 released and slow ocean acidification and warming (see Chapter 26).
Limit development in vulnerable areas.	Reducing beachfront development and ceasing the practice of coral reef harvesting for building materials keeps the areas intact and reduces exposure to land-based pollution.
Prohibit bottom trawling for fish in vulnerable areas and reduce overfishing in general.	Nets dragged across the sea floor can destroy coral reefs and other seabed formations. Using other, safer methods in reef areas will protect the reef from these methods. Keeping populations viable by not overfishing improves an ocean community's ability to withstand perturbations.
Reduce pollution.	Trash, oil spills, and other types of pollution harm sea life. Keeping waters clear, clean, and free of nutrient (fertilizer or sewage) or toxic pollution benefits both ocean organisms and humans. In particular, fungi and bacteria, which thrive when extra nutrients are in the water, can infect and kill coral, so keeping the water free from this type of pollution will reduce infections and keep coral healthy.

↑ This healthy new plate coral in the lagoon on Kanton Island is a sign of hope. In their recovery after being hit by a severe bleaching episode, the coral has grown to a diameter of more than 4 feet.

UNDERSTANDING THE ISSUE

CHECK YOUR UNDERSTANDING

1. **Which of the following is TRUE about ocean acidification?**
   a. CO_2 emissions have reduced the ocean's pH by about 0.1, which corresponds to a 300% increase in ocean water acidity.
   b. Ocean pH has now dropped below a pH of 7.0.
   c. Ocean acidification is harmful, as CO_2 forms carbonic acid when dissolved in water, which eats away at existing calcium-based materials.
   d. In a more acidic ocean, the increase in key nutrients such as nitrogen and iron stabilizes almost all marine ecosystems, counteracting the effects of global warming.

2. **Organisms that live on the ocean floor are said to live in the:**
   a. benthic zone.
   b. abyssopelagic zone.
   c. hadal zone.
   d. pelagic zone.

3. **An estuary is:**
   a. a shallow coastal region affected by the tides, where organisms are adapted to strong currents.
   b. the only ecosystem on Earth that does not depend on photosynthesis.
   c. a pitch-black, cold, high-pressure zone where few organisms live.
   d. a productive marine ecosystem in a region where rivers empty into the ocean, bringing nutrients from land runoff.

4. **Which of the following threats to ocean ecosystems currently has the biggest impact on coral reefs?**
   a. Overfishing
   b. Coastal development
   c. Sediment runoff from land
   d. Oil spills

5. **What is/are the consequence(s) of coral bleaching for coral reefs and other ocean ecosystems?**
   a. Coral reefs attract coral larvae via visual cues, so when reefs lose their color, larvae have difficulty attaching to the top of existing coral skeletons.
   b. Coral bleaching occurs when corals under stress expel their zooxanthellae: Corals can only survive for a short period of time without their photosynthetic algae partners.
   c. Although coral reefs are home to many species, if bleaching occurs, most of these species can move to other marine ecosystems such as estuaries and kelp forests, thus tempering the impact on ocean ecosystems.
   d. All of the above are consequences of coral bleaching for coral reefs and other ocean ecosystems.

6. **The highest percentage of destroyed coral reefs are found in the _____ while the _____ has the lowest percentage of destroyed coral reefs.**
   a. Arctic Ocean; Atlantic Ocean
   b. Indian Ocean; Pacific Ocean
   c. Caribbean Sea; Mediterranean Sea
   d. Middle East; Asian South Pacific

WORK WITH IDEAS

1. What are coral reefs and where are they found? Explain the symbiotic relationship the polyp shares with its resident zooxanthellae.

2. What is ocean acidification? Describe the causes and consequences of this phenomenon.

3. Besides ocean acidification, what threats do coral reefs and other ocean ecosystems face? Describe the causes and consequences of these threats.

4. How can we reduce the threats to coral reefs and other ocean ecosystems? Discuss three strategies.

ANALYZING THE SCIENCE

The data in the table on the following page comes from a recent assessment of coral reefs (*Reefs at Risk Revisited*) prepared by the World Resources Institute.

INTERPRETATION

1. How many different coral reef regions are the global reefs divided into?

2. How are reef areas distributed globally? What patterns do you see in the distribution of reefs and the size of the population that lives within 30 km of reefs? Provide the data to support your conclusions.

3. What are the patterns for local threats to coral reefs relative to population within 30 km of reefs? Do the patterns seem reasonable when considering that local threats consist of overfishing and destructive fishing, marine-based pollution, coastal development, and land-based pollution? Provide the data to support your conclusions.

ADVANCE YOUR THINKING

4. What is thermal stress in coral reefs? What are the patterns in past thermal stress to coral reefs relative to population within 30 km of reefs? Are they the same as with local threats? Why or why not? Provide the data to support your conclusions.

5. How might thermal stress affect reefs in the future? As discussed in the chapter, what is another threat that reefs are facing or will be facing in the future? How are these two stresses related?

6. What challenges do these different threats present for protecting coral reef systems? What is the role of MPAs in protecting coral reefs against these various threats? Based on the information about reefs in the table, which regional reef is most secure and which is least so? Explain your responses.

Integrated threat to coral reefs by region

Region	Reef area (sq km)	Reef area as percent of global	Coastal population (within 30 km of reef), 2000	% of reefs threatened		Reef area in MPAs (%)
				Local threats	Severe thermal stress (1998–2007)	
Atlantic	25,849	10	42,541	75	56	30
Australia	42,315	17	3,509	14	33	75
Indian Ocean	31,543	13	65,152	66	50	19
Middle East	14,399	6	19,041	65	36	12
Pacific	65,972	26	7,487	48	41	13
Southeast Asia	69,637	28	138,156	94	27	17
Global	**249,713**	**100**	**275,886**	**61**	**38**	**27**

EVALUATING NEW INFORMATION

One way for a marine enthusiast to participate in ocean conservation is as a citizen scientist. Citizen science engages individuals or teams of volunteers, many of whom have no special scientific training, in research-related tasks. This allows scientists to accomplish research objectives more easily and also promotes public engagement with science research.

Explore the SciStarter website (scistarter.com).

Evaluate the website and work with the information to answer the following questions:

1. Is this a reliable information source? Does the organization have a clear and transparent agenda?
   a. Who runs this website? Do the organization's credentials make the information presented reliable or unreliable? Explain.
   b. What is the mission of this website? What are its underlying values? How do you know this?

2. From the "Pick a Topic" drop-down menu, select "Ocean & Water".
   a. How many different projects related to oceans and water are there? What sorts of information about each project does the website offer? Provide some examples.
   b. Are there any projects directly related to coral reefs? What are they and how can a citizen scientist be involved?

3. Would you participate in a citizen science project? If so, what type? Do you think that the citizen science model is valid and effective? Explain your responses.

4. How might the citizen science model be useful as a way to influence the public understanding of environmental issues and shape environmental policy?

MAKING CONNECTIONS

DESIGNING MARINE PROTECTED AREAS

Background: The health of the oceans affects our well-being, regardless of whether or not we live on the coast. Historically, we have treated the oceans as an endless resource provider as well as a bottomless waste receptacle. However, numerous scientific studies show that human activities are degrading ocean ecosystems around the world, and these changes are impairing the ocean's ability to provide the ecosystem services we rely on.

Case: One way to protect coastal ocean ecosystems is by creating effective marine protected areas (MPAs) (see Chapter 16 and the NOAA website, www.mpa.gov). You have been selected to be part of a team to evaluate current MPAs and develop guidelines for the future design and management of MPAs. Based on your research and analysis, write a position paper that includes the following:

1. An evaluation of the importance of coastal ocean ecosystems and the challenges of managing them.

2. An analysis of MPAs as a means of protecting coastal ocean ecosystems, including:
   a. an assessment of the ecological and human cost/benefits of establishing MPAs to determine what management priorities and techniques should be.
   b. an evaluation of current MPAs to determine criteria for an effective design.

3. An assessment of the potential challenges to designing and implementing MPAs and a discussion about how to address these challenges.

CORE MESSAGE

Although the oceans are vast, many fisheries are in serious jeopardy due to overfishing. Aquaculture may allow us to raise fish for harvest, taking some pressure off of wild stocks.

GUIDING QUESTIONS

After reading this chapter, you should be able to answer the following questions:

→ How are fish and fisheries like that of the Atlantic cod important for humans?

→ How do technology and the tragedy of the commons interact to jeopardize global fisheries?

→ Why have fisheries declined so precipitously in the last half-century and what is the current status of the world's fisheries?

→ What are some of the ways we are trying to protect our fisheries?

→ What is aquaculture and how might it ease the strain on at-risk fisheries? What trade-offs does aquaculture involve?

Raising sea bass inland in a recirculating aquaculture system is an alternative to marine aquaculture, which can cause problems in coastal ecosystems.

FISH IN A WAREHOUSE?

How one Baltimore fish scientist could change the way we eat

In Frenchman Bay, Maine, eight fishermen huddle around their aquaculturist instructor as he explains how to feed the "fingerling" cod, each a few inches long, in the wired cage beneath the water. The group is part of a new program meant to turn fishers into fish farmers. Instead of harvesting wild fish from the depths of the ocean, with factory ships and huge fishing nets, the fishers will learn how to raise them, from hatchlings to full-sized adults—how to feed them, monitor their health, and ultimately, how to prepare them for sale. Today, the fishers are observing a coastal *net pen*, where fish are raised in a system of stationary, floating nets, usually positioned in coastal waters. Tomorrow they will tour an indoor fish farm where the same types of fish are raised in colossal indoor tanks.

This, say experts, is the future of fishing.

For more than 400 years, the Atlantic cod—a *demersal* fish, or one that feeds along the ocean floor—supported not just a fishing industry but an entire culture down the northeastern coast of North America. Huge **fisheries**—places where fish are caught, harvested, processed, and sold and/or shipped—sprang up in the port towns of Massachusetts, Maine, and Newfoundland, and were sustained for generations almost exclusively by this one fish. Cod became the namesake of one famous Massachusetts cape town, and the source of New England's early wealth. But in the early 1990s, cod catches declined rapidly. By 1992, when the Canadian government closed the Grand Banks area to fishing completely, the annual catch had dropped to just 2% of its historic high. When annual catches fall below 10% of their historic high, scientists call it a **collapsed fishery.** [INFOGRAPHIC 16.1]

Twenty years later, with the moratorium still in place, the fisheries have yet to recover. Cod population estimates still hover at around 30% of what would be needed for the fishery to survive commercial fishing pressure. Meanwhile, some 40,000 people are still out of work and reliant on government support.

How did this happen? The short answer is that nobody owns the ocean. For an individual fisher, the choice is simple: "If I don't take it, someone else will." It doesn't make sense to leave any fish behind, when the immediate value of taking the fish is greater than the immediate cost,

↑ As seen in this photo, cod are big fish and the cod fishery was the economic backbone of colonial New England and Canada; in 1497, John Cabot's crew reported that cod was so abundant that one only had to lower a basket into the water to "fish" for cod.

---

fisheries The industry devoted to commercial fishing or the places where fish are caught, harvested, processed, and sold.

collapsed fishery Annual catches fall below 10% of their historic high; stocks can no longer support a fishery.

Infographic 16.1 | MEET THE COD

ATLANTIC COD, *Gadus mordua*
Conservation status: Vulnerable

↳ Cod are cold-water fish that inhabit coastal waters in the North Atlantic (they are also found in the Pacific). They are usually found in waters 10 to 200 m deep, and live close to the sea floor. Adults eat a wide variety of fish (including other cod) as well as mussels, squid, and crab. Their diet and early life-cycle stages depend on a seabed with a complex structure (lots of hiding places and niches for their prey); therefore, trawling methods that drag a heavy net across the sea floor seriously degrade their habitat and spawning beds. The cod fishery has declined throughout its Atlantic range (blue area), especially in the Grand Banks of the Newfoundland—Labrador Shelf region, where the fishery is seriously depleted.

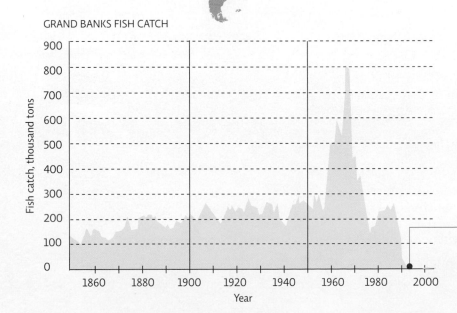

GRAND BANKS

GRAND BANKS FISH CATCH

← The collapse of the Grand Banks fishery was surprising, considering the biology of this fish. Cod are quite large—an adult can be more than 2 meters (6 feet) long and weigh 90 kilograms (200 pounds) or more. When one considers that spawning schools of more than a hundred million fish have actually been observed and that the average female can spawn millions of eggs, it is hard to imagine that our fishing practices could devastate this population. And yet, they did.

1992 Canadian government closes the cod fishery

even if it means that there will be fewer fish to harvest in the future (see Chapter 4 for a discussion of "discounting the future"). As discussed in Chapter 1, this conundrum is a case of *the tragedy of the commons*—any resource to which a group of people have free access (a common) is likely to become overexploited and degraded (the tragedy) as each seeks to maximize personal benefit without consideration for the needs of others.

The longer answer to the problem of fisheries collapse involves a constellation of forces. A trifecta of technological advances—steam engines, flash freezing, and trawler ships that could drag huge nets behind them—enabled fishers to travel further into the ocean, catch more fish, and transport them greater distances than ever before. At the same time, health-conscious eating patterns made fish more popular than any other source of protein. And as technology, rising demand, and the tragedy of the commons conspired, the once-prolific Grand Banks Atlantic cod fishery was decimated. Not only did fish populations plummet, but the increased traffic of fuel-guzzling ships polluted the water. Other marine species that lived in the same waters were lost as **bycatch**, meaning they were

trapped in nets meant to capture cod. And bottom trawler ships did irreparable damage to the sea bed, including the cod's spawning ground.

Bycatch seriously threatens many species today, including small whales and dolphins (more than 300,000 taken per year), sea turtles (more than 250,000 taken per year), and even sea birds (more than 300,000 taken per year). Seals, sea lions, and sea otter populations have declined tremendously (by as much as 85%) in some areas due to bycatch. Other fish are also taken as bycatch, including sharks, rays, and juvenile fish of many species, even cod. Bottom trawlers are particularly damaging—they destroy seabeds (the cod's spawning grounds) and catch or destroy billions of coral, sponges, starfish, and other invertebrates. Bycatch is also a problem with other industrial fishing techniques such as long-line fishing (a main line holds many individual lines with baited hooks) and drift netting (a free floating net entangles fish; this method was banned

---

bycatch Nontarget species that become trapped in fishing nets and are usually discarded. Some methods, like trawling, have very high bycatch levels, and discards often exceed the actual target species catch.

Infographic 16.2 | **BOTTOM TRAWLING**

↓ Bottom trawlers, like those that are used for cod, drag huge nets across the ocean floor, damaging the seabed where cod and other species live and breed.

Other organisms that the fishers cannot use are caught in the large nets and are discarded (unwanted or illegal species) or may drown (turtles and sea mammals). Bycatch can be huge (30–70%) and is the main threat to many whales and marine turtles. Special net modifications can decrease bycatch.

The trawl net has heavy weights to keep it on the seabed and has an opening that can be more than 60 meters (200 feet) wide.

BYCATCH

SEDIMENT PLUME

DAMAGED SEABED

in 1992). Fishing methods that minimize bycatch are available and their use is increasing. [INFOGRAPHIC 16.2]

Such is the story of our last wild food. It begins in the depths of the oceans and ends on our dining room tables. But as fish stocks around the world—not only of Atlantic cod, but also of Chilean sea bass, Alaskan pollock, and Atlantic herring—dwindle to unprecedented lows, scientists and fishers alike are trying to rewrite that story. Their success could mean sparing the world's fisheries from an unthinkable fate; it could also mean that our grandchildren never eat a wild-caught fish.

Humans rely on protein from fish.

Simply put: Humans need fish. We consume more seafood every year than we do beef, pork, and chicken combined. In fact, more than 15% of the world's population rely on fish as their main source of protein. In poorer nations, this preference is driven by the cheaper cost of fish compared with meat and poultry. In wealthier nations like the United States, a preference for seafood emerged as scientists

uncovered an array of health benefits associated with the omega-3 fatty acids found in fish oils—especially the fish oils of cold saltwater species. In recent decades, a bevy of trendy diets have been built around fish consumption.

As technology, rising demand, and the tragedy of the commons conspired, the once-prolific Grand Banks Atlantic cod fishery was decimated.

In 2011, the Food and Drug Administration doubled the recommended amount of fish in its nutritional guidelines, from one serving to two servings per week. The average American now eats about 15 pounds of fish a year. That's twice as much as in the 1950s, and the number is only likely to increase.

And it's not just fish protein we have come to rely on. More than 200 million people around the world earn their living in the fishing industry, which generates some $130

↑ A commercial catch of sand eel fish, which have been overfished, caught on the North Sea, Denmark.

billion annually in global revenue. Developing countries are especially dependent on fish, not only as a source of protein, but also as a source of revenue: More than half of all fish sold in the global market come from developing countries, and fish make up the single biggest developing country export.

But for all the many health and economic benefits this massive industry provides, it has also exacted huge—some would say catastrophic—environmental costs.

Declining fish populations can impact an ecosystem, especially one with a simple food web comprised of few species, like that of the cold waters of the North Atlantic. Unbalanced food webs may be irreversibly changing the ocean ecosystem of the Grand Banks cod. The loss of the higher trophic levels can be particularly disruptive,

and is increasingly occurring. Known as "fishing down the food chain" this loss of apex predators allows some lower-trophic-level species to increase in number. For example, declines in cod and the fish they feed on has increased the population of small jellyfishlike organisms called hydroids. Hydroids make it harder for the cod population to recover by not only feeding on the same food as the very young cod, but by feeding on the juvenile cod themselves. Interestingly, fish lower on the food chain are seen as less desirable for human consumption and some vendors get around this by renaming the fish once it gets to market. For example, there is not much of a market for the invasive Asian carp that is causing problems in Louisiana waters but marketers there have plans to bring it to market as "Silverfin." [INFOGRAPHIC 16.3]

Infographic **16.3** | **FISHING DOWN THE FOOD CHAIN**

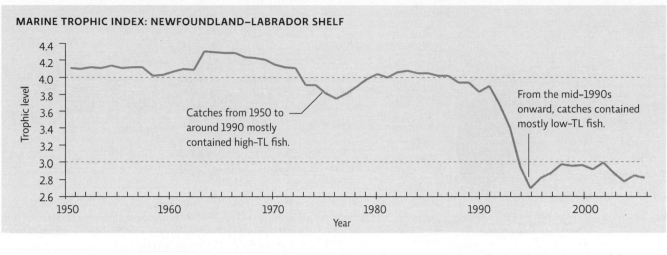

MARINE TROPHIC INDEX: NEWFOUNDLAND–LABRADOR SHELF

Catches from 1950 to around 1990 mostly contained high-TL fish.

From the mid–1990s onward, catches contained mostly low-TL fish.

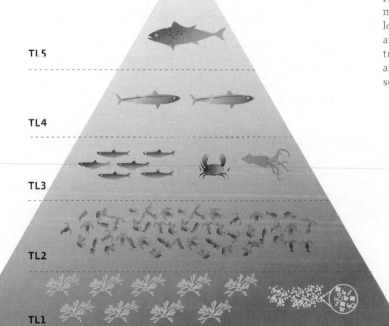

MARINE FOOD PYRAMID

TL5

TL4

TL3

TL2

TL1

↑ The *Marine Trophic Index* (MTI) is a measure of the average trophic level (TL) of fish taken in a given year and is an indicator of the status of a marine ecosystem. Higher-TL species like cod and tuna are typically the more sought-after fish. A catch that contains mostly lower-TL fish suggests that higher-TL fish populations are depleted. To determine the MTI, one identifies the trophic level of all fish taken (how many at TL2, TL3, and so on) and then calculates the average TL represented by all fish taken that year.

← One of the obstacles to the recovery of cod and other large predatory fish is the loss of their own prey. When cod are depleted, fishers pursue the herring, crabs, and shrimp at lower trophic levels, reducing the food supply for the cod, and ultimately jeopardizing the cod recovery.

Today, the oceans—not to mention the fish themselves—are in grave danger. The number of large fish like tuna, cod, and halibut in the ocean is only 10% of what it was in 1950. More than half of the world's fisheries are at their **maximum sustainable yield** (*fully exploited*), the amount that can be harvested without decreasing the yield in future years. Another 32% are **overexploited fisheries**, meaning they are being harvested at an unsustainable level, or **depleted fisheries**, meaning that the population size is very low compared to historic levels and there are not enough fish left to support a fishery.

maximum sustainable yield The amount that can be harvested without decreasing the yield in future years.
overexploited fisheries More fish are taken than is sustainable in the long run, leading to population declines.
depleted fisheries The fish population is well below historic levels and the population's reproductive capacity is low, meaning that recovery will be slow, if at all.

Infographic **16.4** | **STATUS OF MARINE FISHERIES**

STATUS OF GLOBAL MARINE FISHERIES (2010)

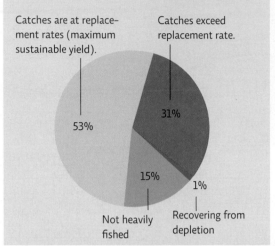

Catches are at replacement rates (maximum sustainable yield).

Catches exceed replacement rate.

53%

31%

15%

1%

Not heavily fished

Recovering from depletion

← Though 68% of global marine fisheries are sustainably fished, the percentage that is overfished (31%) has increased threefold since the 1970s and is a major concern for fishers and fisheries managers.

↓ Fish catches in the Newfoundland–Labrador Shelf area of the North Atlantic show a distinct shift in the species and amounts taken over the years. Fisheries managers called for a 50% reduction in allowable catch of cod in 1988, but political officials only reduced the amount by 10%. The cod fishery was closed in 1992.

FISH CATCH BY COMMERCIAL GROUP: NEWFOUNDLAND–LABRADOR SHELF

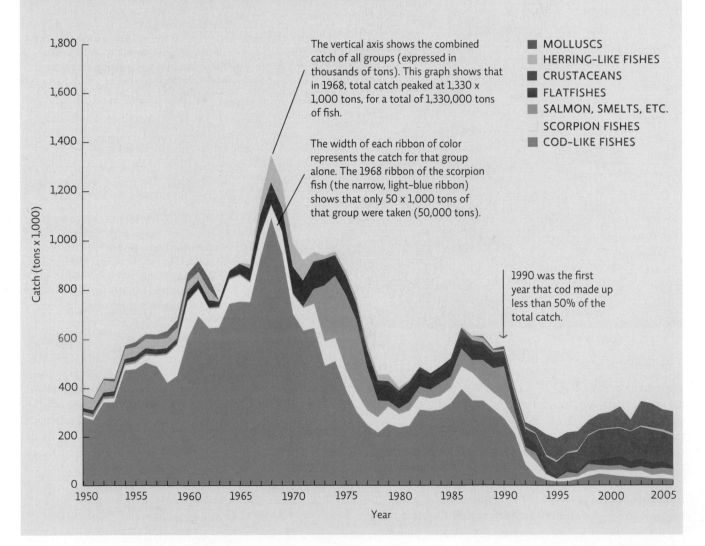

The vertical axis shows the combined catch of all groups (expressed in thousands of tons). This graph shows that in 1968, total catch peaked at 1,330 x 1,000 tons, for a total of 1,330,000 tons of fish.

The width of each ribbon of color represents the catch for that group alone. The 1968 ribbon of the scorpion fish (the narrow, light-blue ribbon) shows that only 50 x 1,000 tons of that group were taken (50,000 tons).

■ MOLLUSCS
■ HERRING-LIKE FISHES
■ CRUSTACEANS
■ FLATFISHES
■ SALMON, SMELTS, ETC.
  SCORPION FISHES
■ COD-LIKE FISHES

1990 was the first year that cod made up less than 50% of the total catch.

Catch (tons x 1,000)

Year

Infographic **16.5** | **PROTECTION FOR MARINE AREAS**

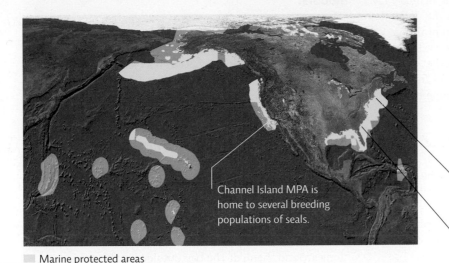

Channel Island MPA is home to several breeding populations of seals.

← Marine protected areas (MPAs) provide varying degrees of protection for different species, depending on the area and the need. Exclusive economic zones (EEZs) extend 200 nautical miles from the coastline of any given nation, giving that nation exclusive rights over marine resources, including fish. Both MPAs and EEZs can help protect species at risk by restricting harvests and use in the area.

— Stellwagen Bank is a highly productive area that supports groundfish (like cod) and whales.

— Gray's Reef supports a diverse community of flora and fauna.

■ Marine protected areas
■ U.S. exclusive economic zone (EEZ)

Unless we change our ways, scientists predict that all commercially valuable wild fish populations may collapse by the middle of the century. [INFOGRAPHIC 16.4]

Laws exist to protect and manage fisheries.

For most of human history, it seemed impossible that we could ever use up the resources in something as vast as the ocean, which takes up 70% of the planet; but in recent decades, that thinking has begun to change. As human populations swell and fish populations plummet, scientists and lawmakers around the world are struggling to reverse the damage that's already been done and to protect our fisheries from further decline.

They started by enacting legislation. For centuries, most nations defined their territorial waters, or exclusive fishing zones, as reaching 12 miles off each coastline. But in the 1960s, it became clear that such zones were no longer adequate. Wide-ranging industrial ships—floating factories, replete with freezing and processing technology—had long freed fishers from the shackles of time and distance. They could now pursue any fishery they liked, no matter how far from their home country; and with no rule of law to constrain them, they could also stay as long as they pleased and take as much as they wanted. As the Atlantic cod would eventually show, no fishery stood a chance against a global army of such ships.

One response to this problem was the creation of **exclusive economic zones (EEZs)**—zones that extend 200 nautical miles from the coastline of any given nation,

where that nation has exclusive rights over marine resources, including fish.

For a long time, most countries, including the United States, simply kicked other nations out of their fisheries while still allowing their own fishers to take as much as they pleased. But with fisheries around the world at or near collapse, fisheries managers have started cracking down. In the U.S. cod fisheries, for example, officials have set strict limits on the amount of all ground (bottom-dwelling) fish of any species that any vessel may take. Once the vessel reaches its predetermined limit, it must stop fishing for the season.

Fisheries managers have also created more than 1,500 **marine protected areas** (MPAs)—discrete regions of ocean that are legally protected from various forms of human exploitation. Many of these MPAs protect only specific elements within the area—certain species of fish, a discrete portion of the habitat (like a spawning ground), or a cultural artifact (like a sunken ship). But at least some of them are fully protected **marine reserves**—"no take" zones, where absolutely nothing may be disturbed by human hands. Sea life is recovering in many MPAs, giving us reason to be encouraged. [INFOGRAPHIC 16.5]

---

exclusive economic zones (EEZs) Zones that extend 200 nautical miles from the coastline of any given nation, where that nation has exclusive rights over marine resources, including fish.

marine protected areas (MPAs) Discrete regions of ocean that are legally protected from various forms of human exploitation.

marine reserves Restricted areas where all fishing is prohibited and absolutely no human disturbance is allowed.

The Magnuson-Stevens Fishery Conservation and Management Act, first passed in 1976, is the primary law governing marine fisheries management in U.S. federal waters. The 1996 amendments focused on rebuilding overfished fisheries, protecting ecosystems, and reducing bycatch. Amendments in 2006 added in market-based programs like *catch-share* ("cap-and-trade" for fish) that appear to be helping with the recovery of some fish stocks.

But managing a resource that makes up 70% of the planet is no small feat, and despite these efforts, fisheries have continued to dwindle. Part of the problem is the enormous amount of illegal, unreported, and unregulated fishing—between $4 and $9 billion worth every year. In some fisheries, 30% or more of the annual catch is harvested with illegal gear, in prohibited areas, or at prohibited times.

Here's another problem: So far, fisheries management has been based on incomplete science. For example, *State of the World's Fisheries and Agriculture*, a biannual report issued by the Food and Agriculture Organization (FAO) that estimates fish stocks and makes recommendations for maintaining fisheries sustainably, bases its conclusions almost exclusively on population studies—research that uses characteristics like a species' rate of reproduction and time to reach adulthood to determine how many fish can be taken without jeopardizing the fishery. Successful management requires more than that. To maintain individual species at sustainable levels, fishery managers need to consider the entire ecosystem in which the species resides—not just the abundance of a given fish, but the relative populations of the fish they feed on, the overall health of their spawning grounds, and a host of other factors.

The Marine Stewardship Council, an international, nonprofit organization that certifies fish food products as sustainable, defines a **sustainable fishery** as one that ensures that fish stocks are maintained at healthy levels, the ecosystem is fully functional, fishing activity does not threaten biological diversity, and the fishery is managed effectively. The Council currently recognizes 104 certified fisheries; another 143 are undergoing assessment for certification.

Of course, no management strategy can reconcile limited supplies with rising demand, which is why, even as net pen operations like the ones in Frenchman Bay, Maine, proliferate around the world, more and more scientists are saying that the real solution to our fish problem won't be found in the ocean at all.

Scientists study the possibility of growing marine fish indoors.

Perched on the edge of Baltimore's Inner Harbor, the University of Maryland's Center of Marine Biotechnology (COMB) looks like a cross between a classic ship and a modern research facility: a sleek glass façade covers one side of the building, a giant sail covers the other. Deep in the lower decks, in the lab of COMB director and marine scientist Yonathan Zohar, land and sea get blurred further still. Pipes, computers, and narrow cylindrical filters buzz and whir above several rows of tanks, each roughly the size, shape, and light-blue shade of a small above-ground swimming pool. Inside the tanks, giant fin fish—mostly sea bass, many weighing as much as 80 pounds—create whirlpools as they retrace the same circumference over and over; they seem blissfully unaware of the machinations around them, or the bay outside, or the ocean beyond.

The operation, known as a *recirculating aquaculture system* (RAS), is Zohar's brainchild. The fish were spawned, hatched, and raised to adulthood in the lab; if all goes to plan, they will be harvested and sold to area restaurants and will eventually end up in the lunches and dinners of local patrons—Mediterranean sea bass, from a Baltimore laboratory. "This is a real leap," says Zohar. "The ability to grow these large predatory fish indoors, away from the ocean, will really change the way we see these animals and the way we think about fish as food."

↑ Dr. Yonathan Zohar works at COMB in Baltimore.

↑ Sea bream fish farm, Shikoku, Japan

Zohar began his career in the 1970s on the banks of the Red Sea, when the environmental movement was in its infancy and global fisheries were just beginning to show signs of distress. Forward-thinking scientists had begun searching for ways to domesticate fish—just as humans had domesticated cattle, poultry, and pork—so that wild stocks might be spared, even as global populations swelled.

Aquaculture had already been used for centuries to grow fresh water fish like tilapia. But most fish farms are family or village operations—just big enough to provide a few dozen people with a steady food supply. Zohar wanted a fish farm that could feed a modern city, or even a country. He also wanted one that could raise large marine species. "With marine aquaculture, we are talking about fish that one, we are running out of, and two, are the most beneficial to human health," he says.

But there was a huge barrier to realizing this vision: most commercially important marine fish—like sea bream, sea bass, and tuna—would not reproduce in captivity. Some scientists had suggested recreating spawning grounds—specific areas where fish come every year to deposit their eggs and sperm—in captive breeding sites.

---

sustainable fishery A fishery that ensures that fish stocks are maintained at healthy levels, the ecosystem is fully functional, and fishing activity does not threaten biological diversity.

aquaculture Fish-farming; the rearing of aquatic species in tanks, ponds, or ocean net pens.

But that seemed wildly impractical to Zohar. "Most of these fish travel hundreds to thousands of miles to reach their spawning grounds," he says. "They move through a wide range of temperatures, water depths, and salinities, and nobody had any clue which of these variables was the key to getting them to reproduce, let alone how to recreate all of that in a finite space." So instead, he tracked several species across the ocean and measured their hormonal changes as they navigated the open waters; as it turned out, one particular hormone was produced in the wild but not in captivity. By developing a hormone supplement for the fish, Zohar and his colleagues tricked the captive fish into breeding as they would in the wild.

But just as that problem was resolved, other problems began creeping up.

Aquaculture presents environmental challenges.

Once scientists could get marine fish to reliably breed in captivity, entrepreneurs began adapting the aquaculture techniques used in freshwater operations to suit larger ocean fish. Before long, elaborate net pen operations began dotting coastlines around the world. This simple technology has enabled fish farms to out-produce traditional fishers. In 2009, aquaculture crossed the threshold of providing more than half of all seafood consumed world wide, with Asia leading the way, and China alone producing some 63% of all farmed fish.

Infographic 16.6 | **NET PEN AND POND AQUACULTURE: PROBLEMS AND POSSIBLE SOLUTIONS**

PROBLEMS	POSSIBLE SOLUTIONS
More diseases and parasites than wild fish	Decrease the density of net pens to minimize the impact. This would decrease concentrated waste (though there would still be some) and decrease/eliminate the need to use antibiotics or pesticides; pursue RAS aquaculture.
Use of antibiotics and pesticides	
Large amounts of waste are released into the environment.	
Fish farms displace mangrove swamps and other wetlands.	Move outdoor fish farms farther from the coast; robotic fish cages that can be navigated through the sea with boat-operated remote controls are being developed.
Farmed fish are fed fish meal made from wild caught fish so still exert pressure on wild stocks	Find alternative foods that do not rely on fully or overexploited wild caught fish.
High-trophic-level fish are typically raised; this requires large amounts of food (2 kilograms [4.5 pounds] of fish for 0.5 kilogram [1 pound] of salmon; 9 kilograms [20 pounds] for 0.5 kilogram [1 pound] of tuna).	Concentrate on fish lower on the trophic chain to increase the efficiency of food conversion.
Escape of nonnative or GMO species	Only use native species if farming in natural waters.

↑ The Aquapod is a submersible net pen for open ocean aquaculture made from recycled materials. Aquapods are designed to be located several miles from the marine coast to lower the environmental impact while optimizing growing conditions for farmed fish.

But, it has also led to a range of environmental problems including depletion of the populations of smaller fish harvested as food for aquaculture species and ecosystem damage from the aquaculture ponds and net pens. For example, hundreds of thousands of acres of tropical mangrove forests have been cleared to make way for shrimp aquaculture ponds. Mangrove trees are an important keystone species that provides habitat and protection for a wide variety of species, including humans, as they protect the coastline from storm surges during hurricanes. Most experts agree that these problems are not insurmountable. But how best to resolve them is a matter of some debate. [INFOGRAPHIC 16.6]

Meanwhile, RAS technology offers its own mix of risks and benefits. "We have to make a distinction between having an impact and being sustainable," says Lorenzo Juarez, deputy manager of the National Oceanic and Atmospheric Administration (NOAA)'s Aquaculture Program. "Anything you do is going to have an impact on the environment. The real question is whether it's an impact the environment can absorb and recover from or one that will do permanent damage."

Indoor fish farming may provide a solution.

Zohar's laboratory starts with water from the same supply that feeds household taps. By adding a precise mix of
salts and trace elements, the lab creates its own seawater,
which it then monitors and regulates by computer as the
seawater cycles through the network of pipes and filters
and swimming pool-sized fish tanks that make up this
miniature, indoor ocean. Everything about the artificial
salt water—from its temperature, salinity, and pH, to
its CO_2 and oxygen concentrations—can be adjusted, on
a tank-by-tank basis, to suit the species of fish and to
mimic the changing conditions the fish would experience
as they moved through both time and distance in the
wild. "The entire system is biosecure," boasts Zohar.
"We don't take a drop of water out of the harbor and we
don't drain a drop of water into the harbor." In fact, the
same several thousand gallons is recycled over and over.
Of course, in areas with water shortages, commandeering
several thousand gallons of water to set up tanks may
not be an option, so this may limit the applicability of
this technology.

For Zohar, recirculating aquaculture represents a lifetime
of research—each detail a carefully crafted response to
the problems associated with existing aquaculture technology. Fish growing in Zohar's tanks require less food
per pound than the same fish grown in a net pen. The
reason for this is simple: Because there are waves, and
because salinity can't be optimized, net pen fish expend
more energy on movement and regulating their internal
salt concentrations (osmoregulation) than they otherwise
might. On top of that, food in the pens is not completely
consumed; a portion of it sinks to the bottom of the pen,
where it can't be recovered or reused. In the RAS setting,
by contrast, aquaculturists can control food intake much
better; and because salinity is always optimal, fish don't
have to invest as much energy in osmoregulation. As a
result, they convert their food into flesh at a much higher ratio.

To further reduce their dependence on smaller marine
fish for food, the researchers are experimenting with a
variety of alternative feeds. The main contender so far
seems to be algae; in a room adjacent to the fish tanks,
long plastic tubes—the size, shape, and color of colossal
lime green freezepops, hang from thin metal racks; each
one is filled with algae that will eventually be converted
into food pellets (along with several other ingredients)
and given to the fish. "All of the optimal ingredients
in marine fish—the omega-3s, etc.—all come from the
base of the food chain," says Zohar. "So why not feed
them algae?"

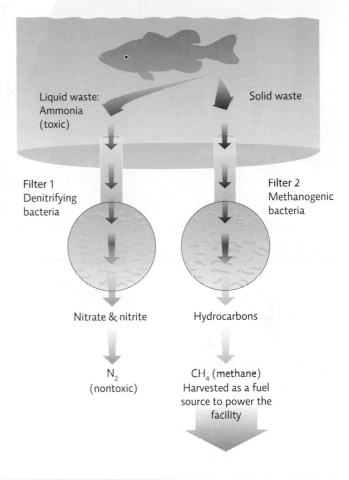

↑ Microbes in the RAS manage the waste in the water. Like
their wild counterparts, these microbes detoxify ammonia
(the same waste product that is so toxic to net pen fish) or
break down solid waste. This system has the added benefit
of providing an energy source for the facility.

To manage the accumulated liquid and solid waste in the
RAS systems, Zohar's team employs carefully calibrated
microbial communities that function much like they do
in the ocean. One community converts ammonia into
nontoxic forms. Other bacteria converts some 96% of the
solid waste into fuel-grade methane. "So on top of all the
other benefits, the system produces energy, about 10 liters
of biogas every day." Zohar explained, on a recent tour of
the facility. That's not enough to power the entire system,
he says, but it certainly offsets some of the energy costs.
[INFOGRAPHIC 16.7]

As good as it all sounds, RAS facilities will have to meet
a number of challenges before they can possibly hope to
replace net pens. Net pens, for one, are much cheaper to

UNDERSTANDING THE ISSUE

CHECK YOUR UNDERSTANDING

1. **Why might aquaculture be considered the "future of fishing?"**
   a. Systems like recirculating aquaculture systems (RAS) are so affordable that more fishers will turn to them.
   b. Today's aquaculture techniques provide fish to consumers without damaging the environment.
   c. It can produce large numbers of fish without depleting wild stocks.
   d. It will raise fish to be released into the ocean to support commercial fisheries.

2. **Why might bottom-trawling fishing techniques make it harder for the cod population to recover even after the ships have left the area?**
   a. No fish are left to breed.
   b. The nets damage the seabed where cod spawn.
   c. The cod leave the area once a trawler has come through.
   d. The ships attract seabirds that eat the cod.

3. **In what way is the RAS waste-treatment plan an example of biomimicry?**
   a. It purifies the water with biodegradable filters.
   b. It depends on bacteria doing what they normally do in nature.
   c. It allows the waste to naturally sink to the bottom of the pool where it remains out of the way.
   d. It focuses on frequent water changes rather than on purification of water.

4. **The collapse of the cod fishery is a tragedy of the commons because:**
   a. some fishers are not allowed to fish in the Grand Banks.
   b. cod was the most common fish in the area.
   c. it was the privatization of the fishery that caused its collapse.
   d. fishers would take as much as possible because if they didn't, someone else would.

5. **A disadvantage of net pen operations is:**
   a. fish that can be successfully raised this way are not popular food items.
   b. these operations are expensive.
   c. they release pollution into the surrounding water.
   d. they cannot be done on a large scale.

6. **"Fishing down the food chain" refers to:**
   a. taking only higher-trophic-level species and throwing back lower-trophic-level species.
   b. making sure equal numbers of individuals are taken from all the trophic levels.
   c. current management techniques designed to restore a depleted fishery.
   d. moving on to take lower-trophic-level species once the higher levels are depleted.

WORK WITH IDEAS

1. Why do we define a fishery in terms of human use? Does that diminish it as an ecological concept?

2. Ecosystems like those found in the extreme environment of the Arctic have simple food webs, with only a few organisms at each trophic level, whereas ecosystems with more moderate climates have more robust food webs with many species. Why is a simple food web more vulnerable to collapse than a more complex one?

3. What are the advantages of RAS like the one Yonathan Zohar is developing and where (in what nations or regions) will they likely be adopted? If adopted in these areas, will they help lower the impact of eating fish? Explain.

4. Consider the triple bottom line and propose (based on current technology) a low-impact way for people in developing nations to meet their need for fish.

ANALYZING THE SCIENCE

The graph to the right comes from *The Ecological Fishprint of Nations* report by Redefining Progress. It tracks the global fishprint (a measure of the ocean area needed to produce the fish catch in hectares per ton of fish, weighted to reflect the ecological productivity of particular biomes) against global biocapacity (a measure of the ocean's ability to supply a steady quantity of fish based on the producer productivity in ocean ecosystems and also expressed in terms of area).

INTERPRETATION

1. In one sentence, explain what has happened over time in terms of the human fishprint and the ability of global oceans to support it.

2. In what year did our fishprint exceed the biocapacity of the ocean?

3. Ecological overshoot is the difference between the fishprint and biocapacity. What was the ecological overshoot in 1980, in global hectares (gh)? In 1990? In 2003?

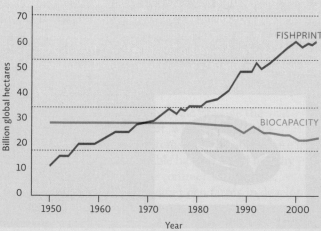

Global Fishprint and Biocapacity 1950–2003

ADVANCE YOUR THINKING

Hint: To answer some of the questions below it might help to access the actual report.

4. Why do you suppose global biocapacity has decreased in recent years?

5. If RAS facilities like the one described in this chapter are widely adopted, what would likely happen to biocapacity and fishprint in the future? Explain why.

6. What would it take to return to a global fishprint and biocapacity as they were in the 1950s or 1960s? Describe three specific strategies using the fishprint as a tool to help accomplish this reversal.

EVALUATING NEW INFORMATION

Should one eat fish? Many of the world fisheries are seriously degraded. But fish are a major protein source for many people around the world and a healthy alternative to other forms of meat. So what is a conscientious consumer to do?

Explore the Seafood Watch section of the Monterey Bay Aquarium website (www.montereybayaquarium.org/cr/seafoodwatch.aspx).

Evaluate the website and work with the information to answer the following questions:

1. Is this a reliable information source? Does it have a clear and transparent agenda?
   a. Who runs this website? Do the organization's credentials make the information presented reliable or unreliable? Explain.
   b. What is the mission of this organization? What are its underlying values? How do you know this?
   c. What data sources does Seafood Watch rely on and what methodology does it employ in calculating its ratings? Are the sources it uses reliable?
   d. Do you agree with its rating system? Do you think that such a rating system is sufficient and useful to help consumers make sustainable choices in selecting fish? Explain your responses.

2. From the Seafood Watch section, select the "Ocean Issues" link.
   a. Do you agree with the aquarium's assessment of the ocean issues? Explain.
   b. Identify a claim that is made and the evidence given in support of this claim. Is it sufficient? Explain.

3. The aquarium offers a seafood recommendation card that identifies safe fish to eat based on health and ecological considerations. Use the site's seafood search option and type in "cod" in the search box. Why are some Atlantic cod identified as fish to avoid and others not?

4. Select the link for pocket guides and choose the one for the area in which you live. Look at the fish listed and find one that you eat frequently. (If you don't eat fish, choose one of the tuna species, a popular fish in the United States.) Is this a good fish to consume? If not, identify an alternate fish you could consume instead. Why are there different pocket guides for different regions of the United States?

5. How might a program like Seafood Watch help the oceans?

MAKING CONNECTIONS

THE FUTURE OF SALMON FISHERIES

Background: According to Seafood Watch, Alaskan wild salmon are an ocean-friendly choice, but other salmon fisheries raise concerns. The management of salmon fisheries is challenging, as salmon require both freshwater and ocean habitats.

One solution is to switch to salmon aquaculture. But while salmon raised in tank systems are environmentally friendly operations, most salmon are farmed in open net pens and should be avoided for their environmental impact. Wastes that pollute water, transmission of disease to wild salmon stocks, and interference with life-cycles of wild fish from farmed fish (which are selectively bred or non-native species of salmon) that escape are the major concerns.

Case: To ensure that the fate that befell the cod fisheries is not repeated for salmon, efforts are being made to develop a sustainable salmon aquaculture. You have been assigned the task of developing a policy for the future of salmon farming based on an evaluation of the following options:

1. The AquAdvantage salmon: a genetically modified salmon that reaches market size in 18 months—half the time of a wild salmon.

2. Integrated multi-trophic aquaculture (IMTA) systems: These blend the farming of fish with the culture of seaweed and shellfish that process the waste the fish generate.

3. Recirculating aquaculture systems (RAS): above-ground, temperature-controlled environments that raise single species of fish without any interaction with the wild.

4. The Arctic char: a salmon-like fish with a smaller footprint.

Research these four options and write a report recommending a course of action to invest in. In your report, include the following:

   a. An analysis of the pros and cons of each proposal including:
      · A discussion of the consequences of each choice from an economic, ecological, food security, and human health perspective.
      · A reflection on the values underlying each proposal.
   b. The best option for the future of salmon aquaculture. Who should be involved in this decision? Provide justifications for your proposal.
   c. A set of guiding principles that can be applied to sustainably managing any fish species as a food source, based on what you now know about fishery management and aquaculture, and their challenges. Discuss the principles you develop and explain why you consider them key for the future of sustainable seafood.

A glass of treated water from the
Groundwater Replenishment System
in Fountain Valley, California.

TOILET TO TAP

A California county is tapping controversial sources for drinking water

CORE MESSAGE

Freshwater is a precious but limited resource, and is essential to life. Some regions consume water faster than it is replenished. And unfortunately, water is not evenly distributed across the globe; many people worldwide lack access to enough clean water. Methods are available to recover and purify otherwise dirty water, but we also need to use water more wisely.

GUIDING QUESTIONS

After reading this chapter, you should be able to answer the following questions:

→ What are the sources of freshwater on Earth and how does water cycle through the environment?

→ What are the causes and consequences of water scarcity?

→ What is an aquifer, how does it receive water, and what problems emerge when too much water is removed?

→ What are some of the ways that our wastewater is treated to make it usable again or safe to release into the environment?

→ How can conservation help us address water scarcity issues?

One of the most exciting moments in Shivaji Deshmukh's career as a water engineer came one bright, sunny day in January 2008. He had gathered with staff from the Orange County Water District (OCWD) in Anaheim, California, to watch for the first time as former sewage water, cleaned using state-of-the-art techniques, was pumped into underground drinking water sources. It was the beginning of a groundbreaking project designed to help save the region from ongoing, and frightening, water shortages.

"It's basically this drought-proof supply of water," says Deshmukh. "Nobody else has done it. Nobody thought a community could support it, because they would be too grossed out by it."

The water that Deshmukh and other engineers watched seep into the region's underground water stores that day in 2008 was treated **wastewater**—including sewage and used water from homes and industrial sites. Understandably, when many residents first heard about the project, they were concerned.

But that same month, Deshmukh and other OCWD staff attended a dedication ceremony at the water treatment plant in Fountain Valley, California, along with hundreds of other people, including various community groups, to honor the massive project. Having that support from the community was key to the project's success, says Deshmukh—but getting it hadn't been easy.

Water is one of the most ubiquitous, yet scarce, resources on Earth.

Even though Earth is covered in more than 1,400 million cubic kilometers of water—about 75% of its surface—only about 1/100 of 1% of that water is usable by humans.

Water provides many important ecosystem services that animals and plants require to live. Up to 75% of the human body, for instance, consists of water. But humans need liquid **freshwater** (which has few dissolved ions

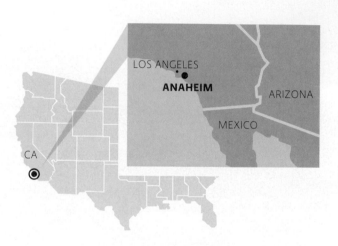

⊚ **WHERE IS ANAHEIM, CALIFORNIA?**

LOS ANGELES
ANAHEIM
ARIZONA
MEXICO
CA

such as salt); ocean water is too salty for human consumption and is toxic in large doses.

Complicating things, nearly 80% of the freshwater on the planet is trapped in ice caps at the poles and glaciers around the world, which contain more than 35 million cubic kilometers—enough to fill 140 billion Olympic-sized swimming pools. [INFOGRAPHIC 17.1]

Wherever there is water, it is constantly moving through the environment via the **water cycle** (hydrologic cycle). Heat from the sun causes water to evaporate from **surface waters** (rivers, lakes, oceans) and land surfaces. At the same time, plant roots pull up water from the soil and then release some into the atmosphere in a process called **transpiration**. Plants with deep roots, like trees, may bring up thousands of gallons of water a year, releasing much of this to the atmosphere. Altogether, the combination of **evaporation** and transpiration—*evapotranspiration*—sends more than 66,000 cubic kilometers of water vapor into the atmosphere every year, equivalent to 17,000 trillion gallons. Once aloft, that water condenses

---

wastewater Used and contaminated water that is released after use by households, industry, or agriculture.

freshwater Water that has few dissolved ions such as salt.

water cycle The movement of water from gaseous to liquid states through various water compartments such as surface waters, soil, and living organisms.

surface water Any body of water found above ground such as oceans, rivers, and lakes.

transpiration The loss of water vapor from plants.

evaporation The conversion of water from a liquid state to a gaseous state.

↑ Crystal-clear purified water from the Groundwater Replenishment System is piped to the Orange County Water District's percolation ponds in Anaheim, California.

Infographic **17.1** | **DISTRIBUTION OF WATER ON EARTH**

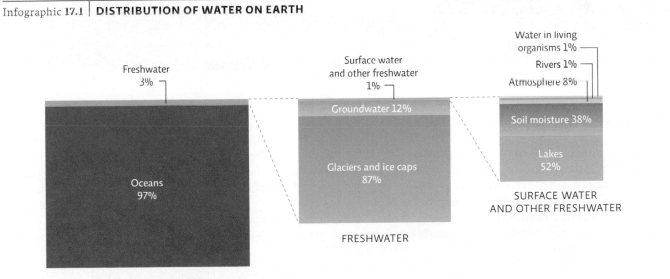

Freshwater
3%

Surface water
and other freshwater
1%

Water in living
organisms 1%
Rivers 1%
Atmosphere 8%

Groundwater 12%

Soil moisture 38%

Oceans
97%

Glaciers and ice caps
87%

Lakes
52%

TOTAL GLOBAL WATER

FRESHWATER

SURFACE WATER
AND OTHER FRESHWATER

↑ Most of the water on Earth is found in the oceans and most of the freshwater is tied up in ice and snow. Only about 0.001% of all of Earth's water is available for us to use, but with more than 350 trillion gallons of water on the planet, that is still a lot of water.

Infographic **17.2** | **THE WATER CYCLE**

↓ Water cycles between liquid and gaseous forms as it moves through space and time. Ocean water (which we cannot use) is converted to freshwater when it evaporates and falls back to Earth as precipitation, refilling freshwater surface and underground water supplies. Liquid freshwater is a renewable resource as long as we don't use it faster than it is naturally replenished.

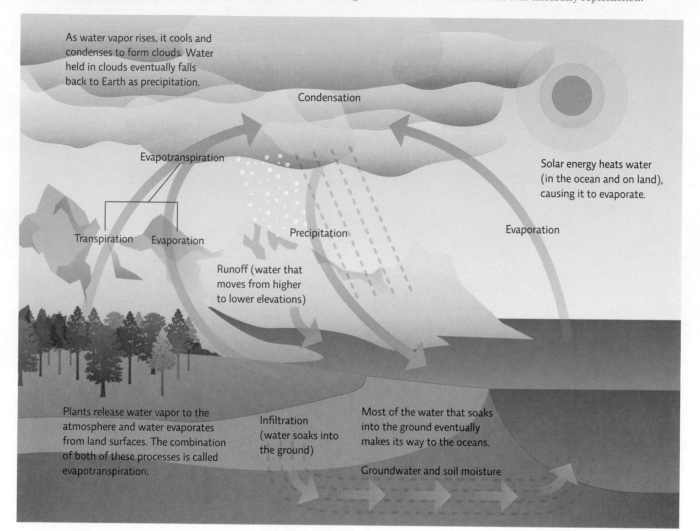

As water vapor rises, it cools and condenses to form clouds. Water held in clouds eventually falls back to Earth as precipitation.

Condensation

Evapotranspiration

Solar energy heats water (in the ocean and on land), causing it to evaporate.

Transpiration   Evaporation

Precipitation

Evaporation

Runoff (water that moves from higher to lower elevations)

Plants release water vapor to the atmosphere and water evaporates from land surfaces. The combination of both of these processes is called evapotranspiration.

Infiltration (water soaks into the ground)

Most of the water that soaks into the ground eventually makes its way to the oceans.

Groundwater and soil moisture

into clouds (**condensation**) and may fall back to Earth as **precipitation** (rain, snow, sleet, etc.). [INFOGRAPHIC 17.2]

Almost all precipitation ends up falling on the oceans, and a tiny remainder falls on land. This latter portion is the part humans can harvest for their own use. Typically, we draw freshwater from **groundwater**. But people don't always live near abundant sources of freshwater, making access a vital issue. Around the world, many areas suffer from **water scarcity**—not having access to enough clean water supplies. In some dry regions, there is simply not enough to meet needs; many arid nations like those of the Middle East, parts of Africa, and much of Australia face water shortages as a way of life. According to the Water Stress Index, compiled by the risk assessment analyst firm Maplecroft, the Middle Eastern nations of Bahrain,

Qatar, Kuwait, and Saudi Arabia are the most water-stressed countries in the world, with the lowest per capita water availability. In other areas, there is enough water but people do not have the money to purchase it or dig wells to access it. By far, per capita domestic (household) use of water in developed nations is much higher than that of developing nations—the United States alone uses more than 400 billion gallons of water a day. A wealthy nation may even be able to invest in costly technology to

condensation Conversion of water from a gaseous state (water vapor) to a liquid state.

precipitation Rain, snow, sleet, or any form of water falling from the atmosphere.

groundwater Water found underground in a region known as an aquifer.

water scarcity Not having access to enough clean water supplies.

Infographic 17.3 | AVAILABILITY AND ACCESS TO WATER WORLDWIDE

↓ Around the world, more than 2 billion people lack access to clean water. An additional 2.6 billion have no access to sanitation.

■ **Physical water scarcity**
Use of water is approaching or exceeding sustainable limits.

■ **Approaching physical water scarcity**
With more than 60% of river flow withdrawn, these areas will face water shortage in the near future.

■ **Economic water scarcity**
Access to water is limited by the ability to pay for it, not by its physical scarcity.

□ **Little or no water scarcity**
Water supplies are abundant and there are no economic constraints to access it.

■ **Not estimated**

↓ Sanitation: Access to adequate sanitation facilities is higher in urban areas than in rural areas in many regions of the world.

↓ Water footprint: Per capita water use is growing faster than our population. Developed countries have a larger per capita "water footprint" than developing nations. In some areas, individuals must make do with only a few gallons a day.

PERCENT OF POPULATION WHO USE IMPROVED SANITATION

■ Urban
■ Rural

Region	Urban	Rural
Sub-Saharan Africa	43	24
South-Central Asia	59	29
East Asia	66	54
Latin America/Caribbean	86	55
Southeast Asia	79	59
Western Asia	93	66
More developed countries	99	92

DOMESTIC WATER USE

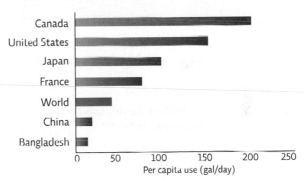

Per capita use (gal/day)

access water (like facilities to remove salt from seawater), but poorer nations do not have this luxury. People in remote regions may have to resort to digging deep wells or installing catchment systems to capture rainwater.

The World Health Organization (WHO) estimates that 1 in 3 people—more than 2 billion—lack sufficient access to clean water; even more lack access to sufficient sanitation facilities (safe disposal of human waste). In developing nations where water and funding for basic sanitation are scarce, people use nearby surface waters to meet their basic cooking, drinking, and washing needs. These waters can be contaminated with raw sewage, which increases the chance for disease transmission. According to the WHO, more than 1 million tons of raw sewage enters the Ganges River of India every minute. In Africa, almost

3,000 people die each day from water-borne diseases like cholera and typhoid fever as a result of poor sanitation and contaminated water.

> The World Health Organization (WHO) estimates that 1 in 3 people—more than 2 billion—lack sufficient access to clean water.

As populations increase, so will scarcity and sanitation issues; according to the United Nations, 2 out of 3 people will face water shortages by 2025. [INFOGRAPHIC 17.3]

↑ The California Aqueduct conveys water past Kettleman City, located halfway between San Franciso and Los Angeles, to Southern California. Local agencies are working to develop a treatment facility that would tap into the aqueduct to provide clean drinking water for Kettleman City.

Like communities around the world, California depends on many sources of water.

Around the world, each region faces its own water challenges. In California, freshwater flows into the northern part of the state when the Sierra Mountain snowpack melts in the spring. But as Earth's climate changes, the state could lose much of its snowpack. Indeed, in 2009, California's precipitation was 80% of its average, and the snowpack only 60% of its average size.

Even in a good snowpack year, the state faces major water issues, explains Deshmukh, now working at the West Basin Municipal Water District in Carson, California. Two-thirds of California's water is located in the northern part of the state, but two-thirds of the state's residents live in the south, he explains. At the moment, the state ships some of the northern water to the south via pipes and canals, which costs money and uses a great deal of electricity. And if there is an earthquake or other natural disaster, that transport system could be cut off.

Many people (not just in California) draw their water from an underground region of soil or porous rock saturated with water, called an **aquifer**. These stores receive water from rainfall and snowmelt that soaks into the ground through **infiltration**. Plant roots take up some of the water along the way, but much of it continues to move downward, filling every available space in the aquifer. As the water trickles down, it becomes naturally filtered by rocks and soil, which trap bacteria and other contaminants as the water passes by. The top of this water-saturated region, referred to as the **water table**, rises and falls due to seasonal weather changes.

The depth of Orange County's groundwater varies, says Deshmukh. At the coast, the aquifer is 6 to 90 meters (20 to 300 feet) deep, but further inland, at its deepest, the

---

aquifer An underground, permeable region of soil or rock that is saturated with water.
infiltration The process of water soaking into the ground.
water table The uppermost water level of the saturated zone of an aquifer.

Infographic **17.4** | **WATER USE**

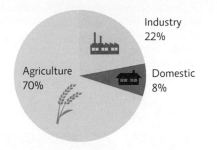

Agriculture 70%

Industry 22%

Domestic 8%

↓ Agriculture remains our biggest user of water and also the sector with the largest amount of waste. There is much room for improvement in how we use water in all sectors. For the average individual in a developed nation, reducing waste is tied to reducing consumption, not just of water but of industrial and agricultural products as well. It may be surprising just how much water goes into some of the products you use.

It takes 2.5 gallons of water to make one plastic water bottle—that is more water than the bottle holds.

GALLONS OF WATER NEEDED TO PRODUCE 1 POUND OF FOOD

BEEF	1,850
PORK	756
CHICKEN	469
BREAD	160
CORN	109
APPLE	84
POTATO	31

GALLONS OF WATER PER PRODUCT

PAIR OF BLUE JEANS	2,900
1 REAM OF PAPER (500 SHEETS)	1,250
COTTON T-SHIRT	766

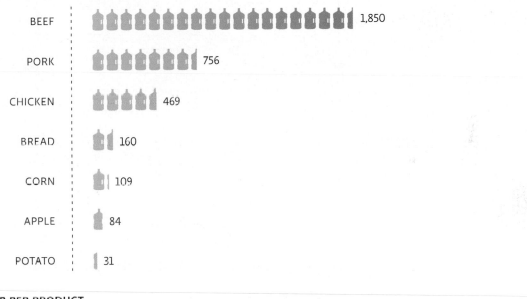

= 100 gallons of water

groundwater extends about 900 meters (3,000 feet) deep. Water quantity is often measured in terms of acrefeet—the amount needed to cover an acre in water that is 1 foot deep, which is equivalent to more than 300,000 gallons. One acrefoot of water is enough for two American families for 1 year. Deshmukh estimates that nearly 5,000 acrefeet of water is accessible from the deepest part of the aquifer. But that deep subterranean water is harder to get, because it costs more to pump it out of the earth than the groundwater closer to the surface.

Aquifers remain dependable sources of water only as long as removal rates don't exceed replenishment rates.

The biggest drain on freshwater supplies isn't the water used to shower, flush the toilet, and wash dishes. In the United States, about 70% of all freshwater withdrawals, from surface or groundwater sources, go to agriculture. Industry is the next biggest consumer of water. [INFO-GRAPHIC 17.4]

In addition, anything that reduces infiltration will reduce the rate at which the aquifer refills and thus decreases the amount of water we can sustainably remove. Infiltration is hampered in urban and suburban settings because of all of the hard surfaces, such as roads and buildings; even the typical suburban lawn is so compacted from the home

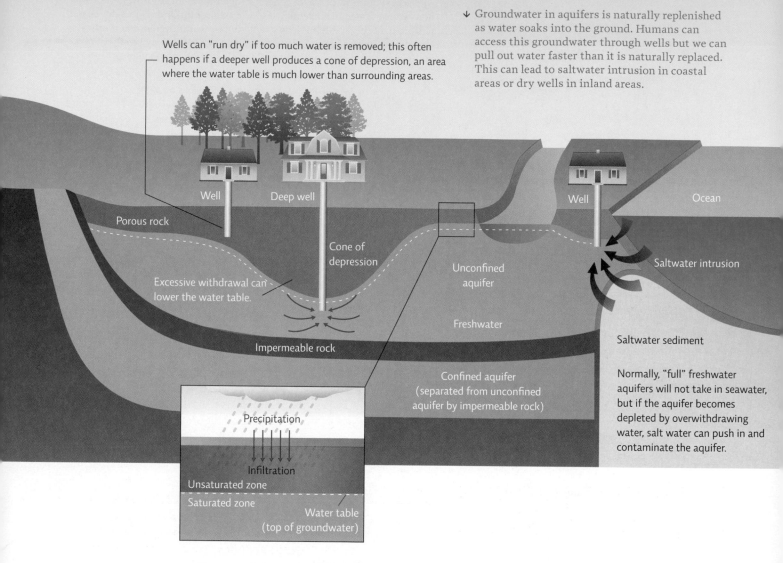

Wells can "run dry" if too much water is removed; this often happens if a deeper well produces a cone of depression, an area where the water table is much lower than surrounding areas.

↓ Groundwater in aquifers is naturally replenished as water soaks into the ground. Humans can access this groundwater through wells but we can pull out water faster than it is naturally replaced. This can lead to saltwater intrusion in coastal areas or dry wells in inland areas.

Well

Deep well

Well

Ocean

Porous rock

Cone of depression

Unconfined aquifer

Saltwater intrusion

Excessive withdrawal can lower the water table.

Freshwater

Saltwater sediment

Impermeable rock

Confined aquifer (separated from unconfined aquifer by impermeable rock)

Normally, "full" freshwater aquifers will not take in seawater, but if the aquifer becomes depleted by overwithdrawing water, salt water can push in and contaminate the aquifer.

Precipitation

Infiltration

Unsaturated zone

Saturated zone

Water table (top of groundwater)

construction process that very little water infiltrates the ground. Urban and suburban designs that provide ways for water to soak into the ground—such as permeable pavement and *rain gardens*—can help refill aquifers as well as help to prevent flooding events after heavy rainfalls. (For more on rain gardens, see Chapter 18.)

Of course, Californians are lucky enough to have plenty of water all along the coast. But removing salt and other minerals from salt water—a process known as **desalination**—is expensive and uses a large amount of energy. Still, there are thousands of desalination plants worldwide operating today that meet some of the water needs of their regions. The largest such facilities in the world are in the Middle East, some of which are processing around 200 million gallons of water per day—about 10 times the volume of the largest U.S. plants (which are in Tampa Bay, Florida, and El Paso, Texas).

But in California, the proximity to salt water has actually threatened the freshwater supply.

Decades ago, Orange County officials discovered to their dismay that salt water was seeping into some of the region's aquifers, putting those precious freshwater stores in jeopardy. Groundwater levels are typically higher than sea level, so salt water doesn't infiltrate aquifers. But as freshwater was pumped out of the county's aquifer inland, the groundwater level dropped, so salty ocean water had started to enter the coastal edge of the aquifer to the west, where it bordered the Pacific. In Orange County, some aquifers are confined by geologic faults, which prevent ocean water from entering at some points—but not everywhere. It was in these unconfined coastal aquifers that **saltwater intrusion** was becoming a problem. [INFOGRAPHIC 17.5]

↑ The Groundwater Replenishment System in Fountain Valley, California, solves two problems at once: water scarcity and dealing with wastewater. This $480 million dollar water treatment system converts the sewage water of Orange County into drinking water, producing 55 million gallons of drinking water every day.

To stem the influx of salt water, in 1975 the Orange County Water District started pumping highly treated (purified) sewage wastewater into infiltrated wells. At 5 million gallons a day, the underground injection created a curtain of freshwater along the California coast that prevented salty ocean water from seeping into the county's aquifer. This water management program was the first to take treated wastewater and pump it into the ground, says Deshmukh.

Most residents had no idea this was going on—even those who thought about water every day. Jack Skinner is a medical doctor and lifelong surfer based in Orange County, whose house in Newport Beach is supplied with water from the county's wells. His second home is in Laguna Beach, surrounded by water; in the morning, he and his wife would watch from their oceanview apartment as the sun would light up a pole stationed about a mile offshore, marking the spot where a nearby sewer treatment plant discharged its **effluent**. "Every morning when we got up, we would see that marker." In the early 1980s, he began experiencing eye infections while distance swimming along the coast. He decided to learn more about the impact of releasing wastewater into the ocean, and how the wastewater was being treated.

Sewage can carry pathogens like viruses and disease-causing bacteria, and surfers get exposed to these when they surf in sewage outflow that is untreated. Public health officials monitor drinking and recreational waters for the presence of **coliform bacteria**. Since many coliforms are intestinal bacteria, their presence could indicate fecal contamination of the water.

As a doctor specializing in internal medicine and cardiology, Skinner was also concerned about potential health problems that could come with exposure over a long period of time to some chemical contaminants that are carried in drinking water. In California, the water sometimes carries traces of toxic chemicals. In the 1990s, officials detected two toxic chemicals: N-Nitrosodimethylamine (NDMA), created by chemical reactions during wastewater treatment with chlorine; and 1,4-dioxane, which came from a computer circuit board manufacturer releasing industrial wastewater into the sewer system.

---

desalination The removal of salt and minerals from seawater to make it suitable for consumption.

saltwater intrusion The inflow of ocean (salt) water into a freshwater aquifer that happens when an aquifer has lost some of its freshwater stores.

effluent Water discharged into the environment.

coliform bacteria Bacteria often found in the intestinal tract of animals; monitored for fecal contamination of water.

Over the years, Skinner had become a self-proclaimed "troublemaker," making sure that wastewater was highly treated before being discharged into the Pacific Ocean, and speaking publicly about water pollution.

One day, the Orange County Sanitation District (OCSD) called Skinner and asked him to come in for a meeting. They wanted to consult with him about a massive new project that they were undertaking that would need the support of the community to succeed. What they said made Skinner very concerned.

There are a variety of approaches to water purification.

In the mid-1990s, Orange County water sanitation engineers and hydrologists found that they were faced with a problem of water scarcity and, at times, the opposite problem—too much water.

As an increasing number of people moved to the region, they used more water, creating excess wastewater. The OCSD already operated a 5-mile "outfall," or underground pipeline that takes treated sewage water past California's beaches and out to the Pacific Ocean. The OCSD also had another mile-long outfall to the Santa Ana River, but it was not enough to handle heavy rainfall which could wash sewage through and out of a facility before it could be adequately treated. About 100 million gallons a day of partially treated sewage water was already flooding the river. Nearly five times that amount would reach the river during a major storm. The county was considering building another pipeline to carry excess water to the Pacific Ocean.

But such an endeavor would be incredibly expensive. So the agency considered a project that would solve both the problem of too much wastewater and too little freshwater. They decided to expand the existing system that used treated wastewater to protect groundwater from infiltration by the ocean. But they knew they would face community opposition.

Since the 1970s, the county had been pumping only small amounts of treated wastewater into the ground—about 5 million gallons per day, and only at the coasts. This new project would expand that amount to 70 million gallons per day, and take place not just at the seawater barrier. The actual amount of treated wastewater that would make its way into people's homes would range, but would ultimately average 15% in north and central Orange County, says Deshmukh. "All of a sudden, it became a significant component of the water supply."

He and the other staff at the OCSD knew they would need to consult the community about the project, known officially as the Groundwater Replenishment System (GWRS)—without the support of community members, it would never get off the ground.

Once the project was conceived, they began intensive community outreach, such as presentations in front of community groups about water scarcity and the need for new sources, and placing representatives at big events. "Any chance we got we would talk about it," says Deshmukh. Part of the multiyear initiative included outreach to community leaders, such as Skinner.

When Skinner initially learned about the project, and the fact that they had been doing a smaller version at the coast for decades, he was uneasy. "Since the late 1970s, my family had unknowingly been drinking that water," Skinner says. "I thought of my daughter, who lives in the area. I thought about my grandson, Robbie. He would be drinking the water, and is probably right now. I wanted to be sure it was safe."

So Skinner set about learning how the wastewater would be treated. The state health department appointed him to serve on a state committee that would review the treatment process for the recycled wastewater, and Skinner began speaking with experts about the techniques.

He learned that the first step of the cleaning process consisted of microfiltration, in which microscopic, strawlike fibers, 1/300 the diameter of a human hair, filter out many suspended solids, bacteria, and other viruses because only water can pass through the center of the fibers.

Then, to render wastewater **potable**—meaning, safe for humans to drink—engineers would perform a crucial step known as reverse osmosis. During this process, they use pressure to force water through a plastic membrane. The pores of this membrane are so tight, explains Deshmukh, that salt and other contaminants (such as pharmaceutical drugs and toxic chemicals) do not pass through, but water does. After reverse osmosis, the water is exposed to ultraviolet (UV) light, which kills any remaining viruses and bacteria. Adding hydrogen peroxide helps convert traces of dioxane and some of the other toxic chemicals that might be present into harmless molecules.

Where some places only clean their wastewater with two or three treatment processes before dumping it, the Orange County water and sanitation districts made the

---

potable Water clean enough for consumption.
wetland An ecosystem that is permanently or seasonally flooded.

decision to spend the money and time to do complete reverse osmosis and complete UV-light disinfection. At completion, the final product is cleaner than state and federal regulations require.

In addition, after the region discovered dioxane and NDMA in the water supply, officials tracked down the sources of those chemicals and rerouted the discharge into wastewater plants that weren't going to recycle the water into the freshwater supply, says Deshmukh.

Engineers also treat wastewater not destined to be potable, says Deshmukh, such as for watering lawns or irrigation, using filtration and disinfection, without reverse osmosis.

The Orange County Water District was Deshmukh's first job after engineering school at the University of California, Los Angeles. His thesis was on reverse osmosis, so he had total confidence in the treatment process. During presentations with community groups, he would explain that he and his parents would be affected by the project, as well. "I live here, my family lives here, this is the water we drink too." During tours of the treatment facility, they would offer visitors a taste of the water on cups labeled "It tastes just like water!" Tasting water that was sewage only 2 hours ago, and seeing its pristine quality, "was very convincing for people," says Deshmukh. In fact, sometimes they had to stop people from taking some home in water bottles, he adds.

Another California community, Arcata, tackled its water purification problems in a different way. Rather than constructing a typical wastewater treatment facility to handle sewage that had been contaminating nearby Humboldt Bay, they repurposed a retired landfill near the coast by converting it to a **wetland**—an ecosystem that is permanently or seasonally flooded. A slow river meanders through the wetland where organisms there purify it; to them, it's not "sewage," it's food. The Arcata facility depends on nature to perform the job of water purification. The water discharged into the ocean is very clean and the health of the bay ecosystem has improved. But while the Arcata facility addresses the need to decontaminate sewage, it does not take the extra steps to produce potable water like Orange County's GWRS. [INFOGRAPHIC 17.6]

↓ This wetland marsh in Arcata, California, is actually part of a wastewater treatment system that uses nature to help purify sewage. The wetland is now an Audubon birding sanctuary.

Infographic **17.6** | **WASTEWATER TREATMENT**

↓ Sewage must be treated before it can be safely released to the environment. Most communities use chemical- and energy-intensive high-tech methods, but systems that mimic nature can also effectively purify water.

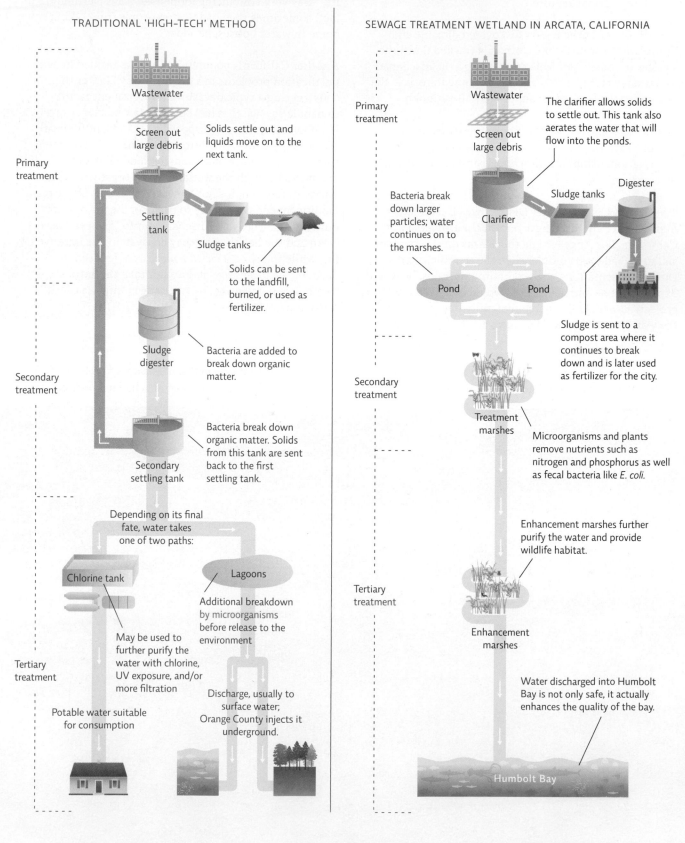

TRADITIONAL 'HIGH-TECH' METHOD

Primary treatment

Wastewater

Screen out large debris

Solids settle out and liquids move on to the next tank.

Settling tank

Sludge tanks

Solids can be sent to the landfill, burned, or used as fertilizer.

Sludge digester

Bacteria are added to break down organic matter.

Secondary treatment

Secondary settling tank

Bacteria break down organic matter. Solids from this tank are sent back to the first settling tank.

Depending on its final fate, water takes one of two paths:

Chlorine tank

Lagoons

May be used to further purify the water with chlorine, UV exposure, and/or more filtration

Additional breakdown by microorganisms before release to the environment

Tertiary treatment

Potable water suitable for consumption

Discharge, usually to surface water; Orange County injects it underground.

SEWAGE TREATMENT WETLAND IN ARCATA, CALIFORNIA

Primary treatment

Wastewater

Screen out large debris

The clarifier allows solids to settle out. This tank also aerates the water that will flow into the ponds.

Clarifier

Sludge tanks

Digester

Bacteria break down larger particles; water continues on to the marshes.

Pond   Pond

Sludge is sent to a compost area where it continues to break down and is later used as fertilizer for the city.

Secondary treatment

Treatment marshes

Microorganisms and plants remove nutrients such as nitrogen and phosphorus as well as fecal bacteria like *E. coli*.

Enhancement marshes further purify the water and provide wildlife habitat.

Tertiary treatment

Enhancement marshes

Water discharged into Humbolt Bay is not only safe, it actually enhances the quality of the bay.

Humbolt Bay

↑ A bird's-eye view of the 1.5-mile-long Three Gorges Dam in Yichang, located in central China's Hubei Province

The GWRS went online in January 2008, and now, months after the water is pumped into the ground, Orange County's water district can tap its artificially replenished groundwater and deliver clean freshwater to people for drinking, irrigation, and other uses. As of fall 2011, it had recycled more than 72 billion gallons of water. Every day, approximately 70 million gallons of recycled water are pumped into wells or percolation basins in Anaheim, where sand and gravel naturally purify the water as it trickles down into the region's aquifers—enough to meet the needs of nearly 600,000 residents.

Solving water shortages is not easy.

A major benefit of the Groundwater Replenishment System, Deshmukh and his colleagues explained to local residents, is that injection into groundwater can address one of the biggest sources of water loss.

The majority of Earth's liquid freshwater is found in lakes. To ensure an ongoing freshwater resource, communities often invest in **dams**. These barriers stop the flow of rivers and create **reservoirs**, large bodies of water that hold freshwater for a variety of uses (freshwater source, flood control, electricity production). In the United States, these often become recreation sites and fishing

resources. While reservoirs are a valuable resource, they lose an enormous amount of water every day through evaporation.

Depending on temperatures, atmospheric conditions, and the surface area of a reservoir or lake, wide-open bodies of water can evaporate thousands of gallons of water a day in desert settings where it's hot and dry. The California Department of Water Resources calculated that fresh-water reservoirs in the South Coast region lost more than 164,000 acrefeet (almost 53.5 billion gallons) to evaporation in 2000 alone. Worldwide, reservoirs lose more water to evaporation than that used for industry and domestic purposes combined.

The construction of dams can also spark political conflicts. In the Middle East, Turkey's plans to build 22 dams that pull water from the Tigris and Euphrates rivers for agriculture and electric power will impact its neighbors Syria and Jordan downstream. With too little water available for too many people, this hotspot may be the site of future conflict.

---

dam Structure that blocks the flow of water in a river or stream.
reservoir Artificial lake formed when a river is impounded by a dam.

Conservation is an important "source" of water.

The GWRS project was expensive: The total price tag to build the system came to about $487 million from federal, state, and local funding.

An easier and cheaper way to maintain water supplies is simply not to waste so much. For example, water-saving irrigation methods limit loss to evaporation and runoff, thus significantly reducing the water that is used. This has the added advantages of protecting surface waters and of preventing soil salinization (the buildup of salt as water evaporates), a common problem in dry climates. Choosing to plant crops more suited to the environment and water availability will also decrease agricultural water use. Many industrial processes are now designed to reuse water rather than discharge it into the environment. Small individual changes in the household can also save a lot of water. [INFOGRAPHIC 17.7]

Infographic 17.7 | **WATER USAGE AND CONSERVATION**

↓ Much of our water usage can be reduced by using new water-efficient technologies and by making behavioral changes that don't waste water. This can be as easy as simply turning off a faucet when not using the water, taking shorter showers, and only using appliances like dishwashers when they are full. Buying less stuff and using less energy also saves water because much water is used to produce both.

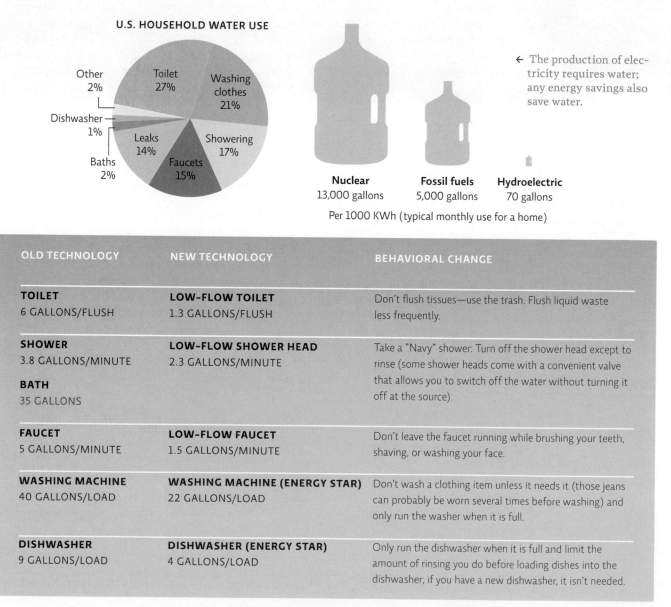

U.S. HOUSEHOLD WATER USE

- Other 2%
- Toilet 27%
- Washing clothes 21%
- Dishwasher 1%
- Leaks 14%
- Showering 17%
- Baths 2%
- Faucets 15%

← The production of electricity requires water; any energy savings also save water.

Nuclear 13,000 gallons **Fossil fuels** 5,000 gallons **Hydroelectric** 70 gallons

Per 1000 KWh (typical monthly use for a home)

OLD TECHNOLOGY	NEW TECHNOLOGY	BEHAVIORAL CHANGE
TOILET 6 GALLONS/FLUSH	**LOW-FLOW TOILET** 1.3 GALLONS/FLUSH	Don't flush tissues—use the trash. Flush liquid waste less frequently.
SHOWER 3.8 GALLONS/MINUTE **BATH** 35 GALLONS	**LOW-FLOW SHOWER HEAD** 2.3 GALLONS/MINUTE	Take a "Navy" shower: Turn off the shower head except to rinse (some shower heads come with a convenient valve that allows you to switch off the water without turning it off at the source).
FAUCET 5 GALLONS/MINUTE	**LOW-FLOW FAUCET** 1.5 GALLONS/MINUTE	Don't leave the faucet running while brushing your teeth, shaving, or washing your face.
WASHING MACHINE 40 GALLONS/LOAD	**WASHING MACHINE (ENERGY STAR)** 22 GALLONS/LOAD	Don't wash a clothing item unless it needs it (those jeans can probably be worn several times before washing) and only run the washer when it is full.
DISHWASHER 9 GALLONS/LOAD	**DISHWASHER (ENERGY STAR)** 4 GALLONS/LOAD	Only run the dishwasher when it is full and limit the amount of rinsing you do before loading dishes into the dishwasher; if you have a new dishwasher, it isn't needed.

In the meantime, supplementing potable water supplies with recycled water is an innovative way to help ameliorate ongoing water issues, says Channah Rock, a water-quality specialist and assistant professor at the University of Arizona. It's rare to find initiatives like Orange County's, she says, but several communities—in Arizona, California, Nevada, and Florida, for instance—are reusing recycled water for nonpotable use, such as for irrigating landscapes and crops, filling fountains and fire hydrants, and flushing toilets. Communities are trying to "match the quality of water with the right use of water," she says. Since people are prohibited from drinking the water that is used for irrigation, for example, says Rock, it's not necessary to subject that water to the same advanced treatment processes as those used for potable water.

Public perception of recycled water projects actually varies, says Rock, depending on how familiar or confident people are with the treatment process and how plagued their community is by water-scarcity issues. In Arizona, which has regions affected by drought, a recent statewide survey found that Arizona residents generally supported most potential uses of recycled water and felt that it was very important that their community use recycled water to help meet its water needs. To many people, she says, what matters is that the water ultimately meets regulatory standards designed to protect public health.

In fact, Rock says, the "toilet to tap" phrase is incredibly misleading, because it leaves out the testing, treatment, and scrutiny that take place in between.

The Groundwater Replenishment Program has received many accolades, including the prestigious 2008 Stockholm Industry Water Award. In 2010, just 10 years after Deshmukh finished graduate school, his work was profiled in *National Geographic*. It's an achievement that fills Deshmukh with pride. "This is water that's normally just wasted in the ocean. For the first time, it was being added to the water basin, cleaner and at a higher quantity than we'd done before."

Once Skinner learned how the water would be treated, spoke to scientists about the effectiveness of the treatment process, and learned the project would have continuous oversight, he was reassured. "I feel they have successfully addressed my concerns," he says. Skinner and Deshmukh even worked together to create videos describing the project. Today, Skinner, his wife, daughter, and grandson consume the recycled water. "I think it's safe for my family to drink."◉

References mentioned in this chapter:
Hoekstra, A., and Chapagain, A. 2007. *Water Resources Management*, 21: 35–48.

BRING IT HOME

❯ PERSONAL CHOICES THAT HELP

Regardless of whether our water comes from an aquifer or a local reservoir, we can make those water sources last longer by taking steps to use our water as efficiently as possible.

Individual Steps
→ If you have an iPhone, download the Waterprint app to determine your total water use; once you have a baseline, see if you can reduce it by 10%.
→ Time your shower and try to reduce it by 1 to 2 minutes.
→ Have a container by the sink or shower to catch water while it warms up—just make sure not to get soap in it. Use this water for watering plants both inside and out.

Group Action
→ Install a rain barrel on your house. Rain barrels allow people to use the rain that falls on the roof of a building to water plants as opposed to letting it run off into the storm drain. If you live in a dorm or an apartment, see if you can get permission to have a rain barrel installed.

Policy Change
→ Do you know where your water comes from? Talk to a city representative to find out where your water comes from, and what steps are being taken to make sure it lasts as long as possible.
→ Encourage local policy makers to ban the watering of lawns, or restrict the use of water for landscaping, to certain days of the week.

UNDERSTANDING THE ISSUE

CHECK YOUR UNDERSTANDING

1. **Approximately 75% of Earth's surface is covered with water. The amount of that water that is drinkable by humans is:**
   a. nearly all of it.
   b. about half of it.
   c. perhaps 1/10 of it.
   d. much less than 1%.

2. **Most humans get their water from:**
   a. underground wells.
   b. rivers with dams.
   c. lakes.
   d. desalination plants.

3. **Infiltration:**
   a. is made easy in urban and suburban areas by all of the lawns.
   b. is made harder in urban and suburban areas by roads, buildings, and lawns.
   c. is the process of removing particulate matter from sewage.
   d. is what happens when seawater enters a freshwater system.

4. **Many people say that desalination (removing the salt from ocean water) is the obvious way around our water shortages. This is:**
   a. becoming relatively easy and inexpensive and will be common soon.
   b. inexpensive now but it is difficult to route the water from the coasts to inland areas.
   c. still very expensive and uses a great deal of energy.
   d. not being done at this time.

5. **If too much water is removed by a well in coastal areas:**
   a. the water table will rise.
   b. a cone of depression will form.
   c. the aquifer might collapse.
   d. salt water can seep into the aquifer.

6. **One creative way that some communities deal with wastewater is:**
   a. creating wetlands that support birds and other wildlife.
   b. using microfiltration, then bottling and selling it as mineral water.
   c. pumping it through a separate water system for people to use for laundry and for yard irrigation.
   d. evaporating it in gigantic reservoirs, creating additional clouds and rain.

WORK WITH IDEAS

1. Why do the problems of water scarcity and unsanitary water conditions often occur together?

2. Discuss at least two of the reasons why having enough potable water is a problem for so many people on Earth.

3. Draw a flow chart of the water cycle. (Don't copy from the book; do your own small drawing.) Follow a single water molecule from a cloud through some portion of the cycle, including a living organism, and back to a cloud.

4. Describe, in your own words, the percentage of total water on Earth that is freshwater, and identify the compartments where it is located.

5. What are some of the things that communities in the United States do to deal with their wastewater?

ANALYZING THE SCIENCE

The World Health Organization's (WHO) *Global Burden of Disease* analysis provides a comprehensive and comparable assessment of mortality and loss of health due to diseases, injuries, and risk factors for all regions of the world. The overall burden of disease is assessed using the disability–adjusted life year (DALY), a time–based measure that combines years of life lost due to premature mortality and time lived in states of less than full health.

INTERPRETATION

1. According to the map, which continent appears to have more problems with the lack of sanitation?

2. What is the DALY status attributable to water and hygiene in most of the world's developed countries?

3. In your own words, describe the DALY status in Eurasia, one of the most populous areas on Earth.

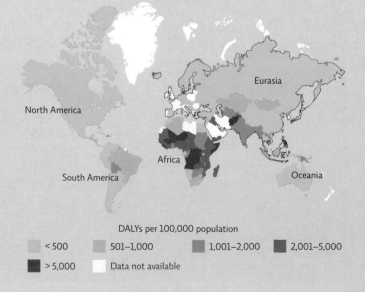

DALYs ATTRIBUTABLE TO WATER, SANITATION, AND HYGIENE (DIARRHEA), 2004

DALYs per 100,000 population

< 500   501–1,000   1,001–2,000   2,001–5,000   > 5,000   Data not available

ADVANCE YOUR THINKING

4. Explain why people in many places in the United States complain of water shortages, while this map indicates that they have access to clean water and sanitation.

5. If much of Africa is grasslands and jungle, why do the inhabitants have so little access to clean water and sanitation?

6. Some countries that border oceans are able to afford desalination plants to improve their access to clean drinking water. If all countries could afford it, would this solve the worldwide problem? Explain.

EVALUATING NEW INFORMATION

In the summer of 1989, Dr. Noah Boaz and his archaeological Earthwatch crews were excavating a site of ancient human habitation along the Semliki River, which runs by the border between Zaire (now the Democratic Republic of Congo) and Uganda. They could not, however, just drink the water from the river, or even swim in it. They had to filter the water a gallon or two at a time and then add chemicals to it in order to remove the waterborne parasites and diseases. Bathing meant wearing shoes and keeping their eyes, nose, and mouth out of the water. The nearby villagers did drink the water, and they had endemic health problems.

More than 2 billion people do not have access to clean drinking water. The results affect all aspects of life in developing countries. According to Water.org, a child dies every 20 seconds from a water-related illness and women in some water-stressed areas spend several hours every day collecting water for their families' basic needs.

There have been many suggestions about ways to improve access to clean water. One of the problems is that in many areas, the lack of access is coupled with a lack of the electricity, developed roads, machines, and equipment necessary to be able to support digging municipal wells and providing pumping stations, reservoirs, and pipelines.

Go to the Global Water website (www.globalwater.org) and explore the links under "Water/Sanitation Facility Projects" and "Technology & Equipment." Then go to the Water.org website (water.org/learn-about-the-water-crisis/facts) and look at some of the featured projects. Finally, go to NEED magazine (www.knowh2o.org/assets/pdf/NEED03_FUTURE.pdf) and look at some of the solutions proposed in the featured article.

Evaluate the websites and work with the information to answer the following questions:

1. Are these authors/sponsoring groups reliable information sources?
   a. Do they give supporting evidence for their claims?
   b. Do they give sources for their evidence, as well as clear explanations?
   c. What is the mission of the organization or of the scientist/author who wrote the article?
   d. Does the author or organization appear to have a workable solution or solutions?

Now go to the bottlelessvancouver website (bottlelessvancouver.wordpress.com/2011/02/25/clean-water-solution-to-third-world-countries) and watch the video about LifeStraw. Then go to the HowStuffWorks website (science.howstuffworks.com/environmental/green-tech/sustainable/playpump.htm) to read about the PlayPump, and to the gizmag website (www.gizmag.com/mobile-bicycle-powered-water-pump/15281) to read about the bicycle-powered water pump.

2. Evaluate the proposed solutions for:
   a. price.
   b. ease of use.
   c. whether they would be portable or stationary.
   d. whether they include pumps for underground water, or items to clean surface water

3. Do any of the proposed solutions stand out as the best possible one for remote or undeveloped areas? Explain your answer.

MAKING CONNECTIONS

RETHINKING TOILETS

Background: In the 1990s, the Arizona State Museum created a major new exhibit featuring the life of the Tarahumara people, who live in the rugged Sierra Madre Occidental mountains in Mexico, in and around the gigantic Barranca del Cobre (Copper Canyon). The museum staff invited a number of the people from this remote area, especially those who had donated their rugs, pottery, carvings, and other artifacts, to come to Tucson to attend the opening. On arriving in Tucson, they declined to stay in the high-rise hotel where museum guests usually stayed. Instead they opted to stay with someone they knew, sleeping in the backyard so they could see familiar stars. When the uses of a bathroom were explained, the group was horrified. Since they lived in the high mountains with only a few streams, water was their most precious resource. The idea of putting human waste into a porcelain bowl of clean, pure water was unthinkable.

Case: People in many areas of the United States are reaching this same conclusion, that clean water should not be used for human waste, or at least that its use should be minimized. Write a position paper for a local magazine, newspaper, or online newsblog encouraging your community to vote for or against the local proposition to reduce the use of clean water for toilets. Include the following in your paper:

a. How much water is used in a standard (pre-1994) toilet? Multiply that amount times 100,000 people (or use your local population number) times an average of three flushes per person per day times 365 days to get an estimate of the amount of water that, literally, goes down the toilet every year in your community.

b. At least three ways to reduce the amount of water used for toilets, including a redesigned toilet (with or without water).

c. Examples of ways other communities are addressing this problem.

CORE MESSAGE

Water pollution decreases our usable water supplies, harms wildlife and human life, and is largely caused by human actions. Some types of pollution may be easier to address than others, but we can decrease pollution by protecting water bodies, restoring forested areas, and limiting the use of potential pollutants.

GUIDING QUESTIONS

After reading this chapter, you should be able to answer the following questions:

→ What is cultural eutrophication and how can water pollution like fertilizer or animal waste end up killing aquatic life?

→ What is water pollution and what are some of the most common types of pollution that enter our surface waters?

→ What is a watershed? What affects the quality of surface water and the quantity of groundwater in a watershed?

→ What are the characteristics of a healthy riparian area and how can this improve water quality?

→ What are some of the best management practices that could reduce nonpoint source pollution in areas like the Chesapeake Bay watershed?

Baltimore's Patapsco River contains raw sewage and other pollutants that it delivers daily to the Chesapeake Bay.

SAVING THE BAY

Teams of researchers try to pin down what's choking the Chesapeake

On a steamy summer day in 1998, microbial ecologist Peter Groffman and some colleagues decided to take a walk. They were in Baltimore, Maryland, for a scientific meeting, but had an entirely separate mission: to figure out what was slowly suffocating the Chesapeake Bay, one of the nation's most important bodies of water. So they chose to walk alongside one of the many streams that feed into the bay, watching the clear water as it bubbled away from them. Then, they saw something that stopped them short: Right next to the stream, a manhole cover line was popped open, and raw sewage was pouring into the stream. They were deep in a large, forested park, with little foot traffic. "This thing could have been flowing for days," he recalls.

One of Groffman's colleagues, a former Baltimore City employee, immediately called the city to fix the leak. That moment sharpened Groffman's determination to find and resolve all the other sources of pollution for the bay. That fall, he helped establish a long-term sampling network that continues today, in which he and his colleagues take samples every week from ten streams around the city to figure out what is getting into the water and where it's coming from.

There is some urgency to the work: Over the decades, the bay's once-productive fisheries have crashed, and thriving oyster beds that once lined the bottom of the bay have disappeared; now, only an estimated 2% of the native population remains. Blue crabs, once common, have declined by nearly 70% since 1990, and are now a rare delicacy. Blooms of algae frequently blanket the water, starving it of oxygen. What is happening?

When oxygen decreases, things go awry.

These problems could be traced back to **water pollution**—the addition of anything that might degrade the quality of the water. The problem is much larger than a few unnoticed open manholes, unfortunately. Over the years, Groffman and others have witnessed firsthand how the Chesapeake Bay is being slowly infiltrated by pollutants that wash off of agricultural lands, suburban lawns, and city streets to contaminate **stormwater runoff** as it flows over the surface of the land after a rainfall. This runoff travels directly from storm drains that capture it, such as those seen on city streets, into the nearest stream.

The list of potential water pollutants is discouragingly long: Industrial chemicals and raw sewage get dumped directly into a body of water; garbage, oil, pesticides, fertilizers, and sediments wash into water from the land; and contaminants such as mercury and acid, air pollutants from fossil fuel combustion, fall back to Earth with the rain, much of it flowing as runoff into rivers, streams, lakes, and seas.

In the Chesapeake area, runoff pollution floods the bay with an excess of nutrients, the first step in a process known as **cultural eutrophication**. Because the nutrients—primarily nitrogen and phosphorus—fuel plant growth, extra amounts trigger explosions of algae, which block sunlight from reaching underwater plants, causing them to die. Although the newly grown algae do emit

⊙ **WHERE IS THE CHESAPEAKE BAY?**

---

water pollution The addition of anything that might degrade the quality of the water.

stormwater runoff Water from precipitation that flows over the surface of the land.

cultural eutrophication A process in which excess nutrients in aquatic ecosystems feed biological productivity, ultimately lowering the oxygen content in the water.

↑ A pipe releasing raw sewage into Stony Run, one of more than 150 rivers and streams that drain into the Chesapeake Bay. The leak discharges up to 20 gallons of contaminated water a minute. At a constant rate, that would be 28,800 gallons a day—a "medium" sewage spill under Maryland Department of the Environment guidelines.

← Unprotected farm fields lose topsoil as well as farm fertilizers and other potential pollutants when heavy rains occur.

Infographic **18.1** | **CULTURAL EUTROPHICATION CREATES DEAD ZONES**

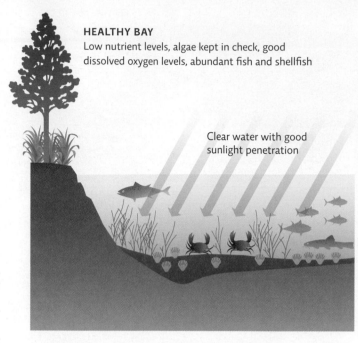

HEALTHY BAY
Low nutrient levels, algae kept in check, good dissolved oxygen levels, abundant fish and shellfish

Clear water with good sunlight penetration

UNHEALTHY BAY
High algae and bacterial growth, low dissolved oxygen, and loss of some aquatic life

Sediment, nitrogen, and phosphorus enter as runoff pollution.

2 Less sunlight can penetrate the algal blooms and sediments.

1 Nutrients cause algal blooms.

Dissolved oxygen levels drop.

3 Underwater photosynthesis decreases, plants die, and oxygen levels drop.

Sediments cloud the water and coat surfaces.

4 As algae die, decomposers (bacteria) increase; biological oxygen demand increases; dissolved oxygen levels drop more.

↑ Healthy bay water is relatively clear with abundant sea life. Oysters and menhaden fish filter out particles keeping water clean, while submerged vegetation produces oxygen to support healthy fish and crab populations.

↑ Sediments that cloud the water, or an influx of nutrients that causes an algal bloom (eutrophication), can prevent sunlight from reaching underwater plants, producing hypoxic regions. An influx of organic material (waste, dead organisms) can also cause hypoxia; growing bacterial populations use oxygen as they decompose the organic matter. Fish and larger, mobile organisms may be able to leave the area; others cannot and will suffer in the hypoxic waters.

oxygen into the water as a by-product of photosynthesis, as a whole, less oxygen is produced when the underwater plants die. As a result, levels of **dissolved oxygen (DO)** in the water plummet—a condition known as **hypoxia**.

Hypoxia is made worse when these nutrients and dead organic matter sink into the deeper waters of the bay, where they are consumed by bacterial decomposers. An influx of food causes these bacteria populations to increase rapidly, which in turn increases the amount of oxygen they consume for cellular respiration, a measure known as the **biological oxygen demand (BOD)** of the water. Water can only hold a limited amount of oxygen, much less than that found in air, so even a small decrease can have immediate effects on other aquatic life.

Another source of bay pollution stems from **sediment pollution**; more than 18 billion pounds of silt, sand, and fragments of clay that pour into the Chesapeake each year. Though sediment also brings nutrients to environmental waters, large amounts cloud the water, making it harder for sunlight to penetrate. This disrupts photosynthesis

in underwater plants, such as the 16 species of seagrasses found in the bay. Sediments can also harm organisms directly by clogging gills and covering the sea (or river) bottom, smothering spawning areas and the nooks and crannies that are habitat to a variety of organisms. [INFOGRAPHIC 18.1]

In the Chesapeake, excess nitrogen, phosphorus, and sediment help explain why the bay is suffering. "There's no argument that those are the three water-quality pollution problems facing the bay, and [they impact] everything else—decimated fisheries, loss of natural filters like seagrasses, and things like this," says Jenn Aiosa, a senior scientist at the Chesapeake Bay Foundation, a nonprofit

dissolved oxygen (DO) The amount of oxygen in the water.
hypoxia A situation in which the level of oxygen in the water is inadequate to support life.
biological oxygen demand (BOD) The amount of oxygen that microbes living in a body of water use.
sediment pollution Eroded soil that is washed into the water through runoff.
estuary A body of water where freshwater rivers meet the sea.

↑ Algae grows on the surface of a pond adjacent to a golf course in the resort town of Mammoth Lakes, California. Normally, algae does not grow in the clear and pristine lakes of mountainous regions, but because of the heavy application of fertilizers on the nearby golf course, runoff into the pond concentrates the plant nutrients and stimulates an algal bloom.

organization that aims to educate people while leading restoration efforts for the bay.

But to Groffman, a researcher from the Cary Institute of Ecosystem Studies in New York, the most interesting culprit in bay pollution is the nutrient nitrogen. His interest in nitrogen was piqued during a basic ecology class he took as an undergraduate student at the University of Virginia in the late 1970s. He became fascinated by the intricate process that allows ecosystems to convert sunlight into energy, driven largely by nitrogen-dependent plant growth. (See Chapter 7 for a look at the nitrogen cycle.) And when that cycle goes awry—say, from too much or too little nitrogen—the whole ecosystem suffers. Groffman was hooked.

The source of pollution can be hard to pinpoint.

The Chesapeake Bay is one of the country's most valuable water bodies, and one of the most stressed by human activities. It therefore serves as the perfect laboratory to understand the effects of those activities, and to figure out how to remedy them.

The Chesapeake Bay is the United States' largest **estuary**, an area where freshwater rivers meet the sea, which in this case is the Atlantic Ocean. Those rivers, in turn, are fed by the water bodies that flow through the densely populated urban and suburban areas of Virginia and Washington, DC; Pennsylvania's Appalachian hills and fields; and Maryland's farmland. It is a shallow bay and the river water that enters it leaves very slowly, so whatever ends up there tends to stay there. With the largest ratio of land to water of any coastal water body in the world, what happens on the land in the Chesapeake has a huge influence on the bay.

Many of the country's other water bodies face similar pressures. In most urban areas, says Groffman, sewer pipes run alongside streams because engineers assume that pipe sewage, if placed alongside the stream, will flow in the same direction and therefore reach treatment plants via gravity. But in many urban areas like Baltimore,

↑ Peter Groffman, fourth from left, shows students a structure that was created to trap trash flowing down an urban stream to prevent it from moving into the Baltimore Inner Harbor area. The structure also directs water to a streamside wetland that provides water-quality improvement.

those pipes are aging. And any time there is a leak—such as the one Groffman caught—that raw sewage flows directly into the stream. "It creates an inherent potential for problems," he says.

In order to mop up the bay's water pollution, we first need to determine where it is coming from. There are two classes of pollution, defined by how they are delivered to the water. Pollution from large discharge pipes of wastewater treatment plants or industrial sites, known as **point source pollution**, is the type that most readily

---

point source pollution Pollution from discharge pipes (or smoke stacks) such as that from wastewater treatment plants or industrial sites.
nonpoint source pollution Runoff that enters the water from overland flow and can come from any area in the watershed.
watershed The land area surrounding a body of water over which water such as rain could flow and potentially enter that body of water.

---

comes to mind. Discharge from sewage treatment facilities contributes about 25% of the nutrient pollution that enters the Chesapeake Bay. Runoff pollution—the same pollution that causes rampant algal bloom in the bay—is known as **nonpoint source pollution** because there is no single discharge point. It enters the water from overland flow and can come from any area in the land that drains into the body of water. Runoff pollution from farmland, suburban lawns, and city streets are all part of nonpoint source pollution.

But water pollution stems from more than just nitrogen and the other nutrients and sediments seen in the Chesapeake Bay. Waterborne pathogens (disease-causing organisms) in raw or partially treated human and animal waste represent a leading cause of sickness and death worldwide. Urban and industrial areas in developed and developing nations also release toxic point source pollution, spewing from industrial discharge pipes or smoke stacks, as well as nonpoint source pollution, like acid runoff from mines.

Even treated wastewater from residential areas can contain enough drug residue from human waste to affect aquatic wildlife. Every drug we ingest gets broken down in our bodies, and the remnants are excreted in our waste. Some of these drug breakdown products include compounds that mimic the hormone estrogen and affect the fish that come into contact with them. In some spots in the Potomac, a majority of male fish now have egg sacs. [INFOGRAPHIC 18.2]

Point sources are often easier to remedy: Identify the source, and make a change. Nonpoint sources, however, demand a more sophisticated, long-term plan. Groffman and others interested in restoring the area needed to find the sources of these pollutants and figure out how they got into the bay.

Excess nutrients flow into groundwater and streams.

When Groffman was thinking about which sites to include in his long-term sampling network, he knew he wanted to represent the diversity of land types that make up the Chesapeake Bay **watershed**—an area of land over which rain and other sources of water drain into a body of water. It was not an easy task: The Chesapeake Bay watershed covers more than 167,000 square kilometers (64,000 square miles) and includes regions in six different states and the District of Columbia, combining highly urban areas, old and new suburban areas, the emerging communities developing outside the suburbs, and agricultural areas. Groffman and his team settled on 10 sites

↳ Water pollution comes from a variety of point and nonpoint sources. Although the main threat to the Chesapeake Bay is excess nutrient and sediment runoff, the EPA identifies pathogens as the leading cause of impaired waters in the United States as a whole. Found in sewage or animal waste, pathogens can come from both point and nonpoint sources.

POINT SOURCES

Some industrial and agricultural sources discharge pollutants directly into a body of water.

NONPOINT SOURCES

A variety of sources contribute pollutants that can run off the surface of the land during rainfall and enter the water; air pollutants can fall directly with the rain.

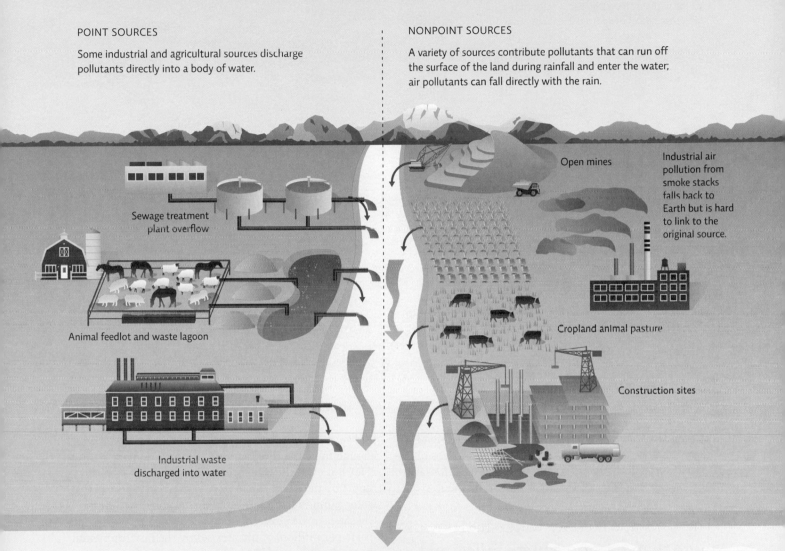

Open mines

Industrial air pollution from smoke stacks falls back to Earth but is hard to link to the original source.

Sewage treatment plant overflow

Animal feedlot and waste lagoon

Cropland animal pasture

Construction sites

Industrial waste discharged into water

LEADING CAUSES OF IMPAIRED SURFACE WATERS IN THE UNITED STATES (2011)

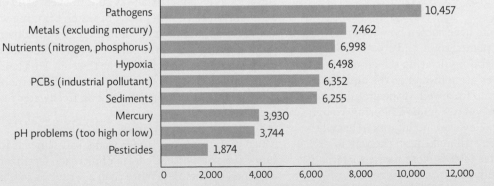

Cause	Number
Pathogens	10,457
Metals (excluding mercury)	7,462
Nutrients (nitrogen, phosphorus)	6,998
Hypoxia	6,498
PCBs (industrial pollutant)	6,352
Sediments	6,255
Mercury	3,930
pH problems (too high or low)	3,744
Pesticides	1,874

Number of impaired bodies of water in the United States

If you could draw a line from hilltop to hilltop around a river and its tributaries (the streams that feed into it), you would be outlining its watershed. Water on the other side of the dotted line flows away from the watershed.

↓ Anything that happens in the watershed can potentially affect the quality of a body of water. This is especially important in terms of nonpoint source pollution that originates on land. Mapping the watershed is an important tool in watershed management. Here the watershed of this river and coastal area is outlined as a black dashed line.

If you find pollution in the water at this point and you have mapped the watershed, you know where to look for the source of the pollution (upstream, in the watershed) and where not to look (downstream, or outside the watershed).

Any rain that falls on the river side of the edge of the watershed could flow downhill into the river or one of its tributary streams.

that capture that diversity, choosing streams located in the dense urban core, older suburbs, newer suburbs of the bay's watershed that were once forested or used for farming, and current agricultural areas. [INFOGRAPHIC 18.3]

The Chesapeake Bay watershed covers more than 64,000 square miles, and includes regions in six different states and the District of Columbia.

A variety of experiments have shown how water flows through a watershed's plants, soils, and streams. Some rainfall soaks into the soil, infiltrating the ground below; some eventually reaches groundwater in the **aquifer**, which people can tap into for a well. If the groundwater is deep enough, infiltration can act as a filtering system that purifies the water. However, pollutants in heavily contaminated water can still make it all the way to groundwater if the water table is close to the surface. For example, nitrate pollution from fertilizer runoff—especially in regions of intense agriculture, like the

midwestern United States—can contaminate well water enough to be life threatening (children are most vulnerable). But even the Chesapeake Bay is at risk: A 2009 report by the Chesapeake Bay Foundation reported that water samples from at least 21% of wells tested in the Chesapeake area exceeded the allowable limit of 10 ppm (parts per million) of nitrates set by the Environmental Protection Agency (EPA).

Any water that doesn't soak into the ground keeps flowing across the land surface as runoff, eventually entering a stream, river, or other body of water, bringing with it whatever pollutants it encounters on its trip downhill. Anything that slows the runoff of water increases the amount of water that infiltrates the soil. This helps keep pollution out of surface waters and also increases groundwater supplies. (For more on freshwater resources, see Chapter 17.)

Much of the excess nitrogen that Groffman was looking for began as a synthetic form of nitrogen fertilizer, first designed by chemists more than a century ago. Plants need nitrogen to grow and survive, so farmers have been relying on this form of fertilizer since the early 1900s to boost their crops. This, along with other sources of nitrogen (for instance, by-products from burning fossil fuels), adds as much nitrogen to the soils as might naturally cycle, essentially doubling the amount of usable nitrogen in the environment.

All that additional nitrogen has to go somewhere; runoff carries the excess fertilizer and animal waste from farms to streams, and eventually to rivers—and sometimes to the bay. Some does soak into the ground where it may be trapped by the soil or find its way into groundwater. But even groundwater flows slowly to the coast, delivering whatever it contains to the bay. And the Chesapeake isn't the only watershed facing this problem: According to the UN, nutrient pollution is the leading type of water pollution worldwide.

aquifer An underground, permeable region of soil or rock that is saturated with water.

Every spring, excess nitrogen flows down the Mississippi River past Louisiana, where it can cause algal blooms and feed microbes in the Gulf of Mexico, snuffing out all underwater life and creating a *dead zone*. Indeed, researchers have found that fertilizers applied in the upper Midwest are responsible for the algal blooms that plague the Gulf Coast. This type of pollution can also hit inland lakes: In Kenya's Lake Naivasha, researchers suspect that recent fish kills may stem from excess fertilizers washing away from flower fields—a new and booming large-scale farming industry on the lake's shores. Worldwide, there are as many as 350 dead zones caused by overfertilization.

This eutrophication has effects: In 2004, a team led by James Hagy, now based at the EPA, evaluated dissolved oxygen levels and nitrogen pollution in the Chesapeake area over a 50-year period, and found that the two were tightly correlated—as nitrogen levels went up, so did hypoxic events.

↓ This summertime satellite image of the Gulf of Mexico depicts water clarity; areas with high sediment loads or phytoplankton (algal) blooms are less clear (more turbid). Sediment can block sunlight and deplete oxygen through diminished photosynthesis. Phytoplankton blooms also result in oxygen depletion and dead zones. Red and orange represent the most turbid water; yellow, green, and aqua are progressively less turbid but still contain enough phytoplankton to trigger oxygen depletion.

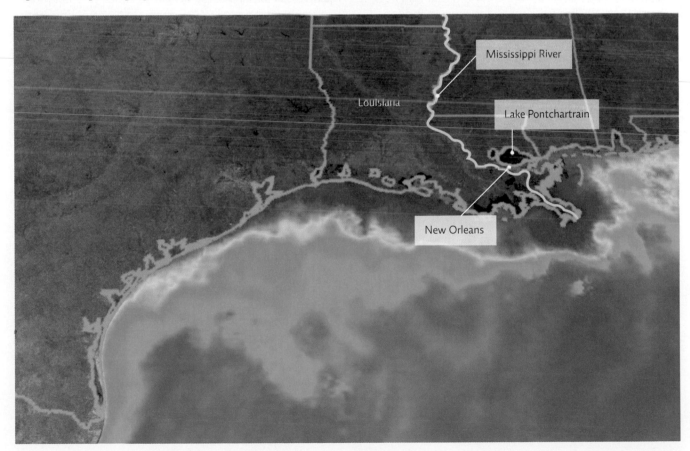

The number and diversity of organisms can indicate the health of a watershed.

Arthur Tuers, a 79-year-old fisher, remembers catching crabs that were 8 inches wide when he was a teenager; now, the largest of these animals might be 4 inches. In fact, a Maryland Department of Natural Resources study of crab catch found that since 1968, the average size of crabs caught there decreased from about 6.2 inches to just over the legal limit of 5 inches in 2000. On top of that, fishers can no longer see the crabs they are aiming to catch through a clear water column. Excess sediment and other pollutants muddy the waters now.

"The best oysters you could find any place in the world were right in this bay," recalls Tuers in a 2009 video taken by the Chesapeake Bay Program, a regional partnership between citizen groups and the local and federal governments, dedicated to cleaning up the bay. "If you wanted some clams, you could go dig up some clams. If you wanted oysters, you could go over here on the pier and take a pair of nippers and get all the oysters you want. But it's not like that anymore."

Historical records from 1880 show that fishers harvested nearly 120 million pounds of oysters each year from the Chesapeake Bay's shores in Virginia and Maryland. A century later, the harvest dropped to only 20 million pounds. The numbers hit a low in 2000, when fishers could gather only about 1 million pounds of oysters in Maryland—the beds were gone.

Oysters are not only an iconic species of the Chesapeake Bay, bringing in millions of dollars for the area, they are also a vitally important species for the bay ecosystem. Oysters are an important food source for a wide variety of organisms at every stage of their life: Sea anemones and sea nettles feed on the free-floating larvae, flatworms and mud crabs eat the juveniles, and larger blue crabs and fish eat first-year oysters. Shore birds like oystercatchers depend on oysters exposed during low tide. Oysters also provide habitat—they form reefs on the muddy seabed that become home to a wide variety of sea life. They stabilize the seabed, reducing the sediment that might otherwise be disturbed and cloud the water. And finally, oysters filter-feed, pumping massive amounts of water through their gills while pulling out pollutants and particles—keeping the water clear, and allowing other animals and plants to thrive.

A major cause of their disappearance has been overharvesting and dredging. But some researchers also believe that eutrophication from excess nutrients in the bay has played a part in the disappearance of the oysters. It turns out that the oysters' decline actually accelerated eutrophication because oysters were no longer removing nutrient pollution. This led to even greater increases in algae and hypoxia—all leading to the loss of more oysters in a *positive feedback loop*. As the oysters were lost, **turbidity** (cloudiness) of the water increased, inhibiting photosynthesis in submerged plants and lowering the dissolved oxygen even further. But it didn't end there. As the submerged plants died, the turbidity increased because plant roots were no longer securing the muddy bottom in shallow areas, which could now be disturbed by surface winds; this reduced photosynthesis even more—another positive feedback loop.

The health of an aquatic ecosystem can be assessed by looking at what lives there—known as a **biological assessment**. The decline of oysters and crabs in the bay is a clear sign of a disturbance, but to identify problem areas, scientists need to assess the streams and rivers that feed the bay. One approach relies on simply netting, identifying, and counting **benthic macroinvertebrates** such as insects and crayfish that live on the stream bottom; if the stream is unhealthy, there won't be many organisms present that are sensitive to pollutants. The abundance and diversity of pollution-tolerant and pollution-sensitive species in the sample can be used to "rate" the stream quality. The Izaak Walton League of America, a nonprofit environmental organization, has a simple sampling protocol that many community volunteers use to monitor their streams regularly—the "Save Our Streams" program. So far, most of the streams in the Chesapeake Bay watershed that have been tested received a rating of fair to very poor; few were rated as good or excellent.

Identifying the types and sources of the pollution is the first step in cleaning up the bay.

The first step in cleaning up the bay, or any body of water for that matter, is to determine which land areas are contributing most to the water pollution, and what factors make some areas more problematic than others. In some of the sites included in their weekly sampling, Groffman and his team take "grab samples," he says, in which they visit the stream and take a water sample by hand once per week. At other sites, they have automated samplers that are programmed to take a sample after a given volume of water passes by. They then take the samples back to the lab, and filter and analyze them for several forms of nitrogen and phosphorus as well as other parameters. When certain samples contain excess amounts, the researchers take detailed samples from the sources feeding that stream, such as suburban lawns and nearby sewer systems.

Infographic 18.4 | LAND USE IMPACTS WATER POLLUTION

↓ According to Groffman's calculations, all areas in the Chesapeake Bay area receive roughly the same amount of nitrogen pollution from air pollution but differ greatly in the introduction of nitrogen from fertilizer use. When the researchers checked area streams for nitrogen pollution, they found that streams in areas that were predominantly agricultural had the most nitrogen in them, followed by suburban streams.

↓ Any actions that increase infiltration in suburban or urban areas, such as more green spaces or the use of permeable pavers that allow water to seep through, reduce the amount of runoff and the delivery of pollutants to streams.

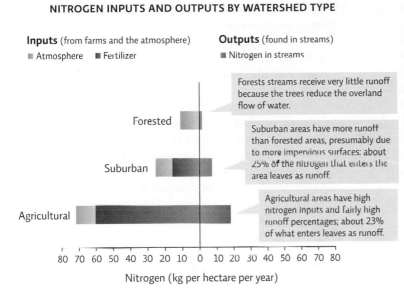

NITROGEN INPUTS AND OUTPUTS BY WATERSHED TYPE

Inputs (from farms and the atmosphere) — Atmosphere, Fertilizer

Outputs (found in streams) — Nitrogen in streams

Forests streams receive very little runoff because the trees reduce the overland flow of water.

Suburban areas have more runoff than forested areas, presumably due to more impervious surfaces; about 25% of the nitrogen that enters the area leaves as runoff.

Agricultural areas have high nitrogen inputs and fairly high runoff percentages; about 23% of what enters leaves as runoff.

Forested / Suburban / Agricultural

80 70 60 50 40 30 20 10 0 10 20 30 40 50 60 70 80

Nitrogen (kg per hectare per year)

The process, he says, is just like analyzing urine samples from a patient. "You can look at the chemistry of the urine to examine the functioning of the patient. You can learn about how a forest or any ecosystem is functioning" by tracking the nitrogen.

Groffman's data show that nitrate levels in urban and suburban areas are higher than levels in natural forest areas, but definitely lower than at agricultural sites. This may seem counterintuitive: Replacing agricultural lands with condominiums leads to better water quality, he explains, because condominiums have less nitrogen runoff, use less fertilizer (though fertilizing lawns is a major nitrogen source in these areas), and pipe their sewage out of the watershed.

Even with lower nitrogen inputs than agricultural areas, urban and suburban settlements still contribute significant runoff pollution because so much of their land surface is covered with cement driveways, asphalt roads, and other impervious surfaces that prevent the infiltration of water into the ground below. [INFOGRAPHIC 18.4]

But these urban and suburban areas still contribute less nitrogen runoff and fertilizer than do agricultural areas. The EPA calculates that on average, agriculture contributes about 60% of the nitrogen to the Chesapeake Bay. In addition, the EPA estimated in April 2011 that more than 160,000 kilometers (100,000 miles) of rivers and streams, and 1 million hectares (2.5 million acres) of lakes, reservoirs, and ponds, across the country had enough nitrogen or phosphorus contamination to be considered "impaired waters."

The nitrogen oxide in emissions from cars and industrial plants that falls as acid rain may be responsible for as much as one-third of the nitrogen that enters the bay.

Reducing nutrient pollution means reducing it at the source and reducing runoff.

Water quality has always been a concern, but perhaps never more so than in 1969. At the time, there were no

turbidity The cloudiness of the water.
biological assessment Sampling an area to see what lives there as a tool to determine how healthy the area is.
benthic macroinvertebrates Easy-to-see (not microscopic) arthropods such as insects that live on the stream bottom.

restrictions on the release of industrial chemical pollutants into bodies of water. That year, the Cuyahoga River in Cleveland, Ohio, made headlines when it caught fire (and not for the first time) because of so much oil and other flammable industrial pollutants floating on the surface of the water. This dramatic event helped spur the passage of the 1972 **Clean Water Act**, which regulates industrial pollution and sets **pollution standards**—allowable levels of a pollutant that can be present in environmental waters or released over a certain time period.

In the Chesapeake, researchers and stakeholders—the state governments and residents of the Chesapeake watershed, as well as federal agencies—have identified a set of **best management practices** to reduce the amount of pollution being delivered to the bay.

These practices differ depending on the way the land in a given area in the watershed is used, explains Margaret Enloe, spokesperson for the Chesapeake Bay Program (CBP). Therefore, the CBP and many regional watershed groups focus their work on making changes on the land that can improve habitats and water quality.

A major component of restoration involves *watershed management*—managing what goes on in the area around the streams and rivers. For example, 25% of the lands in the

---

Clean Water Act The main U.S. federal law that regulates water pollution.
pollution standards Allowable levels of a pollutant that can be released over a certain time period; set by EPA.
best management practices Agreed-upon (or EPA-regulated) actions that minimize pollution problems caused by industrial or land-use impacts.
riparian areas The land area adjacent to a body of water that is affected by the water's presence (for example, water-tolerant plants grow there) and that affects the water itself (for example, provides shade).

watershed are dedicated to agriculture, which Groffman has shown contributes the most excess nutrients to the bay. The CBP works with farmers to manage their use of fertilizers and animal waste and encourages other techniques to prevent excess nutrients and sediment from reaching the bay. To date, they have reached 50% of the targeted reduction.

Another key goal of the CBP is to restore **riparian areas**, the land areas close to the water, by planting vegetated buffer zones that slow runoff and give the rainwater time to soak into the ground. Trees at the water's edge also provide cooling shade, important because cool water can hold more oxygen than warm water. By 2002, CBP partners had restored 2,010 miles of forest buffers; their new goal is 10,000 miles. Widely recognized as an important and effective step at protecting water quality, similar projects are underway across the United States to restore the riparian ares and watersheds of rivers and streams. In the late 1990s, New York City invested billions of dollars to restore and protect areas in the Catskills and Delaware watersheds that supply its water, avoiding the need to construct high-tech and costly filtration systems. [INFOGRAPHIC 18.5]

But these riparian buffer zones and other kinds of treatments that are part of watershed management may only be "a band-aid on the wound," says Tom Jordan, a lead researcher at the Smithsonian Environmental Research Center in Maryland. "Just by changing the diet of the livestock, [farmers] can really improve the situation," he says. Experimenting with different feed to cut back on nitrogen in animal waste or spreading manure at different times of year could be a way to curb excess nutrient runoff from farms.

← Firefighters stand on a bridge over the Cuyahoga River in Cleveland, Ohio, to spray water on a tugboat as a fire—started in an oil slick on the river—moves toward the docks at the Great Lakes Towing Company site. This 1952 blaze, one of 13 fires on the river since the late 1800s, was the most costly, destroying three tugboats, three buildings, and the ship-repair yards.

Plants are nutrient sinks—they store nutrients in their tissues, reducing the amount free to enter the water.

Ground vegetation slows runoff and allows water to soak into the ground. This keeps the water and any pollutants it may carry out of the stream and increases infiltration, which replenishes groundwater.

Shade provided by overhanging trees cools the water; important because cooler water can hold more oxygen than warmer water.

Plant roots stabilize banks and prevent erosion and the deepening of the channel, which can speed water flow and lead to further erosion.

Trees provide food; leaf litter that falls in the water is a main nutrient source for aquatic organisms, including nitrifying bacteria which break down nitrates, reducing the amount sent downstream to the bay.

↑ The area next to a body of water that impacts that water (provides shade and nutrients) and is itself impacted (water-tolerant species live here) is the riparian area. A well-vegetated riparian area reduces the runoff that reaches a body of water by slowing the water's movement across the land so that it soaks into the ground rather than flowing into the stream.

↓ Shown here are the U.S. Department of Agriculture's recommendations for land use in the riparian area. This includes setting aside at least 75 feet of land in managed and undisturbed forest. More may be required for suitable protection in areas with steeper terrain—that is, where runoff flow would be faster.

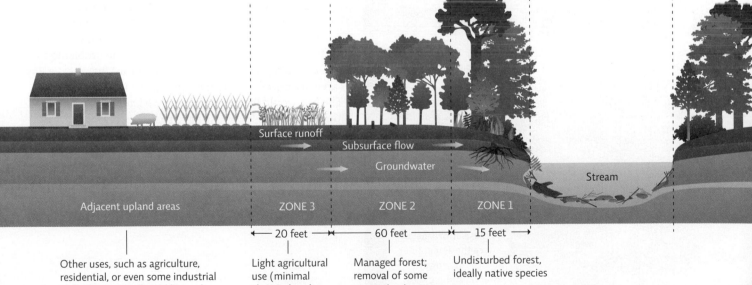

Surface runoff

Subsurface flow

Groundwater

Stream

Adjacent upland areas | ZONE 3 | ZONE 2 | ZONE 1

← 20 feet → ← 60 feet → ← 15 feet →

Other uses, such as agriculture, residential, or even some industrial projects, can be allowed but only with careful handling of toxic chemicals and fertilizer.

Light agricultural use (minimal chemical use); fast-growing grasses trap nutrients that might run off from farms or lawns.

Managed forest; removal of some vegetation increases plant regrowth rates and maximizes nutrient uptake from the soil and subsurface flow.

Undisturbed forest, ideally native species

Infographic 18.6 | INCREASING INFILTRATION OF STORMWATER

↓ Stormwater that doesn't soak into the ground can enter storm drains that flow directly to rivers and streams, or cause floods, especially in heavily built-up urban settings. Anything that increases infiltration can help avoid these stormwater problems.

Trees slow and allow infiltration of runoff.

A green roof has vegetation over a waterproof layer, which can trap some water and reduce roof runoff.

Redirected downspout

Rain barrel captures runoff from roof.

Water soaks into the ground through permeable pavers.

Rain gardens capture gutter and lawn runoff.

Storm drain

Curb cutouts reduce street runoff by diverting it to the ground, where it can infiltrate.

All these efforts help reduce the amount of polluted stormwater entering storm drains and aquatic ecosystems.

In suburban areas, lawns can be major nonpoint sources of nitrogen pollution if homeowners apply too much fertilizer. However, the grass also sequesters nitrogen because of its long growing season, preventing it from flowing to the bay. The CBP encourages homeowners to limit fertilizer use on lawns and to plant native plants and grasses that do not need fertilizer. Planting a **rain garden** of water-tolerant plants in low-lying areas that tend to flood in rain events will also capture water and reduce runoff. [INFOGRAPHIC 18.6]

In urban areas, replacing some hard surfaces with porous surfaces such as green space or permeable pavers reduces runoff by allowing water to seep through; Chicago has already changed a large part of its alley pavement to porous concrete. Green roofs can also help capture some rainwater and slow or prevent its release to the environment (see Chapter 32).

Finally, there is air pollution—a difficult problem to solve since the "airshed" for any body of water is even larger than the watershed itself. CBP partners encourage local authorities to enforce strict pollution standards and to continue to urge residents of the bay watershed to drive less and to maximize their fuel efficiency. [INFOGRAPHIC 18.7]

However, even though legislators and community groups have been trying to save the bay for decades, in years with heavy rainfall, the water quality continues to suffer; sea life is still dying off. The bay has not improved as quickly as we would like, admits Groffman, which has caused some people to question efforts to clean up the bay. It's important to always question whether the approach they're taking is the right one, but it's also important to realize that environmental challenges are often very difficult to solve, he says. "There are just some problems that are really hard, like curing cancer, and the Chesapeake Bay is an example of a really hard problem," says Groffman. "The bay would be much worse if we hadn't done these things," he adds. "We just have to keep working at it."◉

Research articles referenced in this chapter:
Chesapeake Bay Foundation. 2009. *Bad Water 2009: The Impact on Human Health in the Chesapeake Bay Region.*
Groffman, P. et al. 2004. *Ecosystems*, 7:393–403.
Hagy, J.D., et al. 2004. *Estuaries*, 27: 634–658.
Pickett, S., et al. 2008. *BioScience*, 58: 139–150.

rain garden Runoff area that is planted with water-tolerant plants to slow runoff and promote infiltration.

Infographic **18.7** | **BAY RESTORATION**

GOAL	STRATEGY
Reducing pollution at the source	• Agriculture: Test soil and only add needed fertilizers; plant cover crops between seasons to hold soil in place. • Implement better regulation of industrial air and water pollution. • Increase permeable surfaces in urban areas to allow infiltration and decrease runoff; minimize land disturbance when building.

GOAL	STRATEGY
Protecting watersheds	• Establish watershed management plans that protect stream corridors and revegetate watershed areas. • Increase land preservation through donations or purchases by conservation groups to protect sensitive areas.

GOAL	STRATEGY
Restoring habitats	• Protect existing grassbeds in the bay and plant more grassbeds. • Remove non-native species such as nutria (large rodents) and invasive reeds that displace native plants. • Restore wetland habitat where possible, including providing migration routes for fish in area streams.

GOAL	STRATEGY
Managing fisheries	• Establish harvesting limits to help overfished populations recover. • Set aside "no harvest" areas, especially nursery areas, to help depleted populations recover. • Manage the prey species of vulnerable populations so they have enough food.

BRING IT HOME

⊙ PERSONAL CHOICES THAT HELP

We are facing not only shrinking supplies of easily accessible water, but also the potential degradation of this resource due to pollution. By changing products and modifying common practices, we can improve our water quality for years to come.

Individual Steps
→ Read your city's water quality report to see which pollutants are prevalent in your area.
→ Decrease your use of chemicals (fertilizers, pesticides, harsh cleaners, etc.) that will end up in the water supply. For alternatives to traditional yard chemicals, see safelawns.org.
→ Always dispose of pet waste properly. In high quantities it acts as an oxygen-demanding waste and can also spread disease.

Group Action
→ Marking storm drains with "Don't Dump" symbols can remind people not to dump waste liquids down sewer drains. If the drains in your area are not marked, talk to city officials to see if you and other volunteers can mark them.

Policy Change
→ August is National Water Quality month. Take steps every day to reduce your water pollution and help raise awareness in August by writing a letter to your local newspaper outlining simple steps people can take to improve water quality.

UNDERSTANDING THE ISSUE

CHECK YOUR UNDERSTANDING

1. **Water pollution is:**
   a. found only in surface waters near cities.
   b. primarily excess nutrients from lawns, farms, and animal feedlots.
   c. usually from excess carbon being added to the system.
   d. contaminants or excess nutrients in surface waters and in groundwater.

2. **The following items are part of the process of eutrophication.**
   1. Algae quickly reproduce, using up oxygen and blocking sunlight to underwater plants.
   2. Bacteria consume excess wastes and nutrients, using up oxygen.
   3. Underwater plants die.
   4. Excess nutrients enter a body of water.
   What is the correct order for the process?
   a. 1, 2, 3, 4
   b. 4, 1, 3, 2
   c. 3, 4, 2, 1
   d. 2, 4, 1, 3

3. **BOD (biochemical oxygen demand) is:**
   a. the amount of oxygen needed by the largest organism in an area for daily living.
   b. the total amount of oxygen available in a body of water.
   c. the amount of oxygen needed by microbes to decompose nutrients in a body of water.
   d. the amount of dissolved oxygen in a body of water available to organisms such as fish and plants.

4. **Fertilizer from your lawn and motor oil from the leaky oil pan on your car are examples of:**
   a. nonpoint source pollution.
   b. point source pollution.
   c. eutrophication.
   d. pathogenesis.

5. **Three major sources of nitrogen pollution in the Chesapeake, in order from greatest to least, are:**
   a. agriculture, suburban areas, and forests.
   b. agriculture, rainstorms, and urban areas.
   c. urban areas, suburban areas, and agriculture.
   d. logging, agriculture, and urban areas.

6. **A watershed includes:**
   a. only the land that would be underwater during a normal rainfall year.
   b. the surface water and the underground aquifer.
   c. all the uphill land surrounding a river and its streams that can feed water into that river.
   d. all the land downhill from a river that could potentially be flooded.

WORK WITH IDEAS

1. A factory manufactures convenience foods such as frozen lasagna. Its wastes are filtered for large particulates and then fed into the local river. Why might this be a problem for the fish that live in the river?

2. The clams and mussels in your local bay are rapidly declining in number. List the possible water problems, as discussed in the chapter, which might be causing the decline.

3. Compare and contrast three typical point source pollutants and three nonpoint source pollutants from the area where you live.

4. What are potential nonpoint sources of pathogens and drugs entering our waterways and what problems can these cause for people and other organisms?

5. As we saw in Chapter 7, in most natural ecosystems, nitrogen is a limiting factor for the growth of plants, and subsequently for the total biomass of plants and animals. How have humans changed this biogeochemical cycle?

6. You are investigating a die-off of fish in the local mountain lake, Lake Pleasant. It is fed by Hilltop Stream, and the Happy Valley River comes out of the lake and heads down toward Happy Valley. Do any of the following have something to do with the die-off? Justify your answers.
   - The new manufacturing plant just over the ridgetop in the next valley
   - The cattle feedlot on the edge of the Happy Valley River
   - The recent construction of a new subdivision on the eastern slopes above Lake Pleasant
   - The lovely older neighborhood with wide lawns, a golf course, and a large grassy park on the western slope above the lake

ANALYZING THE SCIENCE

The list of the Environmental Protection Agency (EPA) Superfund sites that shows some of the most polluted areas in the United States is on the following page. These are places where chemicals are leaching into the groundwater, polluting the water and possibly causing health problems for residents. Information on these and other sites are given at the EPA scorecard website at scorecard.goodguide.com.

INTERPRETATION

1. How many of the sites on the list are government or military sites? Community sites such as parks or landfills? Mining-related (mines, smelters) sites? Private industrial or manufacturing sites?

2. Identify a particular type of site that seems to be more of a problem than others. Justify your answer.

3. Locate an area of the country that appears to have more problems with groundwater and soil pollution than other areas. What might account for the higher level of pollution in this area?

ADVANCE YOUR THINKING

4. What sorts of pollutants might a landfill contribute to the following water sources?
   a. A nearby stream
   b. The groundwater

5. The Barber Orchard operated for over 80 years on 500 acres, growing apples commercially. The Centredale Manor area in Rhode Island was occupied by a fabric mill. Go to the scorecard website and use the "Search by company, location, or chemical" link. Enter "Barber Orchard" and then "Centredale Manor" to compare the types of pollutants that would be found in groundwater near each of these sites.

6. Enter your zip code in the space provided on the left of the webpage. Which Superfund site is nearest where you live? What problems are found there? Compose a letter to an elected official that expresses your concerns about this site.

Rank	Site	County	Overall site score
1.	Murray Smelter	Salt Lake, UT	86.6
2.	Big River Mine Tailings/St. Joe Minerals	St. Francois, MO	84.91
3.	Lipari Landfill	Gloucester, NJ	75.6
4.	McCormick & Baxter Creosoting Co.	San Joaquin, CA	74.86
5.	Tybouts Corner Landfill	New Castle, DE	73.67
6.	Helen Kramer Landfill	Gloucester, NJ	72.66
7.	Industri-Plex	Middlesex, MA	72.42
8.	Taylor Lumber and Treating	Yamhill, OR	72
9.	Price Landfill	Atlantic, NJ	71.6
10.	Barber Orchard	Haywood, NC	71
10.	Watson Johnson Landfill	Bucks, PA	71
12.	Stoker Company	Imperial, CA	70.94
13.	Pearl Harbor Naval Complex	Honolulu, HI	70.82
14.	Pollution Abatement Services	Oswego, NY	70.8
15.	Asarco, Inc. (Globe Plant)	Adams, CO	70.71
15.	Beede Waste Oil	Rockingham, NH	70.71
15.	Centredale Manor Restoration Project	Providence, RI	70.71
15.	Cherry Point Marine Corps Air Station	Craven, NC	70.71
15.	Circle Smelting Corp	Clinton, IL	70.71
15.	DePue/New Jersey Zinc/Mobil Chem Corp	Bureau, IL	70.71
15.	Eastland Woolen Mill	Penobscot, ME	70.71

EVALUATING NEW INFORMATION

Settling ponds, also known as holding ponds, are locations where contaminated water is allowed to stand so that particulates suspended in the water settle to the bottom of the pond. The water then evaporates or is drawn off, leaving behind the unwanted sediment, which can be from a mine, a quarry, a manure pit at an animal facility, industrial wastewater, stormwater, or other sources.

Go to www.technology.infomine.com/sedimentponds to see an overview of some of the concerns about building and maintaining a settling pond for surface mines. Then visit www.agriculture.gov.sk.ca. Choose an article to read. Summarize that article and evaluate it using all of the items under Question 1.

Evaluate the authors of each website or article and work with the information to answer the following questions:

1. Is this author/sponsoring group a reliable information source?
   a. Does the author or group give supporting evidence for his/its claims?
   b. Does he/it give sources for the evidence?
   c. What is the mission of this organization, or of the scientist/author who wrote the article?

2. Do an Internet search and read two or three news articles about the toxic sludge spill near Kolontár, Hungary.
   a. What is the most recent article you can find and what does it say about the incident?
   b. How similar is the information in the articles?
   c. Do the authors give any sources cited for the "facts" presented?

3. Now do an Internet search using Google Scholar or another scholarly database. Select a peer-reviewed article that you can access in its entirety (not just the abstract—look for a pdf link). Read the article and summarize its purpose and main points.
   a. How does this article compare to the news articles you read? Compare and contrast the types of information regarding scope of coverage, intended audience, clarity of information, and reliability of information (whether reliable sources are cited).
   b. In general, when would you go to a news source for information on a topic like this and when would you go to a scholarly source?

4. What could be done to prevent another incident like what happened at Kolontár from happening again?

MAKING CONNECTIONS

COPPER MINING IN ARIZONA

Background: Copper is a critical element in our modern lives—it's used in our power lines, roofing materials, water pipes, and circuit boards. Over 60% of our nation's copper is mined in Arizona. The owners of the proposed open-pit Rosemont Copper Mine, to be located in the Santa Rita Mountains of southern Arizona, state that it will employ water conservation and recycling techniques never before implemented, to reduce the amount of water it will use. Mines must pump water constantly, both to remove excess water from the mine itself and to prevent some of the toxic compounds used in the mining process from reaching the local aquifer and tainting the water supply. Even after it has removed the valuable ore, a mine must continue to pump in order to keep from polluting the groundwater. In places where pumping has stopped, a pit lake may form—such as the one at the Anaconda Copper Mine in Montana—which then becomes a federal Superfund site that the EPA must clean up.
   The Santa Rita Mountains are mostly public lands, including pristine canyons where thousands of birdwatchers come each year to see species not found elsewhere in the United States, and hikers, photographers, and nature lovers walk the trails and enjoy the unspoiled scenery. Tourism is the major business of the area, with nature preserves, tours, and bed-and-breakfasts active year-round.

Case: You are an investigative reporter, trying to lay out the facts for your readers. Research and write an article explaining:

a. how water is used in the mining process.
b. the types of water pollution that can occur with a copper mine.
c. estimates of the amount of water that will be used annually. Consider both the water used in the mining process and the water pumped away from the hole so that the mine can function.
d. whether the mine and its water use will impact the people who live there.
e. whether the mine will impact the ecology and the groundwater of the Santa Rita Mountains.
f. whether the mine will impact the area's tourism.

Young men work at various levels of a wall at a gold-mining site in Ghana, a country in West Africa.

NO STONE UNTURNED

A bevy of unfamiliar minerals are crucial for our everyday technologies—but they come with a slew of problems

CORE MESSAGE

Modern society is dependent on a limited supply of nonrenewable mineral resources for all kinds of products and industrial processes. However, extracting and processing these resources can cause environmental damage, present occupational hazards for workers, and negatively impact local communities. In addition, some mineral resources are becoming scarce. Sustainable use of mineral resources includes conservation and recycling, as well as substituting some minerals with alternative materials. Meanwhile, the industry must develop less destructive methods and employ environmental safeguards for acquiring and processing these resources.

GUIDING QUESTIONS

After reading this chapter, you should be able to answer the following questions:

→ What are mineral resources and how do we use them in modern society?

→ What geologic forces help shape the planet, the rocks, and the minerals in Earth's crust?

→ How are mineral resources mined and processed?

→ What are the environmental and social impacts of mineral mining and processing?

→ What can be done to reduce the negative impact of mining and processing mineral resources and to address concerns about resource scarcity?

It's hard to imagine a future that's higher tech than the world we live in now, replete as it is with tablets, cell phones, and e-readers. But if you want a glimpse of what such a future might hold, take a peek into Mark Hersam's lab at Northwestern University in Chicago. The engineering professor has developed a carbon-based material that promises to revolutionize the way we produce energy, not to mention the way we make a whole suite of electronic appliances. Imagine flexible solar tents that provide soldiers in the field with a steady, consistent supply of electric power, or solar-powered clothes that double as "wearable electronics," and you'll begin to get a sense of what makes Hersam's work so exciting.

The carbon nanotubes he's developed—tiny, tubular-shaped carbon molecules that resemble rolled-up pieces of chicken wire—are poised to replace indium, a naturally occurring element that is growing in demand but is, like many natural resources, environmentally hazardous to mine. Indium has been the key ingredient in touchscreens and liquid crystal displays (LCDs), and in the solar cells that form the basis of solar power. Its exceeding softness and malleability make it ideal for creating thin transparent semiconductors. But indium is rare in nature and the explosion of LCD technology has threatened to make it

rarer still. Hersam's nanotubes, which are made of carbon, latex, and polystyrene (all cheap and readily available ingredients), and which conduct electricity as well as indium does, could provide an alternative.

Chances are, wherever you're reading this right now, you are surrounded by a bevy of different **minerals**. In addition to indium, **rare earth minerals** are finding their way into more and more modern products. Neodymium, for example, is used to tint your sunglasses and to power the laptop sitting on your desk. Gemstones and precious

◉ WHERE IS INDIUM FOUND IN THE U.S.?

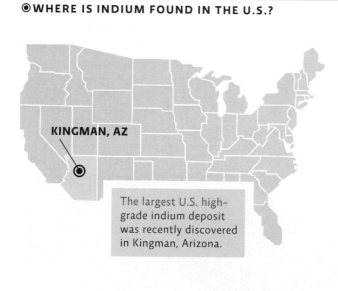

KINGMAN, AZ

The largest U.S. high-grade indium deposit was recently discovered in Kingman, Arizona.

↑ Mark Hersam in his lab at Northwestern University, outside Chicago, Illinois.

Infographic 19.1 | MINERAL RESOURCES ARE A PART OF OUR EVERYDAY LIVES

→ We depend on many mineral resources to make the products we use every day. Rare metals that you've probably never heard of—like indium—and rare earth minerals are increasingly important mineral resources for components of all kinds of modern machines, from cell phones to the motor of a wind turbine. Many minerals you are familiar with also find their way into our homes and the products of our modern age.

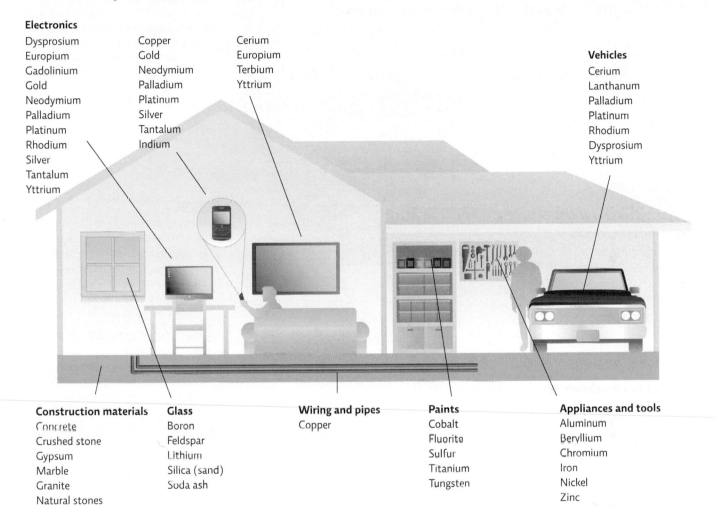

Electronics

Dysprosium	Copper	Cerium
Europium	Gold	Europium
Gadolinium	Neodymium	Terbium
Gold	Palladium	Yttrium
Neodymium	Platinum	
Palladium	Silver	
Platinum	Tantalum	
Rhodium	Indium	
Silver		
Tantalum		
Yttrium		

Vehicles

Cerium
Lanthanum
Palladium
Platinum
Rhodium
Dysprosium
Yttrium

Construction materials

Concrete
Crushed stone
Gypsum
Marble
Granite
Natural stones

Glass

Boron
Feldspar
Lithium
Silica (sand)
Soda ash

Wiring and pipes

Copper

Paints

Cobalt
Fluorite
Sulfur
Titanium
Tungsten

Appliances and tools

Aluminum
Beryllium
Chromium
Iron
Nickel
Zinc

metals like gold and silver are not only in the jewelry you're wearing, but also in most of the electronics you use. In fact, **metals** are an essential component of virtually all electronic devices, thanks largely to their malleability and their ability to conduct electricity.

And those are just the metal minerals. The sandstone, limestone, and gravel that form the buildings, roads, and sidewalks you travel daily? Those are made of minerals too, often containing common nonmetallic minerals like quartz and feldspar, the most abundant minerals on Earth. In fact, minerals are almost everywhere, in almost everything. "They are a deeply ingrained part of everything we know and love," says California Institute of Technology geologist George Rossman. "Human society as we know it would not exist without them." [INFOGRAPHIC 19.1]

mineral A naturally occurring chemical compound that exists as a solid with a predictable, three-dimensional, repeating structure.
rare earth minerals A group of 17 or so elements, similar in chemical structure, that occur close together in nature; they are not necessarily rare but do not occur in concentrated deposits.
metal A malleable substance that can conduct electricity; usually found in nature as part of a mineral compound.

But plucking them from Earth has brought a horrible array of consequences—not just environmental (air and water pollution) but also societal (health hazards and human rights violations). What's more, demand for all types of

↳ Earth is composed of discrete layers. The crust, mantle, and core are distinguished by their distinct chemical composition; regions within these layers that have different physical properties are also recognized. Mineral and fossil fuel deposits are found in the layers of Earth's crust. Powerful geologic forces are constantly but slowly rearranging the face of Earth.

EARTH IS A DYNAMIC PLANET THAT IS CONSTANTLY CHANGING

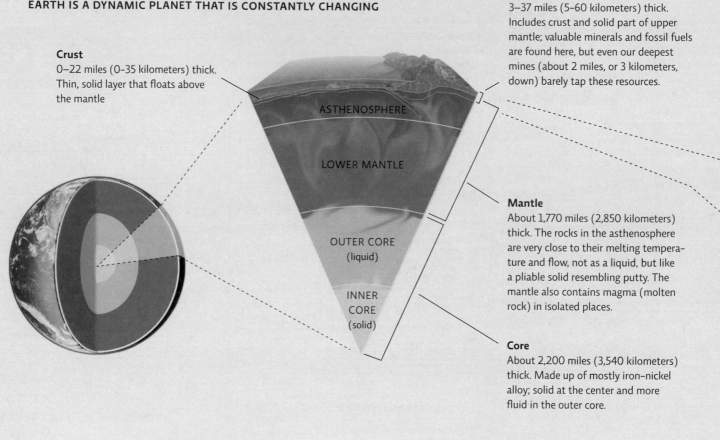

Crust
0–22 miles (0–35 kilometers) thick. Thin, solid layer that floats above the mantle

Lithosphere
3–37 miles (5–60 kilometers) thick. Includes crust and solid part of upper mantle; valuable minerals and fossil fuels are found here, but even our deepest mines (about 2 miles, or 3 kilometers, down) barely tap these resources.

ASTHENOSPHERE

LOWER MANTLE

OUTER CORE (liquid)

INNER CORE (solid)

Mantle
About 1,770 miles (2,850 kilometers) thick. The rocks in the asthenosphere are very close to their melting temperature and flow, not as a liquid, but like a pliable solid resembling putty. The mantle also contains magma (molten rock) in isolated places.

Core
About 2,200 miles (3,540 kilometers) thick. Made up of mostly iron–nickel alloy; solid at the center and more fluid in the outer core.

minerals—from rare earths to precious gems to familiar staples like nickel and copper—is rising, slowly but steadily, as population soars, standards of living increase across the globe, and unprecedented technological advances create a litany of new uses for metals. Hersam's work is just one effort to meet this growing demand. But it may indeed be one of the most promising.

Geologic processes produce mineral resources.

Minerals are naturally occurring solid compounds with a definite crystalline structure and specific chemical composition. They include both metals (like gold, silver, indium, and the rare earths) and nonmetals (like quartz and rock salt), and are classified according to several physical properties, like hardness and luster, that arise from their chemical composition and structure. To make

use of minerals—to find large deposits that we can mine, or to design synthetic compounds with similar properties—we must first understand their basic **geology**, starting with the forces that create them.

Earth is composed of discrete layers of solid and molten rock. The part we mine is the *lithosphere*, which is composed of Earth's solid upper crust and the uppermost portion of the mantle. The lithosphere is broken into **tectonic plates** that float above the *asthenosphere*—a layer of hot, weak rock in the upper mantle that flows slowly and responds to the pressure exerted on it (like putty or molding clay). As molten rock (magma) is forced up from the asthenosphere through *plate boundaries* (places where edges of tectonic plates meet), the plates themselves move, carrying the continents across the face of the planet as they go—ripping them apart and shoving them

↓ The crust is made up of tectonic plates that move slowly, powered by the heat of the innermost parts of Earth. This process changes the face of Earth. It also forms new minerals and redistributes others within Earth's crust. The location of earthquakes and volcanos gives clues about the location of plate boundaries, representing places where plates move against one another or spread apart to release magma from the top part of the asthenosphere.

TECTONIC PLATES

The lithosphere (which includes the solid part of Earth's surface) is broken into tectonic plates that float above the asthenosphere.

At a spreading ridge, plates are moving apart; new crust forms as magma is pushed up and cools. Many important metallic minerals are formed at these ridges and can be mined.

The rocks of the crust contain mineral deposits (like gold).

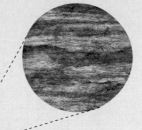

The crust is a very thin layer compared to the size of Earth; the continental crust is thicker than the oceanic crust but the oceanic crust is denser.

Where an oceanic plate meets a continental plate head on, the denser oceanic plate slides below (subducts) the other. This sometimes causes earthquakes, a sudden release of energy, as one plate slides past the other. Trenches may form in this area. When two continental plates collide, mountains can form as the plate edges buckle under the pressure.

together over eons, releasing colossal amounts of energy in the process. [INFOGRAPHIC 19.2]

It is through these processes—through the movement of plates and the release of energy, along with the activity of air and water, that minerals are created. Under such extraordinary heat and pressure, atoms of naturally occurring **elements** combine to form new compounds,

many of which are stable enough to endure for eons. The relative abundance and assortment of elements in any given location determine which chemical reactions will be favored, and thus which compounds will form. Because of this, minerals are not distributed evenly throughout the planet's lithosphere—different minerals are abundant or uncommon in different parts of Earth.

Mineral deposits are almost always found as a component of **rock**. In fact, for the most part, rocks are just conglomerates of minerals—though they occur in a variety of configurations and are constantly formed and transformed via the **rock cycle**. Igneous rocks form where magma emerges and cools at plate boundaries; metamorphic rocks are those that have been altered from their original solid rock state by the heat from magma or the pressure from moving plates; and sedimentary rocks are created by the pressure that builds as sediments (pulverized rocks,

geology The study of the structure of Earth and the processes that have shaped it in the past and shape it today.
tectonic plates Sections of Earth's crust that float above the magma layer.
element A pure chemical substance made up of one type of atom; there are 92 naturally occurring elements; examples include gold and carbon.
rock Conglomerates of one or more minerals that occur in a variety of configurations.
rock cycle The process in which rock is constantly made and destroyed.

Infographic **19.3** | **MINERALS, ROCKS, AND THE ROCK CYCLE**

→ Minerals are the building blocks of rocks. If free of any space restrictions when forming, each mineral would exhibit its characteristic crystalline structure. Constrained by the rock, the crystals are smaller, but each discrete mineral is still visible. Rocks may contain a single mineral (such as rock salt, NaCl) or more than one, such as this granite rock that contains three minerals.

Quartz

Mica

Feldspar

THE ROCK CYCLE CONTANTLY FORMS AND REFORMS ROCKS IN EARTH'S CRUST

→ Rocks form and are transformed when they are subjected to high heat and pressure underground and when they are exposed to wind and water on Earth's surface.

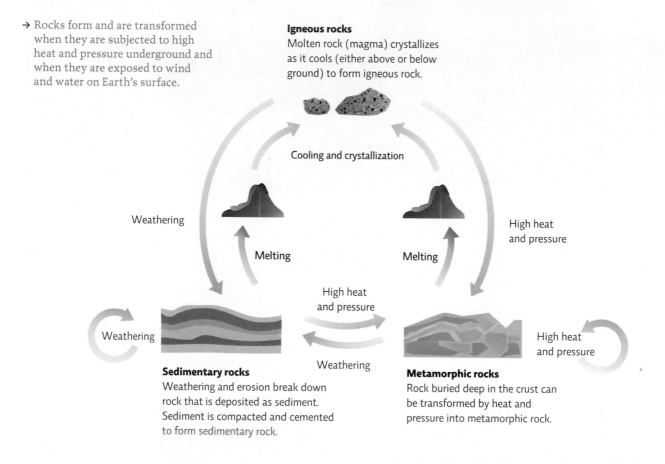

Igneous rocks
Molten rock (magma) crystallizes as it cools (either above or below ground) to form igneous rock.

Cooling and crystallization

Weathering

Melting

Weathering

Melting

High heat and pressure

High heat and pressure

High heat and pressure

Weathering

Weathering

Sedimentary rocks
Weathering and erosion break down rock that is deposited as sediment. Sediment is compacted and cemented to form sedimentary rock.

Metamorphic rocks
Rock buried deep in the crust can be transformed by heat and pressure into metamorphic rock.

shells, or organic detritus) accumulate, piling on top of one another in rivers, streams, and deep basins under the sea. [INFOGRAPHIC 19.3]

Minerals are most often found as components of rocks; **ores** are rock deposits that contain economically valuable amounts of metal minerals—that means not only do the minerals within have practical applications, but they are also concentrated enough to be worth the effort of **mining** and extracting them.

Minerals are considered nonrenewable resources. Yes, they form naturally through the aforementioned Earth processes; but we are using up existing deposits faster than new ones are being created. As with many of Earth's resources, the vast majority of mineral use has occurred in recent decades. In fact, of all of the minerals used in the 20th century, more than half were used between 1975 and 2000; according to one recent study, only 2.5% of all the copper ever mined was used before 1900.

Geologists are not sure how long reserves of any given mineral might last; a mind-boggling array of ever-changing factors must be taken into consideration, and past estimates have been notoriously off. (For example, some groups predicted that indium reserves would dry up by 2000, a date that has since been pushed back to 2020.) The consensus seems to be that for the most important minerals, existing supplies are still orders of magnitude greater than current usage. But not all reserves are created equal: Mining some will be much more environmentally taxing than mining others.

"The real question is not 'do we have enough,'" says Caltech geologist George Rossman. "But rather 'how far are we willing to go to access what we know is there?'"

Extracting and processing mineral resources impacts the environment.

Deep in the Mojave Desert, hidden behind a colossal industrial gate about an hour's drive from the glitz and glitter of Las Vegas, lies Mountain Pass, arguably one of the most important—and controversial—mines in the United States. It was the rare earth minerals harvested from this site—one mineral in particular, called europium—that in the 1960s turned black-and-white television into a Technicolor marvel, and in so doing touched off a technology revolution that has stretched into the present day. Before the Mountain Pass reserves were discovered, human societies had little use for most rare earth minerals. But once europium's usefulness became apparent, and miners started mining for it, they found a treasure trove of other rare earth minerals with properties that made them uniquely suited to a range of tasks.

Gadolinium is used in medical MRI scanners (in fact, it is the very thing that makes MRIs possible). Lanthanum is used in military night-vision goggles and in automobile catalytic converters. And, more recently, neodymium has become the key ingredient in wind turbines and hybrid car batteries. The extreme magnetic ability of this and other rare earth minerals enables them to produce electric currents with very little material. And this, in turn, allows electronic devices to be miniaturized. It's why we can use earbuds—instead of bulky headphones—to listen to our iPods.

To be sure, rare earth minerals are distinct from other minerals in several ways: They are tightly bound to one

---

ore A rock deposit that contains economically valuable amounts of metal minerals.
mining The extraction of natural resources from the ground.

another, tend to be attached to lots of radioactive substrate, and are scattered diffusely through the ores that contain them—all of which presents a string of unique mining challenges. Still, the story of Mountain Pass is like the story of most industrial-scale mines.

> "The real question is not 'do we have enough,' But rather 'how far are we willing to go to access what we know is there?'"
> —George Rossman

It begins with geologists exploring various rock formations to pinpoint mining-worthy deposits—a science in itself, and one that requires remote sensing, exploratory digging or drilling, and chemical testing to evaluate samples. Once a deposit is located, minerals are extracted using one of several mining techniques. For large mining operations like Mountain Pass, virtually all of those techniques require large machinery and explosives to excavate tunnels, expose deposits, and dig out and transport ore. [INFOGRAPHIC 19.4]

After it's been dug out of the earth, the ore is sent through a series of machines known as crushers that grind the rock down into gravel, and then grind it further still—into a fine powder. That fine-crushed powder is then heated to temperatures that, depending on the mineral being extracted, can exceed 1,000°C. During this step, the powder may be mixed with a suite of toxic chemicals—cyanide, mercury, various acids and bases—that bind to the metal, separate it from its nonmineral substrate, and pull it out of the treated sample. This process is known as *smelting*. Additional processing steps follow to purify the mineral in question, producing a form that can be used by manufacturers. [INFOGRAPHIC 19.5]

The same story—search, discover, dig, extract, process—plays out at smaller-scale, "artisanal" mines around the world, where impoverished peasants use crude tools and simpler strategies to achieve the same ends. But as demand for mineral resources has grown, and prices have increased in tandem, large-scale operations like the one at Mountain Pass have come to dominate the landscape.

Such mechanization has increased mine productivity by an order of magnitude—mostly by enabling us to access previously unreachable deposits. But it has also increased the amount of waste produced by an order of magnitude—and in so doing has amplified the environmental and health consequences that mineral mining and processing bring with them. Large industrial mines often require

Infographic **19.4** | **MINING TECHNIQUES**

↓ A variety of mining techniques are available to access mineral resources. The nature of the deposit (depth, concentration, type of ore) all influence the type of mining that will be done. Modern mining depends on heavy equipment and can move tremendous amounts of material in search of the desired mineral.

STRIP MINING is a type of surface mining used in areas where the land is relatively flat and the resource deposit seam is fairly horizontal and close to the surface. This is most commonly used for Wyoming coal deposits and Canadian tar sands (see Chapter 24). Layers of dirt and rock (overburden) are removed to expose the desired resource and the resource is extracted. After it is removed, the overburden can be replaced and the area restored. A variation of this technique (mountaintop removal) is used in Appalachia, but restoration of these areas is more difficult (see Chapter 23).

OPEN-PIT MINING is used when the top of a fairly deep ore deposit is found close to the surface. Drilling and blasting loosen the rock, which is hauled out for processing (if ore) or dumping (if waste rock). The pit is dug out in "benches" that allow trucks and machinery access to the pit. Rare earth ores are usually mined this way, as are aluminum and copper. The largest open-pit mine in the world is the Bingham Canyon Copper Mine in Utah.

PLACER MINING is used when sediments contain heavy metals like gold. Sediments can be dredged from rivers or mixed with water and then sent though a sluice; the heavier material will fall and be trapped, while the water and lighter gravels continue on. Panning for gold is a type of placer mining. Though modern placer methods use heavy equipment to dig and process sediments, smaller operations in remote or less developed areas like this one in Mali still depend on traditional methods.

SUBSURFACE MINING is used when deposits are deep underground or in mountainous areas. Modern equipment is used to excavate mine shafts. Many underground coal mines are in operation around the world. Minerals mined in this way include zinc, silver, gold, uranium, and lead, as well as gems like diamonds and rubies; even salt is sometimes mined this way.

that vast swaths of land be stripped bare, and that an extensive web of roads be built for access and transport. These combined forces—deforestation and road building—contribute to both soil erosion and water pollution as disturbed topsoil bleeds into nearby streams.

Indeed, it is nearly impossible to dig so much, and so deeply, into Earth without affecting the environment. Water traveling through underground mines or piles of waste rock and mill tailings (crushed ore that has been removed of its minerals and discarded) can collect all sorts of contaminants—not only heavy metals like mercury and arsenic that are detrimental to human health, but also exposed sulfur, which lowers the water's pH and produces acid mine drainage that in turns pollutes groundwater and harms aquatic ecosystems. Meanwhile, blasting, digging, crushing, and grinding all create dust,

which pollutes the air with small particles and toxins and in some cases contains radioactive material.

Large mining operations have developed a range of strategies to reduce this dust; they use huge amounts of water when crushing and grinding ore, and they store lightweight waste in ponds to keep it from floating through the air. But these solutions bring their own set of environmental problems—namely, that they generate thousands of gallons of toxic, hazardous waste for every ton of minerals we mine. And because that waste is so concentrated, the threat posed by a spill or leak is tremendous—especially for rare earth mines, where much of the waste is radioactive. [INFOGRAPHIC 19.6]

Trouble started at Mountain Pass when the long, winding pipes that snaked from the mining pits to waste

Infographic 19.5 | **EXTRACTION OF MINERALS FROM ORE REQUIRES MANY PROCESSING STEPS**

↓ Like most forms of ore mining, rare earth mining and processing is energy and water intensive and produces significant waste. Though the exact method differs at every mine, in general the process is the same. Tons of rock must be mined to produce small amounts of useful mineral, generating huge amounts of rock waste. The processing procedure requires multiple steps and produces significant amounts of hazardous waste (toxic and radioactive).

MINERALS MINING: RARE EARTHS

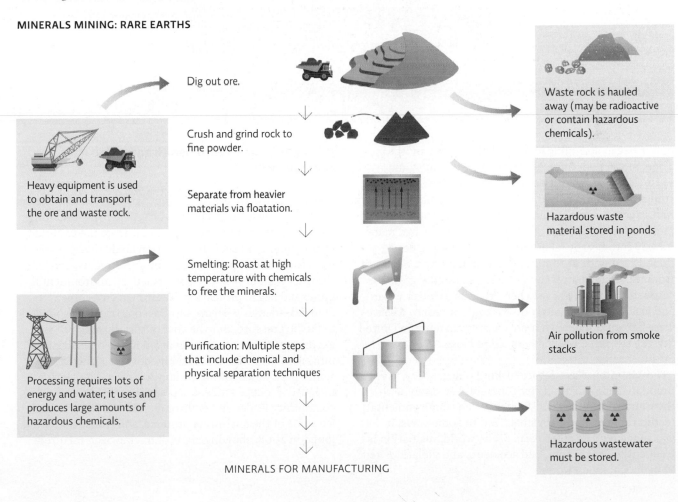

MINERALS FOR MANUFACTURING

Infographic 19.6 | **THE ENVIRONMENTAL IMPACTS OF MINING**

↓ Because it digs up and moves so much earth, there are unavoidable environmental effects of mining.

Habitat damage: Road building, deforestation, water pollution, and loss of soil are all consequences of mining.

Air pollution: Smelters, emissions from heavy equipment, and dust are major sources of air pollution.

Acid mine drainage: The passage of water through exposed sulfur-containing rock at mining sites can acidify the water and damage ecosystems. Heavy-metal contaminants can also be released into the water.

Waste: The amount of waste rock that is removed from mines can be huge, contributing acid drainage as rainwater percolates through the waste piles into soil and groundwater; waste produced from the processing of the ore may also be hazardous.

Carbon footprint: Mining is an energy-intensive industry; energy is used at every stage of the mining, processing, and transport of mineral ores.

Mercury pollution: The use of mercury to extract metals like gold and silver has resulted in substantial mercury contamination in areas around (and downstream from) mines.

ponds began springing leaks. First one; then another. Then another—roughly 60 in all, between 1984 and 1994. The vast majority of them went unreported by the company in charge. By the time federal regulators noticed, investigated, and filed a report, some 600,000 gallons of radioactive, acidic, and otherwise toxic waste had made its way into nearby lakes and the surrounding desert floor. Faced with $100 million in cleanup costs, a lawsuit from the district attorney's office, and rapidly growing competition from China (where, thanks to lower wages and less regulation, newly discovered rare earth reserves were being mined much more cheaply than in the United States), the mine would be forced to close.

"The Mountain Pass debacle is kind of legendary," says Brendan Cummings, an environmental attorney with the Center for Biological Diversity. "The environmental record there was just abysmal." But in some ways, it was just a blip. In other parts of the world, regulation is frequently nonexistent, and accidents and violations tend to be much, much worse.

Mining also comes with significant social consequences.

When a Canadian-based mining giant arrived in Cabanas, El Salvador, wanting to build a large, open-pit gold mine, area residents were wary. They knew from the experience of neighbors that such an operation could divert water from their farms, pollute their air and streams, and threaten their crops. So they formed an environmental group and began protesting. Their own government listened—denying permits to the company in response to public pressure. But the cost of that success proved exorbitant. "[Four] people involved in the opposition to mining in El Salvador [were] killed," says Keith Slack, senior policy advisor for Oxfam America. The victims included a college student, a pregnant mother, and a community leader and environmental activist who had led some of the anti-mine protests, and whose body was found in 2008, showing signs of having been tortured.

↓ Human rights violations associated with mines, such as displacement, violence, and child labor, are perhaps the most troubling by-products of mining for mineral resources.

Use of toxic chemicals: Processing of minerals using hazardous chemicals is done without oversight and proper protection in many small-scale operations in Africa, South America, and Asia.

Child labor: Children may work alongside family members or be forced to work in mines in many areas.

Displaced communities: Mines may displace farmers and entire communities.

Violence: Local residents protesting mining operations have sometimes been met with violence.

Dangerous jobs: Underground mining is a hazardous job (mine collapses, methane explosions, toxic gases, lung disorders).

Polluted environment: When local land and water sources are polluted, residents and the ecosystem on which they depend suffer.

The thread between mining and violence runs long, spans the globe, and involves a range of minerals—from gold and silver to diamonds and emeralds. And it's not just protesters who suffer. Independent diamond miners in Zimbabwe have also been assaulted and threatened—by security guards working for larger-scale mines whose operators want to maintain a monopoly on the region's treasures. Elsewhere in Africa, diamonds harvested from areas controlled by warlords—known as "blood diamonds" or "conflict diamonds"—have been used for ages to support armed conflicts, conflicts that have devastated the lives of millions of people.

Such violence underscores a point that has long been central to any discussion of minerals' true costs: The environment is not the only casualty. Human health and human rights take a bruising, too, especially in developing countries that lack the resources and infrastructure to regulate such behemoth industries. Not only do mines frequently displace communities in these countries—clearing land, polluting water, and filling the air with

noise and dust—but the owners of large industrial mines are notorious human rights violators. They don't generally give area residents a say in the decision to open a mine in a given location. Nor do they typically clean up the pollution after the minerals have all been harvested; nor compensate communities for the loss of land or ecosystem services.

In addition to all of that, some routinely violate child labor laws. The International Labor Organization reports that tens of thousands of children work in gold mines—roughly 25% of them in the Sahel region of Africa. Their small size makes them particularly useful in underground mines; unfortunately, their developing bodies are very vulnerable to the dust and chemicals they breathe in as they work. [INFOGRAPHIC 19.7]

Even for adults, the work is dangerous. Underground mines can collapse, or develop pockets of methane that then cause explosions, or gather high concentrations of other hazardous inhalants, leading to lung disorders and

↑ Since the late 1980s, e-waste from developed countries has been imported to China and broken down in Guiyu. The city, located in Guangdong Province, comprises 21 villages with 5,500 family workshops handling e-waste. According to the local government website, city businesses process 1.5 million tons of e-waste each year.

other illnesses. And the hazards don't end when the ore is brought to the surface. In developing countries especially, smaller-scale "artisanal miners" employ a range of toxic chemicals—mercury and cyanide chief among them—in their own crude smelting processes. These chemicals not only contaminate the soil and water of surrounding communities, but also expose miners and processers to toxic fumes. In Colombia alone, there are an estimated 200,000 artisanal gold miners who mix crushed ore with mercury using their bare hands; mercury concentrations in the air inside such artisanal shops are known to reach 1,000 times the World Health Organization (WHO) limit for exposure.

Safer techniques have long since been developed. But in many areas of Africa, South America, and Asia, large amounts of mercury—and in some cases cyanide—are still used to process ores containing gold and silver. Simply put: Gold is valuable, and mercury and cyanide are cheap. "The thing to remember is that in so many

countries, the people doing the actual mining and processing are poor," says mining analyst John Kaiser. "They're running makeshift, often illegal operations, on a shoestring. They don't have the resources to do it safely."

We can minimize the impacts of mining through better conservation strategies.

If there is a silver lining to the story of mineral resources—resources that fuel life as we know it, resources that we claw into Earth to access, and fight and even die over—it is this: They are eminently reusable. In fact, the greatest reserves yet of gold and silver and neodymium may not be hidden in some desert canyon, but rather sitting in the backs of our own closets and garages, or scattered throughout our landfills as **e-waste** (electronic waste). An average gold mine produces a mere 5 grams of gold per ton of rock—sometimes less, depending on the local geology. A ton of cell phones, on the other

Infographic **19.8** | **ALUMINUM RECYCLING: A SUCCESS STORY**

↓ Recycling metal products to recover the metals for reuse extends their useful life and reduces the need to obtain the metals from mined ores, decreasing the environmental and health impacts associated with mining and processing the ore. A comparison of the two methods for obtaining aluminum to make a new aluminum can—production from the raw material (bauxite ore) or from recycled aluminum cans—highlights the differences between the two processes.

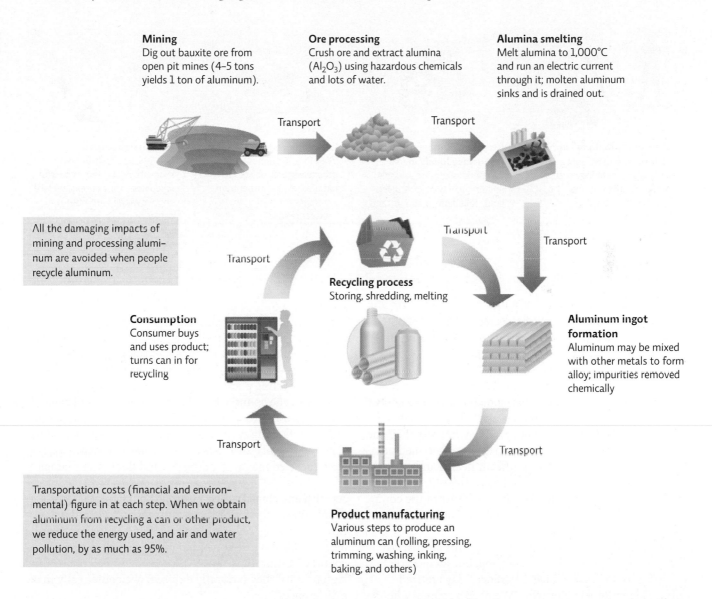

Mining
Dig out bauxite ore from open pit mines (4–5 tons yields 1 ton of aluminum).

Ore processing
Crush ore and extract alumina (Al_2O_3) using hazardous chemicals and lots of water.

Alumina smelting
Melt alumina to 1,000°C and run an electric current through it; molten aluminum sinks and is drained out.

Transport

Transport

All the damaging impacts of mining and processing aluminum are avoided when people recycle aluminum.

Transport

Transport

Transport

Recycling process
Storing, shredding, melting

Consumption
Consumer buys and uses product; turns can in for recycling

Aluminum ingot formation
Aluminum may be mixed with other metals to form alloy; impurities removed chemically

Transport

Transport

Transportation costs (financial and environmental) figure in at each step. When we obtain aluminum from recycling a can or other product, we reduce the energy used, and air and water pollution, by as much as 95%.

Product manufacturing
Various steps to produce an aluminum can (rolling, pressing, trimming, washing, inking, baking, and others)

hand, might contain nearly 200 grams of the precious metal—plus well over 100 kg of copper, 3 kg of silver, and a smattering of neodymium and other rare earth elements. Likewise for our flat-screen televisions, laptops, and iPads: They all contain a bevy of essential and nonrenewable mineral resources.

To be sure, metal recycling is already big business in the United States and around the world. Indeed, the process of recycling scrap metal from cars and other vehicles, not

to mention home appliances, has long been an industry unto itself. And aluminum is so effectively recycled in the United States that on average, the aluminum in any given beverage can is back on the shelf in 2 months, in another can. [INFOGRAPHIC 19.8]

---

e-waste Unwanted computers and other electronic devices such as televisions and cell phones that are discarded; contains valuable metals that can be recovered, but also contains toxic chemicals.

Infographic **19.9** | **ALTERNATIVES THAT REDUCE OUR USE OF MINERAL RESOURCES**

↓ Different choices made by miners, manufacturers, and consumers can reduce the impact of acquiring and processing mineral resources. Some methods reduce overall use (conservation and redesign), while other methods (recycling and use of best practices) reduce the need for, or impact of, mining and processing.

Use "best practices" in mining and processing
Improved technology such as more sensitive sensing equipment and safer processing methods can reduce impact.

Recycling of mineral resources already in use
Cell phones contain valuable metals such as copper, silver, gold, palladium, platinum, and neodymium.

Consumer conservation
Wait for cell phones and other electronic devices to naturally expire, rather than automatically buying the "newest release."

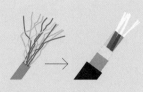

Redesign products
Use less, or substitute with more abundant, less impactful resources—for example, replace copper wiring with fiber-optic cables or use carbon nanotechnology instead of rare minerals in electronics.

It's clear from such successes that recycling can supply needed minerals while eliminating (or at least minimizing) the need for destructive mining practices. But for that to work, it has to be done right. And when it comes to waste, that is still not happening. Most of our discarded cell phones, computers, and flat-screen televisions collect dust in the crevices of our homes and offices, only to be discarded in ordinary trash heaps. Or eventually they end up in a developing-world slum, where impoverished peasants expose themselves to significant health and safety hazards, and decimate their local environment, as they try to extract the various metals contained within the equipment, using the crudest of tools and techniques.

There is, of course, a better way. "Programs similar to the ones we have for plastic recycling, but focused on e-waste, could make a huge difference," says retired mineralogist Robert Housely. "We need a system whereby discarded electronics are picked up from people's homes and delivered to facilities specializing in e-recycling—facilities where lawmakers can mandate—and regulators can enforce—proper health and safety standards."

In other parts of the developed world, an entire industry is being born around that very idea. The growing demand and rising price for elements like indium, gold, and neodymium has made recovering them from hybrid vehicles, electronics, and other products an attractive

endeavor—one that several big-name producers such as Honda, Toyota, and Hitachi are pursuing. In fact, thanks to new technology that has made extracting metals from electronics easier and cheaper, Japan has already opened several recycling plants devoted to electronics (by some estimates, the country has about 300,000 tons of rare earth minerals stored in used electronics). And France is quickly following suit; two new factories are projected to generate roughly 200 tons a year of rare earth minerals from recycled magnets, batteries, and fluorescent lamps.

As with any conservation efforts, though, recycling is only one of the four Rs (see Chapter 30). The other three—refusing, reducing, and reusing—could serve us here as well. "We've gone kind of upgrade mad in this country," Cummings says. "If we used our cell phones and laptops until they naturally expired…we could really make a difference."

Meanwhile, there's another "R" to consider: redesign. Indeed, the Northwestern researchers are not the only ones trying to replace important minerals with less environmentally taxing, more human rights-friendly substitutes. Ceramics (made from sand) are being substituted for some metals. Fiber-optic cables (also made from sand) are increasingly replacing copper wire. And in cases where there is no substitute, process engineers are redesigning products with an eye toward using less material. Aluminum cans are a good example of this: They are thinner today than in the past. In 1975, a pound of aluminum yielded 27 cans; in 2008, the same pound can produce 34 cans.

But most experts agree that neither recycling nor scientific wizardry will replace all that we dig from Earth. Indeed, mining will be necessary for many lifetimes to come. Which is why we need to do it more responsibly, says Cummings. "The key is to minimize the damage as much as possible." We cannot eliminate the negative environmental impacts of mining, but we can reduce them. That means employing *best practices*—those methods recognized as the most efficient and safest available at the time. More-sensitive sensing equipment, for example, reduces the need for exploratory drilling or digging; and more energy-efficient and cleaner-running mine equipment can dramatically reduce the carbon footprint. For his part, Cummings would also like to see would-be miners avoid sensitive ecosystems, like coastal estuaries where many aquatic organisms come to spawn, or streams that house rare, endemic species. The proposed Pebble Mine near Alaska's Bristol Bay, for example, cuts dangerously close to the path traced by wild salmon on their way to inland freshwater streams. "In cases like that, we really have to weigh as a society what our priorities are," he says. [INFOGRAPHIC 19.9]

Today, Mountain Pass is creaking back to life with a $1 billion makeover that the mine's operators have dubbed Project Phoenix. This time around, they say, things will be different. The new facility will recycle wastewater, along with most of the chemicals used to pry the rare earth minerals from their rock substrates; that means no more long, winding, leaking pipes and no more evaporation ponds. So far, critics are optimistic: On one recent tour of the facility, Housely stood on a steel walkway, overlooking a pit that ran at least 500 feet (152 meters) deep, as the mine's lead geologist pointed out some of the company's latest environmental safeguards. "They seemed much more on top of their game," Housely said later. "They may finally get it right this time around."

Meanwhile, back at Northwestern, scientists are betting that environmental sustainability will rank high among those priorities—and that developing smart alternatives now will help us avoid having to scour Earth for minerals like indium later—digging, grinding, smelting, and polluting as we go.◉

References used in this chapter:

Gordon, R.B., Bertram, M., and Graedel, T.E. 2006. *Proceedings of the National Academy of Sciences*, 103: 1209–1214.

Long, K.R., et al. 2010. *U.S. Geological Survey Scientific Investigations Report 2010-5220*. Available at http://pubs.usgs.gov/sir/2010/5220/

Paul, J., and Campbell, G. 2011. *Investigating Rare Earth Element Mine Development in EPA Region 8 and Potential Environmental Impacts*. EPA Document 908R11003.

BRING IT HOME

◒ PERSONAL CHOICES THAT HELP

While mining is necessary to extract many of the minerals and metals needed for constructing the technology that runs our society, we can take action to minimize the negative environmental and health impacts.

Individual Steps

→ Help recycle the rare earth minerals in your electronic goods when they become obsolete. Find a local vendor or electronics recycling event at www.search.earth911.com.

→ Donate your old computers to a charity such as InterConnection, a Seattle-based organization that refurbishes old computers and donates them to nonprofit groups.

Group Action

→ Organize a campuswide or citywide initiative to commit to purchasing electronics from manufacturers who make an effort to get their minerals from conflict-free sources. Find a list of company rankings at www.raisehopeforcongo.org/companyrankings.

→ Sponsor a collection event or permanent site on campus or in your community to collect used cell phones for Cell Phones for Soldiers, a nonprofit organization that provides free phones to veterans and active-duty military members.

Policy Change

→ A device known as a personal dust monitor can tell coal miners exactly how much dust they are being exposed to, helping them moderate their risk of contracting black lung disease. These devices have been available for years, but have never been made mandatory by the Mine Safety and Health Administration (MSHA). Proposals for limiting allowable dust in a mine have also been drafted, but never implemented. Contact the MSHA and the White House regarding the delay in passing these safety measures.

UNDERSTANDING THE ISSUE

CHECK YOUR UNDERSTANDING

1. **Most mineral deposits:**
   a. are evenly distributed around the world.
   b. are short lived and decay quickly after forming.
   c. are found in rocks.
   d. exist as pure seams in Earth's crust.

2. **Society's use of minerals is:**
   a. a luxury rather than a necessity.
   b. mainly of rare earth minerals.
   c. on the increase.
   d. limited to nonmetallic minerals like those found in sand.

3. **Why is an understanding of plate tectonics useful for mineral mining?**
   a. It helps identify likely places to drill into the asthenosphere.
   b. It gives clues as to the likely locations of mineral deposits.
   c. Geologists know that minerals will not be found around plate boundaries.
   d. These geologic conditions can be mimicked in the lab to make minerals.

4. **Human rights violations associated with mines:**
   a. include the use of child labor in many developing countries.
   b. are decreasing around the world because of better oversight.
   c. mainly affect workers; this is a dangerous industry.
   d. are highest in developed countries, where most mines are found.

5. **The aspect of mining that can harm the environment is:**
   a. moving earth and rock to gain access to the mineral deposit.
   b. the initial processing of the rock by crushing and washing.
   c. removing the mineral from rock through smelting and chemical treatment.
   d. All of the above.

6. **Supplies of most mineral resources:**
   a. are abundant and accessible with current technology.
   b. are abundant but can be costly to obtain.
   c. will increase as more are made in the natural rock cycle.
   d. are almost depleted.

7. **The environmental impacts of mining can be reduced by:**
   a. developing methods to more safely process and store waste.
   b. using more energy–efficient mining equipment.
   c. avoiding mining in sensitive environments.
   d. All of the above.

WORK WITH IDEAS

1. Why does the mining of rare earth minerals cause more environmental problems than the mining of more abundant mineral resources such as sand or gravel?

2. List three ways in which you personally use rare earth minerals. What is something that you could do to decrease your rare earth mineral "footprint"?

3. Looking at IG 19.2, what might be an advantage and a disadvantage of mining at the mid–Atlantic ridge?

4. Besides the enormous environmental damage caused by leakage of radioactive waste, what other environmental impacts would you predict that Mountain Pass Mine may have caused?

5. What are some of the negative social impacts that mining may bring about? Do you think those occurred at Mountain Pass Mine? Why or why not?

6. Give an example of how each of the following can reduce the amount of mining needed to meet society's need for minerals: electronics manufacturers, e-waste recyclers, consumers.

ANALYZING THE SCIENCE

Since 2006, China has been decreasing production and export of rare earth minerals. This has precipitated a scramble to establish other sources of these necessary elements of modern technologies. There are now more than 25 new rare earth mining facilities in development outside of China.

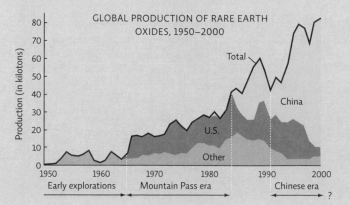

INTERPRETATION

1. How did the quantity of rare earth minerals produced change from 1950 to 2000?

2. Compare the production of rare earth minerals in the United States and China, in the years 1990 and 2000.

3. What caused the decrease in U.S. production of rare earth minerals in 1999–2000?

ADVANCE YOUR THINKING

1. How would you predict this graph will look by the year 2020? Explain your reasoning.

2. Will increased recycling of e-waste cause a decrease in the production of rare earth minerals? Why or why not?

EVALUATING NEW INFORMATION

As cell phone use has skyrocketed, more energy and mineral resources are needed to produce the phones, and more waste is generated in their production and when they are discarded. What can be done to decrease the environmental impact of our increasing use of electronic devices such as cell phones?

Go to the U.S. Environmental Protection Agency (EPA) website on electronic waste recycling (www.epa.gov/osw/conserve/materials/ecycling/index.htm) and review their web pages on "e-cycling."

Evaluate the website and work with the information to answer the following questions:

1. Is the EPA website a reliable information source?
   a. Does it give supporting evidence for its claims?
   b. Does it give sources for its evidence?

2. Select the "Frequent Questions" link. Review the statistics on the relative rates of recycling for computers, TVs, and mobile devices (more information is available at the "Electronics Waste Management" link).
   a. The EPA claims that electronic waste in solid waste landfills is safe for the environment and human health. What evidence does it provide to support this claim? Is it sufficient? Where would you attempt to find more evidence to support or refute this claim?

b. Based on your investigation above, why does the EPA recommend that e-waste be recycled rather than disposed of in landfills? Cite some of the data it uses to support this recommendation. Is it sufficient? Where would you attempt to find more evidence to support or refute this recommendation?

3. What are the specific recommendations the EPA gives for donating a cell phone for reuse, refurbishment, or recycling? Given how easy it appears to be, why do you suppose the recycling rate for cell phones is so much lower than that for computers and TVs?

4. The EPA website links to an external website with a map of e-cycling legislation by state (www.ecycleclearinghouse.org/content.aspx?pageid=10). Find your state and review its current e-cycling legislation. Summarize the legislation—do you support it? Why or why not? How would you modify it to align it more closely with your values and beliefs about e-cycling?

MAKING CONNECTIONS

BLOOD DIAMONDS

Background: The diamond engagement ring is mainly a European and American phenomenon, first documented in the 15th century but uncommon until the 1930s. This popularization of diamonds was brought about in large part by De Beers, a family of companies that is involved in mining, processing, and distributing most of the world's diamonds. De Beers has ensured the continued high price of diamonds by controlling supply and increasing the demand for diamonds through a series of brilliant marketing campaigns.

However, in 1998, the nonprofit organization Global Witness released the first report on "conflict diamonds" or "blood diamonds" used to fund civil war in Angola. The United Nations quickly passed a resolution to stop the sale of such diamonds. In response, the World Diamond Council developed a self-policing certification process rather than submit to governmental oversight: The Kimberley Process has slowed the sale of "conflict diamonds," and it has increased the sale of legal diamonds from poor African countries, helping stabilize those economies and develop infrastructure. However, several key players have recently withdrawn from the Kimberley Process, citing corruption and political pressures that have made it ineffective. This is shown most dramatically in the Kimberley Process certification of Zimbabwean diamonds, in spite of the Zimbabwean dictator's well-documented human rights violations.

Case: It is difficult to make an environmentally conscientious decision about an engagement and/or wedding ring, especially with all of the cultural expectations surrounding this decision. You will develop a "shopper's guide" to present the costs and benefits of engagement and wedding ring options. You should consider:

1. conflict-free diamonds
2. jewels or other stone alternatives to diamonds
3. environmentally responsible gold
4. alternative metals to gold
5. recycled jewels and/or metals
6. other ideas you may have

Research and write the shopper's guide, with enough explanation of each alternative so that a couple can make an informed decision about their engagement and wedding rings. Your report should include:

a. For each option:
   • A discussion of the benefits, including economic, social, and ecological implications
   • A discussion of the possible costs and consequences, including unintended and secondary consequences
   • Information about the availability and pricing of each option

b. A set of guidelines for evaluating new options as they arise—for example, if there are new sources of diamonds, if a cleaner technology is developed for processing gold, or if alternative metals become available. Explain why you consider these guidelines important in choosing an engagement or a wedding ring.

c. A discussion of which of these alternatives, if any, you would choose now that you have completed this research. Explain your reasoning.

A GENE REVOLUTION

Can genetically engineered food help end hunger?

CORE MESSAGE

Although we produce enough food to feed the world's population, nearly 1 billion people are hungry and don't have access to enough nutritious food. The Green Revolution of the 1960s helped fight hunger by increasing global crop yields tremendously but came with a slew of unintended consequences. To feed our growing population, some people support yet another agricultural revolution hinged on genetically modified crops. However, concerns about the safety of growing or eating genetically modified foods trigger strong debate. Opponents recommend that we pursue other agricultural and socioeconomic methods to address global hunger.

GUIDING QUESTIONS

After reading this chapter, you should be able to answer the following questions:

→ What challenges do we face in meeting the nutritional needs of the world's population?

→ What are the basic nutritional requirements for human health and what problems result from poor nutrition?

→ What was the Green Revolution and what were its advantages and disadvantages?

→ What is the focus of the next "green revolution" and how do researchers get new, desired traits into crops?

→ What are the trade-offs of using genetically modified organisms in agriculture, and what are some "low tech" (non-GMO) options for increasing crop production?

Desperate residents plead to workers from the Kenyan Red Cross for food during distribution in Nairobi in 2008. The crowd eventually broke through the gate but was chased back by police with whips.

It was in Burkina Faso—a tiny, landlocked West African country, virtually unknown to the developed world—that a brewing global crisis finally came to a head. On February 22, 2008, riots broke out in the country's two major cities; angry protesters clogged the streets—shouting, throwing rocks, and flipping cars. Soldiers were mobilized to restore order, but the chaos only spread from there. In neighboring Côte D'Ivoire, tear gas was employed and dozens were injured; in Cameroon, some two dozen people were killed; in Egypt, a single boy was shot in the head. Before long, the violence spilled across Africa's borders. Protesters in Yemen torched police stations and blocked roads. In Bangladesh, they smashed cars and buses, vandalized factories, and ultimately injured dozens of bystanders.

What were they rioting over? Food. It had become too expensive. In the 2 preceding years, the cost of rice had risen by 217%; wheat by 136%; and corn and soybeans by 125% and 107% respectively. In the 7 months leading up to the riots, those prices had then doubled. Such increases were not as much of a problem in the United States or Europe, where even the poorest fifth of households spends just 16% of their budget on food. But in the cities that had descended into violence, families were already spending between 50% and 75% of their income on basic staples— rice and beans and bread and milk. Such dramatic and rapid cost increases had pushed those goods out of their reach, and in so doing had nudged too many people toward the brink of starvation. "For countries where food comprises from half to three quarters of [income] consumption," explained World Bank President Robert Zoellick, "there is no margin for survival." In all, he said, some 22 countries were at risk of violent revolts if prices did not soon stabilize.

But what had destabilized them in the first place? Newspaper columnists, politicians, and cable news anchors trotted out the usual suspects. Some blamed global climate change: a drought in Australia, a heat wave in California, and flooding in India had decimated crops of wheat and maize and supplies of beef. Others pointed to the global financial crisis, and to rising fuel prices: Because large, industrial farms rely on fossil fuel—powered machines, the cost of grain is tethered to the cost of oil and gasoline. Others still focused on the growing demand for meat from Asia's expanding middle class: The more grain we devote to cattle rearing, the less we have to sell as food, and the more expensive it thus becomes (see Chapter 21).

But experts said that the trouble with the global food supply goes much further than that—beyond isolated weather events and the slumping global economy, to the very heart of global food policy.

World hunger and malnutrition are decreasing, but are still unacceptably high.

While global food prices have stabilized since the 2008 riots, and grain reserves are currently well stocked, much of the world is still struggling to feed itself. In 2010, the United Nations estimated that some 925 million people (16% of the global population) suffered from **under-nutrition**—meaning that they did not consume enough

◉ **WHERE IS BURKINA FASO?**

ALGERIA

MALI

CÔTE D'IVOIRE

CAMEROON

BURKINA FASO

DR CONGO

UGANDA

South Atlantic Ocean

SOUTH AFRICA

Infographic 20.1 | **WORLD HUNGER**

↓ Though we currently produce enough food to feed the world's population, almost a billion people are undernourished. Of those who are underfed, 98% live in developing nations, but even in wealthy nations there are those who do not have access to enough food or to a balanced diet.

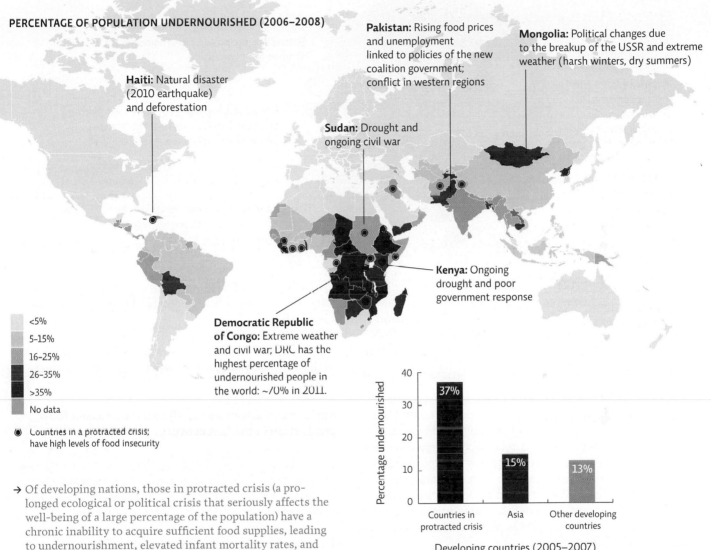

PERCENTAGE OF POPULATION UNDERNOURISHED (2006–2008)

Haiti: Natural disaster (2010 earthquake) and deforestation

Pakistan: Rising food prices and unemployment linked to policies of the new coalition government; conflict in western regions

Mongolia: Political changes due to the breakup of the USSR and extreme weather (harsh winters, dry summers)

Sudan: Drought and ongoing civil war

Kenya: Ongoing drought and poor government response

Democratic Republic of Congo: Extreme weather and civil war; DRC has the highest percentage of undernourished people in the world: ~70% in 2011.

Legend:
- <5%
- 5–15%
- 16–25%
- 26–35%
- >35%
- No data
- ◉ Countries in a protracted crisis; have high levels of food insecurity

→ Of developing nations, those in protracted crisis (a prolonged ecological or political crisis that seriously affects the well-being of a large percentage of the population) have a chronic inability to acquire sufficient food supplies, leading to undernourishment, elevated infant mortality rates, and stunted growth in children.

Bar chart — Percentage undernourished:
- Countries in protracted crisis: 37%
- Asia: 15%
- Other developing countries: 13%

Developing countries (2005–2007)

calories. According to the World Health Organization (WHO), roughly 50,000 people around the world starve to death every single day. [INFOGRAPHIC 20.1]

The United Nations Food and Agricultural Organization (FAO) has set itself the task of eradicating hunger (and extreme poverty along with it). Its goal is to cut the percentage of people around the world who were underfed between 1990 and 1992 by half come 2015, and to one day achieve total food security. **Food security** is defined as all people at all times having physical, social, and economic access to sufficient, safe, and nutritious food. This mission is not just a matter of food quantity, but also of improving food quality. The human diet needs to meet basic nutritional requirements. [INFOGRAPHIC 20.2]

undernutrition Chronic, insufficient calorie intake, resulting in nutrient deficiencies and the inability to meet energy needs.

food security Having physical, social, and economic access to sufficient, safe, and nutritious food.

Infographic 20.2 | **GOOD HEALTH DEPENDS ON GOOD NUTRITION**

↓ A healthy diet contains a variety of foods that provide the proteins, carbohydrates, and fats needed for good health, as well as enough calories to meet daily energy needs. The average adult needs about 1,800 calories a day—more if active, pregnant, lactating, or ill. The food that supplies these calories must also supply micronutrients such as vitamins and minerals. The bulk of the diet should be plant based, but animal products can also be part of a healthy diet.

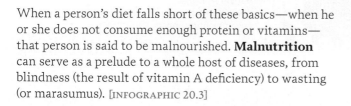

Grains, fruits, and vegetables provide carbohydrates and fiber and are a rich source of many micronutrients, such as the vitamins and minerals needed for good health.

Meat, fish, dairy, legumes, and nuts are good sources of protein as well as of some needed vitamins and minerals.

Fats are also necessary, but some food sources of fat are better than others. The intake of saturated animal fats should be minimized.

There is a little room in the average diet for empty calories (those that provide only energy but no nutrition), but not much!

When a person's diet falls short of these basics—when he or she does not consume enough protein or vitamins—that person is said to be malnourished. **Malnutrition** can serve as a prelude to a whole host of diseases, from blindness (the result of vitamin A deficiency) to wasting (or marasumus). [INFOGRAPHIC 20.3]

Overnutrition, or too many calories, is also a problem. By some estimates, about 1.5 billion people around the world are overnourished, and are thus vulnerable to a whole other set of nutrition-related conditions—namely obesity and type 2 diabetes. Contrary to popular belief, this is also a problem of the poor. It's the cheaper foods that tend to make us fatter. And because those foods are often low in essential nutrients, people who consume too many calories are often just as malnourished as those who consume too few.

Of course, to conquer global hunger, we must grapple with more than just nutrients and calories. Indeed, we must tackle a vast array of problems that lie at the root of the brewing food crisis. Political instability and ecological degradation play significant roles, to be sure. As shown in Infographic 20.1, undernourishment is three times greater in developing nations experiencing a prolonged armed conflict, drought, or natural disaster, than it is in countries not in such crises. But there are other contributors, too—namely, social disempowerment and poverty. Even in the poorest countries (or especially in the poorest countries), there are groups of people who have access to

food and groups of people who do not. To reach the FAO's lofty goals, we must first understand why and how that came to be.

And to do that, we must travel back through recent history, to the last time global hunger was the stuff of headlines.

Agricultural advances significantly increased food production in the last century.

It was the late 1960s. Global population was soaring; food crops were flatlining in some parts of the developing world and plummeting in others. India especially seemed to be hovering on the brink of a massive famine.

To stave off such catastrophe, the international community had launched the **Green Revolution**—a coordinated global effort to eliminate hunger by bringing modern agricultural technology to developing countries in Asia. Working across the globe, scientists, farmers, and world leaders introduced India and China to chemical pesticides, sophisticated irrigation systems, synthetic nitrogen

malnutrition A state of poor health that results from inadequate or unbalanced food intake; includes diets that provide too few or too many calories and/or do not provide the proper nutrients (deficient in one or more nutrients).

Green Revolution Plant-breeding program in the mid-1900s that dramatically increased crop yields and led the way for mechanized, large-scale agriculture.

Infographic 20.3 | **MALNUTRITION**

↓ Malnutrition is defined as a state of poor health that results from inadequate or unbalanced food intake. This includes diets that don't provide enough calories (and thus are nutrient deficient) as well as those that may provide enough (or even too many) calories but are deficient in one or more nutrients.

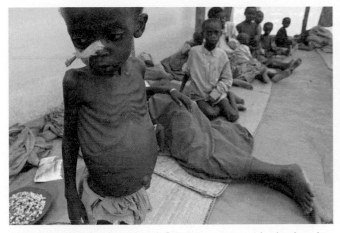

A diet with enough calories but deficient in protein can lead to kwashiorkor (usually in children)—a disorder that produces symptoms such as a bloated belly, loss of muscle mass, and lethargy. The symptoms are a result of the body breaking down its own protein in an attempt to keep vital processes functioning.

Caloric and protein deficiency can result in the condition known as wasting disease, or marasmus—again, children are most likely to suffer from this. Victims are very skinny and frail, very susceptible to infection and disease, and experience stunted growth and developmental problems (including mental retardation).

NUTRIENT	HEALTH PROBLEMS FROM DEFICIENCY	CORRECTIVE MEASURES
Vitamin A: Found in plant-based foods that contain red and orange pigments (carotenoids)	Developmental problems and immune deficiencies; acute deficiencies lead to blindness. More than 100 million children suffer from vitamin A deficiency.	Breast-feeding is helpful since it is a good source of vitamin A; supplements or fortified foods can be used, and golden rice—a variety of rice that is genetically engineered to be high in vitamin A—is also being pursued.
Iron: Meat is a good source of iron; it is also found in spinach, legumes, and whole grains.	Anemia (iron is needed by red blood cells to carry oxygen to cells). As many as 4–5 billion people suffer from iron deficiency—including half of all pregnant women and children under 5; worldwide, 2 billion people are iron deficient enough to be anemic.	Access to foods naturally high in iron or those that have been fortified with extra iron (such as bread and many cereals) helps prevent deficiencies.
Iodine: Found in seafood and in crops grown in regions with iodine in the soil	Thyroid function and fetal brain development are impaired; the WHO calls this the world's most preventable cause of brain damage. Though the situation is improving, almost a billion people worldwide have some level of iodine deficiency.	Iodized salt is an inexpensive solution to this deficiency. Conversely, too much iodine can cause health problems.
Zinc: Found in meat, dairy, and fish	Lowered immunity, stunted growth, and learning disabilities. The WHO estimates that as many as 2 billion people worldwide suffer from zinc deficiencies; it is especially prominent in poor regions where meat or fish intake is low.	Supplements or fortified foods can be used. Plant-breeding programs that increase zinc uptake by crops will help those who live on a largely plant-based diet.

Infographic 20.4 | **THE PERKS AND PROBLEMS OF THE GREEN REVOLUTION**

→ The Green Revolution revolutionized agriculture through selective breeding of the most productive plants, developing high-yield crops (when grown with inputs like fertilizer and pesticides). This increased food supplies worldwide but brought with it a new set of problems.

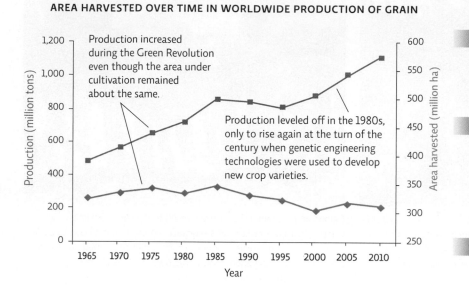

AREA HARVESTED OVER TIME IN WORLDWIDE PRODUCTION OF GRAIN

Production increased during the Green Revolution even though the area under cultivation remained about the same.

Production leveled off in the 1980s, only to rise again at the turn of the century when genetic engineering technologies were used to develop new crop varieties.

NEW PROBLEMS

A loss of crop biodiversity can result when the numerous traditional plant varieties are replaced with a few high-yield varieties. If not planted, we run the risk of losing the traditional varieties for good.

Runoff from farm fields can contaminate surface and ground water with fertilizers and pesticides. Fertilizer and pesticide use can also reduce overall soil fertility; higher productivity drains the soil of its nutrients faster and pesticides can kill needed soil organisms.

Heavy water use can lead to water shortages and soil salinization (salts are left behind when water evaporates), reducing the land's ability to grow crops.

fertilizer, and modern farming equipment—technologies that most industrialized nations had already been using for decades. They also introduced some novel technology—namely, new **high-yield varieties (HYVs)** of staple crops like maize (corn), wheat, and rice. HYVs have been selectively bred to produce more grain than their natural counterparts, usually because they grow faster or larger or are more resistant to crop diseases.

If all that were not enough, developed countries like the United States also began implementing a litany of agricultural policies—tax breaks, government subsidies, and insurance plans—that encouraged their own farmers to plant crops "fencerow to fencerow," and thus added substantially to the world's food supply.

In the short term, the effort was a huge success. The combined force of HYVs, existing technology, and new food policies resulted in a 1,000% increase in global food production and a 20% reduction in **famine** between 1960 and 1990. Today, most experts credit the initiative with the fact that we are now producing enough food to feed every one of the planet's 7 billion or so mouths.

So why are so many people still going hungry?

As it turns out, the Green Revolution has had some unintended consequences that, over time, have threatened some of its successes. For one thing, excessive use of chemical inputs and large-scale monoculture operations have contributed to the twin problems of water pollution and water scarcity, in India especially. For another, heavy reliance on just a few high-yield varieties of each crop has led to a dramatic reduction in agricultural biodiversity. This is no small matter. Without a diverse gene pool, the global food supply is much more vulnerable to pests that can't be controlled by agrochemicals: When all plants in the field are genetically identical, what kills or damages one will probably kill or damage them all. Moreover, as the old varieties fall out of use—as farmers stop producing and conserving the seeds that have been bred through traditional agriculture over thousands of years—many valuable genetic traits become in danger of being lost forever. [INFOGRAPHIC 20.4]

high-yield varieties (HYVs) Strains of staple crops selectively bred to produce more grain than their natural counterparts, usually because they grow faster or larger or are more resistant to crop diseases.
famine A severe shortage of food that leads to widespread hunger.
cash crops Food and fiber crops grown to sell for profit, rather than as food for local families or communities.

↑ Runoff from farmland near the Santa Maria River in California.

↑ A toxic crust of salt and other minerals builds up in heavily irrigated fields in Colorado.

Meanwhile, Africa, a continent plagued by hunger and largely bypassed by the Green Revolution, has nonetheless fallen prey to a different set of the Revolution's unintended consequences: a lack of *food self-sufficiency* (the ability of an individual nation to grow enough food to feed its people), a lack of *food sovereignty* (the ability of an individual nation to control its own food system), and ultimately, a lack of food security.

Here's why: As industrialization and farm subsidies enabled (mostly American) farmers to produce vast surpluses of wheat, corn, and soybeans, the global marketplace was flooded with cheap food from the developed world. Farmers in countries like Burkina Faso could not compete with such cheap and plentiful food imports—plagued as they were by land degradation (drought, soil erosion, water shortages) and armed conflict (violent clashes destroy existing crops and prevent new ones from being planted).

So they converted much of their farmable land to **cash crops** like coffee and cocoa, instead. Rather than feed local populations, these commodities are exported for profit (which is usually held by those in power). Thus a system where grain was locally produced and supplied gave way to a system where it was imported from thousands of miles away. And as they became dependent on food imports, developing countries found themselves at the mercy of forces far beyond their own borders.

And so it was that by 2008, Australian droughts and American agricultural policies had left the people of West Africa to riot in the streets for want of bread. We are indeed making enough food to feed the world. We just aren't getting it to the people who need it most.

Into this morass an additional 3 billion people will soon be born. Global population is expected to reach 10 billion by 2050; experts say that to feed that many mouths we will need to produce twice as much food as we are now producing. That will mean either farming more land or devising more production-boosting technologies. To farm more would mean clearing more forests, and thus destroying more natural habitats, species, and ecosystem services—a dire prescription at a time when the planet is already flirting with ecological disaster. How then, will we make enough food to feed the future?

> Global population is expected to reach 10 billion by 2050; experts say that to feed that many mouths we will need to produce twice as much food as we are now producing.

The next Green Revolution may be a "gene" revolution.

As Robert Paarlberg and his driver made their way up one of Uganda's countless narrow dirt roads, a collection of mud huts encircled by a patchwork of small and midsize fields came into view. The driver stopped and the pair made their way through dry bush, as a young boy tending goats spotted them from the distance and ran toward the huts to inform his kin of impending visitors. When they reached the village, Paarlberg and his driver-translator were greeted by a group of women and children: curious, welcoming, and eager to chat with outsiders.

Paarlberg, a Harvard-based political scientist, had been traveling throughout a northern swath of sub-Saharan Africa—visiting Kenya, Benin, Cameroon, and now Uganda—in an effort to understand the countries' opinions and attitudes about genetically modified crops. The women's hospitality reminded him of the farmers he'd known as a boy growing up in Indiana, but the rest of the scene contrasted sharply with such memories. In Indiana, crops were invariably bountiful: seas of grain billowing beneath an endless sky. In the evening, farmers—most of them hardy, cigar-smoking men—leaned against colossal tractors and other such equipment and bragged about the profits their crops were reaping.

In West Africa, most of the farmers were women and school-aged children (conspicuously not in school). On this particular farm, crops of maize struggled to pop up through parched soil—starved for both water and nutrients. Cassava withered under the strain of a viral infection, and a single cow suffered from untreated sores. Still, the women were warm and chatty. They did not have chemical fertilizers, they told him. Or water pumps. Or any kind of machinery. In fact, there was no electricity. And with the nearest road several kilometers away, everything they bought or sold had to be transported in or out of the tiny farming village on foot. Through the translator and the children who spoke some English, the women fretted about the weather, and crop prices, and how much grain they had in storage. These women knew absolutely nothing of the emerging technology Paarlberg had come to the region to discuss.

As their name suggests, **genetically modified organisms (GMOs)** are organisms that have had their genetic information modified in a way that does not occur naturally. Scientists have been using such organisms for decades to produce insulin, antibiotics, and other important medicines. By inserting certain genes into certain kinds of bacteria (organisms that have been altered in this way are known as **transgenic organisms**), they can coax those microbes into churning out large quantities of medically important compounds—like tiny, living, drug factories.

In the 1990s researchers began applying the same technology to food crops such as corn, rice, wheat, and soybean. By transferring genes for desirable traits (like pest resistance,

Infographic 20.5 | GENETIC ENGINEERING

↓ Genetic material can be transferred from one organism to another. This can give organisms new traits that they haven't naturally evolved but that may prove desirable. For example, the bacterium *Bacillus thuringiensis* (Bt) produces a toxin that kills many types of insect pests. If the gene responsible for that toxin is inserted into a plant's genetic code, that plant will gain the ability to produce the toxin and will be able to ward off pests itself.

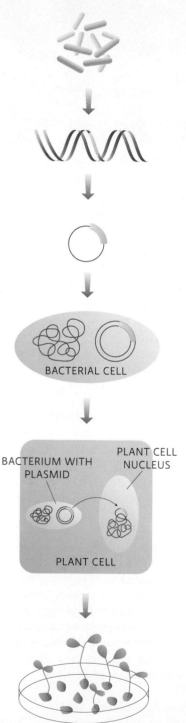

Bacillus thuringiensis (Bt) cells naturally produce pest-killing toxins.

The gene responsible for producing the toxin is isolated and many copies (clones) of the gene are made in a laboratory.

Copies of the gene are inserted into a small piece of circular DNA (called a plasmid) that can be used to deliver the gene to target cells.

BACTERIAL CELL

The plasmid containing the gene is inserted into a different bacterium.

The bacterium delivers the plasmid to a plant cell.

BACTERIUM WITH PLASMID

PLANT CELL NUCLEUS

PLANT CELL

The plasmid travels into the plant cell nucleus and incorporates into the plant's DNA, delivering the new gene.

The plant cells grow into plants and now have the Bt gene, which produces toxins; the seeds from these plants are sold to farmers.

Infographic 20.6 | **EXAMPLES OF GMOs**

TYPE OF GMO	EXAMPLES
HT (herbicide-tolerant) crops are not killed by the herbicide, so the herbicide can be sprayed on the crop/soil directly, where it will kill weeds but not the crop.	Bromoxynil-tolerant canola and cotton Glyphosate- (Roundup-) tolerant corn, cotton, soybeans Imidazolinone-tolerant wheat
Bt crops contain a gene from *Baccillus thuringiensis*, a naturally occurring bacterium that produces a toxin that kills some pests.	Bt corn and potatoes
Nutritionally enhanced food: Genes inserted can increase the amount of a particular nutrient or allow the crop to produce a nutrient it would not normally produce.	Golden rice—produces more vitamin A than average rice
Genetically modified animals are being developed for the food supply.	Salmon that grow faster Pigs that produce more omega-3 fatty acids (a healthy fat)

drought tolerance, and increased nutrient production) from one species to another, they have created a new suite of genetically modified food—plants that can grow more plentifully and thrive in a wider range of habitats than they ordinarily would. [INFOGRAPHIC 20.5]

In the United States, more than 75% of processed food contains GMOs, including 85%–90% of our corn, soybeans, and cotton. Most of these crops have an herbicide-tolerance (HT) gene added that enables them to withstand huge doses of herbicides (chemicals that kill weeds). This trait enables farmers to douse the field with herbicide and kill the weeds without threatening their harvest. Other crops, known as Bt crops, have been engineered to better resist pests; they contain a gene from *Bacillus thuringiensis*, a naturally occurring bacterium that produces a toxin that kills some insect pests. Scientists are also working to add genetically modified animals to our food supply. AquaBounty has developed a genetically modified Atlantic salmon that grows much faster than normal (thanks to genes from the larger Chinook salmon). The company is currently seeking approval from the Food and Drug Administration to bring the fish to market. If approved, this would be the first genetically modified (GM) animal food product sold in the United States. [INFOGRAPHIC 20.6]

genetically modified organisms (GMOs) Organisms that have had their genetic information modified to give them desirable characteristics such as pest or drought resistance.
transgenic organism An organism that contains genes from another species.
Green Revolution 2.0 Programs that focus on the production of genetically modified organisms (GMOs) to increase crop productivity or create new varieties of crops.

Paarlberg and others (including Microsoft founder and famed philanthropist, Bill Gates) believe that GM crops will go a long way toward helping African farmers achieve food self-sufficiency. In fact, GMOs form the basis of what some farmers and scientists like to think of as **Green Revolution 2.0**, or the *Gene* Revolution—the next battle between human hunger and innovation, this time waged in Africa as well as Asia, and against not only hunger but poverty and sickness as well. "When you put the right tools in farmers' hands, the results can be magical," Gates said during a speech in Rome, Italy, in February 2012. "A submergence-tolerant seed capable of surviving in flood waters can grow into a rice surplus that [allows parents to afford to take] a sick child to the doctor....And when that child grows into a healthy and educated young mother, the future is bright."

Of course, for the people of Africa to benefit from this new technology, they would have to first be made aware of its existence and then be persuaded to use it. And as Paarlberg was discovering in his travels, neither of those would happen easily. Not only were most local farmers completely unaware of GM technology, but government officials, who did know about it, were exceedingly wary.

Concerns about GMOs trigger strong debate.

To be sure, genetic engineering is a decades-old technology. But genetically modified food is unlike any of the other products that have been produced with it. With medications, the transgenic organism remains confined to a flask in a lab, and the commercial product is identical (or nearly so) to its naturally produced counterpart.

Genetically modified food is itself the altered item; GMO crops are out in the real world, growing and sharing genes with other organisms, and being eaten by both animals we are raising for food and directly, by humans—many of whom are very uneasy about the prospect.

Since the first genetically modified foods appeared in the early 1990s, a vocal contingent of activists—mostly in the United States and Europe—have rallied against such products—dubbing them "Franken Foods," and strongly contesting their safety. What if humans proved allergic to some of the proteins whose genes were being spliced in, they asked? What if negative health effects took years to manifest? Today, even as GM crops permeate the U.S. food supply, other countries (including China and much of the European Union) continue to fiercely resist the technology.

One enduring concern is that the genes introduced into genetically modified crops could be passed to wild plants through cross-pollination—that is, they could escape into the natural world and be incorporated into other plants for which they were not intended. For example, if a herbicide-tolerance gene were accidently transferred to a weed species, it could enable the weed to grow more aggressively and outcompete other plants, including our crops. This has already happened in the United States. So far, 16 weed species have acquired a gene for herbicide-tolerance. These so-called super weeds, including giant ragweed and pigweed, can be found in 22 states and can tolerate all the herbicide a farmer can spray. They take over entire fields, stop combines, and are tough to clear by hand.

Another fear is that crops engineered to resist pests could inadvertently repel beneficial insects, or that they might give rise to a new population of Bt-resistant pests much more quickly than traditional pesticide spraying would. The rise of secondary pests (those not affected by Bt) is also a concern in some cases. For example, several studies have shown that the amount of pesticides applied to Bt cotton was as great as the amount of pesticide used on traditional cotton, because the elimination of Bt-susceptible pests enabled other Bt-resistant pests to thrive.

But what opponents seem to worry most about is the prospect of putting even more of our food supply under the control of a few multinational corporations. In the United States, companies like Monsanto have been known to tightly guard their GM seeds. Because they are patented, farmers are not permitted to save seeds from one year to the next, but instead must repurchase them year after year. Worse yet, when GM crops grown by one farmer have accidentally (that is, naturally) cross-pollinated another farmer's field, the second farmer has been held liable and forced to pay for the seeds, even if she didn't plant them—even if she doesn't want to grow GM crops. "Food security in private hands is no food security at all," U.S. Senator Tom McGovern has said. "Because corporations are in the business of making money. Not feeding people."

Proponents argue that in Africa especially, where individual countries lack the capacity for large-scale research and development, private investment is essential to agricultural development. "The country programs, agencies, and research centers don't have expertise," Gates said. "And they don't have time to build it from scratch." If we are ever to develop technological solutions to the challenges of developing world agriculture, he says, we will have to rely on the expertise of the private sector. [INFOGRAPHIC 20.7]

Gates and others point out that technological innovation is exactly how developed countries like the United States turned Dust Bowl—era food shortages into vast surpluses in less than a generation. It is unfair, they say, to tell impoverished countries to forego the technologies that we ourselves have benefitted from, especially while the people in those other countries are still starving. "This postmodern resistance to agricultural science makes considerable sense in rich countries, where science has already brought so much productivity to farming that little more seems needed," Paarlberg says. "It becomes dangerous, however, when exported to countries in Africa where farmers remain trapped in poverty."

Meanwhile, in the shadows of this global debate, many African farmers have been devising their own solutions to food insecurity.

It will take a combination of strategies to achieve global food security.

The central plateau of Burkina Faso is marked by poor soils, low crop yields, and very little rain. In many places, the land is so parched that it has crusted over into what local farmers call *zippelle*: hard dry cake. To grow anything here takes a special kind of ingenuity. In the 1980s, with the help of the FAO, farmers revived an ancient strategy known as the *Zaï* pit. During the dry season (between November and May), farmers dig thousands of small pits—12,000 to 25,000 per acre, each about 30 centimeters (12 inches) wide and 15 centimeters (6 inches) deep. Once dug, the pits are filled with roughly a pound of organic matter. After the first rainfall, the organic layer is covered with a thin layer of soil and seeds are placed in the middle of the pit. The top of the pit is then ridged, so that when the rain does fall, it accumulates in the pit and hydrates the seed.

Zaï pits are not the only technology being employed in this landlocked, desert country. Researchers and farmers have also worked together to adapt cutting-edge micro-dose fertilization technology to suit local needs. By applying small and precise quantities of fertilizers close to each seed at planting, they have greatly enhanced fertilizer efficiency: Wheat crops grown in Burkina Faso now require one-tenth of the fertilizer used on their U.S.-grown counterparts; corn crops require one-twentieth.

In addition to Zaï pits and microfertilization, trees have been planted to retain soil and stave off desertification, and livestock manure has been conserved and applied to fields. And cooperative groups have been established throughout the region to manage village cereal banks and community wells. The result has been an increase in grain production that matches some of the first Green Revolution's successes in Asia, and belies the despair of the 2008 food riots. "So much has been written about the disappointments in African agriculture that it is easy to overlook the successes," says Steve Wiggins of the Overseas Development Institute in London. "In some parts of the continent and for particular crops and activities, there have been veritable booms in farming."

Infographic 20.7 | **DO THE TRADE-OFFS OF GMOS MAKE THEM WORTH PURSUING?**

↓ As with other environmental issues, the advantages of using GMOs must be weighed against the disadvantages when deciding whether to use GMO crops (or animals). The conclusion may differ in different areas, depending on available resources and the needs of the people in that region.

GMO TRAIT | **ADVANTAGES** | **DISADVANTAGES**

HERBICIDE TOLERANCE

Weeds are controlled by simply spraying herbicide.

- More herbicides are used since they do not harm the crop (potential water pollutant)
- The farmer is locked into using the herbicide the crop can tolerate.
- The company producing the seed and herbicide has an unchallenged market (no competitors), so prices can be high.

PEST RESISTANCE

Less pesticide may be used because crops can defend themselves from pests.

- The pest-resistance trait could be transferred to weeds.
- The crop may repel beneficial insects like pollinators.
- Pesticide-resistant pest populations may increase.
- Secondary pests (those not affected by the GMO trait) may increase.

CORPORATE FUNDING

Corporations have absorbed the tremendous cost and time of developing useful GMO products, saving taxpayers money.

- The price of the seeds goes up, reflecting the corporation's time and financial investment; farmers must purchase new seeds every year (as opposed to harvesting them for future use).
- The seeds are patented, allowing the corporation to sue farmers whose fields are accidently pollinated by GMO plants.
- If cross-pollination accidently contaminates fields of organic produce, the contaminated crop no longer qualifies as organic.

UNDERSTANDING THE ISSUE

CHECK YOUR UNDERSTANDING

1. What is the main cause of world hunger today?
a. Overpopulation
b. Underproduction of food
c. Inadequate distribution of food to those in need
d. Production of food that is unhealthy

2. Food security is:
a. having sufficient, safe, and nutritious food freely available to everyone in the population.
b. having enough money to buy sufficient, safe, and nutritious food.
c. all people having access to enough calories to survive.
d. all people, at all times, having physical, social, and economic access to sufficient, safe, and nutritious food.

3. Which of the following is not an example of a genetically modified organism (GMO)?
a. Golden rice
b. High-yield variety (HYV) corn
c. Bt cotton
d. AquaBounty salmon

4. Malnutrition is a leading cause of illness and death around the world, and can be caused by:
a. too few calories.
b. too many calories.
c. insufficient micronutrients.
d. All of the above.

5. Proponents of a "Gene Revolution" believe that:
a. humans should be genetically modified to require less food.
b. genetically modified (GM) crop plants are useful, but GM animals are not.
c. GMOs are needed to achieve global food security.
d. GMOs will be useful in developed countries, but not in developing countries.

6. Critics say that GMOs are dangerous because:
a. pesticide- or herbicide-resistant genes may migrate to other species and create "superweeds."
b. beneficial insects as well as crop pests are killed.
c. they place too much power in the hands of large corporations.
d. All of the above.

7. The Green Revolution:
a. increased world food supplies but introduced new problems.
b. was based on traditional, locally adapted crop species.
c. used crop plants that required less fertilizer and pesticides.
d. created food security for the global population.

WORK WITH IDEAS

1. Describe the conditions that brought about the 2008 global food crisis. How did the Green Revolution contribute to the crisis?

2. Why are people still starving, when experts say we currently have enough food for everyone? What are possible solutions to this problem?

3. What is malnutrition and what conditions and diseases result from it?

4. What did the Green Revolution accomplish and how did it do so? Discuss some of the unintended consequences of the methods used in the Green Revolution.

5. Explain how genetic engineering is used to create plants or animals with more desirable traits.

6. Compare the high-tech agriculture methods (HYVs and GMOs) with the low-tech suggestions given in the chapter. What path do you think a nation like Burkina Faso should pursue to achieve food security? Support your answer.

ANALYZING THE SCIENCE

The Food and Agriculture Organization (FAO) of the United Nations monitors world hunger, and has set a goal of decreasing the percentage of underfed people to 10% of the world's population by 2015.

NUMBER AND PROPORTION OF UNDERNOURISHED PEOPLE IN THE WORLD

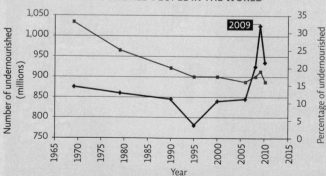

INTERPRETATION

1. Describe in one sentence what the graph shows about change over time of undernourishment. Why are both lines shown here?

2. What accounted for the decline in the number and proportion of undernourished people from 1970 to 2000?

3. What happened to the percentage of underfed people between 2008 and 2009? What do you think led to this change?

ADVANCE YOUR THINKING

4. Do you think the FAO will meet its goal of halving the percentage of underfed people from 1992 to 2015, from 20% to 10%? Why or why not?

5. In looking at both the number and percentage of underfed people, do you think there has been a complete recovery from the 2008 world food crisis? Why or why not?

6. Extrapolate these lines into the future, based on whether the Gene Revolution is successful and whether it is not. Which scenario do you think is more likely? Explain your reasoning.

EVALUATING NEW INFORMATION

Oxfam is an international not-for-profit organization dedicated to addressing issues of poverty and injustice around the world.

Visit the Oxfam website (www.oxfam.org/en).

Evaluate the website and work with the information to answer the following questions:

1. Is the website a reliable information source?
   a. Does Oxfam give supporting evidence for its claims?
   b. Does the organization give sources for its evidence?
   c. Why was Oxfam originally founded back in 1942 and what is its mission today?

2. Select the upper tab entitled "Campaigns" and then select the link on the left for "Agriculture." Select the link for "Global Food Prices in 2011: Questions and Answers."
   a. Do you agree with Oxfam's assessment of the main causes of increasing food prices, and the winners and losers? Explain.

   b. Identify a claim Oxfam makes and the evidence the organization gives in support of this claim. Is it sufficient? Where would you attempt to find more evidence to support or refute its claim?

3. Watch the videos about aid on Oxfam's website: www.oxfam.org/en/video/2010/does-aid-work and www.oxfam.org/en/video/2010/good-aid. Summarize the arguments made about the success of aid. Which of the reasons do you agree with most strongly and why? Which do you find less compelling?

4. Oxfam has proposed a "Robin Hood tax" (www.oxfam.org/en/campaigns/health-education/robin-hood-tax) on big bank transactions that would help support extremely poor communities and people harmed during the 2008 financial crisis. Summarize its proposal and list some of the people supporting it. Do you support it? Why or why not?

5. Do a search on the Oxfam website for GMO foods. What is Oxfam's position on GMOs?

MAKING CONNECTIONS

FOOD INSECURITY CLOSE TO HOME

Background: While food insecurity is most common in developing countries, especially in Asia, there are still many people in the United States living with food insecurity, many of them children. The U.S. Department of Agriculture has defined four classes of food security:

High food security	No reported indications of food-access problems or limitations
Marginal food security	One or two reported indications—typically of anxiety over food sufficiency or shortage of food in the house. Little or no indication of changes in diets or food intake
Low food security	Reports of reduced quality, variety, or desirability of diet. Little or no indication of reduced food intake
Very low food security	Reports of multiple indications of disrupted eating patterns and reduced food intake

Based on these categories, in recent years 11%–12% of American households experienced low or very low food security. However, the 2008 financial crisis caused this number to increase. In 2010 (the most recent data) 9.1% of American households had low food security and 5.4% had very low food security, for a combined total of 14.5%. Of all households, some are more likely to be food insecure: homes with young children (20%), homes with a single mother (35%) or single father (25%), homes with incomes below the poverty line (40%), and African American (25%) and Hispanic (24%) households. Food insecurity is more prevalent in metropolitan areas (17%) and in southern states (16%).

Case: Choose either a household type, an area, or a specific location as your target group. For example, you could focus on homes with young children, or inner-city homes, or your own city or town. Develop a proposal to decrease food insecurity in your target group, using three of the following strategies:

1. Educate and provide logistical support to develop home or community gardens.

2. Work with local retailers to increase fresh food options available to those using Supplemental Nutrition Assistance Program (SNAP; formerly known as food stamps).

3. Establish a public school–community cooperative to broaden healthy food options in school breakfast and lunch programs (perhaps focusing on free and reduced-price lunch programs).

4. Increase business and private involvement in, and donations to, local food banks.

5. Another idea of your own.

Research at least three of these strategies and write a report proposing a course of action to alleviate food insecurity in your target group. Your report should include the following:
a. The current level of food insecurity in your target group, with data in either a graph or table format.
b. An analysis of the advantages and disadvantages of each strategy, including:
   - A discussion of the benefits of each strategy, including economic, social, and health implications
   - A discussion of the possible costs and consequences of each strategy. This could include not only investments of time and money, but unintended social costs.
   - Difficulties in implementing and maintaining the proposed strategy
c. If you could only implement one of these strategies, given limited time and money, which would it be? Explain your reasoning.

Feedlot in the outskirts of Bakersfield, CA.

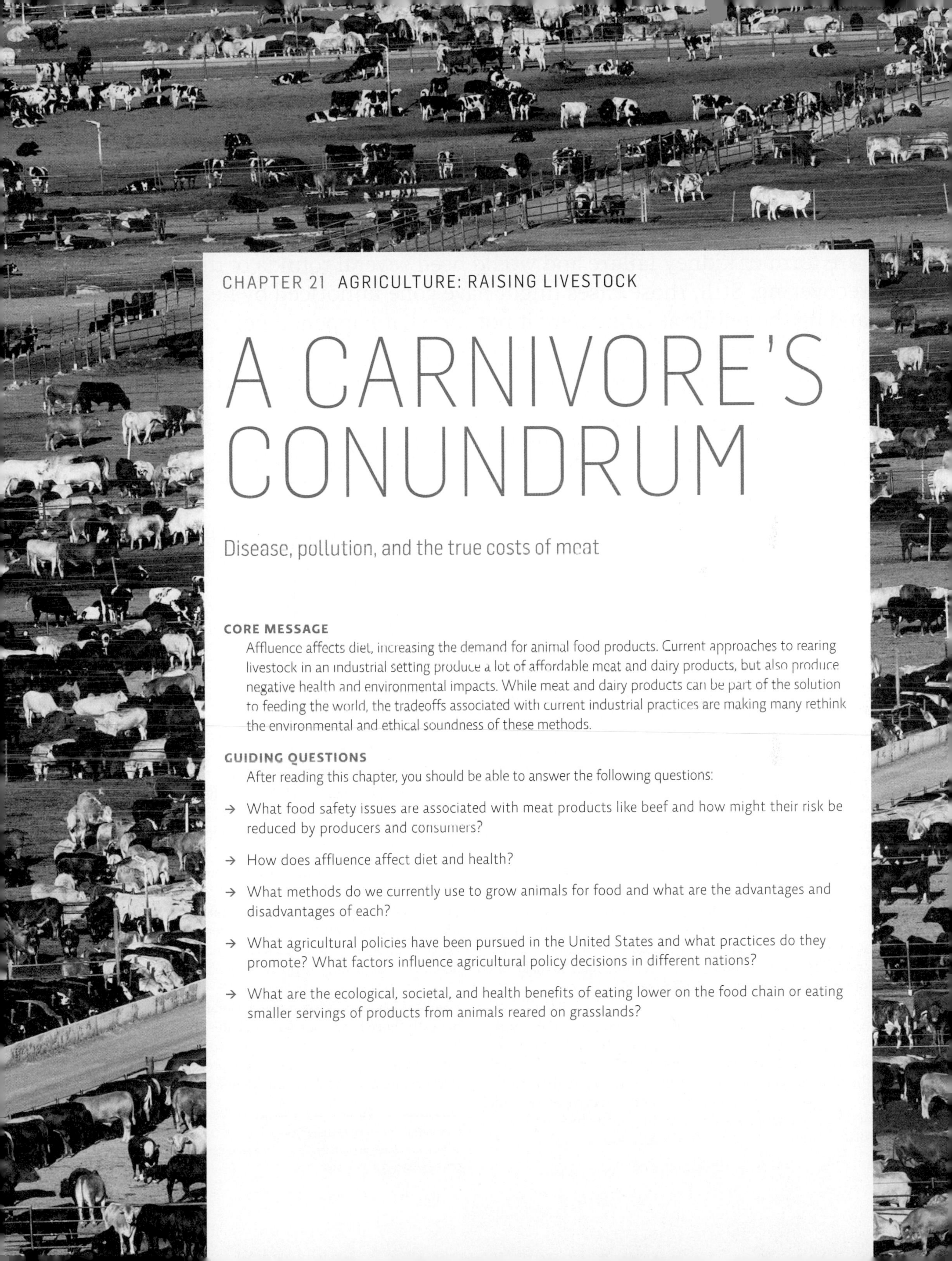

A CARNIVORE'S CONUNDRUM

Disease, pollution, and the true costs of meat

CORE MESSAGE

Affluence affects diet, increasing the demand for animal food products. Current approaches to rearing livestock in an industrial setting produce a lot of affordable meat and dairy products, but also produce negative health and environmental impacts. While meat and dairy products can be part of the solution to feeding the world, the tradeoffs associated with current industrial practices are making many rethink the environmental and ethical soundness of these methods.

GUIDING QUESTIONS

After reading this chapter, you should be able to answer the following questions:

→ What food safety issues are associated with meat products like beef and how might their risk be reduced by producers and consumers?

→ How does affluence affect diet and health?

→ What methods do we currently use to grow animals for food and what are the advantages and disadvantages of each?

→ What agricultural policies have been pursued in the United States and what practices do they promote? What factors influence agricultural policy decisions in different nations?

→ What are the ecological, societal, and health benefits of eating lower on the food chain or eating smaller servings of products from animals reared on grasslands?

On July 5, 2007, an 18-year-old girl in central Pennsylvania fell suddenly and violently ill: diarrhea, abdominal cramps, vomiting. Three days later, a New Jersey man developed the same constellation of symptoms, and in mid-August, so did a 15-year-old girl in Pembroke Pines, Florida. The Florida case was the worst of the three; the patient was hospitalized with a rare form of kidney failure and would need several rounds of dialysis before recovering. Still, those cases might have gone unnoticed by health officials and by the public at large, were it not for what happened next.

On September 7th, when an Albany man also fell ill, epidemiologists working for the New York State Health Department analyzed a hamburger patty recovered from his refrigerator and found that it contained a particularly virulent strain of the bacterium *E. coli*—a strain known as O157:H7.

E. coli O157:H7, or just *157*, as it is called by public health experts, tends to reside in the colon of some farm animals, namely cattle. The cattle themselves are immune to the bacterium, and normally pass it in their feces without incident. But sometimes that feces gets on the animal itself, and sometimes it stays there, even as the animal is slaughtered and made into steak and hamburger that we then eat.

In humans, *157* can cause bloody diarrhea, dehydration, and even kidney failure. The sickness usually clears up within a week in adults. But it can be deadly in people with compromised or underdeveloped immune systems, including infants, young children, and the elderly. In all, 3%–5% of cases prove fatal.

That potential for fatality meant health officials were in a race against the clock. The New York Health Department notified Topps, the company that had processed the meat, who then recalled some 150,600 kilograms (332,000 pounds) of beef from supermarket shelves, restaurant kitchens, and freezers across the country. On September 30th, that recall expanded to include 9.8 million kilograms (21.7 million pounds) of beef—enough to make one hamburger for every single U.S. resident.

As supermarket and restaurant owners across the country began pulling the company's products from their freezer shelves, health officials began the tedious work of tracing the tainted beef back to its origins—trying to discern, from company documents and good detective work, which of dozens of suppliers the offending cattle had come from.

◉ WHERE WAS THE 2007 *E. COLI* OUTBREAK?

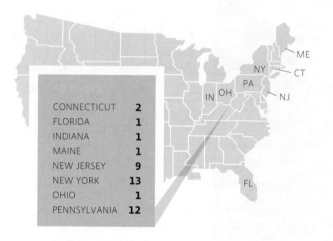

CONNECTICUT	2
FLORIDA	1
INDIANA	1
MAINE	1
NEW JERSEY	9
NEW YORK	13
OHIO	1
PENNSYLVANIA	12

Of course, by then, most of the suspect meat had already been consumed. The final tally: 40 people sick, across eight states, 21 of them hospitalized. "The real numbers were probably larger than that," says Paul Ebner, a professor of animal sciences at Purdue University, "because not every case would be confirmed through proper diagnostics."

Still, those confirmed cases were enough to put Topps, one of the country's largest meat-processing plants, out of business. On October 4th, a class-action suit was filed against the company. On October 5th, embattled by negative publicity and faced with hundreds of millions of dollars in lost profit, the processing plant closed down.

The Beef Recall of 2007 made national headlines, but beneath those headlines were debates that had already been raging for years and continue to this day. Some cattle farmers and beef industry representatives insist that despite a few isolated incidents, our food supply is

---

Concentrated Animal Feeding Operation (CAFO) Many meat or dairy animals are reared in confined spaces, maximizing the number of animals that can be grown in a small area.

safer than it's ever been. Meat is part of a healthy diet, they argue, and industrial-scale production—done in operations known as **Concentrated Animal Feeding Operations (CAFOs)**—allows meat suppliers to raise more beef with fewer animals (more meat per animal); those high-density operations are a more efficient way to rear livestock because they require less grazing land per animal.

Critics, on the other hand, contend that the entire system through which we grow, process, distribute, and consume food is wildly incompatible with human health and environmental sustainability. And food safety issues are just the tip of the iceberg. Our overconsumption of meat has contributed to a vast array of social ills, they say—not only epidemics of obesity and heart disease, but also air and water pollution in regions surrounding crowded animal farms, and biodiversity loss and greenhouse gas emissions across the globe. [INFOGRAPHIC 21.1]

At the heart of the issue are fundamental questions whose answers will impact not only the health of our citizens but also the state of the planet we live on.

Affluence influences diet.

The past few decades have been witness to an unprecedented rise in global affluence. In Africa, a continent still plagued by war, famine, and poverty, consumption

---

Infographic 21.1 | *E. COLI*—**JUST THE TIP OF THE ICEBERG**

↓ Food safety and health issues are real, but are not the only problems associated with the rearing and slaughtering of livestock for food. The industrial method of raising large numbers of animals in crowded conditions has tremendous environmental impacts and raises ethical questions about the way the animals are treated.

Healthy diet
Meat is a good source of protein and can be part of a healthy diet, but the overconsumption of meat is linked to health problems such as obesity and heart disease. However, access to affordable meat could help people in areas where protein-deficiency malnutrition is a problem.

Food safety
Meat can be contaminated during processing; produce can also be exposed to pathogens via contaminated air or water (even ice used in shipping). Contaminated products can end up in processed foods like cookie dough or ice cream.

Air and water pollution
CAFOs and poorly managed pastures are major sources of air and water pollution. If not addressed, this environmental degradation will not only affect our health and that of other species, it will decrease the land's ability to grow our food.

Livestock

Habitat destruction and biodiversity loss
The clearing of land for pasture and cropland (to grow crops fed to animals) is a major cause of habitat destruction worldwide, contributing to species endangerment. This in turn reduces the ability of the area to support the very crops and livestock the land clearing made way for.

Ethical issues
Rearing animals in confined, unhealthy, and unnatural conditions raises ethical questions. Fortunately, you can vote with your dollars—only purchase food raised in a manner that is consistent with your own views on the ethical treatment of animals.

Antibiotic resistance
Most antibiotic use in the United States goes to livestock. This use has been linked to the emergence of antibiotic-resistant strains of bacteria, reducing the effectiveness of those medicines in people.

→ Although hamburgers are more often associated with sickness related to *E. coli* contamination than steak is, fecal matter can contaminate the surface of a steak. Cooking the steak (even to rare temperature) will likely kill the bacteria, but if that same steak is ground up, the *E. coli* is evenly distributed throughout the meat and—if the burger is not cooked completely—the consumer can become infected. This is why the U.S. Department of Agriculture recommends that all ground meat be cooked to an internal temperature of 160°F.

↑ An inspector examines fresh beef sides for potential contaminants and health issues before the meat is cooled for 42 hours at a Cargill meat packing plant in Fort Morgan, Colorado. Cargill is participating in trials of a cattle vaccine against *E. coli, O157:h7.*

of resources has doubled since the start of the 20th century; in developed nations, like the United States, affluence has increased sixfold. As we learned in previous chapters, greater affluence means a greater ecological footprint. Put another way: People with more money consume more resources—not just more material goods, but also more food.

Of course, wealthier people don't just eat more food. They eat more of a certain kind of food—specifically, more animal products, like meat and eggs. This is not entirely a bad thing. In developing nations especially, meat and dairy products mean more and better sources of protein. But while meat and dairy production has increased in both developing and developed nations in recent years, per capita consumption is still much higher in developed nations (as much as 3 or 4 times higher, by some estimates). In fact, most health experts agree that while meat can be part of a healthy diet, Americans especially are eating too much of it—on average, 125 kilograms (276 pounds) per person per year. That overconsumption has been associated in many studies with a wide range of health problems—from cardiovascular disease and diabetes to gout and some cancers. Researchers at the

Harvard School of Public Health have found that as red meat consumption increases, so does one's risk of dying. Replacing some red meat with fish, poultry, or non-meat sources of protein decreased risk by 7%–19%.

It isn't just the overconsumption of meat that is the problem—it's overconsumption in general. A diet high in calories, fat, and sugar, coupled with declines in physical activity, is increasing rates of obesity worldwide. For a recent example of how much affluence influences meat consumption, one need only look to China. In tandem with its economy, the country's annual per capita meat consumption tripled between 1985 and 2005, going from 20 kilograms (44 pounds) per person per year to 60 kilograms (132 pounds) per person per year. [INFOGRAPHIC 21.2]

To accommodate such growing appetites, we have dramatically altered the way we rear livestock. Instead of traditional mixed farms, where crops and livestock are rotated, or grown side by side, we now rear animals in CAFOs, factorylike operations where livestock are densely packed and rapidly grown. Since about the mid-1900s, high-density operations have been the most common method of raising cattle in the United States. Today,

Infographic **21.2** | **AFFLUENCE AFFECTS DIET AND HEALTH**

↘ Worldwide, as income rises, so does the consumption of meat. Eating meat and other animal proteins can improve the health of those who are protein deficient, but a diet high in fat and calories is linked to serious health problems. Though meat is not the only dietary factor linked to these health problems, eating large amounts of meat is believed to be a contributing factor.

PER CAPITA GDP AND MEAT CONSUMPTION BY COUNTRY, 2005

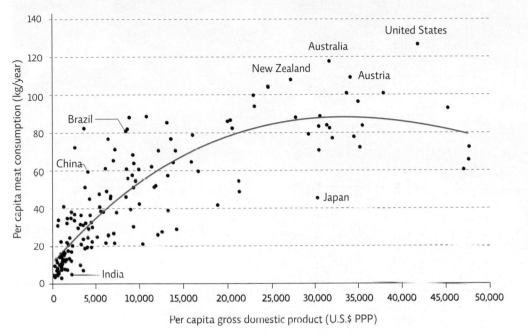

Dollar figures are 2005 U.S. dollars; "PPP" stands for "purchasing power parity," a measure that allows a comparison of income between countries, accounting for different costs of living.

RELATIONSHIP BETWEEN RED MEAT CONSUMPTION AND MORTALITY

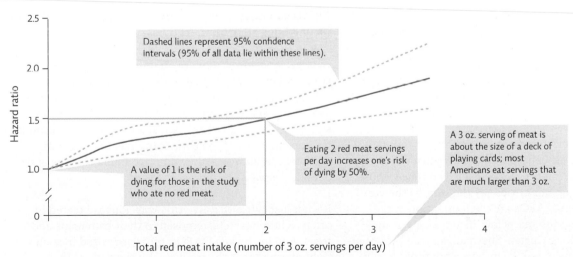

↗ Many studies show strong links between red meat consumption and diabetes, cardiovascular disease, and cancer. An Pan and colleagues from the Harvard School of Public Health evaluated the link between red meat (beef, pork, and lamb) in the diet and mortality (death) in participants enrolled in a long-term (22-year) study that followed the health and diet of more than 50,000 men. Results were adjusted for other variables that affect mortality, such as age, smoking, physical activity, body size, family history of disease, and other dietary choices. The data show that as daily red meat consumption goes up, so does one's risk of death. Similar results were obtained from a similar study which followed more than 100,000 women from 1980–2008.

↑ Cattle herded into a pen in Lubbock, Texas.

poultry (chicken and ducks) and pigs are also predomi-nantly raised in CAFOs. And the practice is spreading to other countries.

So far, CAFOs have enabled us to grow many thousands more food animals than Mother Nature would ever allow.

CAFO managers say that's a good thing. Such dramatic increases in production have not only brought more meat to people of rising affluence, but have also made it cheaper and thus more accessible to all people, even those whose income is not rising. But critics say that that is the very crux of the problem. CAFOs are the foundation of the cheap fast food that too many low-income Americans depend on, says Doug Gurian-Sherman, a senior scientist at the Union of Concerned Scientists, a nonprofit group that lobbies for environmentally responsible food poli-cies. It's these kinds of operations that make McDonald's cheaper than fruits and vegetables. And that economic reality has brought with it serious consequences, not only for human health, but also for environmental safety and animal rights.

CAFOs can raise a large number of animals quickly, but incur a huge environmental cost.

In large, long, low-slung grey buildings on the edge of rural Indiana, tens of thousands of cattle stand nearly motionless, crammed together, mooing and occasionally jostling for space. There are no windows, and thus no sunlight. They moo and they jostle and by many accounts they spend much of the time standing ankle deep in their own feces. What they don't do in many cases is go outside. Or walk around.

It's not just cattle that live this way. Pigs are reared in often tighter quarters, exposed to their own urine and feces. And in chicken CAFOs, birds are raised in such tight quarters that the farmers clip the ends of their beaks off to prevent them from pecking each other to death. Animal rights activists have protested vehemently against the cruelty of such a life, but animal rights are not the priority here. The priority is to get as many animals as possible as fat as possible in as little time as possible. Movement is antithetical to all three. And for cattle, so is a diet of grass.

CAFO cows start their lives on pasture in cow-calf operations. Once delivered to the pens, at around 14 months old, the cattle begin a slow transition from a diet of hay or grass to a diet supplemented with soybeans and grains (including corn, sorghum, and oats). It's important to do this slowly. Because cattle are not naturally adapted to digesting grain, switching their diets too quickly can make them sick. When done properly, the switch to grain (and the lack of movement) enables the animals to gain about 0.9 kilograms (2 pounds) a day. Even then, they may suffer health problems such as liver abscesses or intestinal tract damage. Animals receive medications to combat these diet-imposed health problems.

Some 70% of the cattle that eventually make their way to American dinner plates are "finished"—meaning they spend the last 3 to 5 months of their lives—here, in pens like these.

There are several advantages to rearing livestock in this way—the biggest one by far being the sheer volume of animals that can be produced. According to the Union of Concerned Scientists, CAFOs now produce more than half of all our animal food, even though they account for just 5 percent of all animal farms. The reasons for this are twofold: Not only can a large number of animals be grown on relatively little land, but those animals can also be "turned around" (brought from birth to market) with amazing speed—just under 2 years, compared with several years for the average grass-fed or pen-raised cattle. In fact, if there's one thing both supporters and opponents agree on, it's this: We'd be hard pressed to produce as much meat and chicken and pork, or to produce it as cheaply as we do now, without these types of operations.

The intense mechanization of feedlot farming means that the animals' diet and environment can be better controlled than it would be on more traditional open-pasture or pen farms. And because animals are so concentrated and confined, it is easy to harvest their manure and sell it as fertilizer. The amount of land used directly by the animals—and thus the amount of land that must be cleared of natural vegetation or otherwise denuded—is minimized on a CAFO.

But a growing number of critics say that those advantages in space and production volume come at a huge cost to the environment and to human health.

feed conversion ratio The amount of edible food that is produced per unit of feed input.
water footprint The amount of water consumed by a given group (person, population) or for a process (raising livestock).

As environmentalists are quick to point out, the grain and soybean used to feed all those livestock would feed far more people than the meat we harvest from the animals actually does. The reason for this has to do with the **feed conversion ratio**—how quickly and efficiently any given animal converts the food it eats into body mass. Cattle have a high feed conversion ratio, meaning it takes a lot of feed to produce a gallon of milk and even more to produce a pound of meat. CAFO-reared cattle do have a better feed conversion ratio than do grass-fed cattle; they move around less, gain weight more quickly, and are slaughtered sooner (they burn fewer calories in day-to-day maintenance because they are less active and don't live as long). But they still have a worse feed conversion ratio than either chickens or pigs.

These days, nearly 75% of the corn produced in the United States is used as animal feed—and about half of that goes to cattle. All told, that's about 10 billion bushels of corn, which requires roughly 80 million acres, or one quarter of all U.S. cropland, to grow. Growing all that grain and soybean as feed crops presents its own problems. Though the cattle themselves do not require as much land for grazing, by some estimates, it costs us between 7 and 14 hectares (17–34 acres) worth of grain crop for every ton of beef (about 4 cows) we produce. The food that could be grown on this land would feed 5–10 times more people than the beef it would produce could feed.

These days, nearly 75% of the corn produced in the United States is used as animal feed—and about half of that goes to cattle.

The industrial methods used to raise those feed crops come at a hefty ecological cost that includes habitat loss, water depletion, pollution from field runoff and pesticide application, and soil erosion from cultivated fields (see Chapter 22). The **water footprint** of raising animals this way is also high: We spend, for example, 7 gallons of water to produce a pound of chicken, 23 gallons for a pound of pork, and a whopping 90 gallons for a pound of beef. Much of the water used in a CAFO goes to manure handling. [INFOGRAPHIC 21.3]

Containing both the water and the manure is one of the biggest challenges CAFO operators face. Some manure is collected and sprayed onto cropland, where it acts as a natural fertilizer (though this is risky because the manure can carry pathogens). But most CAFOs produce more manure than can be applied to local fields. Indeed, some of the largest CAFOs produce more waste than a large U.S. city.

Where does it all go? To multi-acre reservoirs known as lagoons. Ebner (from Purdue University) compares these to human wastewater facilities and says that they allow for microbial digestion of the waste. But critics say that they are the equivalent of a city with a population of 3 million people (that's about twice as large as Philadelphia) sending all of its raw sewage to a giant open pit. As the waste breaks down, a variety of gases are produced. When the waste is stirred or agitated, a practice commonly performed before the lagoon is emptied, enough hydrogen sulfide gas can be released to reach toxic levels for nearby workers or animals. And when the lining to one of the lagoons cracks, or when one of the pumps feeding the lagoon springs a leak, the results can be disastrous.

In 1995, when a manure lagoon at a swine operation in North Carolina failed, 95 million liters (25 million gallons) of manure went racing into the nearby New River, killing 10 million fish and closing nearby coastal areas to shell-fishing. Area residents described the stench of manure, urine, and rotting fish—which they endured for weeks—as almost unbearable. In 2007, North Carolina banned the construction of new lagoons at swine CAFOs, favoring technical solutions that process the manure and collect methane from the waste for electricity generation.

Infographic 21.3 | **GROWING LIVESTOCK: FEED AND WATER NEEDS**

↓ The feed conversion ratio shows how much edible food is produced from the food an animal is fed. We never see 100% conversion because the animal has already used most of the energy it has consumed in its life in day-to-day activities; however, some species are better at converting feed to edible food than others are. It may also be surprising how much water is needed to produce meat and dairy products; much of that water goes to grow the feed.

PRODUCTION OF 1 POUND OF MEAT OR DAIRY REQUIRES:

BEEF

6–10 pounds of feed

90 gallons of water

PORK

7 pounds of feed

23 gallons of water

CHICKEN

3 pounds of feed

7 gallons of water

EGGS

4.5 pounds of feed

2.5 gallons of water

MILK

1.1 pounds of feed

30 gallons of water

According to Shalamar Armstrong of Purdue University, and the Indiana Department of Environmental Management, *E. coli* contamination from feedlots has impaired more than 6,000 miles of river. In addition, both nitrogen and phosphorus from manure spills or runoff can cause cultural eutrophication (see Chapters 18 and 22) in lakes, streams, and coastal areas that can result in hypoxic waters; high nitrate levels are also directly toxic both to humans and to aquatic life. Groundwater can be affected as well; wells close to CAFO operations are more likely to be contaminated with dangerous levels of nitrate than those further away.

Air pollution is another big problem; not only are pollutants released at every stage of production, but so many hooves trampling the dirt creates dust, which in dry weather can reach high enough concentrations to compromise human and animal health alike. Odor is a problem, too. "It's like a city's worth of poop," says Kim Ferraro, an environmental lawyer who specializes in CAFO-related issues, and who is representing one Indiana family in a lawsuit against a neighboring CAFO operator. "My clients were literally driven from their homes by the stench."

Another environmental concern is the high **carbon footprint** of CAFOs. Clearing so much land to grow feed crops reduces the amount of carbon that would normally be sequestered by natural vegetation through photosynthesis. Also, fertilizer, equipment, and livestock all emit greenhouse gases, atmospheric gases that trap heat and contribute to climate change. According to the UN Food and Agriculture Organization (FAO), livestock is responsible for some 18% of all anthropogenic greenhouse gas emissions, globally. (For a comparison, transportation is responsible for about 14% of all anthropogenic emissions.) "It's true that CAFOs take up less space than other types of animal operations," says Ferraro. "But once you factor in the hectares of grain crop it takes to raise them in this way, it's obvious that they're still contributing heavily to greenhouse gas emissions."

Disease spreads quickly among living creatures cramped into close and unsanitary quarters; that's as true of animals as it is of humans. So many animals and so much feces cannot help but attract flies and other insects, which are bothersome at best and disease carrying at worst. CAFO farmers are known to dispense large doses of antibiotics in the animals' water and food to combat these vulnerabilities. In fact, 80% of antibiotic use in the United States goes to livestock. While there is no danger of the antibiotics themselves being passed on to consumers in their meat, such prolific use of these valuable medications has contributed to the rise of antibiotic-resistant bacteria which can and do make their way to human hosts

↑ On June 1, 1995, a hog-waste storage lagoon collapsed at Oceanview Farms in Onslow County, North Carolina, sending millions of liters of wastewater into fields and streams. Bulldozers plugged the huge hole with earth to stop the spillage. The state attorney general's office said that the farm responsible for this, the worst agricultural waste spill in North Carolina history, should close because the farm continues to flaunt state regulations.

through the water supply, through other food sources, and through the meat itself. [INFOGRAPHIC 21.4]

For its part, 157 is not a matter of antibiotic resistance. It may simply be a matter of too many animals in too little space. Across the country, scientists and CAFO operators are working to mitigate its spread.

A variety of methods can reduce *E. coli* contamination.

The vast majority of *E. coli* are harmless to both cattle and humans. In many cases they are good—helping us to digest food that we otherwise might not be able to. But some time pretty recently, in the long sweep of evolution, the 157 strain managed to acquire two Shiga toxin genes from a bacterium that causes dysentery. "They're pretty

carbon footprint The amount of carbon dioxide (CO_2) or other greenhouse gas released by an action; a concern because it contributes to climate change.

Infographic **21.4** | **FROM FARM TO YOU**

↘ Beef that comes from CAFOs starts on smaller family farms in cow-calf operations. From birth to slaughterhouse takes about 18 months, shaving 2 or more years off the time it would take to raise a cow to the same size on pasture alone. The advantages of CAFOs are mainly economic, whereas the disadvantages include environmental, health, and ethical concerns.

ADVANTAGES OF CAFOs

CAFO BEEF PRODUCTION

DISADVANTAGES OF CAFOs

Raising cattle for CAFOs

COW-CALF OPERATIONS
Calves are born and kept with their mothers until weaning (about 8 months old and 225 kg (500 lbs)).

Raising feed for CAFO animals

Fossil fuel–derived fertilizers and pesticides are used; fossil fuels are also used to power equipment. Water demands are also high.

• Cow-calf and stocker operations are usually family-run operations that support local economies. They turn grass (food we cannot eat) into meat or dairy (food we can eat).

STOCKER OPERATIONS
Some weaned calves may go directly to a CAFO, but many are sold to stockers who graze the calves on pasture to gain weight (340 kg (~750 lbs)).

GROWING FEED CROPS
An estimated 75% of U.S. corn goes to feed animals; other feed crops include soybeans, barley, and sorghum; CAFO cattle are also fed food and biofuel by-products, reducing the cropland needed to produce feed crops.

• Environmental problems from the use of fossil fuels, pesticides, and fertilizers (water and air pollution, habitat destruction)

• Though CAFO animals use less space, land is still required for growing the animals' feed.

• Farmers are no longer basing the size of the herd on what the local area can support.

• Can grow large numbers of animals in a small space, reducing the need for deforestation

• Better feed conversion ratio per head than pastured animals because the animal is less active and can be fed more calorie-dense feed

• Lower water consumption per animal due to less daily activity

• Manure can be collected to be used as fertilizer, or digested to produce a methane fuel source.

• Higher profits and quicker return on investment for investors and owners

• Penned animals are easier to monitor and treat for health problems.

• Probably the only way to raise the volume of animals Americans consume

CAFO
Cattle arrive around 14 months old and stay 90–180 days. They are fed a diet of forage (hay) and supplemental feed (grain, soybean, etc.) to reach a slaughter weight of 450–635 kilograms (1,000–1,400 lbs). They are usually given antibiotics to ward off disease, and hormones (estrogen, testosterone, or growth hormone implants) to speed growth.

• Ethical issues of rearing animals in close confinement

• Higher water consumption overall by the facility due to the need for cooling and cleaning water

• Large amounts of waste collect in one area, polluting air and water. If not captured, the methane released by cows is a potent greenhouse gas.

• Competition may drive small operators out of business.

• Diseases arise and spread more easily in the close quarters of a CAFO; high antibiotic use linked to emergence of antibiotic-resistant bacteria.

• Meat can be contaminated during slaughter from contact with dirty hides or intestinal contents. The USDA oversees meat processing but allows the slaughterhouse facility itself to perform many of the inspection duties.

• Large slaughterhouses can process huge numbers of animals very quickly.

SLAUGHTERHOUSE

• Inexpensive protein food source for consumer

PROCESSING FACILITY, WHOLESALE, OR RETAIL OUTLET

• Taxpayer money supports government subsidies that decrease the price of feed, distorting the actual cost (health, environmental, and otherwise) of meat that consumers ultimately pay.

potent toxins," says Ebner. "Exposure to just ten colonies or so is enough to get someone very, very sick. You see not only kidney failure but in some of the worst cases, paralysis and even death."

Cattle are not adapted to digest corn, and some scientists have suggested that corn-rich diets are responsible for 157's development: By disturbing the stomach lining, corn makes the gut more acidic and thus facilitates the evolution of acid-tolerant bacteria like 157. But Ebner says the data do not support this idea. For one thing, studies comparing the incidence of 157 in cows fed a high-forage, high-fiber diet to cows fed a typical grain diet have not shown a consistent difference between the two. "It's really hit or miss," he says. "In both groups, we found cows that were positive on Tuesday and then negative on Wednesday and then positive again on Thursday." Ultimately, the gut of a cow is a complex ecosystem and there are likely a variety of factors that contribute to the emergence of 157, making it difficult to pin down the causes.

At any rate, he says, the bigger concern is not how 157 evolved, or even what triggers its growth, but how best to control it. Part of the challenge is that 157 does not affect cows the way it affects humans. In fact, it doesn't affect them much at all. It simply resides in their intestinal tract, near the last section of the colon, doing no harm, eventually passing to the soil in the animal's feces. Even that translocation would not be a problem if the cow were cleaned and butchered properly. But animals living in small, high-density pens tend to get feces on them, and when they aren't cleaned properly, that feces, and the *E. coli* that comes with it, can come in contact with the meat and be passed on to the consumer.

In general, Ebner says, farmers have two basic options for preventing this, and they are evenly spilt over which is likely to be more effective. Preharvest strategies—including antibiotics or a change in diet—aim to kill the bacteria in live cattle, or at least reduce the number living in the animal's colon. Postharvest strategies—like irradiation or intense washing—aim to remove the contamination from slaughtered carcasses.

Beef industry leaders say that they're doing a good job. 2007 was a bad year, to be sure, but those headline-making incidents belie a much less frightening reality: Meat is safer than it's ever been. "Even when you take last year into account, it's nowhere near where it was even 6 years ago," Mohammad Koohmaraie, a former director of the U.S. Department of Agriculture (USDA) Meat Animal Research Center, told *Beef* magazine in 2008. "Before [2007] everyone was bragging on what the beef industry had done." In fact, in 1996, a government-led initiative set out to reduce the number of *E. coli* O157:H7 infections by one-third, come 2010. By 2004, the industry had exceeded that goal. [INFOGRAPHIC 21.5]

But critics say that even those few cases are a few too many, especially given that some have resulted in fatalities.

There are more sustainable ways to grow livestock.

The picture is certainly prettier on a grass-fed farm: Cattle meander leisurely across the meadows, eating grass as they go. They don't reach market weight as quickly as their CAFO-reared counterparts, but neither do they stand in their own feces. Because they are less crowded

---

Infographic **21.5** | *E. COLI* O157:H7 INFECTIONS ARE DECREASING IN THE UNITED STATES

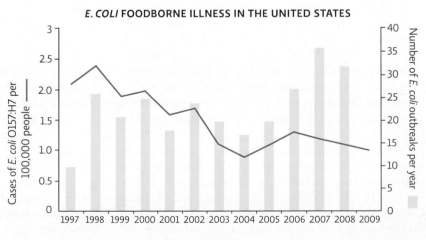

→ The incidence of U.S. cases per 100,000 shows a declining trend (the red line), with the years 2004 and 2009 meeting the 2010 national health objective of no more than 1 case per 100,000 people. (The 2020 health objective of no more than 0.6 cases per 100,000 is the next target.) However, it is interesting to note that while overall case numbers have dropped, the number of outbreaks has increased from 10 in 1997 to 32 in 2008 (the blue bars). There are fewer cases but they are more spread out around the country and are linked to a variety of sources, including beef, spinach, sprouts, and uncooked cookie dough.

E. COLI FOODBORNE ILLNESS IN THE UNITED STATES

↑ Cows at Polyface Farm in Virgina are grass fed and graze using the rotation method. This method is closer to a natural ecosystem than modern farming.

and are eating the food they were adapted, over eons, to eat, sickness is less of a problem and antibiotics less of an urgency.

In the United States in recent years, a niche market has developed for such "grass-fed" beef, which is produced from animals that are raised on pasturelands, rather than in CAFOs. Grass-fed animals don't gain weight nearly as rapidly as feedlot animals do; they eat less-concentrated food (grass, not grain) and spend more time moving around, so they don't get quite as fat. But proponents say that the meat and milk they produce is healthier—it has less saturated fat and more omega-3 fatty acids (a "healthy fat") than CAFO-generated beef. And the animals themselves live better lives—rather than being confined in a tight space, they are free to roam the grasslands and eat the food that they have evolved to digest. In fact, unlike CAFO livestock, grass-fed animals represent a net gain for the human food supply when raised on land that is unsuitable for human crops; they eat grasses that we can't

eat and turn it into beef and dairy products that we can eat (see Chapter 14).

Research shows that raising cattle on pastures would address many of the ecological problems associated with feedlots: Well-managed pastures would reduce erosion and minimize the use of crop inputs like fuel, fertilizer, and pesticides because fewer cows would require less corn. "When we eat from the industrial food system, we are eating oil and spewing greenhouse gases," wrote journalist and sustainable-food advocate Michael Pollan in an article in the *New York Times* in 2008. Smaller, grass-fed operations would replace all that oil with sunlight. Farmers would once again rely on crop diversity and photosynthesis.

One example of how such a system would look can be found on Joel Salatin's polyculture animal farm in Swoope, Virginia. Polyculture farms rear different species of animals (or a mix of animals and plant crops) together—typically in some sort of rotation. Such operations have been touted as a more efficient way to use land resources (see Chapter 22). Salatin's pastures support several different animals, rotated through in a way that maximizes food for each. For example, chickens, released into a pasture after the cows leave, feast on insects congregating on manure. These "free ranging" animals eat a more natural diet and require fewer, if any, synthetic inputs of fertilizer or pesticides. (Consumers need to be careful when they see the term "free range" on food products. The USDA only acknowledges free-range poultry, and the only requirement is that animals have "access to the outside." This can be nothing more than a small space outside a chicken house that the birds rarely use—the animals may not be roaming freely on pasture as they do on Salatin's farm.) According to Salatin, the farm manages to be highly productive even though it uses very little fossil fuel. In one year, his 140 acres produced 30,000 dozen eggs, 10,000–12,000 chickens, 100 beef animals, 250 hogs, 800 turkeys, and 600 rabbits.

But for any of these other options to prevail over the CAFO model, several things must change.

U.S. food policies support industrial agriculture.

To understand our modern food system, we must go back to the 1930s and the Great Depression. It was then that legislators created the **U.S. Farm Bill**. The idea was to provide a safety net for farmers who were so embattled by price volatility and environmental calamity (both inevitable features of a production system that operates at the mercy of weather and global economics) that many of them wanted to stop farming altogether. Programs encouraged farmers to take some land out of production to allow the land to naturally rejuvenate as well as to keep prices high by preventing the market from being flooded. Storage programs for commodity crops (resources that serve as raw materials for other products) like corn and soybean also helped stabilize prices since extra could be stored in good years and supplies could be tapped in bad years. The original Farm Bill also ensured that farmers would be paid at least as much as it cost to produce the food.

But critics contend that the Farm Bill, which is updated every 5 years, has slowly removed these safety nets in favor of new provisions that support behemoth factory farms, including CAFOs, at the expense of small and midsized operations. "There is nothing inherently efficient or economical about raising vast cities of animals in confinement," writes Pollan. He points out that three federal policies in particular, have propped up CAFOs: subsidies that enable feedlot operators to buy grain for

less than it costs to grow; Food and Drug Administration (FDA) approval for the routine use of antibiotics in animal feed (without which he says so many CAFO animals could not survive); and lax regulation of wastewater treatment. "The government does not require CAFOs to treat their wastes as it would require human cities of comparable size to do," he writes.

Two of these "struts" have recently been kicked away. The 2008 global economic crisis drove grain prices up, so that even with subsidies, feedlot operators are now paying at least as much as it cost to grow the grain. And early in 2012, the FDA announced that it would no longer allow farmers to administer antibiotics to their animals without a veterinary prescription. "CAFOs should also be regulated like the factories they are," Pollan writes. They should be "required to clean up their waste like any other industry or municipality." Yes, he admits, such policies will make meat pricier; but that is as it should be. "Paying the real cost of meat, and therefore eating less of it, is a good thing for our health, for the environment, for our dwindling reserves of fresh water, and for the welfare of animals," he writes. "Cheap food is dishonestly priced. It is, in fact, unconscionably expensive."

The UN FAO acknowledges that CAFO livestock may help feed a growing human population, but only if policy changes are made that support more sustainable practices. For example, capturing waste and using it as fertilizer or to generate electricity would turn it into a resource instead of a problem to be solved. Moving CAFOs away from urban centers or smaller livestock operations would prevent the spread of disease between animals and from animals to humans. Yes, sustainably produced meat will cost more: CAFO operators will have to spend money to establish better waste management systems and to adhere to stricter environmental regulations. But that is as it should be. "The price of beef should reflect the true cost of producing it," says Ferraro. "And that includes the environmental costs of pollution, land degradation, and water depletion." [INFOGRAPHIC 21.6]

Consumer choices can increase food supply.

If Pollan's most fervent dreams come to pass—if U.S. agricultural policy changes and CAFOs start giving way to grass-fed, pasture-raised cattle—then we will find ourselves with a lot less meat. What's more, the meat that we

---

U.S. Farm Bill Overarching legislation that deals with many aspects of the production and sale of farm-raised commodity crops, including programs for consumer food assistance, soil conservation, and farm subsidies; updated every 5 years.

Infographic 21.6 | **AGRICULTURAL POLICY MUST CONSIDER TRADE-OFFS**

→ Different regions around the world face different problems and each needs its own solution about the best way to raise livestock. The UN FAO recommends that low-income, developing nations focus more on helping their small farmers earn a living raising livestock and increasing access to affordable food for local citizens. As income increases and the percentage of the nation's income that comes from raising livestock decreases, priorities should shift to improving food safety and reducing the environmental impacts of raising animals.

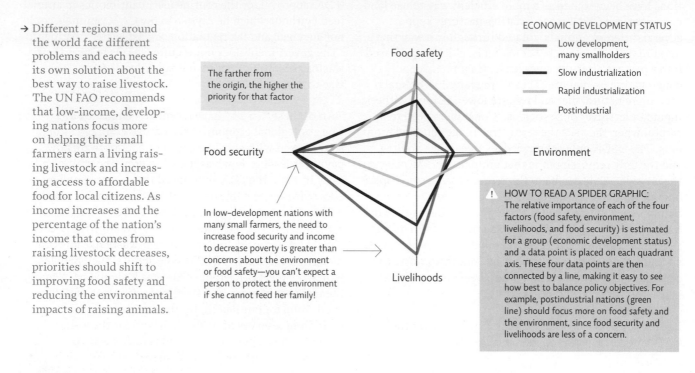

The farther from the origin, the higher the priority for that factor

In low-development nations with many small farmers, the need to increase food security and income to decrease poverty is greater than concerns about the environment or food safety—you can't expect a person to protect the environment if she cannot feed her family!

ECONOMIC DEVELOPMENT STATUS
— Low development, many smallholders
— Slow industrialization
— Rapid industrialization
— Postindustrial

! HOW TO READ A SPIDER GRAPHIC:
The relative importance of each of the four factors (food safety, environment, livelihoods, and food security) is estimated for a group (economic development status) and a data point is placed on each quadrant axis. These four data points are then connected by a line, making it easy to see how best to balance policy objectives. For example, postindustrial nations (green line) should focus more on food safety and the environment, since food security and livelihoods are less of a concern.

do produce will be more expensive. Thus, American consumers will increasingly be faced with what today seems like an impossible choice: spend more money for the same amount of beef that we are now eating, or swap out a large portion of our livestock consumption for more fruits and vegetables, which cost less to grow than meat or poultry, both in dollars and in environmental trade-offs—no manure lagoons, or grain feed crop to worry about.

Experts like to call this *eating lower on the food chain*; many believe it is the key to a sustainable diet. For one thing, it's better for our health. "Diets high in animal fats and red meat confer a high risk of cancer, not to mention heart disease, obesity, and diabetes," says Devra Lee Davis, an epidemiologist and director of the Center for Environmental Oncology at the University of Pittsburgh. "Fat has been called a natural hazardous waste site, because it accumulates fat-loving synthetic organic chemicals, which move up the food chain." Larger mammals accumulate more toxins, she says. And eating lower on the food chain is a good way to avoid ingesting too many of these toxins.

It's also a good way to reduce one's ecological footprint. "By reducing the amount of meat we eat, we can grow and kill fewer animals," writes journalist Mark Bittman in his book *Food Matters*. "That means less environmental damage, including climate change; fewer antibiotics in the

water and food supplies; fewer pesticides and herbicides; reduced cruelty; and so on."

Experts are quick to point out that adhering to our environmental principles need not mean cutting meat out of our diets entirely. In some cases, after all (on certain grasslands or with certain crop-rotation plans), growing livestock really is the best use of a given patch of land at a given time. In fact, one Cornell study concluded that the diet with the lowest footprint for the state of New York was one that included a small amount of meat and dairy. [INFOGRAPHIC 21.7]

Ultimately, the solution will come down to moderation—in the way we grow cattle, and in the way we consume it. Meanwhile, since 2007, an additional 19,032,057 kilograms (41,958,504 pounds) of beef have been recalled from restaurants and supermarkets.◉

Research articles referenced in this chapter:
Armstrong, S.D., *et al.* 2010. In *Genetic Engineering, Biofertilisation, Soil Quality and Organic Farming*, edited by E. Lichtfouse. Dordrecht: Springer.
Callaway, T.R., *et al.* 2009. *Current Issues in Molecular Biology*, 11: 67–80.
Centers for Disease Control. 2011. *Healthy People 2010: Final Review.* www.cdc.gov/nchs/data/hpdata2010/hp2010_final_review.pdf.
FAO. 2009. *The State of Food and Agriculture—Livestock in the Balance.* Rome: Food and Agriculture Organization of the United Nations.
Pan, A., *et al.* 2012. *Archives of Internal Medicine*, 172: 555–563.
Peters, C.J., *et al.* 2007. *Renewable Agriculture and Food Systems*, 22: 145–153.
Russell, J. B., and Rychlik, J.L. 2001. *Science*, 292: 1119–1122.

↓ A 2009 study by Christian Peters of Cornell University compared the number of people that could be supported with locally produced food from farmland in New York State. His research compared diets that varied in meat and fat content ("meat" included meat, milk, and eggs), testing a range that included diets lower and higher in meat and fat than that of the average American, who consumes 6–8 ounces of meat per day, with roughly 40% of calories coming from fat. Though eating more than 4 ounces of meat per day required more land and reduced the number of people the land area could support, diets that contained a small amount of meat (2 ounces per day) were actually the most efficient in terms of carrying capacity, but only when consumption of fat surpassed 35% of the daily diet.

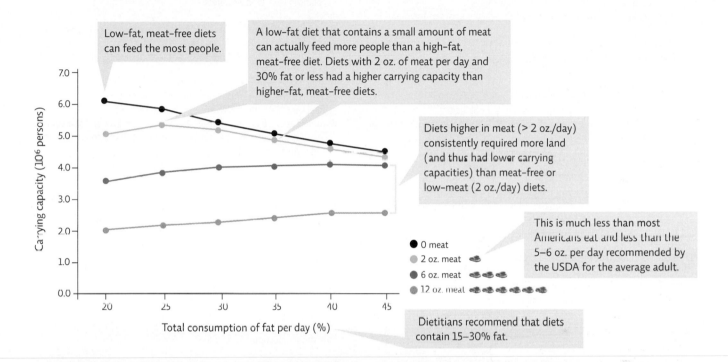

BRING IT HOME

➲ PERSONAL CHOICES THAT HELP

As the demand for meat increases with further economic development, the health and environmental consequences of CAFOs become more and more clear. A careful reconsideration of the dietary choices within affluent countries is the only way to help create a sustainable food production system.

Individual Steps
→ Reducing your own consumption of meat has positive impacts on the environment as well as on your own health. Visit www.meatlessmonday.com for recipes and other resources.

→ Identify brands of meat and dairy products available in grocery stores around you that are labeled as free range, grass-fed, or antibiotic free.
→ Support sustainable-practice farmers who produce food locally. Farmers' markets are a great place to find fresh fruits and vegetables grown within your area.

Group Action
→ Invite local producers to speak to a community group about their practices and products in order to increase public awareness of local food suppliers.
→ Lead a book discussion of Michael Pollan's *The Omnivore's Dilemma* which

brings attention to the problems associated with industrial farming methods as well as some solutions.

Policy Change
→ Organize a community screening of a documentary like *Fresh!* (www.freshthe-movie.com/). Follow it with a discussion of the impacts of industrialized cattle operations on individual health and the environment. Contact local dairy producers for promotional materials or coupons to distribute.

UNDERSTANDING THE ISSUE

CHECK YOUR UNDERSTANDING

1. **In general, as nations become wealthier, what trend is observed regarding meat consumption?**
   a. It declines, because people are more educated and understand the health problems associated with eating meat.
   b. It remains unchanged, because cultural norms, not wealth, dictate diet.
   c. It rises at first and then drops off sharply as people shift to more expensive foods like fish and fresh produce.
   d. It increases, because meat is seen as a desired part of the diet.

2. **What is the most common cause of *E. coli* O157:H7 contamination of meat like beef or pork?**
   a. Poor hygiene at the slaughterhouse; the meat is contaminated with animal feces.
   b. Poor-quality feed given to the animals before slaughter.
   c. The overabundance of antibiotics in animal feed.
   d. Poor handling techniques at the individual grocery store that sells the meat.

3. **In CAFOs:**
   a. animals are grown slowly, grazing on open pasture.
   b. animals are fed a liquid concentrate to speed growth.
   c. animals are crowded together in enclosures and fattened quickly.
   d. animals are exercised and fed constantly to encourage rapid growth.

4. **Some of the problems of CAFOs include:**
   a. fewer animals can be raised per acre of grazing land.
   b. they contribute air and water pollution to the local area.
   c. that CAFOs are not as profitable as traditional ranching.
   d. penned animals are harder to monitor for health problems.

5. **Globally, what are the FAO's livestock-rearing recommendations?**
   a. All nations should phase out the environmentally damaging CAFOs and bring back pasture-raised livestock.
   b. The priorities for developed nations should be food safety issues and protecting farmers' incomes.
   c. Developing nations should focus on methods that increase food security and the income of small farmers.
   d. Livestock rearing in general should be phased out in favor of plant-based food products.

6. **Based on feed conversion ratios, it is more energy efficient to eat chicken than beef because:**
   a. a pound of ground beef has fewer calories than a pound of ground chicken.
   b. chickens are more efficient than cows at converting the food they are fed into body mass, or the food that we eat.
   c. chickens are lower on the food chain than cows.
   d. unlike cows, chickens represent a gain for the human food supply, as they eat items we cannot (such as insects) and convert them into meat we can eat.

WORK WITH IDEAS

1. How does beef get contaminated with fecal bacteria and what can you as a consumer do to protect yourself?

2. What is the evidence that shows that red meat consumption has negative effects on our health? Design a study that would allow you to test the following hypothesis: Red meat consumption in people increases one's risk for heart disease.

3. Compare the trade-offs of rearing cattle in CAFOs or on pasture. Which do you support? Explain.

4. What is the U.S. Farm Bill and how has it changed over time?

5. What is "eating lower on the food chain," and how might it help address some of the problems related to raising animals for food?

ANALYZING THE SCIENCE

These two graphs represent data on meat recalls from 1998–2011, in pounds. Graphs are based on data from USDA Food Safety and Inspection Service (www.fsis.usda.gov).

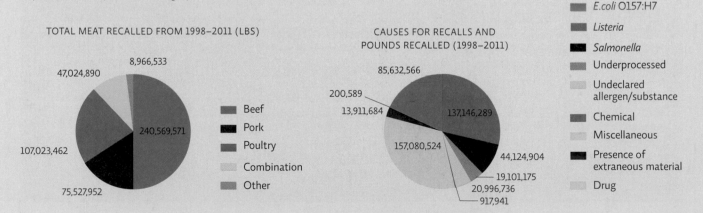

TOTAL MEAT RECALLED FROM 1998–2011 (LBS)

8,966,533
47,024,890
240,569,571
107,023,462
75,527,952

- Beef
- Pork
- Poultry
- Combination
- Other

CAUSES FOR RECALLS AND POUNDS RECALLED (1998–2011)

85,632,566
200,589
13,911,684
137,146,289
157,080,524
44,124,904
19,101,175
20,996,736
917,941

- *E. coli* O157:H7
- *Listeria*
- *Salmonella*
- Underprocessed
- Undeclared allergen/substance
- Chemical
- Miscellaneous
- Presence of extraneous material
- Drug

INTERPRETATION

1. What type of meat was recalled most, and what is the most common reason for recalls?

2. For each type of food recall, calculate its percentage of total food recalls.

ADVANCE YOUR THINKING

3. We are most likely to be exposed to *Listeria* by eating cold cuts like lunch meat. What is it about this type of food that increases our likelihood of getting sick if it is contaminated? Where does *Listeria* rank as a source of contamination?

4. Does the use of CAFOs increase or decrease the likelihood of meat recalls? Explain your answer.

EVALUATING NEW INFORMATION

There are many voices contributing to the discussion about how we raise animals for food (meat and dairy). Information on CAFOs can be found at the Iowa Beef Industry Council website (www.iabeef.org). Check out the link "The Beef Story" and read the article called "Beef from our Farm to your Fork." The Socially Responsible Agricultural Project, at www.sraproject.org, also discusses CAFOs; check out the link "About Factory Farms."

Evaluate the websites and work with the information to answer the following questions:

1. Evaluate the agendas of the two organizations as well as the accuracy of the science behind their positions on CAFOs.
   a. Who runs each website? Do the person's/organization's credentials make the information presented on CAFO issues reliable or unreliable? Explain.
   b. What do you think the mission of each website is? What are the underlying values? How do you know this?
   c. What claims does each website make about CAFO food production and what evidence do they give in support of those claims? Is that evidence credible? Explain.
   d. According to each, how should we be raising food animals?

2. What is the tone (factual, emotional, technical, aggressive...) and intention (to inform, to persuade, to instruct, to entertain...) of each website? Do these websites present objective information or does the information offered seem biased? Explain.

3. Use your information-gathering skills (see Chapter 3 if needed) to locate a web resource that you think is reputable and gives a fairly objective overview of CAFOs. Describe the search process you used, including how long it took you to find this resource. Explain why you think this is a credible and useful source.

4. Based on everything you have learned, what do you feel is the best solution for growing livestock like beef, chicken, and pork in the United States? Support your position.

MAKING CONNECTIONS

EATING LOWER ON THE FOOD CHAIN

Background: In the Cornell study presented at the end of the chapter, researchers looked at diets with little (up to 2 oz. per day) or no meat. These diets would feed more people and could be healthier if they contain the correct balance of nutrients and calories. But the recommendation to "eat lower on the food chain" is not always a welcome one to meat-loving Americans. What about you? Does a diet lower in meat appeal to you? Would you even consider it?

Case: Before you decide, look at your own diet and do some research to find out what a healthy low-meat or meat-free diet would consist of. Your task is to develop a diet plan that includes foods you would actually eat and contains the proper balance of nutrients and calories recommended for someone of your age and activity level.

1. First, analyze your own diet. Write down everything you eat and drink, including portion size, for a typical day (or for a few days). Evaluate that diet:
   a. How much meat do you eat?
   b. How many calories have you consumed? (Try an online calorie calculator or app for this, such as "Calorie Counter & Diet Tracker.")
   c. What percent of this diet is fat? (You may have to calculate this—simply add up fat calories and divide by total caloric intake.)
   d. Go to the USDA website, www.choosemyplate.gov/myplate/index.aspx, to determine what your daily diet should be in terms of calories and servings of protein.

2. Develop either a meat-free diet plan (a "vegan" diet—no animal products of any type) OR a low-meat diet plan (contains 2–3 oz. of meat, dairy, and/or eggs per day). The plan should meet your nutrition needs and be between 20 and 30% fat. The USDA website has suggestions for meat and nonmeat protein sources. Draw up a meal plan for a "sample day" (all your meals and snacks; don't forget beverages), recording total calories and fat calories. Also address the following:
   a. What foods (and how much of each) will you include to meet your caloric needs?
   b. What foods will provide your protein? If you include animal foods, which did you choose and why did you choose these foods to supply your 2–3 oz. (basically 1 small serving) per day?
   c. What percent of this diet is fat?
   d. How likely would you be to actually try out this diet? Explain what appeals to you about this diet and what doesn't. (If you actually try it, write a short reflection on the experience.)

CORE MESSAGE

Achieving food security for all people requires that we build a sustainable food system. Sustainable agriculture methods can help us grow crops within the means of the local ecosystem and without degrading the soil, water, and biodiversity that support that growth. Changes in consumer purchasing practices to support more sustainably produced food will give these foods the economic backing their producers need to succeed, while providing consumers with healthier food.

GUIDING QUESTIONS

After reading this chapter, you should be able to answer the following questions:

→ What is sustainable agriculture and what can such a food system offer us?

→ What are the benefits and drawbacks of industrial agriculture?

→ What kind of sustainable methods can help reduce the environmental impact of industrial farming?

→ What role does the consumer play in helping build a sustainable food system?

→ What are the advantages and disadvantages of sustainable farming and what role will it play in helping us feed the world?

In Japan, the quiet rice paddies of Takao Furuno.

FINE-FEATHERED FARMING

Creative solutions to feeding the world

If there's one thing Greg and Raquel Massa hate, it's weeds—all varieties, but especially the azolla—an insidious, fernlike plant that grows on the surface of water. Each spring, azolla plants invade the couple's rice farm, snaking their way through the dense, muddy paddies that stretch for miles along the Sacramento River near Chico, California. They strangle young rice plants and force Greg into an endless and tedious battle.

The rivalry—Massa versus azolla—has spanned three generations. Greg's grandfather, Manuel Massa, planted the family's (and some of California's) first rice crops in 1916, on the same land that Greg and Raquel now manage. Back then, rice farming was a hard and uncertain life; Manuel was largely powerless against the azolla, which in some seasons claimed his entire crop.

By 1962, when Greg's father, Manuel Jr., took over, American ingenuity and modern science had completely changed the nature of the fight. Heavy doses of chemical herbicides enabled him to obliterate the weed. And specially bred higher-yield rice varieties developed during the Green Revolution (see Chapter 20), along with modern farming equipment and a heavy dose of chemical *fertilizers* (nutrients that boost plant growth), made the family farm both efficient and profitable.

Of course, that modern approach, known as **industrial agriculture**, had its own problems. For one thing, it relied on cheap fossil fuel energy and lots and lots of water, both of which are in much shorter supply these days. For another, it impaired ecosystem services, often gravely, because it involved clearing huge swaths of land to increase production, and focusing all resources on a single crop. On top of that, chemical fertilizers and *pesticides* (such as insecticides that kill insects or herbi-

cides that kill plants) were expensive and not especially good for the land.

Greg and Raquel wanted to find a better, more sustainable way. **Sustainable agriculture** is farming that meets the needs of the farmer and society as a whole without compromising the environment or future productivity. The techniques used will maintain or even enhance the environment. They often do this by mimicking the traits of a sustainable ecosystem—they rely on renewable energy and local matter resources for inputs and depend on biodiversity to trap energy, deal with waste, and control populations. [INFOGRAPHIC 22.1]

So when Greg and Raquel took over in 1997, they converted a portion of their farm to a sustainable farming method known as **organic agriculture**. Instead of synthetic fertilizers and pesticides, organic agriculture employs more natural, or "organic," techniques in the growing of crops—such as using manure as fertilizer and luring in natural predators to control pests. This type of farming uses fewer or no chemicals and in some cases may even produce food that is more nutritious. For example, research by Washington State University soil scientist John Reganold showed that organically grown strawberries had a longer shelf life and a higher level of antioxidants than conventionally grown berries. But organic farming forced Greg and Raquel to battle weeds much as the first Manuel had: with great difficulty.

The trick was to lower water levels enough to kill water-loving weeds, but not so much that the rice crop also died. Each day, Greg would wade into the paddies to see how the rice plants were faring against the azolla. Some weeks,

◉ **WHERE IS CHICO, CALIFORNIA?**

industrial agriculture Farming methods that rely on technology, synthetic chemical inputs, and economies of scale to increase productivity and profits.
sustainable agriculture Farming methods that can be used indefinitely because they do not deplete resources, such as soil and water, faster than they are replaced.
organic agriculture Farming that does not use synthetic fertilizer, pesticides, or other chemical additives like hormones (for animal rearing).
monoculture Farming method in which a single variety of one crop is planted, typically in rows over huge swaths of land, with large inputs of fertilizer, pesticides, and water.

Infographic 22.1 | SUSTAINABLE AGRICULTURE

↓ According to the U.S. Department of Agriculture, sustainable agriculture is farming that uses only limited amounts of nonrenewable resources (like fossil fuels) and does not degrade the environment or the well-being of people or society as a whole.

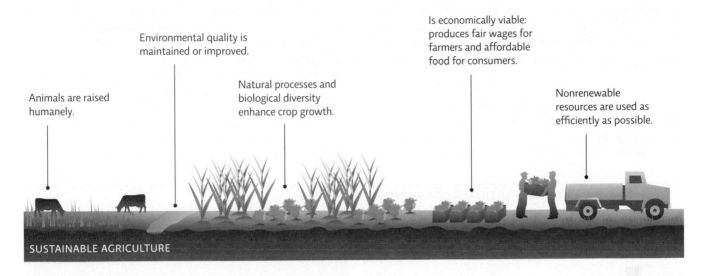

Animals are raised humanely.

Environmental quality is maintained or improved.

Natural processes and biological diversity enhance crop growth.

Is economically viable: produces fair wages for farmers and affordable food for consumers.

Nonrenewable resources are used as efficiently as possible.

SUSTAINABLE AGRICULTURE

he worried the entire crop would die. After a few seasons, the Massas started to despair: How could they make their farm environmentally friendly without losing their livelihood to an army of mangy weeds?

The Massa's story is the story of modern farming; it's the story of how we feed ourselves. And on a planet where population is exploding, climate is shifting, and energy and water resources are running low, it's also a story of constant change.

Modern industrial farming techniques are productive but come with some significant trade-offs.

The changes that helped Greg's father thrive also ushered in a whole new way of growing crops, known as **monoculture** farming. Instead of growing a mix of plants, or growing different crops each season, farmers began growing the same, single crop year after year. Before long, farms that had been populated by a variety of crops morphed into industrial operations, focused on just one crop. Thus, biodiversity was replaced by specialization, and farm ecosystems that more closely resembled nature were replaced by operations completely dependent on technology. "In the 1920s, half of Iowa's farms produced 20 commodities each," says Fred Kirschenmann, a Distinguished Fellow at Iowa State University's Leopold Center for Sustainable Agriculture. "Today 80% of the state's cultivated land is exclusively corn or soybean. Farming systems that were once supported by complexity and diversity of species have now been replaced by

reliance on inputs." Worldwide, 90% of our food comes from just 15 crop species and 8 species of livestock.

The advantages of this new approach were obvious—a single crop was much easier to manage and to mass produce, and with greater ease came not only greater efficiency and greater profits, but also drastically greater amounts of food for a planet that never seemed to have enough.

In recent years, however, the disadvantages of industrial agriculture have become equally apparent. In monoculture farming, the crop that is chosen is not necessarily locally adapted. Instead of choosing the crop best suited to the existing ecosystem, farmers focused on those crops with the highest market demand, and thus the highest dollar value. Because these crops were not locally adapted, and because the volume of plants grown increased exponentially, the average farm became heavily dependent on external inputs—water, pesticides, and fertilizer—added to the farm from outside its own ecosystem.

> Worldwide, 90% of our food comes from just 15 crop species and 8 species of livestock.

Fertilizers boost growth because they provide nutrients needed by plants, such as nitrogen and phosphorus (see Chapter 7 for more on nutrient cycles). And while heavy doses of fertilizer can indeed boost crop production, the excess nutrients (whatever is not used by the plants) can soak into the ground and contaminate the groundwater or be easily washed from fields by rain and modern irrigation practices. The excess, or *runoff*, enters waterways, and ultimately creates hypoxic, or oxygen-poor, regions that threaten aquatic life. This process, known as **cultural eutrophication**, has been a significant problem in the United States, where nutrient runoff from farms in Iowa and other "Corn Belt" states has created a summertime dead zone in the Gulf of Mexico (see Chapter 18). [INFO-GRAPHIC 22.2]

Monoculture crops are also more vulnerable to pests or disease—a single infestation can wipe out the entire crop because what kills one plant will likely kill them all. To deal with this, farmers have turned to pesticides, but this also has proven problematic. To be sure, pesticides kill weed and insect pests and thus dramatically reduce the number of crops lost each year to infestation. But because they are toxic, pesticides also pose a threat to human and ecosystem health. And as scientists quickly discovered, pest populations can develop **pesticide resistance**, an unintentional example of artificial selection (see Chapter 12). Herbicide-resistant weeds and insecticide-resistant insects are cropping up all around the world. This forces us to employ more drastic measures, in the form of higher doses or more toxic chemicals—but as resistance to that next pesticide develops, the cycle repeats itself. It's like an arms race between humans and pests, with the deck stacked in favor of the pests. [INFOGRAPHIC 22.3]

Infographic **22.2** | **THE USE OF FERTILIZER COMES WITH TRADE-OFFS**

ADVANTAGES

INCREASE PRODUCTIVITY

DISADVANTAGES

CULTURAL EUTROPHICATION

TRADE-OFFS

↑ Fertilizer can greatly increase productivity of the soil and is required for many of the high-yield varieties now under industrial cultivation. Vital soil nutrients like nitrogen and phosphorus are often in limited supply (see Chapter 7 for more on nutrient cycles) and plant growth slows if one nutrient starts to run out. The addition of extra nutrients overcomes this deficiency and can boost growth.

↑ Runoff pollution that contains fertilizer can also boost the growth of algae, causing algal blooms that block sunlight and prevent underwater photosynthesis, reducing the addition of oxygen to the water. As organisms die and organic material builds up, decomposing bacteria increase in number, seriously depleting the oxygen content of the water and leading to the death of many more aquatic organisms such as fish.

GROW CROPS IN MARGINAL SOILS

DEVELOP A DEPENDENCE ON FERTILIZER

TRADE-OFFS

↑ Fertilizers help crops grow in areas that may not otherwise be able to support agriculture. This may be the only way to farm in many areas of the world and would help increase local food supplies.

↑ The extra plant growth that fertilizers support can pull other nutrients out of the soil, depleting soils further and requiring even more fertilizer in the future. In addition, some crops may be less nutritious when grown in nutrient-depleted soil, even with the addition of fertilizers (which will not be able to fully restore soil fertility).

Infographic 22.3 | EMERGENCE OF PESTICIDE-RESISTANT PESTS

↓ Exposure to a pesticide will not make an individual pest resistant; it will likely kill it. However, if a few pests survive because they happen to be naturally resistant, they will breed and their offspring (most of which are also pesticide resistant) will make up the next generation. Over time, the original pesticide will no longer be effective and will have to be applied at a higher dose or a different pesticide will have to be used. Application of a pesticide might even increase the size of the pest population by killing the predators that eat the pests.

Pests attack a crop and multiply quickly. There is usually genetic diversity in the population, with some pests being killed more easily than others.

Pesticide is applied to kill the pests.

Only a few survive, including those that are pesticide resistant.

The survivors reproduce.

THE NEXT GENERATION
Survivors reproduce and pass on their traits to offspring.

Pesticide is applied again.

Every time pesticide is applied, it kills vunerable individuals and leaves resistant ones behind to reproduce.

A large, pesticide-resistant population now infests the crop.

Even water inputs have created problems: It turns out that the use of irrigation can result in soil salinization; as water evaporates from the soil, salts are left behind and eventually become so concentrated that they impede crop growth. And if all that weren't enough, the colossal machines used to plow endless fields of corn and soybean have compacted the soil, making it harder for plants to take root and grow.

On top of that, the quality of the food itself can also suffer. Several recent studies have shown a decline in the nutrient content of food grown on soils subjected to years of industrial farming. In one of the first such studies, University of Texas researcher Donald Davis compared the nutritional content of 43 fruits and vegetables grown in 1950 to those grown in 1999, noting a decline in 6 nutrients (protein, calcium, phosphorus, iron, and

vitamins B2 and C). Davis suspects this is due to the impoverished soil in which the crops are grown.

As modern farmers face these enormous challenges, (and global warming and energy and water shortages along with them), the story of how we feed ourselves is changing yet again. This time, we may have to consider not just high-tech solutions but look back to the natural world for answers. In searching for a new weapon against the azolla, Raquel found a Japanese rice farmer who was doing just that.

cultural eutrophication Nutrient enrichment of an aquatic ecosystem that stimulates excess plant growth and disrupts normal energy uptake and matter cycles.
pesticide resistance The ability of a pest to withstand exposure to a given pesticide, the result of natural selection favoring the survivors of an original population that was exposed to the pesticide.

↑ Greg Massa, right, with rice plants in hand.

Mimicking natural ecosystems can make farms more productive and help address some environmental problems.

Takao Furuno was, by most standards, a very successful industrial rice farmer, with annual yields among the highest in southern Japan. But it was a tough grind. Each year he was forced to put all his earnings back into the next year's crop—insecticides, herbicides, irrigation, and fertilizer—so that despite his success, he and his family were left with very little for themselves at the end of each season.

In searching for a better way, he turned, as he often did, to his forebears to see what he could learn from their knowledge. He was surprised to discover that they used to keep ducks in their rice paddies. Like most rice farmers, Furuno considered ducks a pest, albeit a slightly cuter one than the azolla. Adult ducks eat rice seeds before they have a chance to grow, and as they forage, trample young seedlings into the mud. This disturbance creates open patches of water, which in turn, invites only more ducks. "If you're not careful, you end up with a big problem pretty quickly," says Raquel.

But ducklings, Furuno soon realized, were too small to do such damage; for one thing, their bills were not big or strong enough to extract seeds from mud. Instead, they ate bugs and weeds. Azolla was one of their favorites.

Furuno's forbears also grew loaches—a type of fish—in their paddies. The loaches would also eat azolla and could be harvested and sold as food.

Together the ducklings and loaches would keep the weed from strangling the rice crop; but they would not completely eliminate the azolla the way a heavy dose of pesticides would. Furuno quickly discovered that, when kept at this benign concentration, the azolla (which contains symbiotic bacteria that produce nitrogen) actually fertilized the rice. In fact, between the nitrogen from the azolla, and the duck and fish droppings, he soon found that he no longer needed to spend money on synthetic fertilizer.

When raising ducks, fish, and rice crops together, Furuno discovered the root crowns (where the root meets the stem) of rice plants increased to about twice the size that they had been in his old industrial system. A larger root crown meant more rice. "We're not exactly sure why the crowns grew," Furuno told an audience of American farmers at a recent convention in Iowa. "But the ducks seem to actually change the way the rice grows. It's got something to do with the synergy of the whole system." Furuno's operation is an elegant example of biomimicry—a farm operating like a natural ecosystem. This type of farming, known as **agroecology**, considers both ecological concepts (modeling a farm after an ecosystem) and the value of traditional farming knowledge

Infographic **22.4** | **AGROECOLOGY: THE DUCK/RICE FARM**

↓ Takao Furuno's farm is a self-regulating, multiple-species system that naturally meets the needs of the farm ecosystem. All of the species play a role in the system, helping each other and boosting overall production.

THE METHOD

Rice seedlings are planted in flooded rice paddies.

Ducklings are introduced to eat weeds and provide "fertilizer."

Fish are introduced to eat weeds and provide "fertilizer."

Azolla is introduced to add nitrogen; ducks and fish keep the azolla from growing too much.

THE FINAL PRODUCT: AN INTEGRATED SYSTEM

THE HARVEST

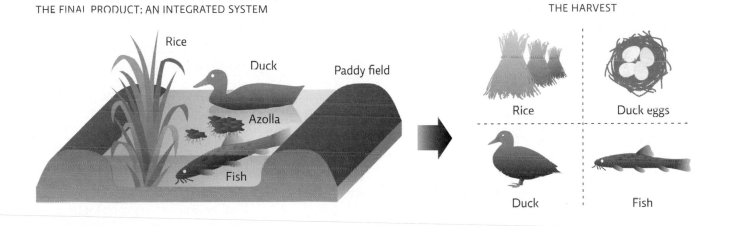

Rice

Duck

Paddy field

Azolla

Fish

Rice

Duck eggs

Duck

Fish

as a rich resource for sustainable and effective farming methods. [INFOGRAPHIC 22.4]

There were other financial gains, too. Duck eggs, duck meat, and fish all fetched a good price in the market. And because he was no longer using pesticides, he could also grow fruit on the edges of his rice field. (He opted for fig trees, as he could harvest the figs annually without having to replant.)

Furuno's farm is an example of **polyculture**—intentionally raising more than one species on a given plot of land. In the decade and a half since Furuno began duck/rice farming, his rice yields have increased by 20—50%, making his among the most productive farms in the world, nearly twice as productive as conventional farms. This kind of success is especially important for farmers in many developing countries, who struggle to produce enough food for current populations. Increasing production with lesser dependence on expensive inputs, and using methods that enhance rather than diminish envi-

ronmental quality, can help communities become more self-sufficient and help them achieve food security.

The Bangladesh Rice Research Institute, which has evaluated Furuno's method and independently verified his success, recommends the technique to Bangladeshi farmers. And by now, some 10,000 Japanese farmers have followed Furuno's lead; his method is also catching on in China, the Philippines...and California.

Furuno's method of duck/rice farming turned out to be a good fit for the Massas. Not only would the rice plants rely on natural fertilizers and natural pest control, but the duck eggs and meat produced would be more humanely grown than those produced by factory farms. The ducks

agroecology Scientific field that considers the area's ecology and indigenous knowledge, and favors methods that protect the environment and meet the needs of local people.
polyculture Farming method in which a mix of different species are grown together in one area.

would grow up in ponds, not crammed together on slats in a barn without access to swimming water. They would get to splash around, and express their "duckiness," as Raquel puts it.

When the ducks first arrived on the Massa farm, they were just 24 hours old, cotton-ball-sized tufts of yellow feathers. The Massa children cared for them in wooden crates in their barn. But as Raquel soon learned, small ducklings grow mighty quickly. In just 2 weeks, they were large enough to turn loose. Furuno had advised stocking about 100 ducks per acre, but Greg and Raquel did not want to sacrifice that much land for this first attempt, so they fenced off just a quarter acre instead. This amount of space provided plenty of room for their 120 ducks to swim and forage in, but as it turned out, not enough food to support them. "They quickly ate all the weeds in the field," says Greg. They also trampled some of the rice plants in their pond because their section of the field was too small. But even as they ran out of weeds to eat, the baby ducks stayed away from the rice plants, just as Furuno had insisted they would.

Integrated pest management is another feature of sustainable agriculture.

Both of Greg and Raquel's azolla-control methods—reducing water levels and employing duck predators—are examples of **integrated pest management (IPM)**, or the use of a variety of methods to help reduce a pest population. The goal of IPM is to successfully control pests while minimizing or eliminating the use of chemical toxins. First, the farmer must examine the lifecycle of the pest and the pest's interactions with the environment to identify the best way to deal with the pest. In general, IPM techniques fall into four different categories: cultural control, biological control, mechanical control, and chemical control.

In sustainable agriculture, farmers use a combination of cultural, mechanical, and biological controls to deal with pest problems, and only resort to chemicals if these methods don't adequately deter the pests. If a pesticide is going to be used, the preference is for natural, biodegradable chemicals that are toxic only to a limited group of organisms. For example, pyrethrum, a compound naturally produced by a flower in the chrysanthemum family, is directly toxic to insects but not to mammals; it is certified for organic agriculture since it breaks down quickly and does not linger in the environment. However, while it does kill pest insects, it is also toxic to *good* insects like honeybees, so its use is avoided during times of pollination. Synthetic pesticides, like those commonly used

in industrial agriculture, are not acceptable for use on certified organic crops but may be part of an IPM plan for conventionally raised crops. [INFOGRAPHIC 22.5]

Sustainable agriculture techniques can protect soil and keep farm productivity high.

Greg and Raquel were well versed in the problems of modern agriculture. Before settling in California to take over the family farm, they had worked as tropical ecologists in Costa Rica, where they learned about traditional, nonindustrial farming methods that help protect the soil and keep productivity high without the use of synthetic fertilizers or pesticides.

Soil is vital to life on Earth—it supports the growth of plants and the animals who feed on those plants. Its formation is very slow (it can take more than 1,000 years to form 2.5 centimeters [1 inch] of topsoil) and depends on myriad soil organisms (see Chapter 14). Modern farming can contribute to a decline in soil fertility and to the direct loss of soil though erosion. However, some traditional (preindustrial) methods can actually help restore or protect soil. For example, methods such as **terracing** and **contour farming** can decrease soil erosion from sloped land. Other techniques that protect the soil include **reduced-tillage** methods, as well as planting a **cover crop** in the off-season to prevent exposed soil from washing away. This last approach has the added advantage of helping to restore fertility. Soil fertility can also be enhanced and pests controlled using **crop rotation** and **strip cropping**. In addition, systems like Furuno's that combine animal and plant rearing return animal waste—a natural fertilizer—to the soil. [INFOGRAPHIC 22.6]

In rice farming, the Massas saw a chance to implement all the concepts and theories they had learned as ecology students, and refined and discussed as scientists, on their

integrated pest management (IPM) The use of a variety of methods to control a pest population, with the goal of minimizing or eliminating the use of chemical toxins.
terracing On steep slopes, land is leveled into steps; reduces soil erosion and runoff down the hillside.
contour farming Farming on hilly land in rows that are planted along the slope, following the lay of the land, rather than oriented downhill.
reduced-tillage Planting crops in soil that is minimally disturbed and that retains some plant residue from the previous planting.
cover crop A crop planted in the off-season to help prevent soil erosion and to return nutrients to the soil.
crop rotation Planting different crops on a given plot of land every few years to maintain soil fertility and reduce pest outbreaks.
strip cropping Alternating different crops in adjacent strips, several rows wide; helps keep pest populations low.

↓ Controlling pests is important for our agricultural yields as well as for our health and for the health of our pets. Rather than using harsh methods in an attempt to completely eliminate the pest (which rarely works anyway), a combination of less hazardous methods can often reduce pest numbers to manageable levels.

INTEGRATED PEST MANAGEMENT HAS SEVERAL STEPS

1. IDENTIFY TRUE PESTS
Not all "bugs" and "weeds" are pests—some plants and animals are innocuous or actually beneficial. By working to control actual pests only, we save time and money, and we help the environment by maintaining diversity and preventing toxic pollution.

2. SET AN ACTION THRESHOLD AND MONITOR PESTS
The pest population size that is unacceptable must be identified. We may have zero tolerance for some pests (e.g., fleas and ticks in our homes) but be able to tolerate small populations of crop pests; we will only act if the action threshold is reached.

3. DEVELOP AN ACTION PLAN
This may include a variety of methods, each of which aids in pest control by excluding, discouraging, or killing pests. The goal is to control the pest while avoiding or minimizing the use of chemical control agents which may be toxic and have unwanted health and environmental effects.

PREVENTION AND CONTROL METHODS

CULTURAL Usually the first method chosen as part of the action plan; involves cultivation techniques that minimize the habitat or food source for the pest so that other control methods can then be used to adequately control the pests.

↑ Strip cropping minimizes potential food for pests that don't disperse over great distances, lessening the chance of an outbreak of pests.

MECHANICAL Relies on methods that physically exclude, trap, or repel pests, such as physical barriers or traps, or removing pests or weeds by hand or with machinery; can be labor intensive but often inexpensive and may be particularly useful in developing countries with plentiful labor but little cash.

↑ Netting can keep out birds and rabbits that would eat the crops. Reducing water levels in rice paddies to kill azolla is an example of mechanical control.

BIOLOGICAL Introducing predators, sterile males, or plants that repel the pest; technique works best if it follows cultural and mechanical steps.

↑ Ladybugs, predatory beetles that eat pests such as this aphid, can actually be purchased and released, or steps can be taken to attract them naturally to the area. The ideal control agent is a specialist whose preferred food is the pest in question, and who does not attack nontarget species.

CHEMICAL Applying chemicals that kill or repel pests; a last resort that is used only if the other three methods cannot control the pests.

↑ To minimize health and environmental concerns, the preferred chemical is the one that can do the job while being the least toxic and the most degradable.

Infographic 22.6 | **SUSTAINABLE SOIL MANAGEMENT PRACTICES**

↓ Many traditional methods are useful for sustainable agriculture because they focus on protecting the soil, the heart of successful farming.

CONTOUR FARMING When farming on hilly land, rows are planted along the slope, following the lay of the land, rather than oriented downhill to reduce the loss of water and soil after a rainfall.

REDUCED TILLAGE Planting crops into soil that is minimally tilled reduces soil erosion and water needs (it reduces water evaporation). It also requires less fuel because of less tractor use.

TERRACE FARMING On steep slopes, the land can be leveled into steps. This reduces soil erosion and allows a crop like rice to stay flooded when needed.

CROP ROTATION Planting different crops on a given plot of land every few years helps maintain soil fertility and reduces pest outbreaks since pests (or their offspring) from the year before will not find a suitable food when they emerge in the new season.

STRIP CROPPING Alternating different crops in strips that are several rows wide keeps pest populations low; it is less likely the pests will travel beyond the edge of a strip and they may not find another row of this crop.

COVER CROPS During the off-season, rather than letting a field stand bare, a crop can be planted that will hold the soil in place. Nitrogen-fixing crops like alfalfa that improve the soil are often chosen.

↑ Ducklings in the Massa Organics rice fields

own land. "We wanted a farm where success was measured not just in crop yield, but in the overall health of the land," Greg says. "A place where we would count profits, but we would also count the number of sandhill cranes and California quail we saw populating the area."

They also wanted to create a farm that was at least partly local (they planned to sell a portion of their rice at local outlets like farmers' markets and food co-ops). More and more consumers are buying food from local farmers; local agriculture supports local economies and provides fresher and thus healthier food to consumers. Because transportation depends on fossil fuels, the more **food miles** a product travels before reaching the consumer, the greater the **carbon footprint** of that food. And much of our food has traveled quite far—about 2,400 kilometers (1,500 miles), on average.

However, transportation is not the main use of fossil fuels when crops are raised industrially. Research by Christopher Weber and Scott Matthews of Carnegie Mellon University determined that about 90% of the carbon footprint for food grown using conventional industrial methods is from the production of the crop (fuel for equipment, raw materials for pesticides and fertilizer production), not its transport. Therefore, buying organically grown produce—even from far away—may reduce the carbon footprint more so than buying locally grown

industrial crops. Of course, the best option is to choose organic foods that are locally grown. Likewise, because beef raised in industrial Concentrated Animal Feeding Operations has one of the highest carbon footprints of all agricultural products (see Chapter 21), one of the best things you can do is to replace at least some of the beef you eat with chicken, pork, fish, or meatless dishes.

Consumers are becoming more aware that things like food miles, and the way food is raised matter for the environment, their communities, and their own health. Organic foods and ethically raised animal products are claiming a bigger share of consumer dollars annually. But this has opened the door to **greenwashing**—making claims about the environmental benefits of sustainably raised or organic foods that are misleading (organic cookies are probably not really healthier than those made with conventional ingredients; "cage free" eggs may still be from chickens living in overcrowded conditions). Consumers need to be diligent about evaluating claims and make informed decisions about what to purchase. [INFOGRAPHIC 22.7]

---

food miles The distance a food travels from its site of production to the consumer.

carbon footprint The amount of carbon released to the atmosphere by a person, company, nation, or activity.

greenwashing Claiming environmental benefits about a product when the benefits are actually minor or nonexistent.

Infographic 22.7 | **CONSUMER CHOICES MATTER**

↓ Because the growing and transport of our food impacts the environment and our own health so much, choosing foods produced in a way that has a lower impact makes a difference. This also supports sustainable agriculture as an economic endeavor, helping the farmers and communities pursuing these methods.

CONSIDER HOW YOUR FOOD IS RAISED...

Industrially grown food is usually cheaper, but has a higher environmental impact and a high carbon footprint (more fossil fuels are used to produce it) than sustainably grown food. The best way to reduce the carbon footprint of the food you buy is to opt for organically grown food when possible (no fossil fuel–derived fertilizers or pesticides).

...AND HOW FAR IT IS SHIPPED

Even though more fossil fuels go into the *production* of industrially grown foods than in shipping it to market, buying food produced closer to home decreases the transportation part of the carbon footprint. For example, in Iowa, the average grocery store apple has traveled 2,779 kilometers (1,726 miles), and the average head of broccoli, 2,971 kilometers (1,846 miles), while locally grown produce has traveled 56 miles on average. So while transportation does not represent the main way fossil fuels are used in agriculture, it still has a significant impact.

PERSONAL FOOD CHOICES

↓ Buying organic food not only reduces the carbon footprint of the food, it is also healthier for you. But it can get expensive, so if your buying dollars are limited, consider focusing your purchases, when you are able, on the "dirty dozen"—the 12 fruits and vegetables most likely to be contaminated with pesticide residue. The "clean 15" are the products least likely to have pesticide residue, so if you can't afford to buy all organic produce, buy these from the regular produce shelf—but always wash all produce well before eating or cooking!

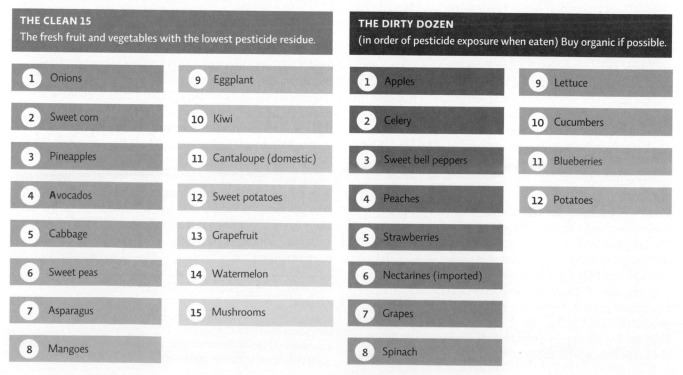

THE CLEAN 15
The fresh fruit and vegetables with the lowest pesticide residue.

1	Onions	9	Eggplant
2	Sweet corn	10	Kiwi
3	Pineapples	11	Cantaloupe (domestic)
4	Avocados	12	Sweet potatoes
5	Cabbage	13	Grapefruit
6	Sweet peas	14	Watermelon
7	Asparagus	15	Mushrooms
8	Mangoes		

THE DIRTY DOZEN
(in order of pesticide exposure when eaten) Buy organic if possible.

1	Apples	9	Lettuce
2	Celery	10	Cucumbers
3	Sweet bell peppers	11	Blueberries
4	Peaches	12	Potatoes
5	Strawberries		
6	Nectarines (imported)		
7	Grapes		
8	Spinach		

Infographic **22.8** | **THE ADVANTAGES AND DISADVANTAGES OF SUSTAINABLE AGRICULTURE**

CONSUMER

ADVANTAGES

• Fresher, tastier, and healthier food (better nutrient profile, no pesticide residue if organic, etc.)
• Satisfaction in making a more ethical and environmentally sound choice

DISADVANTAGES

• Sustainably grown crops may be more expensive.
• Greenwashing can mislead consumers.
• Organic produce may have more blemishes.
• Shelf life of sustainable produce is shorter (not waxed, not picked before ripe)

FARMER & ENVIRONMENT

ADVANTAGES

• Using fewer inputs of water and fossil fuels saves money and causes less environmental damage.
• Soil is not degraded and may be enhanced.
• Less use of toxic chemicals benefits the environment and local communities
• More genetic diversity and species diversity makes it less likely that a pest outbreak or other problem will decimate the entire crop.

DISADVANTAGES

• May be more labor intensive
• Only crops native to the area or suitable for the climate grow well, so a farmer may not be able to grow all crops in all areas.
• Crops grown sustainably may not be as productive per acre as industrially farmed crops (in the short term).
• Fewer government subsidies are available for sustainable agriculture compared to those for industrially grown crops.
• The certification process for getting crops to be labeled as organic takes time and is costly to farmers.

SOCIETY

ADVANTAGES

• Many sustainable methods are less expensive so are suitable for developing nations
• Methods are available that minimize water need—useful in arid areas
• Local production of food can increase food security (see Chapter 20).

DISADVANTAGES

• Productivity per acre is lower for some sustainably produced food (but higher for others).
• Research is needed to identify best methods and crops for a given area.
• Farmers need training to implement these systems (though if indigenous methods are used, it may be the locals who educate the researchers).

It shouldn't be surprising that sustainable and organic agriculture have their own set of trade-offs: While they might be more environmentally friendly than high-input industrial methods, they may also be more expensive and in some cases produce less food per acre of land. But Greg and Raquel were determined to at least try. [INFOGRAPHIC 22.8]

To achieve their goals, the Massas began by installing a recirculation system to reclaim and reuse irrigation water. They planted native oak trees along field borders to serve as a natural windbreak (windbreaks prevent soil from being carried away by wind erosion), and installed nest boxes for wood ducks, barn owls, and bats so that those wild animals would keep area pests in check. "The idea was to restore as much of the natural biodiversity as possible," says Greg, "so that we would not need artificial inputs to run the system." They also took their sustainable ideals to the next level and built a straw house to live in—made of 2 foot thick walls of rice bale, coated on both sides with plastic or stucco—that can withstand the unforgiving heat of a Chico summer and maintain a steady temperature almost entirely by itself.

The house is fireproof, rodent proof, and as Greg likes to joke, bulletproof, too.

A sustainable food future will depend on a variety of methods.

Furuno and the Massas are not the only ones experimenting with polyculture. Indeed, many farmers and scientists across the country are turning to agroecology—working to develop mixed agriculture systems, where instead of planting a single crop, they grow a mix of different species that better replicates the normal ecological community makeup of a given region. Evidence is mounting that such systems can increase a farm's productivity.

Historically, Native Americans planted the "three sisters"—corn, beans, and squash—together in the same plot. Recent studies by plant scientist Mitsuru Tsubo and others confirm a higher productivity per acre in such plots. Though perhaps not a good model for large-scale agriculture, this type of polyculture is recommended for smaller farmers, especially in water-poor areas because it leads to a more efficient use of water by the plants (the more densely planted plots have lower soil temperatures and less water evaporation from soil).

A 2010 report by the International Livestock Research Institute concluded that mixed polyculture farms—ones that, like Furuno's and the Massas', grow both plants and livestock—hold the most promise for intensifying food production worldwide. "It is not big efficient farms on high potential lands but rather one billion small, mixed, family farmers tending rice paddies or cultivating maize and beans while raising a few chickens and pigs, a herd of goats or a cow or two...[who are] likely to play the biggest role in global food security over the next several decades," the institute's executive director, Knut Hove, wrote in the report. "These 'mixed extensive' farms make up the biggest...and most environmentally sustainable agricultural system in the world."

Polyculture operations based on agroecological principles of complex, synergistic systems may do more to free the farmer from expensive corporate inputs than other methods—even large-scale organic farms are industrial operations that have just traded chemical inputs for organic and biological ones. Agroecologist Miguel Altieri argues that all types of monoculture farming, even organic ones, require external inputs and are not a viable option for small farmers or those in developing nations.

Meanwhile, at the Land Institute in Kansas, Wes Jackson is working on a more ambitious plan that he says would correct a mistake made well before the Green Revolution or the rise of factory farms: He wants to replace virtually all of our existing grain crops—which are **annual crops**—with **perennial crop** varieties. Early farmers domesticated annual plants—which grow, produce seeds, and die in a year—because they could manipulate them to produce higher yields from one year to the next (by selecting the best producers and junking the rest, or by crossbreeding plants with good traits). But annuals have to be replanted every season, which requires labor and fossil fuel energy and necessitates tearing up the ground and disrupting the delicate balance of soil ecosystems.

Perennials, on the other hand, can be harvested year after year without disturbing the soil to replant—that means heavy equipment is used less often to manage perennial crops. And the deep roots perennials develop not only hold soil in place, but they tap much farther down into underground aquifers than their annual counterparts, thus dramatically reducing the amount of irrigation needed. Less herbicide is needed for perennial plots as well—weeds do not readily sprout and grow among the established plants. This makes perennials especially attractive for regions with marginal land or an arid climate.

"We hope to advance and enlarge upon the idea that the ecosystem is the necessary conceptual tool for truly sustainable grain agriculture," Jackson told the *Atlantic* in a recent interview. "We believe we can have an agriculture where management by human intervention is greatly reduced."

Saving seeds is a critical step in a sustainable food future.

One unintended consequence of modern monoculture farming is that fewer and fewer varieties of crops are planted. In the past, each region used locally adapted crop varieties developed over hundreds, even thousands, of years by area farmers—crops that were adapted to the local climate, soil conditions, pests, etc. These different varieties were all genetically distinct and represented the genetic diversity of the crop as a whole. But as independent farmers joined the industrialized food production system, their tendency was to plant the single variety of corn or wheat or rice that was currently the highest producer. This has led to an erosion of genetic diversity (the variation among individuals within a species). The problem with this loss is that such genetic raw material allows crops to respond to changes such as the arrival of a new fungal pest, a drier climate, etc. This perhaps has never been as important as it is now, with so many mouths to feed on a rapidly changing planet.

For example, the need to grow rice in arid climates challenges plant breeders to select for plants that can tolerate drier conditions. By going to the myriad wild varieties that exist (known as *wild-type* varieties), breeders can start selecting the individual plants that best tolerate drier conditions, and hopefully produce a variety of rice that can grow under these new conditions. Without a wealth of rice varieties to choose from, plant breeders would have a harder time developing new strains to meet specific needs.

The fear is that, if not planted, many plant varieties could be lost forever. **Seed banks**, such as one located in the high Arctic in Svalbard, Norway, are one solution to

---

annual crops Crops that grow, produce seeds, and die in a year and must be replanted each season.

perennial crops Crops that do not die at the end of the growing season but live for several years, allowing them to be harvested annually without replanting.

seed banks Places where seeds are stored in order to protect the genetic diversity of the world's crops.

this dilemma. There the seeds are kept in cold storage, housed in a tunnel carved out of a mountain; even if the power goes out, these seeds will stay frozen. Managed by Norway and the international Global Crop Diversity Trust, the Svalbard bank holds more than 700,000 varieties of crops from all over the world. The trust also helps developing nations find crop varieties and produce seeds for the seed bank. Local, community seed banks may be of even more immediate use to local farmers, and many are springing up in India and Africa. Most of these programs are run by women's groups, such as those supported by the Indian nonprofit organization, GREEN Foundation (GREEN stands for genetic resource, ecology, energy, and nutrition).

Can sustainable farming methods feed the world?

In the meantime, for the Massas, ducks have proven to be the best possible solution to the challenges of modern rice farming. They ended up with duck meat to sell, and though they didn't take any precise measurements of yields during their first trial run, their rice crops did not appear to suffer at all. "We learned a couple of things," Raquel says. "The ducks trampled some of the rice in their pond, which would not have been a problem if we had used a larger section of the field. I also think we used the wrong breed of duck. They were a little too large to move effectively between the dense plants, and they were not active enough in their foraging activities. These ducks were bred to sit around all day and eat and gain weight quickly for industrial meat production. We are currently researching which breeds to try next."

Still, in the end, the Massas harvested both rice and duck meat. The key to successful duck/rice farming is to remove the ducks before they get big enough to trample rice plants or strong enough to pluck rice seeds from deep within the mud. The hard part isn't knowing when to do this, but having the resolve for what comes next: killing and eating the ducks. The Massa family struggled, but ultimately felt good about the outcome. "I know the conditions in which they were raised were more humane than 99% of the meat ducks in this country," says Raquel. "They had it good and you can taste that in the finished product."◉

Research articles referenced in this chapter:

Altieri, M.A. 2012. *The Scaling Up of Agroecology.* Sociedad Científica Latinoamericana de Agroecología.

Davis, D.R., *et al.* 2004. *Journal of the American College of Nutrition,* 23: 669–682.

Glover, J.D., *et al.* 2010. *Science,* 328: 1638–1639.

Hossain, S.T., *et al.* 2005. *Asian Journal of Agriculture and Development,* 2: 79–86.

Macmillan, S., and Sere, C. 2009. *Back to the Future: Revisiting Mixed Crop-Livestock Systems.* Nairobi, Kenya. International Livestock Research Institute.

Reganold, J.P., *et al.* 2010. *PLoS ONE,* 5: e12346.doi:10.1371.

Tsubo, M., *et al.* 2003. *Water SA,* 29: 381–388.

Weber, C.L. and Matthews, H.S. 2008. *Environmental Science and Technology,* 42: 3508–3513.

BRING IT HOME

◉ PERSONAL CHOICES THAT HELP

While a typical supermarket may seem to present a dizzying array of food choices to the consumer, a look at the ingredient labels betrays our increasing reliance on growing monocultures of common strains of corn, soy, and wheat. These choices will only change if you, the consumer, demand it.

Individual Steps

→ Carefully examine the labels on the food you buy—as your food budget allows, opt for food products that are organically grown and, if available, locally produced.

Group Action

→ Organize a community garden that specializes in heirloom varieties of vegetables that might not be found in the local grocery stores. Start by requesting a seed catalog from www.rareseeds.org.

→ Research specific farming practices that more closely mimic those found in natural ecosystems, such as Joel Salatin's Polyface Farm (http://www.polyfacefarms.com).

→ Subscribe to a community supported agriculture (CSA) farm and receive a weekly supply of sustainably grown produce. A list of CSAs in your area can be found at www.localharvest.org/csa.

Policy Change

→ Identify bodies of water in your area that may be impacted by cultural eutrophication from fertilizer runoff. Meet with local officials and propose ordinances limiting fertilizer use by homeowners, golf courses, or other possible sources of the pollution.

UNDERSTANDING THE ISSUE

CHECK YOUR UNDERSTANDING

1. Techniques used in sustainable agriculture practices include:
a. frequent use of herbicides to keep weeds down.
b. flooding crops frequently with water to reduce pests and increase runoff of excess nutrients.
c. abundant use of fertilizers to allow plants to reach their full capacity.
d. crop rotation to minimize insect and soil pests.

2. Integrated pest management (IPM) includes:
a. growing plants that repel pests alongside the crop.
b. allowing the development of genetic resistance in the pest.
c. killing as many soil and insect organisms as possible with every application of organic pesticides.
d. using pesticides which are persistent in the environment so they will continue to work long after they are applied.

3. The presence of ducks and azolla in rice cultivation has shown that:
a. azolla grows more abundantly when ducks are present.
b. rice yields go up when ducks are removed and azolla is added to the rice paddy.
c. crops such as rice cannot succeed without azolla to provide shade and ducks to provide pollination.
d. restoring some biodiversity to rice paddies reduces pest damage and increases rice yields.

4. Biological controls might include:
a. organic herbicides and pesticides.
b. using species such as fish that eat weeds or pests.
c. taking advantage of natural occurrences such as flooding or drought.
d. All of the above.

5. Soil erosion can be minimized by:
a. slowly and continuously flooding the crops so that the water has time to sink in.
b. plowing crop rows so that they go up and down slopes, allowing irrigation at the top to slowly trickle down.
c. terracing or contoured farming on slopes.
d. leaving the ground unplanted during the nongrowing season so that the soil can replenish itself.

6. According to research, the biggest way to reduce the carbon footprint of the foods you eat is to:
a. eat local produce as much as possible, even if it is not organically grown.
b. eat produce shipped in from countries where it can be grown "in season."
c. eat foods that are grown using organic methods.
d. eat only foods that use conventional industrial methods.

WORK WITH IDEAS

1. Industrial agriculture allows us to grow more food, but many of the technologies used have significant adverse effects. Discuss three specific examples of how growing our food under such an agricultural system may be harmful to us and to the environment.

2. How is the agricultural system illustrated in the duck/rice farm similar to and different from industrial agriculture? Which system do you think is a better way to grow food? Why?

3. List and discuss the major categories of integrated pest management practices. What is the value of pursuing IPM to control pests compared to simply using chemical pesticides alone?

4. What is sustainable agriculture? Give one example each of how we can sustainably raise crops and animals for food. Present an argument, along with three reasons, about why sustainable agriculture is a better way to grow food.

ANALYZING THE SCIENCE

The data in the following graph come from the Farming Systems Trial®(FST) research study conducted by the Rodale Institute. This side-by-side comparison of corn and soybeans grown under organic and industrial agricultural systems was started in 1981 and is one of the longest-running studies of its kind.

INTERPRETATION

1. What does this graph show?
2. Calculate the following for conventional and organic systems: profit per unit of yield; energy input per unit of yield; and greenhouse gas emissions per unit of yield. How do the two systems compare on these three parameters?
3. According to the Food and Agricultural Organization of the United Nations, "Organic agriculture has the potential to secure a global food supply, just as conventional agriculture does today, but with reduced environmental impact." How do the data from the FST study support this statement?

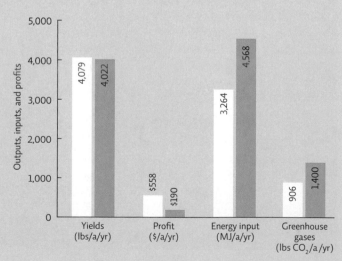

Organic   Conventional   lbs = pounds, a = acre, yr = year, MJ = megajoule

ADVANCE YOUR THINKING

Hint: To answer the following questions, it might be helpful to access the actual Farming Systems Trial report, at www.rodaleinstitute.org/fst30years.

4. According to the FST report, even in drought years, the yields for organic corn were approximately 31% greater than those for conventional (non-drought-resistant) varieties. At the same time, genetically engineered drought-tolerant varieties had yields that were no more than approximately 13% greater than conventional (non-drought-resistant) varieties. Why might this be the case? Why is this an important finding?

5. The FST report also indicates that crops grown using organic methods produced yields equivalent to those of conventional crops, even though they had more weed competition in their fields. Why might this be the case? What makes this an important finding?

6. According to the FST report, the largest energy input in organic systems was diesel fuel, while the largest for conventional systems was nitrogen fertilizer. How could these energy inputs be changed to make both systems more economically and ecologically sustainable?

EVALUATING NEW INFORMATION

Which seed varieties are best? There are groups, such as Renewing America's Food Traditions (RAFT) (www.albc-usa.org/RAFT/), that say we should preserve the original biodiversity and the heirloom varieties that our ancestors grew. Other groups believe that selective breeding and hybridization create superior crops, the downside being that farmers must purchase new seeds every year. Bayer is a huge conglomerate that sells a variety of hybrid seeds and chemicals to enhance growth (bayer-cropscience.us/). Another group, AgBioWorld (www.agbioworld.org), says that it is neutral and reports the news on agricultural practices. Look at the articles and the archives of old newsletters to help you analyze the organizations.

Evaluate the websites and work with the information to answer the following questions:

1. Evaluate the agendas of the three organizations as well as the accuracy of the science behind their positions on heirloom varieties versus hybrid crops.

   a. Who runs each website? Do the person's/organization's credentials make the information presented on food and agriculture issues reliable or unreliable? Explain.

   b. What is the mission of each website? What are the underlying values? How do you know this?

   c. What claims does each website make about the current problems in food production and what the future of agriculture should be? Are their claims reasonable? Explain.

   d. How do the websites compare in providing scientific evidence in support of their assessment of agriculture and their position on the role of heirloom varieties versus hybrid crops? Is the evidence accurate and reliable? Explain.

2. How do the three organizations compare in engaging you, as a citizen, in agricultural policy? Do you think that citizen involvement in policy issues is necessary and effective? Explain your responses.

MAKING CONNECTIONS

WHAT ARE YOU PUTTING IN YOUR MOUTH?

Background: There are many organizations that want to help you make different choices with your food dollars. Some stress convenience, some stress health, some stress reducing the food's environmental footprint, some stress sustainability. How do you decide where to spend your money and what to spend it on?

Case: Research the following new ways to buy food:

- Online grocers: Peapod was probably the first, but there are now many similar services; you order at their online site, they shop and deliver to your door.

- Online purchases in bulk: Zaycon and similar companies claim to have top-quality merchandise, including produce and meat, at wholesale prices. At a scheduled date and time, you drive to a facility and pick it up.

- Organic community-supported agriculture (CSA): You take a subscription, typically for several months. This allows a local farmer, with subscriptions guaranteed, to farm using organic methods. Once a week, you pick up your produce share.

- Subscription wholesale groups: Bountiful Baskets is a good example. You choose each week whether to purchase a regular or an organic basket of fruits and vegetables. On the designated day and time, you go to a central location and pick up your share.

- Rescued produce: On a regular basis, farmers ship produce to a distribution warehouse. However, not all of it is ordered by stores, leaving tons of surplus food. When a warehouse is about to receive a new shipment, the leftover pallets of fruits and vegetables are sent to the local landfill. The 3000 Club in Arizona began Market on the Move (MOM), which picks up surplus produce at warehouses, and takes it to distribution sites. Volunteers staff the sites for 4 hours, and for a $10 donation you receive a produce share (up to 60 pounds of fresh fruits and vegetables).

- Find examples (local if possible) of one or more of these options. Research them and write a report detailing.
  - The type(s) of food offered: all types, fresh produce only, organic, locally grown, etc.
  - The system: an advance subscription, a weekly option, or just show up
  - The cost compared to purchasing the same quality and quantity at a grocery store.

- Write up a proposal to a local group, such as a club or your neighborhood, regarding using one of these options.

BRINGING DOWN THE MOUNTAIN

In the rubble, the true costs of coal

Leveling a mountain for coal in Appalachia. The mountaintop is blown off and dumped into valleys to leave sprawling terraced barren lands in what were once diverse temperate forests.

CORE MESSAGE

Human society runs on energy, and coal continues to be a major reliable energy resource. However, coal mining causes irreversible environmental degradation and the by-products of mining and burning coal pose significant health risks. Despite these drawbacks and the availability of renewable, cleaner energy sources, coal's availability and industrial presence keep it a major energy player. Researchers are developing ways to lessen the impact of burning coal, but as long as we continue to use it, the impact of mining will remain.

GUIDING QUESTIONS

After reading this chapter, you should be able to answer the following questions:

→ How important is coal as an energy source and how is it used to generate electricity?

→ What is coal, how is it formed, and what regions of the world have coal deposits that are accessible?

→ What methods are used to mine coal and what are the advantages and disadvantages of each?

→ What are the advantages and drawbacks of burning coal?

→ What new technologies allow us to mine and burn coal with fewer environmental and health problems?

A thousand feet above the foothills of Central Appalachia, near the Kentucky—West Virginia border, a four-seater plane ducks and sways like a tiny boat on an anxious sea. It's windier than expected, and Chuck Nelson, a retired coal miner seated next to the pilot, grips the door in an effort to steady his nerves. The passengers have come to survey the devastation wrought by mountaintop removal—a form of surface mining that involves blasting off several hundred feet of mountaintop, dumping the rubble into adjacent valleys, and harvesting the thin ribbons of **coal** beneath.

At first, the landscape looks mostly unbroken; mountains made soft and round by eons of erosion roll and dip and rise in every direction, carrying a dense hardwood forest with them to the horizon. But before long, a series of **mountaintop removal** sites come into view. Trucks and heavy equipment crawl like insects across what looks like an apocalyptic moonscape: decapitated peaks and acres of barren sandstone and shale. Smoke curls up from a brush fire as the side of an existing mountain is cleared for demolition. And orange-and-turquoise-colored sediment ponds—designed to filter out heavy metal contaminants before they permeate the water downstream—dot the perimeter.

Here and there, a tiny patch of forest clings to some improbably preserved ridge line. "That's where I live," Nelson says, forgetting his air sickness long enough to point out one such patch. "My God, you would never know it was this bad from the ground." The aerial tour has reached Hobet 21, which, at more than 52 square

↓ The controversial Spruce No. 1 coal mine in West Virginia was originally given a permit authorizing it to dump strip mining waste into 7 miles of creeks and onto 2,000 acres of land. In 2011, the Environmental Protection Agency rejected the permit on the basis of the "irreversible damage" that would be inflicted to the streams, groundwater, and land, including: "the elimination of all fish, killing of birdlife, reduction of habitat value, and risk of human illness."

↑ Seams of coal exposed at a mountaintop removal mining site in Welch, West Virginia

◉ **WHERE IS APPALACHIA?**

HOBET 21 MINE

APPALACHIAN REGION

kilometers (20 square miles), is the region's largest mining operation. So far, sites like this one have claimed nearly a million acres of forested mountain, across just four states: Kentucky, West Virginia, Virginia, and Tennessee. But there is still more coal to mine. And if the coal companies have their way, hundreds of thousands of acres more will be thus obliterated in the coming years.

To stem this tide of destruction, environmental activists have sued the coal industry, the state of West Virginia, and the federal government itself. They argue that mountaintop removal mining destroys biodiversity, pollutes the water beyond recompense, and threatens the health and safety of area residents. And by obliterating the mountains, they say, it also obliterates the culture of Appalachia.

Coal industry reps have countered by decrying the loss of jobs, tax revenue, and business the already impoverished region would suffer if the mines were to close under the weight of too much regulation. They also point out that the culture of Appalachia is as bound to coal mining as it is to the mountains. Both sides count area residents, including miners, among their ranks.

At the heart of the issue is coal itself—our country's dirtiest, and most abundant, energy source—the one most responsible for rising CO_2 levels from electricity production, but also the one we rely on most heavily and the one we are consuming most rapidly. As the Appalachian reserves dwindle, debates raging throughout these decimated foothills are reverberating across our energy-addicted nation.

The world depends on coal for most of its electricity production.

Simply put, coal equals energy. **Energy** is defined as the capacity to do work; like all living things, we humans need it, in a biological sense, to survive. But we also need it to run our societies—to heat and cool our homes; operate our cell phones, lamps, and laptops; fuel our cars; and power our industries. Most of our energy comes from **fossil fuels**—nonrenewable carbon-based resources, namely coal, petroleum, and natural gas—that were formed over millions of years from the remains of dead organisms.

---

coal Fossil fuel formed when plant material is buried in oxygen-poor conditions and subjected to high heat and pressure over a long time.
mountaintop removal Surface mining technique that uses explosives to blast away the top of a mountain to expose the coal seam underneath; the waste rock and rubble is deposited in a nearby valley.
energy The capacity to do work.
fossil fuels Nonrenewable resources like coal, petroleum, and natural gas that were formed over millions of years from the remains of dead organisms.

In the United States alone, we use roughly 1 billion tons of coal per year, the vast majority of it to generate electricity. **Electricity** is a natural form of energy (lightning and nerve impulses are electrical) that we have learned to create on demand, producing it in a central location and sending it out via transmission lines to where we want it to go. More than 40% of electricity generation worldwide, and 45% in the United States, comes from burning coal. (Americans especially love their electricity: In 2008, we used more than 22% of all the kilowatt-hours generated in the world.)

Coal-fired power plants work by feeding pulverized coal into a furnace to generate heat, which then powers a system that produces electricity. It takes roughly 0.5 kilogram (1 pound) of coal to generate 1 kilowatt-hour (kWh) of electricity; that's enough to run ten 100-watt incandescent light bulbs for an hour, an energy efficient refrigerator for 20 hours, or an older, less efficient one for 7 hours. The average U.S. family of four uses about 11,000 kWh of electricity per year. That comes out to

some 5,500 kilograms (12,000 pounds) of coal, or 1,375 kilograms (3,000 pounds) per person.

So, how does coal stack up against other energy sources? On one hand, it produces more air pollution than any other fossil fuel. On the other, it is safer to ship, cheaper to extract, and in the United States at least, more abundant by far. In fact, the United States has 10 times more coal than it does oil and natural gas combined; in 2010 alone, we mined more than 1.1 billion tons. Most of it came from Wyoming, which leads the nation as a coal producer, and Appalachia, which follows behind as a close second. While we exported some of that yield, the vast majority was used to power American households and businesses. [INFOGRAPHIC 23.1]

In terms of net energy, or **energy return on energy investment (EROEI)**—a metric that allows us to compare the amount of energy we get from any individual source to the amount we must expend to obtain, process, and ship it—coal is neither the best nor the worst. It has

Infographic **23.1** | **ELECTRICITY PRODUCTION FROM COAL**

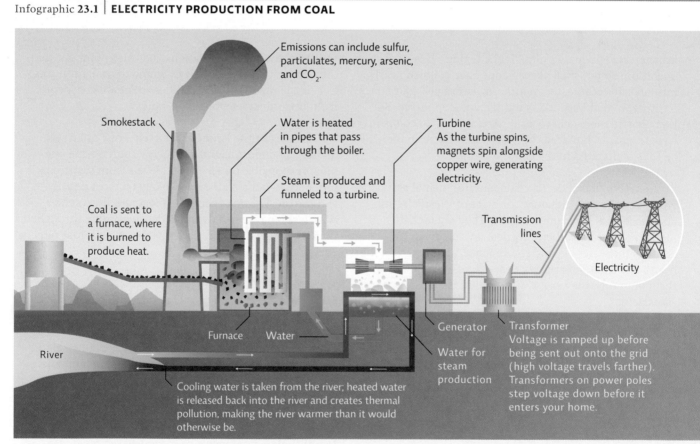

↑ The most common way to generate electricity is to heat water to produce steam; the flow of steam turns a turbine inside a generator to produce electricity. This schematic shows TVA's coal-fired Kingston plant in Tennessee, which generates 10 billion kilowatt-hours a year by burning 14,000 tons of coal a day, supplying electricity to almost 700,000 homes.

↑ A train carrying coal leaves a mountaintop removal mining site and travels through the backyards of homes in Welch, West Virginia.

U.S. ELECTRICITY GENERATION BY FUEL, 2010

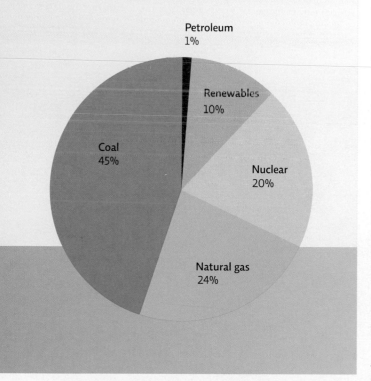

Petroleum
1%

Renewables
10%

Coal
45%

Nuclear
20%

Natural gas
24%

↑ Coal is the main fossil fuel used to produce electricity in the United States and worldwide. Increasing the use of renewable fuels and taking steps to improve energy efficiency and conservation could decrease the role coal plays in energy production.

an EROEI of about 8:1 (8 units of energy produced for every 1 unit consumed for a net production of 7), compared to 15:1 for oil, 6:1 for nuclear, and 20:1 for wind.

There can be no denying the blessings of coal: This sticky black rock has powered several waves of industrialization—first in Great Britain and the United States, now in China—and in so doing has shaped and reshaped the world as we know it. But as time marches on, the costs of those blessings have become all too apparent: They include an ever-growing list of health impacts—from birth defects to black lung disease—and an equally lengthy roster of environmental costs—not only the destruction of Appalachia, but also the pollution of Earth's atmosphere with CO_2 and other greenhouse gases.

This litany of paradoxes has given rise to a deep national ambivalence. While we are consuming more electricity, and burning more coal, than at any time in our history, applications for new coal-fired power plants have been rejected left and right in recent years by determined citizens and local governments from Maryland to Minnesota. "We're caught in a catch-22," says Scott Eggerud, a forest manager with the West Virginia's Department of Environmental Protection. "On one hand, it's like we need the stuff to live; on the other hand, we see that it's kind of killing us." Nowhere is this catch-22 more pronounced than in the foothills of Central and Southern Appalachia.

Coal forms over millions of years.

The Appalachian Mountains were born a few million years before the rise of the dinosaurs, when Greenland, Europe, and North America hovered near the equator—a single giant landmass bathed in a dense tropical swamp. The northwestern bulge of what is now Africa pressed into the easternmost edge of North America, pulverizing the colliding continental margins and forcing a colossal mass of land upward, into a mountain range as high as the Himalayas. It was the last and greatest of three violent clashes that joined all the world's land into a single super-continent, upon which stegosauruses and velociraptors would eventually roam.

The fallout from this cataclysm—the gradual accretion of decomposing swamp vegetation compressed and baked by heat and time—established the Appalachian coal beds. As

electricity The flow of electrons (negatively charged subatomic particles) through a conductive material (such as wire).

energy return on energy investment (EROEI) A measure of the net energy from an energy source (the energy in the source minus the energy required to get it, process it, ship it, and then use it).

Infographic 23.2 | **COAL FORMATION**

↓ Coal is formed over long periods of time as plant matter is buried in an oxygen-poor environment and subjected to high heat and pressure. Places with substantial coal deposits that are retrievable with current technology are called coal reserves.

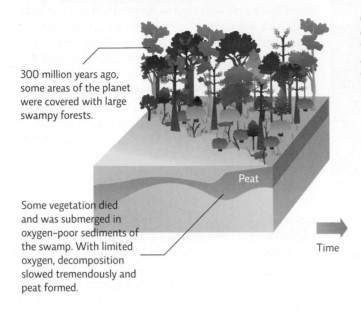

300 million years ago, some areas of the planet were covered with large swampy forests.

Peat

Some vegetation died and was submerged in oxygen-poor sediments of the swamp. With limited oxygen, decomposition slowed tremendously and peat formed.

Time

Over time, pressure built up as more sediment was laid down; increased pressure and heat converted the peat to a soft coal (lignite); in areas with enough pressure and heat, lignite was converted to harder varieties of coal (bituminous and anthracite).

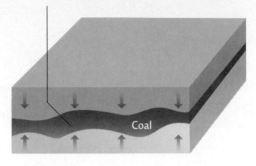

Coal

↓ Seven tons of overburden (the rock, soil, and ecosystem waste) are moved for every ton of coal extracted.

the swamp plants died out, they were buried under a mud so thick it kept oxygen out. Instead of being fully decomposed by bacteria, their remains produced *peat*—a soft mash of partially decayed vegetation. As time passed, and more and more layers of sediment were laid down over the peat, pressure and heat compressed it into the denser rocklike material that we know as coal. [INFOGRAPHIC 23.2]

This same story—tectonic upheaval, followed by deep and rapid burying of organic material, followed by the material's slow compaction into coal—has played out in numerous places across the globe. As a result, coal is found everywhere, though of course, some places have more than others. Europe and Asia hold about 36% of the world's reserves, while the Asian Pacific has about 30%, and North America holds just over 28%. [INFOGRAPHIC 23.3]

The Appalachian beds were once among our most bountiful reserves; seams as tall as a man wound for miles through the mountainside and made for easy harvesting. But after 150 or so years of mining, those reserves have

overburden The rock and dirt removed to uncover a mineral deposit during surface mining.

dwindled noticeably. At current rates of usage, proven coal reserves (those we know are economically feasible to extract) should last about 120 years—longer if deeper reserves can be accessed. And as the layers of minable coal have grown thinner and harder to reach, the coal industry has become both more sophisticated and more destructive in its approach to extracting the coal.

Mining comes with a set of serious trade-offs.

Hobet 21, which has claimed at least 12,000 acres of land, was once the site of several adjacent peaks. To get at the coal beneath those peaks, miners began by clear-cutting the forest above. Next, they drilled holes deep into the side of the mountain (some holes up to 120 meters [400 feet] deep), set dynamite in those holes, and blasted as much as 1,000 feet of mountain into a mass of rubble known as **overburden**. The miners repeated this process several times, until the layers of coal were exposed. Then, using buckets big enough to hold 20 midsized cars, they scooped that rubble into staircase-shaped mounds that filled in entire neighboring valleys from the ground up—all told some 80 million tons of overburden each year. The process obliterated the forest habitat, buried countless

Infographic 23.3 | **MAJOR COAL DEPOSITS OF THE WORLD**

→ Coal reserves are not evenly distributed around the world. The United States has 28% of the world's coal reserves, with much of its best, low-sulfur bituminous coal found in Appalachia.

● Anthracite and bituminous coal ● Lignite

Infographic 23.4 | **MOUNTAINTOP REMOVAL**

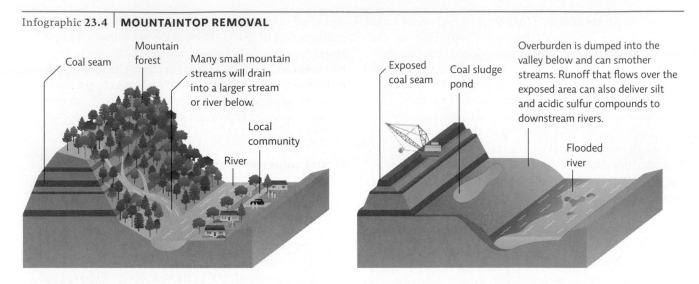

↑ Surface mining techniques are used when coal seams are close to the surface. In Appalachia, the forests are first clear-cut and then explosives are used to blast away part of the mountain. Heavy equipment then digs through debris, dumping the overburden (soil and rock) into the nearby valley, burying streams as the valley is filled in. The exposed coal is dug out and some processing is done on site. Coal sludge left over from processing is stored in ponds on the mining site.

streams, and permanently reordered the land's natural contours. [INFOGRAPHIC 23.4]

This mountaintop removal mining is just one form of **surface mining**. The other type, known as strip mining, employs a similar process: Workers use heavy equipment to remove and set aside overburden so that they can harvest the coal beneath. When they finish mining one strip of land, they return the overburden to the open pit and move on to a new strip. Strip mines are used in areas like Wyoming, where the coal is close to the surface and the ground above is fairly level.

With their reliance on explosives and heavy equipment, surface mines are a far cry from the underground mines, also called **subsurface mines,** that sustained the Nelson family for so many generations. "When our daddies were mining, back in the '40s and '50s, the seams were as tall as full-grown men," says Nelson, who since retiring has become a spokesperson for the anti-mining Ohio Valley Environmental Coalition. "So you could get at 'em the old-fashioned way, with pickaxes and sledgehammers." Those days of plenty are gone, he says. Many of the coal

seams that remain are too thin to be culled by human hands.

To be sure, subsurface mines (which still make up about 60% of all coal mines worldwide, and about 50% of those in the United States) come with their own challenges. Water seeps easily into tunnels, and as it does, toxins leech from the surrounding rocks into the gathering pools. Sulfate in particular produces **acid mine drainage**, which goes on to contaminate soil and streams and has become a major problem with both active and closed mines. In 2011, Duke ecologist Emily Bernhardt and her colleagues reported that not only is acidic water directly toxic to many aquatic plants and animals, it also alters the nutrient cycle of streams in ways that reverberate all the way up the food chain. Subsurface mines also account for 10% of all methane release in the United States; methane is a potent greenhouse gas. [INFOGRAPHIC 23.5]

But subsurface mines also come with some advantages; unlike surface mines, they don't disrupt or permanently alter large surface areas. And because much of the work is still done with workers rather than by machines, they employ more people. In Appalachia, 100,000 mining jobs were lost between 1980 and 1993 as underground mining gave way to mountaintop removal. In Kentucky alone, mining jobs are down 60% in recent years, even though coal production is on an upswing. The loss of traditional mining jobs has bred yet another controversy in the region. Coal industry reps say that the riskier deep-mining jobs are being replaced with higher-paying, safer

surface mining Removing dirt and rock that overlays a mineral deposit close to the surface in order to access that deposit.
subsurface mines Sites where tunnels are dug underground to access mineral resources.
acid mine drainage Water flowing past exposed rock in mines that leaches out sulfates. These sulfates react with the water and oxygen to form acids (low-pH solutions).

Infographic **23.5** | **SUBSURFACE MINING**

→ Subsurface mines are used to access many minerals and ores, including coal. Modern mining depends on powerful machinery to drill out tunnels and remove and transport the materials.

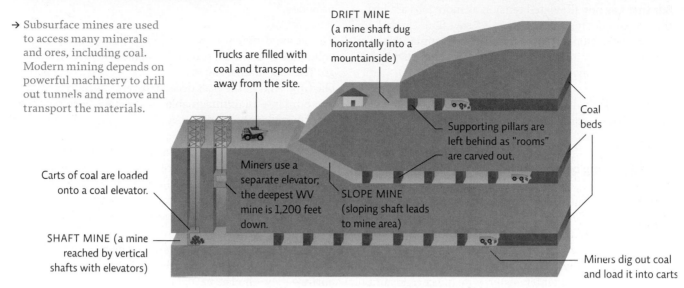

DRIFT MINE (a mine shaft dug horizontally into a mountainside)

Trucks are filled with coal and transported away from the site.

Supporting pillars are left behind as "rooms" are carved out.

Coal beds

Carts of coal are loaded onto a coal elevator.

Miners use a separate elevator; the deepest WV mine is 1,200 feet down.

SLOPE MINE (sloping shaft leads to mine area)

SHAFT MINE (a mine reached by vertical shafts with elevators)

Miners dig out coal and load it into carts

MINING HAZARDS

COAL DUST
Far more miners die from pneumoconiosis (black lung disease) caused by breathing coal dust than from mining accidents. In 2004, the Centers for Disease Control reported 703 coal miner deaths from pneumoconiosis, compared to 28 accidental deaths.

EXPLOSIONS AND MINE COLLAPSE
Methane gas fumes and coal dust are the most common causes of mine explosions; 362 miners died in the West Virginia Monongah coal mine explosion in 1907—the worst mining accident in U.S. history; 29 died after an explosion in the Upper Big Branch mine in 2010.

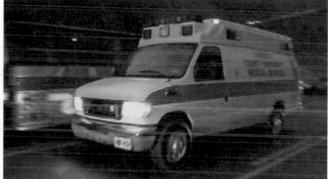

FIRE
Some underground coal mine fires have been burning for hundreds of years; a fire in an abandoned mine in Centralia, Pennsylvania, burning since 1962, has forced the abandonment of the town.

TOXIC FUMES
In 2006, twelve miners died from carbon monoxide poisoning after an explosion trapped them in the Sago Mine in West Virginia.

work—like demolition and heavy equipment operation. But that has not alleviated tension as more jobs are lost than replaced. "It's a double insult," says Tim Landry, a fourth-generation deep miner in West Virginia. "They're not only destroying the land that we love, but they're taking our jobs away, too."

But a loss of jobs is not the community's only—or even its most serious—concern.

Surface mining brings severe environmental impacts.

Bob White is a tiny unincorporated village on the edge of Charleston, West Virginia—just one segment of an endless trickle of trailers and shotgun houses that hug the mountain on each side of long winding valley after long winding valley. Maria Gunnoe, a 40-something waitress, and the daughter, sister, wife, niece, and aunt of coal miners, has lived there, on the same property, all her life. She remembers having free run of the mountain as a child. "When we were kids we used to roam deep in the holler," she says, referring to the mini-valleys that snake through the region's foothills. "We had access to all the

↑ Maria Gunnoe became an environmentalist after the removal of 2,000 acres as the result of mountaintop mining in her area caused flooding to her home and property, poisoned her well water, and made her daughter sick. Gunnoe received the Goldman Environmental Prize for her organizing efforts in her southern West Virginia community.

resources—food, medicine, water—that these mountains provided."

Things are different now. Her own children run into big yellow gates and No Trespassing signs wherever they go. The mountains, she says, have been closed off for blasting. And in the past decade, several million tons of overburden—an unimaginable mass—have been dumped into the valleys around Bob White.

The upheaval has had a noticeable impact on area residents. For one thing, the loss of forest and the compaction of so much soil has increased both the frequency and severity of flooding (without trees, and with the soil so compressed, the ground can't absorb water). "They've filled in like 15 of the valleys around me, and floods are about 3 times more serious than they ever were before," she says. One 2003 flood nearly swallowed her entire valley—house, barn, family and all.

Floods aren't the only problem. In fact, it's the blasting that most scares Gunnoe. It fills the air with tiny particles of coal dust—easily inhaled and full of toxins like mercury and arsenic. Studies show a higher incidence of respiratory illnesses in mining communities. And according to a 2011 study by Melissa Ahern of Washington State University, children in those communities are more likely to suffer a range of serious birth defects, including heart, lung, and central nervous system disorders. "When they were clearing the ridgeline right behind us, we'd get blasted as much as 3 times a day," Gunnoe says. "There were days when we'd have to just stay inside because you couldn't breathe out there. And now, my daughter and I both get nosebleeds all the time."

In addition to filling the air, these toxins also permeate the region's rivers, streams, and groundwater. In a 2005 **environmental impact statement** on mountaintop removal mining, the U.S. Environmental Protection Agency (EPA) reported that selenium levels exceed the allowable limits in 87% of streams located downhill from mining operations. Toxins were as much as 8 times higher in streams near mined areas than in streams near unmined areas. When tests revealed dangerous levels of selenium in the stream behind Gunnoe's house, she and her neighbors started getting their water from town. "We can't trust the water in our own streams anymore," she says. "It's sad, but this is just not the same place that I grew up in."

environmental impact statement A document outlining the positive and negative impacts of any action that has the potential to cause environmental damage; used to help decide whether or not that action will be approved.

↑ In 2008, more than 1 billion gallons of coal fly ash slurry surged into homes and waterways near Harriman, Tennessee, after breaching a coal ash containment pond. It was the largest fly ash release in U.S. history. Many residents have been permanently displaced.

On top of all these immediate threats to human health and safety, area residents like Gunnoe worry about long-term damage to the region's natural environment. They are not alone. Around the world, in fact, mining operations are a major cause of environmental degradation. To be sure, geological resources—those like copper, gold, iron ore, and coal, that are buried deep underground and so must be dug up—are essential to the functioning of modern society. But harvesting them unleashes a dangerous mix of toxic substances that are normally sequestered away from functioning ecosystems, and that living things —including humans—are not generally well adapted to handle.

> "We can't trust the water in our own streams anymore. It's sad, but this is just not the same place that I grew up in."
> —Maria Gunnoe

At surface mines throughout Appalachia, dangerous quantities of iron, aluminum, selenium, and other metals have leached out from the blasted overburden, and may be starting to bioaccumulate and biomagnify up the food chain. Scientists and fishers alike have noted an alarming decline in certain fish that populate area streams and rivers. And recent surveys indicate that biodiversity has decreased in direct proportion to the concentration of such metals in the water. This loss of aquatic life also affects forest life, since many terrestrial animals feed on insects like mayflies and dragonflies that begin their life in the water.

The nutrient cycle in ecosystems surrounding mined areas has also been altered—in some cases, dramatically. Extra sulfates, released from blasted rock, have increased nitrogen and phosphorus availability, and in so doing have led to eutrophication. And as sulfate levels rise, so do populations of sulfate-feeding bacteria. These microbes transform sulfate into hydrogen sulfide, which is toxic to many aquatic plant species.

Meanwhile, overburden has completely buried more than 3,200 kilometers (2,000 miles) of streams, and destroyed an untold range of natural habitats in the process. And destroying habitats invariably threatens local species diversity. Biodiversity in the streams of Appalachia is rivaled only by that in the streams found in rainforests; the diversity of trees is also second only to the tropics. With these ecosystems destroyed on more than 500 mountains in Appalachia, researchers predict the permanent loss of many local plant and animal species.

Slurry impoundments—reservoirs of thick black sludge that accompany each mining operation—are among the

Blasting can damage home foundations and wells, and in rare cases, trigger rockslides. In 2005, a 3-year-old child was killed when a boulder crashed through his house and landed on his bed.

FORMER MOUNTAIN CONTOUR

AFTER MOUNTAINTOP REMOVAL

Typical U.S. Coal Power Plant (500 megawatts)

ANNUAL INPUTS
2.2 billion gallons of water
1.4 million tons of coal (140,000 railroad cars)

ANNUAL POLLUTION PRODUCED INCLUDES
140,000 tons of toxic ash
500 tons of small particles
10,000 tons of nitrogen oxides (NO_x)
10,000 tons of sulfates (SO_2)
170 pounds of mercury
225 pounds of arsenic
3.7 million tons of CO_2
Thermal pollution to nearby waterways

Mountaintop removal permanently damages habitat and streams. In Kentucky alone, more than 2,200 kilometers (1,400 miles) of streams have been damaged or destroyed and 4,000 kilometers (2,500 miles) of streams polluted.

COAL POWER STATION

Home water wells close to reclaimed mines (150 meters [500 feet] or closer) receive significantly higher mine drainage than wells farther away, exceeding EPA allowances for arsenic, lead, iron, and sulfate.

FLY ASH POND

most controversial features of the landscape created by mountaintop removal. They too are a consequence of thinner coal seams. "The thinner seams are messier," says Randal Maggard, a mining supervisor at Argus Energy, a company that has several mining operations throughout Appalachia. "It's 6 inches of coal, 2 inches of rock, 8 inches of coal, 3 inches of rock, and so on. It takes a lot more work to process coal like that." To separate the coal from the rock, miners use a mix of water and magnetite powder known as slurry that good clean coal can float in. Once the sulfur and other impurities have been washed out, the coal is sent for further processing, and the slurry by-product is pumped into artificial holding ponds.

Maggard insists the impoundments are safe. "Before we fill it, we have to do all kinds of drilling to test the bedrock around it," he says. "And then a whole slew of chemical tests on top of that—all to make sure the barrier is impermeable." Still, area residents worry about a breech. With good reason.

In October 2000, the dike holding back the slurry at one of these ponds failed, pouring more than 300 million gallons (30 times the Exxon Valdez spill) of toxic sludge into the Big Sandy River of Martin County, Kentucky. The contamination killed all life in some streams and eventually reached the Ohio River, more than 32 kilometers (20 miles) away. The sludge was 5 feet deep in some places. The EPA would register it as one of the worst environmental disasters in the eastern United States' history. Eleven years later, sludge can still be found a few inches under the river sediments.

↓ The use of coal contributes to environmental and health problems when it is mined or burned. Surface mining, especially mountaintop removal, produces the most environmental damage. Miners and people living in coal mining communities are most at risk for health problems, but air pollution from burning coal affects individuals and ecosystems far removed from the actual power plant.

SUBSURFACE COAL MINE

More than 500 mountains and 1.4 million acres of forest have been damaged or destroyed in Appalachia since mountaintop removal began in the 1970s. More than 200 species are impacted by the practice; 40 or more are already rare or endangered. Habitat loss from mountaintop removal is believed to be the main reason that Cerulean warblers (a songbird) have decreased as much as 80% in some areas of West Virginia.

VALLEY FILL

Piles of mine waste fill valuable plant and animal habitats.

Loss of vegetation and the compaction of soil increases stormwater runoff; stormwater flows over mined or reclaimed areas have been measured at 2–3 times higher than nearby forested areas.

Water and soil contamination from acid mine drainage; 87% of streams tested downstream from mines contained toxic selenium levels that exceeded EPA allowances.

+HOSPITAL

There is a greater incidence of lung, heart, and kidney disease in West Virginia counties with coal mining compared to those without; the incidence of disease was also correlated with the volume of coal mined in the county.

Can coal's emissions be cleaned up?

Of course it's not just the mining and processing of coal that pollutes the environment.

When coal is burned to produce heat energy, it releases a range of toxins that damage the environment and threaten human health: gases (sulfur dioxide, carbon monoxide, nitrogen oxide, and planet-warming carbon dioxide), radioactive material (uranium and thorium), and particulate matter (soot) with particles small enough to irritate lung tissue or even enter the bloodstream if inhaled. Until December 2011, there were no regulations to limit how much of these toxic air emissions power plants could release. The EPA's new Mercury and Air Toxic standards will reduce these emissions by 90% and are predicted to save up to $90 billion in human health costs by 2016 (while only costing $9.6 billion to implement).

Coal-fired power plants also generate tons of toxic fly ash—fine ashen particles made up mostly of silica. Some of that ash is diverted to industry, where it's used in concrete production. But the majority of it is buried in hazardous waste landfills or stored in open ponds like the ones used to contain the slurry waste from mining. In 2008, following a heavy rain event, a 15-meter-tall (50-foot-tall) dike from a pond holding coal fly ash from the TVA Kingston Fossil Plant in Tennessee failed, releasing 1.1 billion gallons of fly ash into nearby rivers and coating vast expanses of riverside land. The EPA estimates that cleanup will cost $1.2 billion—more when property damage and lawsuit settlements are factored in. Federal regulators have since identified more than 100 other ash ponds at risk for breach. [INFOGRAPHIC 23.6]

Infographic **23.7** | **CARBON CAPTURE AND SEQUESTRATION (CCS)**

↓ The capture of CO_2 and subsequent storage that prevents it from reentering the atmosphere will greatly decrease coal's contribution to global climate change. A variety of CCS methods are currently in research and development or are being tested as pilot programs.

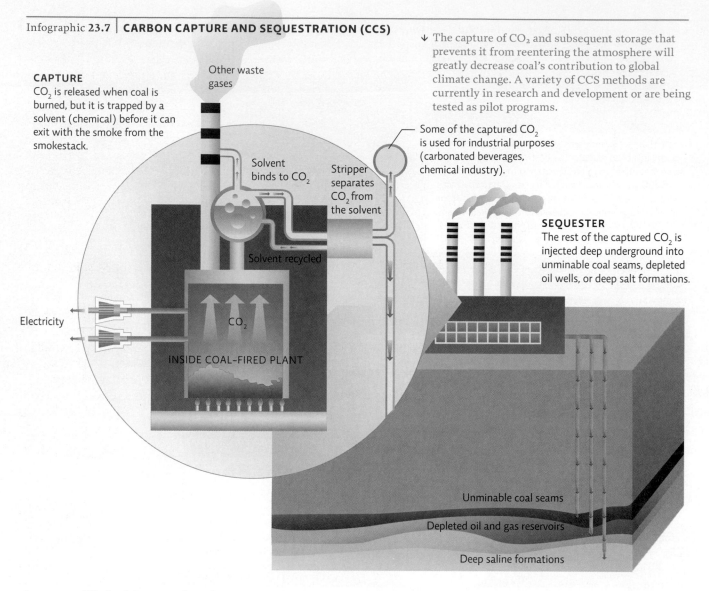

CAPTURE
CO_2 is released when coal is burned, but it is trapped by a solvent (chemical) before it can exit with the smoke from the smokestack.

Other waste gases

Solvent binds to CO_2

Stripper separates CO_2 from the solvent

Solvent recycled

Electricity

CO₂

INSIDE COAL-FIRED PLANT

Some of the captured CO_2 is used for industrial purposes (carbonated beverages, chemical industry).

SEQUESTER
The rest of the captured CO_2 is injected deep underground into unminable coal seams, depleted oil wells, or deep salt formations.

Unminable coal seams

Depleted oil and gas reservoirs

Deep saline formations

In 2011, public health researchers from Harvard University tallied up the costs of mining and using coal. The analysis took into account many (but not all) of the external costs from mining, shipping, burning, and waste production—that is, the costs that are not currently reflected in the market cost of coal. They estimated that coal costs the American public between $300 and $500 billion a year in externalized costs (health, environmental, and property costs). This amounts to an extra $0.10 to $0.26 per kWh in external costs—more than twice what consumers actually pay.

So what do we do? On one hand, we rely on coal to power our society. On the other, the processes of mining and then burning it are hurting us and our environment as much as they are sustaining our way of life.

One potential solution is clean coal technology—technology that minimizes the amount of pollution produced by

coal. For example, scientists and engineers around the world are working on ways to capture the gases emitted from burning coal. We already have the capacity to capture some emissions like particulate matter and sulfur (see Chapter 25).

The next big challenge is **carbon capture and sequestration (CCS)**—the capture and storage of CO_2 in a way that prevents it from reentering the atmosphere. There are many approaches to CCS; some of them promise to capture more than 90% of CO_2 emissions from coal-fired power plants. But most are still in the research and development stage, and so far, progress has been hindered by the cost of capturing the carbon and by a dearth of good

carbon capture and sequestration (CCS) Removing carbon from fuel combustion emissions or other sources and storing it to prevent its release into the atmosphere.

storage options. The extra energy needed to implement CCS must also be considered: Estimates for the additional energy needed range from 25% to 40%—this means 25% to 40% more coal that must be mined, processed, transported, and burned. [INFOGRAPHIC 23.7]

Another emerging clean coal technology involves chemically removing some of coal's contaminants before burning it. While it still contains some toxins, the final product (a liquid or gaseous fuel, depending on the process used), is cleaner than the original coal. Of course, this process requires energy (thus lowering coal's EROEI) and generates hazardous waste (from the toxins found in the coal) that still has to be dealt with.

In fact, as critics are quick to point out, coal can never be truly clean. Each of these "clean coal" technologies produces its own toxic by-products, and none of them eliminate the need to mine coal from deep within Earth's surface.

But if we can't easily abolish coal, or make it totally clean, we can certainly try to use less of it. One way to accomplish this is to design and use more energy-efficient appliances. Less than 40% of the energy from coal is converted to electricity, but the amount of that energy that goes toward powering appliances depends on how efficient those appliances are. We can also reduce our coal consumption through simple conservation efforts like turning off lights and electronics when we're not using them. Every kilowatt-hour of electricity you conserve means roughly 1 pound less of coal being burned. And a growing number of public utilities now offer "green power" programs in which users can choose to buy electricity that has been generated by solar or wind power instead of coal (for more on energy efficiency and conservation, see Chapter 28).

Reclaiming closed mining sites helps repair the area but may not restore the original ecosystem.

In some ways, closed mines—those where all the coal has been harvested—look even more alien than active sites like Hobet 21. Instead of the natural sweep of rolling hills, staircase-shaped mounds covered with what looks like lime-green spray paint join one mountain to the next. Atop some of them, gangly young conifers, evenly spaced,

↓ A valley in West Virginia after mining shows none of the original forest, ridges, or streams that were once found there. The grass is grown from "hydroseed," a seed–chemical mixture sprayed onto surfaces to bind the dirt to prevent erosion. When hydroseed is applied to rock (often done in highway construction), it grows into a plush lawn of grass that dies after a month or two.

Infographic 23.8 | MINE SITE RECLAMATION

↳ Federal law requires that after ceasing operations, the land used for surface coal mines must be restored to close to its original state. This has been accomplished in many cases, but if the area was mountainous, the restored area will never fully recover. After the coal is removed, the area is recontoured to produce a slope and grasses are planted. To replace the mountain streams that were buried by the mining removal process, new channels are constructed to accommodate water flow—but these channels may not support the diverse biological communities that once existed. No matter how much care goes into reclamation, the area will not support a mountain forest community like it once did. However, if done well, a new ecological community may develop.

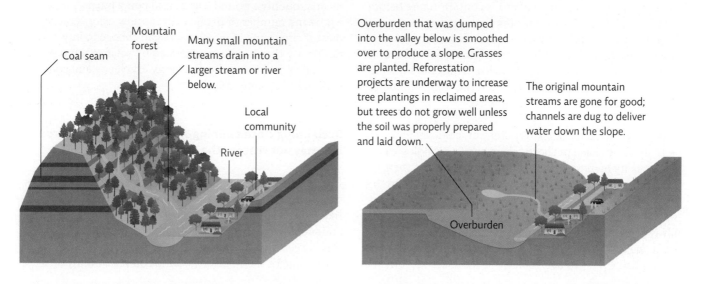

Successfully reclaimed land is suitable for other land uses such as agricultural, residential, commercial, industrial, or is set aside as a natural area. While there have been many successes on smaller-level reclamation sites, more complex reclamation procedures are required to address mountain sites. Successful projects, such as the Kempton Project in Tucker County, West Virginia, that reclaimed 60 acres at a cost of $2.3 million, are rare but show that progress is being made. At Kempton, toxic deposits were removed, native trees and grasses were replanted, streams were relocated, and natural wetlands were installed to neutralize acids from two seeps connected to the former mine site. Though not the mountain it once was, the area is considered a success for mountaintop removal reclamation.

strain toward the Sun. The spray paint is really hydroseed, and most of the gangly trees are loblolly pine, a non-native hybrid that foresters are trying to grow in the region.

Such efforts represent the coal industry's most ardent attempt to honor the U.S. Surface Mining Control and Reclamation Act, which in 1977 mandated that areas that have been surface mined for coal be "reclaimed" once the mine closes. **Reclamation** requires that the area be returned to a state close to its pre-mining condition.

At some surface mines, at least, reclamation is straight-forward (albeit labor-intensive) work: If the mined area was originally fairly flat, the reclamation process includes filling the site with the overburden and contouring the site to match the surrounding land. This relaid rock is then covered with topsoil saved from the original dig. Sometimes, alkaline material such as limestone powder is sprinkled overtop to neutralize acids that have leached into the soil. Vegetation, usually grass, is then planted, leaving other local vegetation to move in on its own.

According to the Mineral Information Institute, more than 2.5 million acres of coal-mined area have been successfully reclaimed using this strategy.

But reclamation has always been a controversial idea. For one thing, the Appalachian forests were created over eons. In fact, in some ways, they were born of the same calamity that laid the coal beneath the mountains. As the supercontinent drifted, carrying the Appalachian range well north of the equator, a temperate hardwood forest replaced the swamps and grew over time into one of the most biologically diverse ecosystems on the planet. In some areas, a single mountainside may host more tree species than can be found in all of Europe, not to mention songbirds, snails, and salamanders that exist nowhere else on Earth.

reclamation Restoring a damaged natural area to a less damaged state.

For another thing, rebuilding a mountain is considerably more difficult than filling in a strip mine where the original land was relatively flat. Critics say that so far, there's no evidence that a site as expansive and as thoroughly destroyed as Hobet 21 could ever be truly reclaimed. In one survey, Rutgers ecologist Steven Handel found that trees from neighboring remnant forests did not readily move in to recolonize mountaintop removal sites, largely because of problems with the soil: At some mines, there was simply not enough of it; at others, it has not been packed densely enough for trees to take root. Those problems have technical solutions, Handel writes, but so far, the seemingly simple act of changing reclamation protocols has been stymied by politics.

And trees are just one facet of reclamation. What about all those streams that were buried? The 1973 Clean Water Act prohibits the discharge of materials that bury a stream or, if unavoidable, requires mitigation practices that return the stream close enough to its original state such that the overall impact on the stream ecosystem is "non-significant." Federal regulations also require that damaged streams be restored. Industry reps argue that they are indeed working to rebuild streams: Once the overburden has been reshaped and smoothed over, they dig drainage ditches and line them with stones in a way that resembles a stream or river. But so far, research shows—and most ecologists agree—that such channels don't perform the ecological functions of a stream. "They may look like streams," says ecologist Margaret Palmer,

from the University of Maryland. "But form is not function. The channels don't hold water on the same seasonal cycle, or support the same aquatic life, or process contaminants out of the water—all things a natural stream does." [INFOGRAPHIC 23.8]

In Appalachia, the arguments over when and where and how to mine for coal are quickly boiling down to a single intractable question: Once it's all gone, how will we clean up the mess we've made? For a story that has played out over geologic time, the question is more immediate than one might think. In West Virginia, coal reserves are expected to last another 50 years, at best. That means no matter what regulations the government imposes, or what methods the coal companies resort to, the day of reckoning will soon be upon us.◉

Research articles referenced in this chapter:
Ahern, M., et al. 2011. *Environmental Research*, 111: 838–846.
Bernhardt, E.S., and Palmer, M.A. 2011. *Annals of the New York Academy of Sciences*, 1223: 39–57.
Epstein, P.R., et al. 2011. *Annals of The New York Academy Of Sciences: Ecological Economics Review*, 1219: 73–98.
Handel, S.N. 2002. *EIS Technical Study Project for Terrestrial Studies*.
Hendryx, M., and Ahern, M. 2008. *American Journal of Public Health*, 98: 669–671.
Pond, G., et al. 2008. *Journal of the North American Benthological Society*. 27: 717–737.
U.S. Environmental Protection Agency. 2005. *Final Programmatic Environmental Impact Statement on Mountaintop Mining/Valley Fills in Appalachia*. Philadelphia, PA: EPA.
Weakland, C.A., and Wood, P.B. 2005. *The Auk*, 122: 497–508.

BRING IT HOME

◉ PERSONAL CHOICES THAT HELP

Although coal is one of our most abundant fossil fuels, its drawbacks are significant. They include CO_2 emissions; the release of air pollutants that cause environmental problems such as acid rain; health problems such as asthma and bronchitis; and massive environmental damage from the mining process. One way to minimize the impact of coal is to reduce consumption of electricity.

Individual Steps
→ Always conserve energy at home and at the workplace.
• Turn off or unplug electronics when not in use.

• Put outside lights on timers or motion detectors so that they only come on when needed.
• Dry clothes outdoors in the sunshine.
• Turn the thermostat up or down a couple of degrees in summer and winter to save energy and money.

Group Action
→ Organize a movie screening of *Coal Country* or *Kilowatt Ours*, which present issues related to coal mining and mountaintop removal from many perspectives.

Policy Change
→ The Appalachian Regional Reforestation Initiative is a great example of how groups, sometimes with very different objectives, can work toward a common goal. Go to arri.osmre.gov to see how this coalition of the coal industry, citizens, and government agencies are working to restore forest habitat on lands used for coal mining.
→ Visit beyondcoal.org to find out about events in your state and for the opportunity to weigh in on the decommissioning of outdated coal power plants and the building of new ones in the United States.

UNDERSTANDING THE ISSUE

CHECK YOUR UNDERSTANDING

1. **Which of the following is NOT true about coal?**
   a. It produces more air pollution than other fossil fuels.
   b. It is difficult to ship.
   c. Extraction is relatively cheap.
   d. The United States has an abundant supply of it.

2. **The purpose of mountaintop removal is to:**
   a. promote housing development.
   b. decrease obstacles in airline flight paths.
   c. gain surface access to coal.
   d. promote the regrowth of prairie habitat.

3. **Emissions from a coal-fired power plant include all of the following EXCEPT:**
   a. oxygen.
   b. carbon dioxide.
   c. mercury.
   d. arsenic.

4. **How do miners separate coal from rock?**
   a. By dissolving the rock with acid
   b. By comparing the colors
   c. By checking to see whether the material floats
   d. By looking for bright, reflective surfaces

5. **How long are coal reserves expected to last in West Virginia?**
   a. 50 years
   b. 100 years
   c. 500 years
   d. Forever

6. **Which of the following is an example of reclamation?**
   a. Reworking coal mines to extract more coal
   b. Replanting the site of mountaintop removal with grass and pine trees
   c. Returning land to the people who originally owned it
   d. Filling in a subsurface mine

WORK WITH IDEAS

1. In your own words, describe the process of coal formation. Why is coal a finite resource?

2. What are the pros and cons of the two types of surface mining? Include in your answer a comparison of the extraction process, the energy return on energy investment (including external costs), and the waste-disposal process.

3. Define and describe the process of carbon capture and sequestration. Why is this process necessary? What are the costs and benefits of the process?

4. What are the environmental consequences of surface mining? Use specific examples from the chapter.

5. Describe the process of subsurface mining. What are the costs and benefits? Why do some miners prefer a subsurface mine to a surface mine, while mining companies prefer surface mines?

6. Define reclamation. Explain why, in your own words, reclamation is a controversial idea. Are there examples in which reclamation has succeeded?

ANALYZING THE SCIENCE

The following graphs depict the fatalities in coal mining and other industries.

INTERPRETATION

1. In Graphs A and B, why are both the absolute number of fatalities and the rate of fatalities presented? Which is the fairest measure to use?

2. Based on Graphs A and B, would you rather work in the coal industry or the metal/nonmetal industry? Why?

3. Why do the numbers in Graph B appear to be larger than those in Graph A? For example, compare the numbers for 1968 in both graphs. What conclusion can you make about graphing data in general?

ADVANCE YOUR THINKING

4. The graphs represent fatality data. Do you know what the fatalities are the result of, based on the graph? What do you assume these fatalities are related to? In your opinion, based solely on these numbers, would you risk being a miner, provided the pay and benefits were good?

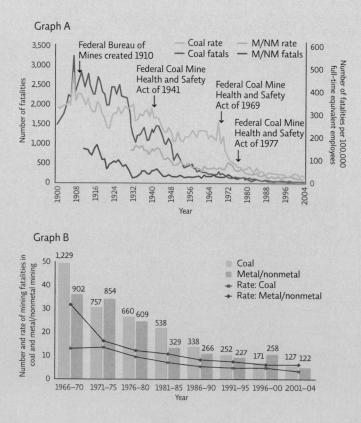

Graph C

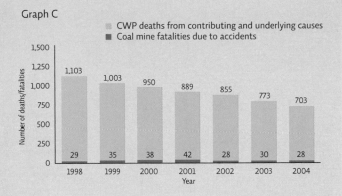

- CWP deaths from contributing and underlying causes
- Coal mine fatalities due to accidents

Number of deaths/fatalities (y-axis): 0, 250, 500, 750, 1,000, 1,250, 1,500

Year	CWP deaths	Accidents
1998	1,103	29
1999	1,003	35
2000	950	38
2001	889	42
2002	855	28
2003	773	30
2004	703	28

5. Now, examine Graph C. It depicts the same data as that shown in Graphs A and B, for the years 1998–2004, but shows the deaths (in numbers) due to accidents AND coal workers' pnuemonconiosis (CWP, also called black lung disease).
   a. What is one difference in the way these data are reported compared to how they are reported in Graphs A and B?
   b. If your only employment choice were to work in a coal mine, what else would you like to know about the data in Graph C?

6. Based on the information in Graph A, what can you say about fatalities and federal regulations? Do you think it would be safe to dispense with federal regulations and allow mine owner/operators to determine safety standards and practices, now that they know the benefits of safety regulations?

EVALUATING NEW INFORMATION

The EPA is responsible, in part, for issuing mining permits, based on the predicted environmental damage from the mines. As with many issues in environmental science, there is often disagreement between government regulators and companies about the extent of environmental damage. Federal regulators argue that some practices are so harmful to the environment that they should be halted; companies argue that halting certain practices will result in severe economic loss.

Go to http://is.gd/ELKPBA. Click on the "Listen to the Story" link, review the photo gallery, and read the story about mountaintop removal.

Evaluate the website and work with the information to answer the following questions:

1. Is this a reliable information source? Does it have a clear and transparent agenda?
   a. Who runs this website? Do this group's credentials make it reliable or unreliable? Explain.
   b. Is the information on the website up-to-date? Explain.

2. When was this story aired and on what program? What type of information is reported in this story?

3. Now, read an ABC News story about mountaintop removal at http://is.gd/NQMsJH.
   a. Who runs this website? Is it comparable to the first website? Why or why not?
   b. Compare the content of the two stories. Which do you think provides more in-depth coverage of the issue?
   c. Now look for links to other stories about mountaintop removal. Do both sites have links to other websites that you could use to learn more if you wished? Are the links more useful or relevant on one site versus the other?

4. One of the links provided on the ABC site is to Massey Energy, which declined to be interviewed for the story.
   a. Why do you think Massey Energy would decline to be interviewed?
   b. Do a brief Internet search for Massey Energy. What types of stories appear? Do a brief review of the stories. Do these stories give you any more insight into Massey Energy's motives for declining an interview?

MAKING CONNECTIONS

DO WE NEED MORE EPA REGULATIONS?

Background: Coal is the most common source of electricity production in the United States and China. Both of these countries are large contributors of greenhouse gases, especially carbon dioxide, which is a product of burning coal. Burning coal also releases other compounds into the air, including pollutants such as mercury and sulfur. A partial solution to this problem is to use clean coal technology.

Case: You are a congressional aide. Part of your job is to provide the congressperson for whom you work with up-to-date information about energy issues. Recently, the EPA unveiled new rules (Mercury and Air Toxics Standards) that set limits on mercury emissions from coal-fired plants. The companies that already produce power with "clean" fuel (that is, not coal) are in favor of the new rules; companies that primarily use coal are not. You've been asked to do research about the new rules to help your congressperson draft a response for the constituents.

1. As part of your research, find out:
   a. What is the purpose (both immediate and long term) of the new rules?
   b. How long have the new rules been under review and revision?
   c. How will the new rules affect the energy companies in your congressperson's voting area? (Assume the area is where you are attending school.)

2. Next, find out what opponents and proponents of the new rules predict about the long-term consequences of the new rules. Which viewpoint do you think is most similar to those your congressperson will have? Explain.

3. As a citizen, independent of your role as a congressional aide, what is your respose to the proponents of the new rules? The opponents? When drafting your response, think about what you know about the history of industry, the EPA, and federal regulations, especially as they relate to coal and current environmental and economic conditions.

CORE MESSAGE

We depend on petroleum and natural gas (nonrenewable fossil fuels) for a range of products, not just to fuel our cars or heat our homes. Our dependence on fossil fuels has led us to become increasingly destructive and extreme in our quest to extract it from the environment.

GUIDING QUESTIONS

After reading this chapter, you should be able to answer the following questions:

→ How are fossil fuels like petroleum formed and why are they considered nonrenewable resources?

→ Where are our current oil reserves found and what is meant by "peak oil"?

→ How is conventional oil extracted? What are some of the environmental consequences of finding, extracting, and using oil?

→ What are the advantages and disadvantages of using natural gas instead of oil?

→ What unconventional sources of petroleum exist? What are the benefits of using them? The costs?

Cleanup crews burn oil from surface waters surrounding the *Deepwater Horizon* oil rig.

CHAPTER 24 PETROLEUM

A DANGEROUS DEPENDENCE

Drilling for answers after an oil rig explosion

↑ John Kessler (left) and his colleague David Valentine in front of ground zero of the *Deepwater Horizon* oil spill in June 2010.

→ Pyrosomes spend their entire life in the open ocean. They are tropical colonial organisms that can grow to 12 feet in length.

As soon as they left shore in Gulfport, Mississippi, John Kessler, an oceanographer then at Texas A&M University in College Station, Texas, started wrinkling his nose. "We could immediately start smelling the oil vapors, even though we were probably 100 miles away from ground zero," Kessler recalls. They could see it, too: sometimes, the surface of the water had a thin sheen, "like when it rains on the driveway and you see that rainbow color," he explains. Other times, the surface was thickly covered in an oil-water mixture that resembled chocolate cake batter. Hundreds of thousands of pyrosomes—marine organisms that resemble small wormlike sea cucumbers—floated dead on the surface. "We'd pull up our equipment and it would have these pyrosomes on them—they were just everywhere," he recalls.

It was June 2010, and Kessler was on a research expedition to find out how the largest oil spill in history was affecting oxygen levels in nearby waters. Two months earlier—ironically, on Earth Day—a huge explosion rocked the ocean 40 miles southeast of the Louisiana coast in the Gulf of Mexico. There, a large bubble of methane gas shot up the column of the oil drilling unit *Deepwater Horizon*, causing a fire that ultimately killed 11 people onboard. Each day following the explosion, the unit gushed upwards of 50,000 gallons of oil into the Gulf waters, despite desperate efforts to stop the flow by BP—the oil company leasing the rig—and the United States government. Because of the leaking oil, in June the National Oceanic and Atmospheric Administration (NOAA) closed 80,228 square miles of the Gulf to fishing.

⊚ WHERE IS THE *DEEPWATER HORIZON*?

Scientists like Kessler sprang into action. The Gulf of Mexico is home to more than 15,000 species of organisms, from miniscule bacteria to huge sperm whales. How would the flood of oil affect them? Kessler was particularly concerned about the effects of excess methane, an organic compound that comprises about half of oil's chemical composition. Methane (CH_4), a simple hydrocarbon made of one carbon atom surrounded by four hydrogen atoms, is a favorite food of many bacteria living deep in the ocean; the problem is, as the bacteria population increases, it depletes the oxygen in the water (using it for cellular respiration) that other marine organisms also need to survive. With an as-yet unknown amount of excess methane polluting the water, the spill had "the potential to deplete oxygen significantly from the Gulf waters and then suffocate life," Kessler says.

Kessler realized he needed to find out just how bad the situation was, so in May, a mere 3 weeks after the explosion on the *Deepwater Horizon*, he and colleague Dave Valentine,

from the University of California at Santa Barbara, applied for a National Science Foundation research grant to study the effects of the oil spill on methane and oxygen concentrations in the Gulf. Two days later, they were funded. "Then we had 2 weeks to get our laboratories mobilized and to the Gulf of Mexico," he recalls, where his plan was to analyze the ocean water around the ongoing spill to see just how much methane was contaminating it and how much oxygen was being depleted.

Oil spills like the *Deepwater Horizon* spill are an unfortunate but potential consequence of oil drilling, and the 2010 disaster is unlikely to be the last of its kind. That's because oil—along with coal and natural gas—are the three principal **fossil fuels** on our planet, and we depend heavily on them for energy and to make the everyday

---

fossil fuel A nonrenewable natural resource formed millions of years ago from dead plant (coal) or animal (petroleum and natural gas) remains.

Infographic 24.1 | **HOW FOSSIL FUELS FORM**

 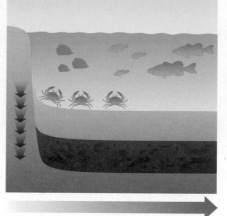 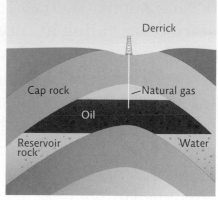

↑ Long ago, some marine organisms died and were buried in sediment. This burial excluded oxygen, and decomposition was greatly slowed down.

↑ As sediments accumulated, the partially decomposed buried biomass was subjected to high heat and pressure. Over the course of millions of years, it was chemically converted to oil or natural gas.

↑ Oil and natural gas move upward in porous reservoir rock until stopped by a layer of dense cap rock. We tap these deposits by drilling into the porous rock reservoirs.

products we use (for more on coal, see Chapter 23). Fossil fuels form over millions of years when organisms die and are buried in sediment before they can decompose. Gradually, as pressure and heat increase, the buried material goes through a chemical transformation into oil, natural gas, or coal. Even though they are produced by natural processes, fossil fuels are considered **nonrenewable** because we are quickly using them up in a fraction of the time it takes for them to form. [INFOGRAPHIC 24.1]

Over the decades, we have come to depend heavily on fossil fuels, especially oil. But disasters like the *Deepwater Horizon* make some scientists and politicians openly question whether our dependence on oil has gone too far. As President George W. Bush famously said in his 2006 State of the Union address, "America is addicted to oil." It's an addiction that cannot be sustained because of the environmental costs of harvesting fossil fuels. We are also quickly running out of fossil fuels.

Petroleum is a valuable resource but it has many drawbacks.

Oil, or **petroleum**, is a liquid fuel made up of hundreds of types of hydrocarbons, organic compounds of hydrogen and carbon. Hydrocarbons take many forms—solid, liquid, or gas—and we have developed methods for extracting them from the depths of the Earth. Oil is found some 300 to 9,000 meters (0.2–5.5 miles) below the surface, where rigs have to dig deep in order to reach it—the *Deepwater Horizon* rig reached a depth of 1,050 meters

(0.6 mile). Surprisingly, though, oil doesn't exist in thick black pools. Instead, if you could look down into an oil reservoir, you'd see rock. **Crude oil** is found as tiny droplets wedged within the open spaces, or "pores," inside rocks, that are so small you'd need a microscope to see them.

World demand for oil rises more than 2% each year, and it is possible we have already passed **peak oil**, the moment in time when oil will reach its highest production levels and then steadily and terminally decline. Some say peak oil occurred in 2002, others in 2006; but still others predict it won't be reached until 2014 or 2020. Whether we just passed it or are about to, passing peak oil will have international economic repercussions as the demand for oil outstrips the supply. Most economists agree that the day of cheap oil is gone. The sooner we prepare for a world

nonrenewable resource A resource that is formed more slowly than it is used, or is present in a finite supply.

oil See *petroleum*

petroleum Liquid fossil fuel useful as a portable fuel or as a raw material for many industrial products such as plastic and pesticides.

crude oil The form of petroleum that is extracted from underground deposits. It is processed into many different products and types of fuel.

peak oil The moment in time when oil will reach its highest production levels and then steadily and terminally decline.

reserves A measure of the amount of a fuel that is economically feasible to extract from a known deposit using current technology.

conventional petroleum reserves Liquid deposits that contain freely flowing oil or oil that can be pumped out.

Infographic 24.2 | PEAK OIL

↓ Peak oil represents the point in time when oil will reach its highest production levels and then steadily and terminally decline. It is likely that the world has reached, or will soon reach, peak oil (red arrow). The dip (black arrow) seen in the early 1980s represented an oil glut; conservation efforts during the oil crisis of the 1970s led to reduced demand, but we have since resumed an upward course of oil consumption.

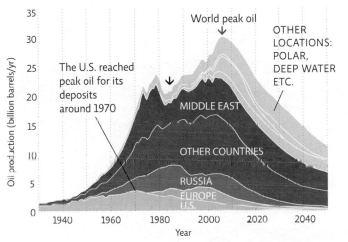

with less oil, the easier the transition to a different energy economy will be. [INFOGRAPHIC 24.2]

Reserves are the amount of a fuel that is economically feasible to extract from a deposit using current technology. **Conventional petroleum reserves** are not evenly distributed around the planet, leading to political problems between countries who have the oil, like those in the Middle East, and those who do not have enough to meet their own needs, like the United States. This distribution problem is compounded by the fact that known reserves are past peak. At current rates of extraction and use, known reserves are expected to last another 40 years (though this number is uncertain because reports of oil reserves tend to be questionable—many nations keep their reserve estimates a secret).

Natural gas reserves are also finite. Russia has the largest amounts of natural gas, with about 25% of the world's reserves; the United States has only 3.4%. Total world reserves of natural gas are expected to last 60—100 years at current rates of use. But like oil, natural gas use is increasing at about 2% per year. At this rate, natural gas could run out much sooner. [INFOGRAPHIC 24.3]

Infographic 24.3 | OIL RESERVES

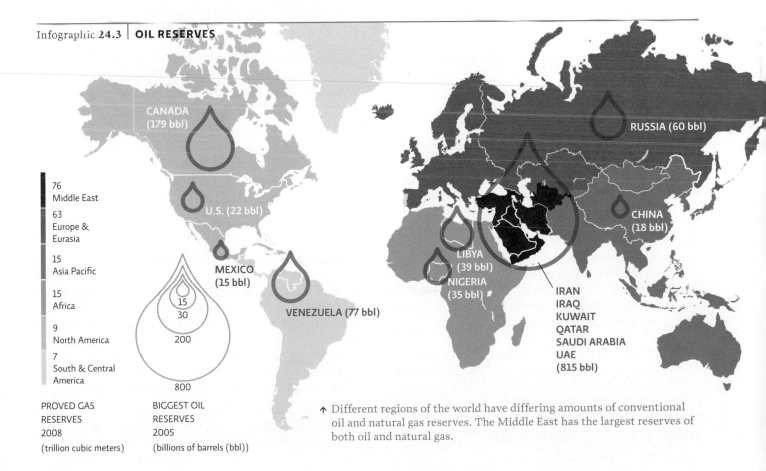

↑ Different regions of the world have differing amounts of conventional oil and natural gas reserves. The Middle East has the largest reserves of both oil and natural gas.

Infographic 24.4 | **HOW IT WORKS: EXTRACTING OIL AND THE OIL SYSTEM**

↓ Oil is obtained by drilling layers of dense rock to reach the reservoir below. At first, oil may easily flow due to the relief of pressure caused by the drill hole. Pumpjacks are used to mechanically pump out more oil when the oil stops flowing freely.

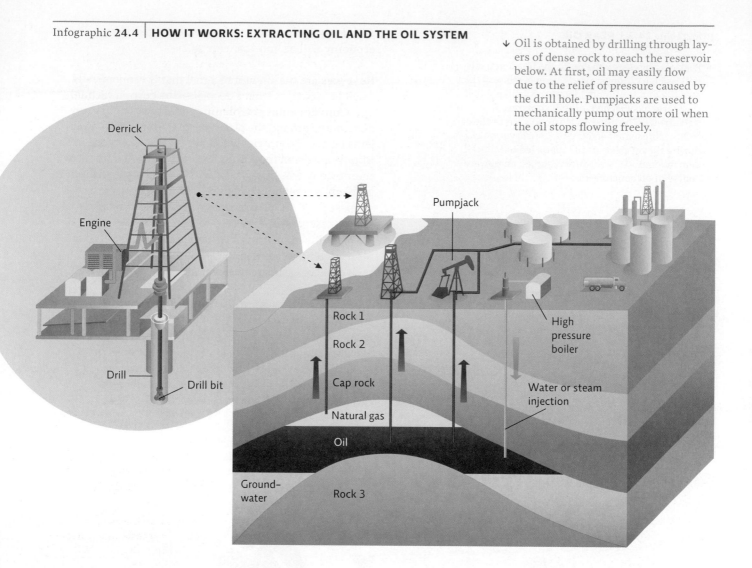

Conventional oil reserves are tapped by drilling wells.

After a new oil well is drilled, nature does most of the work. There is significant pressure on oil deep underground from millions of tons of rocks pressing down and from Earth's heat, which causes gases around the oil and rock to expand. So when a well is first drilled, oil naturally flows upward, escaping like air gushing out of a balloon. This is known as *primary production*, and about 5 to 15% of the oil can be recovered in this first phase. The *Deepwater Horizon* was about to begin its primary production phase when the 2010 spill occurred.

Eventually, the pressure below ground will decrease, so injection wells are drilled nearby and huge hoses pump water through the well into the ground. This increases the pressure below ground and forces more oil to the surface. During this phase of *secondary production*, between 20 and 40% of a reserve's oil can be recovered.

But even after primary and secondary production, there is a lot of oil left in the ground. *Tertiary production* methods like injecting steam or CO_2 into the reservoir allow the recovery of up to 60% of the remaining oil but are costly in both money and energy. [INFOGRAPHIC 24.4]

Processing oil creates fuels and products.

Because it contains so many impurities once it's extracted from the Earth, crude oil must be refined into a usable form. The first refining process, called simple distillation, separates the oil into *fractions*, different hydrocarbon components. To do this, the crude oil is heated and put into a column, or tower. As the oils boil off and rise in the column, the larger compounds condense back to liquid

---

petrochemicals Distillation products from the processing of crude oil that can have many different uses, including as fuel or as industrial raw materials.

form and are collected. The first compounds to boil off include lighter liquids, like liquid petroleum gases and gasoline. Next come jet fuel, kerosene, and diesel fuel. Finally, at temperatures above 600°F, the heaviest products, including grease, wax, asphalt, and residual fuel oil, can be recovered.

Many of the products of distillation, called **petrochemicals**, are used not just as fuel but as raw materials in the production of industrial organic chemicals, pesticides, plastics, soaps, synthetic fibers, explosives, paints, and medicines. The production of a desktop computer, for example, consumes 10 times the computer's weight in fossil fuels, mostly oil. [INFOGRAPHIC 24.5]

Petroleum use comes at a great cost.

Oil extraction is not just financially expensive—it also has environmental consequences at almost every stage. In order to find oil, companies send seismic waves into the ground that bounce back to reveal the location of possible reserves. But doing so disorients marine wildlife; in 2009, ExxonMobil had to abandon its oil exploration in Madagascar because more than 100 whales had beached themselves, presumably as a result of these seismic exploratory methods.

Drilling can affect wildlife, too. Politicians have long debated the merits of drilling in the Arctic National Wildlife Refuge (ANWR), 20 million acres of protected

Infographic 24.5 | **PROCESSING CRUDE OIL**

↓ Crude oil is refined into a variety of chemical products by heating it in a tower. Different-sized carbon compounds "boil off" (vaporize) at different temperatures and heights within the tower and are collected as separate products.

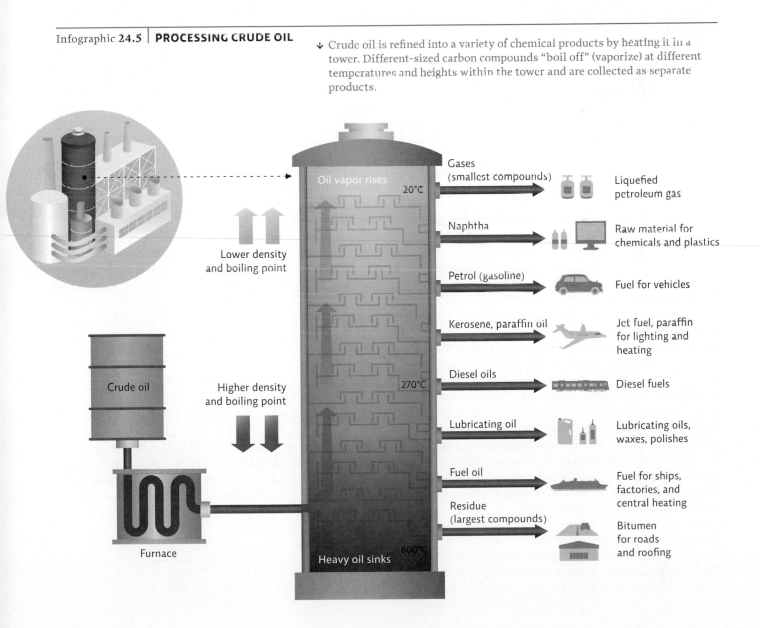

wilderness that was created by Congress under the Alaska National Interest Lands Conservation Act of 1980. The U.S. Geological Survey estimates that parts of ANWR located on the northern Alaskan coast could harbor up to 16 billion barrels of crude oil and natural gas reserves. But drilling there—which could begin in the coming decades—would have serious environmental consequences. The coastal area where drilling would take place is home to arctic foxes, caribou, polar bears, and migratory birds, all of whose habitats could be disturbed by the drilling.

Oil extraction is not just financially expensive—it also has environmental consequences at almost every stage.

Of course, oil spills—an unintended but frequent consequence of oil drilling and transport—can have devastating long-term effects as well. The *Deepwater Horizon* spill threatened the lives of already endangered sea turtles, which feed and live in areas of the Gulf where ocean currents converge—the same places where the spilled oil from the *Deepwater Horizon* tended to accumulate.

Many Gulf birds became drenched in oil, which prevented them from being able to stay afloat and regulate their body temperature. Worse, when the birds tried to clean themselves, many suffered damage to their internal organs, ultimately causing weight loss, anemia, and dehydration. So far, the *Deepwater Horizon* is estimated to have killed more than 2,000 birds. But the long-term effects on bird populations are still unknown. [INFOGRAPHIC 24.6]

During his expedition in the Gulf after the spill, Kessler measured methane and oxygen levels and found that in some areas, methane levels were 100,000 times higher than normal, causing dissolved oxygen levels to drop by as much as 40%. "We were very concerned," he admits. When Kessler and his colleagues went back for additional expeditions in August, "we saw the last remnants of this big methane plume," he recalls—most of it was gone, probably because bacterial populations had exploded in the presence of so much extra food and consumed it rapidly. It was likely, then, that the extra bacterial populations would start to die off and that the depleted oxygen levels would soon start rising again.

In a later *Science* article, University of Georgia marine biologist Samantha Joye and 15 colleagues challenged Kessler's conclusions, suggesting that there was not sufficient evidence to attribute the oxygen decline to

methanogenic bacteria. As noted in Chapter 2, this is one of the strengths of science—scientists constantly evaluate and critique each other's work, leading to better studies and to the accumulation of more evidence.

Assessments made a year after the oil spill revealed mixed results. Research published in the journal *Science* in August 2010, by microbial biologist Terry Hazen, of Lawrence Berkeley National Laboratory in California, found less oil in the water column than expected. His group found a new bacterial strain that was quickly digesting the oil, suggesting that recovery is well underway.

However, other research reported at a scientific conference in February 2011 shows that a great deal of oil still coats the ocean floor, leaving the area a "dead zone:" oil may be out of the water column but it remains in the ecosystem. Marine mammals are still dying at higher-than-normal rates and populations of bottom dwellers like shellfish and crabs are still lower than normal. While the health of the ecosystem as a whole may be better than expected, it is far from recovered a year later and there is still much we don't understand about the event or its aftermath.

On April 19, 2011, the final fishery in the Gulf reopened.

Natural gas is another conventional energy resource.

Natural gas is an important alternative to oil for some purposes, like generating electricity and heat. The United States produces most of the natural gas it needs, importing the rest from Canada. But according to the U.S. Energy Information Administration, which tracks and analyzes energy data for the Department of Energy (DOE), natural gas consumption is expected to increase more than 18% by 2030, an increase with which our current reserves just can't keep up.

The extracting and processing methods for natural gas and oil are very similar. Natural gas sometimes flows freely up wells due to underground pressure, but most natural gas extraction requires some kind of pumping system to force the gas upward. Like oil, natural gas contains impurities and must be refined before it can be used but it is considered the cleanest fossil fuel because it releases much less of the greenhouse gas, carbon dioxide, and other air pollutants than oil or coal when burned.

---

natural gas Gaseous fossil fuel that forms under similar conditions as petroleum but is a simpler hydrocarbon, mostly CH_4 (methane).

Infographic 24.6 | **ENVIRONMENTAL COSTS OF OIL**

↓ Oil has environmental costs at every stage—from exploration for reserves, to extraction, processing, transporting, and burning. If these, often externalized, costs were added to the price of our fuel and petroleum-based goods, the price of the goods would be much higher.

WILDLIFE DISRUPTION

Sonic exploration for oil caused whales to beach near Busselton, Australia.

HABITAT LOSS

An open-pit tar sands mine in Alberta where boreal forest once stood.

EXPLOSIONS

The *Deepwater Horizon* burns.

AIR POLLUTION

Burning petroleum is a major source of air pollution in Mexico City.

OIL SPILL

A severely oiled pelican being rescued in Louisiana. The bird lived.

GROUNDWATER POLLUTION

Oil washed into the marshes of Barataria Bay damaged the marshes themselves and may seep into the aquifers below.

Infographic 24.7 | **FRACKING FOR NATURAL GAS**

↓ Deep deposits of natural gas can be accessed via fracking. Concerns about surface and groundwater contamination make this a contentious resource in populated areas.

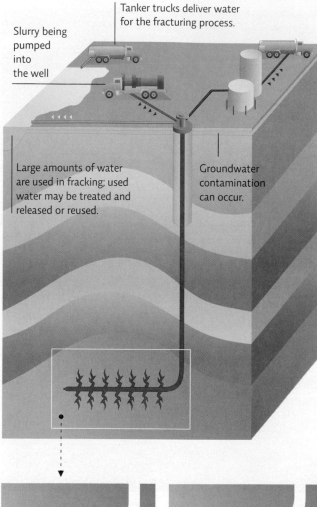

Tanker trucks deliver water for the fracturing process.

Slurry being pumped into the well

Large amounts of water are used in fracking; used water may be treated and released or reused.

Groundwater contamination can occur.

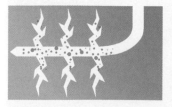

1. A well is drilled down to gas-bearing rock and then extended horizontally.

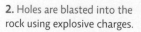

2. Holes are blasted into the rock using explosive charges.

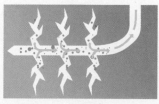

3. A slurry of sand, water, and chemicals is pumped into the fractures.

4. Gas escapes through the fractures and is collected.

As oil and gas reserves shrink and our extraction methods grow ever more harmful, energy companies are looking for alternatives. Another alternative is untapped, deep natural gas reserves. However, some environmental organizations have raised concerns about the safety of the extraction process used to access these deep reserves. For instance, to help release natural gas from the ground after drilling, fluids containing toxic chemicals are injected underground at high pressure in a process known as fracturing or "**fracking**." An estimated 30 to 70% of this liquid resurfaces with the natural gas, bringing the toxic chemicals along with it. These chemicals have the potential to contaminate ground and drinking water, soil, and air. Methane itself may also contaminate groundwater near these wells. (Though areas with methane deposits may normally have some methane in the groundwater, fracking may increase the amount.) [INFOGRAPHIC 24.7]

Unconventional oil resources require different harvesting techniques.

Oil is used for many things in human societies, such as transportation, heating, and making petroleum products like plastics. However, conventional oil supplies won't last forever. Geologists have begun to seek oil in **unconventional reserves**, such as oil shale, and oil and tar sands, most of which require intensive processing to separate oil from other natural elements.

The United States possesses the world's largest concentration of **oil shale**, compressed sedimentary rocks that contain kerogen, an organic compound that is released as an oil-like liquid when the rock is heated. Primarily on federal lands, oil shale contains an estimated 1.23 trillion barrels of oil, more than 50 times the nation's proven oil reserves, according to the U.S. Bureau of Land Management.

Alberta, Canada, is home to the world's largest reservoir of a heavy, black oil known as crude bitumen. The oil, however, is often trapped in sticky, dense conglomerations of sand or clay known as **oil** or **tar sands**. In 2008, Alberta produced 653,000 barrels of synthetic crude oil per day from the tar sands. It is estimated that there are more than 2 trillion barrels of oil in tar sands, though it is not all recoverable. But there are environmental drawbacks to these unconventional oil reserves. [INFOGRAPHIC 24.8]

David Schindler, an ecologist at the University of Alberta in Canada, has spent years studying the effects of tar sand extraction on the water quality of the Athabasca River, which cuts through the heart of one of Alberta's biggest tar sand deposits and mining sites. Research by Schindler

↑ The 800-mile Trans-Alaska Pipeline delivers oil across Alaska to the southern port of Valdez. An even-longer pipeline has been proposed to take oil from Canada's tar sands to Texas.

and his graduate student Erin Kelly shows that concentrations of 13 toxic compounds are higher in the river's tributaries located downstream of oil sand mining than they are upstream, which suggests that the extraction is at least partly responsible for them. Also of concern are the huge lakes of acidic and toxic wastewater that oil companies store at the mining sites as a by-product of the refining process. The pools are so large that they can be seen from outer space.

Tar sand mining is energy intensive and, as a result, produces up to 3.6 times the amount of greenhouse gases than conventional oil and gas production. For any fuel to be worth harvesting, it is important that its energy return on energy investment (EROEI) be favorable—it must require less energy to extract and process than is in the fuel itself.

Unfortunately, damage is not limited to the extraction of fossil fuels. Processing and shipping of fossil fuels generates air and water pollution. Pollution generated when fossil fuels are burned is also a major source of air pollution (see Chapter 25) and a main cause of climate change (see Chapter 26).

Oil consumption drives extraction.

Not all countries use energy equally. The United States, for instance, uses more energy than any other country—its overall primary energy use tripled from 1949 to 2009, according to the Energy Information Administration. Most of this energy comes from fossil fuels—coal, petroleum, and natural gas, with crude oil–based petroleum

fracking A method used to extract natural gas from deep reserves.
unconventional oil reserves Recoverable oil that exists in rock, sand, or clay but whose extraction is economically and environmentally costly.
oil shale Compressed sedimentary rocks that contain kerogen, an organic compound that is released as an oil–like liquid when the rock is heated.
oil or tar sands A heavy, black oil known as crude bitumen that is trapped in sticky, dense conglomerations of sand or clay. It can be mined and processed to produce a substitute for petroleum.

Infographic 24.8 | **UNCONVENTIONAL PETROLEUM RESERVES IN THE UNITED STATES AND CANADA**

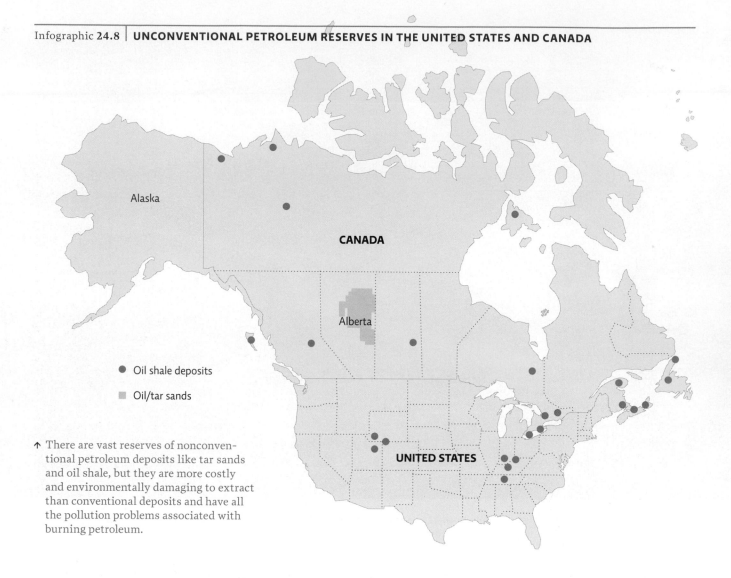

Alaska

CANADA

Alberta

● Oil shale deposits

■ Oil/tar sands

UNITED STATES

↑ There are vast reserves of nonconventional petroleum deposits like tar sands and oil shale, but they are more costly and environmentally damaging to extract than conventional deposits and have all the pollution problems associated with burning petroleum.

↓ Petroleum deposits exist in a variety of forms. Oil shale is known as "rock that burns" (left). Dense, black bitumen can also be found in sand or clay soils such as the tar sands shown here (right).

↑ Tar sands seam where valuable tar sands are extracted with large trucks and shovels, in the Syncrude North Mine, north of Fort McMurray, Alberta, Canada.

as the dominant source. The Houston oilfield services company Baker Hughes Inc. estimates that as of February 2011, the United States had 1,713 rigs like the *Deepwater Horizon* searching the country for oil or gas. Nevertheless, we import most of what we use: In 2009, the United States produced only 11% of the world's petroleum, but consumed 22% of it. Most of the rest was imported from Canada, Venezuela, and Saudi Arabia.

The good news is that in the United States, oil consumption is probably decreasing now, according to a report published in 2009 by IHS Cambridge Energy Research Associates. The report predicts that because of changing demographics and improved vehicle efficiency, U.S oil use peaked in 2005 and is now in decline. But in developing

countries such as China and India—the world's second and fourth largest consumers of oil, respectively, with Japan in third—oil consumption is still rapidly increasing, forcing the United States and other countries to extract as much oil as possible from Earth, regardless of the consequences.

Our economy and lifestyle are dependent upon reliable access to affordable energy, or **energy security**. Yet there are many reasons that our energy supplies might become unreliable or unaffordable. These include dwindling supplies, increasing demand, dependence on energy imports, competition from other countries, and a cartel or monopoly increasing prices or decreasing oil supplies.

In the United States, many of these problems already exist. As supplies dwindle and demand continues to increase, oil becomes harder and harder to extract from the ground, forcing companies to use more powerful—and

energy security Having access to enough reliable and affordable energy sources to meet one's needs.

potentially more dangerous—technologies. But will the risks always be worth the benefits? The *Deepwater Horizon* disaster has led many environmental organizations to demand an end to offshore drilling in the Gulf of Mexico. According to Frances Beinecke, president of the nonprofit Natural Resources Defense Council, "as long as we keep the driving addiction and the oil addiction that we have in this country, we will be jeopardizing really fertile, fragile ecosystems all around the country—onshore and offshore."

The *Deepwater Horizon* spill still threatens many marine species.

On July 15, 2010, after 86 days of oil billowing out of the busted column, workers successfully closed the *Deepwater Horizon*'s wellhead using a 75-ton cap, putting an end to the deluge. When all was said and done, 4.9 million gallons had surged into what the nonprofit environmental group SkyTruth estimates to be 68,000 square miles surrounding the rig.

Infographic **24.9** | **THE GULF OF MEXICO ECOSYSTEM**

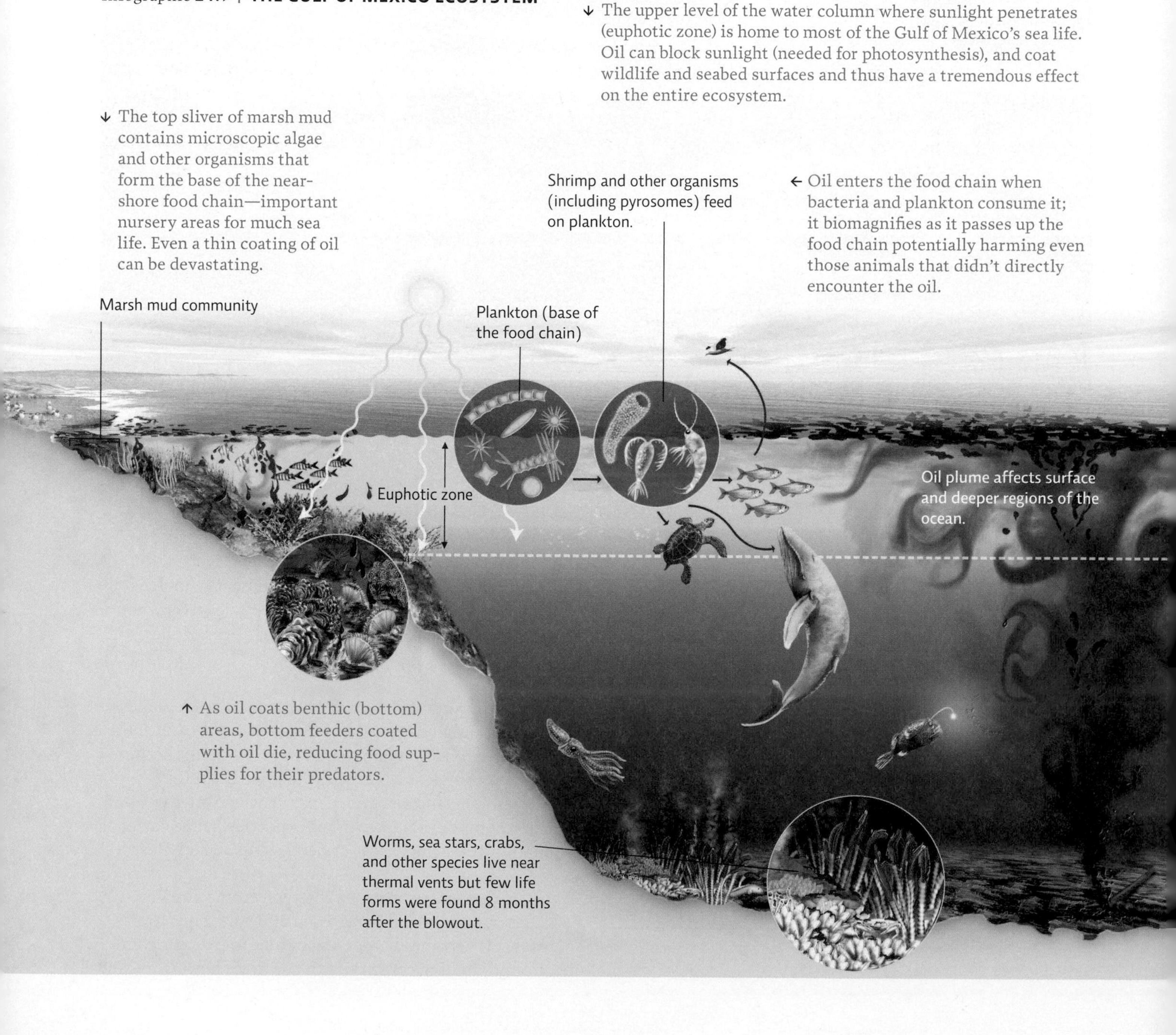

↓ The upper level of the water column where sunlight penetrates (euphotic zone) is home to most of the Gulf of Mexico's sea life. Oil can block sunlight (needed for photosynthesis), and coat wildlife and seabed surfaces and thus have a tremendous effect on the entire ecosystem.

↓ The top sliver of marsh mud contains microscopic algae and other organisms that form the base of the near-shore food chain—important nursery areas for much sea life. Even a thin coating of oil can be devastating.

Shrimp and other organisms (including pyrosomes) feed on plankton.

← Oil enters the food chain when bacteria and plankton consume it; it biomagnifies as it passes up the food chain potentially harming even those animals that didn't directly encounter the oil.

Marsh mud community

Plankton (base of the food chain)

↓ Euphotic zone

Oil plume affects surface and deeper regions of the ocean.

↑ As oil coats benthic (bottom) areas, bottom feeders coated with oil die, reducing food supplies for their predators.

Worms, sea stars, crabs, and other species live near thermal vents but few life forms were found 8 months after the blowout.

The *Deepwater Horizon* spill is still taking its toll. Several species of endangered whales—the blue whale, the sperm whale, and the beaked whale—call the area surrounding the Deepwater Horizon spill home. Estimates of sperm whale populations show numbers have declined since the spill from six whales per 1,000 km in 2007 to only two per 1,000 km. According to NOAA, the sperm whale population is so endangered that only two animals need to die as the result of the oil spill to put the entire Gulf population at risk for extinction. "It's fairly scary," Sidorovskaia says.

In addition, oil could be having other serious impacts that were not obvious at first. The oil from the *Deepwater Horizon* spill is being blamed for killing coral reefs in the Gulf—two species of which are endangered—and threatening the lives of other endangered and threatened species, such as the humpback and blue whale, the gulf sturgeon, and the smalltooth sawfish. The spill has also had a major impact on Gulf fisheries: Although most of the area was subsequently reopened to fishing in October 2010, the short-term gross revenue loss to the fishing industry may have reached $172 million, according to an October 2010 study published by consulting firms IEM and Headwaters Capital. [INFOGRAPHIC 24.9]

So how do we increase energy security and reduce our dependence on fossil fuels so that we can avoid using dangerous extraction methods? There are many possibilities, but the main methods include importing energy from multiple sources while reducing imports overall, exploiting local energy sources, developing alternative energy sources, and reducing energy needs through increased conservation and energy efficiency. Iceland, for instance, is increasing its energy security by focusing on **energy independence**, such that it will meet all of its energy needs without imports by the year 2050.

"America's dependence on oil is one of the most serious threats that our nation has faced," said U.S. president Barack Obama in 2009. But "we will not be put off from action because action is hard." No matter what it takes, America needs to break free of its fossil fuel shackles, and it must do so immediately.◉

---

energy independence Meeting all of one's energy needs without importing any fuel.

Research articles referenced in this chapter:
Hazen, T.C., et al. 2010. *Science*, 330: 204–208.
Joye, S.B., et al. 2011. *Science*, 332: 1033.
Kelly, E., et al. 2010. *Proceedings of the National Academy of Sciences*, 107: 16178–16183.
Kessler, J.D., et al. 2011. *Science*, 331: 312–315.

BRING IT HOME

⊙ PERSONAL CHOICES THAT HELP

The *Deepwater Horizon* incident brought the concerns and dangers of offshore oil drilling to the forefront of public awareness, both in the United States and around the world. Disasters like these are inherently linked to our dependence on a nonrenewable resource. By decreasing our petroleum use, we can reduce the pressure on oil companies to pursue new offshore drilling projects.

Individual Steps

→ Minimize your fuel use when driving by: planning ahead to condense shopping trips and errands and reduce total miles driven; parking as far as you can from your entrance and getting some exercise instead of wasting gas. Not driving? See where you can safely walk, bike, or use public transportation.

→ Reduce your use of disposable plastics like water bottles and single-serve food containers, and always recycle plastics when possible.

→ Turn your thermostat down in the winter to reduce your energy use.

Group Action

→ Organize a carpool system in your community to reduce the number of single-passenger car trips.

Policy Change

→ Work with your school's administration to encourage public transportation and increased bike usage on your campus.

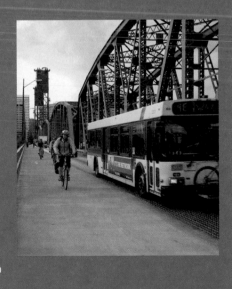

UNDERSTANDING THE ISSUE

CHECK YOUR UNDERSTANDING

1. **What is the correct definition of fossil fuel "reserves"?**
   a. Human-made storage areas for emergency supplies of fossil fuels
   b. Areas where there are known stores of untapped fossil fuels held in reserve for the future
   c. Known sources of untapped fossil fuels that are expected to be both economically and technologically recoverable
   d. Sources of fossil fuels that have not yet been discovered, but which geologists have predicted to exist

2. **To the best of our knowledge, how many years will we be able to rely on oil as a fuel source?**
   a. 10 years
   b. 20 years
   c. 40 years
   d. 80 years

3. **Which of the following describes the process of fracking?**
   a. Piping a mixture of water, sand, and chemicals deep underground to force natural gas to the surface
   b. Blasting cracks, or fractures, in Earth's surface to search for natural gas
   c. Drilling for oil under the ocean's surface
   d. Extracting petroleum products, using a variety of chemicals, from oil shales

4. **Examine Infographic 20.5. According to the data, which of the following has the highest boiling point?**
   a. Jet fuel
   b. Gasoline
   c. Naphtha
   d. Fuel oil

5. **How much more oil is located in the Middle East than in the United States?**
   a. Twice as much
   b. 77 times as much
   c. 200 times as much
   d. 37 times as much

6. **Why do analysts think that oil and gasoline will never be inexpensive again?**
   a. Oil supplies are controlled by political enemies of developed nations.
   b. The world has passed peak oil production.
   c. Steps that prevent oil spills from ever happening again will keep the price of oil high.
   d. As oil companies spend money on research and development of alternative energy sources, those costs will be passed on to consumers through higher fuel costs.

WORK WITH IDEAS

1. Describe the process of conventional oil extraction. Include in your description the following terms: primary production, secondary production, tertiary production, pumpjack, cap rock, and oil reservoir.

2. Compare and contrast the sources of nonconventional oil and natural gas sources in the United States and Canada. What are the advantages and disadvantages of each?

3. Based on what you know about aquatic organisms and communities, make some predictions about the long-term consequences of the *Deepwater Horizon* accident on aquatic life in the Gulf of Mexico.

4. If oil and natural gas are formed from once-living organisms, why are they considered to be nonrenewable resources?

5. What is your interpretation of "energy security"? In your opinion, what is the best path for the United States to take to ensure energy security now and in the future? Should we become energy independent? Why or why not?

6. In addition to driving less and/or driving more fuel-efficient cars, what can you do to reduce your overall oil consumption?

ANALYZING THE SCIENCE

The following graph shows the costs of extracting oil in different parts of the world.

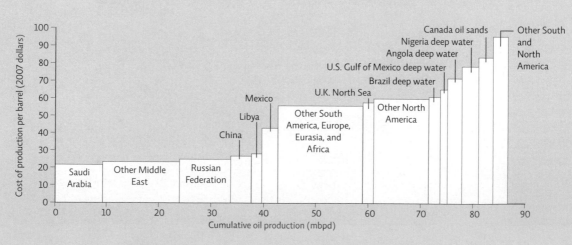

INTERPRETATION

1. Approximately how much oil is produced in "Other North America" compared to Saudi Arabia and "Other Middle East"? If you were the president of Exxon, where would you suggest drilling for oil, given the information in this graph?
2. Why is the bar for "Other Middle East" short and wide while the bar for "Canada oil sands" is tall and narrow? Why is it cheaper to extract oil in Saudi Arabia than in Canada?
3. Calculate the cost per day to produce oil in Saudi Arabia and compare it to the cost in the U.S. Gulf of Mexico deep water. (Hint: Estimate the height of each bar to get the cost to produce a barrel; then estimate the width of each bar to get the total number of barrels in a given region. Divide the cost by the total number of barrels to get the cost to produce oil in that region.)

ADVANCE YOUR THINKING

4. The graph lists only the economic costs associated with oil production. What specifically are the other costs (including environmental, social, and economic)? How do we decide when the costs outweigh the benefits of oil extraction and production?
5. In your opinion, when do the environmental costs (externalities) outweigh the benefits (status quo) of a fossil fuel economy?
6. How did we come to rely so heavily on fossil fuels, and what are the consequences of our dependence? Do you think we will be able to switch to an alternative fuel source before we run out of fossil fuels?

EVALUATING NEW INFORMATION

In recent years, moving the United States toward energy independence from most other nations has become one of the federal government's goals. One path toward this is to increase our use of unconventional fossil fuels, such as tar sands and shales. There are large deposits of tar sands in Canada, and mining has already begun in some areas. The mining products are transported through the transcontinental Keystone pipeline. Oil companies would like to expand the pipeline and have proposed the Keystone XL, which would transport oil from Canada to Texas.

Go to the following websites and read the articles about this issue: http://tinyurl.com/5s6jqxj and http://tinyurl.com/6k9eznl.

Evaluate the websites and work with the information to answer the following questions:

1. Are these reliable information sources? Do they have a clear and transparent agenda?

   a. Who runs these websites? Do the organizations' credentials make them reliable or unreliable? Explain.
   b. Who are the authors? What are their credentials? Do they have the scientific background and expertise to lend credibility to the articles?

2. Briefly summarize what you believe are the most important pros and cons of the Keystone XL pipeline. Based on your reading, what decision would you suggest that the president make regarding this project?

3. What else would you like to know about this project? Do an Internet search to find more information. List at least two more sources, including their urls. In addition, answer questions 1a and 1b for each source. Based on the new information you have gathered, has your opinion about the project changed? Why or why not?

MAKING CONNECTIONS

FOSSIL FUELS AND THE MARCELLUS SHALE

Background: The Marcellus Shale is a geological formation composed of black shale that extends from Ohio and West Virginia across Pennsylvania and lower New York. Geologists have known for over 100 years that large quantities of natural gas are trapped in the shale, but until recently the technology (and the money) to extract the gas did not exist. Given the current situation with conventional fossil fuel sources, and the application of the fracking process to the extraction of natural gas, oil companies are actively drilling for natural gas in new locations, including the Marcellus Shale.

Case: You are a senior member of the city council for a small town in Ohio. The town has been approached by an oil company for permission to lease land in order to drill for natural gas using fracking. A city council meeting has been scheduled before the council votes so that members can learn more about the process and thus make an informed decision about what is right for the community. It will be your job to teach the rest of the council about fracking.

1. Be sure to include the following in your explanation:
   a. A description of the fracking process and where it is currently being used on a regional and local scale

   b. Whether there are possibilities for economic improvement and job creation related to the project. Explain.
   c. Whether there are general environmental concerns related to fracking. Explain.

2. Prior to your recent move to Ohio, you lived in the town of Dimock, Pennsylvania, which had an unpleasant experience with fracking. So, in addition to the information that you already presented to the city council, you decide to prepare and present a detailed case study about Dimock. Be sure to include the following in your presentation:
   a. Stories of residents of Dimock and Damascus, a nearby town, and their experiences with fracking, using videos of residents and the fracking process to illustrate your point
   b. Responses of oil companies to the people affected by fracking
   c. Responses from the EPA and other government agencies to both the oil companies and the people affected by fracking

3. In your summation to the council, provide your opinion of fracking and whether or not it should be permitted in your community.
   a. What is your recommendation to the council? Why?
   b. What other types of information might the council members want before they make a decision? Where would you suggest they go to find the answers to their questions?

Nine-year-old Citron Miller gets a breathing treatment after suffering an asthma attack. Miller missed more than 35 days of school last year because of his asthma. With some 6 million kids affected, asthma has become the most common chronic illness among children in the United States.

THE YOUNGEST SCIENTIST

Kids on the frontlines of asthma research

CORE MESSAGE

Air quality issues span the globe and have serious health effects for humans and other organisms; they also take an economic toll on individuals, industry, and society. Though there are natural sources of air pollution, most is caused by human actions. Pollutants generated in one area can travel great distances to affect other places. Policies that restrict air pollution, and new technology to reduce it at the source, can help us address air pollution problems.

GUIDING QUESTIONS

After reading this chapter, you should be able to answer the following questions:

→ What is air pollution and what is its global impact?

→ What are the main types and sources of outdoor air pollution? What are the common types of air pollutants regulated by the EPA?

→ What are the health, economic, and ecological consequences of air pollution?

→ What are the main sources of indoor air pollution in developed nations such as the United States? What about developing nations? What can be done to reduce these pollutants?

→ What are the economic and societal costs and benefits of mitigating air pollution?

For 10-day stretches during 2003 and 2004, forty-five asthmatic kids from smoggy Los Angeles County, some as young as 9, carried more than books in their backpacks. The kids, from two regions of the county, wore backpacks containing small monitors that sampled the air around them continuously as they went about their daily lives—going to class, playing with friends, having dinner with their families.

Children have a lot to teach us about **asthma**. It's not just that the respiratory disorder more commonly afflicts kids than adults—which it does (1 in 10 American children has asthma, compared to about 1 in 13 adults). Or that research on children with asthma is on the rise. No, in fact, children are actually conducting some of the seminal research on asthma themselves.

And those personal air monitors are far more accurate than measurements taken at local monitoring stations, explains Ralph Delfino, an epidemiologist at the University of California, Irvine, who with his colleagues, recruited the students to help collect the data. "A monitoring station can be many miles from where the subject lives, where they go to school, etc.—so that measurement may not represent their actual exposure very well," he says. The air monitors used in Defino's study detected levels of harmful pollutants in the air surrounding each individual child.

The children selected for the experiment were all currently being treated for mild to moderate asthma and the area they all lived in had significant vehicle air pollution. Each child wore a backpack containing a monitor that would continuously sample the air around the child for 10 consecutive days; the monitor measured the amount of small particles and nitrogen dioxide (NO_2), both of which are commonly present in vehicle emissions and are known to irritate lung tissue. Exposure to these pollutants can trigger asthma symptoms such as wheezing, coughing, or shortness of breath in sensitive individuals.

Ten times a day, the children exhaled into a special bag that assessed their breath for nitric oxide (NO), a chemical marker of airway inflammation—a telltale symptom of asthma.

When the 10 days were over, Delfino compared the types of pollutants to which the kids were exposed with their

asthma A chronic inflammatory respiratory disorder characterized by "attacks" during which the airways narrow, making it hard to breathe; can be fatal.

⊚WHERE IS LOS ANGELES, CALIFORNIA?

LOS ANGELES

ARIZONA

MEXICO

CA

nitric oxide levels at similar points in time. The goal was to paint a picture of the types and amounts of pollutants that exacerbate asthma.

Asthma is a respiratory ailment marked by inflammation and constriction of the narrow airways of the lungs. Delfino's work is important in part because asthma is one of the most common chronic childhood diseases in the United States and other developed nations and a major cause of childhood disability. The United Kingdom has the highest incidence of asthma, with more than 15% of its population diagnosed in 2006 (compared to 10.9% in the United States). Developing nations are also seeing a rise in asthma, especially in urban centers. In all areas, asthma rates are likely underdiagnosed.

In the United States, asthma is the leading cause of school absences, and hence lost revenue for public schools, whose federal funding is based on attendance. The prevalence of childhood asthma more than doubled from 1980 to the mid-1990s, and though it has leveled off in recent years with 7 million children and 17.5 million adults diagnosed, it still remains at historically high levels (because there are more adults than children, overall more adults have asthma, but a larger percentage of children have the disease).

The Los Angeles skyline is obscured by smog.

Researchers believe that **air pollution**—contaminants, either from natural sources or human activities, that cause health or environmental problems—may play a key role in the recent asthma spike. In addition, air pollution has been linked to cancer, infection, and other respiratory diseases—and it not only harms us, but also plants, animals, and even our buildings, bridges, and statues.

Outdoor air pollution has been under examination for many years.

As far back as the 1930s, scientists recognized a link between outdoor air pollution and human illness. In 1930, for instance, 63 people died and 1,000 were sickened in Belgium when a temperature inversion—a situation that occurs when the temperature is higher in upper regions

of the atmosphere than in the lower, causing pollutants to become trapped near Earth's surface—led to a sudden spike in lower atmospheric sulfur levels. And 1948 was the year of the famous London Fog, when acid aerosols trapped in the lower atmosphere in London killed 4,000 people.

In many urban areas and in developed nations today, much of the air pollution comes from vehicle exhaust and industry emissions, including emissions from coal-fired power plants. While less developed nations do have outdoor pollution, their biggest problem is often indoor air pollution from small particles released through burning solid fuels such as charcoal, wood, or animal waste. However, only in the last 25 years have scientists really been able to tease out the link between air pollution and asthma. [INFOGRAPHIC 25.1]

Infographic **25.1** | **AIR POLLUTION IS A WORLDWIDE PROBLEM**

↓ The World Health Organization (WHO) recognizes air pollution as a major threat to human health. Outdoor air pollution in urban areas is responsible for at least 1.3 million deaths annually. Indoor air pollution may be responsible for 2 million or more premature deaths per year, mostly in developing nations; about half of these deaths involve children under the age of 5.

DEATHS DUE TO OUTDOOR AIR POLLUTION, 2008

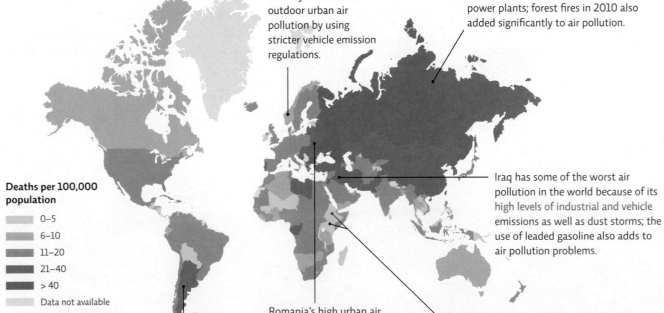

Norway has reduced outdoor urban air pollution by using stricter vehicle emission regulations.

Air quality in many Russian cities is poor, largely because of industry and coal-fired power plants; forest fires in 2010 also added significantly to air pollution.

Deaths per 100,000 population

- 0–5
- 6–10
- 11–20
- 21–40
- > 40
- Data not available

Iraq has some of the worst air pollution in the world because of its high levels of industrial and vehicle emissions as well as dust storms; the use of leaded gasoline also adds to air pollution problems.

Urban areas of Argentina have high levels of outdoor air pollution caused by unregulated vehicle emissions (catalytic converters that reduce emissions are not required).

Romania's high urban air pollution is caused by vehicle and industrial emissions; Romania is moving forward to comply with European Union air quality standards but has far to go.

Though urban air pollution is low, burning charcoal and other solid fuels produces high levels of *indoor* air pollution, giving developing countries like Ethiopia and Kenya some of the highest asthma rates in the world.

↑ Controlled burns of agriculture fields, like this one of an asparagus field in California, help clear land for more planting but release particulate matter (small particles) into the air, contributing to respiratory distress in sensitive individuals.

Outdoor air pollution includes chemicals and small particles in the atmosphere that can be either natural in origin—arising from sandstorms, volcanic eruptions, or wildfires—or come from humans, such as pollution released from factories and vehicles during the combustion of fossil fuels, or from burning any biomass such as wood, crop waste, or garbage. Of these anthropogenic sources, **primary air pollutants** are pollutants released directly from both mobile sources (such as cars) and stationary sources (such as industrial plants). In addition, some primary air pollutants react with one another or with other chemicals in the air to form **secondary air pollutants**. For example, **ground-level ozone** forms when some of the pollutants released during fossil fuel combustion react with atmospheric oxygen in the presence of sunlight. Acid rain is another common secondary pollutant.

Pollution can also move from the troposphere up into the stratosphere, a much thinner layer of the atmosphere that extends from 11 to 50 kilometers (7 to 31 miles) above Earth. This region contains the "ozone layer," an area with high ozone (O_3) concentrations. As we saw in Chapter 2, ozone in the stratosphere is significant because it serves as Earth's sunscreen, blocking some of the dangerous ultraviolet (UV) radiation from the sun. Air pollution can have grave impacts on this layer; for instance, chlorofluorocarbons (CFCs)—compounds that contain carbon, chlorine, and fluorine—can travel up into the stratosphere and destroy ozone.

Don't confuse ground-level ozone with stratospheric ozone depletion. These are two very different problems though they deal with the same molecule—O_3. Ozone in the stratosphere is a good thing—but you don't want to breathe it in as it can directly damage the sensitive tissue of the lungs. Even plants are damaged by the corrosive action of ozone.

air pollution Any material added to the atmosphere (naturally or by humans) that harms living organisms, affects the climate, or impacts structures.

primary air pollutants Air pollutants released directly from both mobile sources (such as cars) and stationary sources (such as industrial and power plants).

secondary air pollutants Air pollutants formed when primary air pollutants react with one another or with other chemicals in the air.

ground-level ozone A secondary pollutant that forms when some of the pollutants released during fossil fuel combustion react with atmospheric oxygen in the presence of sunlight.

↓ There are many sources of outdoor air pollution, both natural and anthropogenic. These sources release primary pollutants, some of which may be converted to different chemicals (secondary pollutants). Prevailing winds transport pollution that reaches the upper troposphere or stratosphere around the globe—no area is immune to air pollution. Agricultural and industrial pollutants have been found in Arctic and Antarctic air, delivered by these prevailing winds.

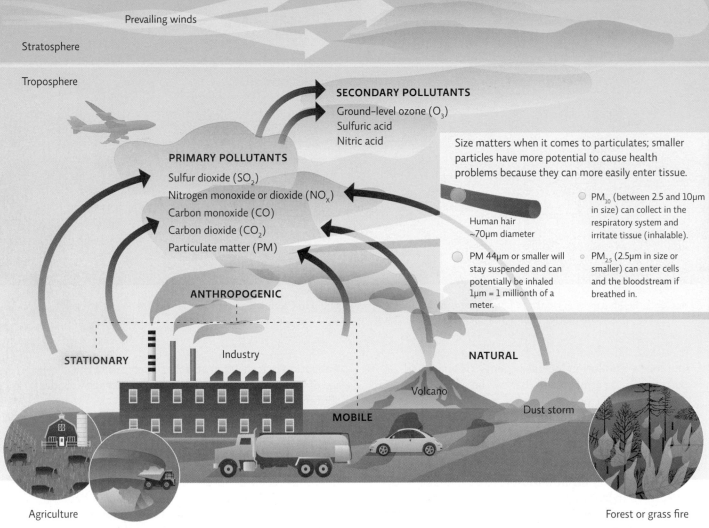

Prevailing winds

Stratosphere

Troposphere

SECONDARY POLLUTANTS
Ground-level ozone (O_3)
Sulfuric acid
Nitric acid

PRIMARY POLLUTANTS
Sulfur dioxide (SO_2)
Nitrogen monoxide or dioxide (NO_x)
Carbon monoxide (CO)
Carbon dioxide (CO_2)
Particulate matter (PM)

Size matters when it comes to particulates; smaller particles have more potential to cause health problems because they can more easily enter tissue.

Human hair ~70μm diameter

PM 44μm or smaller will stay suspended and can potentially be inhaled
1μm = 1 millionth of a meter.

PM_{10} (between 2.5 and 10μm in size) can collect in the respiratory system and irritate tissue (inhalable).

$PM_{2.5}$ (2.5μm in size or smaller) can enter cells and the bloodstream if breathed in.

ANTHROPOGENIC

STATIONARY

Industry

NATURAL

MOBILE

Volcano

Dust storm

Agriculture

Mining

Forest or grass fire

Outdoor air pollution, in all forms, is one of the most dramatic contributors to asthma. When the Environmental Protection Agency (EPA) started regulating pollution back in 1971, the agency did not know the extent to which air pollution could affect health; in fact, the EPA administrator noted at the time that the agency's clean air regulations were "based on investigations conducted at the outer limits of our capability to measure connections between levels of pollution and effects on man." Nevertheless, in 1971, the EPA set standards for the most common but problematic pollutants. Five of these were chemical air pollutants: sulfur oxides, carbon monoxide,

nitrogen oxides, ground-level ozone, and lead; standards were also set for **particulate matter (PM)**—particles or droplets small enough to remain aloft in the air for long periods of time. [INFOGRAPHIC 25.2]

These six "criteria pollutants" are still considered the most problematic and health-threatening pollutants in the United States. A 1993 study published by Harvard University researchers helped establish the link between air pollution and impaired health. A follow-up study estimated that particulate pollution—which includes soot, ash, dust, smoke, pollen, and small, suspended droplets (aerosols)—accounts for 75,000 premature deaths per year. [INFOGRAPHIC 25.3]

Although all particulates reduce visibility, it is the smallest particles—those with a diameter less than 2.5 micrometers (µm), about 1/40th the diameter of a human hair—that aggravate asthma and other chronic lung diseases, and increase one's risk for death.

Outdoor air pollution has many sources.

Where does outdoor air pollution come from? The burning of fossil fuels commercially, industrially, and residentially contributes heavily to outdoor air pollution. Coal and petroleum burning releases emissions that can produce **smog**—a term coined by combining the words "smoke" and "fog"—which is a hazy air pollution that contains a variety of pollutants, including sulfur dioxide (SO_2), nitrogen oxides (NO and NO_2—together expressed as NO_x), particulates, and ground-level ozone.

Factories, incinerators, and mining operations also release pollutants. Industrial pollution releases what is called **point source pollution**, in that it is possible to identify the pollution's exact point of entry into the environment. Hypothetically, point source pollution is easier to monitor and regulate than is **nonpoint source pollution**—pollution from dispersed or mobile sources like vehicles and lawn mowers.

Agriculture is yet another source of nonpoint source outdoor pollution. Toxic pesticides sprayed on crops can become airborne and drift as far as 20 miles; confined animal feeding operations produce significant odor problems and particulate pollution; and animal waste contributes to global warming by releasing the greenhouse gas methane.

In his study in Los Angeles County, Delfino found that particles contained in diesel fuel exhaust were among the worst asthma culprits. In addition, particulate levels were higher in Riverside, one of two regions he tested; the researchers concluded that Riverside had more pollution because it was downwind of the main urban areas in L.A.

In addition to the six criteria pollutants, the EPA also recognizes 187 hazardous air pollutants that can have adverse effects on human health, even in small doses. These toxins may cause cancer, developmental defects, or may damage the central nervous system or other body tissues. They include *volatile organic compounds* (VOCs), a

---

particulate matter (PM) Particles or droplets small enough to remain aloft in the air for long periods of time.

smog Hazy air pollution that contains a variety of pollutants including sulfur dioxide, nitrogen oxides, tropospheric ozone, and particulates.

point source pollution Pollution that enters the air from a readily identifiable source such as a smokestack.

nonpoint source pollution Pollution that enters the air from dispersed or mobile sources.

Infographic **25.3** | **THE HARVARD SIX CITIES STUDY LINKED AIR POLLUTION TO HEALTH PROBLEMS**

→ To see if there was a link between health and small particulate pollution ($PM_{2.5}$; that is, particulate matter smaller than 2.5 micrometers) researchers at Harvard University compared death rates and $PM_{2.5}$ levels in six U.S. cities. The data reveal that as particulate pollution increases, so does the mortality (death) rate. $PM_{2.5}$ pollution and deaths went down in every city between the two sampling periods evaluated—likely a result of stricter pollution limits in the 1990s.

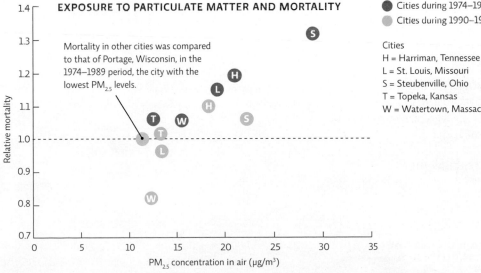

EXPOSURE TO PARTICULATE MATTER AND MORTALITY

Mortality in other cities was compared to that of Portage, Wisconsin, in the 1974–1989 period, the city with the lowest $PM_{2.5}$ levels.

Relative mortality

$PM_{2.5}$ concentration in air (µg/m³)

● Cities during 1974–1989
● Cities during 1990–1998

Cities
H = Harriman, Tennessee
L = St. Louis, Missouri
S = Steubenville, Ohio
T = Topeka, Kansas
W = Watertown, Massachusetts

Infographic 25.4 | SOURCES AND EFFECTS OF REGULATED AIR POLLUTANTS

CRITERIA AIR POLLUTANT	SOURCE	HEALTH/ENVIRONMENTAL EFFECTS
CARBON MONOXIDE (CO) From incomplete combustion of any carbon-based fuel	Vehicles, forest fires, volcanoes	Interferes with red blood cells' ability to carry oxygen; causes headaches and can lead to asphyxia (death)
SULFUR DIOXIDE (SO$_2$) From burning fuel that contains sulfur, such as coal	Industry, volcanoes , dust	Respiratory irritant; harms plant tissue and can be converted to sulfuric acid which damages plants, aquatic creatures, and concrete structures
NITROGEN OXIDES (NO$_x$ = NO AND NO$_2$) From the reaction of nitrogen in fuel or air with oxygen at high temperatures (usually during combustion of a fuel)	Vehicles, industry, nitrification by soil and aquatic bacteria	Respiratory irritant; can increase susceptibility to infection; can be converted to nitric acid, which damages plants, aquatic organisms, and even concrete structures; overfertilizes ecosystems; can cause eutrophication (see Chapter 18)
GROUND-LEVEL OZONE (O$_3$) Formed from reactions between NO$_x$ and VOC with oxygen in the presence of sunlight	Vehicles are the main source of NO$_x$; VOCs can be released from manufactured products or be released directly by industry. Some are also released by trees and other plants.	Respiratory irritant that reduces overall lung function; can reduce photosynthesis in plants
PARTICULATE MATTER (PM) Tiny airborne particles or droplets, smaller than 44 micrometers. The smaller the particle, the more dangerous it is for tissue.	Released during the combustion of any fuel or activity that produces dust; also produced by forest fires, dust storms, and even sea spray	Respiratory irritant; can reduce respiratory and cardiovascular function; reduces visibility; particles can end up in aquatic or terrestrial ecosystem supplying nutrients or acids that can harm organisms that live there
LEAD (Pb) Additive to gasoline, paint, and other solvents; phased out of U.S. gas supply in the 1970s and officially banned in 1996	Lead-based paint in older homes and from other countries; leaded gasoline; soil erosion and volcanoes	Damages nervous, excretory, immune, reproductive, and cardiovascular systems; can accumulate in soils and in the tissues of organisms and can biomagnify up a food chain (see Chapter 3)
OTHER EPA-REGULATED AIR POLLUTANTS	SOURCE	HEALTH/ENVIRONMENTAL EFFECTS
VOLATILE ORGANIC COMPOUNDS (VOCs) Organic molecules (hydrocarbons) that easily evaporate	Solvents, paints, glues, and other organic chemicals; plants naturally release VOCs	Those from human sources can be directly toxic or disruptive to living organisms, including humans; contribute to ground-level ozone formation
MERCURY (Hg) Naturally occurring element	Burning coal; mining and smelting operations; forest fires and volcanoes	A major neurotoxin that can disrupt development in embryos and young children; can bioaccumulate in individuals and biomagnify up the food chain
CARBON DIOXIDE (CO$_2$)	Burning carbon-based fuels such as fossil fuels; forest fires and normal decomposition	Nontoxic so no health effects at normal levels of exposure; greenhouse gas that contributes to climate change, affecting ecosystems worldwide

← Derrick Reliford, 14, in his Bronx home. At 10 years old, he participated in the New York University study to measure how much pollution Bronx children were exposed to. The researchers found that students in the South Bronx were twice as likely to attend a school near a highway as were children in other parts of the city. The South Bronx is home to some of the highest asthma hospitalization rates for children in New York City.

variety of chemicals that readily evaporate but don't dissolve in water. VOCs are released by natural sources such as bogs, and household products including paint, carpets, and cleaners; the main outdoor source is fossil fuel combustion. Another chemical class of concern is *polycyclic aromatic hydrocarbons* (PAHs), which are released as by-products from the combustion of wood, tobacco, coal, or diesel. And in 2007, the EPA ruled that greenhouse gases are air pollutants, giving the agency the authority to regulate carbon dioxide (CO_2) emissions. [INFOGRAPHIC 25.4]

The air we breathe affects our lungs, especially those of children.

If anyone was born to study air pollution, it was Kari Nadeau. Growing up near smoggy Newark, New Jersey, she suffered as a child from terrible asthma and allergies, which she always suspected were related to the pollution surrounding her. Her mom was a public health school nurse and her dad worked for the EPA. "I always had these questions lingering about how much the environment affects people with asthma or allergies," she recalls. Was air pollution a culprit in her respiratory woes? Her instincts told her yes.

And now her research does, too. An assistant professor of pediatric immunology & allergy at Stanford University, Nadeau recently uncovered something surprising: When she looked at the blood collected from kids who came to her clinic from Fresno, California—kids who frequently had terrible asthma—she saw that they had different-looking immune systems than other California kids. Specifically, their immune system's "peacekeeping" functions didn't work as well in keeping asthma-producing inflammation at bay. This observation piqued her curiosity: "I thought, what's different in Fresno? So I went on the Internet and searched and saw that Fresno is the second most polluted city in the country," she says.

Nadeau is now collaborating with scientists in Fresno to study the link between immune system regulatory cells, asthma, and air pollution. To understand the differences among kids from Fresno versus other areas in California, she collected blood from 71 asthmatic children who had spent their entire lives in Fresno, 30 healthy children from Fresno, 40 asthmatic kids from less-polluted Palo Alto, and 40 healthy kids from Palo Alto. She found that all the kids who grew up in Fresno had far higher levels of common pollutants in their blood than did the Palo Alto kids—and the higher their pollutant levels were, the more likely they were to have asthma. Based on her research, Nadeau thinks that air pollutants stifle the activity of genes responsible for maintaining normal immune system function—and that by doing so, they increase the risk for asthma and allergies.

Nadeau's work at Stanford suggests that PAHs are key to understanding the link between air pollution and asthma. "Levels of PAHs and the severity of the asthma were both directly related to the function of immune system cells in our patients," she says, and levels of PAHs were 7 times higher in Fresno, where asthma tended to be more common and severe, than in Palo Alto.

Other studies also support the link between asthma symptoms and air pollution. More children in the South Bronx are hospitalized for asthma than anywhere else in New York State, and since many Bronx children live or attend schools adjacent to congested highways, Bronx congressman José Serrano wondered if the two factors might be related. In 2002, he asked New York University environmental scientist George Thurston if he would be willing to conduct a study to find out. "We thought about it for a nanosecond, and then said, 'sure,'" Thurston recalls. In a study reminiscent of Delfino's, Thurston recruited 40 South Bronx fifth-graders to tote wheeled backpacks containing personal air monitors for a month while rating their respiratory symptoms 3 times a day. "You rolled it, so it wasn't really that heavy," Derrick Reliford, one of the students in the study told the *New York Times*. "They were the rock stars of the class—everybody wanted to help them with the backpacks," Thurston recalls.

The children came from four different schools, two of which were close to a highway and two of which were not. Thurston found that, sure enough, the children who went to schools or lived closer to highways were exposed to more air pollution—in particular, diesel fuel exhaust—and they also had more severe respiratory symptoms.

Low-income or minority areas often have some of the worst air. This raises questions of **environmental justice**—the concept that access to a clean, healthy environment is a basic human right. Polluting industries such as power plants or incinerators are often placed in areas where residents have less ability to fight for their rights—less money, less education, little or no voice in local government. In some cases, even when socioeconomic status is accounted for, minority communities still face more exposure to pollution than average, an example of **environmental racism**. A study conducted in Southern California by Brown University researcher Rachel Morello-Fosche found that one's risk for developing cancer from exposure to polluted air increased as income

decreased. And in general, cancer risk was higher for minorities (Asian, African American, Latino) than for the majority (Caucasians) no matter what the income level.

Children of low-income families are at particular risk: As in the Bronx, their homes and schools are near major roads or factories, and they often come and go to school during rush hour when traffic is heaviest and smog forms.

Traveling pollution has far-reaching impacts.

One of the most problematic characteristics of air pollution is that it moves. Air pollution produced in one city can end up harming humans and other species halfway around the globe. For example, as much as half of the air pollution that falls on the Great Smoky Mountains of Tennessee and North Carolina originates in the Ohio Valley, where it is released by tall smokestacks of coal-burning electrical power plants. Prevailing winds bring the pollution southeast, where the tall mountains in the Southern Appalachians eventually stop it. There, it not only pollutes the air but also produces **acid deposition**—sulfur and nitrogen emissions that react with oxygen and water to form acids that can fall back to ground as acid rain or snow. Acidification of soil due to acid deposition can change the soil chemistry and mobilize toxic metals such as aluminum, hindering the plant's ability to take up water. Acids leach nutrients from the soil, too, reducing the amount of calcium, magnesium, and potassium available to plants in topsoil. Taken together, these impacts can decrease plant growth, weaken the plant so it is more vulnerable to disease or pests, and even kill it. This acid deposition is also a problem throughout the northeastern United States. (See Chapter 15 for more on the pH scale and acidification). [INFOGRAPHIC 25.5]

Proof that air pollution can travel long distances can be found in the far north. Prevailing air currents pick up pollutants from the western United States, conveying them all the way to Lake Laberge in Canada's Yukon Territory, where the moisture condenses, forms clouds, and falls on the lake as rain or snow. Thus, even the most isolated regions on Earth are vulnerable to the effects of air pollution, because atmospheric and hydrologic circulation moves chemical and particulate pollutants around the globe.

Appalachian acid rain, pollution in Lake Laberge, and stratospheric ozone depletion are **transboundary pollution** problems because regions that suffer from the pollution are not necessarily the ones that released the pollutants. This means that even if an area does not produce pollution itself, its air may still be toxic. With

---

environmental justice The concept that access to a clean, healthy environment is a basic human right.

environmental racism Occurs when minority communities face more exposure to pollution than average for the region.

acid deposition Precipitation that contains sulfuric or nitric acid; dry particles may also fall and become acidified once they mix with water.

transboundary pollution Pollution that is produced in one area but falls in a different state or nation.

Infographic 25.5 | **ACID DEPOSITION**

↓ Burning fossil fuels releases sulfur and nitrogen oxides. These compounds react in the atmosphere to form acids. Acid rain, snow, fog, and even dry particles can fall to Earth as acid deposition with the potential to alter the pH of lakes and soil, damaging plant and animal life.

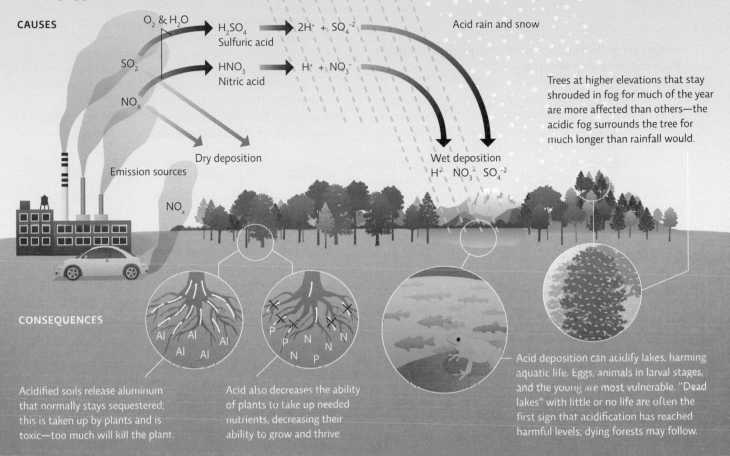

CAUSES

O_2 & H_2O

SO_2

NO_x

H_2SO_4
Sulfuric acid

HNO_3
Nitric acid

$2H^+ + SO_4^{-2}$

$H^+ + NO_3^-$

Acid rain and snow

Trees at higher elevations that stay shrouded in fog for much of the year are more affected than others—the acidic fog surrounds the tree for much longer than rainfall would.

Dry deposition

Emission sources

NO_x

Wet deposition
H^+ NO_3^- SO_4^{-2}

CONSEQUENCES

Al Al Al
Al Al

P N N
P N
N P

Acidified soils release aluminum that normally stays sequestered; this is taken up by plants and is toxic—too much will kill the plant.

Acid also decreases the ability of plants to take up needed nutrients, decreasing their ability to grow and thrive.

Acid deposition can acidify lakes, harming aquatic life. Eggs, animals in larval stages, and the young are most vulnerable. "Dead lakes" with little or no life are often the first sign that acidification has reached harmful levels; dying forests may follow.

ACID DEPOSITION HAS DECREASED

↓ Restrictions imposed by the Clean Air Act have helped decrease acid deposition in the United States. Smokestack scrubbers remove sulfur from coal burning, reducing the SO_2 released. Emission-control technologies on vehicles, such as the catalytic converter, convert dangerous combustion by-products to safer emissions (such as converting NO_x to N_2). This reduces, but doesn't eliminate, these dangerous emissions.

pH OF PRECIPITATION IN THE UNITED STATES

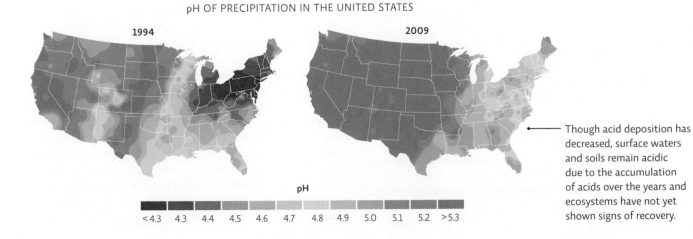

1994

2009

pH

| <4.3 | 4.3 | 4.4 | 4.5 | 4.6 | 4.7 | 4.8 | 4.9 | 5.0 | 5.1 | 5.2 | >5.3 |

Though acid deposition has decreased, surface waters and soils remain acidic due to the accumulation of acids over the years and ecosystems have not yet shown signs of recovery.

the EPA's Air Quality Index available online, people can search for up-to-date air quality reports about any U.S. region. The index also alerts local communities about air quality problems from ground-level ozone and particulate pollution.

Indoor air pollution may pose a bigger health threat than outdoor air pollution.

As Nadeau discovered, outdoor air quality substantially impacts human health. However, we breathe air indoors as well as outdoors, and indoor air quality is a growing concern among public health scientists. In fact, people living in affluent, developed nations may find their greatest exposure to unhealthy air comes from indoors. This is because so much time is spent indoors in homes, schools, or the workplace; these areas contain many potential air pollution sources. For instance, cigarette smoke causes significant health problems, including eye, nose, and mucous membrane irritation; lung damage, which can exacerbate or cause asthma; and lung cancer. Items in our home, like paint, cleaners, and furniture, release VOCs, which can also cause health problems.

Outdoor pollutants can also find their way into our buildings. Radon is a naturally occurring radioactive gas produced from the decay of uranium in rock. It can seep through the foundations of homes and accumulate in basements; exposure to radon can cause lung cancer. Not every area has the type of rock that produces radon, but buildings constructed over areas where soil or groundwater is contaminated with VOCs, such as areas with underground chemical storage tanks, also present an infiltration risk. [INFOGRAPHIC 25.6]

In developing countries, where many people cook and heat with open fires, smoke and soot from burning wood, charcoal, dung, or crop waste are major sources of indoor pollution. Kirk Smith, a professor of environmental health at the University of California, Berkeley, has found that indoor fires increase the risk of pneumonia, tuberculosis, chronic bronchitis, lung cancer, cataracts, and low birth weight in babies born of women who are exposed during pregnancy. "Considering that half the world's households are cooking with solid fuels, this is a big problem," Smith says. "Current estimates are that indoor fires cause the premature death of 1.5 to 2 million women and children per year." Many nonprofit organizations are stepping up to meet this problem with a simple, $50 solar cooker that allows people to cook food without building a fire. This technology has the added advantage of not depleting local biomass resources for fuel.

Air pollution is responsible for myriad health and environmental problems.

The World Health Organization (WHO) estimates that more than 3 million people die prematurely each year as a result of exposure to air pollution. Respiratory ailments are the biggest problem because particulates from soot and smog damage respiratory tissue and increase susceptibility to infection. In addition, as Nadeau's research has shown, asthma rates are higher in people who breathe polluted air—and children are particularly sensitive because they breathe in more air for their size than adults do, and because developing tissue is more vulnerable.

> More than 3 million people die prematurely each year as a result of exposure to air pollution.

Delfino's team of researchers followed up their first study with a second in 2007 that included 53 students with asthma (aged 9–18) with air monitors strapped to their backs. The students also had to breathe into detectors that recorded how much air they were able to blow out from their lungs at once—a measure of lung function. The results suggested that high levels of particulate pollution actually decrease lung function.

Lungs are particularly vulnerable to these small particulates because they get so much exposure (we breathe all the time) and the tissue itself is delicate. Irritants like particles, dust, and pollen can cause the lungs to produce excess mucus in an attempt to trap and expel the irritant. The lining of the airways can become inflamed and swell; in people with asthma, the irritation may trigger muscle contractions which close off the airway completely. Particles smaller than 2.5 μm can actually penetrate cells of the lungs or enter the bloodstream, where they are delivered to other cells of the body. If these particles come from the combustion of fossil fuels or other industrial sources, they may contain toxins, leading to problems associated with toxic exposure.

Since the cardiovascular system depends on the respiratory system to provide oxygen for the body, anything that impairs the lungs also harms the cardiovascular system, which might explain the fact that people living in polluted areas also have higher rates of heart attacks and strokes. Cancer rates are higher in people exposed to air pollution, too—exposure to secondhand smoke and radon are the leading environmental causes of lung cancer, and exposure to smog and vehicle emissions is linked to increased risk for lung and breast cancer.

Infographic 25.6 | **SOURCES OF INDOOR AIR POLLUTION**

↓ For most people, the greatest exposure to air pollution comes from being indoors. There are many sources of air pollution in a home or other building, as these structures tend to trap pollutants, keeping concentrations high. One can reduce exposure by avoiding or limiting the use of carpets, upholstered items, and furniture made with toxic glue and formaldehyde. Safer cleaners and low-VOC paints are readily available. Simple behaviors like taking off your shoes before entering the house and using a vacuum equipped with a HEPA filter will also help—especially important if you have indoor pets. Good ventilation and properly working heating and air conditioning units help keep indoor air pollutants at bay.

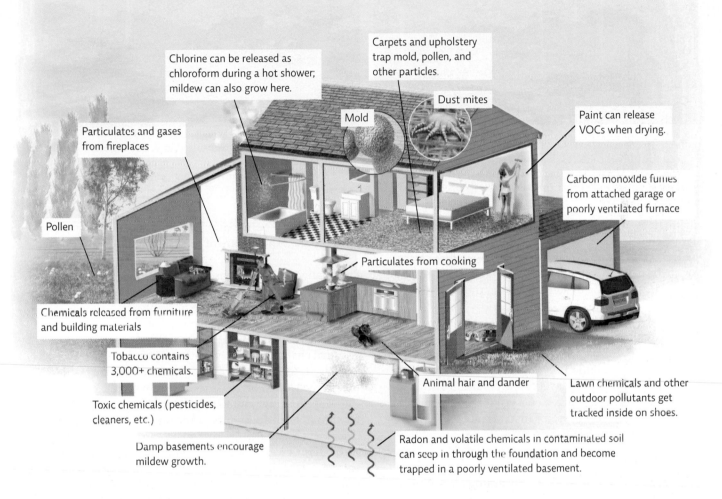

Chlorine can be released as chloroform during a hot shower; mildew can also grow here.

Carpets and upholstery trap mold, pollen, and other particles.

Dust mites

Mold

Particulates and gases from fireplaces

Paint can release VOCs when drying.

Carbon monoxide fumes from attached garage or poorly ventilated furnace

Pollen

Particulates from cooking

Chemicals released from furniture and building materials

Tobacco contains 3,000+ chemicals.

Toxic chemicals (pesticides, cleaners, etc.)

Animal hair and dander

Lawn chemicals and other outdoor pollutants get tracked inside on shoes.

Damp basements encourage mildew growth.

Radon and volatile chemicals in contaminated soil can seep in through the foundation and become trapped in a poorly ventilated basement.

Again, researchers have found that pollution causes babies to be born prematurely and with low birth weight. In addition, air pollution increases the risk that a pregnant woman will suffer from preeclampsia, a form of pregnancy-related hypertension. "We have learned that PAHs end up in the placenta and fetus," explains Beate Ritz, an epidemiologist at the University of California, Los Angeles, School of Public Health.

But humans aren't the only creatures suffering the ill effects of air pollution. Many animals suffer the same respiratory distress as humans—all lung tissue is very vulnerable to air pollution. Invertebrates as a group, especially aquatic ones, seem to be more directly impacted by air pollution (toxic effects or reproductive declines)

than are vertebrates. But aquatic vertebrates like fish and amphibians are certainly feeling severe impacts. Declines in North Atlantic salmon populations have been linked to air pollution–induced water acidity. The higher acidity causes aluminum to build up in water—aluminum is toxic to fish, especially juvenile salmon.

Plant tissues are also vulnerable to pollutants like smog and ozone, as they cause direct damage to sensitive cell membranes. Exposure can damage a leaf's ability to photosynthesize, preventing healthy growth and compromising its survival. Lichens are particularly vulnerable to air pollution such as nitrogen emissions, and their decline is seen as a warning of the potential for damage to forests and crops. Together with changes in soil

↑ Trees inside a FACE (Free-Air CO_2/O_3 Enrichment) area are used to study the effects of air pollution on plants. These trees show damage from the ozone pumped out of the tall vertical pipes (white) that ring the area.

↑ Exposure to acid rain has resulted in yellowing and loss of needles, decreasing overall photosynthesis and stunting the growth of these conifers at high elevations in the Austrian Alps.

chemistry—which can hinder plant growth—pollution damage to plant crops will ultimately cause global crop yields to fall: Estimates of crop yields predict decreases for soybean, wheat, and maize ranging from 4% to 26% by 2030, amounting to dollar losses in the billions.

Finally, pollution damages buildings and monuments. Acid deposition literally eats away at limestone and marble structures; it can etch glass and damage steel and concrete, causing billions of dollars of damage per year. Smog, SO_4, and ground-level ozone pollution also lower visibility by creating haze, a concern for areas that depend on tourism. On hazy days, for instance, it can be impossible to see across the Grand Canyon.

We have several options for addressing air pollution.

Since air pollution often travels to areas that do not produce significant amounts of pollution themselves, regulating air pollution is a particular challenge. How does one country regulate pollution that travels through the atmosphere from another country?

In developed countries, the original approach to dealing with air pollution from human activities was to spread it out—*the solution to pollution is dilution*. Factories, power plants, and other point sources built tall smoke stacks to send emissions high into the atmosphere so that they wouldn't pool at the site of production. The idea was that if dispersed, the amount of pollution in any one area would be too low to cause a problem. But this approach simply doesn't work—industry releases too much

pollution, and air circulation patterns cause some areas to get more than their share of pollution.

Eventually it became clear that regulation would be necessary. The typical approach in the 1970s was **command and control** regulation, a type of regulation that sets national limits on how much pollution can be released into the environment and imposes fines or even brings criminal charges against violators who release more than is allowed. An example of command and control regulation in the United States is the **Clean Air Act**, passed in 1963. It sets a maximum amount, or *air quality standard*, for emissions of pollutants or the presence of pollutants in ambient air. States are responsible for monitoring air quality as well as for developing, implementing, and enforcing compliance plans approved by the EPA. As a result of the Clean Air Act, the United States has seen major reductions in common air pollutants. Removing lead from gasoline, for instance, reduced lead air pollution by 98% from 1970 levels. Sulfur pollution has also been significantly reduced. And the Mercury and Air Toxic standards approved in late 2011, which limit the release of mercury, acid gases, and other pollutants from power plants, are expected to prevent 130,000 cases of serious asthma and as many as 11,000 premature deaths.

The Clean Air Act is, however, now under attack. Because regulation is based on legislation, it is subject to political wrangling; several congressional bills have been introduced that would limit the EPA's ability to regulate air quality. Of particular concern right now is the regulation of carbon dioxide (CO_2). Although it is naturally occurring, and previously considered harmless, CO_2 has now

↑ Statue in Trafalgar Square, in London, England, shows erosion that exceeds normal weathering and is likely due to acid rain.

been strongly linked to climate change (see Chapter 26). In a landmark Supreme Court case, the EPA was given the authority to regulate CO_2 as a pollutant in 2007, but immediately faced political opposition. In 2010, the EPA approved greenhouse gas emission standards (including CO_2, methane, and N_2O) for light-duty vehicles (cars and trucks) that will require new vehicles to produce less greenhouse gas emission; these new regulations will be phased in between 2012 and 2016.

The fact that our air is cleaner today than it was in the 1960s—even with a larger U.S. population and more industry—is evidence that regulations can be effective. Still, these improvements are costly to industry and farmers, and that cost is usually passed on to consumers. For this reason, many individuals and groups oppose such policies, charging that the restrictions are excessive or that the government goes too far in trying to regulate emissions. Some environmentalists worry that if we weaken or dismantle the environmental legislation that protects our air and water (and by extension, our health and ecosystems), we face the return of a highly contaminated environment, compromised health, and diminished ecosystem function and services.

In addition to command and control regulation, there are other ways to curb pollution. One example is **green taxes** on environmentally undesirable actions, such as an extra tax on low-mile-per-gallon vehicles. **Tax credits**, reductions in the amount of tax one pays in exchange for environmentally beneficial actions, fall on the other side of the spectrum. Tax credits encourage consumers to pursue options that might be more expensive than conventional options (such as the purchase of a hybrid automobile); as more people buy the products, the industries that make them can scale up and bring prices down.

Governments also offer **subsidies**, free money or resources intended to promote environmentally friendly

command and control Regulations that set an upper allowable limit of pollution release which is enforced with fines and/or incarceration.

Clean Air Act First passed in 1963 and amended most recently in 1990, this U.S. law authorizes the EPA to set standards for dangerous air pollutants and enforce those standards.

green tax Tax (fee paid to government) assessed on environmentally undesirable activities.

tax credit A reduction in the tax one has to pay in exchange for some desirable action.

subsidies Free government money or resources intended to promote desired activities.

Infographic 25.7 | **APPROACHES TO REDUCING AIR POLLUTION**

↓ We have many approaches that can be used to lessen air pollution, including technology to reduce emissions before a fuel is burned (see Infographic 23.7) and technology to capture emissions after a fuel is burned, as shown below. Some policy tools, such as cap-and-trade, encourage the use of "best available control technologies," that is, the current technology that releases the fewest amount of pollutants. There are economic costs to implementing these changes, but some benefits include new jobs, a competitive advantage for industries that can successfully reduce emissions, and a healthier society and ecosystem.

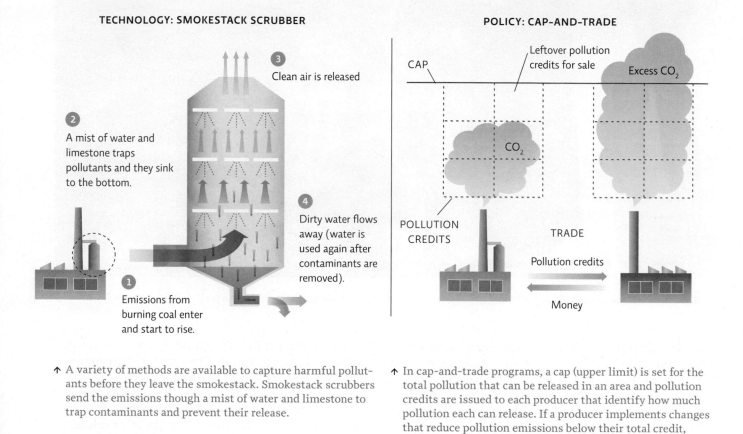

TECHNOLOGY: SMOKESTACK SCRUBBER

3 Clean air is released

2 A mist of water and limestone traps pollutants and they sink to the bottom.

4 Dirty water flows away (water is used again after contaminants are removed).

1 Emissions from burning coal enter and start to rise.

POLICY: CAP-AND-TRADE

CAP

Leftover pollution credits for sale

Excess CO_2

CO_2

POLLUTION CREDITS

TRADE

Pollution credits

Money

↑ A variety of methods are available to capture harmful pollutants before they leave the smokestack. Smokestack scrubbers send the emissions though a mist of water and limestone to trap contaminants and prevent their release.

↑ In cap-and-trade programs, a cap (upper limit) is set for the total pollution that can be released in an area and pollution credits are issued to each producer that identify how much pollution each can release. If a producer implements changes that reduce pollution emissions below their total credit, they can sell their leftover credits to another producer who exceeds their allotment.

activities. And with **cap-and-trade**, also called *permit trading*, a government or regulatory agency sets an upper limit on emissions for a pollutant on a nationwide or regional level, and then gives or sells permits to polluting industries. Users that reduce their pollution emissions below what their permit allows can sell their remaining credits to other users who exceed their allotments. Over time, pollution levels can be reduced as the cap—or limit—is lowered. A cap-and-trade program successfully reduced sulfur pollution from coal-fired power plants in

the United States in the 1990s. A downside to cap-and-trade plans is that pollution can become concentrated in areas where industries choose to buy additional permits rather than reduce emissions.

Technology can also play a big role in improving air quality. To improve indoor air quality we can install air filters and better ventilation systems. To curb pollution emissions from manufacturing, end-of-pipe solutions like scrubbers, filters, electrostatic precipitators, and catalytic converters trap pollutants before they are released. In addition, technology can inspire cleaner methods for extracting energy out of fossil fuels, as is the case with "clean coal" technologies described in Chapter 23. [INFOGRAPHIC 25.7]

cap-and-trade Regulations that set upper limits for pollution release. Producers are issued permits that allow them to release a portion of that amount; if they release less, they can sell their remaining allotment to others who did not reduce their emissions enough.

Mitigating or preventing air pollution costs money, but it is money well spent because it prevents far greater losses down the line—especially in terms of human health. According to a 2003 study published in the *Journal of Allergy and Clinical Immunology*, the average annual cost of care for an asthma patient is $4,912—with 65% of that going to medications, hospital admissions, and nonemergency doctor visits. The remaining 35% goes to indirect costs like lost time at work. Ultimately, asthma costs the United States billions of dollars each year. Nadeau, Delfino, and others hope their research helps policymakers realize just how useful curbing pollution can be. "We're talking about the air we breathe," Thurston says. "There's nothing more communal than that."◉

Research articles referenced in this chapter:

Cisternas, M., *et al.* 2003. *Journal of Allergy and Clinical Immunology*, 111: 1212–1218.

Delfino, R.J., *et al.* 2006. *Environmental Health Perspectives*, 114: 1736–1743.

Delfino, R.J., *et al.* 2008. *Environmental Health Perspectives*, 116: 550–558.

Dockery, D., *et al.* 1993. *New England Journal of Medicine*, 329: 1753–1759.

Laden, F., *et al.* 2006. *American Journal of Respiratory and Critical Care Medicine*, 173: 667–672.

Morello-Frosch, R., *et al.* 2002. *Environmental Health Perspectives*. 110: 149–154.

Nadeau, K., *et al.* 2010. *Journal of Allergy and Clinical Immunology*. 126: 845–852.

Spira-Cohen, A., *et al.* 2011. *Environmental Health Perspectives*. 119: 559–565.

↑ Mass transit options that decrease the number of cars on the road will reduce air pollution. Buses that run on compressed natural gas emit fewer emissions overall but the particulates they release are very small so, while better than a traditional diesel bus, they are still not pollution free.

BRING IT HOME

❍ PERSONAL CHOICES THAT HELP

Individuals can have an effect on air quality by researching the threats to their area, making appropriate behavior changes, and supporting legislation that limits the production of air pollutants.

Individual Steps

→ Reduce your exposure to indoor air pollution by reducing your use of harsh cleaning products, synthetic air fresheners, vinyl products, and petroleum-based candles.

→ Avoid outdoor exercise during poor air quality days. Go to airnow.gov to find the local air quality forecast.

→ Buy a radon detector and carbon monoxide detector for your house to keep you and your family safe.

Group Action

→ Organize a car-free day at your school, community, or workplace, to reduce emissions from vehicles.

→ Work with community leaders and businesses to sponsor a "free public transit" day.

Policy Change

→ If your community does not have public transit, ask community leaders to investigate bringing it to your area.

→ There are many groups working to improve our air quality. Find one in your region and see what issues they are addressing. For a list of national and regional organizations, go to www.inspirationgreen.com/air.

UNDERSTANDING THE ISSUE

CHECK YOUR UNDERSTANDING

1. **In the preindustrial era, what was the primary source of air pollution in cities?**
   a. Volcanic eruptions
   b. Sandstorms
   c. Wood smoke
   d. Pollen

2. **Given the relationship between asthma and air pollution, where would you raise a family to decrease the risk of asthma?**
   a. In an area with low VOCs but moderate to high particulates
   b. In an urban area
   c. Away from major highway systems
   d. In a valley where most people use wood to heat their homes

3. **Air pollution that results when chemicals in the atmosphere react to form a new pollutant is called:**
   a. primary pollution.
   b. secondary pollution.
   c. point source pollution.
   d. particulate pollution.

4. **According to the chapter, more than 24.5 million people in the United States were recently diagnosed with asthma. What is the estimated economic cost related to caring for these patients?**
   a. $1,000 per patient per year
   b. Billions of dollars per year
   c. About $1 million
   d. It is impossible to estimate the cost, due to unknowns such as lost time at work.

5. **Which of the following is NOT a financial incentive to encourage less pollution?**
   a. Green taxes
   b. Cap-and-trade
   c. Tax breaks
   d. Green subsidies

6. **What pollutant did the EPA receive court approval to begin regulating in 2007?**
   a. Ground-level ozone
   b. Sulfur
   c. Lead
   d. Carbon dioxide

WORK WITH IDEAS

1. Using the information presented in Infographic 21.6, explain why indoor air pollution is a cause of growing concern, especially with regard to health problems.

2. Compare and contrast point source and nonpoint source pollution. Provide examples of common pollutants from each source. Explain why industrial pollutants are found in even the most remote places on Earth.

3. List the nine major air pollutants regulated by the EPA. How have the levels of the pollutants in the atmosphere changed since the Clean Air Act was implemented? Based on your answer, would you say that the Clean Air Act has been successful at reducing air pollution?

4. Describe the policy of cap-and-trade. What are some downsides to it? Why do some members of industry dislike this policy as a means of dealing with increasing levels of CO_2?

5. Describe the types of problems that air pollution causes in ecosystems and human health. What is the overall impact in terms of worldwide mortality?

6. Where do you personally receive the most exposure to air pollution? What can you do to lessen its impact?

ANALYZING THE SCIENCE

The graph on the following page indicates levels of ground-level ozone and particulate matter that exceeded national reference levels (the level above which health or ecosystem problems occur) in areas and cities in Canada.

INTERPRETATION

1. What does the *y* axis represent? Choose a city and describe the data for that city.

2. The legend states that $PM_{2.5}$ data is given for 4 years. PM stands for "particulate matter." The number 2.5 represents the size of the particulate. From a health standpoint, why are the $PM_{2.5}$ values reported?

3. How many years of data are graphed for ozone? For particulate matter? Does the difference in the amount of time over which the data have been collected make a difference in your interpretation of these data?

ADVANCE YOUR THINKING

4. Which city would be the worst for your health, based on its levels of pollution? Why?

5. Based on the type of pollution present, what can you predict about the causes of pollution in Sault Ste. Marie versus Montréal? (Hint: Find the two cities on a map and compare the size of each city, their weather, and their primary industries.)

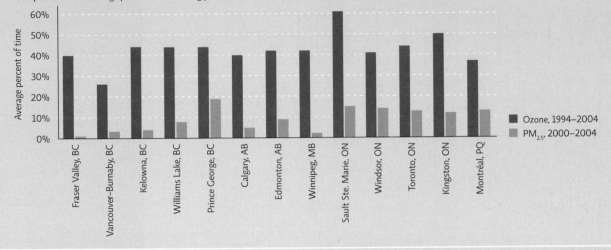

Comparison of the average percent of time smog pollutant reference levels were exceeded in select Canadian cities and the Fraser Valley

EVALUATING NEW INFORMATION

The EPA is tasked with regulating pollutants in the United States. As part of this process, the agency collects and records data for many pollutants, but not all of them. The federal EPA is assisted in this endeavor by state EPAs. However, it is impossible to collect air quality data about every locality in the United States, so most data is collected in and around cities.

Go to www.stateoftheair.org. On the left menu, enter your zip code in the "Report Card" box to get a report about air quality in your area. If no air quality monitoring stations exist in your area, choose your state and look at the data for the county closest to you. Record these data. Then, across the top bar, click on "Key Findings" and read about how the grades were calculated for each county. Finally, click on "Health Risks" and read about the specific health risks associated with both ozone and particulate matter.

Evaluate the website and work with the information to answer the following questions:

1. Is this a reliable information source? Does the organization have a clear and transparent agenda?
   a. Who runs the website? Do this organization's credentials make it reliable or unreliable? Explain.

2. What grade did your area receive for both ozone and particulates?
   a. Based on what you read about how the grade was determined, do you feel the grading system is too lax or too strict?
   b. Why does the American Lung Association advocate for a stricter system?

3. Based on what you know about the levels of pollution in your area and the effects of ozone and particulates on human health, do you believe that the regulations of the Clean Air Act should be loosened, tightened, or remain the same? Should more areas be monitored, or is it sufficient to monitor cities? Why?

MAKING CONNECTIONS

LUNG DISEASE: WHERE YOU LIVE MAKES A DIFFERENCE

Background: Ground-level ozone and suspended particulate matter are two of the most common air pollutants. They are also causes of health problems, especially respiratory illnesses such as asthma. One of the primary sources of ground-level ozone is vehicles. Particulate matter comes from the combustion of organic matter, including wood, fossil fuels, and tobacco.

Case: You are a parent of two children, both of whom have asthma. You are also the primary caregiver for your aging grandmother, who suffers from chronic lung disease. You and your spouse have both been offered jobs in Phoenix, Arizona, and Seattle, Washington. The jobs in each location are equivalent, you have family in both locations, and both locations offer the benefits of large cities. Your decision has come down to which city would be the healthiest for your children and grandmother.

1. Prepare a PowerPoint presentation about each city for your family. In it, include the following:
   a. Pictures of each city
   b. Typical weather for each month of the year, including average ambient air temperature, rainfall, and average number of sunny days
   c. Geography of the area, including altitude, nearby mountain ranges, and bodies of water
   d. Population of each city
   e. Average number of days with air pollution warnings
   f. Typical industries, especially those that would contribute to air pollution
   g. Any natural sources of air pollution
   h. Average incidence of asthma or other lung problems in each city

2. At the end of your presentation, recommend one of the cities to your family, using the evidence you gathered to support your recommendation. Provide the sources you used so that your family can evaluate the thoroughness and reliability of your research.

WHEN THE TREES LEAVE

Scientists grapple with a shifting climate

CORE MESSAGE

One of the biggest environmental problems facing humanity today is climate change. Evidence overwhelmingly points to the fact that climate is changing and that human activity is predominantly responsible for the changes. Climate change is impacting species, ecosystems, and the health and well-being of people around the globe, with more changes to come. Science can help us evaluate the changes that are happening, investigate causes, and also provide information to help make sound policy for dealing with a changing climate.

GUIDING QUESTIONS

After reading this chapter, you should be able to answer the following questions:

→ What is the difference between climate and weather? Why is a change of a few degrees in average global temperatures more concerning than day-to-day weather changes of a few degrees?

→ What is the evidence that climate change is currently occurring? How do scientists determine past and present temperatures and CO_2 concentrations?

→ What natural and anthropogenic factors affect climate, and which are implicated in the climate change we are experiencing now? How might positive feedback loops affect climate?

→ What are the projections for future warming and what are the potential effects of further warming?

→ What actions can we take to respond to a world with a changing climate?

Flames engulf the sky and the ancient forest of the North Woods of Minnesota in a forest fire that occurred unusually early in the season.

Lee Frelich was examining a 700-year-old cedar tree when he first noticed the smoke curling up into the sky over Minnesota's great North Woods. Within an hour, his entire view was filled, so Frelich, an ecologist with the University of Minnesota, and his companions—a photographer and a journalist who had cajoled him into taking them on a tour of the iconic boreal forest—trekked up to the north end of Ham Lake, away from the calamity. They would remain trapped there, amid dense, ancient stands of spruce and fir, for 3 full days, as fire claimed some 75,000 acres around them.

The fire's magnitude was not surprising. Neither, really, was the fact that the flames laid bare a patch of forest that had not burned since 1801.

What was surprising was the timing. It was the first weekend in May—unusually early in the year for such a tremendous fire, especially given that the foot-thick ice had just candled off the lake a few days ago. Frelich's hiking companions, who knew that such fires tended to come in late summer, were surprised. Frelich wasn't. To him it was just one more not-so-subtle reminder that **climate change** was rapidly throwing this ancient landscape into flux.

Other reminders have become commonplace: earlier springs, later winters, some tree species dying off, others popping up in unexpected places. Frelich and his colleagues worry that if current projections hold true, the forests themselves could vanish—converted by stress and time into scrubland or savanna.

Minnesota is not alone. In fact, the great North Woods are but one example of a whole planet in distress. In Africa and the American west, prairies are giving way to deserts (see Chapter 14). In tropical seas around the world, coral reefs are dying off (see Chapter 15). And everywhere, biodiversity is being threatened on a scale not seen since the last mass extinction 65 million years ago (see Chapter 10). Scientists say that all of these changes are occurring as global climate warms with unprecedented speed. What that might mean for the future of our planet is something they are still trying to figure out.

Climate is not the same thing as weather.

Weather refers to the meteorological conditions in a given place on a given day, whereas **climate** refers to long-term patterns or trends. Put another way, the actual temperature on any given day is the weather, while the range of expected values, based on location and time of year, is the climate. We use what we know about a region's climate to predict the weather: Seasonal shifts come at about the same time each year, and winter lows and summer highs generally hover close to expected norms.

Weather can and does vary—sometimes considerably—from one day to the next. But no single weather event—no individual storm, flood, drought, or wildfire (not even one as colossal and ill-timed as the one that trapped Frelich and his friends)—can be attributed to global warming. Global warming is the province of climate. It refers to a rapidly shifting range of temperatures that scientists have measured in myriad locations around the world—not just a few warmer days here and there, but higher high temperatures, more and longer heat waves, earlier springs, and later winters. Climatologists generally require that changes persist over a 30 year time period or longer before being willing to conclude that the variation seen is part of a climate shift, rather than natural variation.

⊚**WHERE ARE THE NORTH WOODS OF MINNESOTA?**

MN

CANADA

THE NORTH WOODS

MN   WI

↑ University of Minnesota ecologist Lee Frelich surveys the damage to the burnt forest immediately following the Ham Lake fire.

To be sure, climate change is not a new phenomenon. In fact, Minnesota's climate has been in flux for thousands of years. Fossil plant and pollen records show that, after the North American ice cap retreated—some 12,000 years ago—the climate warmed so dramatically that tree species' ranges shifted northward at a rate of 50 kilometers per century. Pines, oaks, and other deciduous species replaced the spruce trees that had covered most of the region. As summers became warmer, water levels fell, prairie plants took root, and birches and pine moved north. Then, about 6,000 years ago, the climate cooled a bit, and trees began migrating south and west once again. (Such shifts are sometimes called "tree migration" but the more accurate term is "tree range migration," since they are really changes in a species' range.)

Scientists say things are changing much more quickly now. Instead of thousands of years, organisms that inhabit a given region—not only the boreal forest around Ham Lake, but ecosystems everywhere might have just a few decades to adapt or migrate as climate change makes their current homes uninhabitable.

Evidence of global climate change abounds.

By all accounts, the forests of Northern Minnesota are places of uncommon majesty. Moose and deer meander through an ocean of trees—some of them close to 1,000 years old—that grow out of colossal, hundreds-of-feet-high granite hills whose outcrops reflect the colors of day: pale, soft pink when the Sun rises, and deep, somber rouge when it sets. The serenity belies an unsettling truth: This forest, which has existed for more than 3,000 years—since the early days of the Roman Empire—and even inspired the Wilderness Act of 1964, could vanish within the next century.

To Frelich's well-trained eye, the signs are obvious. He has spent his entire adult life trekking through this

climate change Alteration in the long-term patterns and statistical averages of meteorological events.
weather The meteorological conditions in a given place on a given day.
climate Long-term patterns or trends of meteorological conditions.

ancient landscape, observing it and cataloguing the changes—inch by inch, leaf by leaf. So it's no surprise that when he looks on the placid landscape, he sees a catastrophe unfolding. In one patch of forest, scrawny, young birch trees bud several weeks ahead of schedule. In another, adult birches are dying off rapidly, leaving a graveyard of bony white trunks. "A long growing season is not good for this tree," Frelich says, "because it goes hand in hand with warmer soil, which the paper birch doesn't tolerate well." Meanwhile, red maple, a temperate species that grows as far south as Louisiana, but is far less common up north, is thriving in northeast Minnesota's Sea Gull Lake area. "When you have red maples growing as much as 4 feet in a single year," says Frelich, "you are not talking about a boreal climate anymore."

Regional climate data correlates well with the changes Frelich sees. In the past two decades, springtime has come ever earlier to the region—a week or two sooner than the historical average, according to the Minnesota Department of Natural Resource's climatology office. Eight of the state's 20 warmest years have been recorded since 1981. And for the first time in recorded history, Minnesota logged three mild winters in a row, each with record highs: 1997, 1998, and 1999.

And those data correspond to larger global trends. Overall, 2000 to 2010 is the warmest decade on record, since climatologists started keeping records back in 1850. The seven warmest years have occurred between 1998 and 2010, with 2005 and 2010 tying for warmest individual year. According to the National Oceanic and Atmospheric Administration (NOAA), the global land average temperature increased by 0.96°C (1.73°F) in 2010, compared to the 20th-century average. In higher latitudes, it's even worse: Temperatures have increased by 9°F or more. Sea surface temperatures have also increased, by as much as 5°F in some places.

A few degrees might not seem like much. But even such seemingly small changes in climate can have tremendous impacts on weather, and thus on natural ecosystems and human societies. [INFOGRAPHIC 26.1]

Infographic **26.1** | **A CHANGE IN AVERAGE TEMPERATURE: WHY DO ONLY A FEW DEGREES MATTER?**

↓ A shift of only a few degrees in the average global temperature will likely result in more frequent and extreme heat waves. Compared to the mid-20th-century average, we have increased about 1.0°C (~ 1.7°F). To put this in perspective, at the end of the last ice age 10,000 years ago, Earth's average temperature was only about 3.0°C (5.5°F) colder. At that time there was an ice sheet 1.6 kilometers (1 mile) thick as far south in North America as Chicago.

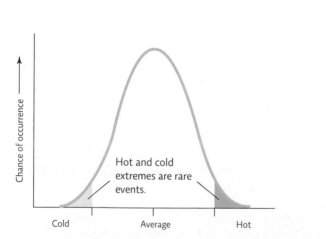

In any stable climate time period (decade, century, millennia), there is some climate variability—in some years, the average temperature is hotter or colder than others but, on average, extremely cold and hot years occur infrequently; most often, temperatures fall somewhere in the middle.

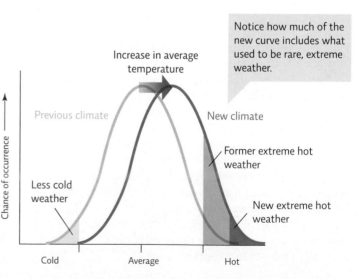

An increase of a few degrees in the "average temperature" shifts the entire curve to the right. This means more years that have the former "extreme" hot weather; it also means the affected area will likely set new records for extreme heat.

↑ Icebergs 60 meters (200 feet) tall, formerly part of the Greenland Ice Sheet, float into the North Atlantic Ocean. Because this ice used to be on land, it contributed to sea level rise when it fell into the ocean.

In fact, we are already seeing some big effects, especially in the Arctic, which is particularly vulnerable to climate change because warming that occurs there causes ice to melt which triggers additional warming. In 2011, researchers at Cambridge University evaluated the weather patterns of Ellismere Island in the Arctic, and found that spring and summer temperatures were 11–16°C (20–29°F) higher than previous years—making the climate similar to a boreal climate 1,000 to 1,400 miles farther south. But as large as that increase was, it was not entirely surprising. Between 2005 and 2007, researchers had measured record losses of sea ice (ice that floats on top of the ocean) in the region. By 2011, they had begun making some dire predictions: Based on the current rate of melting, the entire Arctic could be completely ice free in the summers come 2040.

And it's not just Arctic ice that's melting. In fact, glaciers on the other side of the planet are dwindling just as fast as those in the Arctic, if not faster. Overall, sea ice is decreasing about 3–4% per decade. In Antarctica, the Larsen B Ice Shelf—a colossal sheet of sea ice, more than 210 meters (700 feet) thick and as big as Rhode Island—collapsed in 2002 in just a couple weeks' time. This stunning spectacle, which took climatologists by surprise, was repeated in 2008, when the Wilkins Ice Shelf (also in Antarctica) collapsed, again in 2 weeks' time.

And as ice on land melts, sea levels are rising—by an average of 4 to 8 inches during the last century. About half of this rise is due to land-based ice melt and the other half to thermal expansion—the expansion of water molecules as they heat up. So far, rising sea levels have displaced hundreds of thousands of coastal-dwelling people around the world. [INFOGRAPHIC 26.2]

↳ We know that a variety of factors can alter global temperature, but what is the physical evidence that temperatures have actually increased and that the climate is changing? In other words, what do we predict we would see if warming were occurring and what do we actually see when we test those predictions?

WARMER TEMPERATURES

If climate is indeed warming, we expect to see warmer global temperatures, on average, than in the recent past (more temperature anomalies in the direction of warming).

TEMPERATURE ANOMALIES, AUGUST 2011 (COMPARED TO 1971–2000 BASELINE)

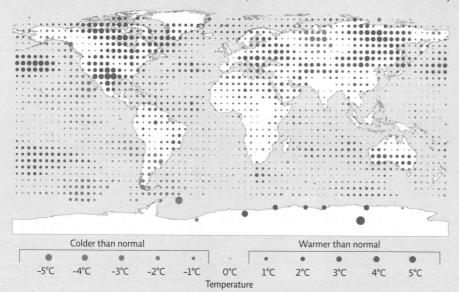

WARMER TEMPERATURES SHOULD LEAD TO

To look for evidence of climate change, scientists look for trends rather than isolated events. For example, NOAA records show that August 2011 was the second-warmest August, behind 1998, since records began in 1880. Taken alone, this event would not suggest climate change, however, this was the 142nd consecutive month that the monthly global land temperature has been above the long-term average and may represent an emerging warming trend.

ANNUAL GLOBAL AVERAGE TEMPERATURE (LAND AND OCEAN)

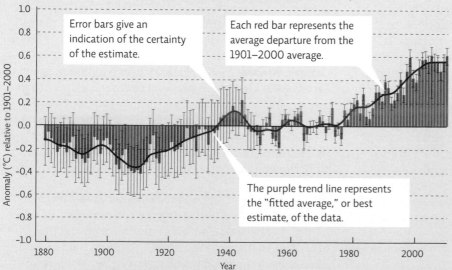

Temperatures were cooler than normal in the first half of the 20th century and have increased since the low seen in 1910. Since the 1980s, temperatures have increasingly exceeded the 20th-century average. The recent plateau may be caused by the release of more air pollution particles that reflect away some incoming solar radiation.

MELTING ICE

If temperatures are warming, we would expect to see more ice melt.

SEA LEVEL RISE

We would also expect to see an increase in sea level as land-based ice melts and as warmer seawater expands.

CUMULATIVE LOSS OF GLACIER MASS

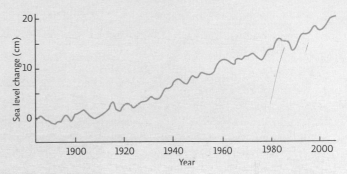

Since the middle of the 20th-century, glaciers around the world have a net loss of ice.

SEA LEVEL CHANGE OVER TIME

Sea level has indeed risen 10–20 cm (4–8 inches) over the last century.

WEATHER EXTREMES

A warmer climate causes more water to evaporate from Earth's surfaces and is expected to produce more extreme weather (heat waves, tropical storms) but also should produce unusual weather as the atmosphere redistributes this heat and moisture around the planet. Areas close to large bodies of water are expected to be wetter (more rain or snow), whereas inner continental areas are expected to be drier as warmer temperatures increase the loss of water from soil, but not enough to fall back as rain.

PRECIPITATION CHANGES Some areas received more precipitation than normal in 2010, as indicated by the size of the circles; others have received less. The departure from normal has increased since 2000, when the National Climatic Data Center first started collecting these data—in other words, the circles have gotten bigger each year since 2000.

INCREASE IN STORMS The intensity of Atlantic hurricanes is measured using the Power Dissipation Index, which takes into account maximum wind speeds and duration of the storm; higher values indicate more intense storms. The increase seen since 1970 tracks the observed changes in sea surface temperature closely.

PRECIPITATION ANOMALIES, JAN 2010 TO DEC 2010 (COMPARED TO A 1961–1990 BASELINE)

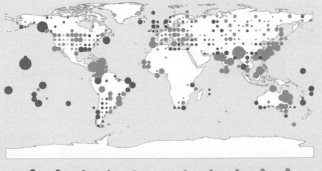

HURRICANE INTENSITY IN THE ATLANTIC OCEAN (YEARLY AVERAGE)

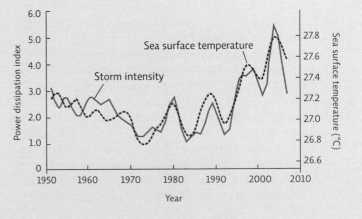

Infographic 26.3 | **THE GREENHOUSE EFFECT**

↓ Life on Earth depends on the ability of greenhouse gases in the atmosphere to trap heat and warm the planet. More greenhouse gases, however, mean more trapped heat and a warmer planet (an enhanced greenhouse effect).

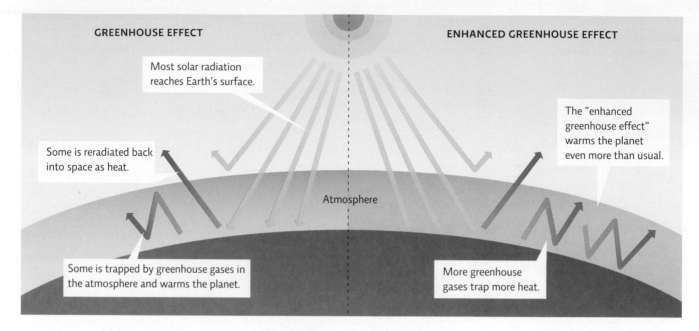

↓ Carbon dioxide (CO_2) is the greenhouse gas released by human actions that currently has the biggest impact on climate. Because different greenhouse gases have different abilities to trap heat, their heat-trapping capacity is expressed as CO_2 equivalents—the amount of CO_2 that would produce the same warming. For example, since a molecule of methane (CH_4) traps 25 times as much heat as CO_2, one methane molecule is equivalent to 25 CO_2 molecules. Much less methane is released in tonnage compared to CO_2, but because each molecule of methane is so much more potent a greenhouse gas than CO_2, it too has a big impact on climate.

GREENHOUSE GASES: GLOBAL EMISSIONS (IN CO_2 EQUIVALENTS), 2004

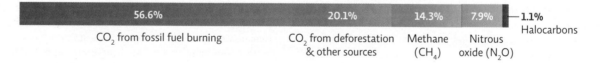

56.6%	20.1%	14.3%	7.9%	1.1% Halocarbons
CO_2 from fossil fuel burning	CO_2 from deforestation & other sources	Methane (CH_4)	Nitrous oxide (N_2O)	

A variety of factors affect climate.

So why is all this happening?

In the past few decades, scientists have discovered that the level of certain gases in Earth's atmosphere is on the rise. These gases, called **greenhouse gases**—which include carbon dioxide, methane, and nitrous oxide—trap heat and help warm Earth, in a process known as the **greenhouse effect**. Normally, the greenhouse effect is a good thing—without it, the average temperature on Earth would be around 0°F—that's about 60° colder than the planet's current average temperature! But starting in the 1980s, scientists began to collect evidence of an enhanced greenhouse effect, which they believe is from the release of greenhouse gases from burning fossil fuels and other

industrial and agricultural practices—in other words, from human activities. [INFOGRAPHIC 26.3]

Greenhouse gases are just one type of **radiative forcer**, or factor that can affect global climate. Another kind of forcer that plays a role in present warming trends is **albedo**, the ability of a surface to reflect away solar radiation. Light-colored surfaces, like glaciers and meadows, have a high albedo—they reflect sunlight, and thus heat, away from the planet's surface. Darker surfaces, like water and dark asphalt, have low albedo—they absorb sunlight, and heat along with it, and then reradiate that heat back to the atmosphere. As surfaces with high albedo are replaced by those with low albedo, not only does the planet warm, but a **positive feedback loop** can be triggered.

Infographic 26.4 | **ALBEDO CHANGES CAN INCREASE WARMING VIA POSITIVE FEEDBACK**

↳ Albedo is a measure of the reflectivity of a surface. The lighter-colored the surface, the higher the albedo. Unreflected (absorbed) light is reradiated as heat, so surfaces with a low albedo release more heat to the atmosphere than high albedo surfaces.

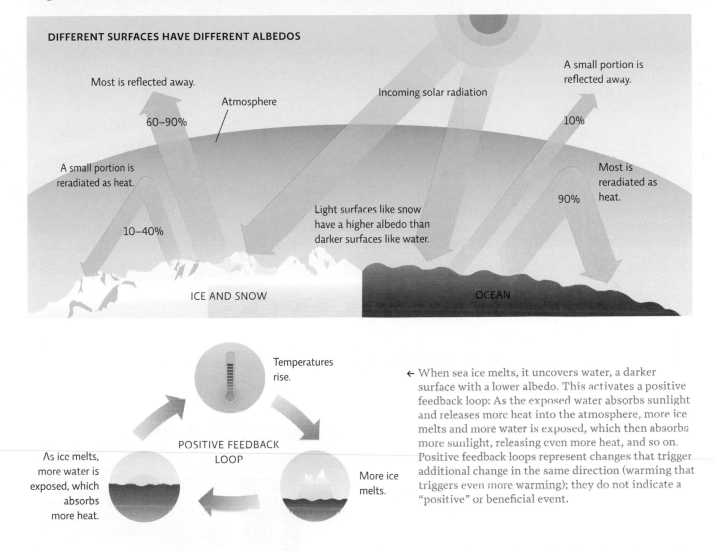

DIFFERENT SURFACES HAVE DIFFERENT ALBEDOS

Most is reflected away.

Atmosphere

Incoming solar radiation

A small portion is reflected away.

60–90%

10%

A small portion is reradiated as heat.

Most is reradiated as heat.

10–40%

90%

Light surfaces like snow have a higher albedo than darker surfaces like water.

ICE AND SNOW

OCEAN

Temperatures rise.

POSITIVE FEEDBACK LOOP

As ice melts, more water is exposed, which absorbs more heat.

More ice melts.

← When sea ice melts, it uncovers water, a darker surface with a lower albedo. This activates a positive feedback loop: As the exposed water absorbs sunlight and releases more heat into the atmosphere, more ice melts and more water is exposed, which then absorbs more sunlight, releasing even more heat, and so on. Positive feedback loops represent changes that trigger additional change in the same direction (warming that triggers even more warming); they do not indicate a "positive" or beneficial event.

Glaciers are a good example of positive feedback: As temperatures rise, glaciers melt and ice (high albedo) gives way to water (low albedo). Because this new watery surface absorbs more heat than the old icy surface, the region warms even faster—replacing even more ice with water, and so on. [INFOGRAPHIC 26.4]

There are other positive feedback loops, too. In the Arctic, for example, the upper levels of permafrost (land that normally remains frozen year-round) are melting during the summer months. When these areas thaw, they release stored carbon, adding more greenhouse gases to the atmosphere. These gases warm the area further, causing even more permafrost melt.

There are also other natural forcers, including clouds. Some have a high albedo and thus work to cool the planet;

others trap reradiated heat from the planet's surface and thus have a warming effect. If warming temperatures cause the formation of more of the high-albedo clouds,

greenhouse gases Molecules in the atmosphere that absorb heat and reradiate it back to Earth.

greenhouse effect The warming of the planet that results when heat is trapped by Earth's atmosphere.

radiative forcer Anything that alters the balance of incoming solar radiation relative to the amount of heat that escapes out into space.

albedo The ability of a surface to reflect away solar radiation.

positive feedback loops Changes caused by an initial event that then accentuate that original event (for example, a warming trend gets even warmer).

→ Climate change research in Alaska monitors CO_2 release from thawing permafrost, along with tundra growth. The project, led by Ted Schuur of University of Florida, has found that in the short term, CO_2 released from melting permafrost leads to increased growth of local tundra vegetation. But in the longer term, the thawing leads to increased atmospheric loading of CO_2 as more is released than can be taken up by the vegetation.

Infographic **26.5** | **CLIMATE FORCERS**

→ A variety of factors can warm or cool the planet. Positive forcers warm and negative forcers cool the climate. Greenhouse gases trap heat in the atmosphere and warm it, while aerosols like sulfate emissions cool it.

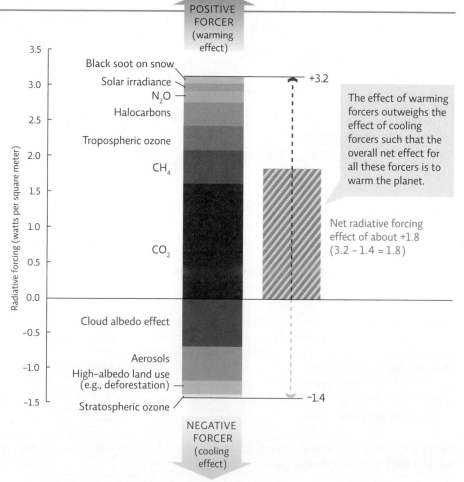

POSITIVE FORCER (warming effect)

Black soot on snow
Solar irradiance
N_2O
Halocarbons
Tropospheric ozone
CH_4
CO_2

+3.2

The effect of warming forcers outweighs the effect of cooling forcers such that the overall net effect for all these forcers is to warm the planet.

Net radiative forcing effect of about +1.8 (3.2 − 1.4 = 1.8)

Radiative forcing (watts per square meter)

Cloud albedo effect
Aerosols
High–albedo land use (e.g., deforestation)
Stratospheric ozone

−1.4

NEGATIVE FORCER (cooling effect)

this could trigger cooling—a **negative feedback loop**. Right now, clouds have a net cooling effect; whether this trend will continue in the future remains to be seen. [INFOGRAPHIC 26.5]

Volcanic eruptions and changes in solar irradiance (like sunspot cycles) are also considered natural forcers, as both have been known to impact climate in the past, though only over short time frames and not as severely as greenhouse gases. Scientists also believe that **Milankovitch cycles** (predictable long-term cycles of Earth's position relative to the Sun) played an important role in earlier climate change events. [INFOGRAPHIC 26.6]

To figure out how much any given forcer or feedback loop is contributing to current warming, or to guess at what future climate might look like based on what we are seeing now, climate scientists must do more than monitor current atmospheric conditions; they must also gather data (on temperature, CO_2 levels, etc.) from the distant past. They do this by studying a wide variety of clues that have been left behind—ice and sediment cores, tree rings, coral reef growth layers, even fossilized mud at the bottom of lakes and rivers (see Chapter 1). These various

sources have consistently corroborated one another: Ice cores paint the same picture as tree rings, and coral reef layers confirm what pollen sediments tell us. This consistency enables us to trust their overall story. [INFOGRAPHIC 26.7]

Climate scientists have used this wealth of historical data to develop what they call *climate models*—computer programs that allow them to make future climate projections by plugging in all the current values (for temperature, CO_2, global air circulation patterns, etc.). These models are used to see how altering the value of certain parameters (say, increasing the amount of atmospheric CO_2) might impact future climate. It may seem ironic that climatologists can predict what the climate will be like 100 years from now, when meteorologists often have a hard time getting the weekly weather forecast right. But scientists say that climate is actually easier to predict

---

negative feedback loop Changes caused by an initial event that trigger events which then reverse the response (for example, warming leads to events that eventually result in cooling).

Milankovitch cycles Predictable variations in Earth's position in space relative to the Sun which affect climate.

Infographic 26.6 | **MILANKOVITCH CYCLES HELP EXPLAIN PAST CLIMATE CHANGE**

↓ Warm periods and ice ages of the past can be attributed in part to Earth's position in space relative to the Sun. Earth has three different cycles that can each have an impact on climate. The current warming we are experiencing cannot be explained by one of these cycles—Earth is currently not in a part of any cycle in which it would have greater warming.

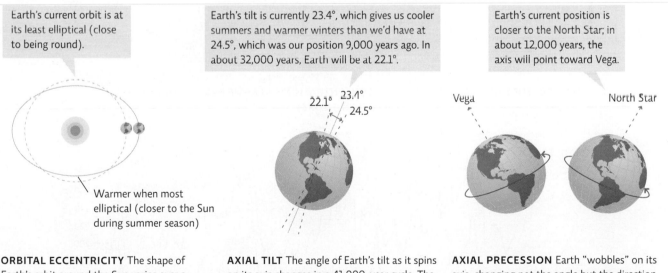

Earth's current orbit is at its least elliptical (close to being round).

Warmer when most elliptical (closer to the Sun during summer season)

Earth's tilt is currently 23.4°, which gives us cooler summers and warmer winters than we'd have at 24.5°, which was our position 9,000 years ago. In about 32,000 years, Earth will be at 22.1°.

22.1° 23.4°
24.5°

Earth's current position is closer to the North Star; in about 12,000 years, the axis will point toward Vega.

Vega   North Star

ORBITAL ECCENTRICITY The shape of Earth's orbit around the Sun varies over a 100,000-year cycle from mostly round to more elliptical.

AXIAL TILT The angle of Earth's tilt as it spins on its axis changes in a 41,000-year cycle. The greater the angle, the greater the extremes between seasons (hotter summers and colder winters).

AXIAL PRECESSION Earth "wobbles" on its axis, changing not the angle but the direction the axis points in a 20,000-year cycle. This changes the orientation of Earth to the Sun and affects the severity of the seasons. When Earth is tilted toward Vega, it is also tilted toward the Sun during summer, making summers hotter in the Northern Hemisphere.

↓ We use a variety of methods to measure temperature and CO$_2$ levels. Current temperatures are monitored directly using instruments such as thermometers and satellites; a variety of analytical equipment is used to measure CO$_2$ levels in the air. Charles Keeling began taking precise CO$_2$ measurements at the Mauna Loa Observatory in 1958, a location chosen because the air there received no local pollution to compromise the data. We can also estimate values from the past indirectly by evaluating physical evidence such as sediment and ice cores, and tree-ring data. Ice cores are particularly helpful; scientists can date the sections by counting the annual layers. They then measure atmospheric gases like CO$_2$ from the bubbles in the ice core sample to determine levels at the time the ice was laid down; measuring the ratio of 2 isotopes of oxygen, O-16 and O-18, gives a very accurate estimate of temperature at the time.

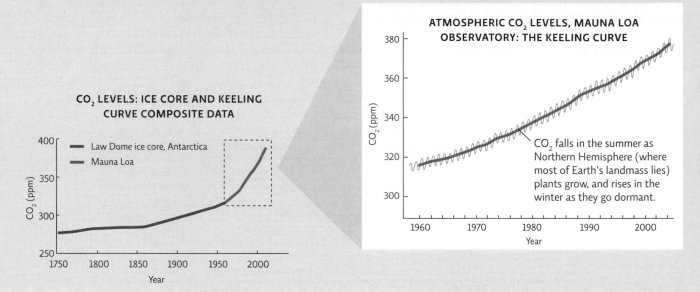

CO$_2$ LEVELS: ICE CORE AND KEELING CURVE COMPOSITE DATA

- Law Dome ice core, Antarctica
- Mauna Loa

ATMOSPHERIC CO$_2$ LEVELS, MAUNA LOA OBSERVATORY: THE KEELING CURVE

CO$_2$ falls in the summer as Northern Hemisphere (where most of Earth's landmass lies) plants grow, and rises in the winter as they go dormant.

↓ A comparison of historic CO$_2$ levels and temperatures, as determined from the Antarctic Vostok ice core, show that the two parameters have been closely aligned over the past 400,000 years. It turns out that the relationship between CO$_2$ and temperature is both one of cause and of effect. In the far past, natural events such as differences in Earth's orbit triggered warming, resulting in the release of more CO$_2$ (an effect), which then caused even more warming (a cause). Today, humans are the source of much of the extra CO$_2$ (and other greenhouse gases) being released. In this case, the CO$_2$ release is preceding the warming—it is the initial cause. The bottom line is that no matter the reason for the release of extra greenhouse gases such as CO$_2$, temperatures change as CO$_2$ increases and decreases.

TEMPERATURE AND ATMOSPHERIC CO$_2$ CONCENTRATION OVER THE PAST 400,000 YEARS (VOSTOK ICE CORE)

- CO$_2$
- Temperature

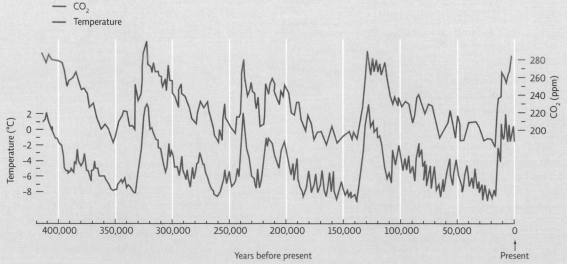

than weather. Climate refers to general trends, while weather is much more specific; it is more like a close-up view rather than a view from afar.

Based on all the clues they have gathered and analyzed, scientists agree that all the natural forcers combined are not enough to account for the rapid climate change that is currently underway. Only when we consider both natural and **anthropogenic** (related to human actions) forcers together do current trends make sense. [INFOGRAPHIC 26.8]

Current climate change has both human and natural causes.

The **Intergovernmental Panel on Climate Change (IPCC)** is an international scientific body, established by the United Nations and the World Meteorological Organization in 1988, that is made up of thousands of scientists from around the world. They evaluate all the

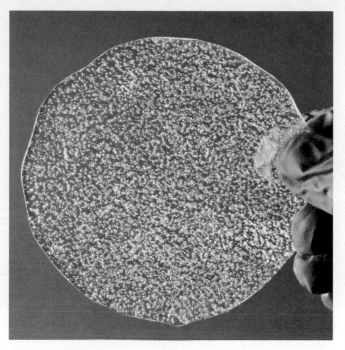

↑ A researcher holds a thin slice of ice from an ice core extracted from Antarctica. The core contains trapped air bubbles that can reveal information about the atmosphere and temperatures of the past.

anthropogenic Caused by or related to human action.

Intergovernmental Panel on Climate Change (IPCC) An international group of scientists who evaluate scientific studies related to any aspect of climate change to give thorough and objective assessment of the data

---

Infographic 26.8 | **WHAT'S CAUSING THE WARMING?**

↓ Climate scientists use computer models (multiple mathematical equations) that take into account the major factors that are known to have affected past climates in order to see what might be responsible for recent warming. Data about natural and anthropogenic factors can be fed into the computer model separately and then together to see which circumstances match up with the warming that has been observed.

COMPUTER MODELS' RECONSTRUCTION OF PAST TEMPERATURES

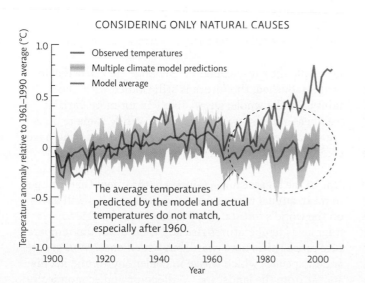

CONSIDERING ONLY NATURAL CAUSES

— Observed temperatures
▓ Multiple climate model predictions
— Model average

The average temperatures predicted by the model and actual temperatures do not match, especially after 1960.

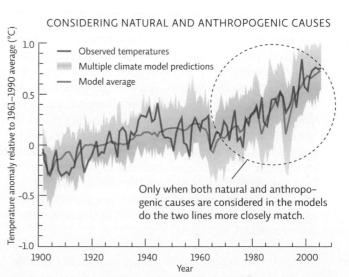

CONSIDERING NATURAL AND ANTHROPOGENIC CAUSES

— Observed temperatures
▓ Multiple climate model predictions
— Model average

Only when both natural and anthropogenic causes are considered in the models do the two lines more closely match.

↑ NASA satellite data reveals that 2011's minimum sea ice extent in the Arctic Ocean, reached on September 9, 2011 as depicted here, was far smaller than the 30-year average (in yellow) and opened up Northwest Passage shipping lanes (in red).

climate science that is published through peer review and compile it into a cohesive series of publications that explain what is understood about the current state of the climate; they also make recommendations that may inform government action.

Here's what everyone agrees on so far: Earth's atmosphere is changing dramatically and with alarming speed. Analysis of CO_2 trapped in ice cores reveals that current levels are higher than at any time in the last 800,000 years and are increasing steadily. In 2011, the atmospheric concentration of CO_2 was 390 ppm (parts per million). This is considerably higher than the preindustrial level of about 280 ppm—a level that had been maintained for millennia.

The majority of this change is due to human activities, especially the burning of fossil fuels, and the consequent release of greenhouse gases like CO_2 into the atmosphere. For most of modern history, the United States has been the biggest emitter of CO_2. In 2007, however, China took the lead: The proliferation of coal-fired power plants that has fueled the country's recent economic and industrial growth has also released copious amounts of CO_2 into the atmosphere. China now releases nearly 30% more CO_2 than the United States (though the United States still releases more per person than any other country in the world, and much of the CO_2 released in the 20th century came from U.S. sources).

Many climate scientists set the upper limit of CO_2 that we should not cross (to avoid substantial negative effects,

like the melting of the Greenland Ice Sheet and other events that we cannot reverse) at 450 ppm (the amount that would result in a 2°C increase in temperature above the average preindustrial temperature—referred to as the "guardrail" temperature). But some scientists place it much lower, at 350 ppm—meaning we must not just reduce the amount of CO_2 we release, we must bring down current atmospheric levels. At our current pace (rapid fossil fuel consumption combined with sluggish efforts to curb our emissions) we will surpass that 450 threshold before the end of this century. If we don't take steps to prevent this, we will likely set in motion a chain of events that could dramatically change the face of the planet: Positive feedback cycles, from glacier and permafrost melt to ocean acidification, to the loss of carbon sinks due to deforestation, could kick into high gear, rapidly accelerating the pace of warming as they do.

But while the causes of climate change are scientifically well established, the future is still riddled with uncertainty. On the whole, we are likely in for more variable weather—not just warmer weather. While some regions are already enduring heat extremes, other areas will have colder winters as weather patterns shift.

One of the biggest uncertainties is the effect that changes in mean annual temperature and precipitation will have on the world's forests. Some species may be able to migrate as temperatures rise and their optimum temperature range shifts northward, but which ones? Will spruce and fir species, which thrive in colder environments, disappear from the landscape? Will iconic and economically important trees like sugar maple and jack pine "move" to Canada? Will important ecological relationships be fractured as some species move or adapt while others (perhaps prey species or pollinators) do not?

Somewhere, buried in the reams of data that scientists like Frelich have spent decades accumulating, lie clues to those very questions.

Some tree species are already migrating north; that doesn't mean they will survive.

Like Frelich, Chris Woodall has spent his entire adult life studying the great forests—both temperate and boreal—that stretch from the northeastern United States well into Canada. Unlike Frelich, Woodall spends most of his time in front of a computer screen, crunching numbers. He works for the U.S. Department of Agriculture's forest inventory program, which maintains roughly 100,000 permanent plots throughout the region—segments of forest where the USDA monitors a host of variables, from temperature and precipitation to tree growth and sapling density. It's a tremendous database, and for the past decade, scientists have used it to develop computerized models of how a warmer climate might change forests in the future. But until recently, no one had looked at whether or not they were already changing.

Woodall's logic was simple: Mature trees tell you where the current range is. Seedlings tell you where that range will be in the future. By comparing the ratio of seedling to mature tree, one should be able to say whether or not any given species is on the move. Using the most recent data collected from 30 states and some 66,000 inventory plots, Woodall compared tree-seedling densities to forest biomass for more than two dozen tree species.

He was astounded by what he found. For most of those species—spruce, jack pine, sugar maple, and several others—the mean location for seedlings proved to be significantly further north than their associated mature trees. "It's like if they had a gravity center, it would be pulling them northward."

Whether they will actually survive in new locations is another story. While studies of the average locations for trees and saplings show northward migration, subsequent studies show contraction, or a loss of trees, at the northern edges. "The obvious question is, 'Well, how does that reconcile with the findings on mean, that they are moving north?'" Woodall says. "The answer is that it's like buffalo rushing off a cliff. As climate warms, you've got all these species gravitating to the northernmost

edges of their traditional ranges. But that doesn't mean they will survive there long term."

Back in the North Woods, Frelich is working to understand why. So far, he's identified several forces that seem to be working in concert. "Warming triggers a whole cascade of events," he says. "Factors that make tree ranges shift northward, and factors that prevent those trees from thriving in their new, more northerly habitats."

> "It's like buffalo rushing off a cliff. As climate warms, you've got all these species gravitating to the northernmost edges of their traditional ranges."
> —Chris Woodall

One of the biggest factors, he says, is deer, which have proliferated like mad in recent years. "In a warmer climate," Frelich says, "you'd expect the maple to advance in the understory, so that as the spruce die off, the maple are ready to take over. Likewise in the south: As maple move northward in response to warming, oak should move in to fill the void." But it turns out that deer like maple much more than they like spruce, and oak even more than maple. "So in places where the deer population is very high, the trees are having a hard time adapting to climate change because the deer are eating up all the early migrators."

And as climate warms, other stresses abound: Snowpack melts earlier, causing more severe water deficits in summer, right when trees need extra water to survive. The whole landscape dries out, creating conditions that favor intense fires and stressing trees so much that they become easy prey for beetle infestation. Pine beetles are a natural part of the life cycle in western forests, but the current outbreak, underway for more than a decade, is unlike anything seen before. "It used to get down to 40 below, every couple years, and that would keep things in check by killing the beetles off," says Frelich. "But that isn't happening anymore."

There are many other indicators that species are responding to climate change. In fact, that they are responding is evidence itself that climate is changing. Because communities are complex assemblages of many species, there are concerns that important community connections will become uncoupled as species respond in different ways to climate change. [INFOGRAPHIC 26.9]

↓ Species have evolved to live and thrive in a certain habitat. If the climate is changing enough to alter ecosystems, we expect to see species responding by changing where they live or the timing of important temperature-dependent biological events. Species' responses such as shifting ranges or earlier blooming and hatching may be the best evidence that climate is actually changing—it is unlikely that these temperature-dependent events would change in this way if the planet were not getting warmer.

SOME SPECIES ARE MOVING TO HIGHER ALTITUDES
A 2008 study of the elevation distribution of 171 forest plants in western Europe looked at where plants were found as well as the optimum elevation for growth. Data were compared for plants from two different time periods: 1905–1985 and 1986–2005. On average, plants shifted their range to higher elevations 29 meters per decade.

SOME IMPORTANT COMMUNITY CONNECTIONS ARE UNCOUPLED
Caterpillars, an important food source that birds feed their young, hatch based on temperature cues; in The Netherlands, the caterpillars are hatching 15 days sooner than they did in 1985. Pied flycatchers, birds who migrate north to their breeding grounds based on day-length cues (and thus have not changed their migration timing), are arriving too late to take advantage of peak caterpillar hatching.

FLOWERING PLANTS ARE BLOOMING EARLIER
Since many biological events are linked to the arrival of spring, we would expect to see earlier blooming, leaf out, and reproduction in species sensitive to temperature cues. One study showed that in Alberta, Canada, aspen, an early spring bloomer, blooms 2 weeks earlier now than normal. Other plants that normally bloom later in spring bloomed 0–6 days earlier.

Climate change has environmental, health, and economic consequences.

The Boundary Waters Canoe Area surrounding Ham Lake—where Frelich and his colleagues were trapped—is the most heavily used chunk of the National Wilderness Preservation System. The boreal forest, its lakes, hiking trails, and breathtaking wildlife—entice some 200,000 visitors every year, providing roughly 18,000 tourism jobs that pay a total of $240 million in wages. Global warming and shifting tree ranges threaten all of that.

"Everyone's worried about losing the forests," Frelich says. "Resort owners, people who own cabins up there, local outfitters that rent camping gear—and the tourists themselves. If the forests burn too much, or if they descend into savanna, or can no longer support the iconic wildlife—like moose, lynx, and boreal owls—that people have come to expect, tourism will dry up. Because no one will want to vacation there."

And it's not just the tourism industry that will suffer in a warmer North Woods. The region as a whole supports about 1,000 forestry and logging jobs, which pay about $50 million in wages. On top of that, some individual species have become industries unto themselves—for example, the sugar maple.

Maple syrup production is heavily dependent on climate. The flow of sweet, sticky tree sap that eventually covers your pancakes is governed by changes in air pressure, which are in turn governed by changes in air temperature. When the temperature drops below freezing, the tree acts as a giant suction system, pulling the sap out of its branches, down into its roots. When the temperature rises above freezing, this action is reversed; the pressure gradient forces sap up from the roots, through the branches and out of any holes—including ones that syrup makers have drilled for taps.

Traditionally, climate in the northeastern United States—from Minnesota to Maine—has provided the optimal

1986–2005

MICHIGAN

1905–1985

SOME SPECIES ARE MOVING TO HIGHER LATITUDES

Several studies report a northward shift in the range of several, but not all, species of birds that have been studied. For example, the blue-gray gnatcatcher has extended its breeding range more than 300 kilometers northward since the 1970s.

In Woodall's study, the sugar maple was one of 11 out of 15 species that showed an average northern shift in range of 21 kilometers. Most seedlings are found in the northern-most regions, a sign that this species is migrating north.

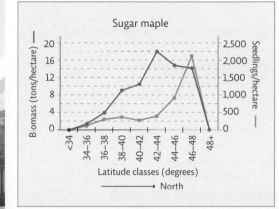

Sugar maple

Biomass (tons/hectare) — 20, 16, 12, 8, 4, 0

Seedlings/hectare — 2,500, 2,000, 1,500, 1,000, 500, 0

Latitude classes (degrees): <34, 34–36, 36–38, 38–40, 40–42, 42–44, 44–46, 46–48, 48+

→ North

SOME SPECIES' POPULATIONS ARE EXPANDING

Many species of insect pests are on the increase, resulting in the decimation of forests around the world. Forests across the Northern Hemisphere are affected by several species of bark beetles. The pest populations normally die back in the winter, helping to keep them in check and allowing the trees to recover. Winter temperatures no longer get cold enough to kill the beetles in some areas, allowing the beetles to thrive year round. The U.S. Forest Service estimates that more than 4 million acres have been affected in the western United States.

freeze-thaw patterns for this process, which syrup makers call "sugaring." But in recent years, the transition from winter to spring has accelerated, leaving fewer freeze-thaw cycles and less sap overall. Meanwhile, across the border in Canada, warmer daytime temperatures have increased the number of freeze-thaw cycles there, and some experts say that Canada is already in the middle of a syrup boom. "If current trends continue," Woodall says, "it's not impossible that the entire industry could one day be lost to Canada."

For Woodall, the stakes are both more basic and more terrifying than the loss of any given industry: Societies that don't protect their forests fail, he says. And it's easy to see why. "Forests stabilize soil and clear water of pollutants," he says. "In fact, the vast majority of Americans drink water that comes from a forested watershed. That means trees are as crucial to our survival as the water we drink." The loss of forests is also another positive feedback loop that threatens to exacerbate climate change: Forests are an important carbon sink—they store trillions of tons of

CO_2 in their plants and soils. When they are burned, or die and decompose, much of that carbon is released into the atmosphere (see Chapter 13 for more on the ecosystem services of forests).

To be sure, some people and places will benefit from climate change. In Greenland, for example, warmer temperatures have enabled farmers to grow a wider variety of crops than they have been able to in the past. Warmer weather during the summer months has also opened the Northwest Passage—a long sought-after shipping route through the Arctic Ocean—which would significantly reduce the transport time for ships that otherwise have to take the southern route through the Panama Canal. High-latitude land in Canada and Siberia will likely become warmer and more habitable—lessening the incidences of cold-related health problems and deaths.

But other areas will suffer more harm than good. Coastal flooding is already affecting low-lying areas, from Bangladesh to New Orleans. Extreme weather has already begun to claim both valuable crops and human lives. And infectious tropical diseases have begun to migrate north of their traditional range (dengue fever, for example, has made its way from Africa and South America into Texas and Florida).

Confronting climate change is challenging.

Jack Rajala's timber company owns 35,000 acres of commercial forest land in Itasca County, Minnesota. In recent years, he's started doing things differently—namely, deliberately thinning out his paper birches in an effort to cultivate more oaks and white pines. That's not to say that the paper birch isn't valuable. But with massive die-offs underway throughout the region, Rajala needs to hedge his bets. "We think we can still facilitate birch," he says. "But it may be an understory tree, not a canopy tree anymore."

Rajala's strategy is called resistance forestry; it includes a handful of techniques aimed at maintaining existing species in their current locations, even as the climate shifts. For example, prescribed burns that mimic historic fire patterns might bolster the ranks of fire-dependent species like paper birch, black spruce, and jack pine, allowing them to spread over a wider area and enhancing their genetic diversity. Planting seeds instead of saplings also helps species hold their ground; natural selection favors the hardiest field-grown seedlings, and so may yield a population better able to survive environmental stresses.

Scientists refer to such efforts, which are intended to minimize the extent or impact of climate change, as **mitigation**.

Mitigation includes any attempt to seriously curb the amount of CO_2 we are releasing into the atmosphere— either by using carbon capture techniques to remove the greenhouse gas from our air and sequester it underground, or by consuming fewer fossil fuels to begin with. In 2004, Princeton University researchers Stephen Pacala and Robert Socolow proposed a "stabilization wedge" strategy—a step-by-step implementation of currently available technology; each step could prevent the release of 1 billion tons of carbon. In 2007, it was estimated that any 8 of the 15 steps, or "wedges," would stabilize CO_2 in the atmosphere at close to 500 ppm in the next 50 years. In 2011, it was estimated that to bring us down to 450 ppm

(the maximum level to keep us from exceeding the 2°C "guardrail") would take 11 wedges—and the longer we wait, the more wedges we'll need. [INFOGRAPHIC 26.10]

On a national or global scale, mitigation efforts can be facilitated in a variety of ways, such as command and control regulations that limit greenhouse gas release; tax breaks; green taxes (in this case, **carbon taxes**); and market-driven programs such as cap and trade (see Chapter 25). Financial incentives that encourage the development and use of noncarbon fuels and more energy-efficient technology is critical (see Chapters 27 and 28).

No matter which strategies we employ, curbing greenhouse gas emissions will take a coordinated global effort, meaning that world superpowers like the European Union, the United States, and China will have to cooperate with the developing nations of the world. So far, efforts have been fraught with obstacles and lack of cooperation. In 1997, an international treaty called the Kyoto Protocol was ratified by every nation except the United States. The treaty set different but specific targets of CO_2 release reduction for various nations; the United States objected because the protocol set much higher reduction requirements for developed countries than it did for developing nations.

Additional criticisms of Kyoto underscore the trouble with confronting such a global problem. Some of those who opposed Kyoto said that it went way too far in curbing greenhouse gas emission; they argued that placing any kind of limit on CO_2 would hurt the economy because it would force industries to spend money updating their infrastructure, limit the amount of work they could do, and place them at a disadvantage compared with countries who had lower reduction targets under the treaty.

Other critics said that Kyoto did not go far enough; given the overwhelming evidence, these critics felt we needed to set much higher reduction targets to make any dent in the problem. They also felt that setting concrete, legally binding reduction targets in developed nations like the United States would actually stimulate the economy because it would force companies in those nations to develop new, cleaner, more efficient technologies that other countries would then buy.

These conflicting views are examples of why climate change is considered a *wicked problem*—one that challenges the bedrock of modern civilization: energy use. Some argue that applying the **precautionary principle** now could help avoid, or at least lessen, some of the most serious outcomes of a changing global climate.

---

mitigation   Efforts intended to minimize the extent or impact of a problem such as climate change.

carbon taxes   Governmental fees imposed on activities that release CO_2 into the atmosphere, usually on fossil fuel use.

precautionary principle   Acting in a way that leaves a safety margin when the data is uncertain or severe consequences are possible.

Infographic **26.10** | **OUR CHOICES WILL AFFECT HOW MUCH THE PLANET WARMS WHICH, IN TURN, WILL INFLUENCE THE SEVERITY OF THE IMPACTS**

↓ The IPCC has several climate change response scenarios that predict how much the climate will change (average temperature increase) based on how the human population responds. These scenarios take into account how quickly fossil fuel use is decreased, how cooperative nations are in sharing technologies and supporting global initiatives to address resource consumption and fossil fuel use. We are currently tracking on a trajectory that is at the upper edge of the worst case scenario (the red line).

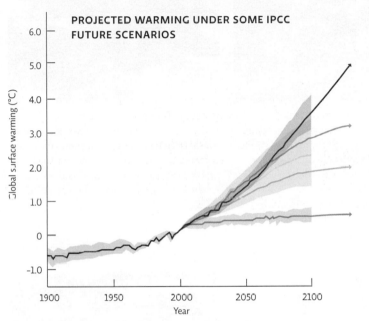

PROJECTED WARMING UNDER SOME IPCC FUTURE SCENARIOS

Different colored lines represent different responses, from no real change in our actions (red) to what we'd expect if greenhouse gases stayed at 2000 levels.

Slow economic growth and spread of technology; continuously increasing resource consumption

Balanced use of a variety of energy sources, including continued use of fossil fuels

Improved technology decreases greenhouse gas emissions and resource consumption; solutions to help economic growth and social justice applied globally

Expected temperatures if greenhouse gases were held to 2000 levels

MITIGATION STRATEGIES HELP REDUCE FACTORS THAT LEAD TO CLIMATE CHANGE

↓ We can take steps to curb climate change by reducing emissions of greenhouse gases and by making better resource and land-use decisions. This will lessen the eventual peak warming we might experience. In 2004, Pacala and Socolow estimated that employing any 7 of 15 potential stabilization wedges, which each reduce CO_2 emissions by 1 gigaton (1 billion tons) per year over the next 50 years, would allow the atmosphere to stabilize close to 500 ppm. By 2007, that had risen to 8 wedges. In 2011, the 2°C "guardrail" target of 450 ppm would require 11 wedges.

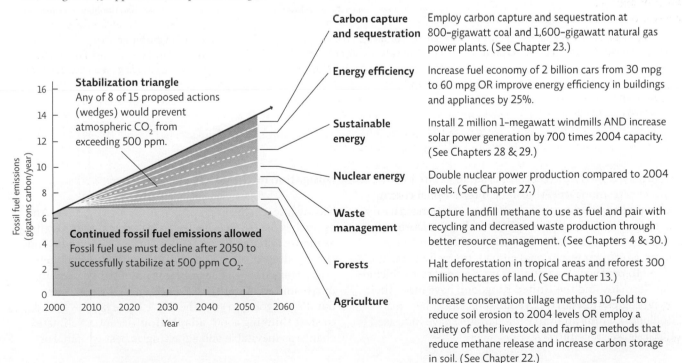

Carbon capture and sequestration Employ carbon capture and sequestration at 800-gigawatt coal and 1,600-gigawatt natural gas power plants. (See Chapter 23.)

Energy efficiency Increase fuel economy of 2 billion cars from 30 mpg to 60 mpg OR improve energy efficiency in buildings and appliances by 25%.

Sustainable energy Install 2 million 1-megawatt windmills AND increase solar power generation by 700 times 2004 capacity. (See Chapters 28 & 29.)

Nuclear energy Double nuclear power production compared to 2004 levels. (See Chapter 27.)

Waste management Capture landfill methane to use as fuel and pair with recycling and decreased waste production through better resource management. (See Chapters 4 & 30.)

Forests Halt deforestation in tropical areas and reforest 300 million hectares of land. (See Chapter 13.)

Agriculture Increase conservation tillage methods 10-fold to reduce soil erosion to 2004 levels OR employ a variety of other livestock and farming methods that reduce methane release and increase carbon storage in soil. (See Chapter 22.)

↓ No matter how hard we try, we will not be able to avoid some future warming, as greenhouse gases emitted in the 20th century will continue to impact climate into the future. Adaptation strategies equip us to adjust to the inevitable warming that will occur.

HEALTH IMPACTS

IMPACT The spread of waterborne pathogens should lead to increased incidence of infectious diseases; an additional 60 million or more people in Africa would be exposed to malaria; fewer cold-related deaths but more heat-related deaths.
ADAPTATION Improve disease surveillance, implement sanitation improvements in flood prone areas, and establish emergency action plans.

CROP PRODUCTIVITY

IMPACT Overall, global crop yields are projected to increase 5–20% with a temperature increase of 1°–3°C. But at temperatures higher than that, overall production is expected to decrease 20–40%; yields will decrease sooner in some areas due to water stress.
ADAPTATION Use erosion-control techniques to improve agricultural productivity; choose crops to fit new conditions.

COASTAL EROSION AND FLOODING

IMPACT 10 to 300 million people will be affected at temperature increases of 2° to 4°C.
ADAPTATION Relocation of some coastal communities may be necessary; construct protective barriers like seawalls and restore wetlands in coastal areas to protect inland areas.

BIODIVERSITY LOSSES

IMPACT Some species may benefit and expand their ranges, but many face extinction. A 2°C increase will likely lead to the extinction of many Arctic species like polar bears; a 3°C increase is predicted to put 20–50% of land species at risk for extinction.
ADAPTATION Wildlife and habitat management to provide migration corridors or relocation assistance; protect vulnerable habitats from further human impact.

DROUGHT

IMPACT The proportion of land area in severe drought is projected to increase from today's levels of 1–3% to 30% by 2090.
ADAPTATION Focus on methods to capture and conserve water, including desalinization in coastal areas; practice pollution prevention to increase and protect water supplies.

FIRE RISK

IMPACT Fire has already increased in some areas; it has more than doubled in boreal North America, and is projected to increase even more in the future.
ADAPTATION Pursue better fire-prevention management, including prescribed burns, thinning of forests to reduce combustible material, and improve fire-response plans.

At any rate, the Kyoto Protocol is set to expire in 2012; so far, despite annual meetings, the international community has had no success in drafting a replacement treaty. The 2011 UN Climate Change Conference in Durban, South Africa, resulted in a legally binding agreement to adopt a future international climate treaty by 2015, but failed to produce any actual treaty. Some nations (including Germany and the United Kingdom) have made their own progress toward CO_2 reductions, but other nations (including China and the United States) have increased CO_2 release since 1997.

Mitigation might not be enough.

Meanwhile, change is coming to the North Woods. And as birches die off and maples try to expand their territory, as moose falter and pine beetles thrive—those who know the woods best say that mitigation will not be enough. "Resisting climate change at this point is like paddling upstream," says Frelich. "It might buy us some time, but it's not going to save the day." So, he says, we need to start thinking about **adaptation**: accepting climate change as inevitable and adjusting as best we can. For human societies at large, this means taking steps to ensure a sufficient water supply in areas where freshwater supplies may dry up; it means planting different crops or shoring up coastlines against rising sea levels; it means preparing for heat waves and cold spells and outbreaks of infectious diseases. [INFOGRAPHIC 26.11]

In the North Woods, it might mean facilitation—moving tree species to entirely new ranges where they don't currently grow, based on the notion that the speed of climate change will make it impossible for natural tree migratory processes, such as seed dispersal, to occur. "The idea is that if we want the forest to adapt, we will have to help it along," says Frelich.

Facilitation has no shortage of critics, many of whom say such tinkering is both dangerous and unnecessary. "Facilitation is my nightmare," says John Almindinger, a forest ecologist with Minnesota's Department of Natural Resources. "That we'll start to believe we're smart enough to figure out how to move things. It's sheer hubris." Besides, he says, many if not most tree species seem to be moving just fine on their own, along traditional forest migration routes. So far, the U.S. Forest Service agrees;

the agency does not allow such bold interventions as planting pines inside the wilderness.

Still, some skeptics are coming around to the idea. "We've changed the landscape through development and agriculture," says Peter Reich, a colleague of Frelich's at the University of Minnesota. "And we've changed the climate, too, with fossil fuel consumption. So we might now need to change the way we manage wild lands to compensate."

On this much, everyone seems to agree: If northern Minnesota is to remain fully forested in the coming century, something will have to be done. "We see it already," says Rajala. "The impact of climate change will be too big to just let nature take its course."◉

Research articles referenced in this chapter:
Beaubien, E., and Hamann, A. 2011. *BioScience*, 61: 514–524.
Both, C., et al. 2009. *Journal of Animal Ecology*, 78: 73–83.
Environmental Protection Agency. 2010. *Climate Change Indicators in the United States*.
Frelich, L., and Reich, P. 2009. *Natural Areas Journal*, 29: 385–393.
Frelich, L., and Reich, P. 2010. *Frontiers in Ecology and the Environment*, 8: 371–378.
Hitch, A., and Leberg, P. 2007. *Conservation Biology*, 21: 534–539.
Lenoir, J., et al. 2008. *Science*, 320: 1768–1771.
NOAA National Climatic Data Center. 2010. *State of the Climate: Global Analysis for Annual 2010*. http://www.ncdc.noaa.gov/sotc/global/2010/13.
Pacala, S., and Socolow, R. 2004. *Science*, 305: 968–972.
Woodall, C.W., et al. 2009. *Forest Ecology and Management*, 257: 1434–1444.

---

adaptation Efforts intended to help deal with a problem that exists, such as climate change.

BRING IT HOME

◉ PERSONAL CHOICES THAT HELP

The effects of climate change are already being felt by humans, other species, and ecosystems around the globe. By changing our actions, we can decrease the greenhouse gases we produce and show policy makers that citizens are interested in preventing global climate change.

Individual Steps
→ Do your part to reduce carbon emissions by conserving energy. Walk or ride a bike instead of driving a car. Share a ride with a coworker rather than driving alone. Negotiate with your employer to telecommute. Live close to where you work or go to school. Reduce your heating and cooling energy use, and always turn off electronics and lights when not in use.

→ If your utility company offers renewable energy, buy it.
→ Reduce the carbon footprint of your food by decreasing the amount of feedlot-produced meat you eat. Buy your food as locally as possible to reduce energy used in transportation.
→ Go to terrapass.com to see how you can offset your CO_2 production from your car, your house, and your airplane travel.

Group Action
→ Volunteer to help build a zero-energy Habitat for Humanity home.
→ Organize a community lecture on climate change with a local university expert or meteorologist as the speaker.

→ Organize an event at your school or community to raise awareness about global climate change and ways to prevent it. Go to 350.org to join a current campaign and for other program ideas.

Policy Change
→ Consider writing, calling, or visiting the office of your legislators to tell them to support funding for research and development of clean and renewable sources of energy. In addition, insist that they support the funding of science, especially those efforts to understand and confront climate change.

UNDERSTANDING THE ISSUE

CHECK YOUR UNDERSTANDING

1. **Which of the following best describes radiative forcers?**
   a. Factors that can alter climate
   b. Events that increase the likelihood of a future change in the same direction as past changes
   c. Long-term patterns in meteorological conditions
   d. Natural factors, such as volcanoes, that affect the amount of sunlight that reaches the surface of Earth

2. **Current atmospheric CO_2 levels are at 390 parts per million (ppm). What is the upper limit of CO_2 predicted by most scientists that we can live with and still avoid substantial negative effects?**
   a. 400 ppm
   b. 560 ppm
   c. 680 ppm
   d. 750 ppm

3. **Scientists have documented that climate change is a normal process. What is the PRIMARY reason that the current warming trend is so alarming to those same scientists?**
   a. It raises the possibility of another ice age.
   b. We have no means of stopping it.
   c. It is solely due to human activities.
   d. It is occurring at a rapid rate.

4. **According to some, the current increase in average global temperature is due to the "wobble" effect. If this were true, which of the following would also be true?**
   a. The axis of Earth would be shifted to point toward the star Vega, resulting in hotter summers in the Northern Hemisphere.
   b. Earth would be rotating around the Sun in an ellipse as opposed to a circle.
   c. Earth would be tilted on its axis at an angle that maximizes the differences between summertime highs and winter lows (temperatures).
   d. The Sun would be rotating around Earth in an ellipse as opposed to a circle.

5. **Which of the following is NOT a mitigation strategy for climate change?**
   a. Green taxes
   b. Cap-and-trade
   c. Carbon-sequestration techniques
   d. Planting different crops

6. **Which of the following factors tend to increase the albedo effect and contribute to cooling?**
   a. Melting glaciers
   b. Sulfate aerosols
   c. Clear, sunny days
   d. Increasing CO_2 levels

WORK WITH IDEAS

1. Compare and contrast the major radiative forcers, including both those that are natural and those that are humanmade. Overall, which forcers are currently having the most effect on global climate? Are they natural or produced by human actions?

2. What is the IPCC? Why is it important?

3. Describe greenhouse gases. What is their function? What human actions have led to an increase in the amount of greenhouse gases in the atmosphere? What has been the result?

4. Describe the types of problems that global climate change causes for ecosystems and human health. Of the impacts given in Infographic 22.11, which do you feel is likely to cause the biggest problem? Why?

5. In the winter of 2010, the northeastern part of the United States had several large snowstorms that resulted in record high snowfall amounts. How does this weather fit in with the notion of global climate change?

6. Describe the process of facilitation and its goal. Why do some scientists claim that facilitation is a bad idea?

ANALYZING THE SCIENCE

The graphs on the following page show 2 of the 15 northern species evaluated in Chris Woodall's study of tree-range migration mentioned in this chapter. The total standing biomass and the total number of seedlings of each species are shown at different latitudes within the study area.

INTERPRETATION

1. What does the purple line on each graph represent? What does the blue line represent?

2. There are 69.2 miles between two adjacent latitude lines. How far apart are the latitude classes shown here? How many miles wide is the study site (from latitude 34° to 48°)?

3. Look at each graph to determine at which latitude class each species shows the most biomass per hectare. Which tree species has more standing biomass at its peak: balsam poplar or American basswood? How can you tell?

ADVANCE YOUR THINKING

4. Is either of these tree populations exhibiting a range migration shift? Present the evidence for your conclusions.

5. Of the two species shown here, which would you expect to see at higher altitudes on a mountainside and which at lower altitudes? Assuming both species could migrate, what do you predict will happen to the populations of these two species if the climate warms a little? If it warms a lot? Explain your answers.

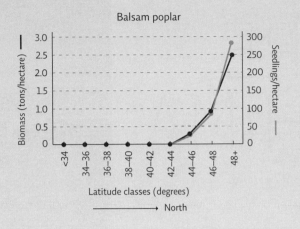

Balsam poplar

American basswood

EVALUATING NEW INFORMATION

Among scientists, there is broad consensus (97%) that climate change is significantly caused by human activity. Yet in a recent poll conducted by the BBC (news.bbc.co.uk/2/hi/science/nature/8500443.stm), 74% of the public are either not convinced or deny that humans are causing global warming. Members of the public get their information from a variety of media sources and information posted on the Internet. How can there be such a large disconnect between scientists and the public?

Go to the Global Warming Hoax page (www.globalwarminghoax.com/news.php?extend.133).

Evaluate the website and work with the information to answer the following questions:

1. Is this a reliable information source? Does it have a clear and transparent agenda?
   a. Who runs the website? Do this person's/group's credentials make the site reliable or unreliable? Explain.
   b. What is the primary message of the website? What evidence is it providing in the short article on Antarctic sea ice?

c. Do you have any questions about the data presented? If so, what are they?

Now go to the Skeptical Science website (www.skepticalscience.com). Click on "Most Used Climate Myths" on the left menu.

2. Is this a reliable information source? Does it have a clear and transparent agenda?
   a. Who runs the website? Do this person's/group's credentials make the site reliable or unreliable? Explain.
   b. What is the primary message of this website? What types of evidence does it provide to support its message?
   c. Click on the "Antarctica is gaining ice" link. Read the article carefully and compare the main point of the article to the article on the Global Warming Hoax site. Based on this article, what are the fundamental flaws of the first article?
   d. Which website do you find more credible? Why?

MAKING CONNECTIONS

IS CLIMATE CHANGE THE BIGGEST CHALLENGE OF OUR TIME?

Background: Many people feel that climate change is the biggest challenge of our generation. While skeptics remain and large portions of the public are unconvinced, the evidence supporting a changing global climate continues to accumulate. Why aren't people convinced by this evidence? Susan Hassol, a climate change communicator, provides a number of explanations. These include:

- The way climate change is presented in the media as a story with two sides.
- A general lack of science literacy among the public.
- A lack of understanding of the scientific process by the public.
- A fundamental misunderstanding between scientists and laypeople involving the use of language—essentially a communication issue.
- A psychological barrier that results when people are presented with something that frightens them.

Case: As a member of your campus environmental organization, you have been asked to participate in a workshop for teachers about the importance of science in education. Your presentation is limited to 30 minutes. Using one of the five preceding explanations as a starting point, and using global climate change as an example, create a presentation that will convince audience members that science is a necessary part of education, especially if we want to solve problems such as climate change.

Make sure that your presentation includes information from experts in the field, presented in both text form and as graphs and figures, and uses language that the average layperson will understand.

CHAPTER 27 NUCLEAR POWER

THE FUTURE OF FUKUSHIMA

Can nuclear energy overcome its bad rep?

The Japanese authorities originally declared a 20-kilometer (12.5-mile) evacuation area around Fukushima, an exclusion zone which may only be entered under government supervision. Four months after the explosion, residents in protective suits are briefed before being escorted to their homes to retrieve a few small items.

CORE MESSAGE

Nuclear energy can create tremendous amounts of power, and with concerns over fossil fuel supplies and climate change, nuclear energy has the potential to be an increasingly important part of the world's energy future. However, there are serious safety concerns with nuclear power, including vulnerability to natural disasters, radioactive waste disposal, and potential for weapons production.

GUIDING QUESTIONS

After reading this chapter, you should be able to answer the following questions:

→ What are radioactive isotopes and why are they important for nuclear power?

→ How do radioactive isotopes decay and what types of radiation are produced? How dangerous is this radiation?

→ How is nuclear energy harnessed to generate electricity in a fission reactor?

→ What problems are associated with the production of nuclear fuel and the disposal of nuclear waste?

→ What are the advantages and disadvantages of nuclear power?

↑ The crippled Fukushima Daiichi Nuclear Power Station 10 months after the disaster.

→ Satellite image of the damaged Fukushima Daiichi Nuclear Power Station on March 14, 2011.

The Fukushima Daiichi Nuclear Power Station is a maze of steel and concrete perched right on Japan's Pacific coast, just 240 kilometers (150 miles) north of Tokyo; its six nuclear reactors supply some 4.7 GW (1 gigawatt = 1 billion watts) of electric power to the country, making it one of the largest nuclear power plants in the world. On March 11, 2011, when a magnitude 9.0 earthquake struck 130 kilometers (80 miles) north of the plant, there were more than 6,000 workers inside. The quake caused a power outage, and in the darkness, chaos ensued: Men and women groped desperately for ground that would not stabilize beneath their hands and feet, shouting in panic as steel and concrete collided around them. When the shaking stopped, emergency lights came on, revealing a cloud of dust. But that was only the beginning of the disaster.

The earthquake had erupted beneath the ocean floor, triggering a tsunami that would arrive at the plant in two distinct waves. The first wave was not big enough to breach the 10-meter-high (33-foot-high) concrete wall that had been built between the plant and the sea. But the second wave, a fearsome mass of water that came 8 minutes later, was. At four stories tall, it bulldozed a string of protective barriers, sent buses and cars and trucks careening into pipes and levers and control panels, and eventually settled, in deep black pools, around the reactors themselves.

They were the strongest earthquake and largest tsunami in the country's long memory. Together, they would claim some 20,000 lives along a 400-kilometer (250-mile) stretch of coast (roughly equal to the distance between Maine and Manhattan). But as the ground steadied and the water subsided, the world's attention would quickly turn to a third disaster, even more precarious and potentially deadly than the first two: the risk of nuclear meltdown at Daiichi.

It's no surprise that the story of nuclear power pivots on calamity. Ever since its potential was first demonstrated, humankind has scurried relentlessly between two competing goals—the desire to harness nuclear energy for our own ends, and the impulse to protect ourselves from its destructive capacity. When the ground trembled beneath Fukushima, concerns over greenhouse gases and global warming had been pushing much of the world—including the United States—toward the former. As the people of Japan scrambled to respond, the world watched closely.

WHERE IS FUKUSHIMA, JAPAN?

Nuclear power harnesses the heat released in nuclear reactions to produce electricity.

In some ways, electricity generated using **nuclear energy** is no different than other forms of thermoelectric power (those that use heat to produce electricity). Just like power plants that run on oil or coal, nuclear plants use heat to boil water and produce steam, which is then used to generate electricity. The difference, really, is in where that heat comes from; coal and oil plants create it by burning fossil fuels. At a nuclear power plant, heat is produced

---

nuclear energy Energy released when an atom is split (fission) or combines with another to form a new atom (fusion); can be tapped to generate electricity.

Infographic **27.1** | **ATOMS AND ISOTOPES**

THE ATOM

→ All of matter is made up of atoms. Each atom is made up of subatomic particles: *protons* and *neutrons* in the nucleus (center) of the atom, make up the mass of the atom. Orbiting around the nucleus are much smaller particles called *electrons*. Different combinations of these subatomic particles produce specific *elements*—a chemical substance made up of only one kind of atom. The number of protons, the *atomic number*, is unique to each element. For example, any atom with only 2 protons is an atom of the element helium. The sum of the number of protons and neutrons gives an element its *mass number*. Helium, shown here, is an element that has 2 protons, 2 neutrons, and 2 electrons.

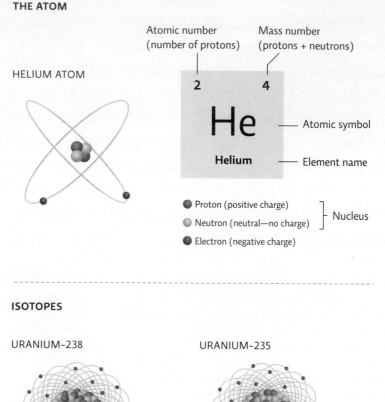

HELIUM ATOM

Atomic number (number of protons) — 2
Mass number (protons + neutrons) — 4

He
Helium

Atomic symbol
Element name

● Proton (positive charge)
○ Neutron (neutral—no charge) } Nucleus
● Electron (negative charge)

ISOTOPES

→ *Isotopes* are atoms that have the same atomic number (number of protons) but a different number of neutrons, and thus a different mass number. An atom with 92 protons is uranium; it can have 146 neutrons, and is called U-238 (92 protons + 146 neutrons = 238). Another uranium isotope, U-235, has 143 neutrons (92 protons + 143 neutrons = 235).

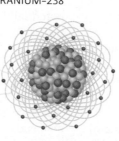
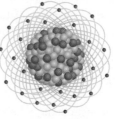

URANIUM-238
● 92 protons ○ 146 neutrons
More neutrons—heavier

URANIUM-235
● 92 protons ○ 143 neutrons
Fewer neutrons—lighter and less stable

through a controlled nuclear reaction—usually a **nuclear fission** reaction.

Nuclear fission reactions are those that result in the splitting of an atom. Their main ingredient is a special type of atom known as a *radioactive isotope*. Some elements can exist in two or more forms—each form has the same number of protons and electrons but a different number of neutrons, and hence, a different atomic mass; these different versions of the atom are called **isotopes**. [INFOGRAPHIC 27.1]

Most isotopes are stable, meaning they do not spontaneously lose protons or neutrons. But some are **radioactive**—they emit subatomic particles and heat energy (radiation) in a process known as radioactive decay. Radioactive decay is measured in half-lives. An isotope's

radioactive half-life is the amount of time it takes for half of the radioactive material in question to decay to a new form. So after one half-life, 50% of the material will decay; in the next half-life, 50% of what's left (or 25% of the original amount) will then decay. After ten half-lives, just 0.1% of the original radioactive material is left. [INFOGRAPHIC 27.2]

nuclear fission Nuclear reaction that occurs when a neutron strikes the nucleus of an atom and breaks it into two or more parts.
isotopes Atoms that have different numbers of neutrons in their nucleus but the same number of protons.
radioactive Atoms that spontaneously emit subatomic particles and/or energy.
radioactive half-life The time it takes for half of the radioactive isotopes in a sample to decay to a new form.

↓ The rate of decay for a given radioactive isotope is predictable and expressed as a *half-life*—the amount of time it takes for half of the original radioactive material (*parent*) to decay to the new *daughter* material (a new isotope, or even a new atom if protons are lost). Radioactive isotopes and their daughter radioactive isotopes continue to decay until they form a stable isotope that no longer loses particles. For instance, U-238 decays initially to thorium-234, which itself will decay over time. The entire decay sequence of U-238 includes progression through at least 13 isotopes (each step with its own half-life that ranges from milliseconds to thousands of years) until the final isotope decays into lead-206, a stable atom.

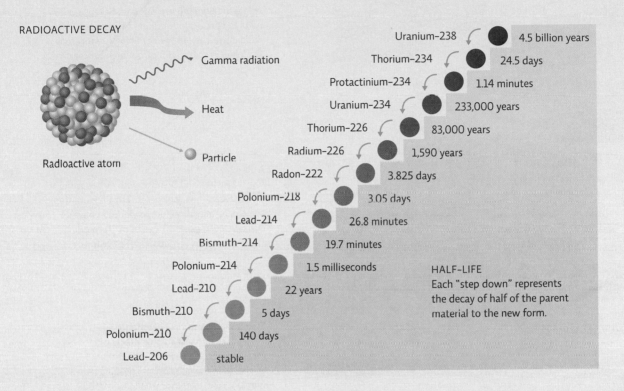

RADIOACTIVE DECAY

Gamma radiation

Heat

Particle

Radioactive atom

Isotope	Half-life
Uranium–238	4.5 billion years
Thorium–234	24.5 days
Protactinium–234	1.14 minutes
Uranium–234	233,000 years
Thorium–226	83,000 years
Radium–226	1,590 years
Radon–222	3.825 days
Polonium–218	3.05 days
Lead–214	26.8 minutes
Bismuth–214	19.7 minutes
Polonium–214	1.5 milliseconds
Lead–210	22 years
Bismuth–210	5 days
Polonium–210	140 days
Lead–206	stable

HALF–LIFE
Each "step down" represents the decay of half of the parent material to the new form.

RADIOACTIVE HALF-LIFE

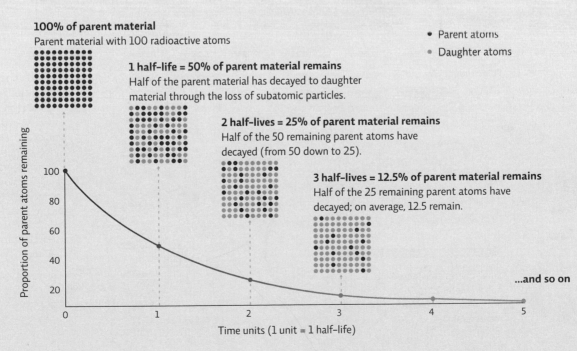

100% of parent material
Parent material with 100 radioactive atoms

• Parent atoms
• Daughter atoms

1 half-life = 50% of parent material remains
Half of the parent material has decayed to daughter material through the loss of subatomic particles.

2 half-lives = 25% of parent material remains
Half of the 50 remaining parent atoms have decayed (from 50 down to 25).

3 half-lives = 12.5% of parent material remains
Half of the 25 remaining parent atoms have decayed; on average, 12.5 remain.

...and so on

Proportion of parent atoms remaining

100
80
60
40
20

0   1   2   3   4   5
Time units (1 unit = 1 half-life)

Infographic **27.3** | **NUCLEAR FUEL PRODUCTION**

↓ Uranium ore (rock that contains uranium) is mined and goes through many stages of processing to produce fuel suitable for a nuclear reactor. The process creates hazardous waste at every step.

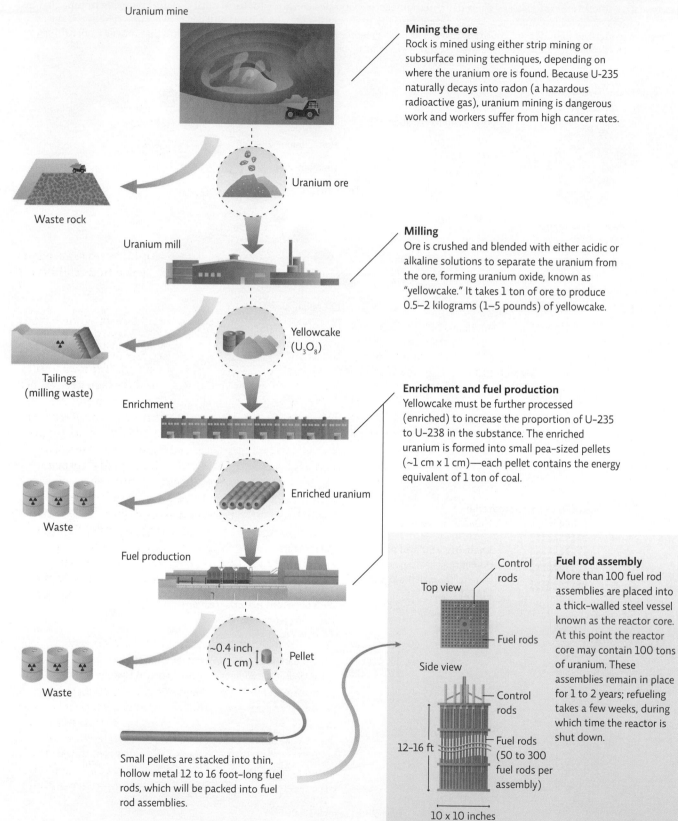

Uranium mine

Mining the ore
Rock is mined using either strip mining or subsurface mining techniques, depending on where the uranium ore is found. Because U-235 naturally decays into radon (a hazardous radioactive gas), uranium mining is dangerous work and workers suffer from high cancer rates.

Uranium ore

Waste rock

Uranium mill

Milling
Ore is crushed and blended with either acidic or alkaline solutions to separate the uranium from the ore, forming uranium oxide, known as "yellowcake." It takes 1 ton of ore to produce 0.5–2 kilograms (1–5 pounds) of yellowcake.

Yellowcake (U_3O_8)

Tailings (milling waste)

Enrichment

Enrichment and fuel production
Yellowcake must be further processed (enriched) to increase the proportion of U-235 to U-238 in the substance. The enriched uranium is formed into small pea-sized pellets (~1 cm x 1 cm)—each pellet contains the energy equivalent of 1 ton of coal.

Enriched uranium

Waste

Fuel production

~0.4 inch (1 cm) Pellet

Waste

Small pellets are stacked into thin, hollow metal 12 to 16 foot-long fuel rods, which will be packed into fuel rod assemblies.

Control rods
Top view
Fuel rods
Side view
Control rods
12-16 ft
Fuel rods (50 to 300 fuel rods per assembly)
10 x 10 inches

Fuel rod assembly
More than 100 fuel rod assemblies are placed into a thick-walled steel vessel known as the reactor core. At this point the reactor core may contain 100 tons of uranium. These assemblies remain in place for 1 to 2 years; refueling takes a few weeks, during which time the reactor is shut down.

Infographic 27.4 | **NUCLEAR FISSION REACTION**

↳ Fission, or the breaking apart of atoms, begins when an atom like U-235 is bombarded with a neutron—this breaks the atom into other smaller atoms and releases free neutrons, which in turn hit other U-235 atoms, causing them to split and release neutrons, and so on. The reaction in the fuel assembly is controlled by the insertion of control rods of nonfissionable material which absorb some of the free neutrons.

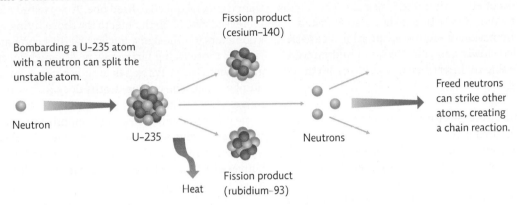

Most nuclear reactors use uranium, which has several isotopes. U-238 is the most stable and is the most abundant form of uranium—it makes up roughly 99% of Earth's total supply. But it's U-235, the most reactive form, that is mined from Earth, processed into nuclear fuel (which, like all mining work, involves both safety and environmental hazards), and packed into the **fuel rods** that are used in facilities like Fukushima. [INFOGRAPHIC 27.3]

A nuclear fission chain reaction begins when U-235 in the fuel rods is deliberately bombarded with a neutron. This bombardment makes the nucleus unstable, causing it to split into a variety of two or more smaller atoms and releasing two or three additional neutrons in the process. These newly released neutrons then hit other U-235 atoms, causing them to split and release even more neutrons, and so on. [INFOGRAPHIC 27.4]

Unlike the type of nuclear reaction at work in a nuclear bomb, which uses much more radioactive material , setting off a chain reaction that is almost instantaneous, the reactions at nuclear power plants are highly controlled. **Control rods**—made of materials such as boron or graphite that absorb neutrons—are placed in the fuel rod assembly between the fuel rods to control the speed of the reaction. They can be added (to slow down or stop the reaction) or removed (to make it go faster).

Even controlled, this chain reaction releases a tremendous amount of heat—10 million times more than is released by burning fossil fuels like coal or oil. The heat is used to boil water, which produces steam, which turns turbines that create electricity.

All types of thermoelectric power take a lot of water—that's why power plants are sited near rivers and oceans; but at the moment, nuclear power requires the most—on average, around 2,460 liters (650 gallons) per megawatt hour (MWh), compared with 1,890 liters (500 gallons) per MWh for coal, and 605 liters (160 gallons) per MWh for natural gas. (A MWh is the production of 1 megawatt—1 million watts—over an hour's time.) That means a typical 1,000-MW nuclear reactor requires 2,460,500 liters (650,000 gallons) of water per minute to flow through the cooling system. Some use much more—each of the two 845-MW reactors at the Calvert Cliffs nuclear power plant in Maryland requires 4,542,500 liters (1,200,000 gallons) of water per minute during operation. Though most of this water (more than 95%) is returned to the source (river or ocean), there are still problems with the release of warmer-than-normal water back into the environment, as well as damage to aquatic life that gets trapped in or against intake filters.

The reason for all that water is simple: With nuclear energy, water is needed not only to produce steam but also to keep spent fuel rods cool and to prevent the reactor from overheating (remember, heat is produced from radioactive decay of fission by-products, which continues even after the reactor is shut down and fission stops; so spent fuel needs constant cooling). Without water to cool them, fuel rods can melt, releasing large amounts of radioactivity; the fuel rod metal casing can also get hot enough to react with steam in a way that produces highly explosive hydrogen gas.

---

fuel rods Hollow metal cylinders filled with uranium fuel pellets for use in fission reactors.

control rods Rods that absorb neutrons and slow the fission chain reaction.

Nuclear energy has a troubled history.

Nuclear energy is the most concentrated source of energy on Earth. Its fearsome power was first demonstrated in 1945, when the U.S. military dropped atomic bombs over the Japanese cities of Hiroshima and Nagasaki. The bombs brought an end to World War II, but they also wreaked havoc on an entire nation of civilians: radiation sickness, cancers that killed slowly and which young children were especially vulnerable to, infertility in some, and birth defects in others.

In 1953, President Eisenhower made his famous "Atoms for Peace" speech, laying out a plan by which this destructive force could be harnessed for good: Instead of building bombs, we would produce cheap, reliable energy. In the years that followed, nuclear physics indeed gave rise to a litany of technologies that benefitted humankind, from radiocarbon dating to X-rays to radiation therapy for cancer.

But while there are now 432 nuclear power plants around the world, nuclear energy itself remains mired in controversy.

Proponents argue that uranium ore (uranium-containing rock) is both more abundant and produces a more efficient fuel than any fossil fuel. For example, 1 pound of uranium produces the same amount of energy as about 100,000 pounds of coal. And the operating costs, per kilowatt-hour, for nuclear power are comparable to those of coal.

Of course, as critics are quick to point out, that estimate does not factor in the great expense of building, maintaining, and then decommissioning actual nuclear plants. In the United States, it costs about $4 billion to build a nuclear reactor (about twice the cost to build a coal power plant), and anywhere from $200 million to $1 trillion to decommission one (most nuclear reactors have a lifespan of about 40 years, after which they need to be disassembled and the radioactive components stored and guarded).

One thing both sides agree on is that nuclear energy is a cleaner way to produce electricity. The processes of generating it emit much less CO_2 than the analogous processes for any fossil fuel, and virtually none of the other problematic combustion by-products, like sulfur dioxide, nitrogen oxides, and particulate matter. According to the Department of Energy, switching from fossil fuels to nuclear energy would be the single most effective way to reduce greenhouse gas emissions in the United States. In the United States alone, existing nuclear power plants already prevent roughly 650 million tons of CO_2 per year

from being released. Worldwide, they prevent close to 2.5 billion tons of CO_2 from entering the atmosphere.

And despite some persistent fears, research suggests that living near a nuclear power plant is actually safer than living near a coal-fired one. A 2011 study found no increased risk of birth defects for those living within 10 kilometers (6 miles) of a nuclear facility compared to the risk for those living farther away. Meanwhile, epidemiologist Javier García-Pérez has found that in Spain, the number of cancer-related deaths does increase as one moves closer to coal-fired power plants. "You still have some environmental hazards, from mining uranium and from radioactive waste and water," says Charles Powers, a professor and nuclear energy scientist at Vanderbilt University "But on balance, nuclear is far cleaner than any fossil fuel."

Still, the debates over safety remain unresolved. Proponents point out that, considering the number of existing plants, and the length of time they have been operating, accidents have been exceedingly few and far between. But opponents say that such safety claims ignore two key points: the vulnerability of nuclear power plants to natural disasters (which, as we will see, can lead to nuclear meltdown and the release of radioactive material into the environment), and the potential for nuclear fuel to be stolen and weaponized. The radioactive waste produced by nuclear power plants is another huge safety issue: It's extremely dangerous; there's a lot of it, and we have yet to come up with a plan for disposing of it safely.

France now generates some 80% of its electricity from nuclear power. But while the United States has the most nuclear reactors of any nation, Americans themselves have been divided over nuclear energy since the 1980s, after two infamous nuclear accidents made global headlines. The first was a partial meltdown at the Three Mile Island plant near Middletown, Pennsylvania, in 1979 (due to an electrical failure followed by a flurry of operator errors). The second was a full nuclear meltdown at the Chernobyl reactor in what is now the Ukraine, in the spring of 1986.

The Three Mile Island incident did not result in any deaths or major public health problems. The Chernobyl meltdown was considerably more severe. An explosion triggered by a test that went awry sent tremendous amounts of radiation wafting over much of Western Russia and Europe. More than a fifth of the surrounding farmland remains unusable to this day, and the World Health Organization estimates that the radiation will ultimately be responsible for some 4,000 deaths when all is said and done. That figure does not include cancer-related deaths, which range from 60,000 (according to one European report) to nearly 1 million (according to a Russian report).

> The radioactive waste produced by nuclear power plants is another huge safety issue: It's extremely dangerous; there's a lot of it, and we have yet to come up with a plan for disposing of it safely.

For their part, the Japanese were terrified of nuclear power after the bombings of Hiroshima and Nagasaki. (The popular Godzilla movies were actually based on a fictional reptile that had been mutated by a nuclear reaction and was coming to exact his revenge on humankind!) But as their country entered its own era of industrialization and economic growth, they were forced to overcome those fears. "Japan had no other natural energy source," says Frank N. von Hippel, nuclear physicist and arms control expert at Princeton University. "Nuclear was their only ticket to becoming a world superpower."

Nuclear accidents can be devastating.

On the day of the quake, each of the three operating reactors at the Fukushima plant held about 25,000 fuel rods, each one about 4 meters (12 feet) long, and filled with pellets of enriched uranium. The power outage had not only plunged the plant into darkness, but also stopped the normal delivery of water to those reactors.

As the world watched, workers at the plant tried everything they could think of to get cold water on the hot fuel. They tried to bring fire trucks and emergency power vehicles in, but the quake and tsunami had rendered the roads impassable. In one desperate attempt, some workers even searched the parking lot for vehicles that might have survived the tsunami with their batteries intact. But it was all to no avail.

Not only were the rods in danger of melting, and of producing hydrogen gas, but without a steady supply of coolant, they were rapidly boiling away all the existing water, thereby creating huge steam pressure in the reactor. To prevent an explosion, the steam would have to be released through a vent. But the design of the Fukushima reactors made this an especially tricky feat.

↓ Police guard a checkpoint at the edge of the exclusion zone leading to the town of Minami Soma, just north of Daiichi. The sign reads "Keep Out."

↳ The most common type of nuclear power plant is a pressurized water reactor (PWR), with 265 in operation around the world. The reactor at Fukushima is a boiling water reactor (BWR), one of 94 in the world. Both designs use fuel assemblies with control rods and water as a cooling and steam source. Temperatures are kept "down" to about 1400°C, well below the melting temperature of the uranium fuel (2800°C) and the metal casing of the fuel rods (2200°C). If the reaction is not kept cool with circulating water, a meltdown can occur. Even if "shut down" by inserting all the control rods to absorb the neutrons and stop the chain reaction, heat will still be produced by the natural decay of the isotopes (they are not being split by bombardment; they are simply spontaneously losing particles).

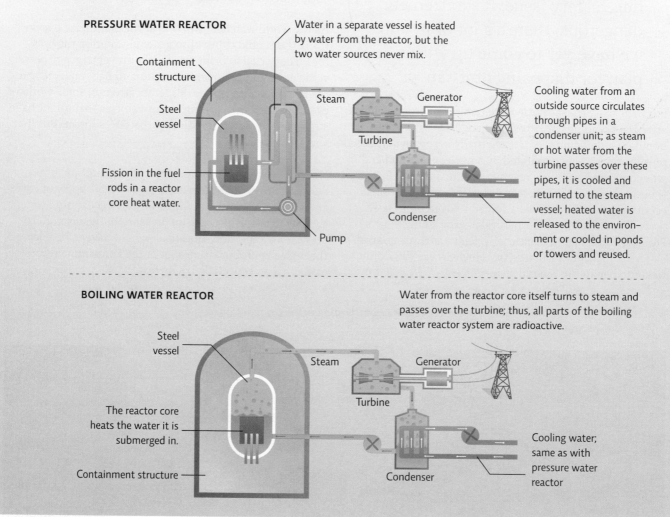

PRESSURE WATER REACTOR

Water in a separate vessel is heated by water from the reactor, but the two water sources never mix.

Containment structure

Steel vessel

Steam

Generator

Turbine

Fission in the fuel rods in a reactor core heat water.

Pump

Condenser

Cooling water from an outside source circulates through pipes in a condenser unit; as steam or hot water from the turbine passes over these pipes, it is cooled and returned to the steam vessel; heated water is released to the environment or cooled in ponds or towers and reused.

BOILING WATER REACTOR

Water from the reactor core itself turns to steam and passes over the turbine; thus, all parts of the boiling water reactor system are radioactive.

Steel vessel

Steam

Generator

Turbine

The reactor core heats the water it is submerged in.

Containment structure

Condenser

Cooling water; same as with pressure water reactor

There are several different types of fission reactors. The most common type worldwide is a *pressurized water reactor* (PWR), where the steam that turns the turbine is not exposed to radiation: Nuclear fission in the core heats water under pressure (like a pressure cooker); that water (which is exposed to radiation) is then piped through, and thus heats, a separate container of water (which is not exposed to radiation); it is the steam from this radiation-free water that turns the turbine. The reactors at Fukushima, however, were *boiling water reactors* (BWR); BWRs produce steam in the reactor core itself. This means that both the steam and the turbine become radioactive in the process. [INFOGRAPHIC 27.5]

It also meant that opening the vent would be akin to pumping radiation straight into the air. Still, not opening it would almost certainly be worse: If any one of the reactors exploded, it would release much, much more radiation than the steam from the vent. "It was this horrible double-edged sword," says von Hippel. "Exactly the kind of situation that we always feared with the BWRs—a lose-lose."

On the morning of March 12th, six workers gathered in the plant's main control room, where each of them swallowed several iodine pills; by loading their thyroid glands with iodine beforehand, they hoped to prevent their bodies from absorbing radioactive iodine from the steam that would pour out of the vent. They donned thick, heavy firefighter uniforms, along with oxygen tanks and masks, in an effort to shield themselves from alpha and beta radiation. Both alpha and beta radiation come from particles (alpha particles are basically a helium nucleus; beta particles are essentially electrons). Neither type can penetrate the skin very well. But if they enter the body through ingestion, both kinds can linger for years, causing organ damage and cancer.

None of the equipment would protect the workers from gamma radiation, which is much more energetic than alpha or beta particles and can easily penetrate walls, skin, and other surfaces. The only way for workers to avoid exposure to gamma rays would be to get in and out as quickly as possible. Dosiometers—mini radiation detectors attached to each fire suit—would let them know, with loud, alarmlike beeps, when they had exceeded the legal limits of exposure. [INFOGRAPHIC 27.6]

As the sun climbed high over the ravaged coastline, the team of six headed into the first reactor to search in darkness for the vent. The plan was to work in teams of two, and to relay their efforts so that no one worker was left hovering over the vent for too long. It took the first team

11 minutes to find the manual gate valve, crank it open a quarter of the way, and retreat. With the vent open partway, the radiation levels climbed so fast that the second team could not even reach it; in just 6 minutes, one of them absorbed a dose of radiation that surpassed the legal limit allowed for 5 years. The third team did not even try.

That afternoon, the building housing reactor No. 1 exploded, sending concrete and steel debris flying through the air and setting off a chaotic, and seemingly irreversible, chain reaction: Radiation levels around the plant climbed exponentially, deterring efforts to vent the other two reactors. In the days that followed, both of them also exploded.

The generation of nuclear waste is a particularly difficult problem to address.

The reactors weren't the only problem. Experts around the world, especially those observing from the United States, were particularly concerned about Fukushima's radioactive waste.

To understand why, it helps to know a little about radioactive waste. In general, there are two kinds: low-level

---

Infographic 27.6 | **RADIOACTIVE ISOTOPES CAN RELEASE ONE OR MORE OF THREE DIFFERENT KINDS OF RADIATION**

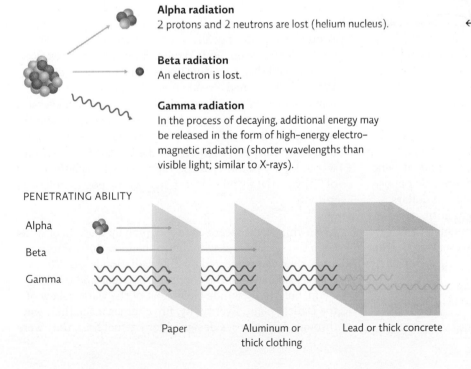

Alpha radiation
2 protons and 2 neutrons are lost (helium nucleus).

Beta radiation
An electron is lost.

Gamma radiation
In the process of decaying, additional energy may be released in the form of high-energy electromagnetic radiation (shorter wavelengths than visible light; similar to X-rays).

PENETRATING ABILITY

Alpha

Beta

Gamma

Paper

Aluminum or thick clothing

Lead or thick concrete

← The different types of radiation have differing abilities to penetrate surfaces. Alpha particles don't travel far and can't even penetrate paper; they do not penetrate skin but can be harmful if inhaled or ingested in food or water. The much smaller beta particles can penetrate the upper layers of the skin but are easily stopped by a thin sheet of aluminum or very heavy clothing. High-energy gamma rays can penetrate the deepest but can be stopped by thick or very dense material such as 20 centimeters (8 inches) of concrete or 2.5 centimeters (1 inch) of lead. Exposure to radiation, especially gamma rays, can damage a wide variety of body organs, leading to radiation sickness; it can also cause mutations that lead to cancer or cause birth defects.

radioactive waste and high-level radioactive waste. **Low-level radioactive waste (LLRW)** is material that has low amounts of radiation relative to its volume, and can usually be safely buried. This includes clothing, gloves, tools, etc., that have been exposed to radioactive material. The United States produces about 100,000 tons (or 2 to 4 million cubic feet) of LLRW per year, all of which is stored at just three sites—in South Carolina, Utah, and Washington.

High-level radioactive waste (HLRW) is another story. As its name suggests, it's more reactive than LLRW; in fact, because so many radioactive by-products are created in fission reactions, HLRW is actually much more radioactive than the original fuel rods. The United States produces about 2,000 tons of HLRW per year, and has some 65,000 metric tons of the stuff in storage right now—that estimate includes both spent fuel rods and waste from the nuclear weapons we've produced.

So far, we have no safe, reliable, long-term storage plan for this waste. Spent fuel rods are stored on site in steel-lined pools, where at least some isotopes—those with short half-lives—can decay to safe levels. Isotopes with longer half-lives require more time to reach this point. Because the isotopes produced in the fission reaction or by the decay of these isotopes are all mixed together, the waste has to be stored for as long as it takes the longest half-life material to decay to safe levels—in some cases that's more than 2 million years.

Almost all of the short-term storage pools in the United States are full; 55 sites are currently storing HLRW in large, steel casks that are surrounded by concrete, waiting for longer-term storage options to become available. So far, we don't have any. "Waste has been one of the biggest sore points in the debate over nuclear energy," says Powers, from Vanderbilt. "It's the one thing we really don't have even a good theoretical solution to."

The United States began construction on a long-term repository at Yucca Mountain in Nevada in 1994, but the project has been repeatedly stymied by opponents who live near the site and activists everywhere who oppose nuclear power. After 15 years and billions of dollars, President Obama halted construction in 2009, a move that fueled only more bickering. Opponents of the Yucca Mountain repository heralded the move as long overdue. But supporters argued that because the Yucca repository

was authorized by an act of Congress, only Congress had the power to cancel the project. The debate promises to continue: Even if a long-term storage option is approved and constructed, we still have the problem of how to safely transport HLRW across the country—a dilemma not likely to be resolved in the near future. [INFOGRAPHIC 27.7]

Meanwhile, back at Fukushima, radioactive waste posed yet another deadly threat. Each of the six reactors had its own swimming pool-like container where used radioactive fuel was stored. The pools were full of years' worth of waste, and they relied exclusively on water to prevent overheating. If just one of those pools ran dry, the consequences would be catastrophic: By some estimates, a worker standing next to a dry pool could receive a fatal dose of radiation in just 16 seconds.

On March 13th, just as experts around the world were contemplating such an event, a fire broke out around the spent fuel pool near reactor No. 4.

Responding to a nuclear accident is difficult and dangerous work.

By March 14th, three days after the earthquake and tsunami, all but a handful of workers had evacuated. The world media would dub the cohort that stayed behind "The Fukushima 50." The radiation levels in the main control room of reactor No. 2 had climbed so high that those remaining workers had to rotate out at regular intervals to avoid being poisoned with radiation.

Around the world, nuclear experts worried aloud over what might happen next: Would there be another explosion? Would the spent-fuel pools run completely dry? Would fuel that had already melted into a heap at the bottom of some reactors now melt through their steel vessels and react with the concrete below? And how far would the wind carry all the radioactive vapors that were escaping into the atmosphere? Experts in the United States said it was at least possible that the vapors could reach the edges of Tokyo—the world's largest city, with a population of 35 million.

Meanwhile, helicopters were scooping up buckets of water from the sea and trying to dump them on the reactors. First, they were deterred by high radiation levels. They bolted lead plates to the choppers' bottoms and tried again, but strong winds blew most of the water askew of the Daiichi plant. Eventually, fire engines made their way through. Using hoses designed for jet-fuel fires, they were

---

low-level radioactive waste (LLRW) Material that has a low level of radiation for its volume.

high-level radioactive waste (HHRW) Spent fuel rods or nuclear weapons production waste that is still highly radioactive.

Infographic 27.7 | **RADIOACTIVE WASTE**

↓ Radioactive waste does not just come from nuclear power plants, but is also generated by industry, the medical field, research laboratories, and weapons production. All of these users produce low-level radioactive waste (LLRW); nuclear power and weapons production is responsible for almost all of the high-level radioactive waste (HLRW). Though much of this material is highly dangerous and remains so for centuries, we currently have no safe way to dispose of it.

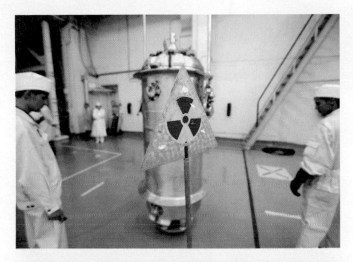

HLRW is a by-product of nuclear fission. The United States lacks a long-term storage facility for HLRW; most spent fuel rods are currently stored onsite at nuclear power plants in steel-lined pools. Some have been moved out of the pools and into dry casks. European and Asian nations are looking into the possibility of joint disposal facilities, though little progress has been made.

← Technicians load highly enriched uranium (HEU) fuel assemblies into casks at the Institute of Nuclear Physics in Kazakhstan.

LLRW includes contaminated items such as clothing, filters, gloves, and other items exposed to radiation. Short-half-life LLRW can be stored until it is no longer radioactive and then disposed of as regular trash. LLRW with a longer half-life is stored in casks; in the United States, LLRW is sent to one of three U.S. storage facilities (four other sites are closed) where it is buried underground; some must be shielded with concrete or lead containment before burial.

← Waste Area G at Los Alamos National Laboratory in Los Alamos, New Mexico. The barrels are being stored before shipment for burial.

MILL TAILINGS Mining and crushing uranium ore produces small particle (sandlike consistency) waste. It contains low levels of radioactive isotopes with long half-lives, such as radium, thorium, and uranium. Piles of mill tailings are stored near milling facilities and covered with clay and rock to prevent the release of radioactive material into the atmosphere—by law, they must remain covered for at least 200 years.

← Trucks haul in and dump uranium tailings, and a bulldozer spreads out the tailings in the disposal cell 48 kilometers (30 miles) north of Moab, Utah.

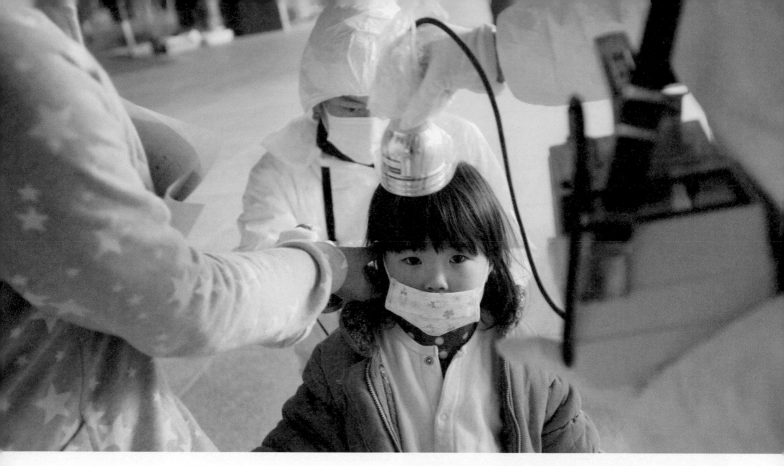

↑ Displaced people who were evacuated from Minamisoma, Futaba, and other towns located near the Fukushima Daiichi Nuclear Power Plant are checked for traces of radiation before they are permitted to enter a sports facility in Fukushima City, where about 1,200 evacuees found temporary shelter.

able to get some water to the places where it was most needed.

The saving move, however, was already in the works by then. "Even as buildings exploded," the *New Yorker* reported, "some of the workers had hooked up a train of fire trucks capable of generating enough pressure to inject water directly into the fuel cores. The drenching continued around the clock, and it went on for months. The process was ungainly, and it produced millions of gallons of radioactive waste that will be dangerous to store, but it probably did more than any other measure to avert a far worse disaster."

The impacts of nuclear accidents can be far reaching.

On March 16th, members of the U.S. Nuclear Regulatory Commission (NRC) determined that anyone more than 80 kilometers (50 miles) from the plant was probably safe from atmospheric radiation emitted by the disaster, based on computer models they had developed.

The finding did not quell public fears. In fact, long after the likely dispersal of radiation had been mapped and made widely known, people in many communities were fiercely divided about whether or not to evacuate. Local newspapers published daily radiation levels alongside weather reports, and average people sorted as best they could through reams of dense, technical information—not all of it reliable—as they tried to make decisions about the health and safety of their families.

For months, it seemed, one new shocking discovery or scandalous revelation followed the next: The Japanese government discovered stores selling beef, spinach, and other food containing small amounts of radiation and scrambled to recall those products. Authorities warned parents not to give local milk to their children because children's cells grow faster and that makes them more vulnerable to the effects of radiation. Distrustful of government reports, and frustrated by the lack of consistent information, average citizens began collecting and testing their own soil and water—sometimes with disturbing results. On October 14, for example, the *New York Times* reported that citizens groups had found 20-odd spots in and around Tokyo that were contaminated with potentially dangerous levels of radioactive cesium.

The impact on the natural environment has been more difficult to determine. "It depends so much on luck," says Powers. "Even if everyone involved in the response does everything perfectly, we're still at the mercy of the wind." At first, the wind at Fukushima Daiichi blew steadily out

↑ The Gösgen Nuclear Plant is located within the town of Däniken, Switzerland. Smoke rising from a cooling tower greets children as they walk home from school.

to sea. But eventually it turned inland, to the northwest, carrying all those radioactive vapors with it. Rain and snow captured some particles, laying them deep in the mountains, forests, and streams around Fukushima. It will take decades before we know the full effect they will have on the ecosystems they have become a part of. But 5 months after the quake, Japan's central government acknowledged that the area within 3 kilometers (1.8 miles) of the plant will likely be uninhabitable for decades.

Will nuclear power play a role in future energy?

The March 2011 tsunami would become the most expensive natural disaster in human history, with losses estimated at $300 billion. In its wake, countries around the world began rethinking their plans to expand their own nuclear energy programs. Germany resolved to phase out all of its nuclear power plants by 2022. Austria, Italy, and Switzerland also reconsidered. In the United States, however, experts fretted in a more undecided way. Critics pointed out that if the Indian Point plant, 56 kilometers (35 miles) north of Times Square in New York City, necessitated the evacuation of an 80-kilometer (a 50-mile) radius (as the Daiichi plant had), it would include some 20 million people. Proponents, meanwhile, held fast to their contention that nuclear power would be an essential,

unavoidable part of weaning the planet from its fossil fuel dependence. Under proper conditions, they insist, it can be a completely safe resource; the U.S. Navy, for example, has been safely using nuclear energy to power vessels since 1954, without incident.

Meanwhile, back in Japan, just 2 months after the quake and tsunami, the prime minister responded to public pressure and global criticism by calling for a temporary shutdown of one nuclear plant in the country's center, where scientists have estimated an 87% chance of a big quake sometime in the next three decades. Other plants that were closed for routine maintenance were told to hold off on reopening, so that, by summer's end, less than a third of Japan's 54 reactors were still running.

The course change was short-lived. Once factory owners began warning that such drastic power cuts would quickly lead to a recession, many of the deactivated power plants sprung back to life by order of the same prime minister who had closed them. Ultimately, the country's leaders would agree to close some of its oldest plants but leave most—36 out of 54, by one estimate—up and running for the foreseeable future. One official told the *New Yorker* that anything else would be "idealistic but very unrealistic." Speaking at an October 2011 forum sponsored by the American Association for the Advancement

Infographic 27.8 | **NUCLEAR POWER: TRADE-OFFS**

↳ What role will nuclear power play in the future? Like all of our energy options, nuclear power has advantages and disadvantages which must be weighed when making this decision.

ADVANTAGES

Operating costs are comparable to those of a fossil fuel power plant; the technology is available now.

No CO_2 is released during operation, so nuclear power does not contribute to climate change (though some CO_2 is released during mining and processing).

Dependable amounts of electricity can be produced—no worries about nighttime or cloudy days (solar power) or windless days (wind power).

Power production can be increased or decreased to meet demand (up to the capacity of the facility).

Uranium supplies should last 80 years but other isotopes can also be used; some reactor designs actually *produce* fuel during the reaction process.

Fuels used for nuclear reactors are energy rich—one small uranium pellet contains the same amount of energy as 1 ton of coal.

DISADVANTAGES

Nuclear power plants are much more expensive to build ($4 billion to build a fission reactor), maintain, and decommission ($200 million to $1 trillion to decommission after a 40- to 60-year lifespan) than fossil fuel power plants.

Mining and processing ore for nuclear fuel produces hazardous waste; surface mining damages the environment and can pollute air and water.

Radioactive waste is very hazardous and we still have no long-term plan for dealing with the waste that is produced from the fission reaction.

Shipping waste (by truck or rail) is also a safety concern and vehemently opposed by those who live on the transport route.

Though accidents are rare in the history of nuclear power, they can have extremely serious consequences and long-term impacts when they occur (Chernobyl and Fukushima).

Some methods of nuclear power production produce radioisotopes that could be used in nuclear weapons production; facilities for the processing of fuel could hide weapons programs.

of Science, Gregory Jaczko, chairman of the NRC, predicted that the impact of Fukushima on the long-term development of nuclear power will be minimal.

Less than a year after the fires had cooled, experts would agree that the Daiichi disaster was something of a draw. "People who support nuclear energy can't argue anymore that the risk is nonexistent," says Powers. "But at the same time, neither can opponents say that a meltdown would automatically result in a zillion immediate casualties."

In a paper published before the disaster at Fukushima, British nuclear engineers Robin Grimes and William Nuttall argued that nuclear power could enjoy a "renaissance" and increase its contribution to electricity production in the future, if only we improved our nuclear technology. They pointed to new "third generation" BWR and PWR reactor designs (and others) available now, and

other technologies ("fourth generation") in development that should be safer and more efficient, producing more electricity with less fuel. Some designs also produce less waste material that could be used to make nuclear weapons, reducing concerns about weapons proliferation.

The third-generation designs in use today are safer, according to safety assessments, and include more fail-safe responses (technical responses that automatically kick in if a problem occurs). But of course, the 2011 earthquake and tsunami have illustrated the need to reevaluate the potential damage of natural disasters. The renaissance would also require better waste-handling options, including the reprocessing of spent fuel (which would decrease the waste produced and increase overall efficiency).

Given the problems associated with fossil fuels, it is likely that nuclear power will continue to have a place in our energy mix. But in making choices about how to pursue our energy future, we must consider the costs and benefits of all our energy options. A "benefit analysis" must evaluate how well a particular energy source meets our energy needs. A "cost analysis" must consider not just the monetary cost of getting kilowatts delivered to our homes; it must also include the environmental and social costs associated with every step of the energy source's life, from acquisition to production to delivery. In addition, as mountaintop removal, oil spills, climate change, and Fukushima demonstrate, there is also a risk assessment that must take place. We must ask two very crucial questions: How risky is the venture (an assessment we can do with at least some degree of accuracy) and how much risk are we willing to take? [INFOGRAPHIC 27.8]

"There is definitely a certain weighing that has to take place," says Powers. "Global warming on one hand, nuclear accidents like Fukushima on the other."

The future of nuclear energy is uncertain.

On August 23, 2011, a magnitude 5.8 earthquake erupted beneath the state of Virginia, causing the North Anna Nuclear Power Plant there to tremble with much greater force than its reactors were designed to withstand: Dry casks, each weighing more than 100 tons and filled with spent fuel rods, shifted inches. It was the region's largest quake in more than a century, and the first time such a calamity had struck an American nuclear power station. Just 5 days later, when a category 5 hurricane by the name of Irene struck the East Coast, workers at three other nuclear power plants noticed that emergency sirens had failed to function properly. And at one plant—the Indian Point Power Plant, the one closest to Manhattan—a discharge canal carrying (nonradioactive) water from the cooling system overflowed due to the high river levels.

On September 9th, the NRC staff suggested ordering power plants to review their ability to survive quakes and floods "without unnecessary delay."◉

Research articles referenced in this chapter:
García-Pérez, J., *et al.* 2009. *Science of the Total Environment*, 407: 2593–2602.
Grimes, R.W. and Nuttall, W.J. 2010. *Science*, 329: 799–803.
Queißer-Luft, A., *et al.* 2011. *Radiation and Environmental Biophysics*, 50: 313–323.

BRING IT HOME

PERSONAL CHOICES THAT HELP

Nuclear energy has been rebranded as "green energy" because it does not emit greenhouse gases. Technology has improved the safety of nuclear facilities; however, there are still safety issues and valid concerns over the long-term storage of nuclear waste. In addition, cost, national security, and uranium supply problems make nuclear power a complicated energy solution.

Individual Steps

→ Use the facility locator on the U.S. Nuclear Regulatory Commission website (nrc.gov) to find out if you have nuclear reactors where you live.

Group Action

As you would expect, the policies endorsed by a particular group will depend on their overall view of nuclear energy. To see two examples, check out the following sites and see how they compare.

→ Visit the Nuclear Energy Institute (nei.org), which is a pronuclear organization, to see proposed legislation regarding increasing our current nuclear energy production.

→ Visit the Nuclear Energy Information Service (neis.org), which is a nonprofit antinuclear organization committed to a nuclear-free future.

Policy Change

→ What is your opinion about the necessity of nuclear energy in the United States? The U.S. Department of Energy website (www.ne.doe.gov) maintains articles, updates, and the latest policy initiatives regarding nuclear energy in America. This resource can help you understand current policy, funding issues, and upcoming legislation.

UNDERSTANDING THE ISSUE

CHECK YOUR UNDERSTANDING

1. **Atoms that have the same number of protons and electrons but different numbers of neutrons are called:**
   a. radioactive.
   b. ions.
   c. isotopes.
   d. subatomic.

2. **After 4 half-lives, about how much radioactive parent material is left?**
   a. 25%
   b. 40%
   c. 0.5%
   d. 6%

3. **What is the expected lifespan of a nuclear reactor?**
   a. 40 years
   b. 100 years
   c. 200 years
   d. 10 years

4. **For what problem related to nuclear power do we not yet have a workable solution?**
   a. Finding a substitute for uranium once it is used up
   b. Safe disposal of HLRW
   c. Producing steam that is not radioactive to turn the turbine
   d. Controlling the fission reaction

5. **Nuclear fission is a reaction that:**
   a. splits an atom, releasing energy.
   b. combines two or more atoms, producing energy.
   c. results in large explosions.
   d. is required to make atoms radioactive.

6. **There are different types of radiation. Which one is most energetic and can therefore penetrate many surfaces, including skin?**
   a. alpha radiation
   b. beta radiation
   c. gamma radiation
   d. particle radiation

WORK WITH IDEAS

1. Describe the process of nuclear fission. Explain how it can lead to a chain reaction in a fission reactor.

2. What are the pros and cons of generating energy from a nuclear source versus a fossil fuel source (coal)? Include in your answer a comparison of the extraction, energy-generating, and waste disposal processes. Based SOLELY on these comparisons, which of these is a better resource to use if we want to minimize environmental degradation?

3. From an economic standpoint, which type of electricity production (nuclear or fossil fuel) is less expensive? Explain. Remember to include both internal and external costs in your answer.

4. Using Infographic 27.4, describe the steps used in mining and processing uranium to the point where it is packed into fuel rods.

5. Suppose a natural disaster that compromises the integrity of a nuclear power plant had occurred in the United States. What might engineers in the United States be able to do immediately that engineers at the Fukushima Daiichi plant in Japan were reluctant to do, based on the differences in design of the plants?

6. In a nuclear fuel assembly, control rods are used to regulate the nuclear reaction. But even if all the control rods are inserted into the fuel assembly, heat is still produced. Explain why.

ANALYZING THE SCIENCE

The following graph depicts the number of accidents that occur in a variety of industries.

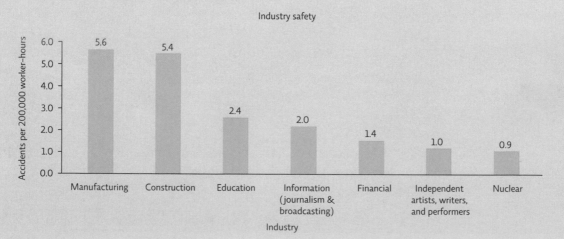

Industry safety

Graph: Accidents per 200,000 worker-hours (y-axis) vs Industry (x-axis)
- Manufacturing: 5.6
- Construction: 5.4
- Education: 2.4
- Information (journalism & broadcasting): 2.0
- Financial: 1.4
- Independent artists, writers, and performers: 1.0
- Nuclear: 0.9

INTERPRETATION

1. What does the height of each bar represent?

2. Based on the graph, which two industries have the highest accident rate?

3. Based on the graph, which two industries have the lowest accident rate?

ADVANCE YOUR THINKING

4. Is there bias in this graph? How would you decide? Think about and answer the following questions:
   a. Which industries are represented? Are they comparable?
   b. Do the data for nuclear workers include accidents that occur as a result of mining, processing, and production?
   c. Which of these industries do you think has the strictest OSHA (Occupational Safety and Health Administration) regulations? Might that account for differences in the number of accidents?

5. If you had access to safety data from all industries, which industries would you show in the graph to obtain a fair comparison to the nuclear industry?

EVALUATING NEW INFORMATION

The Nuclear Regulatory Commission (NRC) is tasked with regulating nuclear power in the United States and "protecting people and the environment." The members of the commission "formulate policies, develop regulations governing nuclear reactor and nuclear material safety, issue orders to licensees, and adjudicate legal matters" (www.nrc.gov/about-nrc/organization/commfuncdesc.html). The NRC maintains a detailed website with information about nuclear power plants and nuclear materials.

Go to the NRC's Facilities page (www.nrc.gov/info-finder/region-state). Find your state in the list of states and territories.

Evaluate the website and work with the information to answer the following questions:

1. Is this a reliable information source? Does the organization have a clear and transparent agenda?
   a. Who runs the website? Do this person's/group's credentials make the information reliable or unreliable?
   b. Is the information on the website up to date? When was the website last updated? Explain.

2. How many nuclear sites are active in your state? How many nuclear sites are being decommissioned?

3. Click on a link for either an active nuclear site or a site undergoing decommissioning.
   a. If there are nuclear power reactors in your state, write down information about their age, intended expiration date, the types of reactor in use (PWR or BWR), and their specific locations. Find the location on a map and record the closest large body of water.
   b. For each nuclear site in your state, describe what the site is, what it contains (in terms of nuclear material), the environmental issues associated with the site (if any), and the plan of action to correct the issues.

4. Consider the information you have gathered, and do a basic risk assessment analysis for one of the sites.
   a. Do you think the facility is a danger to the environment? To human health? Why or why not?
   b. Is the facility located in an area that may have natural disasters? If so, what are they? If a natural disaster were to occur, predict some outcomes.

MAKING CONNECTIONS

WHAT DO WE DO WITH NUCLEAR WASTE PRODUCTS?

Background: One of the problems with nuclear material is that the radiation generated is harmful to life AND it persists in the environment for a long time—from hundreds of thousands to millions of years. To date, we have no long-term solution to the problem of what to do with much of this material. Currently, material is stored in what are described as "temporary" facilities (see Infographic 27.7), although the reality is that they are de facto permanent facilities.

Case: You are member of a citizen's group seeking to prevent the transportation of nuclear waste on the highway that goes through your town. Before joining the group, you were unaware that nuclear waste was being transported by truck to disposal sites. You must research the issue and make a presentation to the members of the group. (The NRC website mentioned above is a good starting place.) Make sure that you include the following:

1. What type of nuclear waste is transported across the country?

2. Where are the three waste storage facilities for low-level nuclear waste? For waste incidental to reprocessing?

3. How is nuclear waste transported? What regulations are in place to provide safe transportation?
   a. Based on your research, what recommendations would you make to your group concerning the current transportation of nuclear waste? For example, is it relatively safe? Have there been releases of radiation to the environment because of transportation? What type of regulations could your locality enact to enhance safety while the material is being transported through your region?
   b. In the event that a long-term disposal site for spent nuclear fuel is approved (perhaps in Yucca Mountain, Nevada), how would you suggest that the material be transported to the site from the nuclear power plants across the United States?

FUELED BY THE SUN

A tiny island makes big strides with renewable energy

Playing soccer under the shade of wind turbines on Samsø Island, Denmark, the first island in the world with 100% renewable energy. Collectively, Samsø's land-based turbines produce about 26 million kilowatt-hours a year, enough to meet the island's demands for electricity.

CORE MESSAGE

In order to become a sustainable society, we need to transition to reliable, renewable energy sources with acceptable environmental and social impacts. No single energy source can replace fossil fuels. Instead, a variety of methods, selected to meet the needs of the population, and availability of local energy sources, will help communities shift to sustainable energy use. Fortunately, we have many good options already at our disposal, with other new methods currently in research and development.

GUIDING QUESTIONS

After reading this chapter, you should be able to answer the following questions:

→ What are the characteristics of a sustainable energy source and what role does renewable energy play in terms of global energy production?

→ How do wind and solar power technologies capture energy? What are the advantages and disadvantages of wind and solar power and how does each compare to fossil fuels in terms of true costs?

→ In what ways can we harness geothermal energy and the power of water and what are some of the trade-offs associated with each resource?

→ What combination of actions did Samsø take to become an energy-positive island and what is the take-home message to other communities who might want to reduce their use of fossil fuels?

→ What roles do conservation and energy efficiency play in helping us meet our energy needs sustainably?

On a typical cold, misty January day in Denmark in 2003, many of the 4,100 residents of a small island gathered together at the beach. Everyone, including the mayor, strained their eyes to see the faint outline of several structures, each over 30 stories tall, located more than 2 miles offshore.

Nestled in the crook of Denmark's mainland, Samsø is home to a small, windswept community of Danish farmers known for their sweet strawberries and tender early potatoes. It is a quiet and serene place. Yet it has been the site of a dramatic revolution—a community transformation that made headlines around the world.

The transformation began on that cold day in 2003. Finally, the mayor pushed a button, and the offshore structures slowly creaked to life. Through the grey mist and rain, people could gradually see the massive blades begin to rotate, converting the power of wind into energy. It was a landmark day in the island of Samsø's ambitious attempt to become the greenest, cleanest, and most energy-independent place on Earth.

"That was a very big moment," recalls Søren Hermansen, a Samsø resident who was key in getting the community behind the project. "Nobody really thought it would happen when we started."

Reliance on renewable energy is a characteristic of a sustainable ecosystem and society.

Samsø used to be just like most communities on Earth: fully dependent on fossil fuels like coal, oil, and gas. As recently as 1998, Samsø imported all of its energy resources from the mainland: Tankers hauled oil into its ports and electricity generated from burning coal was imported via cables.

But in 1997, the government of Denmark decided it was important to promote the idea of **renewable energy**, energy from sources that are replenished over short time scales or that are perpetually available. To do so, they announced a competition: Which local area or island could become self-sufficient on renewable energy? The contest invited applicants to describe which renewable resources were available in their community and how they would be used to replace fossil fuels.

The purpose of the contest was to put communities on track to be more sustainable. To qualify as a **sustainable energy** source, the energy must be renewable, with a low enough environmental impact that it can be used for the

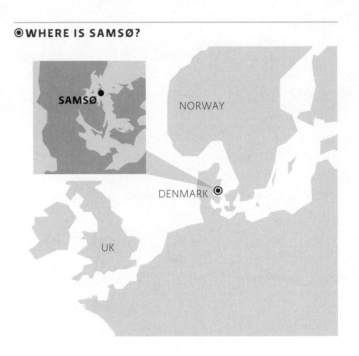

⊚ WHERE IS SAMSØ?

SAMSØ

NORWAY

DENMARK

UK

long term. As a result, any source of energy that causes environmental damage when it is captured or produced, such as generating waste or pollution, is not ideal. For example, experts continue to debate the value of "clean coal"—this type of coal does burn more cleanly than typical coal, but causes environmental damage when it is extracted and requires a lot of energy to process into a cleaner fuel—while still generating toxic waste (for more, see Chapter 23).

Even though Samsø is small—about 26 kilometers (16 miles) long and only 7 kilometers (4 miles) at its widest—it has plenty of the natural resources that are becoming increasingly important sources of sustainable energy. A small firm put together an application for the contest, and the little island won.

↑ Samsø's offshore wind turbines provide much of the electricity used by the island. Since 2005, Samsø has produced more electricity than it uses and exports the extra to mainland Denmark and beyond to neighboring countries.

To become sustainable, Samsø would harness the power of one of its most plentiful natural resources.

A major part of Samsø's transformation was the installation of 11 onshore and 10 offshore wind turbines designed to harness the power of **wind energy**, or energy contained in the motion of air across Earth's surface. There is no shortage of wind on Samsø, and the powerful breezes turn huge blades (up to 40 meters [130 feet] in length) that are connected to a generator, converting mechanical energy into electricity.

Nine of the onshore turbines on Samsø were purchased collectively by groups of farmers who bought shares in their construction. "People on the island are personally invested in this," says Bernd Garbers, a German engineer who lives on the island and is a consultant for the firm that won the energy contest, Samsø Energy Academy. The machines produce 75% of the electricity used by the islanders.

Samsø uses about 500 billion kilojoules (kJ) of energy each year. A barrel of oil produces 6.15 million kJ, so Samsø

renewable energy Energy from sources that are replenished over short time scales or that are perpetually available.
sustainable energy source Energy sources that are renewable and have a low environmental impact.
wind energy Energy contained in the motion of air across Earth's surface.

Infographic **28.1** | **RENEWABLE ENERGY USE**

↘ According to the U.S. Energy Information Administration, as of 2010, renewable sources of energy contributed 3.46 trillion kilowatt-hours (kWh) of total energy production worldwide (about 18% of the total). That amount is projected to more than double to 7.97 trillion kWh by 2035—this will be 22.7% of the total at that time. Of renewable sources, biomass, mostly fuels like wood, charcoal, and animal waste, makes up the largest proportion, followed by energy produced by hydroelectric dams.

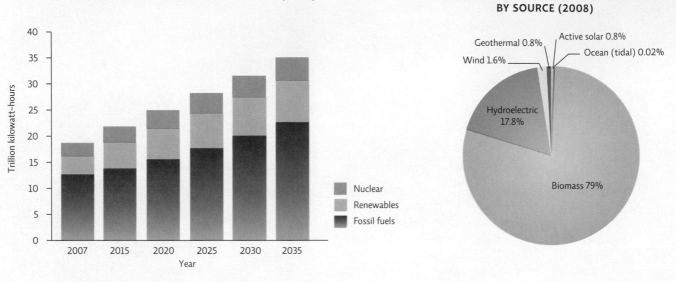

uses 81,300 *barrels of oil equivalents (BOE)* annually (that's 500 billion divided by 6.15 million). The entire human population uses somewhere between 50 and 70 billion BOE per year, and only about 18% of that energy comes from renewable resources. The United States ranks far below that average, fulfilling only about 7% of our energy needs through renewable energy. [INFOGRAPHIC 28.1]

Ideally, Samsø residents would have swapped their vehicles for cars that run on hydrogen or electricity, but those technologies were too expensive and not efficient enough to use on the island. Since Samsø would continue to rely on conventional fuels for transportation, engineers decided to install large offshore wind turbines to produce an equivalent amount of clean energy, offsetting uses by motor vehicles and boats. Islanders are working toward a gradual transition to electric vehicles and those powered by biofuels (see Chapter 29). The island's small size helps make all this feasible since daily "commutes" are short and miles traveled per person per day are lower than in most industrialized nations.

Far off the coast, the offshore turbines are especially efficient at generating energy because wind conditions are better at sea. In some especially windy locations, people

have built strategically placed *wind farms*, which can contain dozens of turbines. In 2010, the world's largest was located in Texas, the leading producer of wind energy in the United States. It has 627 wind turbines that generate enough electricity—781.5 megawatts (MW)—to power 230,000 homes per year. Larger wind farms are under construction, with one in Oregon slated to produce 845 MW. [INFOGRAPHIC 28.2]

But wind energy isn't perfect. First, even for a blustery location like Samsø, wind is intermittent—it stops and starts irregularly, not producing a steady stream of power. And wind turbines are not cheap. Each onshore turbine costs the Samsø islanders the equivalent of $1 million; offshore turbines rang up at $5 million apiece. Beyond cost, wind turbines can create noise, and some people see them as an eyesore. They can also have an impact on the local environment, threatening birds and bats, who are unable to nest near turbines or are killed by rotating turbine blades. As many as 40,000 birds are killed by wind turbines annually in the United States. While that may seem like a large number, it is far less than the number killed by flying into communication towers (40 million) or those killed by domestic cats (hundreds of millions). Still, to decrease the risk from windmills, engineers have

Rotating blades turn a shaft inside the turbine. The shaft is attached to a gear that rotates a higher-speed shaft on the generator.

SHAFT

GEAR BOX

GENERATOR

The spinning generator produces electrical current.

Wind turns the blades.

← Wind turbines can be large or small. All work by the same principle: Spinning blades turn a shaft inside a generator and produce electricity. New designs produce quieter windmills, addressing one of the criticisms of the technology.

↑ Top: Palm Springs, California: Wind turbines generate electricity for Southern California. Bottom: A solar and wind energy-powered house in San Francisco, California.

The scalloped edge of this windmill blade is a new design inspired by the fin of a humpback whale—an example of biomimicry. Founders of the company, WhalePower, have demonstrated that this shape makes the blade more aerodynamic and less likely to stall out at low speeds, allowing it to produce more power than traditional designs.

↑ A solar heating plant near Marup, Samsø Island. Changes in Samsø energy use mean that instead of importing electricity, the island exports it.

tried to avoid placing them in known migratory flight paths or close to areas frequented by birds of prey such as eagles.

The most popular source of sustainable energy is the one that powers the entire planet—the Sun.

Each year, Earth receives a staggering amount of energy from the Sun, more than 4 million exajoules (1 exajoule is 10^{18}, or a million trillion, joules). We use the Sun for many things—to warm our homes, heat our pools, and provide light—but new technologies allow more effective use, and even storage, of the Sun's power. In addition to wind power, the islanders on Samsø decided to tap directly into this amazing natural resource.

Solar energy is energy harnessed from the Sun in the form of heat or light. Wind power is actually an indirect form of solar energy. Wind results from the difference in temperature between different regions of Earth, such as the poles and the equator, causing air to move from cooler regions to warmer regions.

Solar energy can be used in two ways, through active or passive technologies. Around the countryside in Samsø, homes are dotted with **photovoltaic (PV) cells**, also called solar cells. PV cells are **active solar technologies** that convert solar energy directly into electricity. If just 4% of the world's deserts were covered in PV cells, it would supply all of the world's electricity needs. When the Samsø renewable energy project first began, locals were able to buy PV cells for their homes at a low price subsidized by the government.

The islanders also rely on **solar thermal systems**, another active technology that captures solar energy for heating. At the north of the island, rows and rows of solar collectors in a field face the sky, absorbing the Sun's rays and using that energy to heat a massive tank of water to 71°C (160°F). The hot water is then piped into 178 homes in the area for use in heating systems. Another 200 homes on the island, ones farther away from district heating plants, have individual solar collectors to capture solar energy for heat. [INFOGRAPHIC 28.3]

---

solar energy Energy harnessed from the Sun in the form of heat or light.
photovoltaic (PV) cells Also called solar cells, PV cells convert solar energy directly into electricity.
active solar technologies The use of mechanical equipment to capture, convert, and sometimes concentrate solar energy into a more usable form.
solar thermal system An active technology that captures solar energy for heating.

↑ At least 450 residents of Samsø own shares in the onshore turbines, and a roughly equal number own shares in those offshore. Residents are proud of their accomplishments in creating the first island in the world operating on 100% renewable energy.

Infographic 28.3 | **SAMSØ: THE ENERGY POSITIVE ISLAND**

↓ Samsø is pursuing several methods to help it produce enough renewable energy to meet all its needs.

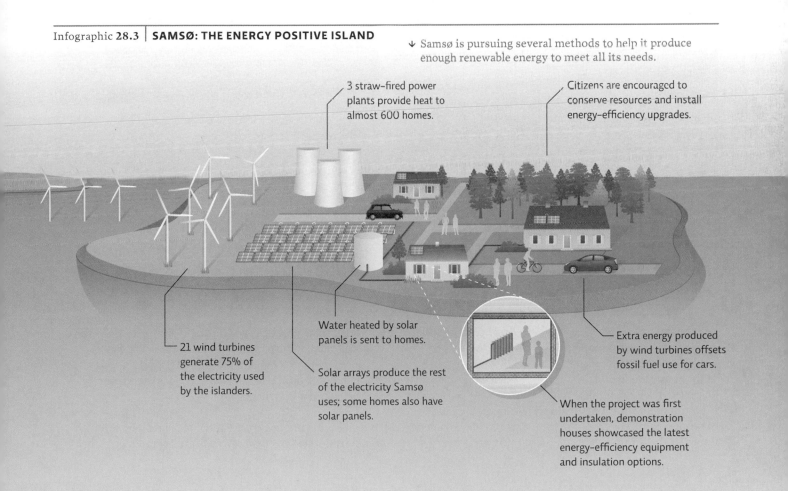

3 straw-fired power plants provide heat to almost 600 homes.

Citizens are encouraged to conserve resources and install energy-efficiency upgrades.

21 wind turbines generate 75% of the electricity used by the islanders.

Water heated by solar panels is sent to homes.

Solar arrays produce the rest of the electricity Samsø uses; some homes also have solar panels.

Extra energy produced by wind turbines offsets fossil fuel use for cars.

When the project was first undertaken, demonstration houses showcased the latest energy-efficiency equipment and insulation options.

Infographic 28.4 | **SOLAR ENERGY TECHNOLOGIES TAKE MANY FORMS**

↓ There are many ways to capture and use solar energy. Passive solar homes are constructed in a way that maximizes solar heating potential.

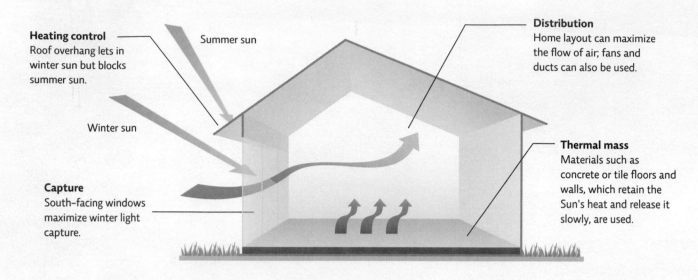

Heating control
Roof overhang lets in winter sun but blocks summer sun.

Summer sun

Winter sun

Capture
South-facing windows maximize winter light capture.

Distribution
Home layout can maximize the flow of air; fans and ducts can also be used.

Thermal mass
Materials such as concrete or tile floors and walls, which retain the Sun's heat and release it slowly, are used.

↓ Active solar technologies capture and convert solar energy for another use such as heating water or producing electricity.

↑ Solar hot-water heaters are the solar technology with the quickest payback for the average homeowner. Sunlight heats a fluid in pipes which then heats water in a tank to be used in the home.

↑ Photovoltaic panels capture sunlight and convert it to electricity that can be used by the homeowner or fed back to the grid and sold to the local utility.

Less expensive alternatives to PV cells or solar thermal systems are **passive solar technologies**. A greenhouse is a simple example of such a system: It captures heat without any electronic or mechanical assistance. Many energy-conscious homes are designed with passive energy in mind, incorporating strategically oriented windows to maximize sunlight in a room and dark-colored walls or floors to absorb that light and heat the home.

Because solar energy is conceptually simple, safe, and clean, with no noise or moving parts, it is the most popular member of the renewable energy club; today, thousands of buildings around the world are powered by PV cells. [INFOGRAPHIC 28.4]

But like wind energy, solar energy is plagued by intermittency and start-up costs. PV cells are becoming cheaper every day, but it can still take a home or business owner many years to recoup the cost of installation. Intermittency is a bigger problem. Sunlight is only available for half of each day, even less in places like Anchorage, Alaska, which see only a few hours of pale sunlight in the winter. Because of this, it is unlikely that any community will rely solely on solar power for its energy needs.

This is a hallmark of a sustainable energy future—because sustainable energy sources have their own strengths and weaknesses, no single source will likely meet the needs of any particular community, certainly not those of the entire world. But together, each can provide its own contribution. For example, solar is productive in daylight hours, whereas wind tends to blow harder at night; thus these two renewable sources complement each other in most places.

But in some regions, it makes more sense to tap other sources of renewable energy.

Energy that causes volcanos to erupt and warms hot springs can also heat our homes.

Unlike Samsø, some communities are fortunate enough to be located near sources of **geothermal energy**. A tremendous amount of heat is produced deep in Earth from radioactive decay of isotopes; temperature increases with depth and the Earth's core may be 5,000°C (9,000°F). It is the same heat that bubbles hot springs and causes geysers to erupt from the ground at Yellowstone National Park in Wyoming. Within the last century, technological advances have enabled us to tap into these vast resources for heat and electricity. "There has been a huge surge in the development of geothermal resources," says Wilfred Elders, co-chief scientist of an ambitious geothermal project known as the Iceland Deep Drilling Project (IDDP). "It is one of the most viable and economically attractive sources of renewable energy."

There are a wide variety of geothermal systems currently in use. **Geothermal heat pumps** (also called ground-source heat pumps) are used in more than half a million

passive solar technologies The capture of solar energy (heat or light) without any electronic or mechanical assistance.
geothermal energy The heat stored underground, contained in either rocks or fluids.
geothermal heat pump A system that actively moves heat from the underground into a house to warm it or removes heat from a house to cool it.
payback time The amount of time it would take to save enough money in operation costs to pay for the equipment.
geothermal power plants Power plants that use the heat of hydrothermal reservoirs to produce steam and turn turbines to generate electricity.

homes around the world. If you have ever been in a cave, you have probably been struck by the cool, constant temperatures there—always around 55°F, whether it's freezing outside or in the middle of a heat wave. No matter what the temperature above, underground remains a constant 55°F. Engineers bury fluid-filled pipes, bringing the fluid to that temperature. They then pump the fluid into homes, essentially providing people with year-round 55°F temperatures. In the winter, the home only needs to be warmed up from 55°F, rather than the outside colder temperatures, and in the summer, this system provides natural cooling. Such pumps are fairly expensive to install, yet have lower monthly energy bills than conventional heating and cooling systems. Areas with more extreme climates save enough money in monthly bills to offset the cost of installation—a metric known as **payback time**—in as little as 5 years.

> Because sustainable energy sources have their own strengths and weaknesses, no single source will likely meet the needs of any particular community.

On the other end of the spectrum are **geothermal power plants**. Generators above geothermal wells use steam released from hydrothermal reservoirs (hot springs) to spin turbines, producing electricity. Hydrothermal reservoirs are areas in Earth's crust that hold heat energy in the form of water, either as steam or liquid. Geysers in Northern California power the world's largest geothermal electrical system, generating more than 1,000 MW of energy—enough for a city as large as San Francisco. Geothermal power plants are reliable and efficient, but their potential is entirely dependent on location. Because drilling is expensive, only sites with enough heat to generate significant electricity are considered for development. These tend to be hot zones, locations like Iceland and the western United States, where the tectonic plates in Earth's crust pull away from or rub against each other, allowing the hot magma to flow upward through cracks in the rock to reach areas closer to the surface.

This type of renewable energy is becoming more popular: Between 2005 and 2010, use of geothermal power worldwide increased by 20%, according to the International Geothermal Association, and is expected to grow even further through 2015. [INFOGRAPHIC 28.5]

Harnessing such a powerful heat source can be difficult, however. In early 2007, an earthquake of magnitude 3.4 on the Richter scale shook the town of Basel, Switzerland.

The quake was attributed to a geothermal mining project in the area, and the project was halted. The Iceland drilling project fell into serious problems when the drill repeatedly got stuck 2 kilometers (1.3 miles) deep into a volcanic crater, leading to months of jammed drill bits and broken pipes. The ultimate goal of the IDDP: Drill into a high-temperature geothermal system (the Krafla volcano, in this case), penetrate twice as deep as conventional geothermal wells (which are usually less than 3,000 meters [10,000 feet] deep), and reach *supercritical fluids,* substances with temperatures above 700°F that are confined by such intense pressure they exist in a limbo state between liquid and gas. The resulting superheated steam could generate 10 times the electricity of most geothermal wells.

Power can be generated from moving water—but at potentially significant environmental and social costs.

The island of Samsø is shaped curiously like a violin. It is also surrounded by one of the most significant sources of renewable energy worldwide: water.

Humans have harnessed the power of falling water for thousands of years, ever since early civilizations used watermills to grind grain into flour. Today, energy produced from moving water—known as **hydropower**—supplies more electricity to the population than any other single renewable resource. Approximately 10% of electricity used in the United States and 20% around the world is generated from hydropower. Large-scale

---

Infographic 28.5 | **GEOTHERMAL ENERGY CAN BE HARNESSED IN A VARIETY OF WAYS**

↓ The high temperatures found underground in some regions can be tapped to generate electricity. Geothermal heat can also be piped directly to communities to provide heat or hot water. Geothermal energy can also be tapped using ground-source heat pumps, reducing the cost to heat and cool individual buildings.

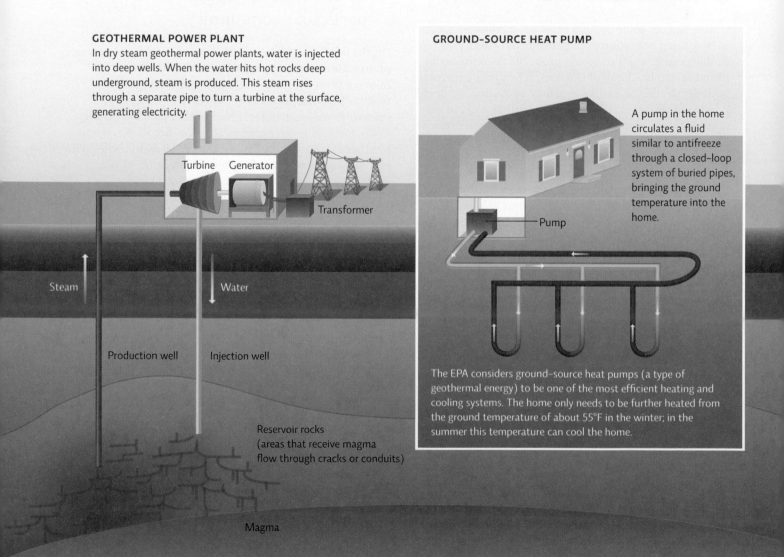

GEOTHERMAL POWER PLANT
In dry steam geothermal power plants, water is injected into deep wells. When the water hits hot rocks deep underground, steam is produced. This steam rises through a separate pipe to turn a turbine at the surface, generating electricity.

Turbine   Generator

Transformer

Steam   Water

Production well   Injection well

Reservoir rocks
(areas that receive magma
flow through cracks or conduits)

Magma

GROUND-SOURCE HEAT PUMP

A pump in the home circulates a fluid similar to antifreeze through a closed-loop system of buried pipes, bringing the ground temperature into the home.

Pump

The EPA considers ground-source heat pumps (a type of geothermal energy) to be one of the most efficient heating and cooling systems. The home only needs to be further heated from the ground temperature of about 55°F in the winter; in the summer this temperature can cool the home.

↑ The geothermal heat produced from the Reykjanes Power Plant in Iceland is mostly used to heat freshwater, which heats 89% of the houses in Iceland.

↑ Installation of a geothermal heating system in a residence. Closed-loop tubing can be installed horizontally in shallow trenches or vertically in much deeper wells. Here, a horizontal loop is being installed.

hydroelectric power plants at giant dams are the source of most of that power.

Like wind energy, hydropower is an indirect form of solar energy. The downward flow of water from mountaintop to ocean is a consequence of the water cycle, a process of evaporation and condensation driven by the Sun's heat. (See Chapter 17 for more information about the water cycle.) Hydropower is abundant, clean, and does not typically produce greenhouse gases.

There are a variety of ways to harness the energy of moving water—from ocean waves and the tides, to capturing energy released from variations in ocean temperatures—but the most common way to generate hydropower is with dams. The Grand Coulee Dam on the Columbia River in Washington State is one of the largest concrete structures on the planet and the third-largest dam. The 12 million cubic yards of concrete that make up the dam could form a 1.2-meter-wide (4-meter-wide) sidewalk wrapped twice around the equator. The dam is the largest electrical power producer in the United States, with a total generating capacity of 6,809 MW. The reservoir created by the dam, an artificial lake that pools behind the structure (called an impoundment), stretches some 240 kilometers (150 miles) back to the Canadian border.

The buildup of water in the reservoir creates an enormous amount of pressure; water diverted from the top of the dam flows through long pipes, called penstocks, to turbines below. As the water rushes past the curved blades

of the turbines, they spin and generate electricity. Three power plants at the Grand Coulee contain 24 generators, which produce enough electricity to power two cities the size of Seattle.

But hydropower from large dams is far from an ideal resource. It wasn't a real option on Samsø, for instance, because the island lacks high mountain peaks whose run-off would feed large rivers. So geography is a key factor. Importantly, hydroelectric systems are also responsible for the loss of major habitats and the displacement of tens of millions of people around the globe. It is the most debated and contentious of any renewable resource.

This is nowhere more apparent than at the Grand Coulee. "The Grand Coulee has had a huge impact for good and for ill," says Michael Garrity, Washington Conservation Director for American Rivers, an organization dedicated to protecting America's rivers. As the final loads of concrete were placed and the first generator at Grand Coulee was switched on in 1941, crowds gathered on the hill above the dam to marvel at the "birth of one of the world's greatest waterfalls," a local paper proclaimed. But nearby, thousands of Native Americans watched in horror as habitats, homes, and livelihoods were irreversibly lost under the pool of water engulfing the land behind the dam.

For nearby Native American communities, the Grand Coulee was a humanmade disaster that sacrificed not only land, but a staple of their economy and culture—salmon. Salmon are migratory fish: They hatch in the freshwater tributaries of the Columbia, migrate downstream into the Pacific Ocean to spend most their lives in salt water, then return to the freshwater environment of their origin

---

hydropower The energy produced from moving water.

↑ A run-of-the-river hydroelectricity station on an estuary of the Hudson River, New York. In such stations, the river still flows freely and is not impounded to produce a lake.

to spawn. But the Grand Coulee Dam blocked the way; salmon and trout runs upstream of the dam were completely wiped out.

Not all hydropower systems have the same impact as a giant dam. A *run-of-the-river hydroelectric* system, often a low stone or concrete wall, doesn't block the water; it merely directs some of the flowing water past a turbine and thus generates electricity from the natural flow and elevation drop of a river, which is less disruptive to the river ecosystem. But energy production is dependent on the flow of a river at any particular time, so this system is only suitable for rivers with dependable flow rates year-round. Such systems are useful in remote areas where electricity can be costly, and they are good candidates to supplement solar power systems, since annual periods of high sunlight often have low water flow and vice versa.

Garrity and other conservation groups still hope that new technologies will someday make existing dams like the Grand Coulee more eco-friendly. "We are advocates for harnessing resources from our rivers," he says, "but in a way that's sustainable."

The true cost of various energy technologies can be difficult to estimate.

One recurring theme with renewable energy technologies is that they are more expensive than traditional (fossil fuel) methods. Even though sunlight, wind, and water

are free, constructing and installing the solar cells, wind turbines, and hydroelectric systems needed to harness their energy are not.

But perhaps we should not be asking why renewables are so expensive, but instead asking why fossil fuels are so cheap. We need to consider *all* the costs—environmental, social, and economic—of these technologies to fairly compare them. Meaning, even if the costs to construct and install a technology are high, those costs are offset by the lower environmental costs of the sustainable technology. This makes comparisons between the energy technologies difficult. Various studies have tried, and so far agree on several points: Fossil fuels, led by coal, have the highest external (environmental) costs. Purely from an economic standpoint, solar energy is the most expensive form of energy. But when external costs are added to production costs to estimate a true cost per kilowatt-hour, wind leads the way as the least expensive method.

It was this logic that helped convince the people of Samsø to personally invest in the ambitious project to convert to 100% sustainable energy. Such a project would not be possible everywhere, says Hermansen. Indeed, he wasn't even sure at first it would succeed, because it needed the buy-in of every one of the island's 4,000-plus residents—literally. This is the only way the project would work, he and others believed—if it was owned by the people. For years, Hermansen and others had to go practically door to door, trying to convince everyone to become part owners in the island's sustainable energy infrastructure. "If we didn't own these wind turbines and other sustainable energy projects on Samsø, a big power company would," says Hermansen. "Eventually, people will get their money back."

At first, the islanders were skeptical. "People were a little scared in the beginning. They were afraid the changes would disturb the landscape and affect tourism," says Garbers. But attitudes slowly changed, thanks largely to Hermansen's efforts.

Hermansen was born on Samsø and grew up there on a family farm. He was an environmental studies teacher with an enthusiasm for all things renewable when he heard about the contest, and it wasn't long before he became the project's first employee. Through extensive community meetings and seminars, he rallied the islanders to the cause. "We are lucky to have him," says Garbers. "If it had been some guy from the mainland promoting the project, maybe the islanders wouldn't have believed him. But Søren was local." This constellation of forces—a small, close-knit community open to the project, community members who personally invested in

↑ The Grand Coulee Dam, located on the Columbia River in central Washington, is the largest concrete structure in the United States. In addition to producing up to 6.5 million kilowatts of power, the dam irrigates more than half a million acres of farmland and provides wildlife and recreation areas. However, its construction eradicated salmon runs and inundated Native American villages and ancient burial sites.

the venture, led by a local individual with a passion for sustainability—helped Samsø achieve its goal.

By now, most of Samsø's residents have invested, starting at around $600 for 1 share in a wind turbine. The logic, says Hermansen, is that all residents share the profits—in a good wind year, the island sells its electricity, and people who purchased 10 shares receive 10 shares' worth of profits. He estimates that each Samsø resident has invested an average of $20,000 over the last 10 years in revolutionizing electricity on the island.

Meeting our energy needs with renewable sources becomes more likely when we pair them with energy conservation measures.

Even though the people of Samsø saw the importance of investing in sustainable energy, they realized that one of the best ways to help achieve energy independence would be to simply use less of it. As energy advisors like to say, *the greenest kilowatt is the one you never use.* Luckily, there are lots of ways to reduce electricity use right now. [INFOGRAPHIC 28.6]

For instance, residents were encouraged to focus on energy efficiency, relying on technology that uses less energy for a given output. In Samsø's cold climate, one of the first conservation efforts revolved around heating homes, mainly by transitioning away from electrical heaters. Energy home audits pinpointed other steps individuals could take, such as using straw, wood chips, and other sources of biomass for heat. Gas stoves were replaced by more efficient electric stoves; people began heating water using electricity generated by the wind turbines or solar panels. Residents were encouraged to apply for grants toward upgrades that made their homes more energy efficient. Similar programs exist in the United States. (See Chapter 32 for more ways consumers can cut energy use.)

Infographic 28.6 | **SAVING ENERGY**

↓ Energy conservation is about wise use: using less and using energy more efficiently. Many of the steps below can be taken immediately, with no investment; others require money and time to implement. Whatever you can do will not only reduce your energy use, it will reduce your energy bill and the pollution generated from producing that energy.

No-cost ways to save energy:	Steps that cost money but have quick payback times:	More expensive options with payback times of greater than 5 years:
• Turn off your computer and monitor when not in use.	• The first step is to have an energy assessment performed to identify where your home is energy inefficient (go to energysavers.gov for more information).	• Replace older windows with double-pane windows suitable for your area—a variety of low emissivity (low-e) window products restrict loss of heating or cooling while allowing in plenty of light.
• Plug home electronics, such as TVs and DVD players, into power strips; turn the power strips off when the equipment is not in use.	• Install attic and wall insulation to meet recommendations for your area.	
• Lower the thermostat on your hot-water heater to 120°F; turn it off if you will be away for several days.	• Weatherstrip and caulk doors and window frames; insulate electrical outlets on outside walls with inexpensive foam cutouts. Check heating ducts for leaks and seal if needed.	• Replace the heating, ventilation, and air cooling (HVAC) system with an energy-efficient model; consider a ground-source heat pump.
• In the winter, open curtains on south-facing windows to let in heat and light during the day; close them at night to reduce heat loss.	• Insulate hot-water pipes and the hot-water heater, especially if the hot-water heater is located in an unheated area of your home.	• Install an on-demand (tankless) hot water heater; this saves energy because water is not constantly being heated, only to cool back down, requiring that it be heated again.
• Be sure to close the fireplace damper when not in use to prevent loss of heat.	• Install a programmable thermostat—you can program it to automatically adjust the temperature when you are not at home or are sleeping.	
• Take short showers instead of baths to reduce hot water use.	• Look for the Energy Star label on home appliances and light bulbs. Energy Star products meet strict efficiency guidelines set by the U.S. Department of Energy and the EPA.	
• Wash only full loads of dishes and clothes; wash clothes with cold water.	• Upgrade to a front-loading washing machine that uses less energy (and less water—which also saves energy).	

Another basic approach to using less energy relies on **conservation**—making choices that result in less energy consumption. Much of conservation involves changing people's behavior; since lighting typically accounts for about 20% of the average home's electric bill, simply turning off the lights when you leave a room can make a big difference. Other options include studying in a room with ample natural light instead of a dark corner that requires artificial light, and lighting your home with compact fluorescent lights (CFLs) or LEDs (light-emitting diodes), which consume less electricity. [INFOGRAPHIC 28.7]

In just a few years, the island of Samsø has transformed itself, capturing the attention of the United States, Japan, and other countries, impressed by how a small farming community became the most energy efficient place on Earth. By 2005, the island was producing more energy from renewable resources than it was using. This transition has the added benefit of cleaner air—Samsø has reduced its nitrogen and sulfur air pollution by 71% and 41%, respectively, and its CO_2 emissions have fallen to below zero, thanks to the fact that the windmills offset more CO_2 than their vehicles emit.

Much of that success stems from community members working together to solve their own local problems, says Hermansen. As he told *Time* magazine in 2008, "People say: 'Think globally and act locally.' But I say you have to think locally and act locally, and the rest will take care of itself."◉

conservation Efforts that reduce waste and increase efficient use of resources.

Research article referenced in this chapter:
Holm, A. *et al.* 2010. *Geothermal Energy International Market Update*. Geothermal Energy Association.

Infographic **28.7** | **ENERGY EFFICIENCY**

↓ The production of electricity from coal is inefficient; most of the embodied energy in the coal is lost as heat. Additional energy is lost transmitting it to your home, so that only about 36% of the original energy from coal arrives at your home in the form of electricity. Making use of energy-efficient devices can reduce overall use of energy.

CONSIDER LIGHT BULBS

A 60-watt incandescent light bulb generates mostly heat and is a very inefficient way to produce light. On the other hand, a 13-watt compact fluorescent light bulb (CFL) produces less heat but the same amount of light as a traditional bulb. Here we compare the two after 8,000 hours of use (the typical lifespan of 1 CFL or 10 incandescent bulbs). Though CFLs actually contain mercury, if both bulbs are illuminated with energy from a coal-fired power plant, using the incandescent would release much more mercury to the environment (and the CFL will only release its 0.4 mg of mercury if it is broken).

ENERGY USED:
(kilowatts)

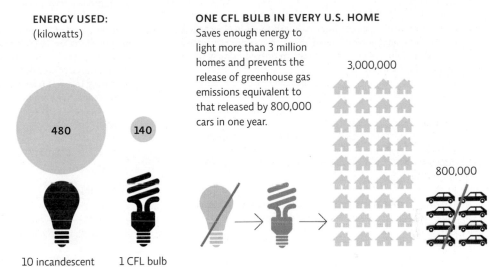

480

140

10 incandescent bulbs

1 CFL bulb

ONE CFL BULB IN EVERY U.S. HOME

Saves enough energy to light more than 3 million homes and prevents the release of greenhouse gas emissions equivalent to that released by 800,000 cars in one year.

3,000,000

800,000

BRING IT HOME

❯ PERSONAL CHOICES THAT HELP

A key component to developing a sustainable society is the use of renewable energies. Renewable energy decreases harmful impacts of mining waste associated with nonrenewable energy, plus reduces the amount of air and water pollution produced when the fuel is processed and burned. The efficiency and availability of renewable energy is rapidly increasing because of the development of new trends and technologies.

Individual Steps
→ Contact your energy provider to see if you can purchase a percentage of your energy from a renewable source.
→ Regardless of your home's energy source, make sure you are using energy efficiently. Review Infographic 28.6 and visit energysavers.gov for more ideas.

→ If you have an iPhone, download the PVme app to see how many solar panels you would need to meet your household energy needs.
→ A major barrier to solar energy for many people is the cost. For information on tax incentives, rebates, and other programs that make using renewable energy easier, check out dsireusa.org or look into the growing trend of leasing solar panels.

Group Action
→ Invest in providing solar energy to lower-income communities through SolarMosaic.com.

Policy Change
→ Contact your state and federal legislators and ask them to support or sponsor legislation that provides financial

incentives for the purchase of renewable technologies.

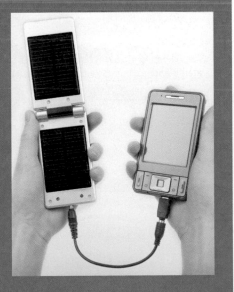

UNDERSTANDING THE ISSUE

CHECK YOUR UNDERSTANDING

1. **Which of the following is a source of renewable energy?**
   a. Natural gas
   b. Wind
   c. Oil
   d. Coal

2. **As a member of the city council in a small town in New Mexico, you must vote on a new source of energy for your growing community. What should be the first step in determining where the energy will come from?**
   a. Rely on historical practices and build a new coal-fired power plant.
   b. Pass a new law that requires all buildings to have solar panels installed.
   c. Conduct an assessment of available energy sources, renewable and nonrenewable.
   d. Provide citizens a tax subsidy for improving the energy efficiency of their homes.

3. **Which of the following renewable energy sources contributes the largest proportion of energy worldwide?**
   a. Wind
   b. Water
   c. Solar
   d. Biomass

4. **Which of the following does the EPA consider to be one of the most efficient cooling and heating systems?**
   a. Ground-source heat pumps
   b. Passive solar
   c. Solar roofing tiles
   d. Electric baseboard heat

5. **What type of energy might you recommend to someone who lives in a remote area with lots of small streams, a rainy season, and a dry season?**
   a. Geothermal heat pump
   b. Large-scale hydroelectric
   c. Solar and run-of-the-river hydropower
   d. Traditional power plant

6. **In some parts of the world, women (and children) walk hours in each direction to collect wood as fuel for heating homes and preparing meals. The long walk is necessary because there is no closer source of wood; it has all been harvested. Which of the following technologies do you think would help ease the burden on women AND benefit the environment?**
   a. Passive solar heating
   b. Solar roofing tiles
   c. Solar cookers
   d. Solar hot-water heaters

WORK WITH IDEAS

1. Describe the characteristics of the five sustainable energy sources presented here. Why are all sources not practical for every location?

2. Compare and contrast the costs and benefits of energy generated from wind, water, sun, geothermal heat, and biological matter.

3. Explain the process of harvesting geothermal energy from supercritical fluids as if you are talking to an older relative unfamiliar with the method. Anticipate in your description questions your relative might have, such as "Why can't we do that here?"

4. Describe the various types of hydropower. In your description, include information about scale, cost, availability, and environmental impact.

ANALYZING THE SCIENCE

A regional power company has proposed building a wind farm in your community. There is plenty of wind, so many people in the community think this is a good idea. Some, however, have cited the environmental impacts of windmills, especially bird and bat mortality. Examine the following graphs and use them to answer the questions that follow.

INTERPRETATION

1. In one or two sentences, summarize the information provided in each graph or table.

2. For each graph or table, provide a title and a legend. What is the take-home message for each graph or table?

ADVANCE YOUR THINKING

3. Based on the data and where you live, what do you think the impact of a wind farm in your area would be on bird and bat mortality? If a wind farm were installed, what recommendations could you make to reduce bat mortality? Make sure to discuss your answer in relation to each of the graphs/tables.

Graph 1

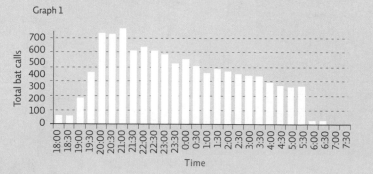

Graph 2

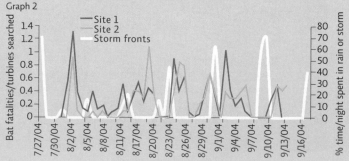

Graph 3

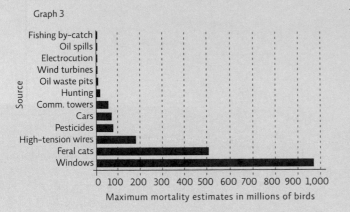

Table 1

Date	Buffalo Ridge, MN	Vansycle, OR	Buffalo Mtn., TN	Stateline OR/WA	Foote Creek Rm, WY	Total bat carcasses	%
May 1–15	0	0	0	–	0	0	0.0
May 16–31	1	0	0	–	1	2	0.4
June 1–15	0	0	0	–	1	1	0.2
June 16–30	3	0	0	–	2	5	0.9
July 1–15	9	0	9	0	2	13	2.8
July 16–31	88	0	0	0	26	110	22.2
Aug 1–15	127	0	10	0	19	151	28.2
Aug 16–31	75	4	0	11	33	128	23.9
Sep 1–15	52	4	8	0	21	81	13.1
Sep 16–30	4	2	–	10	0	20	3.7
Oct 1–15	1	0	0	8	2	11	2.1
Oct 16–31	2	0	0	0	0	2	0.4
Nov 1–15	0	0	0	1	0	1	0.2

EVALUATING NEW INFORMATION

Renewable energy resources have become a priority around the world as we grapple with the cost and limited supply of fossil fuels, not to mention the effect on the environment resulting from their use. How do we know which energy source is appropriate to invest in, with both our time and money? Do we have to give up our current standard of living if we switch to an alternative energy source?

Go to the EPA's Clean Energy website (www.epa.gov/cleanenergy/energy-and-you/how-clean.html). Read the information on the home page, then enter your zip code in the box provided. You will then be asked to select the power company that provides energy to your community (from a drop-down menu). Choose the correct company and click "Next." Two graphs will appear—one that compares the fuel mix (energy sources) of energy production in your community to the national average, and one that compares emissions of specific pollutants in your area to the national average. Read the information provided in the "Make a Difference" section under the graphs.

Evaluate the website and work with the information to answer the following questions:

1. Is this a reliable information source? Does it have a clear and transparent agenda?

   a. Who runs this website? Do this organization's credentials make it reliable/unreliable? Explain.

   b. Who are the authors? What are their credentials? Do they have the scientific background and expertise to lend credibility to the website?

2. In your own words, briefly summarize what you believe are the most important pros and cons of alternative energy resources based on your readings. Do any of these also apply to traditional fossil fuels? Would you be willing to invest your money in research and development of alternative energy sources? Why or why not?

3. What fuel provides most of the energy in your region? Were you aware of this, or are you surprised?

4. How does your community differ from the nation in terms of both fuel sources and emissions?

5. Based on what you know about the weather and geography in your area, what types of alternative renewable fuel sources would you recommend? If you need help with local weather statistics, you can do an Internet search for your region.

MAKING CONNECTIONS

A NEW ENERGY MODEL FOR CENTRAL AMERICA?

Background: The Central American Arc is a chain of volcanoes extending throughout Central America. It is caused by a subduction zone, and massive eruptions have occurred in the past. The countries that the volcanoes are in are all developing nations that are trying to increase their standard of living, which includes providing electricity to all citizens. However, these countries lack fossil fuel resources, so they must import them. In addition, they lack the infrastructure to provide electricity in remote areas.

Case: You are a member of a philanthropic organization that provides expert advice and start-up funds to localities (large and small) in Central America to improve energy facilities and infrastructure. The government of Nicaragua has approached your organization for advice about which types of energy to invest in. Before your organization can make reliable recommendations, you need to gather information about possible sources of energy.

1. Research possible energy fuel sources for Nicaragua. As you conduct your research, answer the following questions:
   a. What are the typical weather patterns?
   b. What is the topography of the country like?
   c. Where do people live?
   d. What types of fuel are available?
   e. What types of fuels can the country afford, now and in the future?

2. Answer these same questions for the country of Iceland.

3. Compare your results for the two countries. Based on your research and comparisons, write a brief report to your organization that details your findings. In your report, make sure to include answers to the following questions:
   a. What recommendations should your organization make to the Nicaraguan government for both short-term and long-term solutions to the problem of energy resources?
   b. Do your recommendations apply to the country as a whole, or do you have region-specific recommendations?

↑ The Cedar Creek Ecosystem Science Reserve in Minnesota has established more than 1,100 long-term experimental plots and 2,300 permanent observational plots distributed across 22 old fields.

→ University of Minnesota ecologist David Tilman (right) and the Nature Conservancy's Joe Fargione (left) study how clearing land for crops releases CO_2.

The rain stopped, and the plants began to die. In 1987, a massive drought struck Minnesota. At the Cedar Creek Ecosystem Science Reserve, David Tilman helplessly watched the grasslands he had been cultivating for more than 5 years—part of a project to test how different amounts of nitrogen and other resources affect growth—wither away.

But while some areas thinned out and became barren, others kept the auburn and green of healthy prairie grasslands. For some reason, certain plots were less affected by the drought. Tilman, a bow tie-sporting ecologist at the University of Minnesota, decided to find out why. Little did he know, his simple plan to study grass in Minnesota would unexpectedly lead him to a new and potentially better type of **biofuel**—a substance that provides the energy needed to power engines—made from a mixture of prairie grasses.

"We weren't thinking about biofuels at all," recalls Tilman. "When you have a new idea, you start imagining what might happen, but you never know what will really happen until you try the experiment."

Biofuels are a potentially important alternative to fossil fuels.

The term "biofuels" makes headlines daily, but the concept is not new—people have been deriving energy from **biomass** for millennia, for instance through burning corn stalks and other crop residue for heat.

⦿WHERE IS THE CEDAR CREEK RESERVE?

THE CEDAR CREEK RESERVE

MINNEAPOLIS

MN   WI

Currently, the world relies primarily on fossil fuels as the main source of energy—but burning them has had a profoundly negative impact on our environment (see Chapters 23–26 for more). And with increasing demand for energy, continued dependence on these nonrenewable fuels is no longer a viable option. Communities are experimenting with ways to produce electricity without fossil fuels (see Chapter 28) and with sustainable alternatives to power motor vehicles.

Biofuels are derived from material from living or recently living organisms (biomass) or their by-products. Fuels derived from biomass are considered renewable since the raw materials can be naturally replenished at least as quickly as they are used. Biofuels also have the advantage of being locally produced and thus they reduce a nation's dependence on other countries for energy.

But biofuels are not a simple solution to our energy problems. For one, their raw materials or **feedstocks**—crops, animal waste, and wood products—require resources to grow. Furthermore, to function as a fuel, these materials must be converted into another form, a process that uses energy.

Today, there is a massive effort to render that process more sustainable. Biofuels currently provide about 4% of the energy used in the United States, more than any other type of renewable resource. According to the U.S. Department of Energy, global biofuel supply and demand is expected to grow more than sixfold, from 12 billion to 83 billion gallons over the next two decades.

But, as Tilman was about to learn, replacing fossil fuels—especially petroleum—is no easy task.

---

biofuels Solids, liquids, or gases that produce energy from biological material.
biomass Material from living or recently living organisms or their by-products.
feedstock Biomass sources used to make biofuels.

Infographic **29.1** | **BIOFUEL SOURCES**

↓ Biomass (material of biological origin) can be burned directly or converted to other forms of fuel, such as bioethanol or biodiesel. Some crops are specifically grown just for fuel production (fuel crops) but waste material can also be used. As long as the feedstocks (biomass sources used in production) are grown and harvested sustainably, biofuels are a sustainable resource.

DIRECT BIOMASS ENERGY

A variety of materials, such as wood, dried manure, and crop waste, can be burned directly. This is the most common energy source in less developed regions of the world but it can lead to problems such as deforestation and air pollution.

Some biomass sources, such as firewood, are intentionally harvested as fuel.

Waste biomass such as cornstalks can be collected and burned; manure can be dried and then burned.

Composite "briquettes" are made from compressing flammable material such as paper, grass, forest, and agricultural waste chips along with shredded plastic (which increases flammability) and are a promising alternative to charcoal.

Biofuels can come from unexpected sources.

Back in Minnesota, Tilman set about testing to see why half his grasslands seemed resistant to drought. After trying "a whole variety of things," Tilman and his team came to a simple conclusion: The plants growing in the most diverse areas—those with up to 20 species of grasses and flowers—were the most able to weather periods of drought. Tilman was baffled. Why should plants competing with each other be more successful during a drought?

He and his team of ecologists decided to study the question empirically. They spread out under a bright blue sky at the Cedar Creek Reserve in Minnesota and planted a smorgasbord of seeds among 152 plots, each plot roughly the size of half a tennis court. The plants began to sprout: gold bunches of junegrass, stately spires of dark blue lupine, bright yellow tufts of goldenrod, and tall, spiky western wheatgrass. Some plots had only 1 species, some 2, others up to 16. Each plant was a **perennial**, meaning it

would grow back every year, differing from **annual** plants such as corn and soybeans, which must be replanted each year. The team used no fertilizer and only watered plots in the first few weeks after planting.

One of the species was switchgrass, a common North American perennial that grows thick and tall (up to 12 feet), with roots penetrating 10 feet into the soil. As Tilman watched his plants grow, he was not the only person thinking about switchgrass. The crop was rapidly becoming a household name for another reason—as a potential source of biofuels.

Often, biofuels are derived from **fuel crops**, those specifically grown to make biofuels. For instance, **bioethanol** can be derived from crops such as corn and sugarcane during a process of fermentation and distillation, similar to that used to make alcoholic beverages. **Biodiesel**, a different kind of biofuel produced from oils, is most often made from high-oil crops like soybeans. [INFOGRAPHIC 29.1]

INDIRECT BIOMASS ENERGY

Biomass can also be converted to other forms and then burned. This produces more energy–rich fuels that are less bulky than the original biomass; these fuels also burn more cleanly (less particulate matter) than the original forms and can fuel engines that currently run on gasoline or diesel fuel.

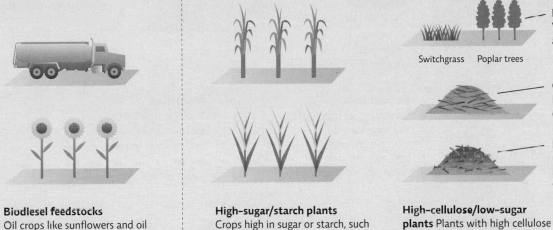

BIODIESEL

Biodiesel feedstocks
Oil crops like sunflowers and oil waste such as restaurant fry grease can be used to make biodiesel.

High–sugar/starch plants
Crops high in sugar or starch, such as corn and sugar cane, are easily fermented to produce ethanol.

BIOETHANOL

Switchgrass   Poplar trees

Fuel crops
Varieties grown just as fuel feedstock

Crop waste
Plant material left over after harvest

Forestry waste
Deadwood or material left over from processing

High–cellulose/low–sugar plants Plants with high cellulose content like grasses, trees, and the nonedible parts of crops are used to make cellulosic ethanol.

Today, ethanol is primarily used as an additive, mixed with gasoline to produce a cleaner burning fuel. It's also more corrosive than gasoline, so mixtures at the gas pump typically contain no more than 10% ethanol (E10 fuels). Some so-called "flex fuel" vehicles are equipped to use up to 85% ethanol mixtures, and thousands of these vehicles are on the road today. Because ethanol has only about two-thirds of the energy found in gasoline, however, it is not suitable for high-demand engines like those in airplanes and large trucks.

There's another convenient source of biofuels—*biowaste*, often in the form of organic leftovers such as crop residues, garbage, or manure. For instance, even though ethanol usually comes from corn and sugarcane, it can also be produced from the stalks, husks, and leaves of plants left over after a crop has been harvested. Any kind of organic material has the potential to be converted to liquid biofuels such as ethanol, or to a gaseous product similar to natural gas known simply as biogas.

The idea of converting waste to energy is appealing—it has the dual benefit of both producing energy and dealing with waste at the same time. For instance, communities are struggling to deal with methane released from decomposing landfill waste, which acts as a potent greenhouse gas, each molecule trapping 25 times more heat in the atmosphere than one of carbon dioxide. Some communities are now working to capture that gas for energy use. Of 2,300 municipal solid waste landfills in the United States, more than 450 have landfill gas projects, according to the Environmental Protection Agency (EPA).

perennial Plants that live for more than a year, growing and producing seed year after year.
annual Plants that live for a year, produce seed, and then die.
fuel crops Crops specifically grown to be used to produce biofuels.
bioethanol An alcohol fuel made from crops like corn and sugarcane in a process of fermentation and distillation.
biodiesel A liquid fuel made from vegetable oil, animal fats, or waste oil that can be used directly in diesel internal combustion machine.

Infographic 29.2 | **WASTE TO ENERGY**

↓ Garbage and agricultural, industrial, and food waste can be used to create a variety of biofuels. Biodiesel can be produced from leftover used restaurant fryer oil, or any oil waste product, such as waste from slaughterhouses. It also has the advantage of being a rather simple production process suitable for large or small scale. Small and midsized operations like Clean Energy Biofuels, located in Atlanta, Georgia, and Knoxville, Tennessee, serve the local community by picking up waste oil from local restaurants, converting it to biodiesel, and then selling it to local users.

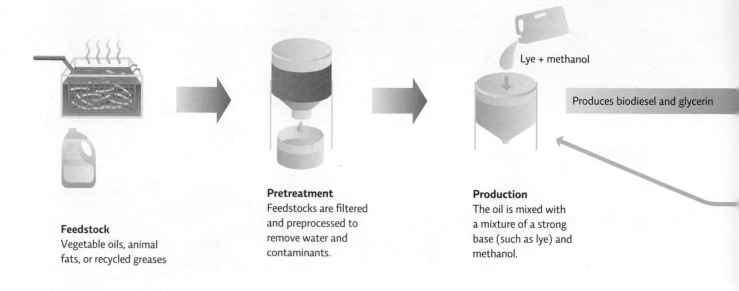

Lye + methanol

Produces biodiesel and glycerin

Feedstock
Vegetable oils, animal fats, or recycled greases

Pretreatment
Feedstocks are filtered and preprocessed to remove water and contaminants.

Production
The oil is mixed with a mixture of a strong base (such as lye) and methanol.

↑ Collection of used vegetable oil from a restaurant to recycle into biodiesel fuel.

Another type of biofuel that can easily be made from waste is biodiesel. This type of biofuel is more energy-rich than ethanol, and can be used directly in diesel engines, with little or no modification. And it can derive from animal fats or waste oil. Indeed, a new industry is emerging around the disposal of used fryer oil from restaurants; once a liability that restaurants paid to have hauled away, this oil can be picked up free of charge by biodiesel entrepreneurs who turn it into a fuel that can be sold to local diesel users.

At the moment, biodiesel comprises only a small portion of the biofuels produced, and cannot be used in gasoline engines. But biodiesel buses and recycling trucks are on the road all around the United States. Currently, biodiesel is the only biofuel to be certified by the EPA, having passed the safety tests required by the Clean Air Act.

Biodiesel can also be produced on a small scale. Across the United States, communities are taking matters into their own hands and forming small biodiesel cooperatives that produce relatively small amounts of biofuel for personal use. In Burlington, North Carolina, T-shirt manufacturer Eric Henry and his friends collect used vegetable oil from 12 local restaurants and pass it through a small biodiesel processor behind Henry's T-shirt plant. "We do it for interests beyond saving money," says Henry in a soft Southern twang. "We do it for environmental reasons, for air quality, and for the local community." [INFOGRAPHIC 29.2]

People have spent the last 100 years trying to use biomass to power engines: Henry Ford's first Model T, rolled out in 1908, was designed to use ethanol as fuel. But it wasn't until the second half of the 20th century that biofuels became big business—and more recently, a hotbed of controversy.

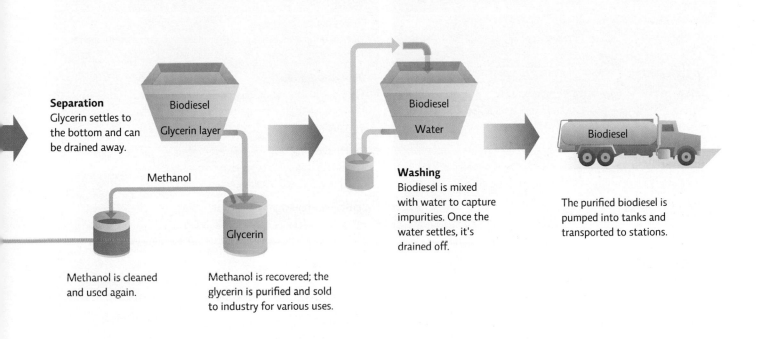

Separation
Glycerin settles to the bottom and can be drained away.

Methanol

Biodiesel

Glycerin layer

Glycerin

Washing
Biodiesel is mixed with water to capture impurities. Once the water settles, it's drained off.

Biodiesel

Water

Biodiesel

Methanol is cleaned and used again.

Methanol is recovered; the glycerin is purified and sold to industry for various uses.

The purified biodiesel is pumped into tanks and transported to stations.

Turning grass into gas is less environmentally friendly than it sounds.

Despite the attraction of putting waste to work, today most biofuels are made from crops grown specifically for energy use. Biodiesel, for instance, typically stems from crops with a high oil content, such as soybeans or rapeseed (canola). For decades, sugar cane has served as the primary source of ethanol in Brazil. But in the United States, corn is the fuel crop of choice.

In the mid-1970s, fuel shortages revived public interest in alternatives to fill cars, fuel stoves, and light homes. As the world's leading producer of corn, the United States had the infrastructure in place to create corn-based ethanol production facilities, which began to pop up in rural farming towns all over the midwestern United States. Between 1999 and 2009, the total number of operating ethanol plants in the United States tripled, growing from 50 to more than 150. The Energy Policy Act, passed by Congress in 2005, required that the country boost its biofuel production to 7.5 billion gallons by 2012. Only 4 years later, in 2009, the country had exceeded that goal, topping 11 billion gallons. Today, nearly 50% of America's gasoline contains some amount of ethanol, according to the American Petroleum Institute.

In the beginning, biofuels were welcomed as an ideal green solution, reducing our dependence on foreign oil and saving the planet along the way. But slowly, that sparkling image began to fade.

Over time, experts realized that the initial attempts to create biofuels weren't much more environmentally friendly than the fossil fuels they were intended to replace. Corn is one of the most energy-intensive crops to grow and harvest, requiring energy inputs like fertilizers and pesticides (made from fossil fuel), as well as fuel needed for the operation of farm machinery; these early corn ethanol projects used as much (or almost as much) energy as they produced. Water was also a concern: One study estimated that a car running on ethanol uses the equivalent of 50 gallons of water for every mile (compared to 2 to 3 gallons of water used to produce a gallon of gasoline).

Another worry is that many natural lands are being converted to farmland for biofuels, endangering native species that live there, such as orangutans in the rainforests of Malaysia and African elephants in Ethiopia. These biofuel crops also displace food crops—farmers who might normally grow corn for food or animal feed are growing it for fuel. This switch drove up the price of corn in 2007, and caused riots in Mexico, Haiti, and

Infographic 29.3 | **BIOFUEL TRADE-OFFS**

ADVANTAGES	DISADVANTAGES
Renewable Can be replaced over short time scales	**Energy content** May not be as high as fossil fuels so more biofuel must be burned to produce the same energy
Versatile Can be used in a variety of ways or converted to fuels that can run our existing machinery; especially important as fuel for mobile sources	**Total energy used** If traditional monoculture crops like corn or soybeans are used that require high fossil fuel inputs, the overall energy return of the biofuel is diminished.
Local Can be grown or collected locally, based on what grows best or is available in a given area, reducing the need to import energy	**Overharvesting** If too much biomass is taken it can reduce forest and grassland contributions to ecosystem services.
Waste to energy Garbage, agricultural, industrial, or food waste can be a feedstock for biofuels.	**Equipment redesign** Some equipment may need to be retooled or redesigned to use biofuels (i.e., vehicles that use ethanol fuel mixtures greater than 10%).
Marginal land Fuel crops can be chosen that will grow on marginal land, which frees up the best land for food production.	**Food issues** Can take over good cropland and reduce the amount of food grown, reducing food supplies; this can drive up the price of food
Pollution Biofuels burn more cleanly and are less toxic than petroleum fuels.	**Pollution** Burning biomass directly (wood, manure, etc.) produces high particulate matter pollution.
Carbon neutral or negative If grown with minimal or no fossil fuel inputs, the carbon released is equal to or less than that taken out of the atmosphere.	**Water** Biofuel crops, especially monoculture annuals like corn and soybeans, consume more water per kWh than is used for traditional fossil fuels (nontraditional oil shale and sands use more water than biofuels).

other nations around the world where corn is a staple food item. And clearing land for biofuel crops releases carbon stored in soil and plants, adding greenhouse gases to the atmosphere as well as disrupting all the benefits those ecosystems provide, such as habitat for a variety of organisms. In addition, the use of crop waste to produce biofuels has its drawbacks—namely the removal of material that normally would have been plowed back into the soil to decompose and add back nutrients. [INFOGRAPHIC 29.3]

Advances in recent years, mostly in increased production per acre, now make corn a better option than it used to be. Even though early attempts to make ethanol from corn often used more energy than they produced, the U.S. Department of Energy recently estimated corn's EROEI (energy return on energy investment) at 1.38, meaning it produced 0.38 units of energy for every unit of energy invested. But even if the energy return could be further improved, should we be using prime farmland to produce

a fuel crop? Is there a way to grow biofuel feedstocks on marginal lands? Along with many other researchers, Tilman was discovering some interesting insights on just how to do that.

Tilman's experiments showed the importance of biodiversity.

Every August for 10 years, as part of their ongoing quest to understand why plants growing in the most diverse areas could more easily weather periods of drought, Tilman's team spread out and harvested small strips of vegetation from each plot. They dried and weighed the plants, determining how much biomass—in this case, plant material—had grown. The results floored them.

"Frankly, we never imagined the effects of diversity would be as large as they turned out to be," says Tilman. The biomass analysis showed that plots with 16 species

↑ Corn grain is processed at Archer Daniels Midland (ADM) ethanol and corn syrup production plant in Decatur, Illinois. Long known as a food company, ADM has invested millions of dollars in biofuel projects.

of plants were an average of 238% more productive than fields planted with only one species (**monoculture**). In a way, this was not a surprise—ecologists have found that plants often grow better together than by themselves because their differences can complement each other. For example, one species may shed its leaves in the late spring, depositing valuable nutrients into the soil that another species takes up in the summer. Ecological communities with high species diversity such as these arose over time as individuals competed for resources. This led to resource partitioning—each species uses particular

resources and is very efficient at accessing and using those resources. Species diversity also leads to mutualistic and commensal relationships, all of which maximize energy uptake and nutrient flow in the community, increasing overall biomass (see Chapter 9).

Tilman kept thinking about the results—specifically, about how they could model biofuel crops after what nature had already figured out: Biodiverse ecosystems that are naturally adapted to their environment are the most productive.

Ethanol is easily made from the starch in corn, but it can also be derived from cellulose, a compound that forms in the cell walls of all plants and makes up the bulk of the fibrous structure of a plant like switchgrass. Like with

---

monoculture Farming method in which one variety of one crop is planted, typically in rows over large areas with large inputs of fertilizer, pesticides, and water.

Infographic 29.4 | **LIHD CROPS OFFER ADVANTAGES OVER TRADITIONAL MONOCULTURE BIOFUEL CROPS**

↳ A life-cycle analysis by David Tilman and colleagues shows that low-input, high-diversity (LIHD) grassland plants offer a higher net energy return and release fewer greenhouse gases than do the traditional monoculture fuel crops like corn and soybeans.

ENERGY RETURN ON ENERGY INVESTMENT COMPARISON

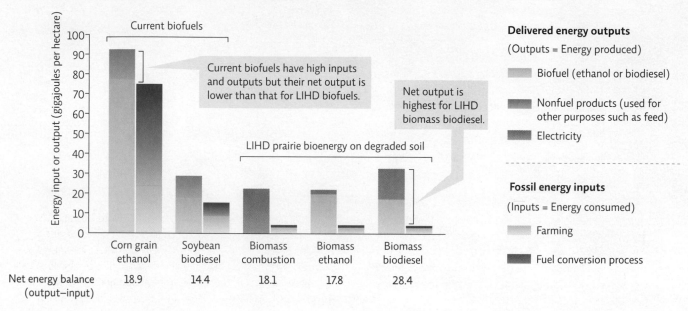

↳ Compared to gasoline, all biofuels produced less greenhouse gas emissions in an EROEI analysis. In particular, fuels from LIHD grasses grown on marginal land produced significantly lower greenhouse gas emissions than higher-input conventional biofuel crops (corn and soybeans).

REDUCTION OF GREENHOUSE GAS (GHG) EMISSIONS BY BIOFUELS

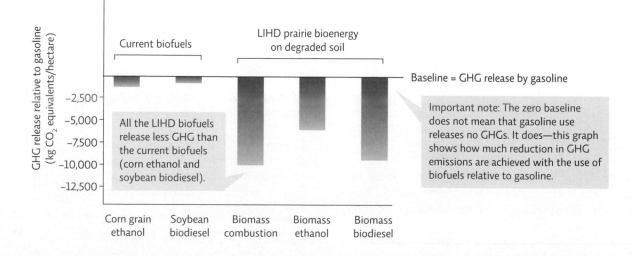

other forms of biofuels, people had been trying to produce fuel from cellulosic material such as wood for decades.

Recently, **cellulosic ethanol** that can be produced from crops such as switchgrass has become a popular alternative to corn ethanol. For starters, it has the potential to circumvent the food controversy, since it won't occupy land that could support food crops. Additionally, switchgrass is resistant to many pests and plant diseases and requires little fertilization, so it consumes less energy to produce. Even more attractive to alternative energy aficionados, switchgrass, like other plants, absorbs CO_2 from the atmosphere and stores carbon in its roots, a process called **carbon sequestration**. By working with a perennial like switchgrass instead of an annual like corn, much of the plant (and its carbon) would remain behind in the deep root system, ready to grow again immediately after harvesting.

However, people who mentioned switchgrass as a source of biofuels "were always talking about monocultures," Tilman remembers, and growing switchgrass as a crop by itself. Tilman realized that by growing switchgrass in high-diversity mixtures of many plant species, a plot could produce a lot more biomass with the same or smaller amounts of irrigation, fertilizer, and management. That meant more biofuel for fewer inputs.

"We kept saying, 'we gotta tell people about this!'" Tilman recalls. To do so, he hired Jason Hill, then an assistant professor of biology at St. Olaf College in Minnesota. Hill collected Tilman's biomass data for each plot over the 10 years, as well as the recorded levels of carbon in the soil and roots of the plants, and analyzed the net energy that could be produced from the grasslands. He compared those data to the amounts of energy harvested from corn and soybeans. The results were overwhelming: The 16-species plots were not only capable of producing more bioenergy than monocultures, they sequestered 14 times more carbon in the soil than the amount released by growing and processing the crop into a biofuel. In other words, they made the grasslands carbon negative—they stored more carbon than was released to grow or use the fuel. Finally, Hill and Tilman noted that diverse grasses can be grown on neglected agricultural lands and in marginal soil, therefore not displacing food production or destroying habitats. The two researchers published their results in a 2006 issue of *Science*, demonstrating for the first time the feasibility

cellulosic ethanol Bioethanol made by breaking down cellulose in plants; a difficult process that has yet to be scaled up to meet large production goals.
carbon sequestration The storage of carbon in a form that does not readily release the carbon to the atmosphere or water.

and potential of biofuels from low-input, high-diversity grasslands, called LIHD biofuels. [INFOGRAPHIC 29.4]

The media went crazy: "Wild flowers could provide a solution to global warming" hailed a Scottish newspaper; "It sounds too good to be true," wrote a journalist at *Discovery News*. Unfortunately, that's exactly what some other scientists thought of Tilman's conclusions—too good to be true.

"When you have a new idea, you start imagining what might happen, but you never know what will really happen until you try the experiment."
—David Tilman

Criticism came swiftly, and from close to home. In the building next door to Tilman's at the University of Minnesota, members of the United States Department of Agriculture (USDA) and professors in the University's agronomy department began to discuss their concerns about Tilman's paper. Michael Russelle, a USDA soil scientist and adjunct professor at the university, suspected that Tilman's group had overestimated the carbon sequestration of the grasslands because of the way they managed the plots—by burning the plots at the end of the season after estimating productivity, they potentially left behind more carbon than harvesting would have done. Because of this, Russelle felt that the energy accounting was "misleading." "Many issues have rapid political implications, and this is one of those. In this case, we felt that the paper wasn't warranted by the data, so we just felt it had to be challenged," recalls Russelle. He approached Tilman and Hill to inform them that he and colleagues were planning to write a rebuttal (an official commentary in response to Tilman's paper). Tilman and Hill were surprised but accommodating. "It's good to have a public dialogue over a controversial paper," says Hill. "You just don't necessarily expect it to come from the building next to you," he adds with a laugh.

The three men met for coffee, but by the end of the meeting, they remained divided. "They didn't convince me," says Russelle. The dissenters published their comment in *Science*, but Tilman and Hill didn't back down, and published a response, backing up their results and study design with other published research. "We essentially said we appreciate your concerns, but we do not think they changed the outcome of the conclusions at all," says Hill.

↑ Bags of algae hang outside Redhawk Power Station near Phoenix, Arizona. Power plant emissions are fed to the algae, which ultimately convert the CO_2 to oils, which are used to make biodiesel—as much as 5,000 gallons per acre each year.

Today, the two groups still maintain opposing viewpoints, but both recognize the importance of the dialogue. "Science is a continual process of putting forth a hypothesis, testing it, evaluating the results, and moving forward," says Russelle.

There is another rising biofuel star: Algae.

In January of 2009, Continental Airlines ran a test flight of a Boeing 737 powered by a blend of normal fuel and a new type of biofuel manufactured from algae.

Like Tillman, algae researcher Stephen Mayfield didn't set out to discover a new type of biofuel. But after spending decades studying and tweaking the genetics of algae, he and his colleagues at the University of California, San Diego, had an epiphany in 2006. At that time, Mayfield was watching TV footage of the glaciers shrinking from climate change, and paying more for gas every time he fueled up. During a meeting to discuss an ongoing research project to engineer algae to produce drugs such

as antibiotics, a new idea arose. "We realized, 'wait a minute, if we can do that, we can produce fuel.'"

Like tiny factories, algae use photosynthesis to convert CO_2 and sunlight into sugars and then convert some of the sugars into oil, which can be harvested and converted to biodiesel. Algae are now making the same kind of headlines as LIHD biofuels, and for good reason.

Algae have some clear advantages over other biofuel sources. Like switchgrass, algae won't compete with food. They grow very quickly and can be raised on salty water, so they don't use up freshwater. Algae are also predicted to generate 30 times more oil per acre than other plants used for biodiesel, such as soybean, oil palms, and rapeseed.

But algae can be finicky—it is difficult to maintain the ideal conditions to maximize oil production—and the facilities are expensive. Plus, unlike corn and other land crops, people haven't been harvesting algae for generations, so scientists have to devise an entirely new infrastructure. Even so, algae-based biofuels are starting to look like big business. There are hundreds of companies now invested in turning algae oil into fuel, according to the National Renewable Energy Laboratory. One company Mayfield cofounded, Sapphire Energy, plans to open the first phase of its 300 acres of algae ponds in a New Mexico facility in the summer of 2012. "That will be a huge milestone for the industry," he says.

Those investments may one day pay off. After Continental Airlines analyzed data from the test flight, they found that the algae biofuel blend resulted in a 1.1% increase in fuel efficiency over regular jet fuel and didn't affect the jet's ability to perform flight maneuvers such as a mid-flight engine shutdown and restart. [INFOGRAPHIC 29.5]

There are many reasons why biofuels have not solved our dependence on fossil fuels.

Despite the hopes of Tilman and others, cellulosic ethanol is not likely to arrive at a pump anytime soon. For one, it's not easy to get the energy out of the plants: While starch from corn is soft, cellulose in grasses is complex and durable, not easily degraded. "There's a reason we make houses and furniture out of cellulosic material such as wood," says Mayfield. "It doesn't like to spontaneously turn into something else." Several attempts use high temperatures, acid, enzymes, and yeast to tear apart the strong bonds—a slow, expensive, and energy-intensive process that has proven difficult to scale up. [INFOGRAPHIC 29.6]

Infographic **29.5** | **BIOFUELS FROM ALGAE**

↓ Algae can be grown as a feedstock for biofuels. Biodiesel is the most common product, but the sugars in algae can also be extracted to produce ethanol, while the solids can even be harvested to produce a high-protein animal feed. Self-contained operations can grow lots of algae in a small space (in tubes or vats) or in outdoor algae ponds—which take up more land area. A New Zealand company, Aquaflow, even plans to harvest wild algae from coastal areas with algal blooms.

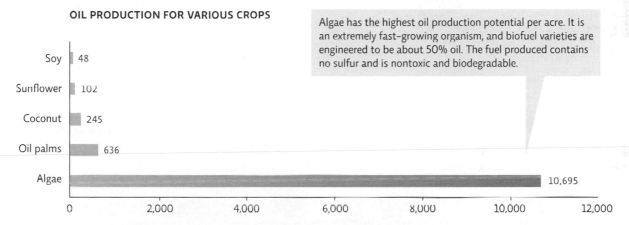

OIL PRODUCTION FOR VARIOUS CROPS

Algae has the highest oil production potential per acre. It is an extremely fast-growing organism, and biofuel varieties are engineered to be about 50% oil. The fuel produced contains no sulfur and is nontoxic and biodegradable.

Crop	Gallons of oil per acre
Soy	48
Sunflower	102
Coconut	245
Oil palms	636
Algae	10,695

Gallons of oil per acre

↑ Researchers test different types of algae to see which produces the best oil to use for biodiesel formation.

↑ Algae pond at Sapphire Energy facility near Las Cruces, New Mexico.

Infographic **29.6** | **BIOETHANOL PRODUCTION**

↓ Ethanol can be produced from the biological fermentation of any plant material. Food sources like grains and sugarcane are easily broken down by yeast (fermentation), whereas plant material high in cellulose is more challenging to break down and requires stronger chemical reactions. Research to improve the breakdown of cellulose and make the process more efficient is underway.

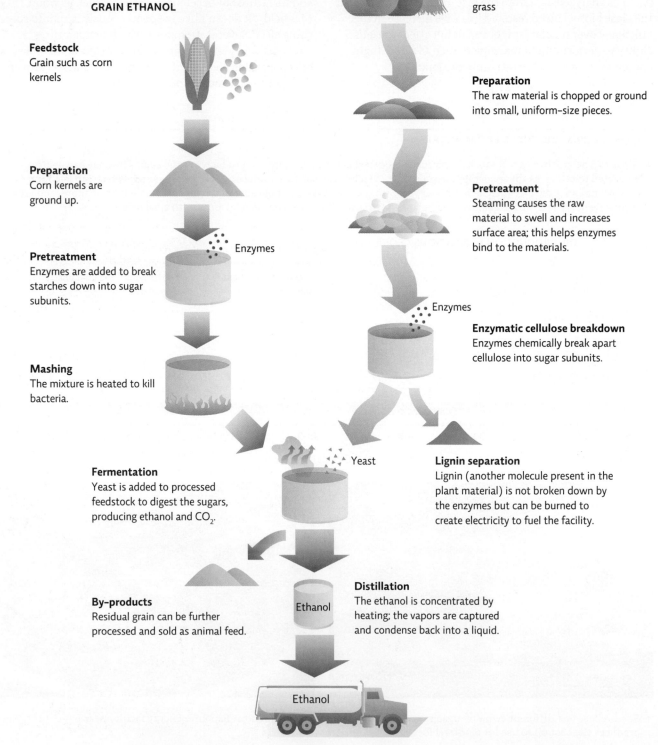

GRAIN ETHANOL

Feedstock
Grain such as corn kernels

Preparation
Corn kernels are ground up.

Pretreatment
Enzymes are added to break starches down into sugar subunits.

Enzymes

Mashing
The mixture is heated to kill bacteria.

Fermentation
Yeast is added to processed feedstock to digest the sugars, producing ethanol and CO_2.

Yeast

By-products
Residual grain can be further processed and sold as animal feed.

CELLULOSIC ETHANOL

Feedstock
Plant material such as switch-grass

Preparation
The raw material is chopped or ground into small, uniform-size pieces.

Pretreatment
Steaming causes the raw material to swell and increases surface area; this helps enzymes bind to the materials.

Enzymes

Enzymatic cellulose breakdown
Enzymes chemically break apart cellulose into sugar subunits.

Lignin separation
Lignin (another molecule present in the plant material) is not broken down by the enzymes but can be burned to create electricity to fuel the facility.

Ethanol

Distillation
The ethanol is concentrated by heating; the vapors are captured and condense back into a liquid.

Ethanol

← Don Walton, switchgrass area specialist for the University of Tennessee Extension, loads bales of switchgrass grown for the Tennessee Biofuels Initiative onto a truck on the farm of Dwaine and Randall Peters in Vonore, Tennessee.

There is currently no large-scale commercial facility online for converting cellulose-rich biomass to ethanol, whether from switchgrass monocultures or LIHD grasslands, but there are some smaller prototype facilities, and more are under construction around the world. One began operation in Vonore, Tennessee, in 2010 as part of the $70.5 million Biofuels Initiative of the University of Tennessee. The goal is to create a template for the logistics of growing, handling, storing, and transporting the thousands of bales of switchgrass that would be processed daily at a commercial facility. "It's the holy grail," says Michael Palmer, a professor of botany at Oklahoma State University who tracks biofuel research. "Millions and millions of dollars from the government and private sectors are being pumped into the technology of cellulosic conversion." Verenium Biofuels Corporation, a Massachusetts-based firm that has proprietary rights for special enzymes used in the breakdown of cellulose, recently sold its business and demonstration facility in Louisiana to BP for $98 million. The goal is to produce low-cost cellulosic ethanol from grasses, wood chips, agricultural waste, or even garbage.

Algae, too, will take several years before it can be scaled up, predicts Mayfield. It's relatively easy to convert algae to diesel fuel, he says, but building the infrastructure to do this on a large scale—such as massive ponds—delays progress. He predicts that algae will become widely used as a biofuel source before switchgrass, however, simply because it has a bigger customer: the military, which can pay a premium for renewable diesel to fly its jets. Furthermore, there already is a major source of ethanol—corn. "I think algae are more likely to be commercially deployed in the next several years."

Multiple solutions will be needed to help replace fossil fuels.

Like all strategies to replace fossil fuels, biofuels will never have just one golden ticket winner. Instead, a patchwork of feedstocks grown and processed across the country will help provide our fuel. The good news is that we have a wide variety of options, which enables us to focus on the types of biofuels that work best for different needs or in different regions.

Whatever the future, one thing is sure—to be truly environmentally friendly, we may simply have to stop using so much fuel. This means increasing energy efficiency to decrease waste, and basic conservation—simply using less.

The energy efficiency of vehicles can be improved with lighter materials, more aerodynamic shapes, more energy-efficient engines, and hybrid technologies that pair electric motors with internal combustion engines. Total electric vehicles would lessen the demand for liquid fuels. Drivers can also increase the fuel efficiency of whatever vehicle they drive by adopting better driving habits and of course, by driving less overall. [INFOGRAPHIC 29.7]

In 2007, Congress passed the Energy Independence and Security Act, which requires the United States to produce 36 billion gallons of renewable and alternative fuels per year by 2022 (up from 11 billion gallons produced in 2009)—equivalent to about one-quarter of the entire country's yearly use of gasoline.

We are behind schedule—the economic downturn and obstacles to scaling up cellulosic ethanol have stalled progress in recent years. The goal for 2010 was

Infographic 29.7 | **ENERGY EFFICIENCY AND CONSERVATION ARE PART OF THE SOLUTION**

↓ Choosing the vehicle that meets your needs but also gets good gas mileage is an important part of the solution to our energy problems. Go to www.fueleconomy.gov to compare different makes and models of cars for fuel efficiency. This can also save you money—a car that is 10 MPG more efficient than another will save the driver almost $1,000 in fuel costs per year (assumes 15,000 miles driving per year with gasoline at $3.65/gal). All the major (and some smaller) car companies have hybrid and/or electric vehicles on the road or in planning. Though hybrids generally get better gas mileage than other cars, many traditional (and less expensive) models get 30 to 40 MPG.

PLUG-IN HYBRID VEHICLES

Gasoline–hybrid vehicles have a battery system, charged when the car is braking or coasting, that helps run the motor. The car goes from electric to gasoline mode seamlessly. Plug-in hybrids can be plugged into a household 120V AC outlet to add charge to the battery and extend the battery range.

ELECTRIC VEHICLES

The Nissan Leaf is a totally electric vehicle with no gasoline engine backup. It will go 75–100 miles per charge (about $3 worth of electricity), with a public charging station infrastructure being developed; owners also install a charging station at home.

↓ Driving habits that maximize mileage will make our biofuels go farther.

HOW CAN YOU INCREASE YOUR MPG?

DRIVE MORE EFFICIENTLY

Don't be an aggressive driver
Avoid "jackrabbit" quick starts; avoid speeding up to a stop.

Don't speed
Most cars get their best gas mileage between 35 and 55 miles per hour; when possible, try to drive in your top gear (cruising gear).

Use cruise control
By keeping a more constant speed, fuel efficiency increases.

Avoid idling
Turn off your car if you will be parked, even briefly. Starting the car back up takes less fuel than idling for more than a few seconds. Hybrid cars that automatically turn off the gas engine save idling gas waste.

TAKE CARE OF THE CAR

Tune it up
Keep your vehicle in tune to maximize fuel efficiency—don't skip maintenance visits.

Keep tires properly inflated
Under- or overinflated tires decrease mileage and make for unsafe driving; check tire inflation regularly, especially when the seasons change—air pressure increases in warm weather and decreases when it gets cold.

Unload the trunk
Avoid carrying excess weight in the vehicle—the heavier the load, the lower the mileage.

PLAN YOUR OUTINGS

Combine trips and plan your route
A cold car gets worse gas mileage; combining trips while driving a warm car—even starting and stopping several times as you drive around town—will save gas; plan your route to avoid backtracking.

Commuting
Drive the most fuel-efficient car available when commuting; consider working from home when possible and carpool with coworkers if you can; using high-occupancy vehicle (HOV) lanes can improve your mileage and get you to work more quickly.

250 million gallons of cellulosic ethanol; that year, we produced only 25 million gallons—off by a factor of 10. Uncertainties about whether tax credits and subsidies will remain available make cellulosic ethanol a risky investment, and many venture capitalists have pulled out of projects.

And yet, we are moving ahead. Many U.S. states offer rebates or tax exemptions on fuels for commercial vehicles that operate on biodiesel blends or E-85 fuels, and financial incentives like tax credits or grants are offered to biodiesel producers or distributors. In August of 2011, the U.S. federal government launched a $510 million program, in the form of matching funds, to help finance biofuel projects directed toward military transportation applications.

Biofuels are also growing in popularity around the world. Brazil, a world leader in ethanol biofuel, has no light vehicles that run on 100% gasoline, and Brazilian car manufactures have developed vehicles that run on ethanol blends or 100% ethanol. China and India have national policies mandating the increase of the ethanol component of gasoline. In addition, China, India, and other developing countries are supporting biodiesel production from locally grown crops such as the succulent plant jatropha. Meanwhile, China is pursing the development of technologies to produce biofuels from municipal garbage. Smaller-scale programs are also underway in many nations around the world.

carbon debt The amount of carbon released during the first 50 years of land clearing for biofuel feedstocks.

Despite ongoing controversies and setbacks, the future of biofuels looks bright.

Working with Nature Conservancy scientist, Joseph Fargione, Tilman and Hill set about to determine exactly how much CO_2 is released by clearing land for crop biofuels, which involves burning vegetation and plowing carbon-rich soil. The results were shocking: The team found it could take decades—or even centuries—of biofuel production to make up for the amount of CO_2 released during land conversion, what they called the **carbon debt**. For instance, it will take 93 years to make up for the carbon debt resulting from digging up central grasslands in the United States to produce corn ethanol. To repay the carbon debt for clearing peatland rainforest for palm biodiesel in Indonesia and Malaysia: 423 years. The only biofuel with zero carbon debt? Bioethanol from prairie biomass grown on marginal cropland.

The results solidified Tilman's hopes for a new type of biofuel that circumvents the problems of corn ethanol, such as fuel from switchgrass or algae. "Biofuels, if used properly, can help us balance our need for food, energy, and a habitable and sustainable environment," Tilman wrote to the country in a 2007 opinion piece in the *Washington Post*. "We have the knowledge and the technology to start solving these problems."◉

Research articles referenced in this chapter:
Fargione, J., et al. 2008. Science, 319: 1235–1238.
Russelle, M.P., et al. 2007. Science, 316: 1567b.
Tilman, D., et al. 2006. Nature, 441: 629–632.
Tilman, D., et al. 2006. Science, 314: 1598–1600.
Tilman, D., et al. 2007. Science, 316: 1567c.

BRING IT HOME

❍ PERSONAL CHOICES THAT HELP

Biofuels represent a potential replacement for fossil fuels and might be an especially important fuel for transportation. Despite the promise of biofuel use, there are also trade-offs such as the fossil fuel inputs needed to grow fuel crops. Developing new biofuel technologies may help us meet the energy needs of the future.

Individual Steps
→ Visit ecogeek.com to find information on experiments and new advances in biofuel technology.

→ Review Infographic 29.7 and take steps to make your vehicle and driving more fuel efficient; make the investment in a hybrid or electric vehicle if you can.

Group Action
→ Host a movie night to watch *FUEL*, an award-winning film that looks at the history of biofuels as well as possible solutions for the future.
→ Follow the Veggie Van Organization on Facebook and Twitter to see which schools they are visiting and to see how they use their biofuel vehicles for education.

Policy Change
→ Use the Advanced Biofuels USA site (advancedbiofuelsusa.info) to learn about sustainable biofuel options and stay current on proposed biofuel legislation.

UNDERSTANDING THE ISSUE

CHECK YOUR UNDERSTANDING

1. **Which of the following best describes a fuel crop?**
   a. Crops that are specifically grown to make biofuels
   b. Waste biomass from food crops that is converted to fuel
   c. Waste oil from restaurants used to make biodiesel
   d. Manure patties remaining in a field grazed by herbivores

2. **What is a LIHD biofuel?**
   a. Biofuel with a high-density lipid content
   b. Biofuel produced from waste oil, with a low impact on the environment
   c. Biofuel produced from algae grown in high density in fertilizer–rich liquid
   d. Biofuel produced from grasslands with high species diversity, grown with few inputs

3. **Which of the following produces the most oil per acre of crop?**
   a. Soybeans
   b. Algae
   c. Sunflowers
   d. Oil palms

4. **How do plants sequester carbon?**
   a. During photosynthesis, carbon is stored in the leaves of plants.
   b. Plants take in atmospheric CO_2 and store carbon in their roots.
   c. Plants convert CO_2 into oxygen, replacing CO_2 in the atmosphere.
   d. Sugar is converted to CO_2 and released during respiration.

5. **Compared to gasoline, which of the following produces the smallest amount of greenhouse gases?**
   a. Corn grain ethanol
   b. Soybean biodiesel
   c. LIHD biodiesel
   d. LIHD ethanol

6. **Direct sources of biomass energy include wood, corn stalks, grass bales, and _____.**
   a. sugar cane
   b. used restaurant oil
   c. dried manure
   d. biodiesel

WORK WITH IDEAS

1. Describe the environmental issues associated with biofuels. Include issues specifically related to the use of corn.

2. You are an investor looking to get into the biofuels market, and you plan to choose one source for generating biofuel. Based on the information in this chapter, which source would you choose and why?

3. List and describe the current sources of fuel (both direct and indirect) that come from biomass. Include in your description the sources of biofuel and the feasibility of producing large quantities of biofuel from these sources (to the best of our knowledge at the present time).

4. Compare and contrast corn and switchgrass as sources of biofuels. What are the benefits of each? What are the costs?

5. Describe the process of biodiesel production using leftover restaurant oil. Is this possible on a small scale?

6. Explain, in your own words, why biofuels (especially cellulose and algae) have not replaced fossil fuels.

ANALYZING THE SCIENCE

The following graph shows the total environmental impact of biofuels compared to greenhouse gas emissions. Each data point is plotted relative to the emissions and impact of gasoline; fuels that fall in the white area of the graph are considered to be better choices.

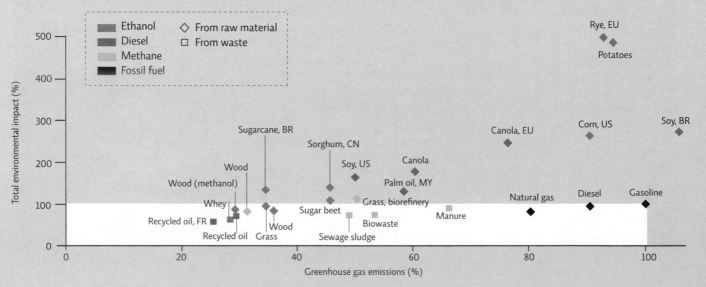

INTERPRETATION

1. What does each axis of this graph represent?

2. What types of fuel are being compared, and what are their original sources?

3. Locate each of the following on the graph: U.S. corn ethanol, Brazilian (BR) sugarcane ethanol, Brazilian soy diesel, Malaysian (MY) palm oil diesel, and gasoline. Note that the first four are the most economically important biofuel sources.
   a. Rank all five in terms of their total environmental impact, from highest to lowest. Which has the highest environmental impact and which has the lowest? Assuming that all were equally abundant, which would be the most environmentally friendly choice, based on total environmental impact?

ADVANCE YOUR THINKING

4. What kinds of things does the measure "total environmental impact" likely include? (Hint: Remember to look at Infographic 29.3.)

5. Why is the environmental impact for Brazilian soy diesel high?

6. Assuming that the scientists who evaluated these data are correct, what type of fuel contributes the least amount of greenhouse gases and has the lowest total environmental impact? Why do you think its impact is low relative to other fuel sources?

7. According to the graph, which biofuel would make the most *environmental* sense to use in the United States, given our current standard of living and the resources we have access to?

EVALUATING NEW INFORMATION

Biofuels are increasingly available to consumers at petroleum retailers across the nation. The National Biodiesel Board maintains a website about biodiesel and its use. Go to that site (http://www.biodiesel.org/Home) and explore it by opening each of the links at the top of the page.

Evaluate the website and work with the information to answer the following questions:

1. What type of information does the site provide under each link?

2. Is this a reliable information source? Does it have a clear and transparent agenda?
   a. Who runs the website? Do this person's/organization's credentials make the site reliable or unreliable?
   b. Is the information on the website up to date? When was the last time the website was updated? Explain.

3. Click on the "Using Biodiesel" tab; then select the "Finding Biodiesel" link to find retailers in your state. How many are there? Where are these retailers located? If you wanted to start using biodiesel (and properly converted your engine), would you be able to consistently find a source of biodiesel?

4. Under "Using Biodiesel" at the top of the page, open the link to "Market Segments."
   a. Explore this page. What types of users have contributed stories about their use of biodiesel?
   b. Choose the "Market Segment" link entitled "General Interest" and read about biodiesel. Choose 3 of the claims made there and read the paragraph that accompanies them. What evidence is given in support of each claim. In your opinion, is this evidence sufficient? Explain.
   c. If you drove a vehicle that could use biodiesel, would you use it? Why or why not?

MAKING CONNECTIONS

THE FUTURE OF ENERGY

Background: Environmental science is complex because there are no "magic bullets" that solve environmental problems. Each potential response comes with trade-offs. A major goal of environmental science is to identify sustainable choices that reduce our impact in a manner that is socially, culturally, and economically acceptable. Such choices occur in every sphere of society and are everyone's responsibility.

Case: As an environmental science student, you have learned about the pros and cons of sources of energy from fossil fuels to renewable energy. You have been assigned to draft a proposal for future energy policy on university campuses in your state that will be sent to the state governing body of the university system.

1. Be sure to include the following in your proposal:
   a. The number and size of public institutions in your state.
   b. The types of energy that are currently being used on each campus. For example, at one of your big state schools, what type of fuel does the bus system use? What is the source for electricity production in your state? Do any of the schools have mechanisms in place to either reduce energy consumption or to generate their own power?

2. Based on your research, what recommendations would you make to the governing body to decrease the quantity of energy consumption and to increase the production of sustainable energy on the state's campuses? In your recommendations, include options for the institutions as a whole and for individuals in those institutions. Remember to consider the location of each institution in your state, and assume there will be money to fund such an initiative.

3. Conclude your report with a section on benefits to the institution and to members of the institution (faculty, staff, and students). In addition to becoming more energy efficient, how will this initiative help your campus?

Debris is displayed on the deck of SEAPLEX's research ship *New Horizon* in the summer of 2009.

A PLASTIC SURF

Are the oceans teeming with trash?

CORE MESSAGE

We are often unaware of the waste we produce, but it profoundly affects the environment. Waste that cannot be reused or recycled simply accumulates; some of it is toxic or otherwise disruptive to living things. We can address its impact by minimizing the waste we generate in the first place, recovering and recycling waste materials, and disposing of all waste safely.

GUIDING QUESTIONS

After reading this chapter, you should be able to answer the following questions:

→ What kinds of trash do humans generate, particularly in developed nations like the United States?

→ What options do we have for dealing with solid waste and what are the trade-offs for each option?

→ What is hazardous waste and what can the average person do to reduce his or her production of it?

→ What are some of the environmental consequences of our waste production and disposal methods?

→ How can industry and individuals reduce the amount of waste that they produce?

From the deck of the *SSV Corwith Cramer*, the surface waters of the North Atlantic looked like smooth, dark glass. The 134-foot oceanographic research vessel had set out from Bermuda just 24 hours before on a month-long expedition aimed at tracking human garbage across the deep sea. It was a cool mid-June evening, the sun was setting, and the crew had just launched its first "neuston tow" of the trip. Giora Proskurowski, the expedition's chief scientist, stood watching as the tangle of mesh skidded along the water's surface, or *neuston layer*, keeping pace with the ship as it went. It was hard to imagine that any garbage would be found in such a flat, serene landscape.

After 30 minutes, crew members pulled the net—dripping with seagrass and stained red with jellyfish—from the water. Sure enough, when Proskurowski got closer, he could make out scores of tiny bits of plastic glistening in the mesh. The crew members counted 110 pieces, each one smaller than a pencil eraser and no heavier than a paper clip. Based on the size of the net and the area they had dragged it over, 110 pieces came out to nearly 100,000 bits of plastic per square kilometer.

Despite the water's pristine look, the yield was hardly surprising. In the past 20 years, undergraduate students on expeditions like this one had handpicked, counted, and measured more than 64,000 pieces of plastic from some 6,000 net tows. That might not sound like much, given the vastness of the Atlantic and the smallness of the plastic. But the tiny bits were gathering in very specific areas, known as *gyres*. Gyres are regions of the world's oceans where strong currents circle around areas with very weak,

or even no, currents. Lightweight material, like plastic, that is delivered to a gyre by ocean currents becomes trapped, and cannot escape the stronger circling currents. When scientists first evaluated all the data from the Atlantic Gyre, they discovered surprisingly dense patches of plastic—more than 100,000 pieces per square kilometer—across a surprisingly large "high-concentration zone." The press and general public would come to know this region as the "Great Atlantic Garbage Patch," cousin to a "Great Pacific Garbage Patch" discovered around the same time, and, they suspected, to several other patches scattered about the ocean.

Scientists knew where the plastic was coming from (open landfills and litter-polluted gutters around the world). And they understood why it was being trapped in the gyres (wind patterns and ocean currents). But other questions remained. No one could say how big the Atlantic Patch actually was, or how all that plastic was affecting ocean ecosystems. Were the toxins found in plastic accumulating up the food chain, making their way into fish, bird, or even mammal diets? Were all those tiny bits of solid surface providing transport for invasive species? And why wasn't there more of it? Despite a five-fold increase in global plastic production and a fourfold increase in the amount of plastic discarded by the United States, the concentration of plastic in North Atlantic surface waters had remained fairly steady across the 22 years for which data existed.

To answer these questions, Proskurowski, his team, and their captain would sail the *Corwith Cramer* all the way out to the Mid-Atlantic Ridge, an underwater mountain range which lies about 1,000 miles farther east than any previous plastics expedition had ever gone. "That's far enough from Bermuda that getting back will be a challenge," he wrote on the ship's blog as land faded from view. "The same forces that drew plastic into this

◉ WHERE IS BERMUDA?

MID-ATLANTIC RIDGE

BERMUDA

ATLANTIC OCEAN

EXPEDITION ROUTE

↑ Team members ready a neuston net for deployment.

↓ Giora Proskurowski on the *Thomas G. Thompson*

particular part of the ocean—namely low and variable winds and currents—will make operating a tall ship sailing vessel tricky, to say the least." But if all went according to plan, Proskurowski thought, he and his colleagues would be the first to reach the so-called garbage patch's easternmost edge.

Waste is a uniquely human invention, generated by uniquely human activities.

In natural ecosystems, there is no such thing as waste. Matter expelled by one organism is taken up by another organism and used again. This natural recycling is consistent with the **law of conservation of matter**, which states that matter is never created or destroyed; it only changes form. Forms of matter that are dangerous to living things (think arsenic and mercury) tend to stay

---

law of conservation of matter Matter can neither be created nor destroyed; it only changes form.

Infographic 30.1 | U.S. MUNICIPAL SOLID WASTE STREAM

↓ The United States produces more trash, per person, than any other nation. Though municipal solid waste makes up a small part of the total waste produced (~2%), it is the type of waste we can most directly impact with our day-to-day choices. Generation of MSW was 4% lower in 2009 than in 2008; this is attributed to the economic downturn rather than to increased recycling (which also decreased 3.3% in the same time period).

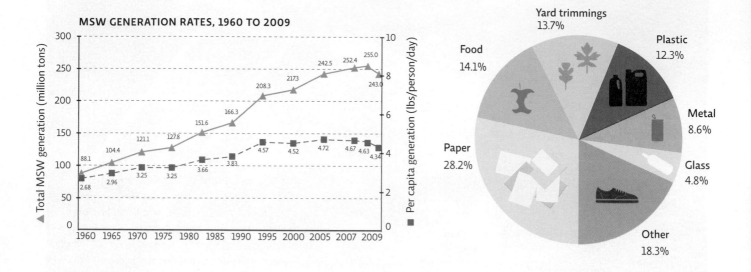

buried, deep underground, and are released only during extreme events like volcanic eruptions.

Human ecosystems are another story. By taking matter out of the reach of organisms that can use it, we continually disrupt this natural cycle. We do this by converting usable matter into synthetic chemicals that can't easily be broken down, and by burying readily degradable things in places and under conditions where natural processes can't run their course.

For example, paper and cardboard break down easily in a compost bin, thanks to the microbes that feed on them. But we typically keep this type of waste in landfills, where the lack of water, oxygen, and microbes force it to decompose much more slowly than it otherwise might.

Thus matter—the physical substance of which the universe is made—becomes **waste**—a uniquely human term used to describe all the things we throw away. Waste that can be broken down by microbes is considered **biodegradable**. Waste that can be broken down by chemical

and physical reactions is considered degradable, even if that degradation takes a long time. Some waste—mostly synthetic molecules like the pesticide DDT and CFCs once found in aerosols—is considered **nondegradable**. These molecules are chemically stable and don't degrade in normal atmospheric conditions. And because they haven't been around for very long, no organism has yet developed (through mutation or genetic recombination) the enzymes to use them as food.

Almost any human activity you can think of generates some form of waste. Processes that produce food and consumer goods generate *agricultural* and *industrial waste*, which in the United States account for about 54% of all garbage. The harvesting of coal and precious metals like gold and copper generates *mining waste*, which can pollute air, water, and soil, and makes up an additional 33% of U.S. waste. And the increasingly complicated act of living—in houses, apartments, dormitories, and small businesses around the world—produces its own steady stream of trash, referred to as **municipal solid waste (MSW)**, or at the community level, an MSW stream. [INFOGRAPHIC 30.1]

In the United States, MSW streams make up just 2% of total waste. But because Americans create more garbage per person than any other nation in the world, 2% is still a lot of trash. In 2009 alone, each American produced about 1.97 kilograms (4.34 pounds) of solid waste per

---

waste Any material that humans deem to be unwanted.
biodegradable Capable of being broken down by living organisms.
nondegradable Incapable of being broken down under normal conditions.
municipal solid waste (MSW) Everyday garbage or trash (solid waste) produced by individuals or small businesses.

↑ Five major gyres are found in the world's oceans and there are floating bits of plastic in all of them. These "garbage patches" are not floating islands and aren't even necessarily visible when gazing at the water. Much of the debris is very small and lies just below the surface. The highest density captured in one of the 6,100 sampling tows was 520,000 pieces per square mile. It is hard to estimate the size of any garbage patch since the material is so spread out and may reach down to depths of 20 meters (65 feet).

day, up from about 1.22 kilograms (2.7 pounds) per day in the 1960s. With more than 300 million people living in the United States, this adds up to about 243 million tons of household trash per year; that's twice the per capita amount produced by the European Union and as much as 10 times the amount produced by most developing countries.

The vast majority of this garbage comes from a familiar array of goods: paper, wood, glass, rubber, leather, textiles, and of course plastic—cheap enough to have become a staple of both advanced and developing societies, light enough to float, and durable enough to persist for hundreds of thousands of years across thousands of miles of ocean.

How big is the Atlantic Garbage Patch and is it growing?

The first garbage patch was discovered in 1997 by Captain Charles Moore, a veteran seafarer who was crossing the Pacific on his way back from an international yacht race.

As the press caught wind of what sounded like a giant plastic island in the middle of the ocean, a media frenzy ensued. Some reports said the patch was twice the size of Texas; others said that in the densest portions, plastic outweighed plankton (plankton consists of all the organisms floating in the ocean's upper reaches—from microbes to jellyfish). Still others claimed that the patch was growing exponentially.

But none of those claims was accurate. In fact, most of them were completely unfounded.

Part of the trouble was semantic. "An oceanographer understands the term 'patch' to mean 'an uneven distribution,'" says Kara Law, a scientist at the Woods Hole Oceanographic Institute who has led several expeditions on the *Corwith Cramer*. "We say the upper ocean is 'patchy' because organisms are often observed in clumps, separated by regions with sparse populations. But when reporters and laypeople hear 'patch,' they incorrectly think of an 'island' or a 'continent' of trash."

In 2009 alone, each American produced about 1.97 kilograms (4.34 pounds) of solid waste per day.

A bigger problem was the data itself. MSW and global plastic production data indicated that more plastic was being produced, used, and discarded throughout the world. And according to one scientific study, the amount of plastic being ingested by subarctic sea birds had nearly doubled between 1975 and 1985. But it was impossible to tell whether the patches themselves were growing or not. Data from the Pacific Patch suggested that the concentration of plastic had risen by an order of magnitude between the 1980s and 1990s. But that comparison was of limited value because sampling methods differed widely from one expedition to the next. In the Atlantic, where sampling methods had remained consistent across the years, no such increase could be detected.

Some scientists had suggested that the nets used on most plastics expeditions were too porous to trap the smallest of particles and that with finer nets, much more plastic would be found at the surface. Others thought that plastic was hiding below the surface, throughout the mixed layer—a layer of uniform density at the top of the water column that is saturated with sunlight, low in nutrients, and easily mixed by wind. Part of the *Corwith Cramer*'s mission was to try and figure out which hypothesis might be correct.

By July 3rd, Proskurowski's team had gleaned at least part of the answer. In the ship's daily blog, he wrote:

The results from the past several days of tucker trawls (which collect water samples from discrete depths within the mixed layer) have been very interesting. While we've seen fairly low numbers of plastic at the surface, we have observed an almost equal amount of plastic from one meter's water depth, slightly less at ten meters and no plastic below the mixed layer. These tows show that plastic is undoubtedly being mixed down into the water column, and what we measure at the surface is, in many cases, not the sum total of what is out here. Like many environmental problems, the closer you look at the system, the worse it appears.

It would be a few more weeks before they found out just how much worse.

How we handle waste determines where it ends up.

"It is with startling accuracy that so many tiny particles of plastic end up in such obscure but well-defined stretches of ocean," Proskurowski wrote one morning, as the *Corwith Cramer* floated under the early summer sun near the mid-Atlantic ridge. "Especially given how long and convoluted the journey they took to get here was."

In fact, all of our solid waste makes a similar journey, one that begins when we toss something we no longer need into a trash bag that is then carried from building to curb to garbage truck, before making its way to one of several kinds of waste facilities.

Open dumps are places where trash—both **hazardous** (waste that presents a health hazard) and nonhazardous— is simply piled up. Because they are one of the cheapest ways to get rid of human trash, they are common in undeveloped countries, where entire communities often spring up around the dumps, and people survive on what

---

open dumps Places where trash, both hazardous and nonhazardous, is simply piled up.

hazardous waste Waste that is toxic, flammable, corrosive, explosive, or radioactive.

leachate Water that carries dissolved substances (often contaminated) that can percolate through soil.

sanitary landfills Disposal sites that seal in trash at the top and bottom to prevent its release into the atmosphere; the sites are lined on the bottom, and trash is dumped in and covered with soil daily.

they scavenge from the waste piles. Open dumps attract pests such as flies and rats, which can be a human health hazard. Open dumps also contribute to water pollution: Rain either washes pollutants away from the dump to surrounding areas or pulls it along as it soaks into the ground. If this contaminated water, called **leachate**, continues to travel downward, it can contaminate the soil and groundwater.

Sanitary landfills, more common in developed countries, seal in trash at the top and bottom in an attempt to prevent its release into the environment. Several protective layers of gravel, soil, and thick plastic prevent leachate from depositing toxins into groundwater below the land-fill. The trash is covered regularly with a layer of soil that reduces unpleasant odors, thus attracting fewer pests.

But there is a downside to these, too. The compacting of trash under a layer of soil excludes oxygen and water so

↑ At Phnom Pen, Cambodia's municipal garbage dump, people work around the clock collecting plastic, metals, wood, cloth, and paper, which they sell to recyclers.

↑ Processing electronic waste in China: The recovery process is not done safely and exposes the workers and community at large to toxins. Though officially banned, export of electronic waste to China still continues and other Asian and African countries continue to accept the waste.

well that the aerobic bacteria (those that require oxygen to live) and other detritovores that normally decompose at least some of the waste can't survive. Newspaper that would degrade in a matter of weeks is preserved in landfills for decades. Anaerobic microbes (those that live in oxygen-poor environments) pick up some of the slack.

But they are much slower and produce lots of methane, a combustible greenhouse gas that is 20 times more potent than CO_2. As a result, landfills are a significant anthropogenic contributor of methane in the United States. [INFOGRAPHIC 30.2]

Infographic 30.2 | **MUNICIPAL SOLID WASTE DISPOSAL**

↓ As indicated in the Environmental Protection Agency's solid waste hierarchy, reducing waste at its source (homes and small businesses) is the top choice in waste management, with land-filling as the last choice. However, more than 50% of our solid waste ends up in a landfill.

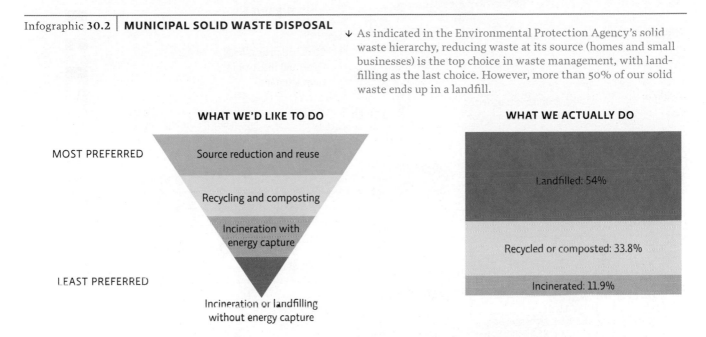

WHAT WE'D LIKE TO DO

MOST PREFERRED

Source reduction and reuse

Recycling and composting

Incineration with energy capture

LEAST PREFERRED

Incineration or landfilling without energy capture

WHAT WE ACTUALLY DO

Landfilled: 54%

Recycled or composted: 33.8%

Incinerated: 11.9%

↓ In a sanitary landfill, an area is dug out, lined to prevent groundwater contamination from leachate; trash is dumped and covered with dirt frequently (this dirt may take up to 20% of the landfill area.) Newer landfills have a leachate-collection system built in; older landfills can be retrofitted to collect leachate. Leachate from holding ponds is treated before being released into the environment.

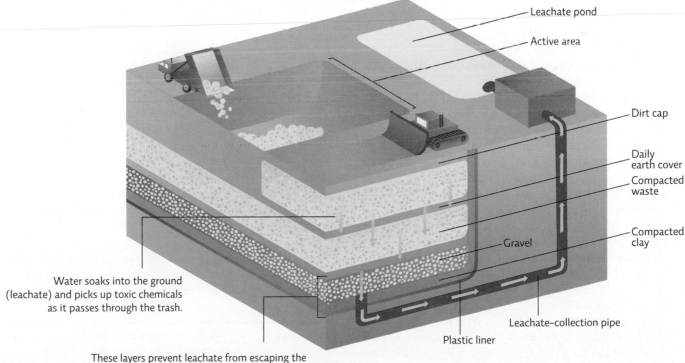

Leachate pond

Active area

Dirt cap

Daily earth cover

Compacted waste

Compacted clay

Gravel

Leachate-collection pipe

Plastic liner

Water soaks into the ground (leachate) and picks up toxic chemicals as it passes through the trash.

These layers prevent leachate from escaping the landfill area and reaching groundwater below.

Infographic **30.3** | **HOW IT WORKS: AN INCINERATOR**

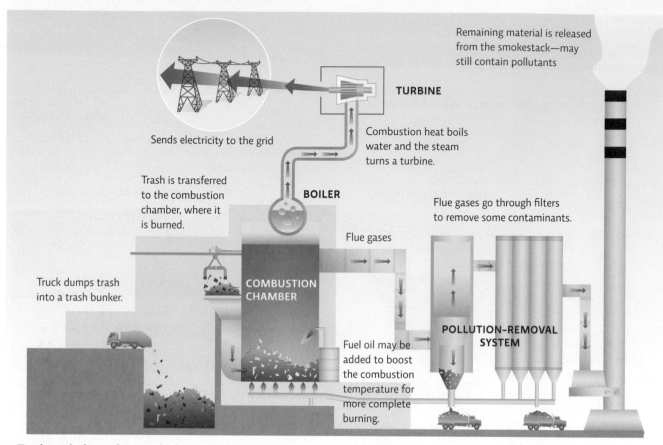

Remaining material is released from the smokestack—may still contain pollutants

TURBINE

Sends electricity to the grid

Combustion heat boils water and the steam turns a turbine.

Trash is transferred to the combustion chamber, where it is burned.

BOILER

Flue gases go through filters to remove some contaminants.

Flue gases

Truck dumps trash into a trash bunker.

COMBUSTION CHAMBER

POLLUTION-REMOVAL SYSTEM

Fuel oil may be added to boost the combustion temperature for more complete burning.

↑ Trash can be burned at very high temperatures in incinerators (most of which are designed to also generate electricity) but in some facilities, fuel oil must be added to the mix for more complete combustion. Cleaning systems remove particulates, sulfur, and nitrogen pollutants, as well as toxic pollutants like mercury and dioxins. The ash is considered toxic waste and must be buried in hazardous waste landfills. Municipal solid waste, medical waste, and some hazardous waste is incinerated in the United States.

A lot of trash—thousands of tons per day—also winds up in specially designed **incinerators**. Burning waste in this way reduces its volume dramatically—by about 80-90%, in fact. But it also pollutes the air and water, and produces ash, some of which is toxic, and must therefore be disposed of in a separate, specially designed landfill. Incinerators are also extraordinarily expensive to build, and tipping fees (fees charged to drop off trash) are usually much higher at an incinerator than at a landfill. [INFOGRAPHIC 30.3]

As dumpsters and landfills fill up, cities and towns begin shuffling their waste from state to state and country to country. New York City, for example, sends its trash to landfills or incinerators in New Jersey, Pennsylvania, Ohio, Virginia, and South Carolina. Other cities ship trash overseas on garbage barges. And the United States as a whole sends its old electronic devices, such as computers, cell phones, and televisions, collectively known as **e-waste**, to some countries in Asia and Africa. There,

impoverished villagers do their best to extract the precious metals within. It's dangerous work. In addition to gold, copper, and zinc, e-waste contains a suite of toxic metals such as lead, mercury, and chromium. When they are released by unsafe and poor extraction methods, these toxins cause a wide range of medical conditions—from birth defects to brain, lung, and kidney damage to cancer. The soil, air, and water surrounding e-waste disposal sites don't fare much better, as toxic chemicals are released into all three during the extraction process (see Chapter 19 for more on e-waste).

Amid this trash transfer, too much of our waste, especially our plastic, is escaping to the open sea. Some of it is blown there by aberrant winds from the tops of open landfills.

incinerators Facilities that burn trash at high temperatures.
e-waste Unwanted computers and other electronic devices such as televisions and cell phones that are discarded.

Some is carried through faulty sewage systems, or in the trickling currents of litter-polluted gutters. To be sure, not all of it is plastic. But other types of waste—textiles, glass, wood, and rubber—sink or degrade relatively quickly. Plastic just floats along. Eventually time, saltwater, and sunlight break it down from its recognizable, everyday forms—combs, candy bar wrappers, CD cases—into fragments so tiny that even thousands of them together can't be seen by a naked eye trained upon a calm sea.

And as far as it has traveled, the plastic in the garbage patch still has a long way to go. It will take decades, maybe even centuries, for those fingernail-sized fragments to degrade into smaller molecules. And even then, they will not cease to pollute. In one study of the North Pacific Gyre, virtually all water samples, even those that were free of plastic debris, contained traces of polystyrene, a common plastic used in a wide range of consumer goods.

Improperly handled waste threatens all living things.

The consequences of mismanaging our trash are manifold. When disposed of improperly, chemical waste can wreak havoc on plant and animal life (even some household trash is considered hazardous and should not be disposed of in a landfill or municipal incinerator); incinerators create small particle air pollution, and landfills produce methane. All of this threatens the balance of life and the health of ecosystems. [INFOGRAPHIC 30.4]

Aquatic life is especially vulnerable. Sea mammals get tangled in everything from discarded fishing nets to plastic six-pack rings, often with fatal consequences. On top of that, many fish and nearly half of all seabirds eat plastic, often by mistake (plastic bags floating in the open ocean look a lot like jellyfish). Some of these animals choke on the plastic or are poisoned to death by its toxins. [INFOGRAPHIC 30.5]

Other sea animals live long enough to be consumed by predators, including humans. That's no small matter. BPA, an organic compound used in plastic, has been shown to interfere with reproductive systems, and styrene monomers, the subunits of polystyrene, are a suspected carcinogen. Plastic also absorbs fat-soluble pollutants such as PCBs and pesticides like DDT. These toxins are known to accumulate in the tissue of marine organisms, bio-magnify up the food chain, and find their way into the foods we eat. (For more on bioaccumulation see Chapter 3).

Researchers suspect that floating bits of plastic can also serve as an attachment point for fish eggs, barnacles, and many types of larval and juvenile organisms. Thus each tiny bit of plastic could potentially transport harmful, invasive, or exotic species to new locales. "I think one of the most underrated impacts of these so-called garbage patches is the introduction of hard surfaces to an ecosystem that naturally has very few of them," says Miriam Goldstein, a PhD candidate at Scripps Institute of Oceanography who studies the Pacific patch. "Organisms that live on hard surfaces are very different than those that float freely in the ocean. And adding all that plastic is providing habitat that would not naturally exist out there."

However, knowing that these things can happen is not the same as proving that they are happening in the patches.

Infographic **30.4** | **HOUSEHOLD HAZARDOUS WASTES**

↓ Hazardous wastes are those that are toxic, flammable, explosive, or corrosive (like acids). Many chemicals that enter your home are actually hazardous and should not be discarded in the regular household trash. The EPA recommends that you contact your local solid waste agency for information on disposing hazardous materials. You can also call 1-877-EARTH-911.

To protect your health and that of your family and the environment, avoid or reduce your use of hazardous chemicals such as:

Drain openers
Oven cleaners
Automotive oil and fuel additives
Grease and rust removers
Glue
Bug and weed killers
Mold and mildew removers
Paint thinners, strippers, and removers

Other materials that are also considered hazardous and may need to be disposed of as hazardous waste include:

Batteries
Fluorescent light bulbs
Mercury thermometers

Infographic **30.5** | **PLASTIC TRASH AFFECTS WILDLIFE**

COMPARISON OF FOOD INGESTED BY LAYSAN ALBATROSS CHICKS IN TWO REGIONS OF THE PACIFIC

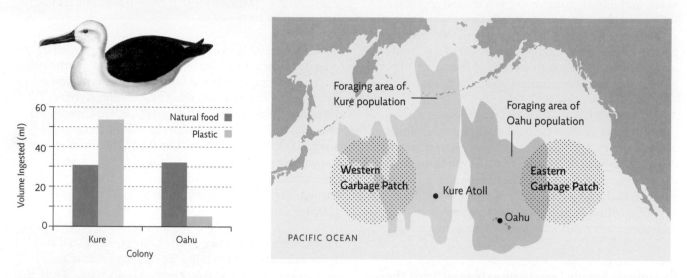

↑ Research by Lindsay Young and her colleagues compared the food ingested by Laysan albatross chicks in two populations, more than 1,200 miles apart in the Pacific. Their data show that while both populations consumed roughly the same amount of actual food, chicks in the western Pacific near Kure ingested 10 times more plastic than chicks near Oahu. In addition, the Kure chicks had 4 times as many plastic pieces and these pieces were, on average, twice as large as those of the Oahu chicks. A comparison of the foraging areas for these two populations gives a clue as to why the Kure birds eat so much more plastic than the Oahu birds.

In general, as Goldstein points out, gyres tend to be areas of very low productivity, which there are very few large fish there. It is not yet clear to scientists whether large numbers or important species of fish (or sea birds) are ingesting plastic from the gyre, or if toxins from the plastic are accumulating in their tissues.

On June 21st, three weeks into their journey, Proskurowski wrote on the ship's blog:

At 0930 this morning, we sampled what I predict will be the largest amount of plastic [the program] has ever recorded in 25 years of sampling. When the tow started, I could distinctly see a few pieces in the upper layer of water and it looked like it was going to be a "good one"—one where we got enough plastic to keep the lab busy. A couple minutes into the tow, suddenly we started to see more and more macro debris—a toilet seat, white plastic bags, oil jugs, a few bread bag fasteners, Styrofoam cups, several shoes, a few foot insoles, a loofah sponge, sunscreen, and liquor bottles—sometimes appearing to form loosely organized windrows (indicative of a special type of turbulence in upper ocean called Langmuir circulation). While everyone was commenting on all the debris, a red 5-gallon bucket drifted by the port bow with a school of about 20 fish underneath it. The bucket got caught in the net. Fearing it would tear the net, we ended the tow early and pulled the bucket with two of its associated schools of grey triggerfish and thousands of tiny

fragments of plastic on deck. Team scientist Skye Moret quickly dissected one fish and found 46 pieces of plastic in its guts. This school of fish, a coastal species that typically lives on reefs, was thousands of miles from land with plastic-filled stomachs.

He was right. At 23,000 pieces of plastic—more than 26 million pieces per square kilometer—the June 21st tow was the largest in the research program's 25-year history. It would take two lab members (rotating every hour or so) more than 14 hours of continuous work to process all of it.

When it comes to managing waste, the best solutions mimic nature.

While nondegradable trash like plastics present a challenge for disposal, much of our waste is biodegradable and we can apply the concept of biomimicry—emulating nature—to teach us how to better deal with this part of our waste stream. **Composting**—allowing waste to biologically decompose in the presence of oxygen and water—can turn some forms of trash into a soil-like mulch that can be used for gardening and landscaping.

composting Providing good conditions for the decomposition of biodegradable waste, producing a soil-like mulch.

Infographic **30.6** | **COMPOSTING**　↓ Composting can reduce a household's trash tremendously. A simple compost pile can be started in the backyard, or a compost bin can be built or purchased.

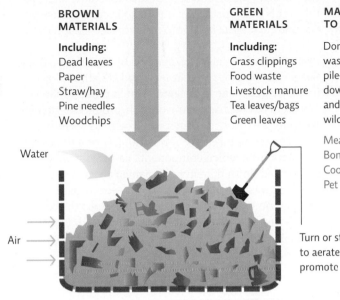

BROWN MATERIALS

Including:
Dead leaves
Paper
Straw/hay
Pine needles
Woodchips

GREEN MATERIALS

Including:
Grass clippings
Food waste
Livestock manure
Tea leaves/bags
Green leaves

MATERIALS TO AVOID

Don't put these wastes in your pile–they won't break down at the same rate and will attract wildlife and pests:

Meat scraps
Bones
Cooking oil
Pet waste

Water

Air

Turn or stir the pile regularly to aerate the pile and promote decomposition.

↑ A variety of compost bins can be used. Even small tubs can be used indoors for those living in apartments.

↓ Municipal composting facilities like this one in Sevier County, Tennessee, use large digesters to process household waste after recyclables have been removed. The digesters produce a mulch that residents can pick up at no cost.

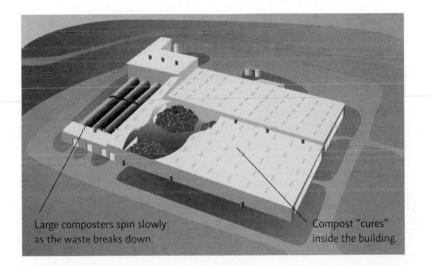

Large composters spin slowly as the waste breaks down.

Compost "cures" inside the building.

↑ Household plant-based food scraps can be added, but meat products should not be included.

↑ The end product is a rich, soil-like mulch that can be used in gardens.

Composting can be done on a small scale (in an apartment) or on a large one (in municipal "digesters"). But this only works for organic waste, such as paper, kitchen scraps, and lawn debris. [INFOGRAPHIC 30.6]

Another promising solution to some of our waste woes involves converting garbage or its by-products into usable energy. For example, the heat produced during incineration might be converted into steam energy or electricity. And the methane produced during anaerobic decomposition can also be harnessed as an energy source. Engineers and scientists are working out ways to capture the methane from landfills, and to use methane-producing microorganisms to process trash in vats or digesters.

In fact, the concept of spinning waste into energy that could then light our homes or fuel our cars is so appealing that it has even extended to waste found in the open ocean. Not long after the Atlantic and Pacific patches were found, a San Francisco-based environmental group tried devising ways to turn the plastic bits into diesel fuel. That plan proved impossible. "There is no way to remove all those millions of tiny pieces without also filtering out all the millions of tiny creatures that inhabit the same ecosystem," says Kara Law. "So once it's out there, all we can really do is hope that nature eventually breaks it all the way down." That could take hundreds to thousands of years. In the meantime, she says, we need to stop adding to the patches that already exist.

Life cycle analysis and better design can help reduce waste.

The real trick to keeping garbage out of the ocean is to go back to the beginning, before the things we throw away are even made. Environmentally friendly products continue to grow in popularity, and manufacturers have given a name to the tactics used to create these products: lifecycle analysis. By assessing the environmental impact of every stage of a product's life—from production, to use, to disposal—an increasing number of companies are trying to reduce the amount of waste generated by the things they design, make, and sell. Cradle-to-cradle analysis takes this even further as it tries to increase reuse potential and turn *waste* back into *resource* (see Chapter 4).

Part of this shift has been spurred by legislation. Some European countries, and at least 19 U.S. states, have implemented "take back laws," which require manufacturers to take back some of their products—namely, computers—after consumers are finished with them. This creates an incentive for manufacturers to design products from which components can easily be salvaged and reused. Indeed, many companies are trying to cut down on waste by *de-manufacturing*—or disassembling equipment, machinery, and appliances into the component parts so that those parts can be reused. BMW, for example, designs its vehicles so that at least 97% of the components can be reused or recycled.

Infographic **30.7** | **INDUSTRIAL ECOLOGY**

↓ The industrial park in Kalundborg, Denmark, is a prime example of industrial ecology. Here, by design, there are 24 different connections between various industries and local farms, such that the waste of one becomes the resource for another.

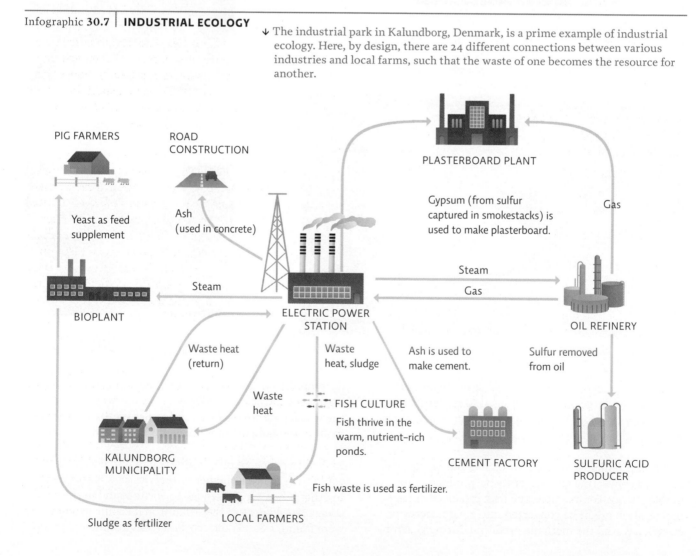

↑ The Foster family of Stowe, Ohio, with their polymer-based possessions.

On a grander scale, in **eco-industrial parks**, industries are positioned so that they can use each other's waste in the same industrial area; one company's trash becomes another's raw material. This type of strategic thinking is known as *industrial ecology*; in addition to "waste-to-feed" exchanges, participants in eco-industrial parks also decrease their ecological footprint by coordinating activities and sharing some spaces, like warehouses and shipping facilities. This form of waste reduction can occur within a single company, between two companies, or between a company and a community. For example, one UK company sells the soapy water produced by its shampoo manufacturing plants to area car washes, rather than throwing it away. The same company also uses its ethanol waste—a by-product of its manufacturing processes—to heat some of its facilities. [INFOGRAPHIC 30.7]

Consumers have a role to play, too.

Advertising—a virtual staple of advanced societies—bombards us from every corner of modern life: not just on the televisions in our living rooms, but on taxicab computer screens, billboard-laden subway cars, the pop-up ads that invade our laptops. The message is surprisingly uniform: To live a happy, more fulfilling life, we simply must have more "stuff." A good deal of that stuff—though certainly not all of it—is made of plastic. "Plastics are incredibly useful and have made possible many of the greatest breakthroughs in technology and standard of living," Proskurowski says. "But they have also made our lives lazier. We now buy a bottle of water rather than refill a canteen. We buy individually wrapped bags of mini carrots, instead of buying carrots that are straight from the ground and have to be washed. There are countless other examples, and as we learn with every net tow, there are significant costs to the planet for those choices."

So how do we start making different choices? As any good environmentalist will tell you, it comes down to the "4 Rs": refuse, reduce, reuse, recycle.

The first thing we can do is simply **refuse** to use things that we don't really need, especially if they are harmful to the environment. This may be as minor as declining to take a plastic bag for a few items purchased at the drug store or as major as choosing a bicycle or walking over taking the car to work. The logic is simple: When we save a resource by refusing to use it, that resource lasts longer, which in turn means that less pollution will be generated disposing of it and producing replacements, and more will be available for future uses. "Refusal doesn't mean never using the resource," says David Bruno, founder of the 100 Things Challenge, a popular movement to pare down our worldy possession to 100 items or fewer. "It just means using it at a more sustainable rate."

If we can't completely refuse a given commodity, we can still try to **reduce** our consumption of it, or minimize our overall ecological footprint by making careful purchases. People who must drive to work can minimize their fossil fuel consumption by choosing a more fuel-efficient vehicle. Those who cannot drink tap water might purchase a specialized faucet or pitcher filter instead of relying on bottled water. And all consumers can greatly reduce the amount of waste they generate by paying special attention to packaging, which accounts for one-third of all U.S. trash, and roughly half of all paper used.

And if we can't avoid using a product, our next best choice is to **reuse**—the third "R"—something consumers can do with just a little effort, by choosing durable products over disposable ones. "Products produced for limited use are really just made to be trash," Bruno says. "They pull resources out of the environment and produce pollution at every step of production, shipping, and disposal." He advises considering use and reuse each time we head to the store; whether you are purchasing clothing, razors, cups, or plates, ask yourself: How long will this last and for what other purpose might it be used?

---

eco-industrial parks Industrial parks in which industries are physically positioned near each other for "waste-to-feed" exchanges (the waste of one becomes the raw material for another).

refuse The first of the waste-reduction "4 Rs": Choose NOT to use or buy a product if you can do without it.

reduce The second of the waste-reduction "4 R's": Make choices that allow you to use less of a resource by, for instance, purchasing durable goods that will last or can be repaired.

reuse The third of the waste-reduction "4 R's": Use a product more than once for its original purpose or another purpose.

Infographic 30.8 | **THE FOUR R'S**

REFUSE: DON'T USE IT

- Avoid disposables (they are made to be trash!).

- Opt out of junkmail at www.optoutprescreen.com and www. dmachoice.org.

- Choose goods with no packaging.

- Bring your own cloth bags when shopping or don't accept a bag if you don't need one.

REDUCE: USE LESS

- Choose goods with minimal packaging.

- Buy in bulk (but only if you will use all that you buy).

- Buy durable, well-made, and repairable goods.

- Use both sides of paper at work or school.

- Post notices on a central bulletin board or send them via email to reduce copies.

- Buy local, fresh food; it comes with less packaging.

- Compost your kitchen and yard waste.

REUSE: USE IT AGAIN

- Use products—such as shopping bags, food containers, and plasticware—multiple times.

- Rent, borrow, or lend items.

- Reuse products in different ways: Use yogurt containers to hold screws, scrap paper for a note pad, and so on.

- Repair broken equipment, tools, furniture, and toys (this becomes possible when you choose items that can be repaired).

- Buy and sell old clothes and household goods or donate them to charities.

- Choose reusable containers for leftovers rather than plastic bags or wrap.

- Bring your own coffee mug to work.

RECYCLE: RETURN IT FOR REPROCESSING

- Check with your local waste management service to see which recyclables it takes (just because a product or its packaging says it is recyclable doesn't mean you can recycle it in your area).

- Buy products made from recycled material (close the loop).

- Encourage family, friends, and coworkers to recycle as well.

- Help start a recycling program at your workplace or in your community.

Reusing also applies to industry. TerraCycle is a U.S. company that produces worm compost fertilizer and packages it in used soda and water bottles. The company also collects hard-to-recycle packaging like candy bar wrappers and juice pouches and turns them into new products like backpacks.

Once we've refused, reduced, and/or reused a given commodity as much as possible, we are left with the final R, **recycling**, the reprocessing of waste into new products. Recycling has several advantages. By reclaiming raw materials from an item that we can no longer use, we limit the amount of raw materials that must be harvested, mined, or cut down to make new items. In most cases, this helps conserve limited resources, not only trees and precious metals, but also energy. To execute this step properly, we must first have purchased items that can be recycled. We must also *close the loop* by purchasing items that are made of recycled materials to encourage manufacturers to make products from recycled materials. [INFOGRAPHIC 30.8]

Of course, even recycling comes with trade-offs. For example, while less energy and water is used making paper from paper than is used making paper from trees, there are also energy and environmental costs in collecting used paper, storing it, and shipping it to recycling plants. Once these costs are factored in, sustainably harvesting a forest of fast-growing trees may prove to be more environmentally friendly than recycling used paper. [INFOGRAPHIC 30.9]

By journey's end, the crew of the *Corwith Cramer* had spent a total of 34 days at sea, traveled 3,817 nautical miles, conducted 128 net tows, and counted 48,571 pieces of plastic along the way. Crew members found the northern and southern boundaries of the Atlantic Patch, at latitudes near Virginia and Cuba, respectively. But despite making it as far as the Mid-Atlantic Ridge, they never found the eastern edge.

As the coast of Bermuda came back into view, on July 14, 2010, Proskurowski wrote his last blog post:

It is easy to brush off the topic of plastic pollution in the ocean. It occurs thousands of miles from land, in regions of the planet that are rarely visited by humans, and relatively sparsely populated by marine life. It is also easy to pass off responsibility, to say 'I recycle.' or, 'It must come from some other (developing) country,' or, 'it is all fishing or marine industry waste.' While there may be kernels of truth in all those arguments, the reality of the situation is that open ocean plastic pollution

---

recycle The fourth of the waste-reduction "4 R's": Reprocessing items to make new products.

occurs over incredibly large regions of the Earth, has widely distributed point sources, and—because the oceans connect the whole globe—has far-reaching consequences.◉

Research articles referenced in this chapter:

Law, K. L. *et al.* 2010. *Science* Vol. 329 no. 5996: pp. 1185–1188.

U.S. Environmental Protection Agency. 2010. *Municipal Solid Waste in the United States: 2009 Facts and Figures* (EPA 530-R-10-012).

Young, L.C., *et al.* 2009. *PLoS ONE*, 4: e7623.

Infographic **30.9** | **OPTIONS FOR PRODUCT DISPOSAL**

↓ We have better options than simply throwing away many products. An item like a plastic bottle can be recovered and reused, as the innovative company TerraCycle does, or the bottle may be recycled into another product.

RECYCLING REQUIRES 3 STEPS: Consumers and industry must turn in or collect materials for recycling, the material must be used to make new products, and the products must be bought by consumers.

Collection

Purchase   Production

RECYCLING PLASTIC: The number code on a plastic item indicates the type of resin it is made with. In many areas, #1 and #2 are the only plastics that can be recycled; in others, all types are taken by recyclers. Check with your community recyclers to see what is taken in your area.

| 1 | 2 | 3 | 4 | 5 | 6 | 7 |
| PETE | HDPE | V | LDPE | PP | PS | OTHER |

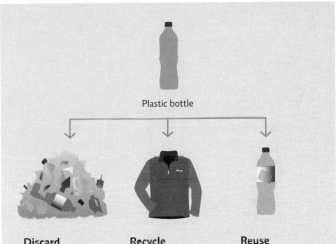

Plastic bottle

Discard
The bottle may end up in the environment such as the Atlantic Garbage Patch.

Recycle
Consumers turn in used plastic bottles to be recycled into new products like this Patagonia fleece jacket.

Reuse
Companies such as TerraCycle collect used bottles and package products such as fertilizer in them.

BRING IT HOME

❂ PERSONAL CHOICES THAT HELP

How much solid waste you produce is under your control. By reducing the amount of waste you produce, you reduce how much money we spend on waste disposal as a whole, and at the same time place less pressure on the resources used to produce consumer goods. Reducing your solid waste is very easy and can save you money in the process.

Individual Steps
→ Track your trash. Record what you throw out for a week by category and weight. How could you reduce your total trash weight by a quarter? By half?

→ Use the information in Infographic 30.8 (strategies to refuse, reduce, reuse, and recycle) to identify five changes you can make to reduce your solid waste.

Group Action
→ Start recycling unusual items in your community. TerraCycle is a company that takes items that usually end up in the garbage, like candy wrappers, corks, and chip bags, and recycles them into new products, like purses or backpacks.
→ Talk to friends and family about having "no gift" or "low gift" celebrations. Instead of buying lots of presents, treat

friends to a dinner or a fun activity. For large families, use a grab bag or draw names and buy for only specific people.

Policy Change
→ Talk to community leaders to discuss the possibility of starting a community-wide composting program.
→ Research recycling rates of participation in your community. Advocate for recycling education and curbside recycling programs.

UNDERSTANDING THE ISSUE

CHECK YOUR UNDERSTANDING

1. **All of the following are consistent with the law of conservation of matter EXCEPT:**
   a. there is no such thing as waste.
   b. matter is neither created nor destroyed, but it can change form.
   c. any matter that is dangerous to living things is quickly biodegraded into a harmless form.
   d. one organism's waste matter is taken up by another organism and reused.

2. **Garbage that you generate in your home, at work, or at school would be called:**
   a. municipal solid waste.
   b. hazardous waste.
   c. industrial waste.
   d. nonbiodegradable waste.

3. **In sanitary landfills, the primary safeguard against groundwater contamination is to:**
   a. do nothing, as groundwater contamination is not a problem with landfills.
   b. line the landfill with impermeable clay and plastic.
   c. design the landfill to use the natural slope of the land so water runs downhill without being contaminated.
   d. locate landfills on industrial sites where groundwater is not used for drinking.

4. **Which of the following statements about incinerators is FALSE?**
   a. Incinerators reduce the volume of material that needs to be landfilled.
   b. Incinerators are expensive to build and operate.
   c. The burning of waste reduces the toxic and hazardous materials in the waste stream, as the heat neutralizes the toxins.
   d. Burning solid waste in incinerators can be used to produce energy.

5. **The best solutions to managing waste would include:**
   a. source reduction and application of natural decomposition processes.
   b. supporting the siting of a landfill in your community to reduce waste transportation costs.
   c. disposing of hazardous waste in oceans, as the water dilutes the waste.
   d. shipping electronic waste to poor countries to provide jobs there in extracting precious metals from the electronic devices.

6. **An industrial ecologist's goals would NOT include:**
   a. increasing resource-use efficiency by altering manufacturing processes.
   b. designing products that are durable and can be de-manufactured.
   c. identifying points in the product life cycle at which waste products can be used in other processes.
   d. identifying points in the product life cycle at which waste products can be landfilled.

WORK WITH IDEAS

1. Why is waste considered to be a human invention? Furthermore, why do we distinguish between different types of waste?

2. Compare and contrast landfilling and incineration. What are the trade-offs for each option?

3. What is biomimicry? Describe two ways this concept can be applied to reducing and managing waste.

4. What is industrial ecology? How can it be applied in managing waste? What is the consumer role in this process?

ANALYZING THE SCIENCE

The data in the following table come from a report by Greenpeace based on research published on plastic debris in the world's oceans between 1990 and 2005.

INTERPRETATION

1. How many different species are included in these data and how many are affected?

2. For seabirds:
   a. What percentage of seabird species had entanglement records? Which two groups had the highest rate of entanglements?
   b. What percentage of seabird species had ingestion records? Which two groups of seabirds had the highest rate of ingestions?

3. For marine mammals:
   a. What percentage of marine mammal species had entanglement records? Which two groups had the highest rate of entanglements?
   b. What percentage of marine mammal species had ingestion records? Which two groups had the highest rate of ingestions?

Species group	Total number of species worldwide	Number and percentage of species with entanglement records	Number and percentage of species with ingestion records
Sea turtles	7	6 (86%)	6 (86%)
Seabirds	312	51 (16%)	111 (36%)
Penguins	16	6 (38%)	1 (6%)
Grebes	19	2 (10%)	0
Albatrosses	99	10 (10%)	62 (63%)
Pelicans and cormorants	51	11 (22%)	8 (16%)
Gulls and terns	122	22 (18%)	40 (33%)
Marine mammals	115	32 (28%)	26 (23%)
Baleen whales	10	6 (60%)	2 (20%)
Toothed whales	65	5 (8%)	21 (32%)
Fur seals and sea lions	14	11 (79%)	1 (7%)
True seals	19	8 (42%)	1 (5%)
Manatees and dugongs	4	1 (25%)	1 (25%)
Sea otters	1	1 (100%)	0

ADVANCE YOUR THINKING

Hint: Access the actual report at http://bit.ly/tdQ7Kr.

4. Scientists suspect that entanglement is a significant cause of population decline for many species, but they consider the reported entanglement rates to be conservative.
   a. Based on the data in the table, which group (sea turtles, seabirds, or marine mammals) is likely to be most affected by entanglement? Why?
   b. Why might reported entanglement rates underestimate the real problem?

5. Ingestion refers to animals eating plastic. While many species of marine mammals, seabirds, and sea turtles ingest plastic, some groups ingest more than others. What might explain these differences in ingestion rates among species?

6. Not much is known about the specific consequences of aquatic species ingesting plastics. In the process of science, observation leads to more questions. List three questions about ingestion of plastics that should be studied. How would you go about answering one of these questions?

EVALUATING NEW INFORMATION

Curbside recycling programs are a very convenient system for recycling metals, paper, glass, and plastic. However, they are available to only about half the population in the United States today, and what can be recycled is limited. So where do you go if you do not have curbside recycling, or if you want to recycle or safely dispose of such things as electronics, batteries, or books? One source of help is the Earth911 website, which has information on how to recycle a vast range of items as well as information on the latest recycling laws, ideas for living a green lifestyle, and feature stories on people and companies who are working to make a difference.

Go to the Earth911 website (earth911.com).

Evaluate the website and work with the information to answer the following questions:

1. Is this a reliable information source? Does the organization have a clear and transparent agenda?
   a. Who runs this website? Do the organization's credentials make the information presented reliable or unreliable? Explain.
   b. What is the mission of this website? What are its underlying values? How do you know this?

   c. What data sources does Earth911 rely on for its information and what is its policy on what the organization puts on its website? Are the sources it uses reliable?
   d. Do you agree with the organization's assessment of the problems with and concerns about waste and recycling? What about its solutions? Explain.

2. On the home page, scroll down to the "Search with Earth911" section and click on the icon for Electronics. If the listing that opens is not for your location, type in your location at the top of the page. What options exist in your community to recycle electronics? Describe two.

3. Click on the "Articles" tab at the top of the home page to get a list of articles that relate to electronics. What types of articles are offered? Who writes these articles? Does the content seem useful and credible? Explain your response.

4. How useful is a website like Earth911? How can a website like this influence societal understanding of waste issues and facilitate a change in behavior on the part of individuals and businesses?

MAKING CONNECTIONS

IS IT TIME TO SKIP THE UBIQUITOUS PLASTIC BAG?

Background: In the United States, we use approximately 100 billion plastic bags a year. After their short use, these bags often end up as litter and many eventually make it to the ocean. Concerns over the impacts of plastic bags on marine wildlife have led to a movement to restrict or ban plastic bags. Ireland and China have instituted fees on single-use plastic bags, and in January 2011, Italy became the first country to implement a nationwide ban on them. In the United States, plastic bag use in the city of San Francisco has dropped by 5 million per month since it instituted a ban on plastic bags in March 2007.

Case: Your community is considering some sort of restriction on plastic bags. You have been assigned the task of evaluating alternative strategies to reduce plastic bag pollution, including:

1. implementing an outright ban on single-use plastic bags.

2. instituting a fee for consumers to use plastic bags. The fee would be used for waste education and litter clean-up efforts.

3. establishing a reusable bag credit for consumers. This credit would be paid by the businesses, as it would save them having to purchase plastic bags.

Research these (and possibly other) options and write a report recommending a course of action for your community. In your report include the following:
   a. An analysis of the pros and cons of each proposal, including:
      · a discussion of the consequences of each choice from economic, ecological, and convenience/practical perspectives.
      · a reflection on the values underlying each proposal.
   b. Based on the information at hand, what is the best option for your community? Who should be involved in this decision? Provide justifications for your proposal.
   c. From what you now know about the consequences of plastic pollution and the challenges of managing plastic waste, develop a set of guiding principles that could be applied to addressing the issue of plastic packaging. Discuss the principles you develop and explain why you consider them to be key to the future of reducing and managing plastic waste.

CORE MESSAGE

Society seeks to protect the natural environment and public health by establishing environmental policies that define what is acceptable behavior for individuals, groups, or nations with respect to the environment. National and international policies are needed when environmental problems extend across state or national boundaries. Because environmental problems are complex, policies are often compromises between various stakeholders; getting agreement on national or international policies can be difficult.

GUIDING QUESTIONS

After reading this chapter, you should be able to answer the following questions:

→ Why are environmental policies sometimes needed at a national or even international level?

→ How are U.S. environmental laws established and how does lobbying influence the process? What are some of the major U.S. environmental laws?

→ What policy tools can be used to implement and enforce environmental policy?

→ How are international policies established and enforced?

→ What international policies have been established to deal with climate change and where do we currently stand on this issue?

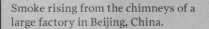

Smoke rising from the chimneys of a large factory in Beijing, China.

COUNTERFEIT COOLING

In the global efforts to thwart climate change, some lessons are learned after the fact

In the spring of 2006, when he was still a law student at Stanford University, Michael Wara had a Eureka! moment—a discovery that would eventually confirm scientists' and policy makers' worst fears. Wara, a former climate-change scientist who was now focused on climate law, had set out to assess a particularly controversial global environmental policy— one that had been implemented a few years earlier in an effort to slow the rise of greenhouse gas emissions.

The *emissions trading* policy known broadly as **cap and trade** was based on a deceptively simple-sounding idea: A *cap*, or upper limit, is set on the amount of any given greenhouse gas that a company is allowed to emit each year. If the company reduces its emissions below that upper limit, it earns credits for the difference. The company can then sell, or *trade*, those "carbon credits" in the global marketplace—often for a hefty sum, usually to other companies that have yet to meet their own emission reduction targets. By purchasing the credits, companies that can't manage to reduce their own emissions can still contribute to reduction initiatives elsewhere, in other industries or other parts of the world (see Chapter 25).

Straightforward as it all sounded, though, policy makers and environmentalists had been sharply divided since the program's inception on one major question: Would the policy actually result in emission reductions? With the emissions trading program now well established, it was time to find out. Wara had deployed an army of formulas across a mountain of spreadsheets, over many, many weeks.

The final tallies were more than disheartening: Somehow, this program that had been designed to reduce greenhouse gas emissions was actually contributing to their increase. "It really shocked us," he recalls. "Nobody could believe it at first."

It looked like one group in particular was gaming the system: The coolant makers of India and China, whose gaseous products (called HFCs) are used to keep air conditioners and refrigerators nice and cold. These companies were eliminating a waste gas produced as a by-product of the coolant manufacturing process and earning thousands of credits (that they sold for tens of millions of dollars to developed countries) to do so. Both the coolant and this by-product waste gas were well-known, potent, greenhouse gases. And the companies were churning out both in spades. "They were producing twice to three times as much coolant gas as they needed to meet market demand," says Wara. "It didn't matter if the coolant sold or not, because they were making a fortune off the credits they were earning by destroying all the by-product."

Carbon and greenhouse gas regulation—including the cap-and-trade system created for it—had come amid a protracted and impassioned global conversation. Political leaders, policy makers, scientists, and environmental activists all over the world had converged on the question of how best to curb emissions and thus stave off global warming. How, then, with all that chatter and debate, did it come to this?

The answer to that question contains a much broader lesson about the challenges of protecting the environment through treaties and legislation.

Public policies aim to improve life at the social level.

Environmental policies give us guidelines meant to restore or protect the natural environment—sometimes by repairing damaged ecosystems, other times by reducing or mitigating the impact we humans have on our planet. As we learned in the first chapter of this book, environmental problems are often "wicked problems," meaning that they tend to be very complex; they have multiple causes and consequences, along with multiple stakeholders. Most of the biggest environmental issues that we now face—like pollution, species endangerment, and climate change—are also **transboundary problems**; they occur across state and national boundaries. That

---

cap-and-trade An emissions trading plan that set upper limits for pollution release (the cap); producers are issued permits that allow them to release a portion of that amount and they can sell or trade permits they do not need to use.

environmental policy A course of action adopted by a government or organization that is intended to improve the natural environment and public health or reduce human impact on the environment.

transboundary problem A problem that extends across state and national boundaries; pollution that is produced in one area but falls in or reaches other states or nations.

adaptive management A plan that allows room for altering strategies as new information comes in or as the situation itself changes.

Infographic 31.1 | POLICIES GUIDE OUR CHOICES AND ACTIONS

↓ National and international policies are needed to address global environmental problems because what happens in one area can affect another. Human impact, such as the transport of non-native species, can have far-reaching consequences. In addition, the atmosphere, rivers, and oceans can very effectively deliver pollution from one area to the next. For example, greenhouse gases released in any part of the world impact global climate. Environmental issues that affect the entire planet require a global response, guided by policies we all agree to follow.

Overexploitation of common resources like ocean fish stocks affects everyone.

Air currents transport pollutants throughout the atmosphere.

Rivers transport species, nutrients, and pollution, which eventually reach the sea.

Ocean currents transport pollutants to nearby and distant areas.

Rising sea level caused by ice melt and thermal expansion affects low-lying areas first.

Damming or pulling too much water out of rivers can cause them to run dry in other regions downstream.

→ Water currents

→ Air currents

means that solving them requires the cooperation of individual states and countries around the world.

That makes environmental policy tricky. From the start, lawmakers must juggle a handful of potentially daunting factors: effectiveness, or whether the policy can attain the desired goal; the negative trade-offs that might result from the policy; who will absorb the cost burden (external and internal costs) of the policy, for both its enactment

and its aftereffects; and whether the policy is flexible enough to accommodate changes—the task of **adaptive management**. Even once all those hurdles are cleared, another remains, and it's often the largest of the bunch: public support. Policies that may be effective, affordable, and flexible can still die before they ever have a chance to be enacted if voters and/or lawmakers don't agree that they are necessary. [INFOGRAPHIC 31.1]

Within the United States, policy can be set at three basic levels: local, state, and national. Before the 1960s, environmental issues mostly dealt with how best to use resources (see Chapter 1). Addressing pollution or environmental damage was not a key objective. And environmental issues were primarily handled at the state level. To be sure, there were a few environmentally focused federal laws on the books (for example, the Oil Pollution Act of 1924 banned the release of oil into coastal waters). But in general, federal environmental regulations were considered an intrusion on state sovereignty. In fact, most environmental problems were only addressed after the fact, through litigation—an arrangement that too often favored the polluters: It was even more difficult back then than it is today to prove that toxins from a factory or dump that had seeped into the water or permeated the air were killing livestock or causing human illnesses.

Eventually, though, things began to change. Industry grew, and so did pollution. As it did, environmental problems began slipping across state lines, so that water and air pollution from one state affected another. In the 1960s and '70s, federal legislators, prompted by a massive national outcry, realized that more regulation was needed, and so devised a new set of policies that could function across state lines: **performance standards**. By determining how much pollution could be released in the first place, environmental regulation shifted from after-the-fact litigation to prevention. The shift proved effective. It's much easier to prove that an industry's emissions violate a legally binding federal standard than it is to prove that those emissions caused cancer rates to skyrocket; and if the standards are sound, the harm should be averted.

By 1969, the era of modern environmental policy had begun. That year, with performance standards gaining a foothold, the **National Environmental Policy Act (NEPA)** was codified into law. NEPA established environmental protection as a guiding policy for the nation, mandating that the federal government take the environment into consideration before taking any action that might affect it. It also established a process that remains central to environmental regulation—where the need for legislation is determined by available scientific evidence, and where various solutions are analyzed and compared, in excruciating detail, before a decision is made.

NEPA's signature feature has been the **environmental impact statement (EIS)**—a report that details the likely effect of a proposed action, such as building a road or upgrading a nuclear facility. The goal of an EIS is to identify problems before they occur so that stakeholders can choose the most acceptable course of action (maybe move that road 10 miles south to avoid disturbing that forest; maybe don't build the road at all). To keep the process transparent, the findings are made available to everyone—citizens, policy makers, and special interest groups—and everyone is given a chance to respond (through letters and public hearings). [INFOGRAPHIC 31.2]

In NEPA's wake came a wave of iconic legislation, most of it passed with overwhelming bipartisan support. Many of the environmental laws passed in the 1970s have a mechanism that allows individual citizens (or groups—including state governments) to demand enforcement via the **citizen suit provision**. Violations can be reported; if they aren't dealt with in a satisfactory manner, the citizen or group can file a lawsuit against the violator (individual,

---

Infographic **31.2** | **POLICY DECISION MAKING**

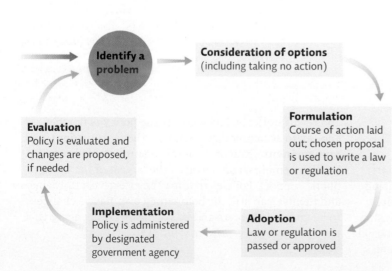

← Policies are created and revised using some basic steps that allow policy makers to systematically evaluate the situation and possible responses. The process starts with identifying the problem, considering available options for responding, and evaluating the costs and benefits. A policy is then drafted, further evaluated by interested parties, and if found acceptable, formally adopted. In the United States, bills must be passed by the House of Representatives and the Senate and then signed by the president to become law. Regulations based on those laws are proposed and administered by regulatory agencies (like the Environmental Protection Agency) in a similar manner. The process itself is responsive and allows for adaptive management—reentering the policy cycle for revision as new or changing information comes to light.

Identify a problem

Consideration of options (including taking no action)

Formulation Course of action laid out; chosen proposal is used to write a law or regulation

Evaluation Policy is evaluated and changes are proposed, if needed

Implementation Policy is administered by designated government agency

Adoption Law or regulation is passed or approved

↓ In the United States, environmental protection and regulation were originally seen as state issues, but by the middle of the 20th century, it became apparent that many environmental problems crossed state lines and would be best handled through federal legislation. These landmark environmental laws were passed, beginning in the 1960s, during a period of tremendous bipartisan cooperation and support for taking steps to ensure a clean and healthy environment. They have been amended many times to deal with changing or new environmental problems.

National Environmental Policy Act (NEPA), 1969	Established environmental protection as a guiding policy for the nation. It mandates that the federal government take the environment into consideration by completing an environmental assessment before taking action that might have an environmental impact.
Clean Air Act (CAA), 1970 (originally passed in 1963)	Regulates air pollutants that are hazardous to human health by setting standards about the amount of pollutants that can be present or released into the air. Greenhouse gases such as CO_2 were not originally covered by the CAA since they are not toxic, but in 2009 the Supreme Court gave the EPA the authority under the CAA to regulate greenhouse gases (since they directly impact climate change).
Clean Water Act (CWA), 1972	Regulates water quality by setting standards for the release or presence of specified toxic or hazardous water pollutants
Endangered Species Act (ESA), 1973	Protects and aids in the recovery of endangered and threatened species of fish, wildlife, and plants in the United States
Toxic Substances Control Act (TOSCA), 1976	Regulates the production and distribution of designated toxic chemicals
Comprehensive Environmental Response, Compensation, and Liability Act (CERCLA), 1980	Commonly called "Superfund," CERCLA requires that responsible parties clean up sites contaminated by hazardous materials and holds them liable for the costs and damages. It also provides funding to decontaminate sites when the owners cannot be found or cannot afford the cost of clean-up.

private company, or the government—even the regulatory agency mandated to enforce regulation) that has allegedly failed to uphold the existing law. [INFOGRAPHIC 31.3]

To implement and enforce all these new federal environmental laws, Congress established the **Environmental Protection Agency (EPA)** in 1970. The EPA is a regulatory agency that establishes rules and regulations to support each environmental law as it is passed. EPA officials set the standards that ensure the goals of any given law are met. They are also tasked with holding individual states and corporations accountable. If a given entity fails to comply with a given rule, the EPA has the authority to step in and mandate changes. They can, for example, force a power plant to make upgrades that decrease pollution, or close down a factory for repeated violations, or fine an individual state for failing to curb its vehicle-generated air pollution. They can also force entities to pay for clean-up costs and in certain cases can revoke operating permits.

Of course, the EPA's reach extends only to our country's borders. And, as the case of the coolant factories shows, most environmental problems tend to stretch way beyond those.

performance standards Targets set that specify acceptable levels of pollution that can be released or exist in ambient (outdoor) air; industries must act to meet these standards.

National Environmental Policy Act (NEPA) A 1969 U.S. law that established environmental protection as a guiding policy for the nation and required that the federal government take the environment into consideration before taking action that might affect it.

environmental impact statement (EIS) A document that outlines the positive and negative impacts of a proposed action (including alternative actions and the option of taking no action); used to help decide whether or not that action will be approved.

citizen suit provision Allows a private citizen to sue, in federal court, a perceived violator of certain U.S. environmental laws, such as the Clean Air Act, in order to force compliance.

Environmental Protection Agency (EPA) An independent federal agency responsible for the implementation and enforcement of environmental laws in the United States.

Infographic **31.6** | **SETTING INTERNATIONAL POLICIES**

↘ Establishing policies at the international level takes time, a lot of work, and compromise. Effective policies are generally those that provide benefits across sectors (not just for one group or region), address the causes as well as the consequences, identify specific targets, and include the flexibility of science-based adaptive management to allow for revision. To succeed, international policies must have buy-in at multiple levels—government, citizens, industry, and among nations—and have effective enforcement.

← Brazilian President Fernando Collor de Mello, center, is applauded as he signs the Framework Convention on Climate Change at the 1992 Earth Summit in Rio de Janeiro, Brazil.

International meetings like this are held to address pressing issues and may result in treaties that identify the course of action that signatory parties agree should be pursued. Meetings typically include presentations by policy makers, scientists, and even individual citizens, and sessions that assess the situation. Negotiations follow and, if successful, conclude with the drafting of a treaty that specifies what the nations/parties agree to do.

← Chairperson Raúl Estrada-Oyuela shakes hands with an official after the Kyoto Protocol was adopted at the Kyoto International Conference Center in Japan, December 11, 1997.

Broad treaties like the 1992 UNFCCC are just the beginning and require additional agreements that lay out exactly what will be done. In this case the Kyoto Protocol provided that direction. A protocol is a treaty that specifically indicates what will be done—precise goals or targets are set. The protocol also indicates how compliance will be assessed and enforced.

← Canadian Environment Minster Peter Kent's notes for his announcement that Canada was withdrawing from the Kyoto Protocol in December, 2011.

For international treaties like the Kyoto Protocol, compliance depends on the cooperation of the signatory parties to do what they said they would do when they signed the treaty. Beyond this, there are few international avenues available to enforce compliance. Other nations may put pressure on the violator by imposing economic sanctions (reducing or cutting off trade) but this does not guarantee compliance and is often seen as a last resort.

That treaty was only the beginning of a long, slow, deliberative process that involved myriad entities, and a seemingly endless stream of meetings, debates, negotiations, and votes. Agreeing that something needs to be done is one thing; coming to a consensus on what to do is another. Policy making, especially at the international level, is a painfully slow business. In fact, it took several years for the UNFCCC to establish a formal statement, essentially saying "we should do something about climate change." [INFOGRAPHIC 31.6]

After that came the **Kyoto Protocol** (1997), an international treaty that lays out exactly what that *something* is: specific greenhouse gas reduction targets, with deadlines. But agreeing on these targets was not easy. Debates raged for months and months, making international headlines each step of the way and frequently pitting the business sector against the environmentalist movement, scientists against policy makers, and countries against countries. At issue was how, exactly, to curb greenhouse gas emissions.

Ultimately, countries like the United States and Great Britain that had historically released the most greenhouse gases were given much bigger reduction targets than other countries—in all, 37 developed nations and the European Union itself (called Annex 1 parties) were given specific reduction targets relative to their emissions in the baseline year of 1990. Because developing countries had not released as much in the past, and did not necessarily have the resources to reduce current emissions, they were not given any reduction targets.

Developed nations—the United States especially—argued that this was unfair: If developing nations could burn more fossil fuels—and China and India were two developing nations that were burning lots of fossil fuels—they would gain an unfair advantage in the global marketplace; lucrative industries would desert us and flock to them. But developing nations were adamant: Emission caps of any kind would stunt their economic growth. Why should they have to bear even a fraction of the burden of reducing pollution that was caused almost exclusively by the developed, industrialized world? [INFOGRAPHIC 31.7]

Enter the **Clean Development Mechanism (CDM)**, created by the Kyoto Protocol in 2003 to broker a compromise that would encourage both developed and developing countries to curb their emissions. Here's what they came up with: Developed, industrialized nations would pay for emissions-curbing technology and projects in developing, unindustrialized nations, by purchasing "carbon offset credits" from them. "For example," Wara explained in a recent *Nature* paper, "rather than build an ineffectual but cheap coal-fired plant, a Chinese utility might instead build a more efficient gas-fired plant that emits less CO_2; the difference in potential carbon emissions between the coal plant and the gas plant can, after monitoring and certification, be converted into CDM credits that can be sold to an industrialized nation. The revenue from the credits enables the utility to afford the more expensive gas plant."

While the United States still declined to ratify Kyoto (the lack of specific reduction targets for all nations proved unacceptable to the members of the U.S. Senate, who rejected the Kyoto Protocol in a 97–0 vote), every other developed nation did sign on. Eventually, the CDM and its cap-and-trade system gave rise to an entire industry built around carbon credits. But it didn't happen overnight. Like most international environmental policies, Kyoto was designed to allow a gradual shift so that countries would have time to respond, and the economy would not take too big a hit; and also so that adaptive technology could be developed that would make the tasks at hand easier to accomplish. This is a common approach: Treaties and protocols like Kyoto identify a target, and then outline a robust time frame, replete with interim targets and deadlines, all in service to the ultimate goal—in this case, zero carbon emissions.

But as Wara and others know all too well, when policies are slow to move forward, problems are also slow to emerge.

Policies sometimes have unintended consequences.

The numbers on the computer screen in Wara's Stanford office are digital avatars; their real-life counterparts—invisible gaseous molecules that trap heat and thus warm our atmosphere—exist a world away. They can be found in the large, fluffy plumes of smoke that curl up from the pipes of aged coolant factories in regions like the state of Gujarat in western India, then billow and disperse across low-slung, densely packed metropolises. Those factories produce two gases in particular: HCFC-22, used in refrigeration and air conditioning; and HFC-23, which is merely a by-product of creating HCFC-22. Because both gases are potent global warmers, both are regulated under the Kyoto Protocol.

There are six Kyoto gases in all: carbon dioxide (CO_2), methane, nitrous oxide, hydrofluorocarbons (HFCs),

---

Kyoto Protocol The 1997 amendment to the UNFCCC that set legally binding specific goals for reductions in greenhouse gas emissions for certain nations that ratified the treaty.

Clean Development Mechanism (CDM) A UN program that allows a country with a greenhouse gas reduction commitment to implement emission-reduction projects in developing countries.

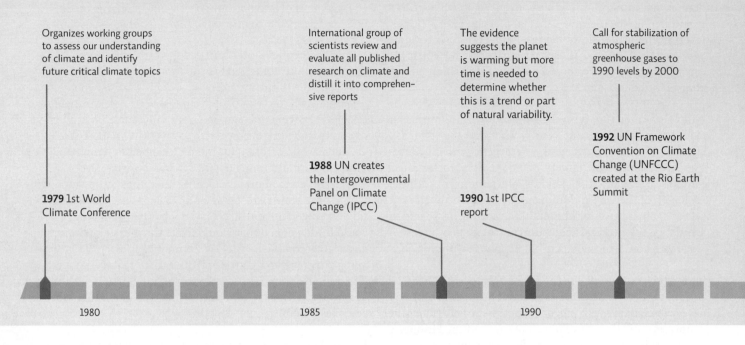

Organizes working groups to assess our understanding of climate and identify future critical climate topics

1979 1st World Climate Conference

International group of scientists review and evaluate all published research on climate and distill it into comprehensive reports

1988 UN creates the Intergovernmental Panel on Climate Change (IPCC)

The evidence suggests the planet is warming but more time is needed to determine whether this is a trend or part of natural variability.

1990 1st IPCC report

Call for stabilization of atmospheric greenhouse gases to 1990 levels by 2000

1992 UN Framework Convention on Climate Change (UNFCCC) created at the Rio Earth Summit

1980   1985   1990

perfluorocarbons (PFCs), and sulfur hexafluoride. For the purposes of crediting, each is converted to its *equivalent* in carbon dioxide, widely agreed to be the most abundant and most massively emitted of the six. So carbon dioxide is assigned a value of "1" (meaning that a single credit represents permission to emit 1 ton of CO_2 per year), and the remaining six are valued in relation to that, based on how potent they are as warmers and how long they can be expected to remain in the atmosphere. Methane is valued at 21 (1 ton of it equals 21 tons of CO_2, and thus 21 carbon credits), nitrous oxide at 310, and so on.

To be sure, the Kyoto Protocol has achieved some successes. The United Kingdom, for example, met its 2012 Kyoto target—a 12.5% reduction in greenhouse gas emissions relative to the 1990 baseline—in 2000, more than a decade early. By the actual deadline, it had cut emissions by a whopping 23%—or double its Kyoto target. Germany also exceeded its Kyoto targets by nearly 10% by the 2012 deadline.

The United States also took steps to monitor and reduce emissions as part of its UNFCCC commitment. Though U.S. greenhouse gas emissions are still higher than 1990 levels, thanks to energy efficiency improvements and land-use changes (like reforestation projects) they are starting to fall—dropping 6% between 2005 and 2010. Improving fuel efficiency in vehicles is also part of the

U.S. approach. New targets for **corporate average fuel efficiency (CAFE) standards** of vehicles have been set; the 2011 average for cars and light trucks of 27.3 miles per gallon (MPG) will incrementally increase to 54.5 MPG by 2025. [INFOGRAPHIC 31.8]

The coolant factories, though, stand in stark contrast to these achievements, in large part because of the way gases were assigned value under Kyoto. HFC-23, which is a potent greenhouse gas and has an atmospheric lifetime of 270 years, has been valued at 11,700. That means eliminating just 1 ton of this by-product gas earns a company 11,700 carbon credits. And that means the gas—a mere by-product with no commercial value of its own—is very valuable. When approved as a CDM project, industries in developing countries earn carbon-offset credits by destroying these gases; they then sell these credits to Annex 1 parties who are themselves trying to meet Kyoto targets. According to Wara's calculations, in some years—when carbon credits were selling for a lot and coolant was selling for just a little—the companies made more than twice as much from the credits as they did from the coolant itself. "That's a profound distortion of the market," he

corporate average fuel efficiency (CAFE) standards A target of the minimum fuel efficiency (MPG) that manufacturers must meet; evaluated as a weighted average of all the cars and light trucks each manufacturer produces.

Set reduction targets for greenhouse gas emissions by developed nations; these countries could offset some emissions through mitigation actions in other countries (e.g., reforestation projects).

Both conferences reconfirm that reductions in greenhouse gas emissions are crucial and identify the need for programs to help less developed nations participate, but no binding targets are set.

Set a deadline of 2015 to establish binding targets to replace Kyoto; extend Kyoto compliance to 2020 with new targets to be set in the near future for developed and developing nations

The evidence suggests human impact is contributing to global warming.

Determine that scientific evidence for warming is significant and action should be taken to avoid serious consequences

Agree that global temps should not exceed 2°C above preindustrial temps; agree to include developing nations in future target pledges

2011 Durban Platform for Enhanced Action, UNFCCC

1995 2nd IPCC report

1998 Kyoto Protocol to the UNFCCC

2001 3rd IPCC report

2007 Bali Road Map, UNFCCC

2009 Copenhagen Summit, UFNCCC

2010 Cancun Agreements, UNFCCC

2005 Kyoto Protocol takes effect

95 2000 2005 2010

Infographic **31.8** | **EMISSION TRENDS**

↓ The International Energy Agency tracks CO_2 emissions from fossil fuel combustion, the leading contributor of greenhouse gas emissions. Data comparing 2009 emissions to the baseline year of 1990 are shown here. While some nations did have a significant drop in CO_2 emissions, others actually increased their emissions. Overall, the Annex 1 nations of the Kyoto Protocol (developed nations who committed to emission targets) reduced emissions and exceeded their Kyoto target of a 4.7% reduction by a full 10%. Much of this reduction was due to the economic restructuring of many of the countries of the former Soviet Union and Eastern bloc. The United States, which did not ratify the protocol and was not bound by its target of a 7% reduction, actually increased CO_2 emissions by 6.7%. Developing nations were not given any reduction targets and many had an increase in emissions. Overall, compared to 1990 emission levels, the world saw a net increase of 38.3%.

PERCENT CHANGE IN CO_2 EMISSIONS FROM 1990 TO 2009

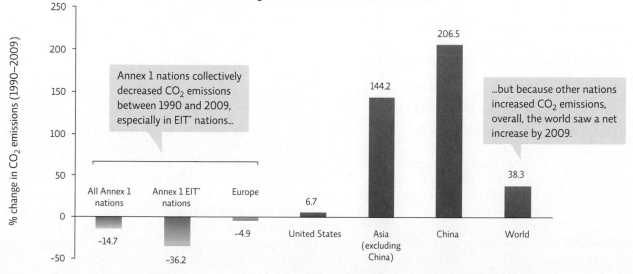

Annex 1 nations collectively decreased CO_2 emissions between 1990 and 2009, especially in EIT* nations...

...but because other nations increased CO_2 emissions, overall, the world saw a net increase by 2009.

*Economies in transition (EIT) include the Russian Federation, the Baltic states, and several central and eastern European states.

Infographic **31.9** | **A CARBON CREDITING SYSTEM GETS SIDETRACKED**

CDM EMISSION CREDITS ISSUED, BY TYPE

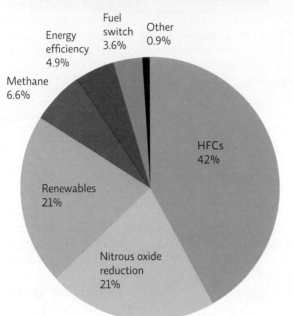

← The main intention of the Clean Development Mechanism was to fund low-carbon energy-production facilities in developing nations in order to reduce global greenhouse gas emissions. Because it is cheaper to build a new power plant from scratch in a developing country (that would be building one anyway) than to prematurely close down or retrofit an older polluting power plant in a developed country, this was seen as a way to reduce global emissions in a more cost-effective way. But rather than funding renewable energy projects and other methods of reducing greenhouse gas emissions, the bulk of CDM money has gone to HFC projects.

says. "Once that balance shifts, you're not in the coolant business anymore, you're in the carbon business."

At least some factories resorted to dubious measures to maintain that business: Data from Wara and others revealed that during crediting periods, they would deliberately use less efficient manufacturing processes in order to generate as much HFC-23 as possible, so that they could maximize their profits by destroying the gas. They also overproduced HCFC-22 "above levels that otherwise would be produced in response to HCFC-22 market demand," according to a report by the NRDC, "simply in order to maximize HFC-23 for destruction."

> "Once that balance shifts, you're not in the coolant business anymore, you're in the carbon business."—Michael Wara

"It's just such a far cry from what policy makers envisioned happening when they set this whole thing up," says Andersen. "The idea was that the money paid for the credits would fund renewable energy projects in the developing world—things like solar and wind power." Instead, the vast majority of that money—billions of dollars—has been given to the refrigeration industry. The

CDM reports that since the UN program began, more than 40% of all credits have been given to just 19 HCFC-22-producing coolant factories—the majority of them in China and India. Countries in sub-Saharan Africa, which were initially expected to be big beneficiaries of the CDM program, have been left out in the cold. [INFOGRAPHIC 31.9]

Adapting policies is necessary but difficult.

So far, undoing the flawed CDM program has been as challenging as getting it into place was—a lesson, say critics, in the importance of adaptive management. Reforming the system would be much easier if more flexibility or fail-safe options had been built into the original plan.

These days, at any rate, there are no shortages of solutions being proposed. Some experts advocate a more intense focus on short-lived greenhouse gases like methane and black carbon (or soot). For his part, Wara advocates going in the opposite direction and making the global carbon market a market for CO_2 rather than for all six gases. "That's the most important one," he says. "It's emitted in the most prodigious quantities, and has a very long atmospheric life." Most CO_2 comes from the energy sector, so focusing on that gas alone would be a good way to ensure that the carbon market pushes humanity away from fossil fuels toward more sustainable energy sources.

China and India have fiercely resisted such changes, and critics charge that politics and political lobbying have had undue influence on reform efforts. When European delegates at one global climate conference suggested that any payments for incineration of HFC-23 should go into an international fund to help factories retool or phase out both the by-product and its underlying coolant product altogether, the Chinese government blocked the initiative, insisting that the money go directly to its own clean development fund.

Likewise, when several Kyoto countries objected to awarding credits to natural gas-burning plants (proponents argued that natural gas emits less CO_2 than a coal plant would; opponents argued that such projects stray too far from the stated goals of the CDM—to fund and support clean, non-fossil fuel-based energy projects), the Chinese factions within the council overrode them.

In the few short years since the CDM was launched, the coolant manufacturers on the panel have amassed both power and influence. As Martin Hession, a past chair of the CDM told the *New York Times* recently, even raising the possibility of trimming future payments was "politically hard." "China and India both have representatives on the panel," the paper reported. "And the new chairman, Maosheng Duan, is Chinese." Some policy makers have worried that if the coolant makers aren't paid to destroy the HFC-23, they will simply release it into the atmosphere. That, says Wara, would be catastrophic.

In 2010, though, the European Union finally put its foot down. And so did the United Nations. Responding to public pressure, both groups began dramatically altering the way they value and pay out HFC-23 emission-reduction credits. The European Union has announced that as of 2013 it will no longer accept HFC-23 credits from companies in its carbon trading system (which happens to be the largest in the world, by a long shot). The United Nations is refusing to award HFC-23 credits to any new factories, and in the fall of 2011 revised downward the percentage of coolant gas that would be eligible for the HFC-23 reduction credit. As Hession told the *New York Times*, the United Nations believes that such measures will eliminate the incentive to overproduce coolant gases.

Others are not so sure. Only time will tell.

In the meantime, many economists still agree that emissions trading schemes are the best hope we have for curbing greenhouse gas emissions. Even Wara remains hopeful. "I am still enamored of market-based approaches to these problems," he says. "For better or worse, those are the incentives that people respond to." Besides, he adds, "As messy a story as Kyoto is, it's still one we can draw valuable lessons from. In California we have a cap-and-trade program spinning into action now. They've learned a lot of lessons from the CDM and I think their program is much stronger because of it."◉

Research articles referenced in this chapter:
Andersen, S.O., and Sarma, K.M. 2010. *Making Climate Change and Ozone Treaties Work Together to Curb HFC-23 and Other "Super Greenhouse Gases".* NRDC Issue Paper.
International Energy Agency. 2012. *CO₂ Emissions from Fuel Combustion: Highlights.*
Wara, M. 2007. *Nature* 445(7128): 595-596

BRING IT HOME

❂ PERSONAL CHOICES THAT HELP

The process of writing and revising policy, proposing it, voting on amendments, then finally enacting it as law is a complex and often messy one. The legislative process is often referred to as "sausage making" because of all the steps, input, and the fact that the final product often looks much different than the original.

Individual Steps
→ Find out who and what is influencing your elected politicians. The website www.opensecrets.org allows you to look up the top individuals and industries that contribute to any candidate's campaign.

Group Action
→ When an important issue is not adequately addressed, concerned citizens often form petition drives. Signatures are collected and delivered to politicians, who can propose new legislation. Form a group to petition for an important issue that you feel is being overlooked and present the collected signatures to any politician who can propose new policy.

Policy Change
→ Actions speak louder than words. Visit www.votesmart.org and find out the voting record of your representative. How does he or she vote on environmental issues like climate change? If you do not feel your representative's record is moving the country forward, volunteer for another candidate whose policies you support during the next election cycle.

UNDERSTANDING THE ISSUE

CHECK YOUR UNDERSTANDING

1. **Why are international laws and policies necessary to address some environmental issues?**
   a. Legislation at the national level does not address important environmental problems.
   b. National legislation cannot address environmental problems that cross national boundaries.
   c. National legislation is not effective because people won't vote for pro-environment laws.
   d. International laws and policies are easier to enforce than national laws and policies are.

2. **Modern U.S. environmental policy:**
   a. requires that environmental impacts be evaluated before federal action is taken, based on an analysis of possible outcomes.
   b. allows input only from stakeholders, e.g., land owners or local citizens.
   c. is focused on damage mitigation rather than on prevention.
   d. is mainly found at the state or local level.

3. **Which of the following can be used to implement or enforce environmental policy?**
   a. Taxes or tax credits
   b. Penalties such as fines or jail time
   c. Cap and trade
   d. All of the above.

4. **Effective international environmental policies:**
   a. are simpler to implement than national policies are.
   b. benefit only a small number of nations and interest groups.
   c. allow for revision based on science or changing needs.
   d. focus on the causes, not the consequences, of environmental issues.

5. **International policies may be enforced:**
   a. through a multitude of international laws.
   b. by international bodies such as the United Nations.
   c. more easily and effectively than national laws and policies.
   d. mainly through voluntary compliance of the nations involved.

6. **International climate change policy:**
   a. has established new binding targets to replace Kyoto targets, which expired in 2012.
   b. requires that developing countries reduce carbon emissions.
   c. has relied heavily on market solutions such as emission caps and carbon credit trading.
   d. has decreased global carbon emissions to below 1990 levels.

7. **Since the 1997 Kyoto Protocol:**
   a. some nations have decreased their carbon emissions.
   b. the United States, even though it was not a Kyoto signatory, met its Kyoto target of a 6% reduction below 1990 levels.
   c. global carbon emissions have fallen below 1990 levels.
   d. All of the above.

WORK WITH IDEAS

1. Give one advantage and one disadvantage of addressing environmental issues at the international level.

2. Describe how U.S. environmental policy has changed from the 1960s to the present day.

3. Give an example of a market-driven approach to solving environmental problems. How does this differ from command and control regulation of environmentally damaging behavior?

4. How does political lobbying affect national and international environmental policy? Give two examples of lobbying groups involved in environmental policy.

5. Which treaty would you identify as the foundation of international policy on climate change, the UN Framework Convention on Climate Change (UNFCCC) or the Kyoto Protocol? Explain your reasoning.

6. Given the current status of the Kyoto Protocol, identify one success and one failure of this treaty.

ANALYZING THE SCIENCE

The U.S. Energy Information Administration (EIA) was established by law to be an independent and impartial energy authority. It provides statistical and analytical information on energy issues, including pollution and climate change, to support public understanding and policy making. The EIA's data on carbon dioxide (CO_2) emissions and atmospheric concentrations are shown in the figure on the next page.

INTERPRETATION

1. Describe in one or two sentences how atmospheric concentrations of carbon dioxide have changed since about 1750.

2. How have anthropogenic (human-caused) emissions of carbon dioxide changed since about 1860? Why do you think that there are no data on anthropogenic emissions before 1860?

3. What relationship do you see between anthropogenic emissions of carbon dioxide and atmospheric concentrations of carbon dioxide?

ADVANCE YOUR THINKING

4. How would you predict this graph would look if the Kyoto Protocol is successful? If it is not successful? Explain your reasoning.

5. About 75% of anthropogenic emissions of carbon dioxide are created during the burning of fossil fuels and about one-third of this is from vehicle emissions. Suppose we started using electric cars. Identify the circumstances under which this might or might not lead to decreased atmospheric carbon dioxide.

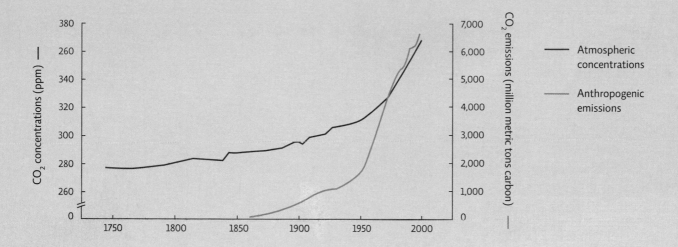

EVALUATING NEW INFORMATION

Climate change can feel like an overwhelming problem, far beyond one person's ability to influence. How much does it really affect the environment if you drive a truck instead of a car, or keep your air conditioner set at 70° instead of 72°? Well, the Environmental Protection Agency (EPA) has created a calculator to answer that question.

Evaluate the website and work with the information to answer the following questions:

1. Visit the Environmental Protection Agency website (www.epa.gov).
    a. What is the mission of the EPA?
    b. Is the EPA website up to date? Does it appear to be accurate? Reliable? Explain.
    c. How long has the EPA been a part of the U.S. government? Does it appear to have been effective? Explain.

2. Go to the EPA emissions calculator (click on the link on the home page, or go to www.epa.gov/climatechange/ghgemissions/ind-calculator.html).
    a. Complete the Household Carbon Footprint Calculator.
    b. What is your estimated annual level of greenhouse gas emissions (personal or family)? What is the largest source of your greenhouse gas emissions?
    c. Do these estimates seem accurate to you? Why or why not?
    d. What actions did the calculator identify that you could use to decrease your emissions? How much emissions could you reasonably save if you followed the recommendations?
    e. Identify one action that you would be most likely to do. What is it and how much impact would it have?
    f. What action would you be least likely to do? Why?
    g. If individuals were to follow the EPA recommendations, what impact do you think this would have on global carbon emissions? Explain your reasoning.

MAKING CONNECTIONS

MEETING NEW CAFE STANDARDS: INCREASING FUEL EFFICIENCY AND DECREASING GREENHOUSE GAS EMISSIONS IN AMERICAN CARS

Background: Cars in the United States have the lowest average fuel efficiency among developed countries, in spite of the corporate average fuel efficiency (CAFE) regulations enacted by Congress in 1975. This is due in part to the American fondness for small trucks, which under CAFE have lower performance standards than cars. In August 2012, President Obama announced new CAFE standards that, if met, will double average fuel efficiency of American cars by 2025 (to 54.5 mpg from 27.3 mpg). Besides reducing American dependence on foreign oil, the revised standards are expected to significantly decrease U.S. greenhouse gas emissions.

Case: There are multiple ways for a car manufacturer to increase its fleet's average fuel efficiency. You are in the strategic planning department of a major American automobile manufacturer and you have been charged with developing a proposal for a major modification to the fleet that will allow the company to achieve CAFE standards. You will evaluate the following four options:

1. Introduce a new technology that saves fuel and reduces emissions (e.g., electric air conditioning or solar battery–charging panels).

2. Create a new hybrid or all–electric car to add to the fleet.

3. Create a new car or modify a current car to use alternative energy sources (e.g., biodiesel, natural gas, algae).

4. Modify a current car to be more fuel efficient.

After researching these options, write a proposal that includes the following:
    a. A review of the advantages and disadvantages of each option, including environmental, economic, political, and social considerations
    b. Your assessment of the best option, based on the advantages and disadvantages discussed
    c. A proposal for remediating at least one of the disadvantages you identified for your "best option"

Of the options you researched, which car would you be most likely to purchase? Explain why you would choose that type of car.

Clay Garden, built on the site of a burned-down home by a resident across the street, provides urban farming opportunities for local residents. In the background are the Webster Morrisania public housing projects, which provide homes for some of the poorest people in the Bronx.

THE GHETTO GOES GREEN

In the Bronx, building a better backyard

CORE MESSAGE

Cities can be both an environmental blessing and a curse. Using green strategies to plan or retrofit cities can benefit citizens, business, and the environment—not to mention reduce environmentally related health problems and degradation to natural resources.

GUIDING QUESTIONS

After reading this chapter, you should be able to answer the following questions:

→ What is the pattern of global urbanization and megacity growth in recent decades?

→ What are the trade-offs associated with cities or urban areas?

→ What is environmental justice? How does urban flight contribute to and result from environmental justice problems?

→ What environmental problems does suburban sprawl generate?

→ How can we create cities that are environmentally sustainable and promote a good quality of life for the residents?

As she walked her dog Xena—a scruffy puppy she had found tied to a tree in her South Bronx neighborhood—Majora Carter considered her options. The 32-year-old aspiring filmmaker had moved back home to save money while she attended graduate school. Initially, she had wanted as little to do with the decaying neighborhood as possible. But then she'd gotten involved in a local artists' group and taken work at a community development center. Now, a colleague at the city parks department was offering her a $10,000 grant to come up with a waterfront development project for her neighborhood. Carter was balking.

At the moment, she and her neighbors were busy fighting a mammoth waste facility that the city was trying to move from Staten Island to the East River waterfront. With 30 transfer stations in the South Bronx, their tiny parcel of New York already handled 40% of the entire city's commercial waste, not to mention four power plants, two sludge processing plants, and the largest food distribution center in the world. All told, some 60,000 diesel trucks passed through the neighborhood every week. In exchange for this burden, area residents boasted the highest asthma and obesity rates in the country, along with some of the poorest air quality. Another waste facility would only make matters worse. Consumed with this battle, Carter wasn't sure she had the time or energy to take on a development project.

Besides, the idea of developing waterfront property in the South Bronx seemed a bit naïve to her. Like most of her neighbors, Carter had lived in the neighborhood most of her life; she knew full well how inaccessible the surrounding river was to residents.

The waterfront—all of it—had long been claimed by industry. There was simply nothing left to develop, she thought, as she and Xena made their way along their usual route—past the transfer station, roaring with diesel-powered, garbage-filled 18-wheelers, along a winding string of garages filled with auto glass shops, metal work, and produce shipments.

And then Xena began pulling her toward an abandoned lot, one they had passed a million times without bothering to notice. After a futile effort to resist, Carter allowed herself to be led down a garbage-strewn path, through a ramble of towering weeds—the kind of place one would never venture alone at night. There at the end, sparkling in the early morning light, was the East River. Carter stood in awe. How many other forgotten patches of waterfront were there, she wondered? Maybe the river wasn't so inaccessible after all.

More people live in cities than ever before.

For the first time in human history, more than half the world's population lives in **urban areas**—densely populated regions that include both cities and the suburbs that invariably surround them. In the United States the proportion is even higher: 80% of Americans are urban dwellers. This migration of more and more people toward urban living is called **urbanization** and it's happening around the world at an unprecedented rate. As global population swells, rural lands are morphing into urban and suburban ones, and ordinary cities are growing into *megacities*—those with at least 10 million residents. With more than 18 million inhabitants, the New York City metropolitan area qualifies as the largest city in the United States and one of the world's 25 megacities. [INFOGRAPHIC 32.1]

⊙ WHERE IS THE BRONX, NEW YORK?

---

urban areas Densely populated regions that include cities and the suburbs that surround them.

urbanization The migration of people to large cities.

Concrete Plant Park, a 7-acre park on the site of a former concrete batch mix plant, is one of the milestones in the creation of the Bronx River Greenway, an environmental effort to transform the Bronx River. New York City officials and local activists re-established salt marshes on the riverbank once strewn with trash and tires and opened it up to recreation and water activities.

Infographic **32.1** | **URBANIZATION AND THE GROWTH OF MEGACITIES**

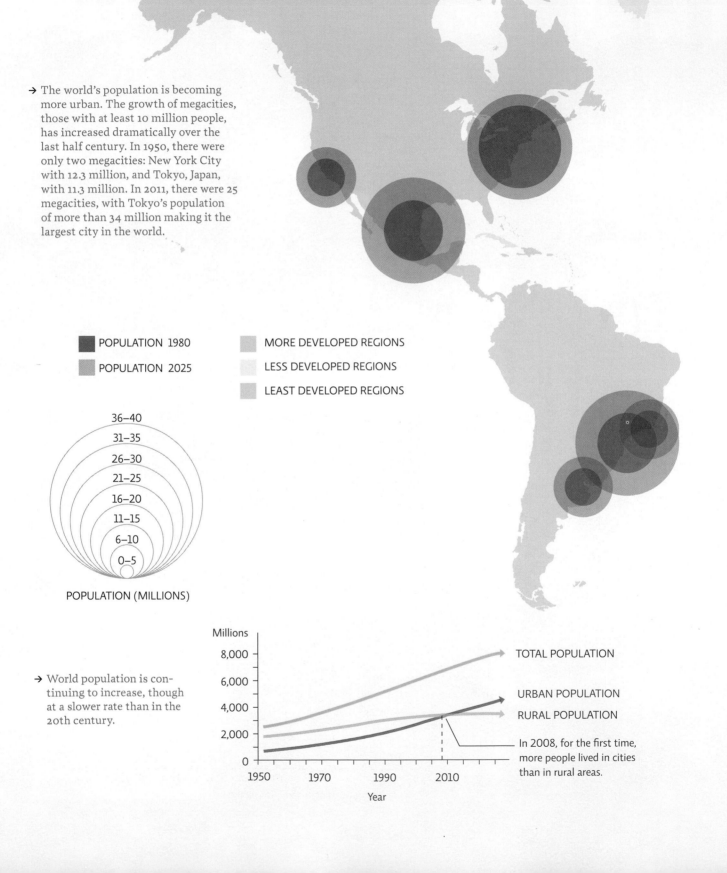

→ The world's population is becoming more urban. The growth of megacities, those with at least 10 million people, has increased dramatically over the last half century. In 1950, there were only two megacities: New York City with 12.3 million, and Tokyo, Japan, with 11.3 million. In 2011, there were 25 megacities, with Tokyo's population of more than 34 million making it the largest city in the world.

POPULATION 1980

POPULATION 2025

MORE DEVELOPED REGIONS

LESS DEVELOPED REGIONS

LEAST DEVELOPED REGIONS

36–40
31–35
26–30
21–25
16–20
11–15
6–10
0–5

POPULATION (MILLIONS)

→ World population is continuing to increase, though at a slower rate than in the 20th century.

Millions

8,000 — TOTAL POPULATION

6,000 —

4,000 — URBAN POPULATION

RURAL POPULATION

2,000 —

0 —

1950   1970   1990   2010

Year

In 2008, for the first time, more people lived in cities than in rural areas.

→ The United Nations predicts that by 2025, there will be 29 cities with populations over 10 million. Most of those cities will be in Asia.

POPULATION OF THE 10 METROPOLITAN AREAS PROJECTED TO HAVE THE LARGEST POPULATIONS IN 2025

■ 2025
■ 1980

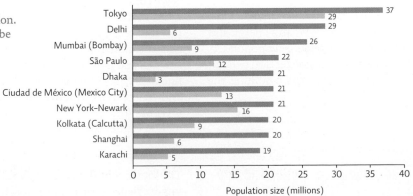

City	1980	2025
Tokyo	29	37
Delhi	6	29
Mumbai (Bombay)	9	26
São Paulo	12	22
Dhaka	3	21
Ciudad de México (Mexico City)	13	21
New York–Newark	16	21
Kolkata (Calcutta)	9	20
Shanghai	6	20
Karachi	5	19

Population size (millions)

Infographic 32.2 | **MANY URBAN AREAS HAVE LOWER PER CAPITA FOOTPRINTS THAN AVERAGE**

→ Due to higher population densities, less personal vehicle travel, smaller homes, and efficiencies of scale, people living in large urban areas typically have a lower ecological footprint than those in suburban areas (but not necessarily lower than rural areas). A 2009 study by geographer David Dodman found that, of the cities analyzed, the only two urban areas with higher footprints than the national average were in China—Shanghai and Beijing (only Shanghai is shown here). However, a look at the data shows that their footprints were not high compared to other large cities but rather, the national footprint in China is very low due to the large rural, low-income population.

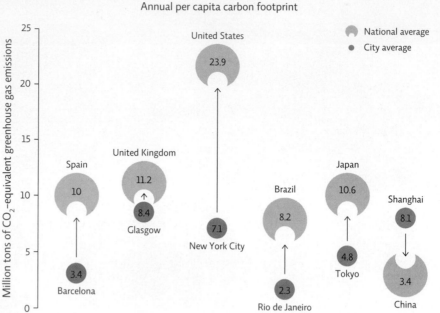

Annual per capita carbon footprint

National average
City average

United States 23.9
Spain 10 / Barcelona 3.4
United Kingdom 11.2 / Glasgow 8.4
New York City 7.1
Brazil 8.2 / Rio de Janeiro 2.3
Japan 10.6 / Tokyo 4.8
Shanghai 8.1 / China 3.4

Million tons of CO$_2$-equivalent greenhouse gas emissions

To be sure, cities bring some obvious advantages to their inhabitants: more job opportunities, better access to education and health care, and more cultural amenities, to name a few. But as far as the environment is concerned, urbanization is both a blessing and a curse. On the plus side, concentrating people in smaller areas (building up rather than out) can reduce the development of outlying agricultural and wild spaces and thus protect existing farms and ecosystems. Higher population densities also make some environmentally friendly practices more cost effective. For example, it's easier to implement recycling and mass transit programs in cities because there are more people to share the cost of these services. Living in smaller homes that are closer to needed amenities and having access to mass transit also decreases the energy use—and the carbon footprint—of urban dwellers compared to those who live in suburban areas. [INFOGRAPHIC 32.2]

On the minus side, cities are locally unsustainable: They require the import of resources like food and energy and the export of waste. Because they are densely populated, most cities are also hotbeds of traffic congestion (which pollutes the air) and sewage overflow (which pollutes the water).

Another problem stems from the way cities are designed and built—namely, the replacement of vegetation with pavement. Plants absorb water, filter air, and regulate area temperatures; pavement and concrete do not. In fact, the blacktop that covers most cities prevents rainwater from being absorbed into the ground, which in turn diminishes groundwater supplies and can lead to flooding (see Chapter 18). Cities also require an abundance of energy. This trifecta—too few plants, too much pavement, and high energy use—conspire to trap solar heat absorbed, and put off, by buildings, making most cities warmer than their surrounding countrysides. This phenomenon is known as the **urban heat island effect**. [INFOGRAPHIC 32.3]

All urban dwellers are vulnerable to the health effects associated with pollutants. But in most cities, the pros and cons of city living are unevenly realized. For example, New York City is one of the wealthiest, most populous cities in the world, but most of the cultural amenities, top-notch health-care facilities, and job opportunities are concentrated in Manhattan, while most of the garbage, sewage, and power plants are located in the Bronx. This imbalance has spawned a whole new area of activism known as **environmental justice**, based on the idea that no community should be saddled with more environmental burdens and less environmental benefits than any other.

The movement is particularly relevant in the most impoverished cities in the world. In Mumbai, the largest

---

urban heat island effect The phenomenon in which urban areas are warmer than the surrounding countryside due to pavement, dark surfaces, closed-in spaces, and high energy use.
environmental justice The concept that access to a clean, healthy environment is a basic human right.

583

Infographic **32.3** | **TRADE-OFFS OF URBANIZATION**

ADVANTAGES

Lower impact per person due to smaller homes and less travelling.

Higher energy efficiency in stacked housing than in freestanding buildings.

More transportation options lessen the need for personal vehicles; closer proximity to destinations make walking and mass transit viable options.

Zoning ordinances are easier to implement.

More job opportunities; local collaboration from a diverse community fosters innovation and ingenuity.

More services for citizens, including more educational and cultural opportunities and better health-care options.

DISADVANTAGES

Concentrated wastes that have to be transported away

Urban heat island effect increases energy needs and can have health consequences.

Dependence on food and resource inputs from outside the city

Disease and violence may be higher in concentrated inner-city areas.

Traffic congestion and its associated air pollution

Less green space leads to stormwater problems.

↓ A volunteer gardener at Finca Del Sur, a garden in the South Bronx, tends the corn stalks while a passenger train goes by in the background. The garden was created on an empty plot of land bordered by a highway exit ramp and a commuter train line.

Infographic 32.4 | **URBAN FLIGHT IS OFTEN "WHITE FLIGHT"**

↑ Urban flight, the movement of people out of inner city areas, is often driven by the decay of urban areas ("flight from blight") and the influx of lower-income groups who are often minorities.

city in India, with 20.5 million people, almost 7 million are slum dwellers who live in horrid, overcrowded conditions—as many as 18,000 people per acre—without adequate sanitation or running water; one report showed only one toilet per 1,440 residents. Globally, more than 1 billion of the world's population live in slums, mostly in large cities in developing countries.

With 90% of future population growth predicted to occur in large cities, urban planners are desperately searching for ways to create cities where the basic needs of their residents are met and where the environmental benefits outweigh the environmental costs. The story of how the South Bronx waterfront was lost and then reclaimed provides important lessons about how to do this.

Suburban sprawl consumes open space and wastes resources.

In the late 1940s when Carter's father, a Pullman porter and the son of a slave, first bought the house Carter would grow up in, the South Bronx was a mostly white working class suburb of Manhattan. But as more blacks and

Hispanics moved to the area, seeking their share of the American Dream, whites fled to nearby commuter towns. The process of people leaving a city center for surrounding areas, originally made possible by the automobile and later by mass transit, is known as **urban flight**. While in many cities—especially those in developing nations—immigration exceeds emigration, urban flight today is triggered by a variety of forces, including over-crowding, noise and air pollution, and in some cases the high cost of city living. [INFOGRAPHIC 32.4]

No matter what the cause, urban flight results in **suburban sprawl**—a slow conversion of rural areas into suburban and exurban ones. **Exurbs** are more sparsely populated towns beyond the immediate suburbs whose residents also commute into the city for work.

As its name suggests, suburban sprawl tends to spread out over long corridors in an unplanned and often inefficient manner. By covering ever-greater swaths of terrain with concrete and pavement, sprawl reduces the amount of land available for farming, wildlife, and ecosystem services. Because of the haphazard way in which they develop, the resulting communities are heavily dependent on driving; unlike cities, which are densely populated

Infographic 32.5 | **SUBURBAN SPRAWL**

↳ Suburban development often leads to suburban sprawl—low population density in developments that appear outside of a city. Homes typically get larger the farther they are from the city and residents have a larger ecological footprint (larger homes and more time spent driving). The suburbs now have their own suburbs—the exurbs, which are commuter towns that are beyond the traditional suburbs but whose residents still commute into the city, often an hour or more each way. Both suburbs and exurbs often displace farmland and wildlands.

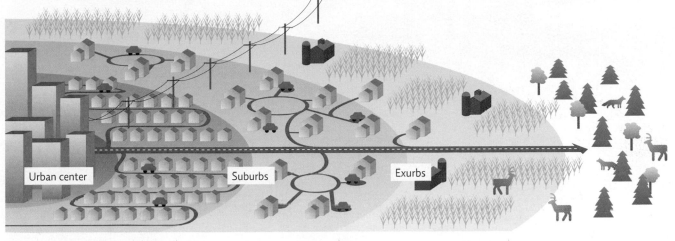

Urban center

Suburbs

Exurbs

Homes get larger (and less energy efficient) with distance from urban center. Single-family homes are more common and take up more space (on bigger plots of land). Commuting time increases.

Loss of arable land as developers purchase and subdivide fertile fields

Loss of species habitat

and can accommodate mass transit systems, suburbs and exurbs force residents to drive almost everywhere they need to go. And because suburban homes are typically larger than urban ones (and exurban homes are often even larger than suburban ones), they tend to have a greater ecological footprint. [INFOGRAPHIC 32.5]

Urban planners of today are well aware of these perils and often work to mitigate them. But in the 1960s, prolific **urban planner** Robert Moses was all too happy to accommodate urban flight and the sprawl that came with it. It was Moses who commissioned the Cross Bronx Expressway, a highway that enabled commuters from suburban Westchester County in the north to completely bypass the Bronx as they traveled in and out of Manhattan each day. But the expressway also displaced 600,000 Bronx residents and further segregated the ailing borough from the rest of New York City. "The South Bronx was utterly cut off," says Marta Rodriguez, a lifelong Bronx resident and a colleague of Carter's. "We didn't stand a chance."

To make matters worse, in the Bronx and elsewhere in the 1960s, urban flight led to redlining—the process whereby

banks rule certain sections of a city off-limits to any type of investment. Bronx landlords quickly discovered that if their neighborhood was redlined, torching their buildings and collecting the insurance money would yield greater profits than renting or selling. By the 1970s, the burning of tenement houses had spiraled out of control, so much so that at one Yankees game, an addled sportscaster famously declared, "Ladies and gentlemen, the Bronx is burning." And as the shopping centers and apartment buildings were shuttered or burned, other industries took their place—namely, the garbage disposal operations and auto parts manufacturers that had been shunned by wealthier enclaves. Before long, the South Bronx had been transformed into an industrial wasteland. In most cities, *zoning laws*—laws that restrict the type of development allowed in a given area—create a buffer between

urban flight The process of people leaving an inner city area to live in surrounding areas.
suburban sprawl Low-population-density developments that are built outside of a city.
exurbs Towns beyond the immediate suburbs whose residents commute into the city for work.
urban planner A person who develops land-use plans in and around cities.

↑ Majora Carter received a MacArthur "genius" grant for her work with Sustainable South Bronx. She stresses that the environmental movement is not just one of the middle-class majority who can afford to buy organic food, drive hybrid cars, and live in areas with little pollution. Low-income families also deserve a clean and healthy environment.

commercial and residential areas so that factories aren't wedged between houses. But in the Bronx, such laws were routinely ignored. Without other options, area residents were forced to accept factories and warehouses built on the ashes of apartment complexes. And because these new industrial neighbors preferentially hired commuters from outside the Bronx, unemployment rates skyrocketed—along with crime, poverty, and asthma.

It turned out that the patch of waterfront Carter and Xena had stumbled upon was a relic of the Moses-era highway expansion, sitting as it did beneath the Sheridan Expressway, a stretch of highway originally meant to cut across the entire northeast Bronx. The Sheridan was abandoned when planners realized it would run through the Bronx Zoo, a popular tourist destination. But by then the damage was done. Hunts Point—Carter's neighborhood—had been isolated and the surrounding waterfront condemned to wasteland.

Carter knew that replacing the abandoned lot with a park would be a big first step toward righting some of the wrongs that her community had endured. The trees and plants would trap pollutants from the air, preventing them from infiltrating people's lungs. The grass and soil would absorb rainwater so that it could no longer carry trash and detritus from the streets into the river. And

claiming even a small patch of waterfront for themselves would give Carter and her neighbors a sense of ownership, not to mention a connection to nature and a place to stretch their legs.

Indeed, studies by urban planning expert Reid Ewing and others have shown that parks improve both the physical and psychological health of people who live near them. And cities themselves—cities that, like the Bronx, were once plagued by drug trafficking and rampant gun violence—have shown that more green space can also mean less crime. In Bogotá, Colombia, in the late 1990s, for example, a particularly environmentally conscious mayor noticed that while his city was designed to accommodate heavy automobile traffic, the vast majority of his electorate did not drive. So he narrowed municipal thoroughfares from five lanes to three, expanded bike lanes and pedestrian walkways and established a string of parks and public plazas throughout the city. The result? People stopped littering. Crimes rates dropped. And slowly but surely, city residents reclaimed their streets.

Bogotá was not so different from the South Bronx, Carter thought. If that city could go green on a third-world budget, surely she and her colleagues could raise enough money to do the same. Starting with the $10,000 seed grant from the city parks department, they leveraged a small fortune in additional grants, donations, and private investment, until they had finalized plans to build a $3 million park, complete with gardens, grassy knolls, and East River kayaking. Hunts Point Riverside Park—the spot that Carter stumbled upon—would be the borough's first waterfront park in more than 60 years. But that was just the beginning.

> Carter knew that replacing the abandoned lot with a park would be a big first step toward righting some of the wrongs that her community had endured.

Environmental justice requires engaged citizens.

Energized by their successful riverside park project, Carter and her neighbors formed a nonprofit called the Sustainable South Bronx (SSBx). The group immediately set its sights on an even grander vision: They would create a greenbelt around the entire community—1.5 miles of waterfront greenway, 12 acres of new waterfront open space, and 8.5 miles of green streets (those with landscaped medians)—all connected by an interlinking system of bike and pedestrian pathways that stretched from the Hunts Point Riverside Park, around the South Bronx's winding edges, all the way to the existing 400-acre park on Randall's Island. They would also disassemble the Sheridan Expressway and turn it into 28

acres of additional parkland, some of which they would designate as *conservation easements*—tracts of land that the city would sign a legally binding agreement not to develop.

It was an ambitious agenda indeed—an expensive one, too—and would require the support of administrators and elected officials from the Bronx to Manhattan to the state capital in Albany. "There is a big fear that environmental justice is fiscally irresponsible," says Carter. "People running the city think 'how can we spend money on parks when we're coming up short on schools, and clinics, and job training, and healthcare?' What they don't realize is that parks can actually help with those things, too." Parks not only increase community pride, but also create green jobs and improve health.

Convincing community members of these benefits would prove as difficult as convincing legislators. Getting them to come out and oppose a landfill was one thing; area residents knew all too well what another trash heap would do to their neighborhood. But getting them to support a park? They had more pressing concerns. Theirs was the poorest congressional district in the city; at the moment, more than 20% of residents were unemployed. And their neighborhood hadn't had a waterfront park in more than 60 years, let alone an entire greenway. Why bother now? "We'd ask people, 'What would you like to see in your neighborhood?' and they really didn't have an answer," Rodriguez says. "They'd never been asked that question before. It was as if having parks was too far in the future for them."

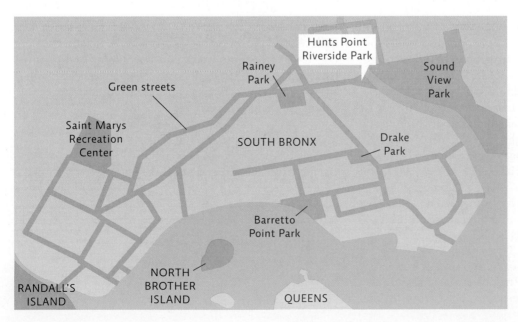

← The proposed South Bronx Greenway will connect the Hunts Point Riverside Park to Randall's Island with pedestrian and bike paths, providing a much-needed recreational area and a viable alternative means of transportation while improving air quality and offering opportunities for economic development.

In fact, the SSBx vision folded readily into a growing movement aimed at making cities more environmentally friendly and socially equitable. **New Urbanism**, as the movement is called, maintains that cities (both now and in the future) have the capacity to reduce our per-person ecological footprint, even as they improve the quality of life for people, provided they are designed properly. The City University of New York Institute for Research on the City Environment estimates that if a city is designed and built with an eye toward sustainability, the ecological footprint of any given urban dweller could be trimmed to about half that of the average American.

On top of the carbon savings, the consensus emerging from environmentalists, sociologists, and economists is that the future lies in cities—that's where most people will live and perhaps is where most people should live. Cities promote interaction among a diverse group of people. This promotes the exchange of ideas and lessens cultural and economic barriers.

The future depends on making large cities sustainable.

Sustainable cities are those where the environmental pros outweigh the cons—where sprawl is minimized, walkability is maximized, and the needs of inhabitants are met locally. In recent years, urban planners have come up with a wide range of strategies for accomplishing these goals. To achieve self-sufficiency, for example, a sustainable city might maintain a mixture of open and agricultural land along its outskirts. Such land could provide a large part of the local food, fiber, and fuel crops, along with recreational opportunities and ecological services. Waste and recycling facilities could also be located nearby, along with other enterprises aimed at producing resources needed by area residents. To stave off sprawl, the same city might establish urban growth boundaries—outer city limits beyond which major development would be prohibited. Keeping any outward growth that does occur as close to mass transit as possible minimizes the impacts of transportation, just as building "up" (parking garage) rather than "out" (huge, expansive parking lot), minimizes the amount of land used. To encourage more walking and less driving, zoning laws might allow for mixed land uses, where residential areas are located reasonably close to commercial and light industrial ones.

Of course, building an ideal city from scratch is easy compared with the task of overhauling an existing city, especially when that city is as densely populated and ever expanding as New York. Upgrading decaying infrastructure like roads, public places, and sewage and water lines can be more expensive than new construction and the process is disruptive to residents. Urban retrofits are certainly possible but sometimes their very success raises property values to the point that the original residents can no longer afford to live in their own neighborhoods. Even so, there are plenty of ways that American cities can push themselves into the environmental plus column. For example, **infill development**—the development of empty lots within a city—can significantly reduce suburban sprawl. And even the most car-friendly of cities has a range of options for reducing traffic congestion and the air pollution that comes with it: reliable public transportation, car sharing programs that allow residents to use cars when needed for a monthly fee, and sidewalks and overhead passageways that allow pedestrians to safely cross busy roads. The cumulative effect of strategies like these, which help create walkable communities with lower ecological footprints, is known as **smart growth**. [INFOGRAPHIC 32.6]

Persuading people to support smart growth, as Carter and her colleagues soon discovered, was a matter of showing them that the benefits could be economic as well as environmental. "You need to show them what we call the triple bottom line," says James Chase, vice president of SSBx (and Carter's husband), referring to the economic, social, and environmental impacts of any decision. "Developers, government, and residents all need some tangible, positive return." A major park project would surely be a boon for all three. Developers would be guaranteed millions in waterfront development contracts. Residents could look forward to cleaner air and water, a prettier neighborhood, and better health as a result. The state and city government would save a bundle in health-care costs. The greenbelt would also spur the local economy—such a vast stretch of public space would attract street vendors, food stands, bicycle shops, and sporting goods stores.

It would also require a green workforce. Some of the undeveloped property that Carter and her neighbors hoped to convert into parkland was contaminated with hazardous waste. These sites are called *brownfields*, and they require a special type of cleanup, or remediation, before they can be developed.

The surrounding wetlands, suffering from decades of neglect, would also need to be restored. And maintaining the new trees, plants, and parks they hoped to create

---

New Urbanism A movement that promotes the creation of compact, mixed-use communities with all of the amenities for day-to-day living close-by and accessible.
infill development The development of empty lots within a city.
smart growth Strategies that help create walkable communities with lower ecological footprints.

Infographic 32.6 | SUSTAINABLE CITIES AND SMART GROWTH

↓ Smart growth can be applied to large cities or to smaller communities. It employs strategies that make efficient use of land to create pleasant livable communities with a lower ecological footprint than current suburban areas.

Take advantage of compact building design and incorporate environmentally friendly technologies.

Create a range of housing opportunities and choices.

Renovate and develop existing communities (rather than building outside the city).

Foster distinctive, attractive communities with a strong sense of place.

Encourage community and stakeholder collaboration; make development decisions that are fair and cost effective.

Mix land uses to place residential and commercial areas together.

Provide urban green space; preserve farmland and critical environmental areas.

Create walkable and bike-friendly neighborhoods.

Provide a variety of "clean" transportation choices into and around the city.

Infographic 32.7 | **GREEN BUILDING**

↓ There are many steps that can be taken to build or retrofit a building so that it has less environmental impact and is a healthier environment for those who live, work, or go to school there. The nonprofit group, Green Building Council, certifies buildings through its LEED program (Leadership in Energy and Environmental Design). Buildings receive a standard, silver, gold, or platinum rating, based on a variety of criteria that include energy efficiency, sustainable building material use, and innovative design.

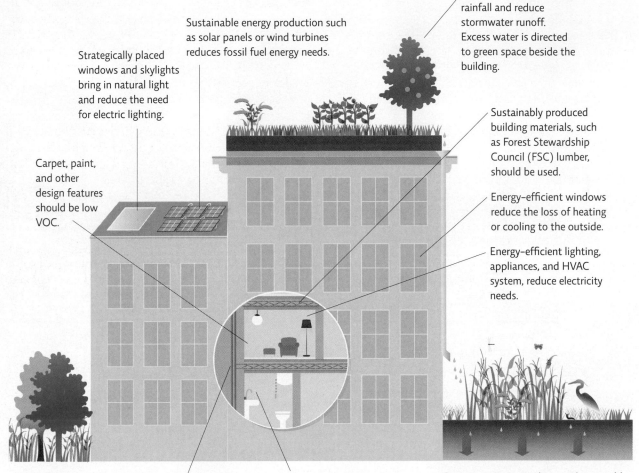

Green roofs help absorb rainfall and reduce stormwater runoff. Excess water is directed to green space beside the building.

Sustainable energy production such as solar panels or wind turbines reduces fossil fuel energy needs.

Strategically placed windows and skylights bring in natural light and reduce the need for electric lighting.

Carpet, paint, and other design features should be low VOC.

Sustainably produced building materials, such as Forest Stewardship Council (FSC) lumber, should be used.

Energy–efficient windows reduce the loss of heating or cooling to the outside.

Energy–efficient lighting, appliances, and HVAC system, reduce electricity needs.

Energy-efficient heating systems might recycle heat from equipment (industrial or commercial settings) or use more efficient designs such as radiant floor heat.

Water-saving devices such as waterless urinals and motion-sensitive low–flow faucets are encouraged.

Green space, rain gardens, and permeable or grass pavers allow stormwater infiltration and offer opportunities for building users to enjoy nature.

would require a workforce trained in urban forestry. Anxious to claim these emerging professions—all of which promised job security and living wages—for their own community, Carter and her neighbors launched the Bronx Environmental Stewardship Training—a green job training program that teaches South Bronx residents the principles of urban forestry, brownfield remediation, and wetland restoration. Program participants—many of

them ex-convicts and high school dropouts facing prison time—also learn how to install solar panels and retrofit older buildings to make them energy efficient. So far, 82% of the participants have found jobs in the green economy and 15% have gone on to college.

The green economy includes a movement known as **green building**, that is, the construction of buildings that are better for the environment and the health of those who use them. The best green buildings are awarded a Leadership in Energy and Environmental Design (LEED) certification. The Bronx Library Center is a silver-certified

green building Construction and operational designs that promote resource and energy efficiency, and provide a better environment for occupants.

LEED building. It earned the silver certification by recycling 90% of the waste materials created during the construction of the building, using architectural design and efficient heating and cooling systems to save 20% of energy costs, and using sustainably grown wood in 80% of the construction lumber.

Carter and her team also launched Smart Roofs, LLC, a green-roof and green-wall installation company. Green roofs are one type of rain garden—an area seeded with plants that can tolerate wet conditions (see Chapter 18 for more on rain gardens). A 2004 study by the New York City Department of Design and Construction found that consumers could save more than $5 million in annual cooling costs if green roofs were installed on just 5% of the city's buildings. According to a study by Columbia University, the same amount of green roofing could achieve an annual reduction of 350,000 tons of greenhouse gases. And Riverkeep, an environmental nonprofit, found that green roofs can retain 810 gallons of stormwater for every $1,000 of investment—easing pressure on the city's overburdened sewer systems and mitigating water pollution from storm runoff. Green roofs also lessen the urban heat island effect. [INFOGRAPHIC 32.7]

Green roofs require factory-made artificial soil (ordinary topsoil is too heavy and can clog drainage systems). And to prevent leaks, a specially designed membrane system must be installed. Producing the artificial soil, creating and installing the membranes, and planting the right mix of species to trap water requires labor. Skilled labor. "Once we figured out the employment factor, we had a win-win-win," says Chase. "There's all these jobs—good jobs—that are going to take off in the next decade, but that not a lot of people know how to do right now. It was a clear opportunity for us." To help the company take off, SSBx secured tax credits from the state legislature. Building owners who install green roofs on at least 50% of their available rooftop now receive a 1 year property tax credit of up to $100,000. Carter offered up her own roof as the first test case.

By the time the Hunts Point Riverside Park opened, dozens of cities across the country—from Madison, Wisconsin, to Miami, Florida—had taken up the mantle of sustainability and smart growth.◉

Research articles referenced in this chapter:
Beckett, K., and Godoy, A. 2010. *Urban Studies*, 47: 277–301.
Dodman, D. 2009. *Environment and Urbanization*, 21: 185–201.
Ewing, R., et al. 2003. *American Journal of Health Promotion*, 18: 47–57.
United Nations Population Division. 2008. *An Overview Of Urbanization, Internal Migration, Population Distribution And Development In The World.* New York: United Nations.

BRING IT HOME

◔ PERSONAL CHOICES THAT HELP

A sustainable community is one that promotes economic and environmental health and social equity. It is one in which the health and well-being of all citizens are considered, while those citizens help implement and maintain the community.

Individual Steps
→ Investigate and support sustainable businesses in your area (sustainablebusiness.com). Research products before you purchase them to understand the impact of your consumption choices (goodguide.com).
→ If you have a balcony or yard, plant flowers, vegetables, or trees.
→ Support local businesses by shopping and dining close to home.

Group Action
→ Join neighborhood cleanup days. If you can't find one, organize one.
→ Start a petition to get more bike lanes in your city. Ride public transit more often.
→ Find out how colleges and universities are working toward sustainable practices at AASHE.org.

Policy Change
→ Attend a meeting of your city council or county commission and ask members to look into smart growth opportunities.
→ See how well you can plan for a sustainable community. Play the new PC strategy game Fate of the World and see how policies you put in place impact global climate change, rainforest preservation, and resource use.

UNDERSTANDING THE ISSUE

CHECK YOUR UNDERSTANDING

1. **In 2025, where will most of the world's megacities be found?**
   a. North America
   b. Europe
   c. Africa
   d. Asia

2. **In response to redlining in the 1960s, landlords in the Bronx:**
   a. increased rents, forcing working-class people to move out of the borough.
   b. subdivided apartments into smaller units to increase the number of renters.
   c. burned their buildings to collect insurance money.
   d. constructed walkways for tenants to access the nearest subway.

3. **A city that promotes smart growth:**
   a. encourages development at the city edges.
   b. provides tax incentives for people who own more than one car.
   c. allows vacant lots to accumulate in the city for a more open look.
   d. mixes land uses to place residential and commercial areas together.

4. **Which of the following is an example of environmental injustice?**
   a. Locating industries away from where people live
   b. Building garbage dumps in high-poverty, low-income areas
   c. People chaining themselves to trees to prevent the trees from being cut down
   d. Preventing the construction of a dam to save an endangered species of fish

5. **What unexpected impact did Bogotá experience as a result of building more green space?**
   a. Decline of mass transit use
   b. Lower crime rates
   c. Lower property values
   d. More athletic injuries

6. **Which of the following describes a brownfield?**
   a. An area that requires cleanup of pollution and hazardous waste before it can be developed
   b. A field of dormant grasses in the fall
   c. An area where treated sewage is spread to dry and create compost
   d. In a city, a vacant lot where buildings once stood

WORK WITH IDEAS

1. Explain the triple bottom line, using the Bronx waterfront restoration as an example.

2. Describe a green roof system. What are the costs and benefits? Why might green roofs be especially beneficial in cities?

3. You are taking a road trip with friends, driving from New York to Los Angeles in July via the interstate highway system. Along the way, your friends begin to notice that the temperatures are more bearable driving through the country than the city. Explain why this happens.

4. Describe a LEED-certified building. What criteria are used to evaluate a building for LEED certification?

ANALYZING THE SCIENCE

A new LEED-certified science building has been proposed for your campus. The school is excited because the new building will generate good publicity and save money on energy costs. As part of the new building, current energy use will be displayed visibly for anyone to see. An example of the output is provided below using August 2009 data from an office building constructed to LEED gold standards.

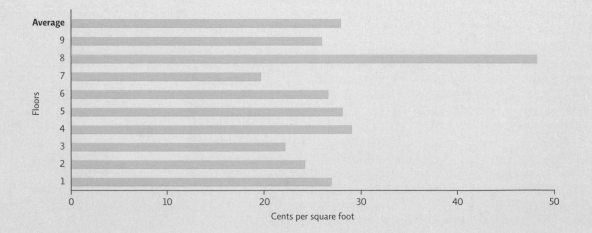

INTERPRETATION

1. In one or two sentences, summarize the information provided in the graph.

2. Provide a title and a legend, including the take-home message for the graph.

ADVANCE YOUR THINKING

3. Based on the data and what you know about LEED-certified buildings, what recommendation would you make to the future inhabitants of the new building on your campus?

4. It is relatively straightforward to design and construct a new building to meet LEED standards. But even existing structures can be renovated to meet LEED standards. Think about areas of homes that are often renovated. What can individual consumers do to improve the sustainability of their existing homes?

EVALUATING NEW INFORMATION

More and more people around the world live in cities, and the number of megacities is increasing. There are costs and benefits to living in cities, and in an effort to increase the benefits, there is a growing movement toward "greening" cities. Environmental and social scientists have published a number of studies documenting the effects of greener cities on environmental and human health.

Go to the website Green Cities: Good Health (depts.washington.edu/hhwb/Top_Introduction.html). Read the introduction to the site.

Evaluate the website and work with the information to answer the following questions:

1. Is this a reliable information source? Does it have a clear and transparent agenda?
   a. Who runs this website? Do the organization's credentials make it reliable/unreliable? Explain.
   b. Who are the authors? What are their credentials? Do they have the scientific background and expertise to lend credibility to the website?

On the menu bar, click on the "Research Themes" link. Then, on the left menu, choose one of the following links: "Crime & Fear," "Active Living," or "Mental Health & Function."

2. What type of information is provided on the page? What is the source of the information?

3. List a couple of the "Fast Facts" provided. Then scroll down the page and read each section. For the topic you chose, what is the primary claim? What data is provided to support the claim?

4. Do you find the data convincing? Why or why not? How does what you read relate to your own life? Give specific examples.

MAKING CONNECTIONS

BUILDING GREEN

Background: In 2011, a record number of damaging tornados occurred in the midwestern and southern parts of the United States. The town of Joplin, Missouri, suffered devastating losses; so many buildings were destroyed that the amount of debris was greater than that generated by the destruction of the World Trade Center in Manhattan. Most residents have remained and the process of rebuilding is ongoing.

Case: You are a resident of Greensburg, Kansas. Greensburg suffered a similar loss in 2007. It is a much smaller town than Joplin, but faced the same problems of rebuilding. As a member of the town who participated in reconstruction, you have been invited to Joplin to describe what happened in Greensburg and the subsequent rebuilding strategy.

Research Greensburg, Kansas, and prepare a three-part presentation for the citizens of Joplin.

1. In Part 1, include:
   a. a brief history of the town, including its main attractions, population density, and location.
   b. the type of people who live in the town. Include information about how they make a living, their religious beliefs, and their overall political views.
   c. a brief description of the tornado and subsequent damage.

2. In Part 2, include:
   a. an explanation of why the community decided to rebuild Greensburg as a "green city".
   b. a description of the plan they adopted and how they set about accomplishing their goals.
   c. the results of the rebuilding effort to date.

3. In Part 3, present your conclusions and recommendations for the town of Joplin. Base these both on your knowledge of what happened in Greensburg and on your understanding of the benefits of greening cities.

APPENDIX 1

BASIC MATH SKILLS

Math skills are needed to evaluate data and even to understand much of the information in news reports. Here we present a review of some basic skills that will be useful in this class and in other science classes.

AVERAGES (MEANS)

To calculate an average, add all the numbers in the data set and divide by the number of numbers.

Example: The sum of these numbers is 10,547; 10,547 / 8 = 1318.375. You could round this off for an average or mean of 1,318.

Data set	
1,004	766
2,349	988
456	1,203
1,882	1,899

WORK WITH AVERAGES

Problem 1: If these were your grades on exams, what would be your exam average?
84, 73, 93, 95, 79, 86

PERCENTAGES/FREQUENCIES

To **convert a fraction to a percentage**, divide the numerator (top number) by the denominator (bottom number) and multiply by 100.
Example: To express the fraction $2/_5$ as a decimal, divide 2 by 5, which equals 0.4. Multiply this by 100 for your answer: 40%.
To **convert a decimal to a percentage**, multiply by 100; a shortcut for this is simply to move the decimal over two places to the right.
Example: $0.08 \times 100 = 8\%$

WORK WITH PERCENTAGES

Problem 2: If 8 out of 32 frogs in a pond have deformities, what percentage of frogs have deformities?

Problem 3: In a pond, 25% of the frogs have leg deformities. If there are 100 frogs in the pond, how many have deformities?

Problem 4: If there are 68 frogs in the pond and 25% have deformities, how many have deformities? (First, make an estimate based on your answer to Problem 3—will it be a higher or lower number? This will help you decide if the answer you calculate is reasonable or whether you might need to recalculate.)

SCIENTIFIC NOTATION

In science, we often use very large or very small numbers. To make these easier to present, scientists use scientific notation, which multiplies a number (called the coefficient) by 10 (the base) raised to a given power (the exponent). If the coefficient is 1, we can leave it off and simply show the base and exponent (e.g. $1 \times 10^2 = 10^2$). The exponent tells us how many orders of magnitude larger or smaller to make the number. In other words, the exponent is telling us how many zeros the number will have: $10^2 = 100$; $10^3 = 1,000$, and so on. Negative exponents represent decimals; for example: $10^{-2} = 0.01$; $10^{-3} = 0.001$, and so on.

Here is a simple shorthand way to evaluate numbers given in scientific notation. Move the decimal place to the right if 10 has a positive exponent, and to the left if the exponent is negative. The number of spaces the decimal place is moved is equal to the exponent. For example, 10^2 tells us to move the decimal place 2 spaces to the right; 10^{-2} means we move it 2 spaces to the left.

By convention, we always designate the coefficient as a whole number (2) or a decimal, with the decimal point at the "10" position (2.3). In other words we would write 2.3×10^5, not 23×10^4. Both are technically correct but the first is the preferred format.

Examples:
$2 \times 10^6 = 2,000,000$
$2.36 \times 10^5 = 236,000$
$4.99 \times 10^{-4} = 0.000499$

Some typical values you might run across include:
$10^6 = 1$ million
$10^9 = 1$ billion
$10^{12} = 1$ trillion

MEASUREMENTS AND UNITS OF MEASURE

There are many handy conversion calculators on the Internet, but it is still useful to have a general idea of how large various units of measure are and how metric and English systems of measurement compare.

LENGTH

Metric
1 kilometer (km) = 1,000 meters (10^3)
1 meter (m) = 100 centimeters
1 centimeter (cm) = 10 millimeters
1 millimeter (mm) = 0.000001 meters (10^{-6})
1 micrometer (μm) = 0.000000001 meters (10^{-9})
1 nanometer (nm) = 0.000000000001 meters (10^{-12})

English
1 mile (mi) = 5,280 feet
1 yard (yd) = 36 inches (in) or 3 feet
1 foot (ft) = 12 inches

Conversions
1 km = 0.621 mi
1 m = 39.4 in
1 cm = 0.394 in

1 mi = 1.609 km
1 yd = 0.914 m
1 in = 2.54 cm

MASS

Metric
1 metric ton (mt) = 1,000 kilograms
1 kilogram (kg) = 1,000 grams
1 gram (g) = 1,000 milligrams
1 milligram (mg) = 0.001 grams (10^{-3})

English
1 ton (t) = 2,000 pounds
1 pound (lb) = 16 ounces (oz)

Conversions
1 mt = 2,200 lb
1 kg = 2.2 lb
1 lb = 454 g
1 lb = 0.454 kg
1 g = 0.035 oz

VOLUME

Metric
1 liter (L) = 1,000 milliliters
1 milliliter (ml) = 0.001 liters

English
1 gallon (gal) = 4 quarts
1 quart (qt) = 2 pints or 4 cups
1 pint (pt) = 16 fluid oz

Conversions
1 L = 0.265 gal or 1.06 qt
1 gal = 3.79 L

AREA

Metric
1 hectare (ha) = 10,000 square km

English
1 acre (ac) = 4,840 square yards (yd)

Conversions
1 ha = 2.47 ac
1 ac = 0.405 ha

CONCENTRATIONS

Metric
1 part per million (ppm) = 1 mg/L
1 part per billion (ppb) = 1 μg/L
1 part per trillion (ppt) = 1 ng/L

TEMPERATURE CONVERSIONS

Fahrenheit (°F) to Celsius (°C): °C = (°F − 32) × $\frac{5}{9}$
Celsius (°C) to Fahrenheit (°F): °F = (°C × $\frac{5}{9}$) + 32

In general:
1°C = 1.8°F
1°F = 0.56°C

Answers to problems:
1. 85   **2.** $\frac{8}{32}$ = 0.25 = 25%   **3.** 100 × 0.25 = 25   **4.** 68 × 0.25 = 17

APPENDIX 2

DATA-HANDLING AND GRAPHING SKILLS

This tutorial offers a quick look at the basics of working with data and graphing.

Scientists gather data to learn about the natural world. Data can be organized into graphs, which are "pictures" or visual representations of the data. Because they can condense and organize large amounts of information, graphs are often easier to interpret than a simple list of numbers. They show relationships between two or more variables that help us determine whether the variables are correlated in any way and allow us to look for trends or patterns that might emerge. To be effective, graphs should be constructed according to conventions, and must be accurately plotted and properly labeled. Certain types of graphs are more suitable than others to show particular types of data, so it is important to choose the correct graph for your data.

The following sections describe variables found in graphs; data tables; and the types of graphs commonly used in environmental science.

VARIABLES

The **independent variable** is the parameter the experimenter manipulates—it could be whether or not a group is exposed to a treatment (given a medicine, exposed to a particular wavelength of light), is part of a distinct group (trees at specified distances from a stream), or is a group followed over a period of time (monitored daily, yearly, etc.). If you were setting up a data table in which to record the data your experiment would produce, you would be able to fill in the values for the independent variable *before beginning* the actual experiment.

The **dependent variable** is the response being measured in the experiment—the responding variable. The experiment is being conducted to see if this variable is "dependent on" the independent variable. In other words, when you change the independent variable, does the dependent variable change as a result? If you were setting up a data table in which to record data, you would be able to include a column heading for the dependent variable, but you would not be able to enter the values until the experiment was complete. There may be more than one dependent variable being tested.

DATA TABLES

Data tables have a conventional format. The independent variable is shown in the left-hand column and the data for the dependent variable or variables are shown in columns to the right of that. To be useful, the data table needs to have a descriptive title (what data are we looking at?) and the units of measure must be included.

ANNUAL ATLANTIC HERRING CATCH

Independent variable →

Dependent variable →

Year	Herring catch (1,000 tons)
1965	731
1970	580
1975	382
1980	270
1985	180
1990	150
1995	45
2000	20
2005	26

Units of measure are given—here we multiply each value by 1,000 to find the number of tons of catch that year; using the 1,000-fold conversion in the units allows us to use numbers that are easier to interpret.

TYPES OF GRAPHS

A. LINE GRAPHS

In science, researchers often test the effect of one variable on another. Line graphs are used when the independent variable is represented by a numerical sequence (1, 2, 3...) rather than discrete categories (red, yellow, blue...). The dependent variable is always a numerical sequence.

Steps to producing a line graph

1. **Determine the *x* axis and the *y* axis.** The independent variable is usually shown on the *x* axis (the horizontal axis). In a line graph, this variable is one that changes in a predictable, numerical sequence, such as the passing of time, increasing concentrations of a solution being tested, habitat distance from the seashore, etc. The data being collected (the response being observed) represent the dependent variable, which is shown on the *y* axis (the vertical axis). The axes require a descriptive title that indicates exactly what each axis represents, along with units of measure, if needed.

2. **Set up the axes.** Set up each axis so that the largest data value for that variable is close to the end of the axis, leaving as much of the available space for your graph as possible. Aim for 5 to 15 "ticks" (the small dividing marks) on any given axis—don't overload it with 50 tiny ticks or have so few that it is hard to place data points. It is essential to evenly space the ticks on a given axis, keeping the increments the same numerical size. On the sample graph, all the x axis ticks are 10 years apart; all the y axis ticks are 100 units apart. The increments will depend on your particular data; however, be sure they are presented in the same size. Give the graph a descriptive title.

3. **Plot the points.** Plot each point by finding the x-value on the x axis and moving up until you reach the y-value position across from the y axis. If you are graphing more than one set of data, draw data points as different shapes or colors and provide a legend to identify each data set.

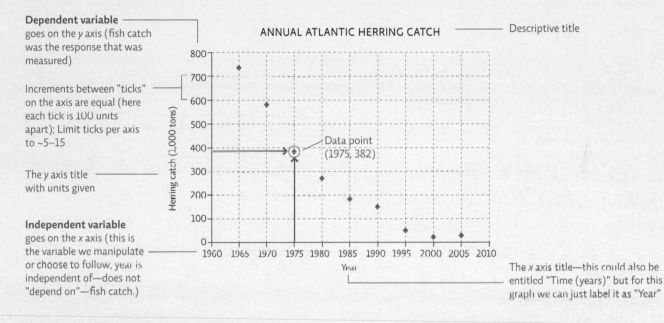

Dependent variable goes on the y axis (fish catch was the response that was measured)

Increments between "ticks" on the axis are equal (here each tick is 100 units apart); Limit ticks per axis to ~5–15

The y axis title with units given

Independent variable goes on the x axis (this is the variable we manipulate or choose to follow, year is independent of—does not "depend on"—fish catch.)

Descriptive title

Data point (1975, 382)

The x axis title—this could also be entitled "Time (years)" but for this graph we can just label it as "Year"

4. **Draw the line.** Once data points are in place, you can draw your line, but don't simply connect the dots unless they all line up exactly. Step back and visualize what kind of trend the data are showing and draw a line that approximates that trend. These trend lines can be mathematically determined but can also be fairly accurately estimated by simply drawing in a line that goes through the center of the data—about as many points will be above as below the line. You can draw a straight line or you may elect to draw a curve to accommodate shifts in the trend.

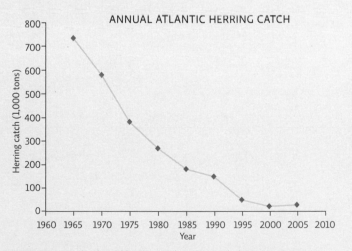

"Connecting the dots" like this implies that each data point is perfectly accurate and that this exactly represents the relationship between the two variables.

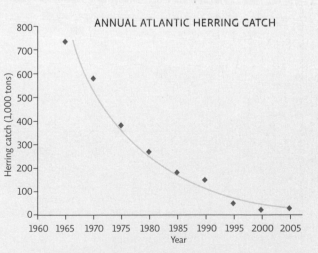

Drawing in a "trend line" that floats through the cloud of points is a more accurate estimation of the actual relationship seen between the two variables. We could draw a straight trend line for this data, but since it seems that the rate of decline lessens as times goes by, the curve seen here may better represent the relationship.

Interpreting the data

Once the graph is made, we can evaluate the data and draw conclusions. The first step is to simply describe the relationship seen—this is a statement of the *results* (observations). Here we see that between 1965 and 1980, herring catches dropped off dramatically and thereafter continued to drop, but more slowly. Now that we understand the relationship between the two variables, we can draw *conclusions*—make some inferences: What might have caused this relationship? What else may be true because this relationship exists? We could infer from these data that the herring population size also decreased in this time frame. If we know that the same number of fishers were fishing for herring the same number of days each year, the slower decline after 1980 might represent the fact that the fish are more difficult to catch because the population size is smaller. We might also conclude that it has not been as profitable to fish commercially for herring since 1980 as it was in the 1960s and 1970s. These last three statements are conclusions (inferences) based on the results of the study (observations); they are not observations themselves.

Interpolation and extrapolation (projections)

Plotting line graphs also allows us to estimate values of *y* or *x* within the range of our data set, values that we did not actually measure (*interpolation*). We can also create a projection of data points beyond our data set (*extrapolation*) by extending the line. This assumes the same trend will hold at higher or lower *x* axis values, which may or may not be true; therefore, extrapolations are not likely to be as accurate as interpolations. Extrapolations, also called projections, are often shown as dashed lines.

WORLD POPULATION

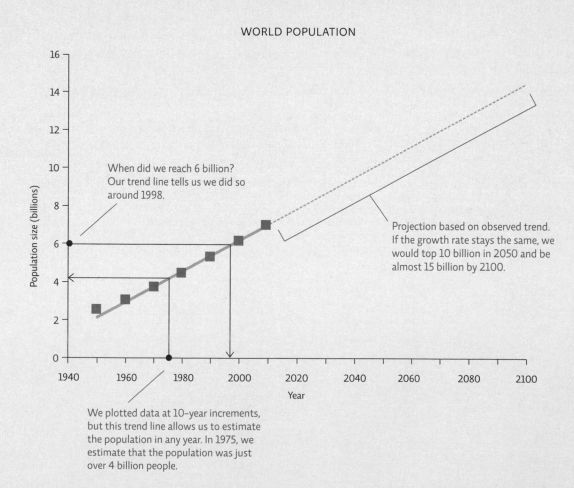

When did we reach 6 billion? Our trend line tells us we did so around 1998.

Projection based on observed trend. If the growth rate stays the same, we would top 10 billion in 2050 and be almost 15 billion by 2100.

We plotted data at 10-year increments, but this trend line allows us to estimate the population in any year. In 1975, we estimate that the population was just over 4 billion people.

B. SCATTER PLOTS

Scatter plots (with or without a trend line) are used when any x-value could have multiple y-values. For instance, in the second graph shown here, data were collected from various countries. Girls in four of the countries surveyed receive, on average, 4 years of schooling; therefore, there are 4 data points over the x axis value of 4. But each of those countries had different y-values (total fertility rate). Here it would make no sense to "connect the dots." The resulting line would be impossible to follow.

It is more appropriate to construct a line that passes through the cloud of points and shows the "trend," just as we did with the line graph. Data points can be entered into a computer graphing program to calculate a "best fit" line, but you can also estimate the path yourself. To pick the best fit line, draw a line (straight or curved) that passes centrally through the cloud of data points, with about as many points above as below the line—the occasional point far away from the others won't impact the line significantly. When completed, your line may or may not be straight, but it should not connect the dots.

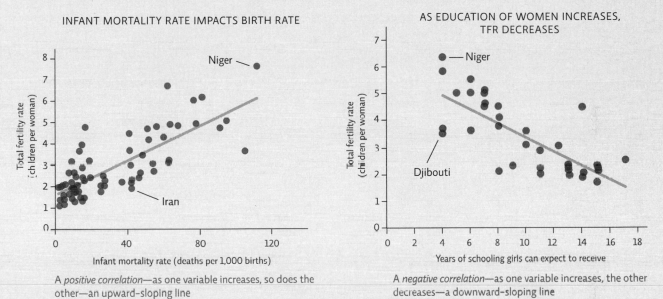

A *positive correlation*—as one variable increases, so does the other—an upward-sloping line

A *negative correlation*—as one variable increases, the other decreases—a downward-sloping line

C. PIE CHARTS

Pie charts are useful when the groups represented by the independent variable are all discrete categories (e.g. red, yellow, blue. . .) rather than a numerical sequence (e.g. 1,2,3. . .). In addition, the categories also represent all the subsets of a whole—all the category values add up to 100%. In other words, you have the entire pie! Data values and/or category titles can be shown either inside each "slice" or outside the pie. The data could also be shown as a bar graph (see Section D on the next page), but showing it as a pie chart instead allows one to more easily compare the size of each group to the other groups and to the whole.

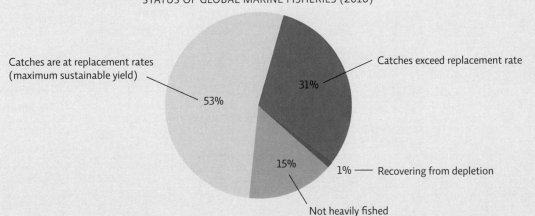

STATUS OF GLOBAL MARINE FISHERIES (2010)

D. BAR GRAPHS

Bar graphs are appropriate in some cases. As with a pie chart, the key consideration is whether the independent variable (the variable you are manipulating in your experiment) is part of a numerical sequence (in which case, a line graph or scatter plot would be used) or represents discrete, separate groups. When you have discrete, separate groups, a bar graph can be used—it would make no sense to connect the data from one group to the next in a line. As with a line graph or scatter plot, the independent variable usually goes on the x axis and the dependent variable is shown on the y axis.

For example, researchers examined the stomach contents of birds from two different colonies. "Colony" is the independent variable because the researchers chose to see if where a bird lived would affect the type of food it ate. The volume of each food type ingested is the dependent variable—it is the data that the researcher set out to find and it may change according to colony location.

FOOD AND PLASTIC EATEN BY TWO ALBATROSS COLONIES

Y axis values represent a numerical sequence; ticks are placed in equal increments, with the largest number slightly higher than the largest data value.

Different groupings within a category are shown as different colors (or patterns, if black and white) and a legend is provided.

Bar height corresponds to the y axis value for that particular group.

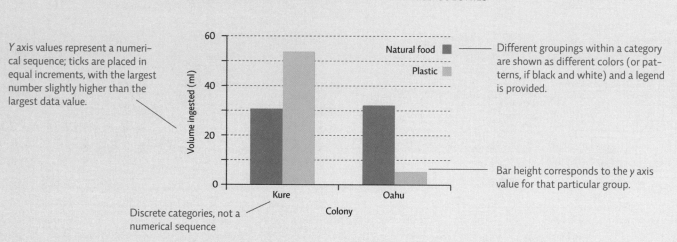

Discrete categories, not a numerical sequence

Sometimes it is easier to place the independent variable on the y axis if the labels themselves are long. This prevents the need to place labels sideways, making them harder to read.

The graph below, which shows the percent of women who would like to use birth control but have no access to it (unmet need), displays the independent variable (region) on the y axis and the dependent variable (percent with unmet need) on the x axis.

UNMET NEED FOR BIRTH CONTROL

The dependent data value can be shown on the bar graph for quick viewing.

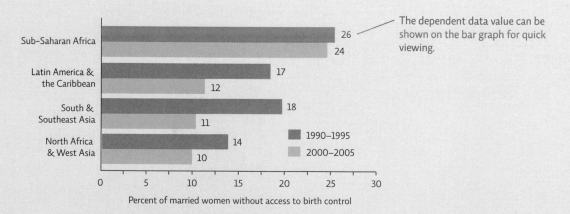

E. AREA GRAPHS

Another useful graph that allows us to view the relative proportion of all the groups being compared is an area graph. It is used when we have a line graph (the independent variable on the *x* axis is a numerical sequence) showing multiple lines. Each data set (line) is part of a larger group—here we show total fish catch broken down by type of fish. Each line is graphed "on top" of the other, and the space between the lines is filled in with a different color. This is useful because at any given *x* axis point (say, the year 1968) we can see what the total fish catch was as well as how much each type of fish contributed to the total catch. The width of the "ribbon" for each fish type at that point represents its *y* axis value—in this case its catch in 1,000 tons. (The *y* axis value opposite the ribbon represents the total for all groups.)

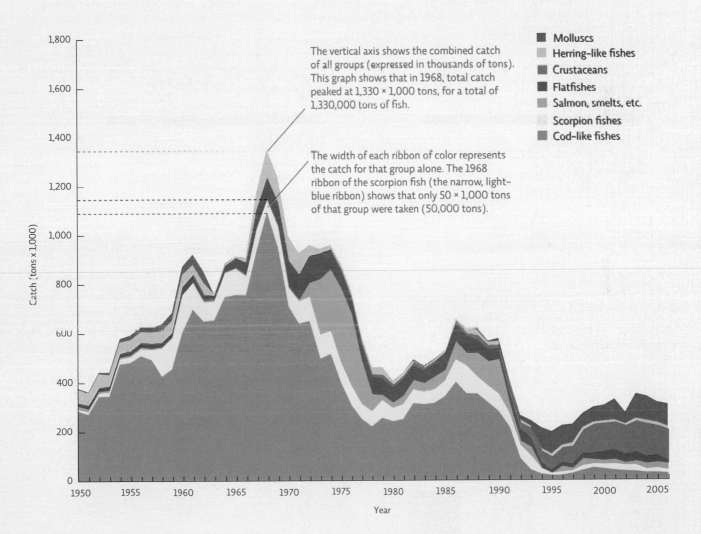

FISH CATCH BY COMMERCIAL GROUP: NEWFOUNDLAND–LABRADOR SHELF

The vertical axis shows the combined catch of all groups (expressed in thousands of tons). This graph shows that in 1968, total catch peaked at 1,330 × 1,000 tons, for a total of 1,330,000 tons of fish.

The width of each ribbon of color represents the catch for that group alone. The 1968 ribbon of the scorpion fish (the narrow, light-blue ribbon) shows that only 50 × 1,000 tons of that group were taken (50,000 tons).

Legend:
- Molluscs
- Herring-like fishes
- Crustaceans
- Flatfishes
- Salmon, smelts, etc.
- Scorpion fishes
- Cod-like fishes

STATISTICAL ANALYSIS

DESCRIPTIVE STATISTICS

In science it is not enough to simply collect data and graph it in order to draw conclusions. Suppose we see a difference between data collected for different groups. How different must the data sets be in order to conclude that the groups are different from each other? And how can we determine whether the treatment we applied—say, growing plants with a new fertilizer—really affected growth? We turn to statistical analysis.

Let's look at an example. We are growing two sets of plants, identical in every way except that one is grown without any fertilizer (the control group) and the other is grown with fertilizer (the test group). To draw conclusions, examine the values in the data table.

We begin with descriptive statistics—what are the characteristics of our data set? We calculate useful statistical values for each data set such as the *mean* (the average), the *range* (the highest value minus the lowest value), and the *sample size* (the number of subjects in each group). We might also calculate other values (with the help of any number of readily available online or calculator-based programs) that help describe the data set, such as *standard deviation* (the average amount of variation of each data value from the mean) and *standard error* (a measure that gives us an idea of how accurate our calculated mean really is, based on the standard deviation). Standard error bars are often shown with data as ± values (for example, 14.9 ± 1.4) or as error bars on a graph.

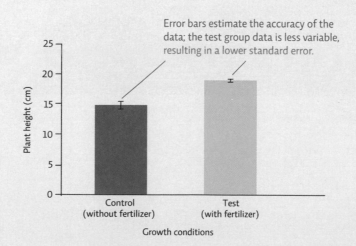

Mean 14.9 cm

Mean 19 cm

**CONTROL GROUP—
GROWN WITHOUT FERTILIZER**

	Height (cm)
Sample size: 10	
Range: 15	5
Mean: 14.9	11
Std deviation: 4.3	14
Std error: 1.4	15
	15
	16
	17
	18
	18
	20

**TEST GROUP—
GROWN WITH FERTILIZER**

	Height (cm)
Sample size: 10	
Range: 4	17
Mean: 19	17
Std deviation: 1.2	18
Std error: 0.38	18
	19
	20
	20
	20
	20
	21

**AVERAGE HEIGHT OF PLANTS GROWN WITH AND
WITHOUT FERTILIZER**

Error bars estimate the accuracy of the data; the test group data is less variable, resulting in a lower standard error.

INFERENTIAL STATISTICS

We can take our analysis further and evaluate the data with an inferential statistical test to determine how likely it is that the data we obtained from the two groups in our experiment actually represent different responses or whether our two groups are both just subsets of a single, larger group. The statistical test gives us a *p-value*—a number that tells us how much overlap there is between the data sets. In science, we generally require that there be no more than a 5% overlap between the two data sets. If the high end of one set (the control group here) overlaps just a little with the low end of the other data set (our test group), and this overlap is no more than 5%, we can conclude that the two groups most likely represent two distinct populations, a result of the treatment we applied—in this case, the addition of fertilizer. If the overlap had been more than 5% (a p-value > 0.05), we would not have sufficient evidence to conclude that they were indeed two different groups, but instead we would say they were likely to be a single group that varied widely.

The data showed this:

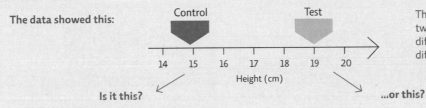

The average height of the plants in the two groups is different—but are they different enough to be considered two different populations?

Is it this?   **...or this?**

THE DATA POINTS COULD ALL BE PART OF ONE POPULATION

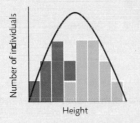

If the two groups are really part of one population, we might have inadvertently put faster-growing individuals in the test group and/or slower-growing ones in the control group. If that is true, retesting or using a larger sample size should produce some short control and some tall test plants.

THE DATA POINTS COULD REPRESENT 2 SEPARATE POPULATIONS

If the two groups really did respond differently to the treatment (fertilizer use), retesting or using a larger sample size should still produce this same trend, with most test plants being larger than control plants.

For this data set, a *t-test* (a simple statistical test) yields a p-value of 0.035—our data sets overlap 3.5%. Therefore, we can conclude that the two groups are different at the 0.05 level. Because our experimental design eliminated other variables that might have affected growth (the only difference between the two groups was whether plants received fertilizer), it is reasonable to conclude that the greater growth was caused by the fertilizer. As you read about studies and evaluate the authors' conclusions, look for the p-value given with the analysis of the data—if the calculated p-value is larger than 0.05, the author will probably conclude that there is not sufficient evidence to conclude that the variable tested had an effect.

OTHER FACTORS AFFECT RESULTS

VARIABILITY

Data sets with a lot of variability are less likely to show significant differences, even if the means are different—there is more of a chance that the two data sets overlap so much that they must be considered part of the same population.

LITTLE VARIABILITY

LOTS OF VARIABILITY

SAMPLE SIZE IS IMPORTANT

Small sample sizes are less reliable because, due to sampling error, we may have inadvertently sampled mostly unusual subjects; this would give us an incorrect picture of the entire population.

Mean

This smaller sample overestimates the mean.

SMALLER CONTROL GROUP

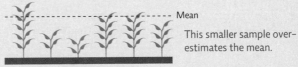

Mean

LARGER CONTROL GROUP

This larger sample suggests that larger plants are not the norm.

ANSWERS TO *CHECK YOUR UNDERSTANDING* QUESTIONS

Go online for the rest of the answers to the chapter problems! → bcs.whfreeman.com/saesextended

CHAPTER 1: ENVIRONMENTAL LITERACY
1. b
2. a
3. b
4. c
5. d
6. a

CHAPTER 2: SCIENCE LITERACY AND THE PROCESS OF SCIENCE
1. a
2. c
3. b
4. d
5. c
6. c

CHAPTER 3: INFORMATION LITERACY
1. a
2. d
3. a
4. c
5. b
6. d

CHAPTER 4: ENVIRONMENTAL ECONOMICS AND CONSUMPTION
1. b
2. a
3. d
4. d
5. c
6. a

CHAPTER 5: HUMAN POPULATIONS
1. a
2. c
3. c
4. b
5. d
6. b

CHAPTER 6: ENVIRONMENTAL HEALTH
1. b
2. c
3. b
4. d
5. d
6. a
7. a
8. c

CHAPTER 7: ECOSYSTEMS AND NUTRIENT CYCLING
1. d
2. b
3. a
4. c
5. a
6. b

CHAPTER 8: POPULATION ECOLOGY
1. d
2. d
3. a
4. c
5. d
6. b

CHAPTER 9: COMMUNITY ECOLOGY
1. d
2. c
3. a
4. b
5. c
6. b

CHAPTER 10: BIODIVERSITY
1. a
2. b
3. c
4. a
5. b
6. d

CHAPTER 11: PRESERVING BIODIVERSITY
1. d
2. d
3. c
4. a
5. b
6. c
7. d

CHAPTER 12: EVOLUTION AND EXTINCTION
1. c
2. d
3. a
4. a
5. a
6. c

CHAPTER 13: FORESTS
1. b
2. d
3. a
4. c
5. c
6. b

CHAPTER 14: GRASSLANDS
1. a
2. b
3. b
4. a
5. b
6. a

CHAPTER 15: MARINE ECOSYSTEMS
1. c
2. a
3. d
4. a
5. b
6. b

CHAPTER 16: FISHERIES AND AQUACULTURE
1. c
2. b
3. b
4. d
5. c
6. d

CHAPTER 17: FRESHWATER RESOURCES
1. d
2. a
3. b
4. c
5. d
6. a

CHAPTER 18: WATER POLLUTION
1. d
2. b
3. c
4. a
5. a
6. c

CHAPTER 19: MINERAL RESOURCES
1. c
2. c
3. b
4. a
5. d
6. b
7. d

CHAPTER 20: FEEDING THE WORLD
1. c
2. d
3. b
4. d
5. c
6. d
7. a

CHAPTER 21: AGRICULTURE: RAISING LIVESTOCK
1. d
2. a
3. c
4. b
5. c
6. b

CHAPTER 22: AGRICULTURE: RAISING CROPS
1. d
2. a
3. d
4. b
5. c
6. c

CHAPTER 23: COAL
1. b
2. c
3. a
4. c
5. a
6. b

CHAPTER 24: PETROLEUM
1. c
2. c
3. a
4. d
5. c
6. b

CHAPTER 25: AIR POLLUTION
1. c
2. c
3. b
4. b
5. b
6. d

CHAPTER 26: CLIMATE CHANGE
1. a
2. b
3. d
4. a
5. d
6. b

CHAPTER 27: NUCLEAR POWER
1. c
2. d
3. a
4. b
5. a
6. c

CHAPTER 28: SUN, WIND, AND WATER ENERGY
1. b
2. c
3. d
4. a
5. c
6. c

CHAPTER 29: BIOFUELS
1. a
2. d
3. b
4. b
5. c
6. c

CHAPTER 30: MANAGING SOLID WASTE
1. c
2. a
3. b
4. c
5. a
6. d

CHAPTER 31: ENVIRONMENTAL POLICY
1. b
2. a
3. d
4. c
5. d
6. c
7. a

CHAPTER 32: URBANIZATION AND SUSTAINABLE COMMUNITIES
1. d
2. c
3. d
4. b
5. b
6. a

GLOSSARY

A

abiotic The nonliving components of an ecosystem, such as rainfall and mineral composition of the soil. (Chapter 7)

acid deposition Precipitation that contains sulfuric or nitric acid; dry particles may also fall and become acidified once they mix with water. (Chapter 25)

acidification The lowering of the pH of a solution. (Chapter 15)

acid mine drainage Water flowing past exposed rock in mines that leaches out sulfates. These sulfates react with the water and oxygen to form acids (low-pH solutions). (Chapter 23)

active solar technologies The use of mechanical equipment to capture, convert, and sometimes concentrate solar energy into a more usable form. (Chapter 28)

adaptation A trait that helps an individual survive or reproduce (Chapter 12); efforts intended to help us deal with a problem that exists, such as climate change. (Chapter 26)

adaptive management A plan that allows room for altering strategies as new information comes in or as the situation itself changes. (Chapters 2 and 31)

additive effects Exposure to two or more chemicals has an effect equivalent to the sum of their individual effects. (Chapter 3)

age structure The part of a population pyramid that shows what percentage of the population is distributed into various age groups of males and females. (Chapter 5)

age structure diagram A graphic that displays the size of various age groups, with males shown on one side of the graphic and females on the other. (Chapter 5)

agroecology Scientific field that considers the area's ecology and indigenous knowledge, and favors methods that protect the environment and meet the needs of local people. (Chapter 22)

air pollution Any material added to the atmosphere (naturally or by humans) that harms living organisms, affects the climate, or impacts structures. (Chapter 25)

albedo The ability of a surface to reflect away solar radiation. (Chapter 26)

annual Plants that live for a year, produce seed, and then die. (Chapter 29)

annual crops Crops that grow, produce seeds, and die in a year and must be replanted each season. (Chapter 22)

antagonistic effects Exposure to two or more chemicals has a lesser effect than the sum of their individual effects would predict. (Chapter 3)

anthropocentric worldview A human-centered view that assigns intrinsic value only to humans. (Chapter 1)

anthropogenic Caused by or related to human action. (Chapters 1 and 26)

applied science Research whose findings are used to help solve practical problems. (Chapter 1)

aquaculture Fish-farming; the rearing of aquatic species in tanks, ponds, or ocean net pens. (Chapter 16)

aquifer An underground, permeable region of soil or rock that is saturated with water. (Chapters 17 and 18)

artificial selection Humans decide which individuals breed and which do not in an attempt to produce a population with desired traits. (Chapter 12)

asthma A chronic inflammatory respiratory disorder characterized by "attacks" during which the airways narrow, making it hard to breathe; can be fatal. (Chapter 25)

atmosphere Blanket of gases that surrounds Earth and other planets. (Chapter 2)

B

background rate of extinction The average rate of extinction that occurred before the appearance of humans or occurs outside of mass extinction events. (Chapter 12)

benthic macroinvertebrates Easy-to-see (not microscopic) arthropods such as insects that live on the stream bottom. (Chapter 18)

best management practices Agreed-upon (or EPA-regulated) actions that minimize pollution problems caused by industrial or land-use impacts. (Chapter 18)

bioaccumulation The buildup of fat-soluble substances in the tissue of an organism over the course of its lifetime. (Chapter 3)

biocentric worldview A life-centered approach that views all life as having intrinsic value, regardless of its usefulness to humans. (Chapter 1)

biodegradable Capable of being broken down by living organisms. (Chapter 30)

biodiesel A liquid fuel made from vegetable oil, animal fats, or waste oil that can be used directly in diesel internal combustion machines. (Chapter 29)

biodiversity The variety of life on Earth; it includes species, genetic, and ecological diversity. (Chapters 1 and 10)

biodiversity hotspot An area that contains a large number of endemic but threatened species. (Chapter 10)

bioethanol An alcohol fuel made from crops like corn and sugarcane in a process of fermentation and distillation. (Chapter 29)

biofuels Solids, liquids, or gases that produce energy from biological material. (Chapter 29)

biological assessment Sampling an area to see what lives there as a tool to determine how healthy the area is. (Chapter 18)

biological oxygen demand (BOD) The amount of oxygen that microbes living in a body of water use. (Chapter 18)

biomagnification The increased levels of fat–soluble substances in the tissue of predatory animals that have consumed organisms that have bioaccumulated toxins. (Chapter 3)

biomass The sum of all organic material—plant and animal matter—that make up an ecosystem, including material from living or recently living organisms or their by-products. (Chapters 7 and 29)

biome One of many distinctive types of ecosystems determined by climate and identified by the predominant vegetation and organisms that have adapted to live there. (Chapter 7)

biosphere The sum total of all of Earth's ecosystems. (Chapter 7)

biotic The living (organic) components of an ecosystem, such as the plants and animals and their waste (dead leaves, feces). (Chapter 7)

biotic potential (*r*) The maximum rate at which the population can grow due to births if each member of the population survives and reproduces. (Chapter 8)

birth rate The number of births per 1,000 individuals per year. (Chapter 8)

boom–and–bust cycles Fluctuations in population size that produce a very large population followed by a crash that lowers the population size drastically, followed again by an increase to a large size and a subsequent crash. (Chapter 8)

boreal forests Coniferous forests found at high latitudes and altitudes characterized by low temperatures and low annual precipitation. (Chapter 13)

bottleneck effect The effect of population size being drastically reduced; leads to the loss of some genetic variants and results in a less diverse population. (Chapter 12)

bycatch Nontarget species that become trapped in fishing nets and are usually discarded. Some methods, like trawling, have very high bycatch levels, and discards often exceed the actual target species catch. (Chapter 16)

C

canopy Upper layer of a forest formed where the crowns (tops) of the majority of the tallest trees meet. (Chapter 13)

cap–and–trade Regulations that set upper limits for pollution release. Producers are issued permits that allow them to release a portion of that amount; if they release less, they can sell their remaining allotment to others who did not reduce their emissions enough. (Chapters 25 and 31)

carbon capture and sequestration (CCS) Removing carbon from fuel combustion emissions or other sources and preventing its release into the atmosphere. (Chapter 23)

carbon cycle Movement of carbon through biotic and abiotic parts of an ecosystem. Carbon cycles via photosynthesis and cellular respiration as well as in and out of other reservoirs such as the oceans and soil. It is also released by human actions such as fossil fuel burning. (Chapter 7)

carbon debt The amount of carbon released during the first 50 years of land clearing for biofuel feedstocks. (Chapter 29)

carbon footprint The amount of carbon dioxide (CO_2) released to the atmosphere by a person, company, nation, or activity; a concern because it contributes to climate change. (Chapters 21 and 22)

carbon sequestration The storage of carbon in a form that does not readily release the carbon to the atmosphere or water. (Chapter 29)

carbon sinks Places such as forests, ocean sediments, and soil, where accumulated carbon does not readily reenter the carbon cycle. (Chapter 13)

carbon taxes Governmental fees imposed on activities that release CO_2 into the atmosphere, usually on fossil fuel use. (Chapter 26)

carrying capacity (*K*) The population size that an area can support for the long term; it depends on resource availability and the rate of per capita resource use by the population. (Chapters 1, 5, and 8)

cash crops Food and fiber crops grown to sell for profit, rather than as food for local families or communities. (Chapter 20)

cause–and–effect relationship An association between two variables that identifies one (the effect) occurring as a result of or in response to the other (the cause). (Chapter 2)

cellular respiration The process in which all organisms break down sugar to release its energy, using oxygen and giving off CO_2 as a waste product. (Chapter 7)

cellulosic ethanol Bioethanol made by breaking down cellulose in plants; a difficult process that has yet to be scaled up to meet large production goals. (Chapter 29)

citizen suit provision Allows a private citizen to sue, in federal court, a perceived violator of certain U.S. environmental laws, such as the Clean Air Act, in order to force compliance. (Chapter 31)

Clean Air Act First passed in 1963 and amended most recently in 1990, this U.S. law authorizes the EPA to set standards for dangerous air pollutants and enforce those standards. (Chapter 25)

Clean Development Mechanism (CDM) A UN program that allows a country with a greenhouse gas reduction commitment to implement emission–reduction projects in developing countries. (Chapter 31)

Clean Water Act The main U.S. federal law that regulates water pollution. (Chapter 18)

climate Long–term patterns or trends of meteorological conditions. (Chapter 26)

climate change Alteration in the long-term patterns and statistical averages of meteorological events. (Chapter 26)

climax community The end stage of ecological succession in which the conditions created by the climax species are suitable for the plants that created them so they can persist as long as their environment remains unchanged. (Chapter 9)

climax species Species that move into an area at later stages of ecological succession. (Chapter 9)

closed-loop system A production system in which the product is folded back into the resource stream when consumers are finished with it, or is disposed of in such a way that nature can decompose it. (Chapter 4)

clumped distribution Individuals are found in groups or patches within the habitat. (Chapter 8)

coal Fossil fuel formed when plant material is buried in oxygen-poor conditions and subjected to high heat and pressure over a long time. (Chapter 23)

coevolution A process of evolution in which two species each provide the selective pressure that determines which traits are favored by natural selection in the other. (Chapter 12)

coliform bacteria Bacteria often found in the intestinal tract of animals; monitored for fecal contamination of water. (Chapter 17)

collapsed fishery Annual catches fall below 10% of their historic high; stocks can no longer support a fishery. (Chapter 16)

command and control regulation Regulations that set an upper allowable limit of pollution release which is enforced with fines and/or incarceration. (Chapters 25 and 31)

commensalism A symbiotic relationship between individuals of two species in which one benefits from the presence of the other but the other is unaffected. (Chapter 9)

community ecology The study of all the populations (plants, animals, and other species) living and interacting in an area. (Chapter 9)

competition Species interaction in which individuals are vying for limited resources. (Chapter 9)

composting Providing good conditions for the decomposition of biodegradable waste, producing a soil-like mulch. (Chapter 30)

Concentrated Animal Feeding Operation (CAFO) Many meat or dairy animals are reared in confined spaces, maximizing the number of animals that can be grown in a small area. (Chapter 21)

condensation Conversion of water from a gaseous state (water vapor) to a liquid state. (Chapter 17)

conservation Efforts that reduce waste and increase efficient use of resources. (Chapter 28)

conservation biology The science concerned with preserving biodiversity. (Chapter 11)

conservation genetics Scientific field that relies on species' genetics to inform conservation efforts. (Chapter 11)

conservation reserve program Farmers and ranchers are paid to keep damaged land out of production to promote recovery. (Chapter 14)

consumer An organism that obtains energy and nutrients by feeding on another organism; includes animals, fungi, and most bacteria. (Chapters 7 and 9)

contour farming Farming on hilly land in rows that are planted along the slope, following the lay of the land, rather than oriented downhill. (Chapter 22)

control group The group in an experimental study that the test group's results are compared to; ideally, the control group will differ from the test group in only one way. (Chapter 2)

control rods Rods that absorb neutrons and slow the fission chain reaction. (Chapter 27)

conventional petroleum reserves Liquid deposits that contain freely flowing oil or oil that can be pumped out. (Chapter 24)

convention on biological diversity An international treaty that promotes sustainable use of ecosystems and biodiversity. (Chapter 11)

Convention on International Trade in Endangered Species of Wild Fauna and Flora (CITES) An international treaty that regulates global trade of selected species. (Chapter 11)

coral reef Colonies of tiny animals (coral) that produce a calcium carbonate exoskeleton that over time build up to form large underwater structures (the reef) in shallow, warm, tropical seas. (Chapter 15)

core species Species that prefer inner areas of a habitat—areas deep within the habitat, away from the edge. (Chapter 9)

corporate average fuel efficiency (CAFE) standards A target of the minimum fuel efficiency (MPG) that manufacturers must meet; evaluated as a weighted average of all the cars and light trucks each manufacturer produces. (Chapter 31)

correlation Two things occur together—but it doesn't necessarily mean that one caused the other. (Chapter 2)

cover crop A crop planted in the off-season to help prevent soil erosion and to return nutrients to the soil. (Chapter 22)

cradle to cradle Management of a resource that considers the impact of its use at every stage of the process. (Chapter 4)

critical thinking Skills that enable individuals to logically assess the information they find, reflect on that information, and reach their own conclusions. (Chapter 3)

crop rotation Planting different crops on a given plot of land every few years to maintain soil fertility and reduce pest outbreaks. (Chapter 22).

crude birth rate The number of offspring per 1,000 individuals per year. (Chapter 5)

crude death rate The number of deaths per 1,000 individuals per year. (Chapter 5)

crude oil The form of petroleum that is extracted from underground deposits. It is processed into many different products and types of fuel. (Chapter 24)

cultural eutrophication A process in which excess nutrients in aquatic ecosystems stimulate excess plant growth and disrupt normal energy uptake and matter cycles. (Chapters 18 and 22)

D

dam Structure that blocks the flow of water in a river or stream. (Chapter 17)

death rate The number of deaths per 1,000 individuals per year. (Chapter 8)

debt-for-nature swap A wealthy nation forgives the debt of a developing nation in return for a pledge to protect natural areas in that developing nation. (Chapter 11)

decomposers Organisms such as bacteria and fungi that break down organic matter all the way down to constituent atoms or molecules in a form that plants can take back up. (Chapter 9)

deforestation Net loss of trees in a forested area (Chapter 13)

demographic factors Population characteristics such as birth rate or life expectancy that influence how a population changes in size and composition. (Chapter 5)

demographic transition Theoretical model that describes the expected drop in once-high population growth rates as economic conditions improve the quality of life in a population. (Chapter 5)

density dependent Factors, such as predation or disease, whose impact on the population increases as population size goes up. (Chapter 8)

density independent Factors, such as a storm or an avalanche, whose impact on the population is not related to population size. (Chapter 8)

dependent variable The variable in an experiment that is evaluated to see if it changes due to the conditions of the experiment. (Chapter 2)

depleted fisheries The fish population is well below historic levels and the population's reproductive capacity is low, meaning that recovery will be slow, if at all. (Chapter 16)

desalination The removal of salt and minerals from seawater to make it suitable for consumption. (Chapter 17)

desertification The process that transforms once-fertile land into desert. (Chapter 14)

detritivores Consumers (including worms, insects, crabs, etc.) who eat dead organic material. (Chapter 9)

developed country A country that has a moderate to high standard of living on average and an established market economy. (Chapter 5)

developing country A country that has a lower standard of living than a developed country, and has a weak economy; may have high poverty. (Chapter 5)

discount future value To give more weight to short-term benefits and costs than to long-term ones. (Chapter 4)

dissolved oxygen (DO) The amount of oxygen in the water. (Chapter 18)

dose-response curve A graph of the effects of a substance at different concentrations or levels of exposures. (Chapter 3)

E

ecocentric worldview A system-centered view that values intact ecosystems, not just the individual parts. (Chapter 1)

eco-industrial parks Industrial parks in which industries are physically positioned near each other for "waste-to-feed" exchanges (the waste of one becomes the raw material for another). (Chapter 30)

ecolabeling Providing information about how a product is made and where it comes from; allows consumers to make more sustainable choices and support sustainable products and the businesses that produce them. (Chapter 4)

ecological diversity The variety within an ecosystem's structure, including many communities, habitats, niches, and trophic levels. (Chapter 10)

ecological footprint The land needed to provide the resources for, and assimilate the waste of, a person or population. (Chapters 1 and 4)

ecological succession Progressive replacement of plant (and then animal) species in a community over time due to the changing conditions that the plants themselves create (more soil, shade, etc.). (Chapter 9)

economics The social science that deals with how we allocate scarce resources. (Chapter 4)

ecosystem All of the organisms in a given area plus the physical environment in which they interact. (Chapters 7 and 9)

ecosystem approach A conservation strategy that focuses on protecting the ecosystem as a whole in an effort to protect the species that live there. (Chapter 11)

ecosystem restoration The repair of natural habitats back to (or close to) their original state. (Chapter 11)

ecosystem services Essential ecological processes that make life on Earth possible (Chapter 4); benefits provided by functional ecosystems that are important to all life (including humans); includes such things as nutrient cycles, air and water purification, and ecosystem goods such as food and fuel. (Chapter 10)

ecotones Regions of distinctly different physical areas that serve as boundaries between different communities. (Chapter 9)

ecotourism Low-impact travel to natural areas that contributes to the protection of the environment and respects the local people. (Chapters 11 and 13)

edge effects The different physical makeup of the ecotone which creates different conditions that either attract or repel certain species (for instance, it is drier, warmer, and more open at the edge of a forest and field than it is further in the forest). (Chapter 9)

edge species Species that prefer to live close to the area where two different habitats meet (ecotone areas). (Chapter 9)

effluent Water discharged into the environment. (Chapter 17)

electricity The flow of electrons (negatively charged subatomic particles) through a conductive material (such as wire). (Chapter 23)

element A pure chemical substance made up of one type of atom; there are 92 naturally occurring elements; examples include gold and carbon. (Chapter 19)

emergent layer The region where a tree that is taller than the canopy trees rises above the canopy layer. (Chapter 13)

emerging infectious diseases Infectious diseases that are new to humans or that have recently increased significantly in incidence, in some cases by spreading to new ranges. (Chapter 6)

emigration The movement of people out of a given population. (Chapter 5)

empirical evidence Information gathered via observation of physical phenomena. (Chapter 2)

empirical science A scientific approach that investigates the natural world through systematic observation and experimentation. (Chapter 1)

Endangered Species Act The primary law under which biodiversity is protected in the United States. (Chapter 11)

endemic Describes a species that is native to a particular area and is not naturally found elsewhere. (Chapters 10 and 12)

endocrine disruptor A molecule that interferes with the endocrine system, typically by mimicking a hormone or preventing a hormone from having an effect. (Chapter 3)

energy The capacity to do work. (Chapter 23)

energy flow The one-way passage of energy through an ecosystem. (Chapter 7)

energy independence Meeting all of one's energy needs without importing any fuel. (Chapter 24)

energy return on energy investment (EROEI) A measure of the net energy from an energy source (the energy in the source minus the energy required to get it, process it, ship it, and then use it). (Chapter 23)

energy security Having access to enough reliable and affordable energy sources to meet one's needs. (Chapter 24)

environment The biological and physical surroundings in which any given living organism exists. (Chapter 1)

environmental/ecological economics New theories of economics that consider the long-term impact of our choices on people and the environment. (Chapter 4)

environmental ethic The personal philosophy that influences how a person interacts with his or her natural environment and thus affects how one responds to environmental problems. (Chapter 1)

environmental health The branch of public health that focuses on factors in the natural world and the human-built environment that impact the health of populations. (Chapter 6)

environmental impact statement (EIS) An evaluation of the positive and negative impacts of any proposed action on the environment, including alternative actions that could be pursued; used to help decide whether or not that action will be approved. (Chapters 8, 23, and 31)

environmental justice The concept that access to a clean, healthy environment is a basic human right. (Chapters 25 and 32)

environmental literacy A basic understanding of how ecosystems function and of the impact of our choices on the environment. (Chapter 1)

environmental policy A course of action adopted by a government or organization that is intended to improve the natural environment and public health or reduce human impact on the environment. (Chapter 31)

Environmental Protection Agency (EPA) The federal agency responsible for setting policy and enforcing U.S. environmental laws. (Chapters 3 and 31)

environmental racism Occurs when minority communities face more exposure to pollution than average for the region. (Chapter 25)

environmental science An interdisciplinary field of research that draws on the natural and social sciences and the humanities in order to understand the natural world and our relationship to it. (Chapter 1)

epidemiologist A scientist who studies the causes and patterns of disease in human populations. (Chapters 3 and 6)

estuary The region where freshwater rivers meet the sea. (Chapters 15 and 18)

evaporation The conversion of water from a liquid state to a gaseous state. (Chapter 17)

evolution Differences in the gene frequencies within a population from one generation to the next. (Chapter 12)

e-waste Unwanted computers and other electronic devices such as televisions and cell phones that are discarded. (Chapters 19 and 30)

exclusive economic zones (EEZs) Zones that extend 200 nautical miles from the coastline of any given nation, where that nation has exclusive rights over marine resources, including fish. (Chapter 16)

experimental study Research that manipulates a variable in a test group and compares the response to that of a control group that was not exposed to the same variable. (Chapter 2)

exponential growth Population size becomes progressively larger each breeding cycle; each female produces a J curve when plotted over time. (Chapter 8)

external cost Costs that are not taken into account when a price is assigned to a product or service. (Chapter 4)

extinct/extinction The complete loss of a species from an area; may be local (gone from an area) or global (gone for good). (Chapter 12)

extirpation Local extinction of a species in one or more areas, though some individuals exist in other areas. (Chapters 8, 10, and 12)

exurbs Towns beyond the immediate suburbs whose residents commute into the city for work. (Chapter 32)

F

falsifiable An idea or a prediction that can be proved wrong by evidence. (Chapter 2)

famine A severe shortage of food that leads to widespread hunger. (Chapter 20)

feed conversion ratio The amount of edible food that is produced per unit of feed input. (Chapter 21)

feedstock Biomass sources used to make biofuels. (Chapter 29)

fisheries The industry devoted to commercial fishing or the places where fish are caught, harvested, processed, and sold. (Chapter 16)

flagship species The focus of public awareness campaigns aimed at generating interest in conservation in general; usually an interesting or charismatic species, such as the giant panda or tiger. (Chapter 11)

food chain A simple, linear path starting with a plant (or other photosynthetic organism) that identifies what each organism in the path eats. (Chapter 9)

food miles The distance a food travels from its site of production to the consumer. (Chapter 22)

food security Having physical, social, and economic access to sufficient, safe, and nutritious food. (Chapter 20)

food web A linkage of all the food chains together that shows the many connections in the community. (Chapter 9)

Forest Ecosystem Management (FEM) Focuses on managing the forest as a whole, rather than on maximizing yields of a specific product. (Chapter 13)

forest floor The lowest level of the forest, containing herbaceous plants, fungi, leaf litter, and soil. (Chapter 13)

fossil fuel A nonrenewable natural resource formed millions of years ago from dead plant (coal) or animal (petroleum and natural gas) remains. (Chapters 23 and 24)

fossil record The total collection of fossils (remains, impressions, traces of ancient organisms) found on Earth. (Chapter 12)

founder effect The effect when a small group with only a subset of the larger population's genetic diversity becomes isolated; the small group evolves into a different population, missing some of the traits of the original population. (Chapter 12)

fracking A method used to extract natural gas from deep reserves. (Chapter 24)

freshwater Water that has few dissolved ions such as salt. (Chapter 17)

fuel crops Crops specifically grown to be used to produce biofuels. (Chapter 29)

fuel rods Hollow metal cylinders filled with uranium pellets for use in fission reactors. (Chapter 27)

G

gendercide The systematic killing of a specific gender (male or female). (Chapter 5)

gene frequencies The assortment and abundance of particular variants of genes (alleles) relative to each other within a population. (Chapter 12)

genes Stretches of DNA, the cell's hereditary material, that each direct the production of a particular molecule (usually a protein) and influence an individual's traits. (Chapter 12)

genetic diversity The heritable variation among individuals of a single population or within the species as a whole. (Chapters 10 and 12)

genetic drift The change in gene frequencies of a population over time due to random mating that results in the loss of some gene variants. (Chapter 12)

genetically modified organism (GMO) Organism that has had its genetic information modified to give it desirable characteristics such as pest or drought resistance. (Chapter 20)

geology The study of the structure of Earth and the processes that have shaped it in the past and shape it today. (Chapter 19)

geothermal energy The heat stored underground, contained in either rocks or fluids. (Chapter 28)

geothermal heat pump A system that actively moves heat from the underground into a house to warm it or removes heat from a house to cool it. (Chapter 28)

geothermal power plants Power plants that use the heat of hydrothermal reservoirs to produce steam and turn turbines to generate electricity. (Chapter 28)

grasslands A biome that is predominately grasses, due to low rainfall, grazing animals, and/or fire. (Chapter 14)

green building Construction and operational designs that promote resource and energy efficiency, and provide a better environment for occupants. (Chapter 32)

green business Doing business in a way that is good for people and the environment. (Chapter 4)

greenhouse effect The warming of the planet that results when heat is trapped by Earth's atmosphere. (Chapter 26)

greenhouse gases Molecules in the atmosphere that absorb heat and reradiate it back to Earth. (Chapter 26)

Green Revolution Plant-breeding program in the mid-1900s that dramatically increased crop yields and led the way for mechanized, large-scale agriculture. (Chapter 20)

Green Revolution 2.0 Programs that focus on the production of genetically modified organisms (GMOs) to increase crop productivity or create new varieties of crops. (Chapter 20)

green tax Tax (fee paid to government) assessed on environmentally undesirable activities. (Chapters 25 and 31)

greenwashing Claiming environmental benefits about a product when the benefits are actually minor or nonexistent. (Chapter 22)

gross primary productivity A measure of the total amount of energy captured via photosynthesis and transferred to organic molecules in an ecosystem. (Chapter 9)

ground-level ozone A secondary pollutant that forms when some of the pollutants released during fossil fuel combustion react with atmospheric oxygen in the presence of sunlight. (Chapter 25)

groundwater Water found underground in a region known as an aquifer. (Chapter 17)

growth rate The percent increase of population size over time; affected by births, deaths, and the number of people moving into or out of a regional population. (Chapter 5)

H

habitat The physical environment in which individuals of a particular species can be found. (Chapters 7 and 9)

hazardous waste Waste that is toxic, flammable, corrosive, explosive, or radioactive. (Chapter 30)

hectare (ha) Metric unit of measure for area; 1 ha = 2.5 acres (ac). (Chapter 13)

herbivore An animal that feeds on plants. (Chapter 14)

high-level radioactive waste (HHRW) Spent fuel rods or nuclear weapons production waste that is still highly radioactive. (Chapter 27)

high-yield varieties (HYVs) Strains of staple crops selectively bred to produce more grain than their natural counterparts, usually because they grow faster or larger or are more resistant to crop diseases. (Chapter 20)

hormone A molecule released by the body that directs cellular activity and produces changes in how the body functions. (Chapter 3)

hydropower The energy produced from moving water. (Chapter 28)

hypothesis A possible explanation for what we have observed that is based on some previous knowledge. (Chapter 2)

hypoxia A situation in which the level of oxygen in the water is inadequate to support life. (Chapter 18)

I

immigration The movement of people into a given population. (Chapter 5)

incinerators Facilities that burn trash at high temperatures. (Chapter 30)

independent variable The variable in an experiment that the researcher manipulates or changes to see if it produces an effect. (Chapter 2)

indicator species The species that are particularly vulnerable to ecosystem perturbations, and that, when we monitor them, can give us advance warning of a problem. (Chapters 9 and 11)

industrial agriculture Farming methods that rely on technology, synthetic chemical inputs, and economies of scale to increase productivity and profits. (Chapter 22)

infant mortality rate The number of infants who die in their first year of life per every 1,000 live births in that year. (Chapter 5)

infectious diseases Illnesses caused by an invading pathogen such as a bacterium or virus. (Chapter 6)

inferences Conclusions we draw based on observations. (Chapter 2)

infill development The development of empty lots within a city. (Chapter 32)

infiltration The process of water soaking into the ground. (Chapter 17)

information literacy The ability to find and evaluate the quality of information. (Chapter 3)

instrumental value The value or worth of an object, organism, or species is based on its usefulness to humans. (Chapters 1 and 10)

integrated pest management (IPM) The use of a variety of methods to control a pest population, with the goal of minimizing or eliminating the use of chemical toxins. (Chapter 22)

Intergovernmental Panel on Climate Change (IPCC) An international group of scientists who evaluate scientific studies related to any aspect of climate change to give a thorough and objective assessment of the data. (Chapter 26)

internal cost Those costs—such as manufacturing costs, labor, taxes, utilities, insurance, and rent—that are accounted for when a product or service is evaluated for pricing. (Chapter 4)

interspecific competition Competition among individuals of different species. (Chapter 9)

intraspecific competition Competition among members of the same species. (Chapter 9)

intrinsic value The value or worth of an object, organism, or species is based on its mere existence; it has an inherent right to exist. (Chapters 1 and 10)

invasive species A non-native species (a species outside of its range) whose introduction causes or is likely to cause economic or environmental harm or harm to human health. (Chapter 12)

in vitro **study** Research that studies the effects of experimental treatment cells in culture dishes rather than in intact organisms. (Chapter 3)

in vivo study Research that studies the effects of an experimental treatment in intact organisms. (Chapter 3)

IPAT model An equation (I = P x A x T) that identifies 3 factors that increase human impact (I) directly: population size (P), affluence (A), and technology (T). (Chapter 4)

isotopes Atoms that have different numbers of neutrons in their nucleus but the same number of protons. (Chapter 27)

K

keystone species A species that impacts its community more than its mere abundance would predict. (Chapters 9 and 11)

K-selected species Species that have a low biotic potential and that share characteristics such as long lifespan, late maturity, and low fecundity; generally show logistic population growth (Chapter 8)

Kyoto Protocol A 1997 amendment to the UNFCCC that set legally binding specific goals for reductions in greenhouse gas emissions for certain nations that ratified the treaty. (Chapter 31)

L

landscape conservation An ecosystem conservation strategy that specifically identifies a suite of species, chosen because they use all the vital areas within an ecosystem; meeting the needs of these species will keep the ecosystem fully functional, thus meeting the needs of all species that live there. (Chapter 11)

law of conservation of matter Matter can neither be created nor destroyed; it only changes form. (Chapter 30)

LD50 (lethal dose 50%) The dose of a substance that would kill 50% of the test population. (Chapter 3)

leachate Water that carries dissolved substances (often contaminated) that can percolate through soil. (Chapter 30)

life expectancy The number of years an individual is expected to live. (Chapter 5)

limiting factor The critical resource whose supply determines the population size of a given species in a given biome. (Chapter 7)

logical fallacies Arguments which attempt to sway the reader without using actual evidence. (Chapter 3)

logistic growth The kind of growth in which population size increases rapidly at first but then slows down as the population becomes larger; produces an S curve when plotted over time. (Chapter 8)

low-level radioactive waste (LLRW) Material that has a low level of radiation for its volume. (Chapter 27)

M

malnutrition A state of poor health that results from inadequate or unbalanced food intake; includes diets that provide too few or too many calories and/or do not provide the proper nutrients (deficient in one or more nutrients). (Chapter 20)

marine protected areas (MPAs) Discrete regions of ocean that are legally protected from various forms of human exploitation. (Chapter 16)

marine reserves Restricted areas where all fishing is prohibited and absolutely no human disturbance is allowed. (Chapter 16)

maximum sustainable yield The amount that can be harvested without decreasing the yield in future years. (Chapters 13 and 16)

metal A malleable substance that can conduct electricity; usually found in nature as a part of a mineral compound. (Chapter 19)

Milankovitch cycles Predictable variations in Earth's position in space relative to the Sun which affect climate. (Chapter 26)

mineral A naturally occurring chemical compound that exists as a solid with a predictable, three-dimensional, repeating structure. (Chapter 19)

minimum viable population The smallest number of individuals that would still allow a population to be able to persist or grow, ensuring long-term survival. (Chapter 8)

mining The extraction of natural resources from the ground. (Chapter 19)

mitigation Efforts intended to minimize the extent or impact of a problem such as climate change. (Chapter 26)

monoculture Farming method in which one variety of one crop is planted, typically in rows over huge swaths of land, with large inputs of fertilizer, pesticides, and water. (Chapters 22 and 29)

Montréal Protocol An international treaty that laid out plans to phase out ozone-depleting chemicals like CFC. (Chapter 2)

mountaintop removal Surface mining technique that uses explosives to blast away the top of a mountain to expose the coal seam underneath; the waste rock and rubble is deposited in a nearby valley. (Chapter 23)

Multiple-Use Sustained-Yield Act U.S. legislation (1960) mandating that national forests be managed in a way that balances a variety of uses. (Chapter 13)

municipal solid waste (MSW) Everyday garbage or trash (solid waste) produced by individuals or small businesses. (Chapter 30)

mutualism A symbiotic relationship between individuals of two species in which both parties benefit. (Chapter 9)

N

National Environmental Policy Act (NEPA) A 1969 U.S. law that established environmental protection as a guiding policy for the nation and required that the federal government take the environment into consideration before taking action that might affect it. (Chapter 31)

natural capital The wealth of resources on Earth. (Chapter 4)

natural gas Gaseous fossil fuel that forms under similar conditions as petroleum but is a simpler hydrocarbon, mostly CH_4 (methane). (Chapter 24)

natural interest Readily produced resources that we could use and still leave enough natural capital behind to replace what we took. (Chapter 4)

natural selection The process by which organisms best adapted to the environment (the fittest) survive to reproduce, leaving more offspring than less well-adapted individuals. (Chapter 12)

negative feedback Changes caused by an initial event trigger events that then reverse the response (i.e., changes brought on by warming lead to cooling). (Chapters 15 and 26)

net primary productivity (NPP) A measure of the amount of energy captured via photosynthesis and actually stored in the photosynthetic organism. (Chapter 9)

New Urbanism A movement that promotes the creation of compact, mixed-use communities with all of the amenities for day-to-day living close by and accessible. (Chapter 32)

niche The role a species plays in its community, including things like how it gets its energy and nutrients, what habitat requirements it has, and which other species and parts of the ecosystem it interacts with. (Chapters 7 and 9)

nitrogen cycle Continuous series of natural processes by which nitrogen passes from the air to the soil, to organisms, and then returns back to the air or soil through decay or denitrification. (Chapter 7)

nitrogen fixation Conversion of atmospheric nitrogen into a biologically usable form, carried out by bacteria found in soil or via lightning. (Chapter 7)

noncommunicable diseases (NCDs) Illnesses that are not transmissible between people; not infectious. (Chapter 6)

nondegradable Incapable of being broken down under normal conditions. (Chapter 30)

nonpoint source pollution Pollution that enters the air from dispersed or mobile sources or enters the water from overland flow. (Chapters 18 and 25)

nonrenewable resources Resources whose supply is finite or not replenished in a timely fashion. (Chapters 1 and 24)

nuclear energy Energy released when an atom is split (fission) or combines with another to form a new atom (fusion); can be tapped to generate electricity. (Chapter 27)

nuclear fission Nuclear reaction that occurs when a neutron strikes the nucleus of an atom and breaks it into two or more parts. (Chapter 27)

nutrient cycles Movement of life's essential chemicals or nutrients through an ecosystem (also known as biogeochemical cycles). (Chapter 7)

O

observational study Research that gathers data in a real-world setting without intentionally manipulating any variable. (Chapter 2)

observations Information detected with the senses—or with equipment that extends our senses. (Chapter 2)

oil See *petroleum*. (Chapter 24)

oil or tar sands A heavy, black oil known as crude bitumen that is trapped in sticky, dense conglomerations of sand or clay. It can be mined and processed to produce a substitute for petroleum. (Chapter 24)

oil shale Compressed sedimentary rocks that contain kerogen, an organic compound that is released as an oil-like liquid when the rock is heated. (Chapter 24)

open dumps Places where trash, both hazardous and nonhazardous, is simply piled up. (Chapter 30)

ore A rock deposit that contains economically valuable amounts of metal and minerals. (Chapter 19)

organic agriculture Farming that does not use synthetic fertilizer, pesticides, or other chemical additives like hormones (for animal rearing). (Chapter 22)

overburden The rock and dirt removed to uncover a mineral deposit during surface mining. (Chapter 23)

overexploited fisheries More fish are taken than is sustainable in the long run, leading to population declines. (Chapter 16)

overgrazing Too many herbivores feeding in an area, eating the plants faster than they can regrow. (Chapter 14)

overpopulation More people living in an area than its natural and human resources can support. (Chapter 5)

ozone Molecule with 3 oxygen atoms that absorbs UV radiation in the stratosphere. (Chapter 2)

P

parasitism A symbiotic relationship between individuals of two species in which one benefits and the other is negatively affected (a form of predation). (Chapter 9)

particulate matter (PM) Particles or droplets small enough to remain aloft in the air for long periods of time. (Chapter 25)

passive solar technologies The capture of solar energy (heat or light) without any electronic or mechanical assistance. (Chapter 28)

pastoralists Individuals who herd and care for livestock as a way of life. (Chapter 14)

pathogen An infectious agent that causes illness or disease. (Chapter 6)

payback time The amount of time it would take to save enough money in operation costs to pay for the equipment. (Chapter 28)

peak oil The moment in time when oil will reach its highest production levels and then steadily and terminally decline. (Chapter 24)

peer review Researchers submit a report of their work to a group of outside experts who evaluate the study's design and results of the study to determine whether it is of high-enough quality to publish. (Chapters 2 and 3)

perennial Plants that live for more than a year, growing and producing seed year after year. (Chapter 29)

perennial crops Crops that do not die at the end of the growing season but live for several years, allowing them to be harvested annually without replanting. (Chapter 22)

performance standards Targets set that specify acceptable levels of pollution that can be released or exist in ambient (outdoor) air; industries must act to meet these standards. (Chapter 31)

persistence The length of time it takes a substance to break down in the environment. (Chapter 3)

persistent chemicals Chemicals that don't readily degrade over time. (Chapter 3)

pesticide resistance The ability of a pest to withstand exposure to a given pesticide, the result of natural selection favoring the survivors of an original population that was exposed to the pesticide. (Chapter 22)

petrochemicals Distillation products from the processing of crude oil that can have many different uses, including as fuel or as industrial raw materials. (Chapter 24)

petroleum Liquid fossil fuel useful as a portable fuel or as a raw material for many industrial products such as plastic and pesticides. (Chapter 24)

phosphorus cycle Series of natural processes by which the nutrient phosphorus moves from rock to soil or water, to living organisms, and back to the soil. (Chapter 7)

photovoltaic (PV) cells Also called solar cells, PV cells are silicone–based devices that convert solar energy directly into electricity. (Chapter 28)

pioneer species Plant species that move into an area during early stages of succession; these are often *r* species and may be annuals, species that live one year, leave behind seeds, and then die. (Chapter 9)

point source pollution Pollution that enters the air or water from readily identifiable sources such as discharge pipes or smoke stacks. (Chapters 18 and 25)

policy A formalized plan that addresses a desired outcome or goal. (Chapter 2)

policy tools Methods that can be used to enforce or implement regulations or achieve desired outcomes. (Chapter 31)

political lobbying Contacting elected officials in support of a particular position; some professional lobbyists are highly organized, with substantial financial backing. (Chapter 31)

pollution standards Allowable levels of a pollutant that can be released over a certain time period; set by EPA. (Chapter 18)

polyculture Farming method in which a mix of different species are grown together in one area. (Chapter 22)

population All the individuals of a species that live in the same geographic area and are able to interact and interbreed. (Chapter 8)

population density The number of individuals per unit area. (Chapters 5 and 8)

population distribution The location and spacing of individuals within their range. (Chapter 8)

population dynamics The changes over time of population size and composition. (Chapter 8)

population growth rate The change in population size over time (births minus deaths over a specific time period). (Chapter 8)

population momentum The tendency of a young population to continue to grow even after birth rates drop to "replacement rates" (2 children per couple). (Chapter 5)

positive feedback Changes caused by an initial event accentuate that original event (i.e., changes brought on by warming lead to even more warming). (Chapters 15 and 26)

potable Water clean enough for consumption. (Chapter 17)

precautionary principle A rule of thumb that calls for leaving a safety margin when the data about a particular substance's potential for harm are uncertain and where the substance may cause unexpected or unpredictable effects. (Chapters 2, 3, 26, and 31)

precipitation Rain, snow, sleet, or any form of water falling from the atmosphere. (Chapter 17)

prediction A statement that identifies what is expected to happen in a given situation. (Chapter 2)

primary air pollutants Air pollutants released directly from both mobile sources (such as cars) and stationary sources (such as industrial and power plants). (Chapter 25)

primary sources Sources of information that present new and original data or information, including novel scientific experiments or observations and first-hand accounts of any given event. (Chapter 3)

primary succession Ecological succession that occurs in an area where no ecosystem existed before (for example, on bare rock with no soil). (Chapter 9)

producer A photosynthetic organism that converts solar energy to chemical energy via photosynthesis. (Chapters 7 and 9)

protected areas Geographic spaces on land or at sea recognized, dedicated, and managed to achieve the long–term conservation of nature. (Chapter 11)

public health The science that deals with the health of human populations. (Chapter 6)

R

radiative forcer Anything that alters the balance of incoming solar radiation relative to the amount of heat that escapes out into space. (Chapter 26)

radioactive Atoms that spontaneously emit subatomic particles and/or energy. (Chapter 27)

radioactive half-life The time it takes for half of the radioactive isotopes in a sample to decay to a new form. (Chapter 27)

rain garden Runoff area that is planted with water-tolerant plants to slow runoff and promote infiltration. (Chapter 18)

random distribution Individuals are spread out over the environment irregularly with no discernable pattern. (Chapter 8)

rangeland Grassland used for grazing of livestock. (Chapter 14)

range of tolerance The range, within upper and lower limits, of a limiting factor that allows a species to survive and reproduce. (Chapter 7)

rare earth minerals A group of 17 or so elements, similar in chemical structure, that occur close together in nature; they are not necessarily rare but do not occur In concentrated deposits. (Chapter 19)

receptor A structure on or inside a cell that binds a hormone, thus allowing the hormone to affect the cell. (Chapter 3)

reclamation Restoring a damaged natural area to a less damaged state. (Chapter 23)

recycle The fourth of the waste-reduction "4 R's": Reprocess items to make new products. (Chapter 30)

reduce The second of the waste-reduction "4 R's": Make choices that allow you to use less of a resource by, for instance, purchasing durable goods that will last or can be repaired. (Chapter 30)

reduced tillage Planting crops in soil that is minimally disturbed and that retains some plant residue from the previous planting. (Chapter 22)

refuse The first of the waste-reduction "4 Rs": Choose NOT to use or buy a product if you can do without it. (Chapter 30)

remediation Restoration that focuses on the cleanup of pollution in a natural area. (Chapter 11)

renewable energy Energy that comes from an infinitely available or easily replenished source. (Chapters 1 and 28)

replacement fertility rate The rate at which people must be born to replace those dying in the population. (Chapter 5)

reproductive strategies How quickly a population can potentially increase, reflecting the biology of the species (life span, fecundity, maturity rate, etc.). (Chapter 8)

reserves A measure of the amount of a fuel that is economically feasible to extract from a known deposit using current technology. (Chapter 24)

reservoir (or sink) Abiotic or biotic component of the environment that serves as a storage place for cycling nutrients (Chapter 7); artificial lake formed when a river is impounded by a dam. (Chapter 17)

resilience The ability of an ecosystem to recover when it is damaged or perturbed. (Chapter 9)

resource partitioning When different species use different parts or aspects of a resource, rather than competing directly for exactly the same resource. (Chapter 9)

restoration ecology The science that deals with the repair of damaged or disturbed ecosystems. (Chapter 9)

reuse The third of the waste-reduction "4 R's": Use a product more than once for its original purpose or another purpose. (Chapter 30)

riparian areas The land area adjacent to a body of water that is affected by the water's presence (for example, water-tolerant plants grow there) and that affects the water itself (for example, provides shade). (Chapter 18)

risk assessment Weighing the risks and benefits of a particular action in order to decide how to proceed. (Chapter 3)

rock Conglomerates of one or more minerals that occur In a variety of configurations. (Chapter 19)

rock cycle The process in which rock Is constantly made and destroyed. (Chapter 19)

rotational grazing Moving animals from one pasture to the next in a predetermined sequence to prevent overgrazing. (Chapter 14)

r-selected species Species that have a high biotic potential and that share other characteristics such as short lifespan, early maturity, and high fecundity. (Chapter 8)

runoff Water that flows downhill across the land surface, usually after a rainfall (Chapter 13); see also *stormwater runoff.*

S

saltwater intrusion The inflow of ocean (salt) water into a freshwater aquifer that happens when an aquifer has lost some of its freshwater stores. (Chapter 17)

sanitary landfills Disposal sites that seal in trash at the top and bottom to prevent its release into the atmosphere; the sites are lined on the bottom, and trash is dumped in and covered with soil daily. (Chapter 30)

science A body of knowledge (facts and explanations) about the natural world, and the process used to get that knowledge. (Chapter 2)

scientific method Procedure scientists use to empirically test a hypothesis. (Chapter 2)

secondary air pollutants Air pollutants formed when primary air pollutants react with one another or with other chemicals in the air. (Chapter 25)

secondary sources Sources of information that present and interpret information from primary sources. Secondary sources include newspapers, magazines, books, and most information from the Internet. (Chapter 3)

secondary succession Ecological succession that occurs in an ecosystem that has been disturbed; occurs more quickly than primary succession because soil is present. (Chapter 9)

sediment pollution Eroded soil that is washed into the water through runoff. (Chapter 18)

seed banks Places where seeds are stored in order to protect the genetic diversity of the world's crops. (Chapter 22)

selective pressure A nonrandom influence affecting who survives or reproduces. (Chapter 12)

service economy A business model whose focus is on leasing and caring for a product in the customer's possession rather than on selling the product itself (selling the *service* that the product provides). (Chapter 4)

sex ratio The relative number of males to females in a population; calculated by dividing the number of males by the number of females. (Chapter 5)

single species approach A conservation strategy that focuses on protecting one particular species. (Chapter 11)

sliding reinforcer Actions that are beneficial at first but that change conditions such that their benefit declines over time. (Chapter 1)

smart growth Strategies that help create walkable communities with lower ecological footprints. (Chapter 32)

smog Hazy air pollution that contains a variety of pollutants, including sulfur dioxide, nitrogen oxides, tropospheric ozone, and particulates. (Chapter 25)

social traps Decisions by individuals or groups that seem good at the time and produce a short-term benefit, but that hurt society in the long run. (Chapter 1)

soil erosion The removal of soil by wind and water that exceeds the soil's natural replacement. (Chapter 14)

solar energy Energy harnessed from the Sun in the form of heat or light. (Chapter 28)

solar thermal system An active technology that captures solar energy for heating. (Chapter 28)

solubility The ability of a substance to dissolve in a liquid or gas. (Chapter 3)

species A group of plants or animals that have a high degree of similarity and can generally only interbreed among themselves. (Chapter 7)

species diversity The variety of species in an area, including how many are present (richness) and their abundance relative to each other (evenness). (Chapters 9 and 10)

species evenness The relative abundance of each species in a community. (Chapter 9)

species richness The total number of different species in a community. (Chapter 9)

statistics The mathematical evaluation of experimental data to determine how likely it is that any difference observed is due to the variable being tested. (Chapter 2)

stormwater runoff Water from precipitation that flows downhill over the surface of the land. (Chapter 18)

stratosphere Region of the atmosphere that starts at the top of the troposphere and extends up to about 31 miles; contains the ozone layer. (Chapter 2)

strip cropping Alternating different crops in adjacent strips, several rows wide; helps keep pest populations low. (Chapter 22)

subsidies Free government money or resources intended to promote desired activities. (Chapters 25 and 31)

subsurface mines Sites where tunnels are dug underground to access mineral resources. (Chapter 23)

suburban sprawl Low-population-density developments that are built outside of a city. (Chapter 32)

surface mining Removing dirt and rock that overlays a mineral deposit close to the surface in order to access that deposit. (Chapter 23)

surface water Any body of water found above ground such as oceans, rivers, and lakes. (Chapter 17)

sustainable A method of using resources in such a way that we can continue to use them indefinitely; capable of being continued without degrading the environment. (Chapters 1 and 4)

sustainable agriculture Farming methods that can be used indefinitely because they do not deplete resources, such as soil and water, faster than they are replaced. (Chapter 22)

sustainable development Economic and social development that meets present needs without preventing future generations from meeting their needs. (Chapters 1 and 4)

sustainable energy source Energy sources that are renewable and have a low environmental impact. (Chapter 28)

sustainable fishery A fishery that ensures that fish stocks are maintained at healthy levels, the ecosystem is fully functional, and fishing activity does not threaten biological diversity. (Chapter 16)

sustainable grazing Practices that allow animals to graze in a way that keeps pastures healthy and allows grasses to recover. (Chapter 14)

symbiosis A close biological or ecological relationship between two species. (Chapter 9)

synergistic effects Exposure to two or more chemicals has a greater effect than the sum of their individual effects would predict. (Chapter 3)

T

tax credit A reduction in the tax one has to pay, in exchange for some desirable action. (Chapter 25)

tectonic plates Sections of Earth's crust that float above the magma layer. (Chapter 19)

temperate forests Found in areas with four seasons and a moderate climate, receive 30–60 inches of precipitation per year, may include conifers and/or hardwood deciduous trees (lose their leaves in the winter). (Chapter 13)

terracing On steep slopes, land is leveled into steps; reduces soil erosion and runoff down the hillside. (Chapter 22)

tertiary sources Sources of information that present and interpret information from secondary sources. (Chapter 3)

testable A possible explanation that generates predictions for which empirical evidence can be collected to verify or refute the hypothesis. (Chapter 2)

test group The group in an experimental study that is manipulated somehow such that it differs from the control group in only one way. (Chapter 2)

theory A widely accepted explanation of a natural phenomenon that has been extensively and rigorously tested scientifically. (Chapter 2)

threatened species Those that are at risk for extinction; various threat levels have been identified ranging from "least concern" to "extinct." (Chapter 11)

time delay Actions that produce a benefit today set into motion events that cause problems later on. (Chapter 1)

total fertility rate (TFR) The number of children the average woman has in her lifetime. (Chapter 5)

toxicologists Scientists who study the specific properties of any given potential toxin. (Chapter 3)

toxins Chemicals that cause direct damage upon exposure. (Chapter 3)

trade-offs The imperfect and sometimes problematic responses that we must at times choose between when addressing complex problems. (Chapter 1)

tragedy of the commons The tendency of an individual to abuse commonly held resources in order to maximize his or her own personal interest. (Chapter 1)

transboundary pollution Pollution that is produced in one area but falls in a different state or nation. (Chapter 25)

transboundary problems Pollution that is produced in one area but falls in or reaches other states or nations. (Chapter 31)

transgenic organism An organism that contains genes from another species. (Chapter 20)

transpiration The loss of water vapor from plants. (Chapter 17)

triple bottom line Considering the environmental, social, and economic impacts of our choices. (Chapters 1 and 4)

trophic levels Feeding levels in a food chain. (Chapter 9)

tropical forests Found in equatorial areas with warm temperatures year-round and high rainfall; some have distinct wet and dry seasons but none has a winter season. (Chapter 13)

troposphere Region of the atmosphere that starts at ground level and extends upward about 7 miles. (Chapter 2)

true cost Including both internal and external costs when setting a price for a good or service. (Chapter 4)

turbidity The cloudiness of the water. (Chapter 18)

U

ultraviolet (UV) radiation Short-wavelength electromagnetic energy emitted by the Sun. (Chapter 2)

unconventional oil reserves Recoverable oil that exists in rock, sand, or clay but whose extraction is economically and environmentally costly. (Chapter 24)

undernutrition Chronic, insufficient calorie intake, resulting in nutrient deficiencies and the inability to meet energy needs. (Chapter 20)

understory The smaller trees, shrubs, and saplings that live in the shade of the forest canopy. (Chapter 13)

uniform distribution Individuals are spaced evenly, perhaps due to territorial behavior or mechanisms for suppressing the growth of nearby individuals. (Chapter 8)

United Nations Framework Convention on Climate Change (UNFCCC) A 1992 international treaty that formally recognized that climate change was an emerging problem and that precautions should be taken to prevent dangerous anthropogenic interference with Earth's climate system. (Chapter 31)

urban areas Densely populated regions that include cities and the suburbs that surround them. (Chapter 32)

urban flight The process of people leaving an inner city area to live in surrounding areas. (Chapter 32)

urban heat island effect The phenomenon in which urban areas are warmer than the surrounding countryside due to pavement, dark surfaces, closed-in spaces, and high energy use. (Chapter 32)

urbanization The migration of people to large cities. (Chapter 32)

urban planner A person who develops land-use plans in and around cities. (Chapter 32)

U.S. Farm Bill Overarching legislation that deals with many aspects of the production and sale of farm-raised commodity crops, including programs for consumer food assistance, soil conservation, and farm subsidies; updated every 5 years. (Chapter 21)

V

vector-borne disease An infectious disease acquired from organisms that transmit a pathogen from one host to another. (Chapter 6)

W

waste Any material that humans deem to be unwanted. (Chapter 30)

wastewater Used and contaminated water that is released after use by households, industry, or agriculture. (Chapter 17)

waterborne disease An infectious disease acquired through contact with contaminated water. (Chapter 6)

water cycle The movement of water from gaseous to liquid states through various water compartments such as surface waters, soil, and living organisms. (Chapter 17) water footprint The amount of water consumed by a given group (person, population) or for a process (raising livestock). (Chapter 21)

water pollution The addition of anything that might degrade the quality of the water. (Chapter 18)

water scarcity Not having access to enough clean water supplies. (Chapter 17)

watershed The land area surrounding a body of water over which water such as rain could flow and potentially enter that body of water. (Chapter 18)

water table The uppermost water level of the saturated zone of an aquifer. (Chapter 17)

weather The meteorological conditions in a given place on a given day. (Chapter 26)

wetland An ecosystem that is permanently or seasonally flooded. (Chapter 17)

wind energy Energy contained in the motion of air across Earth's surface. (Chapter 28)

worldviews The window through which one views one's world and existence. (Chapter 1)

Z

zero population growth The absence of population growth; occurs when birth rates equal death rates. (Chapter 5)

zoonotic disease A disease that is spread between infected animals (not merely a vector that transmits the pathogen but another host that harbors the pathogen through its life cycle) and humans. (Chapter 6)

CREDITS/SOURCES

INFOGRAPHIC SOURCES AND REFERENCES

CHAPTER 1:
IG 1.7 Adapted from UN Human Development Indices, 2008 (http://is.gd/7UqClm)

CHAPTER 2:
IG 2.6 Adapted from UN Vital Ozone Graphics, 2007 (http://is.gd/vMwgpH)
IG 2.7 Satellite images, Source: NASA Graphs adapted from ozone graph from "Refrigerant Reclaim Australia," 2009 Annual Report (https://www.refrigerantreclaim.com.au/AR06/) and (http://is.gd/eW5qbl)

CHAPTER 3:
IG 3.3 Adapted from:
Nagel, S. C., *et al.* (1997). *Environmental Health Perspectives*. 105: 70–76.
Ishido, M. & J. Suzuki (2010). *Journal of Health Science*. 56: 175–181.
Lang, I. A., *et al.* (2008). *Journal of the American Medical Association*. 300: 1303–1310.

CHAPTER 4:
IG 4.1 Adapted from Costanza, R., *et al.* (1997). *Nature*. 387:253–260.
IG 4.2 Data from Global Footprint Network

CHAPTER 5:
IG 5.1 Data from *UN World Population Prospects, the 2010 Revision*.
IG 5.2 Bar graph adapted from Population Reference Bureau, *2010 World Population Data Sheet*; map adapted from World Resources Institute (http://is.gd/t2A8lx)
IG 5.3 Table from Population Reference Bureau, *2010 World Population Data Sheet*; graph adapted from the United Nations Department of Economic and Social Affairs
IG 5.5 Data from U.S. Census Bureau
IG 5.6 Infant mortality adapted from CIA Factbook, 2011 estimates; Desired Fertility adapted from Pritchett, L. (1994). *Population and Development Review*. 20: 1–55; Education adapted from Pew Research Center, *The Future of the Global Muslim–Population*, 2011;

Family planning adapted from Sedgh, G., *et al.* (2007). *Women With an Unmet Need for Contraception in Developing Countries and Their Reasons for Not Using a Method*, Guttmacher Institute.
IG 5.7 Adapted from Population Action International
IG 5.8 Adapted from *The Ecological Footprint Atlas 2010*

CHAPTER 6:
IG 6.3 Graph adapted from World Health Organization (2006) *Preventing Disease Through Healthy Environments* http://www.who.int/quantifying_ehimpacts/publications/preventingdisease/en/index.html
IG 6.5 Image adapted from the CDC (www.dpd.cdc.gov/dpdx/HTML/Dracunculiasis.htm) and CNN (http://www.cnn.com/2010/HEALTH/04/05/guinea.worm.lifecycle/index.html)
IG 6.6 Map adapted from World Health Organization (2006) *Preventing Disease Through Healthy Environments* (page 8) and WHO Global Health Observatory Data: Total Environment—Total Burden of Disease (2004), from http://apps.who.int/ghodata/; Top Ten Causes of Death adapted from World Health Organization, *Deaths across the globe: an overview*, http://www.who.int/mediacentre/factsheets/fs310/en/index4.html
IG 6.7 Map adapted from Emory University (2011) *End game for Guinea worm disease is near* at http://www.emory.edu/EMORY_REPORT/stories/2011/03/report_from_carter_center_guinea_worm_disease.html

CHAPTER 7:
IG 7.4 Map adapted from http://is.gd/Kh0ni4 Biome graph adapted from R. Ricklefs, (2000) *The Economy of Nature*, W.H. Freeman, New York

CHAPTER 8:
IG 8.6 Predator–prey graph adapted from George Dovel, 2008. *The Outdoorsman*, Number 30.

CHAPTER 9:
IG 9.3 Adapted from Florida Fish and Wildlife Conservation Commission
IG 9.7 Maps adapted from Comprehensive Everglades Restoration Plan website (http://is.gd/l6XXjK)

CHAPTER 10:
IG 10.2 Data from *Number of Living Species in Australia and the World Report*, 2009
IG 10.5 Adapted from Conservation International (PDF: http://is.gd/PsXTqP)

CHAPTER 11:
IG 11.1 Graph adapted from Baillie, J.E.M, *et al.* 2010. *Evolution Lost: Status and Trends of the World's Vertebrates*, Zoological Society of London
IG 11.2 ICUN categories from: The International Union for Conservation of Nature (Lead Author); Mark McGinley (Topic Editor) "IUCN Red List Criteria for Endangered". In: Encyclopedia of Earth. Eds. Cutler J. Cleveland. [First published in the Encyclopedia of Earth October 12, 2009; Last revised Date October 12, 2009]
IG 11.7 Map adapted from UNEP (2012) Global Environmental Outlook–5 Report, http://unep.org/geo/pdfs/geo5/GEO5_report_full_en.pdf.; Graph adapted from Center for Biological Diversity (extinction data) and Statistics Sweden, Environmental Accounts and Natural Resources (global protected areas).

CHAPTER 12:
IG 12.2 Based on research by Nachman, M.W. 2005. *Genetica*, 123: 125–136
IG 12.4 Map adapted from Savidge, J.A. (1987) *Ecology*, 68: 660–668

CHAPTER 13:
IG 13.1 Map adapted from http://is.gd/Kh0ni4
IG 13.3 Adapted from Costanza, R., *et al.* (1997). *Nature*. 387:253–260.
IG 13.4 Map adapted from UN Food and Agriculture Organization (PDF: http://is.gd/ma67Ps)

CHAPTER 14:
IG 14.1 Map adapted from http://is.gd/KhOni4
IG 14.3 Source: U.S. Department of Agriculture (http://is.gd/pFRKaX)

CHAPTER 15:
IG 15.1 Graph adapted from NOAA (http://is.gd/pAuD5o); maps adapted from Cao, L. & K. Caldeira. (2010) Climatic Change.99, 1–2.
IG 15.3 Map and graphs adapted from UN Environmental Programme (http://is.gd/VI5oz2)

CHAPTER 16:
IG 16.1 Map adapted from Aotearoa (http://en.wikipedia.org/wiki/Atlantic_cod); graph adapted from UN Millennium Ecosystem Assessment (http://is.gd/25KrJl)
IG 16.3 Graph and fish catch adapted from The Sea Around Us Project
IG 16.4 Data from UN Food and Agriculture Organization
IG 16.5 Map adapted from NOAA: National Marine Protected Areas Center (http://www.mpa.gov/dataanalysis/maps/)

CHAPTER 17:
IG 17.3 Map adapted from UN Food and Agriculture Organization (http://is.gd/1TgsbR); graph adapted from the Population Reference Bureau; Domestic Water Use Graph adapted from Hoekstra, A. Y. and A. K. Chapagain (2007). *Water Resources Management*, 21:35–48.
IG 17.4 Adapted from World Water Assessment Programme (WWAP); gallons of water per food or product from *National Geographic* (http://is.gd/MCeBGJ)
IG 17.7 Pie chart adapted from University of Georgia Cooperative Extension; data for water use per 1,000 kwh from Institute of Electrical and Electronics Engineers

CHAPTER 18:
IG 18.2 Source: EPA
IG 18.4 Adapted from Groffman, P., *et al.*, (2004). *Ecosystems*, 7:393–403.
IG 18.5 Source: U.S. Department of Agriculture
IG 18.7 Source: Chesapeake Bay Program

CHAPTER 20:
IG 20.1 Map adapted from http://www.fao.org/hunger/en/; Bar graph from FAO (2010) State of Food Insecurity in the World http://www.fao.org/docrep/013/i1683e/i1683e.pdf
IG 20.4 Data for graph from NUEweb, website of the Department of Plant and Soil Sciences, Oklahoma State University, http://www.nue.okstate.edu/Crop_Information/World_Wheat_Production.htm

CHAPTER 21:
IG 21.2 Upper graph adapted from: FAO. 2009. *The State of Food and Agriculture—Livestock in the Balance*. Rome: Food and Agriculture Organization of the United Nations; Lower graph adapted from: Pan, A., *et al.* 2012. *Archives of Internal Medicine*, 172: 555–563.
IG 21.3 Adapted from Smil, V. 2001 *Feeding the world: A challenge for the twenty-first century* Cambridge, MA: MIT Press.
IG 21.5 Data from CDC at http://wonder.cdc.gov/data2010/.
IG 21.6 Source: FAO. 2009. *The State of Food and Agriculture—Livestock in the Balance*. Rome: Food and Agriculture Organization of the United Nations
IG 21.7 Graph adapted from Peters, C.J., *et al.* 2007. *Renewable Agriculture and Food Systems*. 22(2):145–153.

CHAPTER 22:
IG 22.4 From: Furuno, Takao. *The Power of the Duck*. Tasmania: Takari Publications, 2001

CHAPTER 23:
IG 23.1 Source: Tennessee Valley Authority; pie chart data from EIA Annual Energy Outlook, 2011 (EIA = Energy Information Administration)
IG 23.3 Map adapted from Britannica Encyclopedia (http://is.gd/MUgevK)
IG 23.5 Adapted from Kentucky Geological Survey (http://is.gd/nyyEbh)
IG 23.7 Adapted from World Coal Association

CHAPTER 24:
IG 24.2 Adapted from Campbell, C.J. 2004. The Association for the Study of Peak Oil and Gas.
IG 24.3 Source: U.S. Energy Information Association, *Oil & Gas Journal*
IG 24.7 Adapted from Technovelgy (http://is.gd/LerVhk)
IG 24.8 Map sources from the USGS (http://is.gd/EtlzWr) and (http://is.gd/UEcFpO)

CHAPTER 25:
IG 25.1 Source: the World Health Organization
IG 25.2 Source: EPA
IG 25.3 Adapted from Laden, Francine, *et al.*, (2006). *American Journal of Respiratory and Critical Care Medicine*. 173:667–672.
IG 25.4 Adapted from PhysicalGeography.net; maps adapted from National Atmospheric Deposition Program
IG 25.5 Source: EPA
IG 25.6 Smokestack scrubber adapted from Encyclopedia Britannica; cap-and-trade adapted from the *Washington Post*

CHAPTER 26:
IG 26.1 Source: adapted from IPCC, 4th Assessment Report, Working Group 1 Report: *The Physical Science Basis*
IG 26.2 Temperature anomalies source: National Climatic Data Center/NOAA; annual global average temperature (land and ocean) source: National Climatic Data Center/NOAA; Cumulative loss of glacier ice source: United Nations Environmental Programme; sea level change over time source: National Climatic Data Center/NOAA; precipitation changes from: National Climatic Data Center/NOAA; Increase in storms from: EPA
IG 26.3 Bar graph adapted from Intergovernmental Panel on Climate Change, 4th Assessment Report, Working Group 1 Report: *The Physical Science Basis*
IG 26.5 Adapted from Intergovernmental Panel on Climate Change, 4th Assessment Report, Working Group 1 Report: *The Physical Science Basis*
IG 26.7 Upper graphs source: adapted from NASA; lower graph source: adapted from United States Global Change Research Program
IG 26.8 Adapted from Intergovernmental Panel on Climate Change, 4th Assessment Report, Working Group 1 Report: *The Physical Science Basis*

IG 26.10 Top graph adapted from Intergovernmental Panel on Climate Change, 4th Assessment Report, Working Group I Report: *The Physical Science Basis*; bottom graph source: Pacala, S. & R. Socolow (2004). *Science*, 305:968–972.

IG 26.11 Adapted from Intergovernmental Panel on Climate Change, 4th Assessment Report, Working Group II Report: *Impacts, Adaptation and Vulnerability*

CHAPTER 27:

IG 27.1 Adapted from aboutnuclear.org

IG 27.2 Decay chain: adapted from Department of Energy; bottom graph adapted from GeoKansas

IG 27.3 Flow chart adapted from World Information Service On Energy (WISE) Uranium Project; Fuel assembly adapted from climateandfuel.com

IG 27.4 Adapted from atomicarchive.com

IG 27.5 Adapted from Nuclear Regulatory Commission and the Union of Concerned Scientists

IG 27.6 Image adapted from Stannered (http://is.gd/PV4jfL)

CHAPTER 28:

IG 28.1 Bar graph source: U.S. Energy Information Administration, *International Energy Statistics Database*; pie chart source: U.S. Energy Information Administration

IG 28.2 Adapted from US. Department of the Interior, *Wind Energy Development Programmatic Environmental Impact Statement*

IG 28.3 Adapted from *Power and Energy* (http://is.gd/5B34FF)

IG 28.4 Source: U.S. Department of Energy

IG 28.5 Source: U.S. Department of Energy (http://is.gd/WECJHW); ground source heat pump adapted from Tennessee Valley Authority

IG 28.7 Source: EnergyStar.gov

CHAPTER 29:

IG 25.3 Adapted from Tilman, D., *et al.* (2006) *Science* 314:1598–1600.

IG 29.5 Adapted from oilgae.com (http://is.gd/yQDsh0)

IG 29.6 Adapted from alternate-energy-sources.com

IG 29.7 Source: U.S. Department of Energy

CHAPTER 30:

IG 30.1 Source: EPA

IG 30.3 Source: EPA

IG 30.5 From: Young, L.C., *et al.* (2009). PLoS ONE 4(10): e7623

CHAPTER 31:

IG 31.7 Adapted from UNEP, 2005, *Vital Climate Change Graphics Update* at http://www.grida.no/graphicslib/detail/kyoto-protocol-timeline-and-history_76cf#

IG 31.8 Data from International Energy Agency (2011) *CO$_2$ Emissions from Fuel Combustion*: Highlights at www.iea.org/co2highlights.

IG 31.9 Data from CDM pipeline: UNEP Risoe Centre, *CDM/JI Pipeline*, accessed October 2012 from http://cdmpipeline.org/cdm-projects-type.htm

CHAPTER 32:

IG 32.1 Map source: UN Population Division; CIA World Factbook; Urban vs. rural line graph adapted from United Nations, Population Division; megacities bar graph adapted from United Nations, Population Division

IG 32.2 Adapted from *National Geographic*

IG 32.4 Adapted from socialexplorer.com

IG 32.6 Source: EPA

ANALYZING THE SCIENCE DATA SOURCES

CHAPTER 1:

Modified from Brown, L.R., "World on the Edge—Food and Agriculture Data—Livestock and Fish." pp. 7, 15. Earth Policy Institute. http://www.earthpolicy.org/datacenter/pdf/book_wote_livestock.pdf.

CHAPTER 2:

Modified from http://undsci.berkeley.edu/article/ozone_depletion_01. Figure 28 – A Plot of Chlorine Monoxide and Ozone Concentrations from Ozone Depletion: Uncovering the hidden hazard of hairspray. Originally from: Anderson, J.G., *et al.* 1989. Ozone destruction by chlorine radicals within the Antarctic vortex: the spatial and temporal evolution of ClO–O$_3$ anticorrelation based on in situ ER-2 data. *Journal of Geophysical Research* 94:11465–11479.

CHAPTER 3:

Adapted from http://pollutioninpeople.org/results/report/chapter-3/metals_1 (Figure 3: Mercury levels in participant hair; http://pollutioninpeople.org/results/report/chapter-6/ddt_pcb_2) (Figure 7: DDT exposure was measured as the breakdown product p,p'DDE in blood serum).

CHAPTER 4:

Modified from Brown, L.R., "Learning from China: Why the Existing Economic Model Will Fail." Earth Policy Institute. 8 Sep 2011. http://www.earthpolicy.org/data_highlights/2011/highlights18.

CHAPTER 5:

Adapted from United Nations, Department of Economic and Social Affairs, Population Division (2011): World Population Prospects: The 2010 Revision. New York. (Updated: 15 April 2011). http://esa.un.org/unpd/wpp/Analytical-Figures/htm/fig_1.htm (Graph A); (Updated: 5 July 2011). http://esa.un.org/unpd/wpp/Analytical-Figures/htm/fig_13.htm (Graph B).

CHAPTER 6:

Adapted from Fan, V., 2012. *Malaria Estimate Sausages by WHO and IHME.* Center for

Global development. http://blogs.cgdev.org/globalhealth/2012/02/malaria-estimate-sausages-by-who-and-ihme.php

CHAPTER 7:

Modified from: http://www.globalchange.umich.edu/globalchange1/current/lectures/kling/rainforest/rainforest_table.html. Original data from: J. Terborgh 1992. Diversity and the tropical rain forest. Scientific American Library, W. H. Freeman, New York, xii + 242 pages.

CHAPTER 8:

Modified from "Kaibab Plateau Deer Population: 1907–1940" http://www.hhh.umn.edu/centers/stpp/pdf/KaibabPlateauExercise.pdf (Redrawn from Leopold, A)

CHAPTER 9:

Adapted from Rogers, J.D., 2008. *Journal of geotechnical and geoenvironmental engineering* http://web.mst.edu/~rogersda/levees/History%20New%20Orleans%20Flood%20Control-Rogers.pdf. Page 6, Fig. 7 (adapted from Kolb and Saucier 1982).

CHAPTER 10:

Modified from Cincotta, R.P., *et al*, 2000. *Nature*. VOL 404. 27

CHAPTER 11:

Modified from IUCN; Figure 3 from *Summary Statistics*, www.iucnredlist.org/about/summary-statistics

CHAPTER 12:

Adapted from "OVERPOPULATION: A Key Factor in Species Extinction." Center for Biological Diversity. http://www.biologicaldiversity.org/campaigns/overpopulation/.

CHAPTER 13:

Adapted from "FAO Forestry Paper 163: Global Forest Resources Assessment 2010: Main report." Food and Agriculture Organization of the United Nations, Rome 2010. http://www.fao.org/docrep/013/i1757e/i1757e00.htm Chapter 6 - Protective functions of forest resources.

CHAPTER 14:

Adapted from Heidenreich, B. "What are Global Temperate Grasslands Worth? A Case for their Protection." Figure 1, Habitat Conversion and Protection in the World's 13 Terrestrial Biomes. The World Temperate Grasslands Conservation Initiative, International Union for Conservation of Nature. July 2009. http://www.iucn.org/about/union/commissions/wcpa/wcpa_puball/wcpa_pubsubject/wcpa_grasslandspub/?4266/What-are-Global-Temperate-Grasslands-worth-A-case-for-their-protection
Originally from: Hoekstra J.M., *et al*, 2005. "Confronting a biome crisis: global disparities of habitat loss and protection." *Ecology Letters*. 8:23–29.

CHAPTER 15:

Modified from Burke, L., *et al*, "Reefs at Risk Revisited." February 2011. Page 42, Table 4.1 http://www.wri.org/publication/reefs-at-risk-revisited.

CHAPTER 16:

Modified from Talberth, J., *et al*, "The Ecological Fishprint of Nations: Measuring Humanity's Impact on Marine Ecosystems." http://www.wwf.dk/dk/Service/Bibliotek/Hav+og+fiskeri/Rapporter+mv./FishprintofNations 2006. Page 8, Figure 3.

CHAPTER 17:

Adapted from Figure "DALYs attributable to water, sanitation and hygiene (diarrhea), 2004." World Health Organization. Public Health Information and Geographic Information Systems (GIS). 2011. http://gamapserver.who.int/mapLibrary/Files/Maps/Global_wsh_daly_2004.png

CHAPTER 18:

Adapted from Scorecard: The Pollution Information Site. GoodGuide. Environmental Protection Agency. http://scorecard.goodguide.com/env-releases/land/rank-sites.tcl

CHAPTER 19:

Modified from USGS Fact Sheet 087-02, Figure 1, http://pubs.usgs.gov/fs/2002/fs087-02/

CHAPTER 20:

Adapted from FAO, *The State of Food Insecurity in the World, 2010*, figures 1 and 2 on page 9. http://www.fao.org/docrep/013/i1683e/i1683e.pdf

CHAPTER 21:

Graphs based on data from USDA Food Safety and Inspection Service (www.fsis.usda.gov)
http://www.sraproject.org/wp-content/themes/default/images//Recall_Total_by_Type_Lbs.pdf
http://www.sraproject.org/wp-content/themes/default/images//Recall_Total_by_Cause_Lbs.pdf

CHAPTER 22:

Modified from Rodale Institute. "The Farming Systems Trial." Figure "Comparison of FST Organic and Conventional Systems." http://www.rodaleinstitute.org/fst30years

CHAPTER 23:

Modified from National Institute for Occupational Safety and Health, 2007 World Report. http://www.cdc.gov/niosh/programs/mining/risks.html (Graphs A and B)
Table 2-1 as reported by the National Center for Health Statistics, CDC. http://blogs.wvgazette.com/coaltattoo/files/2010/10/blacklungdeathchart1.jpg (Graph C)

CHAPTER 24:

Adapted from Murphy, D., *Business Insider*. http://www.businessinsider.com/does-peak-oil-even-matter-2010-12. Figure 1. Estimates of the cost of production for oil production from various locations. Data from Cera.[4]

CHAPTER 25:

Modified from Environment Canada, National Air Pollution Surveillance (NAPS) Network, Ottawa, 2004. http://www.ecoinfo.org/env_ind/region/smog/smog_e.cfm.

CHAPTER 26:

Modified from University of Florida, School of Natural Resources and Environment. *The SNRE Source*, Vol. 1 Issue 1, Fall 2005. http://snre.ufl.edu/pubsevents/source/fall05/stormy.htm. Calculated from http://hurricane.csc.noaa.gov/hurricanes/.

CHAPTER 27:
Modified from Clean Energy Insight. http://www.cleanenergyinsight.org/wp-content/uploads/2009/08/comparingindustrysafety_graph.jpg. U.S. Bureau of Labor Statistics, 2007.

CHAPTER 28:
Adapted from Sibley, David. "Causes of Bird Mortality." November 18th, 2010. Sibley Guides: Identification of North American birds and trees. http://www.sibleyguides.com/conservation/causes-of-bird-mortality/. (Graph 1)
Adapted from Erickson, W., *et al.* (2001). "Avian Collisions with Wind Turbines: A summary of existing studies and comparisons to other sources of Avian Collision Mortality in the United States." West Inc., prepared for the National Wind Coordinating Committee. http://www.duke.edu/web/nicholas/bio217/ptb4/batdata.html (Table 1)
Adapted from Arnett, E., *et al.* (2005). "Relationships Between Bats and Wind Turbines in Pennsylvania and West Virginia: An assessment of fatality search protocols, patterns of fatality, and behavioral interactions with wind turbines." Report prepared for Bats and Wind Energy Cooperative. (Graphs 2 and 3)

CHAPTER 29:
Modified from: http://jcwinnie.biz/wordpress/imageSnag/GHG_from_various_transportation_fuels.png
Originally from Scharlemann and Laurance, 2008. *Science*. 43–44.

CHAPTER 30:
Adapted from "Plastic Debris in the World's Oceans." Table 2.1 Number and Percentage of Marine Species Worldwide with Documented Entanglement and Ingestion Records. Greenpeace. www.unep.org/regionalseas/marinelitter/.../plastic_ocean_report.pdf

CHAPTER 31:
Adapted from Energy Information Agency, *What are Greenhouse Gases?*, Figure 1. www.eia.gov/oiaf/1605/ggccebro/chapter1.html)

CHAPTER 32:
Modified from Appelbaum, A. *Fast Company*, 6 Jan 2011.

PHOTOS

FRONT COVER:
Tim Griffith

BACK COVER:
David Maisel/INSTITUTE

TABLE OF CONTENTS:
p. II Alex McLean
p. IV James Balog/Aurora Photos
p. VII NSIDC courtesy Ted Scambos and Rob Bauer
p. IX Galen Rowell/Mountain Light/Aurora
p. XI Béatrice Jaud, from the film Severn the voice of our children—J+B Séquences
p. XIV Peter Essick/Aurora
p. XVI Nina Berman/Noor/Redux
p. XXX (top) Jorgen Caris/Hollandse Hoogte/Redux; (bottom) Mitchell Kanashkevich/The Image Bank/Getty Images
p. XXXV Alessandro Grassani/Invision/Aurora Photos

CHAPTER 1:
pp. 0–1, p. 3 Wolfgang Kaehler/Picade; p. 4 (left) Ashley Cooper/ HG/Aurora Photos; (right) Rafn Sigurbjornsson/Verkis Consulting Engineers; p. 6 NSIDC Courtesy of Ted Scambos and Rob Bauer; p. 8 (left to right) Mads Nissen/Panos Pictures; Rex Features via Associated Press Images; Martin Roemers/Panos Pictures; p. 11 Gabrielle & Michel Therin-Weise; p. 15 Andreas Strauss/Getty Images; p. 16–17 (left to right) Portrait by Charles Willson Peale, 1791, National Park Service; Theodore Roosevelt Collection, Harvard College Library (560.51 1903-115); Rachel Carson Council/ ZUMA Press/Newscom; Courtesy Martin Rowe/Green Belt Movement.

CHAPTER 2:
pp. 20–21 Courtesy Linnea Avallone/Concordiasi Team; p. 23 (top) George Steinmetz/National Geographic Stock; (bottom) Courtesy Dr. Susan Solomon; p. 29 David Hay Jones/Photo Researchers, Inc.; p. 32 Bettmann/Corbis; p. 34 NASA (x3); p. 35 Bart Coenders/iStockphoto.

CHAPTER 3:
pp. 38–39 ULTRA.F/Getty Images; p. 41 Bettmann/Corbis; p. 42 Roger Winstead;

Gnatcatcher); Raymond Gehman/National Geographic Stock (red maple); Kent Wood (pine beetle); p. 478 (clockwise from top left) John-Michael Maas/AFP/Getty Images/Newscom (Health Impacts); Laurent Weyl/Collectif Argos (Crop Productivity); Cedric Faimali/Collectif Argos (Coastal Erosion); Layne Kennedy (Fire); Svan Torfinn/Panos (Drought); Creative Commons (Biodiversity).

CHAPTER 27:
pp. 482–483 AP Photo/David Guttenfelder; p. 484 (top) AP Photo/David Guttenfelder; (bottom) Digital Globe/Getty Images; p. 491 AP Photo/David Guttenfelder; p. 495 (top) Justin Jin/Panos; (middle) Peter Essick/Aurora Photos; (bottom) Stuart Johnson/Deseret News; p. 496 Christoph Bangert/Stern/laif/Redux; p. 497 Mark Henley/Panos; p. 499 Andy Duback/Greenpeace.

CHAPTER 28:
pp. 502–503 Alessandro Grassani/Invision/Aurora Photos; p. 505 Andrew Henderson/National Geographic Stock; p. 507 (top) ZUMA Press/Newscom; (bottom) Stefan Falke/laif/Redux, pp. 508–509 Alessandro Grassani/Invision/Aurora Photos, p. 510 (left) Zuma Press/Newscom; (right) Gilles Rolle/RFA/Redux; p. 513 (left) Pep Bonet/Noor/Redux; (right) Perrot Oliver /SIPA; p. 514 Stephanie Roland; p. 515 Kirkendall & Spring Photography; p. 517 iStockphoto/Thinkstock.

CHAPTER 29:
pp. 520–521 Ben Lowy/Getty Images; p. 522 (top) Courtesy David Tillman; (bottom) Tim Rummelhoff; p. 524 Courtesy Robert Williams; p. 526 Everett Kennedy Brown/EPA/Newscom; p. 529 Ben Lowy/Getty Images; p. 532 Robert Clark/Institute for Artist Management; p. 533 (left) AP Photo/Sherri Barber/Coloradoan; (right) Courtesy Sapphire Energy; p. 535 Emily Spence/Knoxville Daily News; p. 536 (left) Yuriko Nakao/Reuters; (right) AP Photo/Rick Bowmer; p. 537 Courtesy FUEL the movie.

CHAPTER 30:
pp. 540–541 Scripps Institution of Oceanography, UC San Diego; p. 543 (top) Sea Education Association; (bottom) Courtesy Giora Proskurowski; p. 545 Sea Education Association; p. 546 (top) Nigel Dickinson/Leader Photos; (bottom) Stuart Isett/Polaris; p. 551 (top) Alan Marsh/Wave/Corbis; (middle) Chris Price/iStockphoto; (bottom) David Hoffman; p. 553 Sarah Leen/National Geographic Stock.

CHAPTER 31:
pp. 558–559 TAO Images Limited/Getty Images; p. 564 courtesy NASA; p. 567 Jamey Stillings; p. 568 (top to bottom) Eduardo DiBaia/Associated Press; Kyodo/ Associated Press; Chris Wattie/Reuters.

CHAPTER 32:
pp. 576–577 Nina Berman/Noor/Redux; p. 579 Nina Berman/Noor/Redux; p. 583 Nina Berman/Noor/Redux; p. 586 Nina Berman/Noor/Redux; p. 591 Brian Holmes.

INFORMATION GRAPHICS
MGMT. design

RENDERED ILLUSTRATIONS
Nicolle Rager Fuller, Sayo-Art LLC

INDEX

WORLD MAP

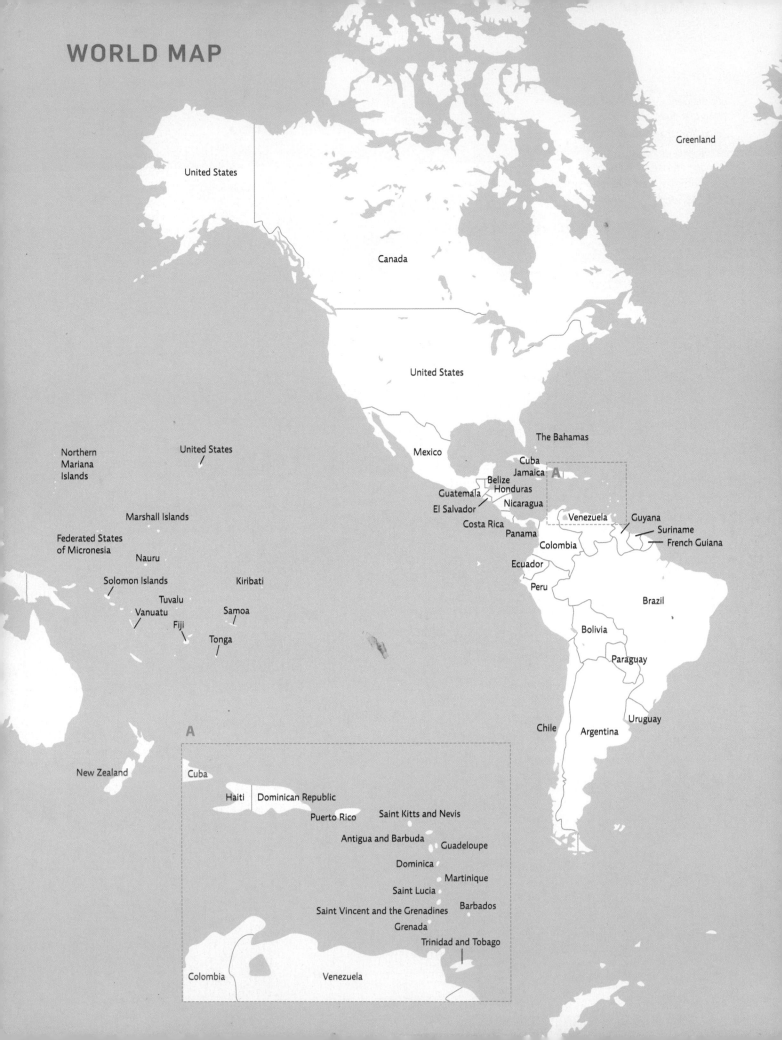

Greenland

United States

Canada

United States

Northern
Mariana
Islands

United States

Mexico

The Bahamas

Cuba
Jamaica

Belize
Guatemala   Honduras
El Salvador   Nicaragua
Costa Rica
Panama

A

Venezuela

Guyana
Suriname
French Guiana

Colombia

Marshall Islands

Ecuador

Federated States
of Micronesia

Peru

Nauru

Solomon Islands

Brazil

Tuvalu

Vanuatu

Kiribati

Bolivia

Fiji

Samoa

Paraguay

Tonga

Uruguay

Chile   Argentina

New Zealand

A

Cuba

Haiti   Dominican Republic

Puerto Rico   Saint Kitts and Nevis

Antigua and Barbuda
Guadeloupe

Dominica

Martinique

Saint Lucia

Saint Vincent and the Grenadines   Barbados

Grenada

Trinidad and Tobago

Colombia   Venezuela